SEDIMENTARY DEPOSITION IN RIFT AND FORELAND BASINS IN FRANCE AND SPAIN

(Paleogene and Lower Neogene)

Sedimentary Deposition in Rift and Foreland Basins in France and Spain

(Paleogene and Lower Neogene)

Georges Busson
and B. Charlotte Schreiber
EDITORS

Columbia University Press
NEW YORK

Columbia University Press
Publishers Since 1893
New York Chichester, West Sussex

Library of Congress Cataloging-in-Publication Data
Sedimentary deposition in rift and foreland basins in France and Spain (Paleogene and lower Neogene) / Georges Busson and B. Charlotte Schreiber, editors.
p. cm.
Includes bibliographical references (p. —) and index.
ISBN 0-231-06786-0 (alk. paper)
1. Evaporites—France. 2. Evaporites—Spain. 3. Sedimentary basins—France. 4. Sedimentary basins—Spain. 5. Geology, Stratigraphic—Paleogene. 6. Geology, Stratigraphic—Miocene. I. Busson, Georges. II. Schreiber, B. Charlotte.
QE471.15.E8S43 1997
553.2′82—dc20 96-32223

♾ Casebound editions of Columbia University Press books are printed on permanent and durable acid-free paper.
Printed in the United States of America
c 10 9 8 7 6 5 4 3 2 1
p 10 9 8 7 6 5 4 3 2 1

Contents

Acknowledgments

Since 1977 the National Center for Scientific Research of France (CNRS) has supported a major research project concerning confined facies in France, with Dr. Georges Busson as the originator and coordinator. The formation of this research group engendered the title "Nature and Genesis of Confined Facies" and successively gave rise to a series of cooperative research programs (RCP 459 from 1977–1981), plus a coordinated research group entitled GRECO 52 (from 1982–1985), and finally a united association, UA 1209 (1986–1988). One important part of the work of this group has been concerned with the various segments of the Paleogene rift basins of France. In this project the studies of Yves Guillevin, Didier Masson, Martine Keiner, and Dominique Couilloud have contributed greatly to our knowledge: many of their results have been integrated into and aided the general development of the recent work presented in this volume.

A number of the contributions to this volume arise from thesis studies by their authors. The work of Robert Moretto concerning the geology of the Bresse basin, is taken from the body of his thesis (1982–86) carried out at the University of Nancy, co-directed by Jean Hilly and Georges Busson, under a research grant from CNRS (1986–88). Alain Curial at the University of Lyon was favored by a grant from the Geological Society of France, and was financially supported, in part, by Schlumberger (1982–86), plus a research contract with CNRS (1986–88), in his studies of the geology of the Bresse basin; he implemented the study through the application of borehole logs (directed by Georges Busson). Since 1981, the work of Marie-Madeleine Blanc-Valleron has been centered on the Paleogene basins of upper Alsace, the subject of her doctoral thesis (thèse d'Etat) under the direction of Georges Busson, which she began in 1983, and which has been integrated with the projects cited previously. Daniel Dumas, after an earlier study concerning the Paleogene evaporites of the Narbonne-Sigean area (diplôme d'etude approfondie, 1983), began a doctoral thesis concerning the Paleogene evaporites of the Valence basin at the University of Lyon in 1984. This work was completed in 1988 under the direction of Georges Truc and Georges Busson. Based on this thesis, Gilles Dromart graciously wrote the article for this volume concerning the Valence basin, with the aid and concurrence of Daniel Dumas.

Jean-Marie Rouchy, a researcher within the same study group within CNRS, has synthesized the structural and geological framework for the various basins in France that are the segments of the Paleogene rift. Marc Helman, a post-doctoral research fellow at Queens College (CUNY), has contributed to this introductory chapter by typifying the subsidence behavior of these various rift basins, based on the data and work provided by Jean-Marie Rouchy, Daniel Dumas, Alain Curial, and Marie-Madeleine Blanc-Valleron.

The same research groups under CNRS supported the studies of actual modern sedimentary environments in the marine salt ponds of the western Mediterranean, that have been the subject of two separate volumes (Busson et al., 1982; Ortí et al., 1984). Among these researchers, Dominique Geisler-Cussey simulaneously carried out studies of the modern settings and the Triassic evaporites of eastern France (1986, University of Nancy: in her thesis (thèse d'Etat), which was supervised by Jean Hilly and Georges Busson). Selected and significant results from this study are reported here and serve as the sedimentologic background to the Paleogene evaporites.

All of this work was dependant on the continuous financial support, over a period of twelve years, by the Centre National de la Recherche Scientifique (CNRS), and all of this research took place through their aid. In particular, we wish to acknowlege the support given by the General Delegatation of the Scientific and Technical Research Committee for the framework of the project entitled "Potential of Subsurface Resources" (assistance no. 77–7–1516 and 81—D–1100). The National Society of Elf-Aquitaine financed and aided very greatly the study of the modern evaporite environments from 1980–83 (contract #3845, "Actual Sedimentation and Diagenesis of Hypersaline Environments"). The Mines de Potasse d'Alsace supported the research of Marie-Madeleine Blanc-Valleron through a contract with the Louis Pasteur University of Strasbourg (1981–85) and the United Association #1209 has also supported her research project "The Study of the Potassic Evaporite Basin of Mulhouse (Southern Alsace)" since 1985.

The National Agency for Radioactive Waste Management has extended considerable aid through their support contract #1240 (1986–88) "Geologic and Well-Log Study of the Paleogene Salt Formations of Bresse" upon which two of the researchers were dependant for a period of two years. In addition to this, many other researchers benefited from the support of the laboratories in which they were located: The Geological Laboratory of the National Museum of Natural History (Marie-Madeleine Blanc-Valleron, Georges Busson, Jean-Marie Rouchy), the Laboratory of Sedimentary Facies Associations of the University of Nancy (Dominique Geisler-Cussey and Robert Moretto), the Department of Earth Sciences of the University Claude Bernard of Lyon (Alain Curial, Gilles Dromart, and Daniel Dumas).

In addition to the researchers mentioned above, other French researchers who were associated with this project, but on a less structured basis, also received significant support for their research given by CNRS in France, including Georges Truc (Lyon) and Denise Nury (Laboratory of Structural and Applied Geology, Aix-Marseilles).

Our research group has worked with the Laboratory of Mechanics of Solids of the Polytechnic School of Paris in multidisciplinary collaboration that was both amicable and fruitful. The work of the Organic Geochemical Laboratory of the French Institute of Petroleum (IFP) permitted aquisition of the analysis of the organic material utilized in our work, and this data is sythesized in this book. The collaboration of the specialists of Schlumberger and the ENSPM (Ecole Nationale Supérieure du Pétrole et des Moteurs de l'IFP) has permitted the detailed interpretation of both the sedimentologic and well log data.

The results presented in this volume are derived from data gained from subsurface core samples and records. Access to these subsurface records was aided considerably by the Service for Registry of Oilfield Reservoirs (Rueil-Malmaison) which functioned very efficiently and at no cost for academic research up until 1987, the year in which we put together these presentations. The Mines de Potasse d'Alsace have put at our disposition the numerous wells, cores, mines, and observational reports that have been accumulated over the past few decades. Gaz de France has lent us their cores, well-logs, drilling reports, collection of photographs and other records obtained in the areas of Etrez (Bresse) and Tersanne (Valence). We wish to thank the various societies and organizations such as Géostock, Rhône Poulenc, National Society of Elf-Aquitaine, Societé des Petroles de Valence (SPV), Compagnie Française des Pétroles Total, and the regional offices of Bureau de Recherches Géologiques et Minieres (BRGM) as well as the city hall of the City of Valence which provided the many logs, cores, and seismic profiles for this academic research.

In January 1987, Georges Busson invited B. Charlotte Schreiber to examine the results of the studies from these evaporitic basins of France. After studying these documents, and with the concurrance of the members of the working group, she concluded that it was the appropriate time to create a synthesis of this work. Thus began a prolonged period of work in which she needed to remain in Europe for extended spans of time in 1987, 1988, and 1989. In order to carry out this work she received financial support from the National Museum of Natural History (as associate professor in 1987 and 1989) and from NATO in 1988. She has also been supported in this work by the Research Foundation of the City of New York. Facilities for editing this book have been provided by Oxford University (1990) and by Lamont-Doherty Earth Observatory (Columbia University) from 1987.

Sedimentary Deposition in Rift and Foreland Basins in France and Spain

(Paleogene and Lower Neogene)

PART 1

Introduction

GEORGES BUSSON AND B. CHARLOTTE SCHREIBER

The sedimentary filling in the numerous Paleogene basins of southeastern France and northern Spain is characterized by the presence of thick depositional layers that have formed under diverse but confined conditions. Most of these basins were strongly evaporative and some of the basins in France and Spain—those that subsided very rapidly—have been filled by salt (mostly halite but also potassic salts). In France these salt areas are typically the segments of the Western European rift that sundered the European continent from the North Sea to the Mediterranean during the Paleogene. Numerous other basins in France received thinner, less saline, deposits and were entirely contemporaneous with the deeper basins also formed during the upper Ludien and Rupelian (Stampian), and were probably also related to the general period of rift tectonics. In the Ebro Basin of northern Spain the evaporite basins, formed by Pyrenean crustal loading, were longer-lived, persisting from the base of the Cenozoic until the Miocene. They formed in the region just to the south of the Pyrenees that was strongly affected by subsidence created in response to the advance of the thrust sheets that moved from north to south (i.e., crustal loading).

These various confined sedimentary basins in both France and Spain were spread over great distances and developed within paleogeographic and structural contexts that were very different from one another, some in association with ancient massifs (the Vosges, the Black Forest, the Massif Central, and Castille), others either in conjunction with the evolution of the embryonic mountain chains (the Alps and Pyrenees), or in relation to the distensive birth of the Mediterranean. All evolved within an increasingly arid climate and were exquisitely sensitive to both local and worldwide variation in sea level.

These basins are interesting for the depositional sequences that formed within them because, within a small area, a multiplicity of sedimentary facies can be found. These Paleogene basins are often nearly undeformed and have not been deeply buried; in most cases the sediments have not been significantly altered and are well-preserved.

Generally speaking, these buried sediments (evaporites, evaporative carbonates and argillites, and intercalated layers of organic matter) are those for which modern analogs are not readily available, hence their interpretation is somewhat difficult and can be controversial. The well-preserved segments of the intracontinental rifts examined in this study are of great economic importance and offer an ideal setting for the study of synrift sedimentation. Additionally, in many instances these sediments are good source rocks and may have significant hydrocarbon potential.

The Quality of the Available Data

Because of the economic potential of these basins, or of parts of them, copious data of good quality are required for evaluation and development: gravity maps, seismic profiles, and borehole data (logs and cores). For this reason, abundant drilling has been carried out, above all specialized boreholes for the investigation of many of the evaporites: boreholes for the discovery and mapping of potassium salts, for the dissolution of halite, and for the creation of subterranean cavities for liquid hydrocarbon storage are among the special research projects from which we have drawn our data. The boreholes especially made for the study of the evaporites were the main object of the logs adapted for these facies. They are devices of exceptional efficacy in the determination of the mineralogical, petrographic, and geochemical characteristics of the sections. Data sufficient to establish the sedimentologic evolution of the basins that are both continuous in time and space are based on information collected within each borehole, and from one borehole to the other. But the abundance of cores constitutes the most remarkable and exceptional record of these salt-bearing basins. Also, the basins of Haute Alsace, Catalonia, and Navarre have revealed a precious wealth of information in the vast amount of data collected from drilling for mines and their exploitation.

The results presented here are from French, Spanish, and Anglo-American study groups and rely on research that has been carried out on diverse scales of investigation and as multidisciplinary studies. We think that the results demonstrate that the interest in the depositional products produced within modern confined sedimentary environments required intensive study by several groups simultaneously. The sedimentary rocks that filled the Paleogene rift basins also require multidisciplinary study. Use of seismic reflection records, borehole logs, and the geochemical, petrographic, and sedimentary analysis that now exist in great abundance are so detailed that they often can be extended as far as the fine scale of nanofacies. This investigation concludes with an effort of synthesis and reflection that systematically reaches as far as the scale of the basin and the paleogeographic context of the basinal assemblage.

Presentation

In the first paper Geisler-Cussey reviews modern evaporites that may be either marine or continental in origin, setting the stage for the diverse sedimentary facies to follow. The next paper by Rouchy attempts to set the tectonic stage for the entire study presented in this book, particularly for the French Paleogene. This is followed by papers covering the major Paleogene evaporite areas of France, Haut-Alsace (Blanc-Valleron and Schuller), Bourg-en-Bresse (Curial and Moretto), Valence (Dromart and Dumas), and Southern Provence (Nury and Schreiber).

The discussion of the Paleocene to Miocene evaporative deposits of Spain begins with a detailed structural introduction (Santanach) and a detailed stratigraphic and sedimentologic overview of the entire sequence (Ortí). The eight following papers deal with various areas of the Ebro Basin in order, through time: the first continental evaporitic phase in the central area (Tremp gypsum, upper Paleocene) (García Veigas); the first marine evaporitic phase in the south Pyrenean foredeep (Beuda gypsum—middle Eocene autochthonous-allochthonous zone) (Ortí and Rosell); the second marine evaporitic phase in the south-Pyrenean foredeep (the Priabonian potash basin (upper Eocene; autochthonous-allochthonous zone)) (Rosell and Pueyo); lacustrine evaporitic deposits in the south Pyrenean foreland basin (Barbastro and Puente La Reina gypsum—Eocene-Oligocene) (Sáez and Salvany); continental evaporitic sedimentation along the SE border of the Ebro Basin (Catalanides mountains) (Paleocene—lower Oligocene) (Ortí); continental evaporitic sedimentation in Navarra during the Oligocene to lower Miocene (Falces and Lerín formations) (Salvany); Glauberite deposits of the Lerín formation (upper Oligocene: Alcandre zone, LaRioja-Navarra) (Salvany and Ortí); continental sedimentation in the Miocene (Ortí and Salvany).

The volume ends with a summary and general overview on what has been learned, presented by Busson.

1

Modern Depositional Facies Developed in Evaporative Environments (Marine, Mixed, and Nonmarine)

DOMINIQUE GEISLER-CUSSEY

Overview

The aim of this work is to present some specifically useful examples of modern evaporite sedimentation that will help in the understanding of the Paleogene deposits outlined in this study. Marine, marine-marginal, and nonmarine environments (table 1.1) all have to be considered in order to develop the concepts of what happened sedimentologically in terms of deposition in both the center and at the edges of these evaporative basins. In addition there are already a number of general sedimentological reviews in the literature concerning the deposition of modern evaporative facies (Busson et al. 1982; Kendall 1984; Ortí, Busson et al. 1984; Schreiber 1978, 1986, 1988). While it is evident that there are additional microfacies formed in deeper water (Lowenstein and Hardie 1985), modern analogs are not well studied. Additionally, efflorescent salt crusts are also known, but are only occasionally preserved (Smoot and Castens-Seidell 1994).

Marine subaqueous sedimentation that takes place in hypersaline water bodies fed from the sea has been known for thousands of years. The ancient Egyptians, Greeks, and Romans controlled and harvested salt in these environments and undoubtedly earlier civilizations did the same. Modern hydrochemical studies in these settings were first made by Usiglio (1849) who recognized the similarity between the natural accumulation of sulfates and halite and ancient deposits; he sought to duplicate and quantify this process to create a working model for the understanding of the various salt and potash deposits of France and Germany. Only recently, however, have general studies of these saline environments been made from the actual viewpoints of the chemical, biological, and sedimentological character of the deposits. The marine-fed depositional examples described here (table 1.1), are located on the western coast of the Mediterranean (figure 1.1). Special emphasis is given to these marine subaqueous evaporative environments as analogs throughout this volume.

The sedimentologic framework of the Paleogene deposits presented here is within a rather siliciclastic matrix, and the evaporative sedimentation on the sandy coast of Baja California (Mexico) seems to be an even more appropriate modern depositional example. However, marine-marginal sedimentation is well known in the sabkha environment developed on the largely carbonate substratum of the Trucial Coast in the Persian Gulf (Shearman 1966; Kinsman 1966, 1969; Butler 1970; Purser 1973).

In this volume the saline lakes of Spain are presented as a suitable model of nonmarine sedimenta-

TABLE 1.1.

Examples, with references, used in this section to illustrate marine subaqueous, mixed, and continental evaporitic environments.

Water Sources	Environments	Examples	Hydrochemistry	Deposits
Marine subaqueous	Controlled salt ponds	Secovlje (Yugoslavia) Salin-de-Giraud (France) Santa Pola (Spain)	Herrmann et al. 1973 Landry and Jaccard 1982 Gouleau 1982 Landry and Jaccard 1984 Gouleau 1984	Geisler 1982-1986 Thomas 1982 Ortí, Pueyo, Geisler-Cussey and Dulau 1984
Mixed	Marine marginal	Baja California(Mexico)	Pierre 1982	Holser 1966 Pierre 1982-1983-1984 Lowenstein + Hardie 1985
Continental	Lakes	Zaragossa-Lerida (Spain)	Pueyo 1978-79 Pueyo and Ingles-Urpinell 1986	Pueyo 1978-79 Pueyo and Ingles - Urpinell 1986
	Rifts	Baringo and Bogoria (Kenya)	Tiercelin et al. 1986 Renaut + Tiercelin 1994	Tiercelin et al. 1986 Renaut + Tiercelin 1994

TABLE 1.2.

Summary of the main hydrochemical and sedimentological characteristics of the marine, mixed, and continental evaporitic environments presented in this paper.

ENVIRONMENTS			Marine	Mixed	Continental	
MODELS			Controlled salt ponds	Marine marginal	Saline lake	Rift lake
SUBSTRATUM			Variable	Siliciclastic + gypsum-halite	Evaporitic	Volcanic
HYDROGEOCHEMISTRY			Sea water: Cl^-, SO_4^{--}, Na^+, K^+, Ca^{++}, Mg^{++}	Cl^-, SO_4^{--}, Na^+, Ca^{++} enriched seawater	Cl^-; SO_4^{--}; Na^+, Mg^{++}	CO_3^{--}; $CO_3^-H^-$; Na^+
MINERALS	Aragonite	$CaCO_3$	■			
	Calcite	$CaCO_3$	■			■
	Gypsum	$Ca\,SO_4 \cdot 2H_2O$	■	■	■	
	Halite	$NaCl$	■	■	■	
	Dolomite	$Ca.Mg\,(CO_3)$		■		
	Glauberite	$Na_2Ca\,(SO_4)$			■	
	Mirabilite	$Na_2SO_4 \cdot 10H_2O$		■		
	Magnesite	$Mg\,CO_3$		■		
	Celestite	$Sr\,SO_4$			■	
	Polyhalite	$K_2MgCa\,(SO_4) \cdot 2H_2O$			■	
	Thenardite	Na_2SO_4			■	
	Bloedite Na	$Mg(SO_4) \cdot 4H_2O$			■	
	Gaylussite	$Na_2\,Ca(CO_3)_2 \cdot 3H_2O$				■
	Trona	$Na_2H\,(CO_3)_2 \cdot 2H_2O$				■
	Nacohlite	$Na\,H\,CO_3$				■
	Magadiite	$Na\,Si_7O_3(OH)_3 \cdot 3H_2O$				■
SEDIMENTARY STRUCTURES	Laminae		■			
	Layers		■	■		
	Efflorescence crusts				■	■
	Nodules		■	■	■	■
	Scattered crystals			■	■	■
	Concretions					■
ORGANISMS	Cyanobacteria		■	■	■	■
	Diatoms		■			■

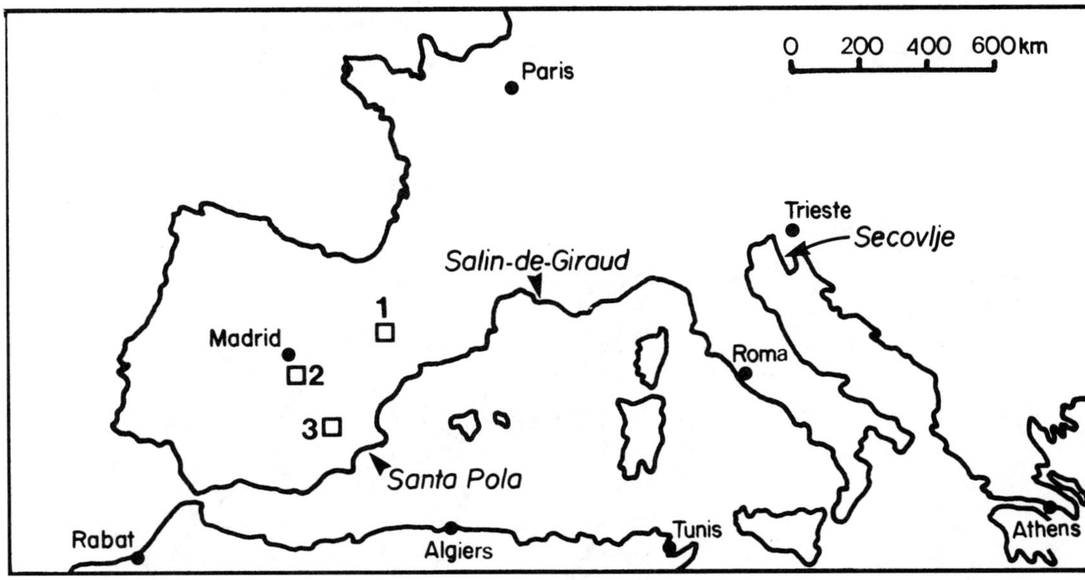

Fig. 1.1. Location map of solar salt works on the western Mediterranean Coast (Secovlje, Croatia; Salin-de-Giraud, France; Santa Pola, Spain) and of areas with saline lakes in Spain presented in this study (1 = Zaragossa-Lerida; 2 and 3 = La Mancha).

tion. Because continental ground waters feeding the lakes originate from the weathering of an assemblage of Triassic evaporites, they are mineralized. Similar outcrops also were present during the Paleogene and the deposition of those older evaporites originated from evaporated meteoric ground waters flowing into those basins. Of course, consideration of the rift valley framework of the Paleogene evaporites may also logically include a consideration of the sedimentation in saline lakes of the east African rift system. Other nonmarine sedimentation in continental lakes has been discussed by Eugster and Hardie (1978); Hardie, Smoot, and Eugster (1978).

For each of the modern evaporative settings outlined in this review, the behavior of usual major elements (Ca^{+2}, SO_4^{-2}, Na^+, Cl^-, Mg^{+2}, K^+) in the brines will be given first so as to follow the changing composition of water during a normal course of evaporation. The analytical results generally are taken from and apply to the surficial brine body, but also, when data are available, information is given about the interstitial brines to reach a better understanding of the chemical framework of early diagenetic processes.

Special attention will be given to the behavior of Mg^{+2} leading to dolomitization and to genesis of magnesium-rich clays in the first stages of diagenesis. Some consideration will be made of minor elements such as bromine, which is an interesting salinity marker, and silica, because of the almost permanent relationship between siliceous diagenetic features and fossil evaporites.

Marine Subaqueous Environments: Controlled Salt Pond Model

The climate on the western coast of the Mediterranean sea is characterized by a low annual average rainfall (500 to 300 mm) and a high evaporation rate (around 2000 mm) causing a significant net water loss and therefore generating hypersaline brines throughout the year in the south of Spain or just during summer in the northern parts (Perthuisot 1982; Ortí, Pueyo, and Truc 1984). This favorable climatic background explains the common presence of numerous salt ponds in this area (controlled commercial lakes).

The solar salt works of Secovlje near Portoroz in Croatia, Salin-de-Giraud on the Rhône delta in southern France, and Santa Pola, south of Alicante

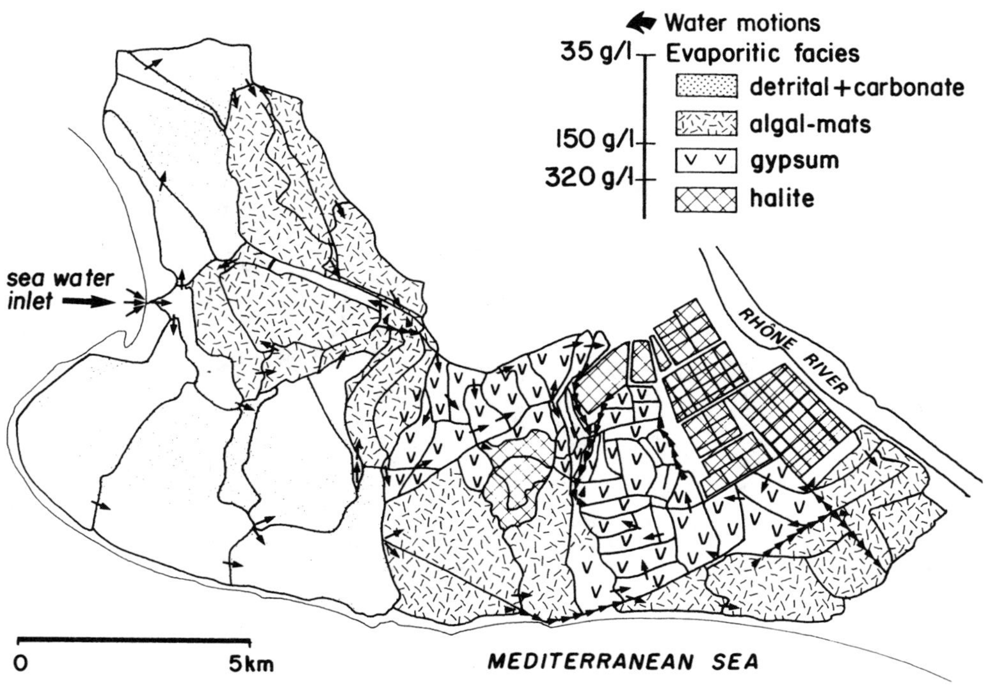

FIG. 1.2. Map of the solar salt works of Salin-de-Giraud, France, showing the water flow pattern and the distribution of evaporitic facies with increasing salinity (after Perthuisot 1982 and Geisler-Cussey 1986).

in Spain, will be considered in this study (figure 1.1).

HYDROCHEMISTRY OF MAJOR ELEMENTS

The chemistry of modern sea water and brines was first studied in a systematic way in the solar salt works of Secovlje by Herrmann et al. (1973). Many authors use these data as a reference to illustrate the behavior of major elements during sea water evaporation. To confirm this chemical evolution during evaporation the salt works of Salin-de-Giraud and Santa Pola have recently been studied from a hydrochemical viewpoint that considered the surficial brines and also the pore brines (Landry and Jaccard 1982, 1984; Gouleau 1982, 1984).

In the salina of Salin-de-Giraud the water is pumped into the salt works directly from the sea and flows over long distances (about 50 km) from one pond to the next, to allow maximum evaporation before reaching the ultimate ponds in which the salt crystallization process takes place (figure 1.2). During this transit the salinity of the brines is controlled, either remaining constant in each specific pond throughout the evaporation period during spring and summer (the ponds being emptied in winter), or undergoing seasonal fluctuations when remaining inundated all year long.

In the salinas of Santa Pola, the ponds are not emptied during winter. The sea water pumped into the salt works also flows over some distance and through several concentrating ponds until it reaches saturation with respect to halite (figure 1.3). In some instances high salinity brines are reinjected in the distribution network generating anomalous circuits, in order to provide special salts, but these circuits will not be taken into account here.

1—Surficial brines.

The behavior of the usual major elements is very similar in all salt works observed, including those from Secovlje (Herrmann et al. 1973), Salin-de-

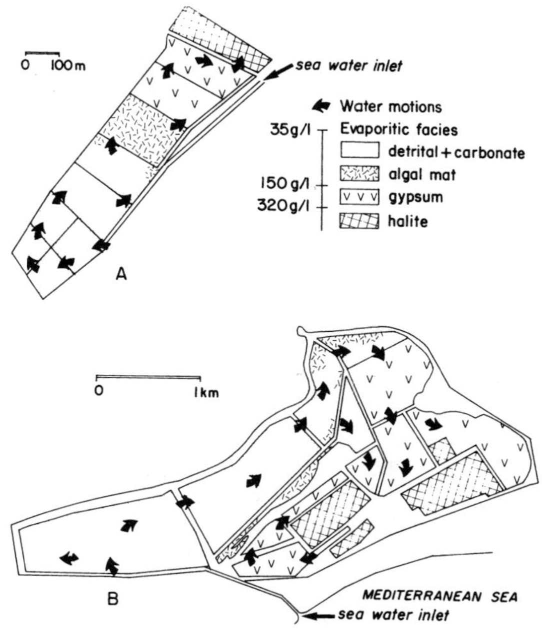

FIG. 1.3. Map of the two solar salt works of Santa Pola, Spain: Bonmati (A) and Bras del Port (B), showing the water flow pattern and the distribution of evaporitic facies with increasing salinity (after Geisler-Cussey 1986).

Giraud (Landry and Jaccard 1982) and Santa Pola (Landry and Jaccard 1984) and the results will be presented together in this review (figure 1.4).

(a) Magnesium shows a linear increase throughout the whole evaporation process but with a higher slope, after halite precipitation, because of a marked removal of sodium and chlorine from the brine and therefore a higher relative concentration of the remaining elements. This evolution in concentration is in agreement with the fact that no magnesium-bearing mineral forms until after halite precipitation, and thus gives magnesium the characteristic of being a good salinity marker in the evaporation process of sea water.

(b) Potassium increases along with sea water evaporation and shows the same trends in concentration as magnesium.

(c) Calcium first increases until reaching the first gypsum occurrence and afterwards decreases due to gypsum precipitation, which utilizes the available Ca^{+2} ions of the brines. Chemical precipitation of calcium carbonate seems very limited in these environments and does not affect the calcium content of the brines to any appreciable degree. The curve illustrating the calcium content of the brine is however the only one showing a notable scatter in the plotted analysis. Such statistical spread probably is related to the dissolution of some biodetrital carbonate particles such as shell fragments of *Cardium* in the first ponds after the entrance of sea water, and also to early gypsum destruction by bacteria causing localized enrichments in calcium of the brines.

(d) Sulfate shows a linear increase with increasing salinity. The slope of the curve is relatively constant until reaching halite precipitation and then becomes steeper, as already seen in the case of magnesium and calcium (because of massive halite precipitation). In fact, the large excess of SO_4^{-2} ions in relation to the available Ca^{+2} ions leads only to a slight flattening of the sulfate curve in the area of gypsum precipitation because only a portion of the available SO_4^{-2} ions is actually incorporated within the gypsum.

(e) Sodium displays a linear increase with increasing salinity until the first halite precipitation. Afterwards the sodium curve shows a straight decrease because of the massive incorporation of Na^+ ions into the halite crystal lattice.

(f) Chlorine shows the same linear increase as the sodium plot until halite precipitation (in which Cl^- ions are of course involved) leading thereafter to a rather constant chlorine content in the brine. The elevated post-halite chlorinity is linked to a slight

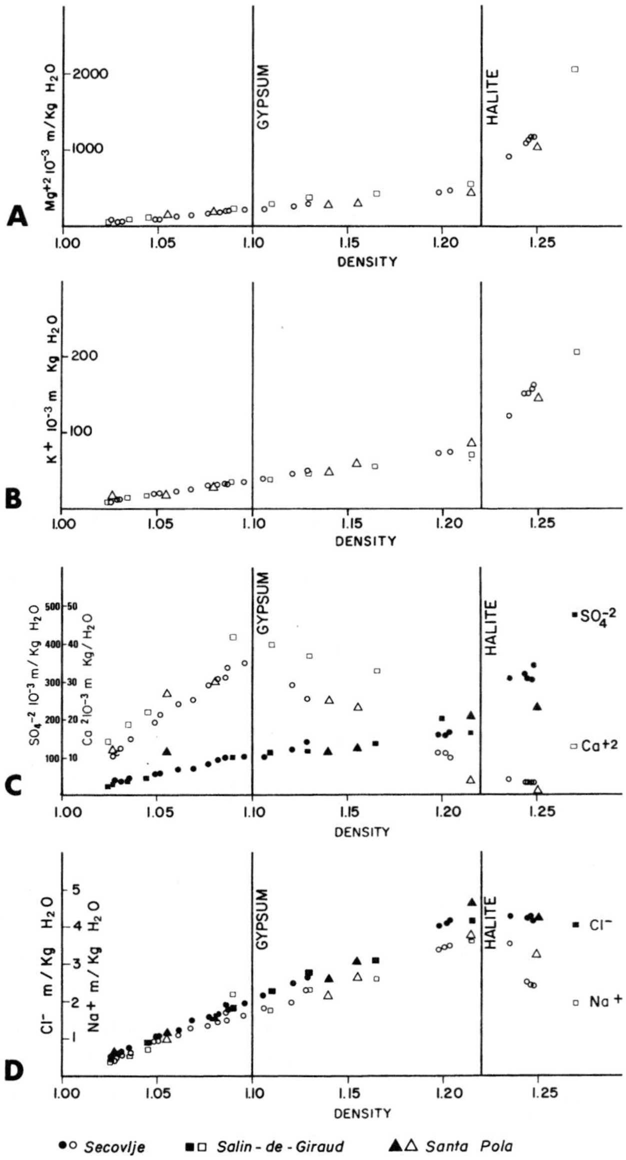

FIG. 1.4. Changes in major elements in the surficial brines with increasing density of three solar salt works located on the western Mediterranean coast: Secovlje, Croatia (after Herrmann et al. 1973); Salin-de-Giraud, France (after Landry and Jaccard 1982); Santa Pola, Spain (after Landry and Jaccard 1984).

excess of Cl^- ions in relation to Na^+ ions within the original water.

In the three salt works considered in this study the brines show a progressive salinity increase as a function of progressive sea water evaporation from the entrance of normal sea water to the salt pans where halite precipitation takes place. These natural but controlled environments show very similar behavior patterns for the major usual elements; however the harvesting of salt each year avoids prolonged contact between halite deposits and residual brines, preventing salt recycling and the study of its effects on the "back reactions" more usual in nature.

In general the solute behavior in concentrating sea water follows several different pathways. In the considered salinity range, Mg^{+2} and K^+ are conservative elements used as concentration indicators. SO_4^{-2} and Ca^{+2}, which combine in the precipitation of gypsum, and Cl^- and Na^+, precipitating as halite, show the same fractionation mechanisms because of mineral precipitation.

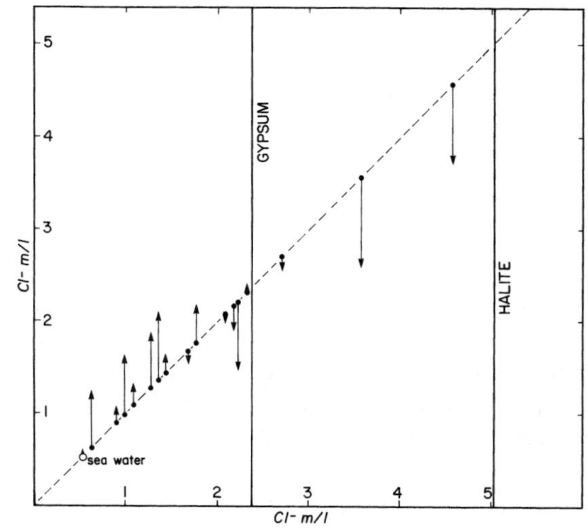

FIG. 1.5. Chlorine content in the interstitial brines versus chlorine content in the surficial brines of the pond studied (in 1980 and 1981) in the solar salt works of Salin-de-Giraud, France (modified from Gouleau 1982). Plotted values correspond to the chlorine content in the surficial brines; arrows indicate relative increase and decrease with depth (down to 10 cm) of the chlorine content in the interstitial brines.

2—Interstitial brines.

Data concerning pore brines of sediments from the salinas are only available from the salt works of Salin-de-Giraud (Gouleau 1982) and Santa Pola (Gouleau 1984).

Gouleau has thus afforded us an insight into the differences between the surface water and the pore water.

Salin-de-Giraud. Analysis of the pore brines within the sediments allows the identification of two different behavioral trends within the major pore-water elements; the evolution of chlorine content is given as an example (figure 1.5), but the general trend is the same for the other elements. In the first ponds, before gypsum precipitation, the major element content within the interstitial brines increases with depth. The reverse evolution, i.e., a general decrease of content with depth, is observed in the ponds in which gypsum precipitation takes place.

This peculiar behavior of major elements in the interstitial brines may be explained based on the way the salt works is operated. Concentrated brines are stored in ponds located close to the entrance of sea water through the winter period. These dense, most evaporated brines difuse downward and become pervasive throughout the existing sediment. This salinity trend is preserved during summer because of the introduction of the less concentrated superficial brines, which have a lower density and do not displace the winter storage brines. On the contrary, the ponds precipitating gypsum in summer are emptied in winter, undergo winter emersion and therefore are subjected to meteoric waters that produce a salinity decrease with depth into the sediment, which then is preserved during summer.

The behavior of major elements in the interstitial brines at Salin-de-Giraud emphasizes the importance of winter emersion that controls the composition of pore brines. More surprisingly, there are no traces of early diagenesis in this rather unstable chemical system.

Santa Pola. In the Santa Pola salt works, the ponds are filled with brines all year long. In the upper 50 centimeters of sediments, including part of the pre-evaporite substratum, the interstitial brines show, for example, a chlorine content very close to the content analyzed in superficial brines (figure 1.6). The same general behavior pattern is observed for most of the other major elements, indicating very few fluctuations in the brines with depth and therefore a general stability of the chemical system in space and time, with almost the same analytical results for the samples taken during two field studies (1981 and 1983). Some fresh water may mix with

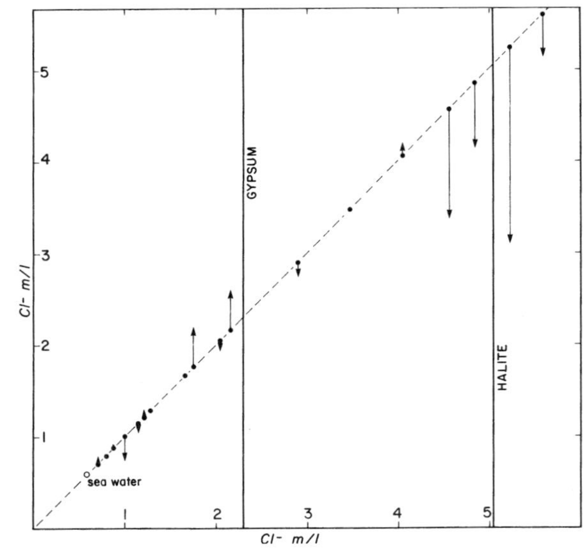

FIG. 1.6. Chlorine content in the interstitial brines versus chlorine content in the surficial brines of the ponds studied (in 1981 and 1983) in the solar salt works of Santa Pola, Spain (modified from Gouleau 1984). Plotted values correspond to the chlorine content in the surficial brines; arrows indicate relative increase and decrease with depth (down to 50 cm) of the chlorine content in the interstitial brines.

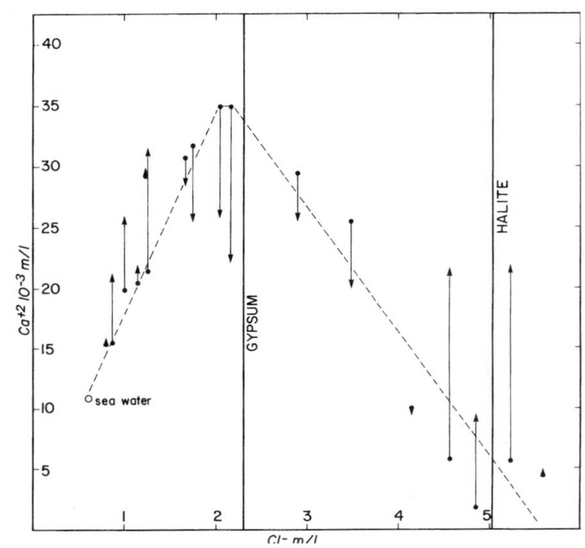

FIG. 1.7. Calcium content in the interstitial brines versus chlorine content in the surficial brines of the ponds studied (in 1981 and 1983) in the solar salt works of Santa Pola, Spain (modified from Gouleau 1984). Plotted values correspond to the calcium content in the surficial brines; arrows indicate relative increase and decrease with depth (down to 50 cm) of calcium content in the interstitial brines.

the brines on the edges of the salt works, introducing locally interstitial dilute brine, without any effect on the salinity of surficial brines. This does happen in the salt ponds considered here.

Two of the major elements, Mg^{+2} and Ca^{+2}, show some variations from the behavior of the other elements. The Mg^{+2} content in the interstitial brines increases slightly with depth when algal mats start to develop within the buried sediments, because of the dissolution of Mg^{+2}-rich calcite located in the substratum below. The behavior of Ca^{+2} is more complex (figure 1.7). The content of Ca^{+2} increases with depth due to dissolution of carbonates of the substratum in the first ponds, close to the entrance of sea water, but in subsequent ponds there is a decrease with depth. The latter is the result of early diagenetic processes which first generate thin layers of calcite; after reaching higher concentrations of Ca^{+2}, gypsum is formed as nodules composed of tiny lenticular crystals.

The Santa Pola salt works are a good example from which to follow the various geochemical pathways which lead to early diagenetic processes that develop in a subaqueous marine environment.

The modern subaqueous marine environments presented in this review form within relatively shal-

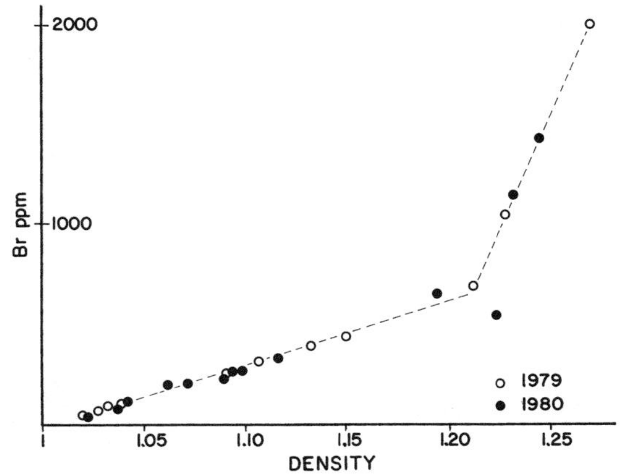

FIG. 1.8. Bromine content versus density in the surficial brines of the solar salt works of Salin-de-Giraud, France (after Geisler 1982).

low water bodies (some decimeters to meters in depth) and generally have no observed brine stratification and limited interaction with the underlying sediment. However, they do demonstrate that the evolution of the ionic content and concentration determines the precipitation of the main evaporative minerals (carbonate, gypsum, halite) as observed in fossil evaporites. This evolution is typical of a nearly pure marine environment.

CHEMISTRY OF MINOR ELEMENTS

1—Bromine.

The behavior of bromine during sea water evaporation is well known through the experimental work of German researchers (Boeke 1908; Braitsch and Herrmann 1963). Bromine does not precipitate as a mineral species during normal evaporation of sea water; but its content shows a progressive linear increase and the curve becomes steeper after halite precipitation takes place. This behavior is very clearly demonstrated with the data obtained in the salt works of Salin-de-Giraud (figure 1.8) (Geisler 1982), and is very similar to that observed for magnesium and potassium in the same settings. Therefore bromine is also an excellent indicator of salinity and salinity fluctuations. However, bromine is incorporated into the crystalline lattice of halite and also of other chlorides as a substitute for chlorine. This feature permits calculation of a distribution coefficient corresponding to the ratio of bromine content in the crystal to bromine content in the brines. For halite the experimental value of the distribution coefficient is 0.14; if the brine is magnesium-rich, the value is lowered to 0.07 (Braitsch and Herrmann 1963). The values obtained in the salt works of Secovlje (Herrmann et al. 1973), Salin-de-Giraud (Geisler 1982), and Aigues-Mortes (Sichere 1976) in southern France average around 0.12, which is in very good agreement with experimental data. This very low value for the distribution coefficient does not significantly affect the increase of bromine in evaporating sea water.

If fluid inclusions are abundant in halite crystals, the bromine content will be elevated because of the high bromine content in the trapped brines rather than in the content in the crystalline lattice itself. Therefore, the analytical results have to be considered critically and supported by careful petrographic study.

The first halite precipitating from sea water evaporation shows an average bromine content of 70 ppm. Lowered or even very low bromine values are the result of halite recycling by fresh water (Holser 1966a). However the recycling of potassium-bearing salts with a higher bromine content may lead to halite having an elevated bromine content (Hardie 1984). It follows from this that absolute values for bromine content, as a function of the bromine content of the first halite precipitation, are sometimes difficult to interpret, depending on the type of water. Fresh, marine, more or less concentrated brines that recycle halite or other salt deposits are very significant. However the variations in

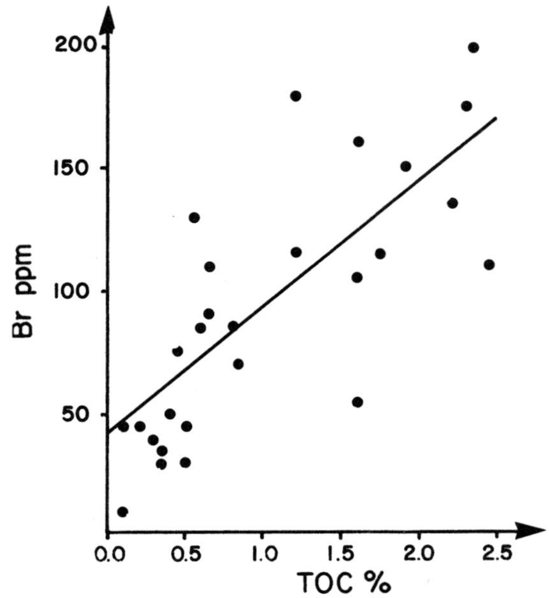

Fig. 1.9. Bromine content versus total organic carbon (TOC) content from bulk sediment analysis of samples from the solar salt works of Salin-de-Giraud, France (after Thomas 1982). Correlation coefficient = 0.764.

bromine content in any specific depositional sequence may be considered as good indicators of relative salinity fluctuations.

Another significant feature is that bromine and total organic carbon contents show a close correlation in the salt works of Salin-de-Giraud (figure 1.9) (M. Thomas 1982). Bromine is probably complexed with biopolymers in the sediment and constitutes a kind of ionic reservoir available for future chemical exchange during diagenesis.

2—Magnesium.

Magnesium-bearing clays such as smectite or interlayered illite-smectite show a close correlation with total organic carbon content in the salt works of Salin-de-Giraud (figure 1.10) (M. Thomas 1982). Moreover, experimental work on algal mat layers shows a 3- to 4-fold increase in magnesium content within the algal-rich layers compared to that in the surrounding brine, suggesting the existence of a possible magnesium complex within the organic matter itself (Gebelein et al. 1973). In this case the organic matter again acts as a reservoir for ions that will react later during early diagenesis with the host sediment. The released magnesium may either be adsorbed by clays that develop interlayer gaps in hypersaline environments or may be substituted for calcium in carbonates. Magnesium-rich clays seem more linked to the presence of organic matter than to the highest salinities. The formation of magnesium complexes with organic matter leads to a layer-by-layer diagenesis in older deposits, causing the formation of thin, laminated rocks that consist of alternating magnesium-rich and other clay layers or dolomite and other carbonate layers, depending on the nature of the host sediment.

3—Silica.

The silica content in surficial and interstitial brines has been studied in the salt works of Salin-de-Giraud (Gouleau 1982; Gouleau and Noël 1984) and Santa Pola (Gouleau 1984) and the behavior of silica is very different in each salt work considered.

Salin-de-Giraud. In the surficial brines the silica content is constant, around 10 mmoles/l, until the point of gypsum precipitation when it increases as a function of sea water evaporation (figure 1.11). In the pond areas before gypsum precipitation, the interstitial brines show a marked increase in silica content with depth, sometimes rising to over 450 mmoles/l, after which the silica has a rather constant content, close to that measured in surficial brines.

The behavior of silica after gypsum precipitation

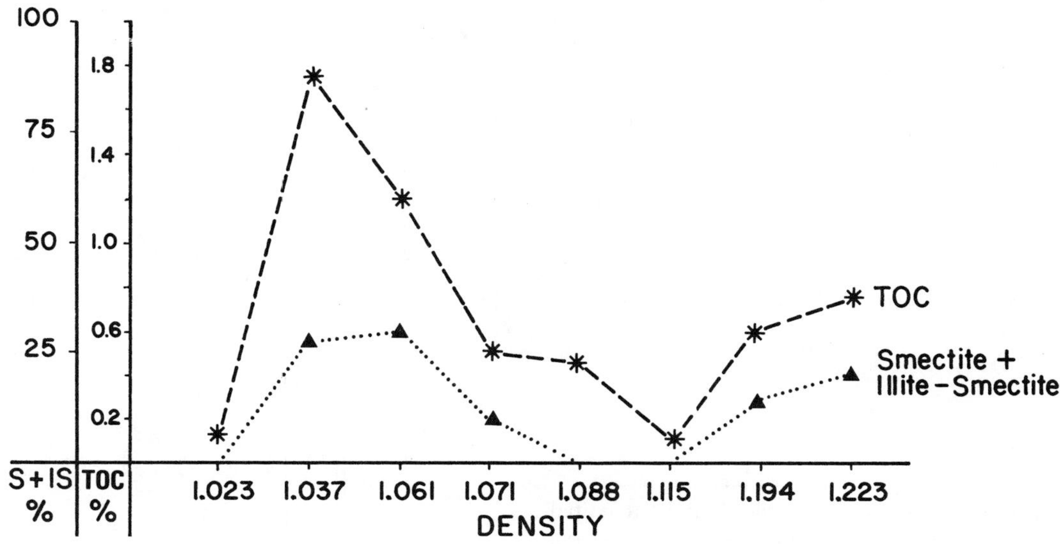

FIG. 1.10. Total organic carbon (TOC) content and clay (smectite, illite-smectite) content versus density from bulk sediment analysis of surficial samples from the solar salt works of Salin-de-Giraud, France (after Thomas 1982).

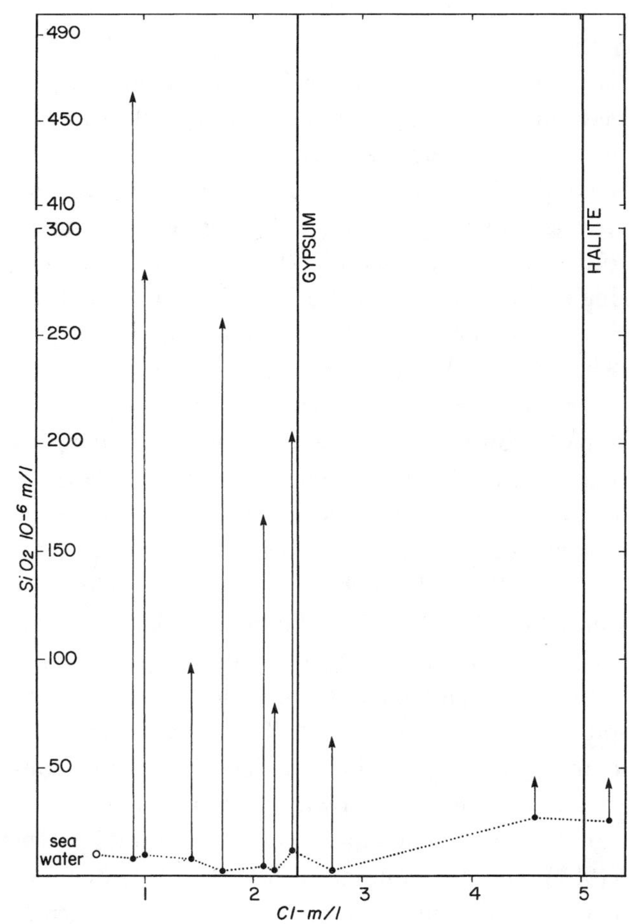

FIG. 1.11. Silica content in the surficial and interstitial brines versus chlorine content in the surficial brines of the ponds studied (in 1980 and 1981) in the solar salt works of Salin-de-Giraud, France (modified from Gouleau 1982). Plotted values correspond to the silica content in the surficial brines. Arrows indicate relative increase with depth (down to 10 cm) of silica content in the interstitial brines.

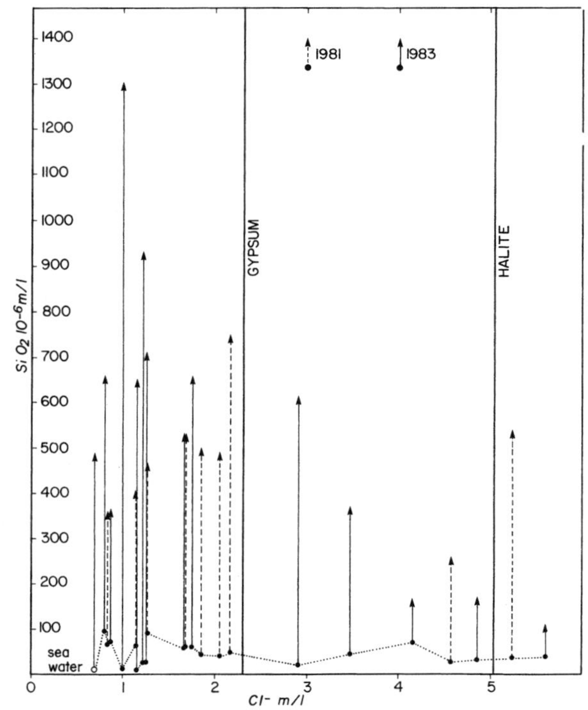

FIG. 1.12. Silica content in the surficial and interstitial brines versus chlorine content in the surficial brines of the ponds studied (in 1981 and 1983) in the solar salt works of Santa Pola, Spain (modified from Gouleau 1984). Plotted values correspond to the silica content in the surficial brines. Arrows indicate relative increase with depth (down to 50 cm) of silica content in the interstitial brines.

is similar to that of ions such as magnesium or potassium which show a simple increase with sea water evaporation. Before gypsum precipitation, the silica content in the surficial brines is very low and kept constant by the growth and development of numerous diatoms that use all the available silica. The high silica content in the interstitial brines is the result of rapid diatom dissolution with depth. The silica-enriched interstitial brines recycle back to the surface and bring new silica supplies that are rapidly consumed by new diatoms. Such silica enrichment of the interstitial brines may explain the early siliceous diagenesis very often encountered in association with fossil evaporites.

Santa Pola. The surficial brines show fluctuating silica content in the ponds before gypsum precipitation, dependent on the sampling season, and a rather constant content (40 mmoles/l) after gypsum precipitation (figure 1.12). The fluctuating silica content in the water of the first ponds is mainly controlled by diatom growth. For instance, when sampled after maximum phytoplanktonic bloom (1981) the silica content in the brines was high because dissolution of diatoms had already occurred. On the other hand, samples taken from the brines during phytoplankton bloom (1983) showed lower silica content because this period coincided with maximum use and most available silica was stored in the diatoms tests. After reaching gypsum saturation the silica content stays almost constant, with a slight decrease when salinity increases, indicating no direct salinity control.

In the interstitial brines, beginning at the top of the sediment, the silica content increases notably in comparison with the surficial brines (figure 1.12). On the edges of the salt works where meteoric groundwaters seep through the host sediment the silica content becomes very high, up to 1600 mmoles/l. In the areas of highest salinity in the surficial brines (150 to 200 g/l) the rate of silica enrichment in the pore brines is rather low and in areas of moderate salinity it fluctuates without any dependence upon salinity. The uncertain fluctuations of silica content in the interstitial brines is probably the result of sedimentological control related to silicate diagenesis that is quite common in evaporative sediments.

These salt works illustrate the behavior of silica

in two different frameworks. In Salin-de-Giraud the silica content is controlled by organic activity and salinity, probably because of the period of winter emersion that interrupts the natural cycle.

In Santa Pola the behavior of silica is more complicated and difficult to explain because the brines are always present in the host sediment and react with it.

EVAPORATIVE FACIES

An analysis of the sedimentation within the salt works of Salin-de-Giraud (Geisler 1982; Geisler-Cussey 1986) and of Santa Pola (Ortí, Pueyo, Geisler-Cussey, and Dulau 1984) reveals the depositional mechanisms that are taking place in these hypersaline, marine subaqueous environments.

The different facies encountered present a lateral distribution linked to increasing salinity from the sea water entrance to the salt ponds. The first deposits of these marine-fed settings are carbonates, followed by nearly pure organic accumulations, gypsum, halite, and finally the evolution of "mother liquors" which have very high densities. Because these high density waters are not conserved in the solar salt works and are discarded or otherwise processed for the extraction of useful elements, most of the possible potassic, magnesium, and other salts are never observed.

1—Carbonate deposition.

The deposition of carbonates has been observed in the Santa Pola salt ponds in the salinity range between normal sea water (35 g/l) and 80–85 g/l. Carbonates occur first as a soft unstructured mud and, where the water reaches higher salinities, as very thin laminae generally associated with algal mats.

Carbonate mud with a peloidal texture occurs in the first ponds of the salina where sedimentation develops as a soft calcareous mud directly overlying the clayey substratum. This sediment accumulates at about 1 cm/year. Magnesian calcite and aragonite are present in the muds in subordinate amounts. SEM observation of the mud shows the presence of poorly differentiated crystals of all three carbonates.

This mud contains a network of flexible organic tubes arranged vertically in the micritic sediment. This mixture covers the bottom of the first ponds of the salina system. Kironome larvae frequently occupy the interior of these tubes, which they probably produce through a mucoid secretion. The larvae ingest the carbonate mud and are responsible for the ultimate peloidal texture of the mud. Fecal pellets may be observed either as isolated components (150 to 350 mm) or as aggregates (up to 500 mm) (figure 1.13A).

Bioturbation produced by birds and fish is also an important factor in the final texture of this soft sediment. Scattered patches of green algae (Cladophora) and grasses such as *Ruppia maritima* commonly colonize the bottom of these ponds. Ostracoda, diatoms, foraminifera, and bivalve shell fragments (mainly *Cardium*) are found in this mud. In Salin-de-Giraud, *Cardium* shells participate, for salinities up to 55 g/l, in the carbonate sedimentation as important accumulations of shell fragments.

The total carbonate content of the mud is high, up to 85–90%, and is mixed variably with silty quartz, clay minerals, and organic matter. The distribution of the different constituents shows a good correlation between percent detrital particles and aragonite or calcite; this is possibly related to the presence of shell fragments. Magnesian calcite has a different distribution, suggesting another origin, probably associated with the photosynthetic activity of green algae or bacteria.

Millimetric and inframillimetric lithified laminae: Sediment containing millimetric laminae occur in Santa Pola primarily as fragments commonly associated with algal mats in the last ponds of dominantly carbonate sedimentation. They are mainly composed of magnesian calcite and a lesser amount of calcite and aragonite. SEM observation shows that they are composed of fusiform crystals with poorly differentiated crystalline faces.

Some filaments, probably of algal origin, crosscut the laminae, representing the very beginning of algal settlement on the bottom of the ponds after carbonate precipitation. Lithification is assumed to

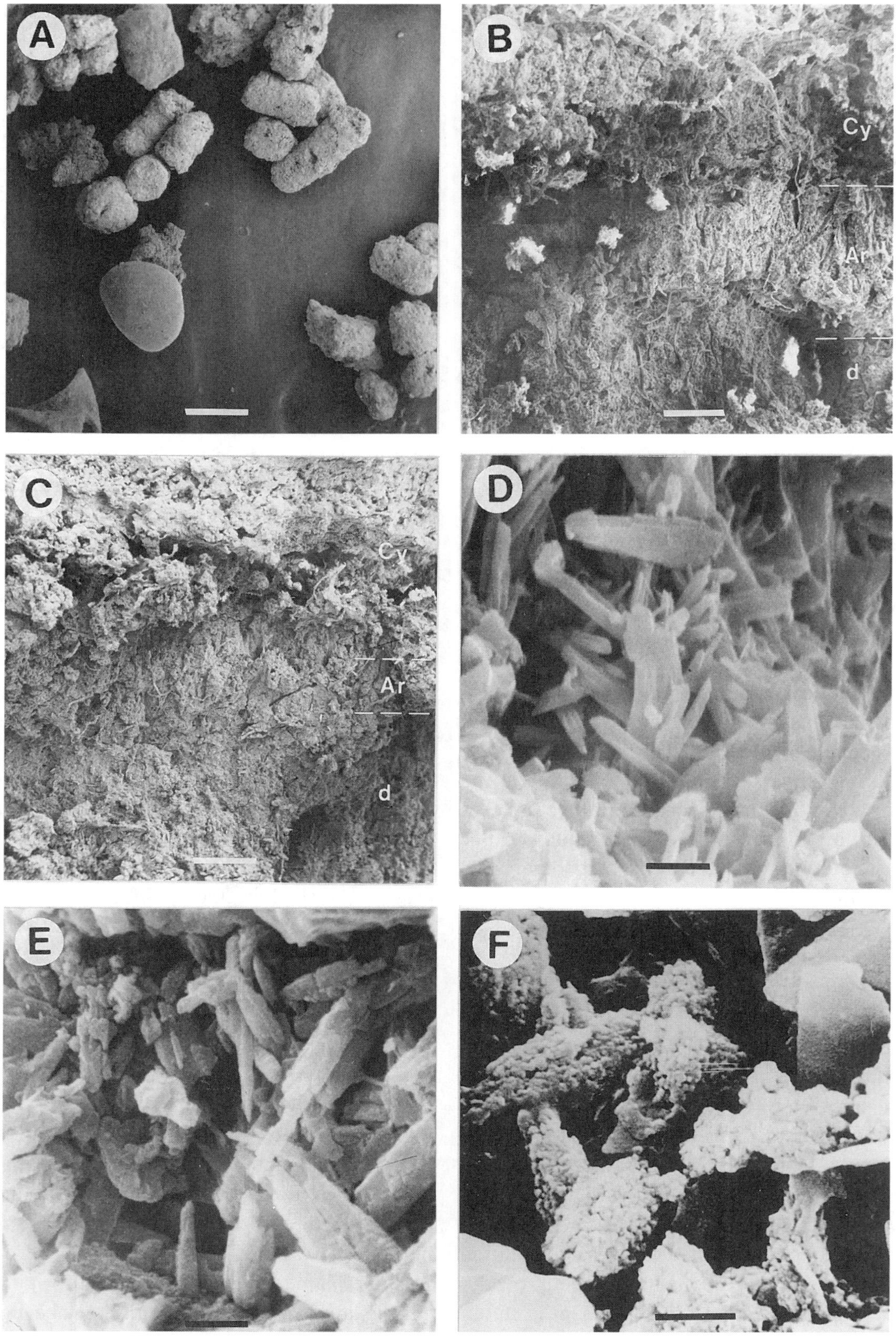

Fig. 1.13

be an early diagenetic event caused by the alteration of the earlier aragonite laminae.

Inframillimetric carbonate laminae were observed in the first ponds of algal mat development at Salin-de-Giraud. They alternate regularly with algal and detrital laminae and are cross-cut by algal filaments (figure 1.13B and C). Close to the surface they are entirely composed of aragonite needles (figure 1.13D), but a few millimeters down into the sediment, they rapidly become altered to fusiform crystals with poorly differentiated crystalline faces (figures 1.13E, and F), looking very similar to the magnesian calcite of the lithified laminae. In this microfacies it is possible to follow the very early diagenetic transformation of aragonite into magnesian calcite.

Except for the biodetrital carbonate particles, the carbonate sedimentation is due to both primary biochemical and chemical precipitation. In the first ponds the carbonate deposits are the main sediment, but they are progressively preserved and the morphology is carried over from the progressive development of the algal filaments with increasing salinity.

2—Laminated algal-mat deposition.

The laminated algal mats are encountered in the salinity range of 80 g/l and continue to dominate the sedimentation up to 150 g/l. For the lowest salinities they occupy the margins of the ponds and with increasing salinity they colonize the entire bottom of the ponds. If they undergo emersion, they develop polygonal desiccation cracks. The mats are commonly associated with carbonates but may alternate with gypsum as well.

The algal mats display a very thin millimetric lamination made up of the alternation of dark organic and light detrital laminae (figures 1.14A), sometimes with associated layers of direct chemical origin (i.e., carbonate or gypsum precipitates).

Microscopic as well as SEM study allows observation of the different laminar types and their very regular alternation within the algal mats (figures 1.14C and D). The organic laminae show a network of cyanobacterial filaments alternating with detrital laminae composed of silty quartz, clays, and detrital and biodetrital carbonate particles.

The laminae of chemical origin are composed of aragonite for the lowest salinities (40 to 60 g/l) and gypsum for the highest salinities (140 to 150 g/l) (figure 1.14E); these occur only in these specific salinity ranges. SEM study permits very clear observation of the aragonitic laminae with their needle-shaped crystals and their progressive transformation to fusiform crystals with increasing depth (figures 1.13D, E and F) in the uppermost layers. In the gypsum layers the crystals constitute a fine-grained gypsum mush (figure 1.14F).

Both the chemical laminae and the detrital lami-

Fig. 1.13. Carbonate deposition in marine solar ponds.

A. Fecal pellets composed of micrite make up most of the carbonate mud sediment. Santa Pola. Scale bar, 0.5 mm (photo courtesy of J. J. Pueyo).

B and C. SEM photomicrograph of a vertical section of the uppermost layers in a laminated algal mat showing (from top to bottom) cyanobacterial (cy), aragonitic (ar), and detritial (d) laminae which alternate regularly downward. The cyanobacterial filaments vertically crosscut the aragonitic and detrital laminae. Salin-de-Giraud. Scale bar, 300 microns.

D. SEM photomicrograph of aragonitic needles in the uppermost aragonitic layer of a laminated algal mat displaying well developed, regular faces. Salin-de-Giraud. Scale bar, 2.5 microns.

E. SEM photomicrograph of aragonitic needles with an etched aspect, located a few centimeters below the uppermost part of a laminated algal mat. The crystals show irregular faces, but the needle shape is still recognizable. Salin-de-Giraud. Scale bar, 2.5 microns.

F. SEM photomicrograph of aggregates of small carbonate crystals in a laminated algal mat buried under a sandy level and a surficial algal mat. The elongated shape of these aggregates may be linked to an aragonitic precurser. Salin-de-Giraud. Scale bar, 1 micron.

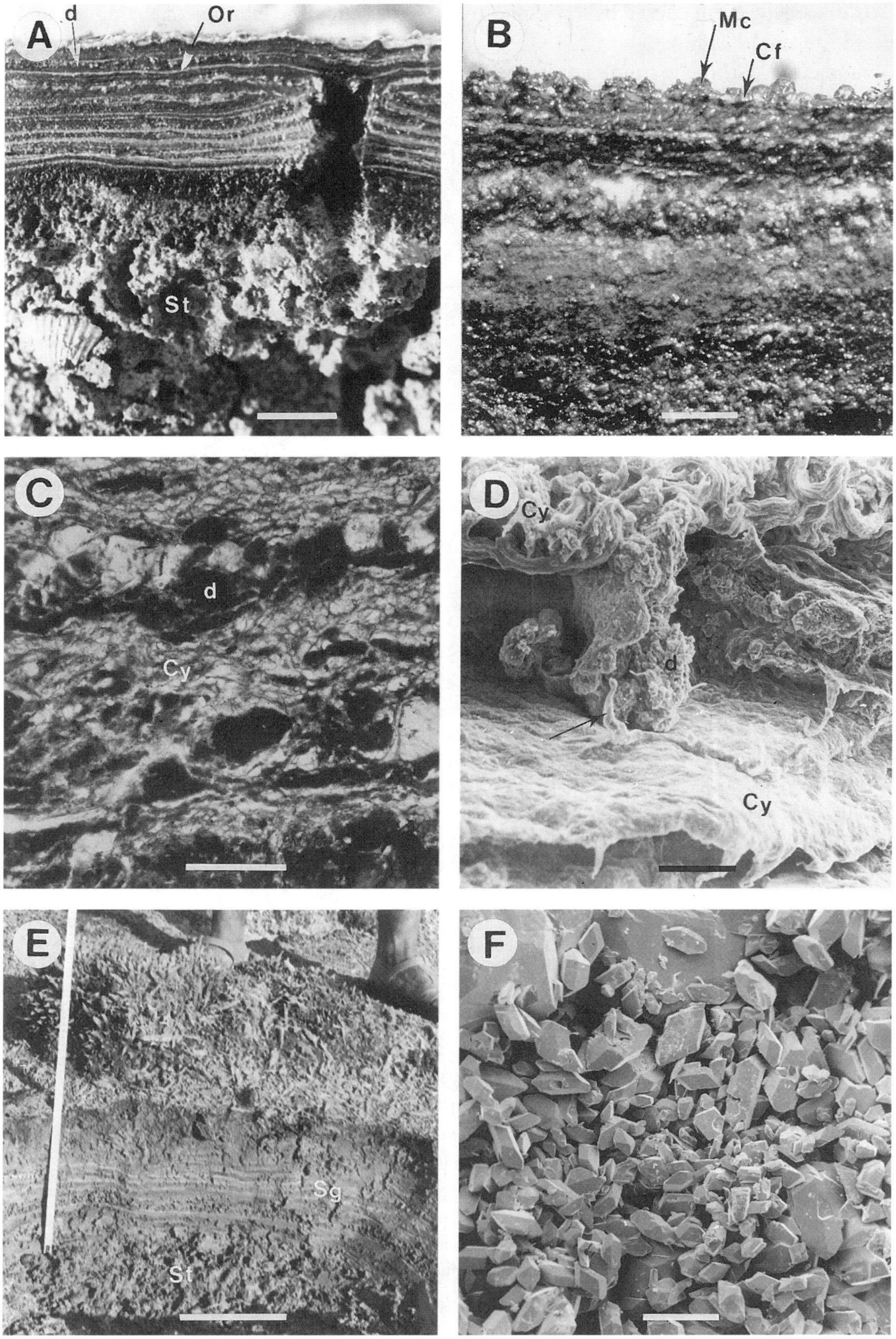

FIG. 1.14

nae are commonly cross-cut vertically by cyanobacterial filaments on a very fine scale.

The regular alternation of cyanobacterial and detrital laminae, cross-cut vertically by cyanobacterial filaments, speaks in favor of episodic growth of cyanobacteria interrupted by the influx of detrital material. The episodic character is not directly seasonal, being uneven, but is tied to new water supplies entering the ponds.

In the range of the lowest salinities the influx of detrital material provided by the influx of new water is followed by reconcentration of the water, and then by renewed precipitation of carbonate. Upon reaching the range of the highest salinities, precipitation of gypsum takes place, and the environment becomes less favorable to the development of cyanobacteria.

After each new water influx with its detrital supply (whether or not associated with carbonate or gypsum precipitation), the cyanobacteria form filaments that cross-cut the detritically and chemically formed laminae and recolonize the entire bottom of the ponds.

3—Organic material.

The dominant benthonic cyanobacterial population is *Microcoleus chthonoplastes,* which is observed in association with other cyanobacterial populations, depending on the ambient salinity (J. C. Thomas and Geisler 1982; J. C. Thomas 1984).

Within the lowest part of the salinity range in Salin-de-Giraud, *Microcoleus* filaments may be covered by a *Lyngbya aestuarii* layer in each of the organic laminae. *Lyngbya aestuarii* also occurs as a fluffy mat which is easy loosened and that probably has very little chance to be preserved in the sedimentary record.

Microcoleus is the main constituent of the algal mats in Salin-de-Giraud, while some *Chloroflexaceae* occur in association with it in Santa Pola (figure 1.14B). In the organic laminae the *Chloroflexaceae* forms an orange-colored layer at the bottom and simultaneously a green layer of *Microcoleus* develops above. It is noteworthy that the uppermost organic lamina usually traps various sedimentary particles together with a large number of diatoms (figure 1.15A). A transverse section of a *Microcoleus* filament shows trichomes in cross section surrounded by their common sheath (figure 1.15B).

The bacterial population is closely associated with the other organic material (Cornée 1982, 1984). The main bacterial groups encountered in this setting belong to the following types: heterotrophic aerobic and anaerobic halophilic forms—some of which produce calcium carbonate in the labora-

FIG. 1.14. Laminated algal mats developed in marine solar ponds.
A. Vertical cross-section of a thinly laminated algal mat with alternating dark organic (or) and light detrital (d) laminae developed on a clayey substratum (St). Santa Pola. Scale bar, 1 cm (photo courtesy of J. J. Pueyo).
B. Stratified cyanobacterial population in the organic layers of a laminated algal mat. The uppermost tubercular green layer is composed of *Microcoleus* (Mc) and the orange layer below contains Chloroflexaceae (Cf). The organic layers are black as a result of reducing conditions. Santa Pola. Scale bar, 1 cm (photo courtesy of J. C. Thomas).
C. Photomicrograph (plane light) of alternating cyanobacterial laminae (cy), forming a sheet of linked filaments, and detrital laminae (d) composed of quartz grains and carbonate peloids. Santa Pola. Scale bar, 0.5 mm.
D. SEM photomicrograph of the upper layers of a laminated algal mat. The uppermost cyanobacterial lamina (cy) shows linked *Microcoleus* filaments; the next one beneath is more compacted because of the lack of trapped detrital particles (d). Some cyanobacterial filaments (arrow) crosscut the intermediate detrital lamina. Salin-de-Giraud. Scale bar, 150 microns.
E. Laminated organic sandy-gypsum deposit (sg) on a clayey substratum (st). Alicante. Scale bar, 20 cm (photo courtesy of J. J. Pueyo).
F. SEM photomicrograph of an extremely fine interlocking gypsum mush forming a gypsum lamina alternating with organic cyanobacterial laminae. Santa Pola. Scale bar, 100 microns (photo courtesy of J. J. Pueyo).

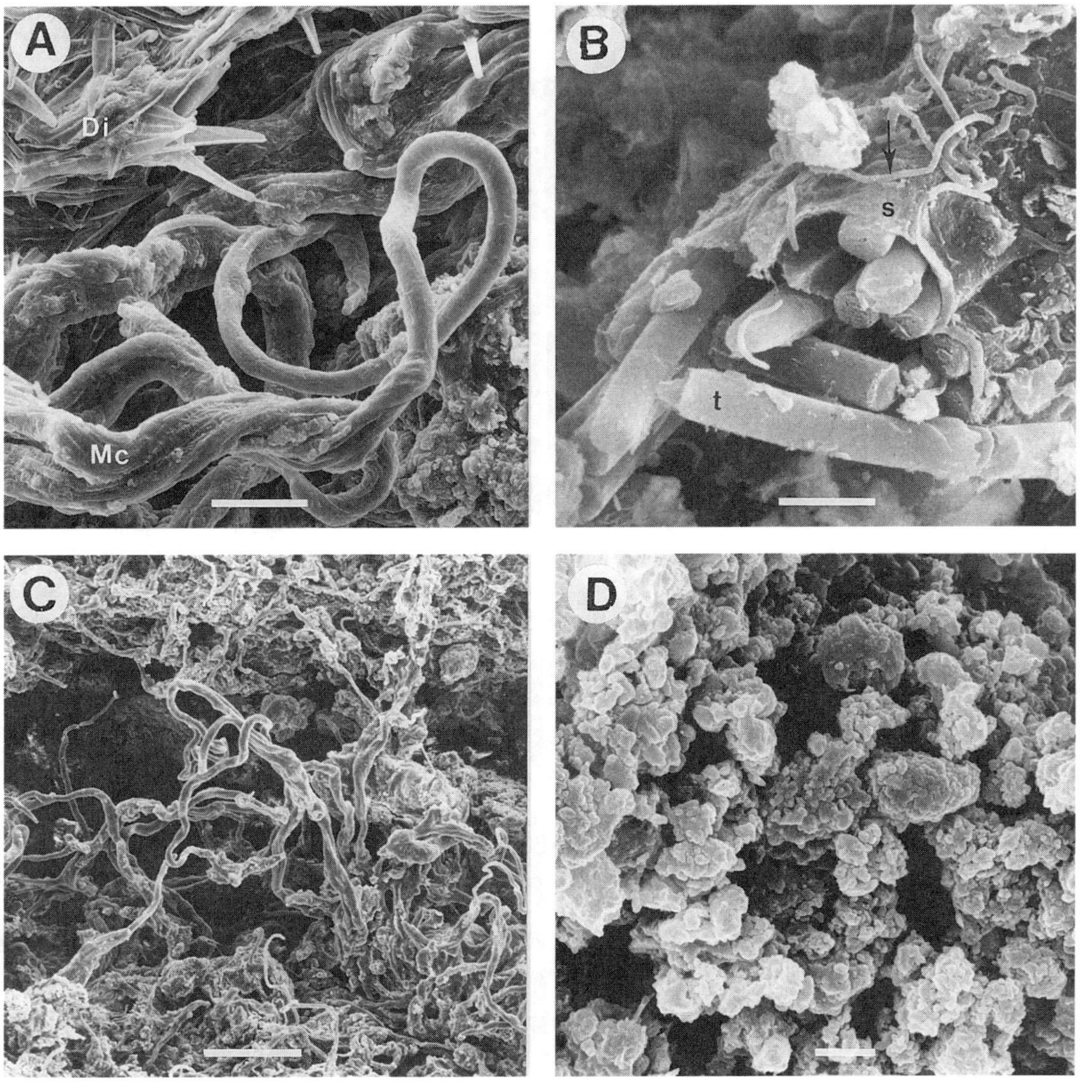

Fig. 1.15. Organic material in algal mats as seen in marine solar ponds.

A. SEM photomicrograph of the uppermost layer of a laminated algal mat in which filaments of *Microcoleus* (Mc) trap detrital particles and many diatoms (Di) to the top of the organic layer. Salin-de-Giraud. Scale bar, 40 microns.

B. SEM photomicrograph of a cross-section of a *Microcoleus* filament showing the trichomes (t) surrounded by their common sheath (s). Some bacteria (arrow) develop on the filament indicating a close biological association. Salin-de-Giraud. Scale bar, 5 microns (photo courtesy of A. Cornée).

C. SEM view of preserved cyanobacterial filaments a few millimeters above the surface, showing a well spaced organization. Santa Pola. Scale bar, 200 microns.

D. SEM view of well spaced detrital particles with some holes suggesting a former cyanobacterial layer having been totally destroyed upon burial. Santa Pola. Scale bar, 4 microns.

tory—and anaerobic sulfate-reducing and anaerobic photosynthetic forms.

Bacteria develop on cyanobacterial filaments because here they find cellular extracts, produced by the cyanobacteria, that permit their growth (figure 1.15B). Their distribution in the sediment is irregular and they commonly gather in clusters, where the bacteria send out filaments toward algal organic material.

Cyanobacterial filaments may persist for a few millimeters or centimeters beneath the surface (figure 1.15C). Most filaments disappear rapidly, leaving elongate holes in the entrapped sediment (figure 1.15D). If sediment trapping is very poor, the filaments become compacted and remain as very thin undifferentiated organic laminae before being destroyed further down in the burial history of the algal mat (figure 1.14D).

The diatoms observed on the top of the uppermost cyanobacterial lamina dissolve very rapidly upon burial and provide a silica source for early diagenesis of the sediment.

The bacterial population shows a decreasing assemblage downwards, probably in relation to the scarcity of cyanobacterial filaments and therefore a very limited nutrient supply or perhaps also in response to changes in Eh and pH.

4—Diagenetic gypsum.

Diagenetic gypsum occurs mainly in Santa Pola under a permanent water layer as nodules or lenses developing syndiagenetically at the expense of the algal mats (figure 1.16B) sometimes pervading the entire algal mat, and leaving just a few relics of the original deposit (figure 1.16A). A nodular mosaic structure may form after extreme alteration. At the contact between primary sediment and the very early diagenetic gypsum, growing gypsum crystals can be seen in the process of incorporating detrital particles and/or algal filaments.

The gypsum crystals are lenticular shaped (figure 1.16C), but sometimes they may be tabular (figure 1.16D). The lenticular shape is induced by the presence of traces of specific types of organic material (Cody 1979) and provides a good example of this morphological control.

This early gypsum forms in part because of the elevated salinity of the interstitial water. Since these early diagenetic gypsum nodules or lenses are generated within a subaqueous environment the presence of nodular gypsum cannot be a criteria for suPratidal evaporite deposition.

5—Gypsum deposition.

Gypsum occurs as the principal deposit forming in the salinity range of 150 g/l to 320 g/l, and continues to a much lesser extent when massive halite precipitation begins. Gypsum develops first as scattered millimetric crystals, commonly limited to the margins of the ponds. With increasing salinity, gypsum grows into continuous crusts covering the entire bottom of the pond. The largest crystals occur below the surface on the top of small topographic highs in the ponds. If they undergo periods of emersion, the gypsum crusts develop polygonal desiccation cracks, as do the carbonates.

With increasing salinity, gypsum deposition may display different morphological aspects:

a—Transition zone between algal mats and gypsum precipitation (140 to 150 g/l).

Orange mucoid *Aphanothece* mats develop in the ponds of first gypsum precipitation in Santa Pola and bind scattered, millimetric, but very rounded gypsum crystals (figure 1.17A). The rounded morphology of these gypsum crystals may be caused by minor episodes of dissolution when new water supplies enter in the pond and lower the salinity.

b—Crusts of porous gypsum (150 to 200 g/l).

The gypsum deposits display a very porous texture and the crystals are mixed or covered with orange *Aphanothece* mats. The sediments may be structureless, but in other instances contain poorly laminated horizons built up by layers of organic material. The crusts of porous gypsum are usually formed by small (5 mm) gypsum nodules composed of small (0.5 mm) more or less interlocking, lenticular gypsum crystals. This lenticular morphology of the gypsum crystals seems again to be associated

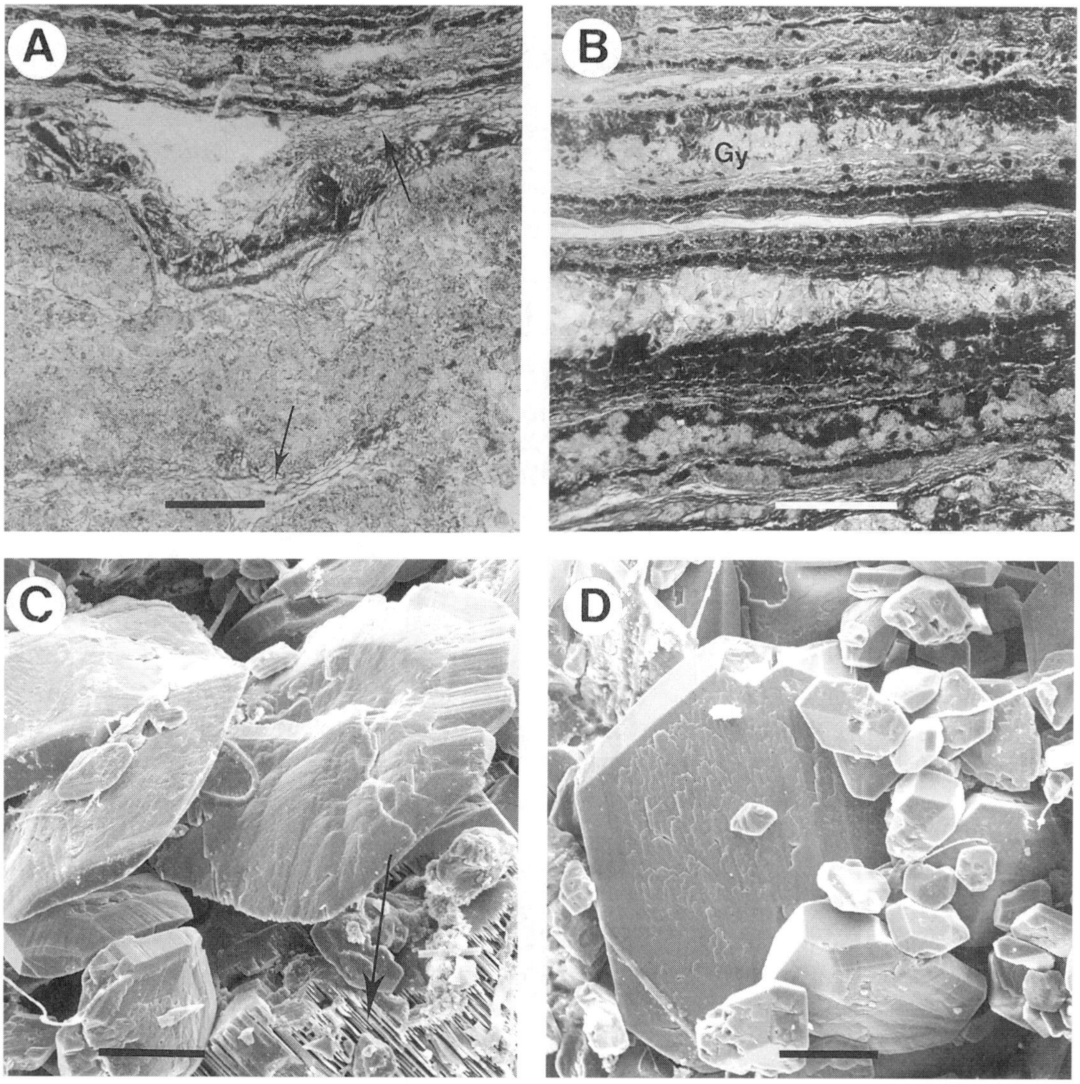

Fig. 1.16. Gypsum associated with algal mats, as seen in marine solar ponds.
A. Photomicrograph (plane light) of a totally gypsified algal mat showing some relics of organic laminae (arrows). Santa Pola. Scale bar, 2 mm.
B. Photomicrograph (plane light) of a laminated algal mat overgrown with diagenetic gypsum (Gy) forming lenses and nodules more or less parallel to the layering. Santa Pola. Scale bar, 2 mm.
C. SEM view of lenticular gypsum crystals developed in a laminated algal mat, displaying irregular faces. One of the crystals shows the easy cleavage of gypsum (arrow). Santa Pola. Scale bar, 100 microns.
D. SEM view of tabular gypsum crystals observed in a laminated algal mat. Santa Pola. Scale bar, 100 microns.

with a high organic content, and the porosity of the crusts probably is caused by the decay and destruction of the earlier orange mucoid *Aphanothece* mats.

c—Layered gypsum crusts (200 to 320 g/l).

Two types of layered crusts can be distinguished based on the morphology of the gypsum crystals and their relation to salinity.

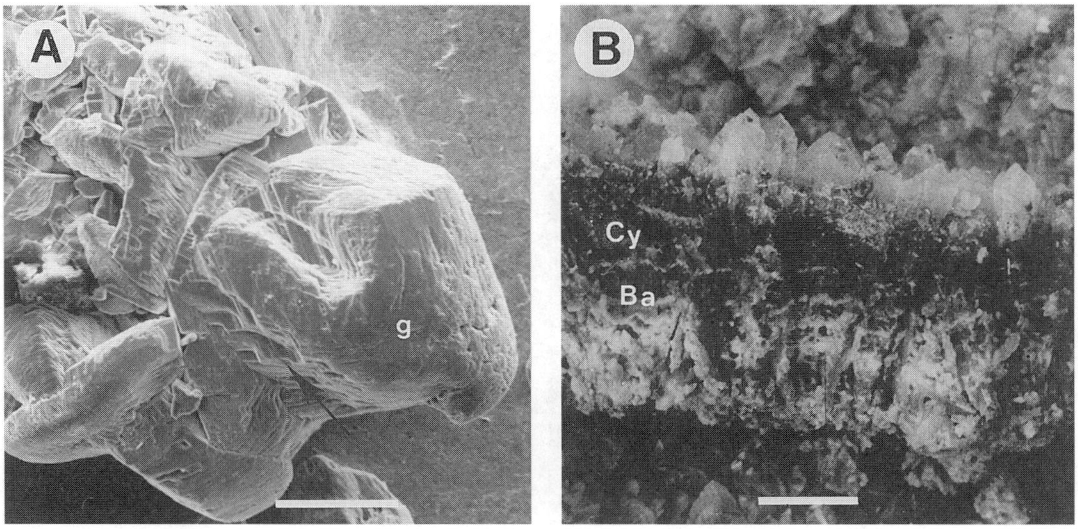

Fig. 1.17. Gypsum and associated organic material in marine solar ponds.
A. SEM photomicrograph of rounded gypsum crystals (g) associated with orange mucoid cyanobacteria, *Aphanothece* (arrow). This peculiar shape forms during periods of dissolution. Santa Pola. Scale bar, 200 microns.
B. Organic populations associated with a gypsum crust of selenite forming three distinct layers from top to bottom: orange mucoid coccoid cyanobacteria, green filamentous cyanobacteria (cy), and purple photosynthetic bacteria (ba). Santa Pola. Scale bar, 2 cm (photo courtesy of J. C. Thomas).

Crusts of selenitic gypsum (200 to 300 g/l) with well developed crystals, are actually made up of superposed "basic" crusts (figure 1.18A). Each layer begins at the bottom with inframillimetric gypsum crystals, some of which grow larger and may extend vertically through a few subsequent layers of crusts (figure 1.18B). These superposed crusts also have horizontally aligned primary voids in the upper portion of the layer (figure 1.18A). Cyanobacterial and bacterial populations develop as three distinct organic layers toward the top of these crusts (figure 1.17B). The uppermost displays a more or less continuous orange mucoid layer of coccoïd cyanobacteria with a dominant *Aphanothece* population. A green layer of filamentous cyanobacteria develops underneath the orange one, and is in turn underlain by a purple layer of photosynthetic bacteria such as *Chromatium* (J. C. Thomas 1984). These bacteria take SO_4^{-2} ions out of the brine for their life processes, and it is also possible that they utilize the sulfate of the gypsum crystals; therefore they could participate actively in early gypsum destruction and the development of loose sandy residual gypsum commonly encountered in some salt works, and termed gypsite (Cornée 1984).

The intimate association of gypsum with specific organisms is a very important feature which explains the layered structure of many gypsum crusts. Periods of lowered salinity caused by new water supplies in the ponds permit the development of cyanobacterial sheets on top of the uppermost gypsum crusts and may disrupt the continued vertical crystal growth of the gypsum crystals. Thus new gypsum nucleation takes place above the cyanobacterial sheet, forming the next depositional layer. The presence of voids observed in the upper portion of each layer is the result of decay of the earlier organic material (figure 1.18A). At higher salinities cyanobacterial growth becomes less important but organic remains are preserved as inclusions within the gypsum crystals. The crystals grow without interruption but display a zoned pattern of cyanobacterial remains parallel to progressive growth faces in the crystals.

FIG. 1.18. Layered gypsum crusts in marine solar ponds.
A. Vertical cross-section of a layered gypsum crust of selenitic crystals. The bottom (b) of each crust is continuous and the top (t) shows primary horizontal cavities. Santa Pola. Scale bar, 1 cm.
B. Photomicrograph (polarized light) of a thin layered gypsum crust showing the superposition of millimetric crusts with some through-growing crystals. Salin-de-Giraud. Scale bar, 2 mm (photo courtesy of N. Dulau).
C. Vertical cross-section of a layered gypsum crust of acicular crystals. Each basic crust is continuous at the bottom (b) and becomes very porous toward the top (t) with more or less orthogonal cavities (arrow). Santa Pola. Scale bar, 1 cm.
D. SEM view of rosette-like acicular gypsum crystals. Santa Pola. Scale bar, 0.5 mm (photo courtesy of J. J. Pueyo).

Crusts of acicular gypsum (300 to 320 g/l) have been observed in Santa Pola in the zones of highest salinity in the gypsum precipitation range. Acicular gypsum crystals form in sheaves or rosettes and are organized into basic crusts in which primary cavities are present near the upper part of the layer in orthogonal shapes (figure 1.18C, D). Crystalline growth is episodically interrupted by the development of halite when salinity rises in summer. Subsequent salinity decrease allows new gypsum precipitation and formation of a new crust, following or associated with dissolution of halite.

In short, the layered gypsum crusts result from sporadic inhibition of gypsum growth caused by changing cyanobacterial populations during periods of lowered salinity or halite deposition during periods of higher salinity.

6—Halite deposition.

Halite deposition occurs in the salinity range above 320 g/l. The highest salinities mesured in the controlled salt ponds reach 370 g/l. As observed for the other deposits, halite appears first on the margins of the ponds and spreads to the entire bottom with increasing salinity. In the ponds in which halite precipitates, the brines are red, as a result of the presence of hyperhalophilic bacteria such as *Halobacterium* and also algae such as *Dunaliella salina.* Some filamentous cyanobacteria occur that also coat halite and gypsum crystals (figure 1.19A).

The shape of halite crystals varies with increasing salinity, forming cubes, chevrons, and cubic hoppers (with skeletal faces).

a—Crusts composed of halite cubes (320 g/l to 325 g/l).

Crusts of cubic halite occur in the ponds of earliest halite precipitation and correspond generally to annual deposition, as they undergo partial dissolution and truncation during winter. The halite crust shows small crystals at the base, some of which become larger going toward the top (figure 1.19B). Halite cubes are generally perfect and display a milky inclusion-rich center indicating very rapid growth at first that then slowed down, generating a limpid outer rim. Gypsum precipitation may also have occured as small crystals accumulated between the halite cubes in the overlapping range of coprecipitation. This organization of halite crystals is the result of competitive growth in open water.

b—Crusts composed of chevron halite (325 g/l to 370 g/l).

These are generated in the salt ponds in which salt harvest takes place. The crusts are made up of beds of inverted halite pyramids marked by thin gypsum laminae (figure 1.19C). Microscopic observation shows that very thin gypsum laminae cut through halite chevrons or interrupt their growth after localized dissolution (figure 1.19D). In the second case a new halite layer usually is generated from small crystalline nuclei which undergo competitive growth.

These halite crusts are the result of halite growth which is periodically interrupted by phases of dilution, permiting formation of the gypsum. After dissolution of the uppermost halite deposit, the resulting gypsum lamina is more or less discontinuous and halite may continue to grow without interruption. If a continuous gypsum lamina does grow, it introduces a marked layering in the halite crust. The halite crusts, on the whole, are very porous and are marked by vertically elongated dissolution cavities probably generated during the dilution stages (figure 1.19D). The annual salt harvest interrupts the futher evolution of the halite crust structure; early diagenetic cementation by halite precipitation from interstitial brines is only known from some natural deposits (Shearman 1970).

c—Crusts composed of skeletal hoppers (over 370 g/l).

These are cubic halite crystals having depressed faces (skeletal growth) occuring in ponds in which the mother brines of the salt, with salinities greater than 370 g/l, are collected. The difference in morphology is probably due to greater concentration of ions in the water (high levels of supersaturation), the resulting rapid growth rate of the salt, and also to the composition of the water.

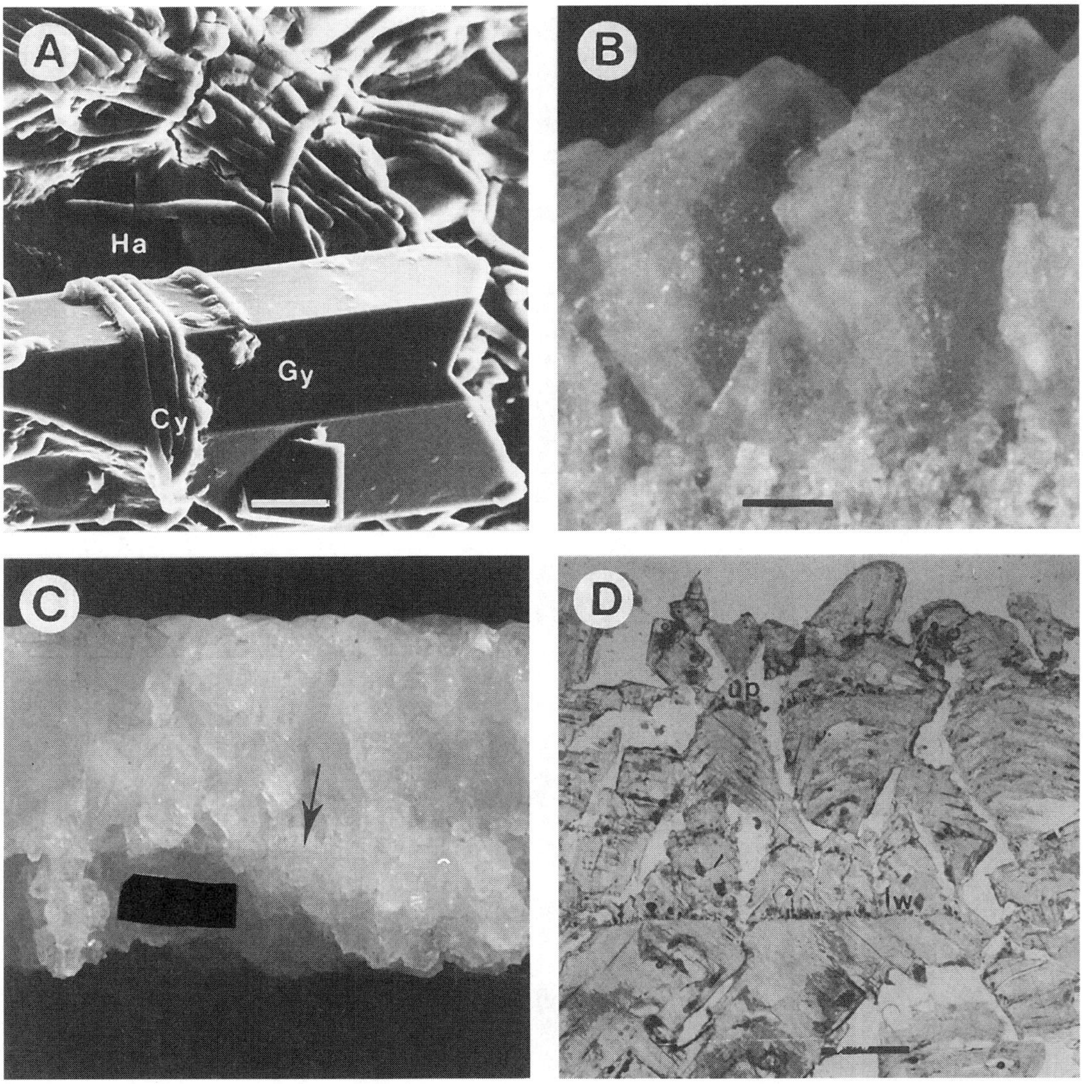

Fig. 1.19. Halite deposition in marine solar ponds.

A. SEM view of a large halite crystal (Ha) and a small gypsum crystal (Gy) in the foreground covered with the filamentous cyanobacteria, *Phormidium* (Cy). Salin-de-Giraud. Scale bar, 20 microns (photo courtesy of A. Cornée).

B. Vertical cross-section of a halite crust composed of cubes pointing an apex or an edge (coign) upward. Small crystals develop on the bottom and some expand upward. Salin-de-Giraud. Scale bar, 2 mm.

C. Vertical cross-section of a layered halite crust composed of chevrons. Layering is due to dissolution surfaces marked by thin gypsum laminae (arrow). Santa Pola. Crust is 3 cm thick.

D. Photomicrograph (plane light) of a cross-section of a halite crust composed of chevrons. The lower gypsum lamina (lw) corresponds to a dissolution period followed by gypsum precipitation before new halite crystals are generated. The upper gypsum lamina (up) is not continuous enough to interrupt halite growth which proceeds upwards in crystallographic continuity. Santa Pola. Scale bar, 2 mm.

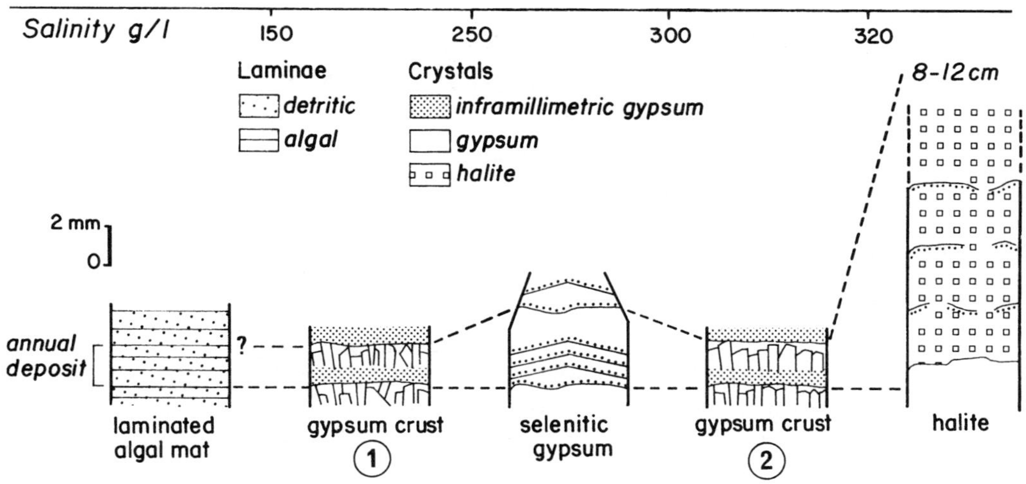

FIG. 1.20. Lateral evolution with increasing salinity of annual depositional sequences in the solar salt works of Salin-de-Giraud, France (after Geisler-Cussey 1986). Layering of gypsum crust: 1 = due to ephemeral cyanobacteria; 2 = related to dissolution of previous halite.

SUMMARY

The evaporative sediments in these controlled salt ponds are characterized by a laminated or layered structure. The alternation of millimeter- or centimeter-thick layers results mainly from variations of salinity producing a lateral shifting of the related depositional environments with their specific facies. A similar layering of the deposit is typical of many fossil evaporites.

The winter emersion in Salin-de-Giraud salt works permits the determination of the annual quantity of precipitated evaporites produced by sea water evaporation (figure 1.20). Sedimentation of a few laminae, each less than 1 mm thick, in the algal mats, a gypsum crust up to a few millimeters thick, and 8 to 12 cm of halite, are lateral annual depositional equivalents. Except for the scale, the relative thickness is close to that observed in ancient evaporative basins.

The evaporative sequence in the ponds described here is controlled by man, but it contains examples of sedimentation that may develop in the marginal setting of a natural evaporative coastal lagoon (figure 1.21). The aim here is to show lateral evolution of facies with increasing salinity. The result is an assemblage very similar to that which might be encountered in ancient marine evaporites.

The Mixed Environment: Marine Marginal Model

The marine-marginal environment is located at the boundary between marine and continental domains where sea water and continental meteoric water are commonly mixed. Evaporites develop in areas of limited sea water influx and high evaporation rates, generating a chemically confined environment. This arrangement is achieved in the marginal sabkhas of the Trucial coast in the Persian Gulf and was studied in such detail that it became the classical model for this kind of environment (Shearman 1966; Kinsman 1966, 1969; Butler 1969, 1970). However the substratum in this area is essentially calcareous, while the Paleogene evaporites are generated in a largely silicilastic framework. Similar marine marginal environments exist on the sandy Pacific coast of Baja California in northern Mexico and seem to be a more appropriate model for the Paleogene deposits.

Baja California, Mexico, provides good examples of natural coastal depressions that are periodically, once a year or more, flooded by sea water and are also affected by storms that introduce meteoric water. The fact that the sediments are preserved permits study of the effects of early diagenesis con-

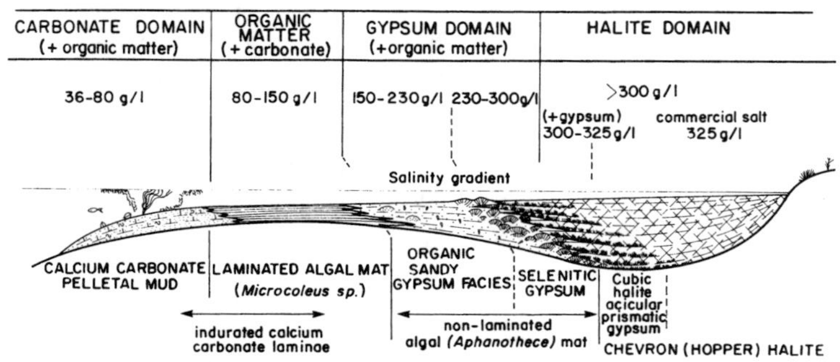

Fig. 1.21. Lateral evolution with increasing salinity of evaporitic facies, idealized as an horizontal sequence, from the data of the solar salt works of Santa Pola, Spain. Relative thickness of each facies is comparable to that present in the salina, all facies forming over the same time interval (after Ortí et al. 1984).

trolled by mixing of seawater-derived brines and continental groundwaters. Two settings will be considered successively: Ojo de Liebre lagoon and Salina Ometepec.

OJO DE LIEBRE LAGOON

The Ojo de Liebre lagoon in Baja California (figure 1.22) has been extensively studied by Holser (1966b), Pierre (1982, 1983), and Pierre et al. (1984). This lagoon communicates with the open sea through a 4 km wide channel and covers over 400 km^2 under a very dry climate. The margins of the lagoon are clearly defined, well controlled, and represent the largest salt works in the world.

Since some areas are not controlled, and the sediment-brine relationship is still the natural one (figure 1.22). The uncontrolled areas include part of the supratidal margins of the lagoon that where studied in three settings: location A, where algal mats develop, location B at the southeastern end, and location C at the southern end of the lagoon.

The chemical behavior of major elements has been studied in the surficial brines of the lagoon subjected to progressive evaporation; the behavior of these elements has also been examined in the interstitial brines of the studied areas (figures 1.23, and 1.24). The data from the marine-fed salt works of Secovlje in Croatia are repeated in each diagram as a comparison with a single-source evaporating sea water sequence in order to provide a reference for the determination of the effects of continental ground water supplies in the interstitial environment (Pierre 1982).

a—Brines of the lagoon.

In the brines of the lagoon the major elements (Mg^{+2}, Ca^{+2}, Na^{+}, SO^{-2}_{4}, Cl^{-}) show nearly the same evolution as in the salt works of Secovlje, except for K^{+}, which shows a slightly higher content. The general behavior of the ions is in good agreement with progressive evaporation of normal sea water.

The bromine content in the lagoon brines is twice as high as in the brines of the salt works of Secovlje. The higher bromine content seems related to the decay of cyanobacteria in the algal mats fringing the lagoon and the consequent release of the stored bromine. This again illustrates the importance of organic matter as an ion reservoir in any sedimentary system.

b—Supratidal algal mat zone.

This setting is composed on the landward side by supratidal evaporative flats cross-cut by tidal channels. The lower part of the flats, forming a 500 m

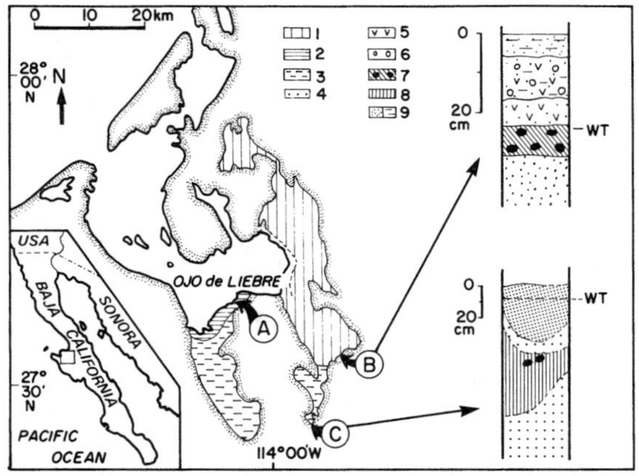

Fig. 1.22. Location map of Ojo de Liebre lagoon. Vertical cross-section of the deposits is given for locations B and C (modified from Pierre 1982, 1983, Pierre et al. 1984). Number key: 1 = controlled salt ponds, 2 = algal mats area, 3 = evaporitic deposits, 4 = halite, 5 = gypsum, 6 = nodular polyhalite, 7 = massive polyhalite, 8 = dolomite, 9 = sand and clays.

wide tongue, is covered with algal mats displaying polygonal desiccation cracks. The algal mats, a few centimeters thick, are associated with detrital and carbonate deposits, mainly pellets and aragonitic needles. Inland of the algal zone there are small gypsum dunes reworked by the wind.

Most of the major elements analyzed in the interstitial brines—magnesium, potassium, sulfate, and chlorine—show the same behavior as observed for evaporating sea water (figure 1.22, location A). The concentration of these interstitial brines stays below the domain of gypsum precipitation. The calcium content is lower in some places because of localized calcium carbonate precipitation and higher in others (the edges of this chemical system) because of calcium carbonate dissolution caused by fresh water supplies causing brine dilution and/or pH fluctuations. Sodium content shows a slight depletion probably related to adsorbtion by clays or absorption by halophytic plants.

c—Southeastern end of the basin.

The data presented here concern measured waters from a small satellite pond where interstitial brines were sampled within a substratum composed of a variably-sized siliciclastic silt (figure 1.22, location B). The density increases from the edges to the center of the pond. The behavior of major elements differs from that of simple evaporation of sea water. Calcium and chlorine are depleted from the normal sea water concentration curve; sulfate and sodium are in excess; magnesium and potassium are in excess in the center of the pond and depleted on its edges. Comparison of relative contents shows landward enrichment in Ca^{+2}, SO_4^{-2}, and Cl^- in the marine brines derived from the main lagoon as a result of mixing with continental groundwaters which have previously dissolved gypsum and halite. A massive supply of SO_4^{-2} ions brought in by these continental groundwaters creates a sulfate-rich environment where Mg^{+2} and K^+ ions are amply available.

During three successive field trips (1979, 1980, and 1981) important modifications of the sediment appeared. In 1979, the evaporative succession in the center of the basin was mainly composed of a gypsiferous layer reaching a maximum thickness of 30 cm, topped with a millimeter-thick halite crust. A few centimetres below the surface, the gypsum sediment was partially or totally replaced by millimeter-size, white polyhalite nodules ($K_2 Mg Ca_2 (SO_4)_4.2H_2O$). Minor amounts of glauberite ($Na_2 Ca(SO_4)_2$) as white crystals that pseudomorph gypsum crystals and of celestite also were present as millimetric white nodules scattered within the sediment. In 1980, the bottom of the gypsum layer that was in contact with the interstitial brines was totally transformed into polyhalite with a significant volume increase, filling the cavities and moving apart the detrital host sediment. The earlier upper polyhalite layer was also enlarged. In 1981, both polyhalite layers reached a thickness of 10 cm each.

This example shows how very rapidly early dia-

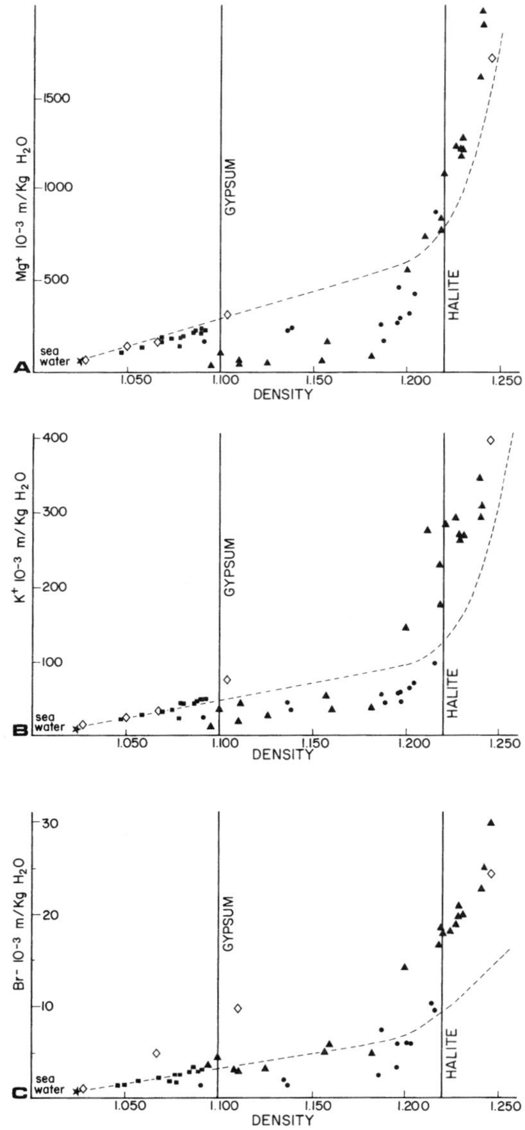

Fig. 1.23. Behavior with increasing density of magnesium, potassium, and bromine in the surficial brines of the lagoon and in the interstitial brines of the three areas studied in Ojo de Liebre lagoon (after Pierre et al. 1984). The dashed line is the curve from the solar salt works of Secovlje, Croatia (after Herrmann et al. 1973). Lozenges = brines of the lagoon, squares = location A, triangles = location B, circles = location C.

genesis occurs as a result of mixing between marine-derived brines and continental groundwaters which previously dissolved halite and gypsum. The SO_4^{-2} enrichment in the presence of K^+ and Mg^{+2} ions within the interstitial environment is responsible for the genesis of polyhalite, glauberite, and celestite (Pierre 1983).

d—Southern end of the basin.

A well-studied lateral pond of the main estuary, also in a siliciclastic sediment setting, is taken as an example (figure 1.22, location C). Here the density of the interstitial brines also increases from the edges to the center. Once again, the behavior of major elements in the interstitial brines is different when compared to simple evaporating sea water. Sodium and chlorine are present in excess as a result of halite dissolution by inflowing meteoric groundwaters. Magnesium and potassium are slightly depleted, and sulfate shows a large deficit. The behavior of these ions also depends upon meteoric groundwater supplies. The marine-derived interstitial brines seeping through the sediment from the north mix

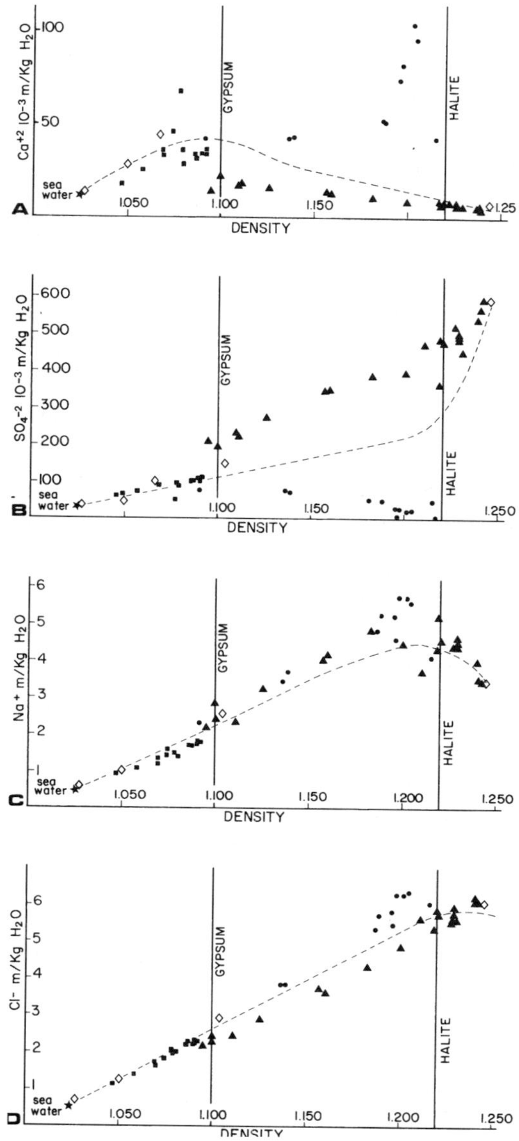

Fig. 1.24. Behavior with increasing density of calcium, sulfate, sodium, and chlorine in the surficial brines of the lagoon and in the interstitial brines in the three areas studied in Ojo de Liebre lagoon (after Pierre et al. 1984). The dashed line is the curve from the solar salt works of Secovlje, Croatia (after Herrmann et al. 1973). Lozenges = brines of the lagoon, squares = location A, triangles = location B, circles = location C.

with continental groundwaters and undergo dilution causing dissolution of carbonate shell fragments and therefore show Ca^{+2} enrichment. The Ca^{+2} excess is responsible for the precipitation of almost all the available sulfate as gypsum. This SO_4^{-2}-poor environment is characterized by the mixing of Mg^{+2}-rich, marine-derived brines and calcium bicarbonate continental waters, modified by capillary evaporation.

In the central part of the basin the deposits are composed of a 30 cm thick halite crust, disappearing laterally, overlying fine siliceous sands containing scattered lenticular gypsum crystals. The quantity of gypsum becomes greater toward the edges of the basin. These deposits are underlain by a fine, light green sediment, partially made out of green siliceous sands (figure 1.22, location C). In addition there are inclusions of light green plastic sediment composed of numerous little rhombohedrons of dolomite (5 to 15 mm) either isolated or in aggregates. Forming beneath the brine water table, it reaches a maximum thickness under the area where the halite crust develops. The dolomitic sediment shows a rapid lateral thickness decrease away from the halite area. In this setting dolomite precipitates directly from the interstitial brines without any carbonate precursor, as the host sediment is a siliceous sand. This mechanism is made possible by the low SO_4^{-2} content in the interstitial brines due to previous depletion by gypsum precipitation.

The mechanism of dolomite formation during early diagenesis in a carbonate host sediment is well known from the studies of the Trucial coast of the Persian Gulf (Shearman 1966). The evaporative complex of Ojo de Liebre provides an example of the genesis of primary interstitial dolomite in a siliciclastic host sediment (Pierre et al. 1984).

The evolutionary trends in the two sub-basins reflect the complexity of the mechanisms involved in the mixing zone between marine and continental environments. They depend upon (1) the origin of the water, marine or nonmarine, and the composition of the various brines flowing into the basin; (2) capillary evaporation concentrating the brines; and (3) precipitation and dissolution of various mineral species. These recent marine marginal environments permit an understanding of what has happened on the edges of large ancient evaporite basins at the contact between marine and continental domains.

SALINA OMETEPEC

One of the main points of interest in the study of this setting is to illustrate the morphologies within preserved halite deposits. Shearman (1970) has described and interpreted them and more recently Lowenstein and Hardie (1985) went back to these deposits for a detailed study.

The sediment displays a bedded structure linked to the alternation of halite and detrital layers, the latter being silty mud (clay) or gypsiferous. The halite layers show various morphologic aspects and may be made up of: halite rafts cemented by clear halite; inclusion-rich chevron structures with lateral syntaxial growth of clear halite and dissolution surfaces truncating chevrons; vertically or horizontally elongated cavities filled with internal muddy sediment on the bottom; and clear halite that fills the remaining porosity. The genesis of these halite deposits may be described as a distinctive saline pan cycle (Lowenstein and Hardie 1985) characterized by a succession made up of three developmental stages: flooding, evaporative concentration (leading to a saline lake), and desiccation.

The flooding stage corresponds to the influx of water loaded with detrital particles and undersaturated with respect to halite. The top of the previous halite layer suffers dissolution that sharply truncates the chevron structure of the existing halite. The dissolution process affects also the bulk sediment, generating vertical dissolution pipes along discontinuities such as crystalline faces or between inclusion-rich layers, resulting in skeletal halite spines and horizontally elongated cavities along depositional discontinuities. Even gypsum crystals within the halite may be rounded and corroded and then moved along to form detrital gypsum layers. The detrital load of the floodwaters will settle on the bottom of the pan and within the dissolution pipes as an internal sediment.

The saline lake stage is the result of renewed brine concentration by evaporation and halite dissolution of the salt from the previous cycle. Halite crystallizes first at the brine-air interface as millimeter- to centimeter-size pyramidal hoppers and rectangular and square-shaped plates. After a few hours of continued evaporation, supersaturation causes rapid and preferential growth of the edges of halite crystals and also formation of halite with numerous fluid inclusions. Crystals may coalesce as rafts and sink to the bottom very rapidly where they constitute "cumulates" and act as nuclei from which halite crystals will develop on the bottom by syntaxial growth in a few days or weeks. Competitive upward growth produces vertically oriented and elongated halite crystals that are markedly zoned by inclusion-rich and inclusion-free bands related to variations in crystal growth rates. Depending on the orientation of the parent halite crystal, the resulting overgrowth will be a chevron with an upward pointing edge or a cornet with an upward pointing face. The halite bed generated in this way shows an accumulation of hoppers and plates at the bottom and an upward syntaxial overgrowth of chevrons and cornets. The resulting halite crust is very porous.

The evolution of the desiccation stage depends upon the thickness of the surficial water layer, the evaporation rates, and the importance of additional fresh water supplies. Total desiccation will happen more or less rapidly after several weeks and diagenetic growth of halite, precipitating from interstitial brines, may start. Early diagenesis of halite occurs as cementation of dissolution cavities and intercrystalline voids by clear halite without specific orientation. Displacive clear halite also grows in the detrital layers in contact with or under the brine water table and often incorporates host sediment along growing faces. Cementation by halite produces a gradual volume increase and eventually the halite crust undergoes polygonal fracturing, creating tepee-like structures that are rarely preserved because they are mostly dissolved or swept over by the next flood phase. The polygonal structure however is frequently preserved and even emphasized by dissolution features and mud infill.

The Continental Environment: Saline Lake Models

A review of numerous saline lakes has led to a better knowledge of their depositional environment (Hardie, Smoot, and Eugster 1978). Saline lakes in Spain and in the Eastern African rift system serve as examples of evaporative sedimentation in a continental environment.

SALINE LAKES OF SPAIN

Numerous saline lakes have developed in Spain in the Zaragoza-Lerida area along the Ebro valley and have been extensively studied (Pueyo 1978/79; Pueyo and Ingles Urpinel 1986). The modified Mediterranean climate in this region is almost semiarid (200 to 400 mm/year rainfall) because of the rain shadow effect of the Pyrenees, the high sunshine ratio, and prevailing dry NW winds. These saline lake sites form in depressions generated by local dissolution of Tertiary evaporites in the substratum and by lateral facies changes (from clays to sandstones) forming paleochannels that commonly create lateral limits to the depressions. Erosion is essentially eolian, leading to the accumulation of detrital particles (quartz and clays) on the southeastern edges of these saline lakes.

The run-off of meteoric water feeds the lakes, which are largely ephemeral. The depressions are hydrologically closed, and evaporation is greater than inflowing water supplies, leading to the genesis of saline lakes. The relatively limited size, up to a few hundred meters, of these saline lakes and the low relief on the edges just permits the development of an outer mud flat and an inner, mostly ephemeral, lake with a saline mud flat and a saline pan (figure 1.25).

HYDROCHEMISTRY OF MAJOR ELEMENTS

Surficial brines: The meteoric water seeps through Tertiary evaporites and becomes a $Cl - SO_4 - Na - (Mg)$-rich brine as shown on the $Na + K - Mg - Ca$ (figure 1.26A) and $Cl - SO_4 - HCO_3 + CO_3$ (figure 1.26B) diagrams of Hardie, Smoot, and Eugster (1978). The evolution of brine composition depends upon concentration by evaporation, precipitation, and dissolution of earlier mineral species.

Enrichment in SO_4^{-2} and Cl^- ions is related to dissolution of gypsum and halite in the existing Tertiary evaporites. Calcium carbonate precipitation explains the low $HCO_3 + CO_3^{-2}$ content in the brines, and gypsum precipitation lowers the Ca^{+2} content even further. The Na^+, Cl^-, and SO_4^{-2} content may fluctuate depending on the precipitation of halite, mirabilite ($Na_2SO_4 \cdot 10H_2O$) and bloedite ($Na_2SO_4 \cdot MgSO_4 \cdot 4H_2O$). This setting is not a perfectly closed chemical system as some of these minerals may dissolve with the next rain and others may be taken away by the next strong wind storm since they are commonly surficial deposits.

The behavior of Mg^{+2} and K^+ ions is similar to that observed in evaporating sea water, with a linear increase of the concentration curve showing a higher slope after first halite precipitation. As they

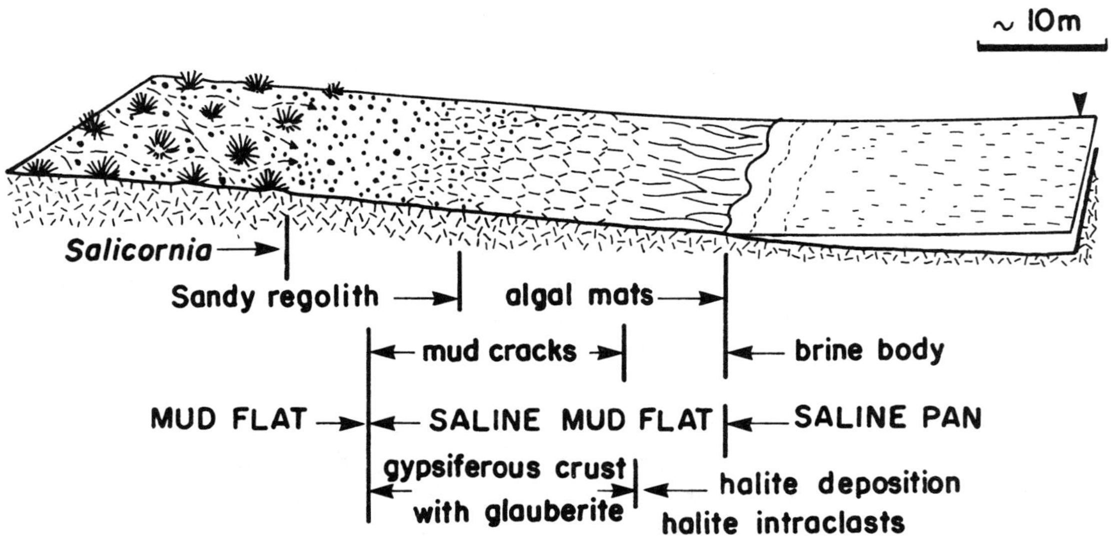

FIG. 1.25. Schematic distribution of the sedimentary environments and related evaporitic deposits encountered in continental saline lakes of the Zaragossa-Lerida area, Spain (after Pueyo 1978/79).

do not precipitate as minerals, the Mg^{+2} and K^{+} constitute good references to follow salinity fluctuations.

Interstitial brines: The pore brine content is higher on the western edges of these lakes where clays accumulate preferentially as a result of the northwestern winds. The analytical results are obtained from the upper 40 cm of the lake sediments and show an average salinity of around 200 g/l (100 to 300 g/l), allowing gypsum precipitation. Because of capillary evaporation the salinity increases from bottom to top, but is affected to a lesser extent in places where superficial brines stay longer. The Na/Cl and K/Na ratios are higher than in surficial waters because of adsorption of Na and K by clays. This kind of exchange is favored by CO_2 release when organic matter decays.

Other Spanish lakes in the region of La Mancha, southeast of Madrid were also studied (Marfil et al. 1975; de la Peña et al. 1982). The only available geochemical data present the behavior of the surficial brines, which show a very similar pattern to that of the brines of the saline lakes in the Zaragossa-Lerida area (figure 1.26). The meteoric groundwater seeps through Triassic and Tertiary evaporites, resulting in Na − Mg − Cl − SO_4-rich brines. However the plotted analysis diverges slightly from the edges of the Na + K − Mg − Ca diagram, indicating a higher Ca content which is probably related to additional contributions from the dissolution of associated Jurassic and Cretaceous limestones which enter into the groundwater system.

The saline lakes of Spain are a good example of what happens when meteoric waters dissolve fossil evaporative deposits, as well as showing that the behavior of major elements is different from that in evaporating sea water. These waters generate, besides gypsum and halite, other minerals which do not occur in the marine depositional sequence such as mirabilite, bloedite, glauberite, thenardite, epsomite, and hexahydrite.

EVAPORATIVE FACIES

Two types of deposits are encountered in these settings. Surficial sediments form by precipitation from surficial brines, by capillary evaporation generating surface efflorescence, and by wind reworking of previously deposited sediments. Early diagenetic minerals form within the host sediment from interstitial brines. Before discussing these de-

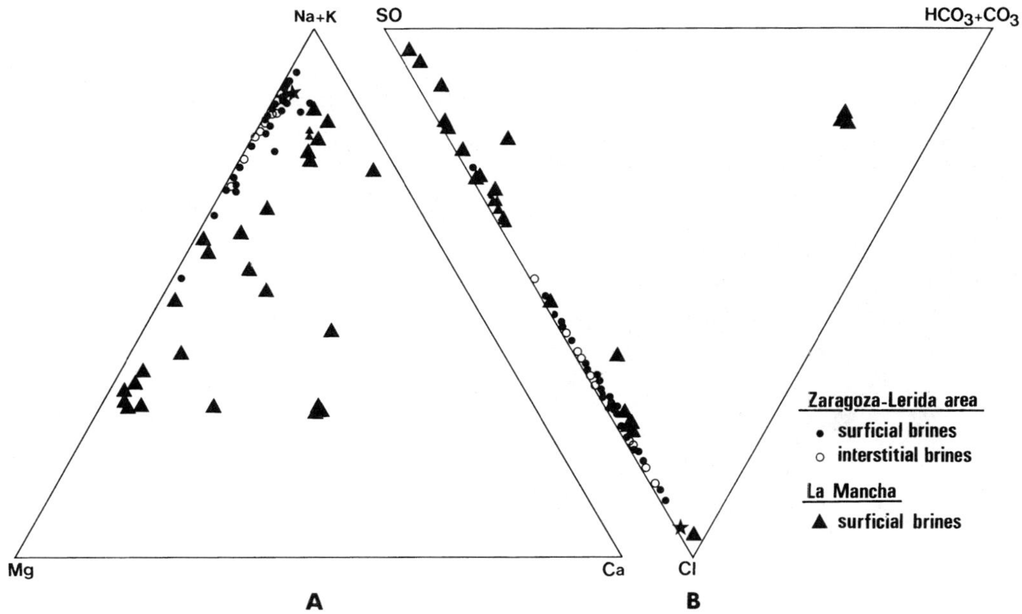

FIG. 1.26. Composition of surficial and interstitial brines in continental saline lakes utilizing Na + K − Mg − Ca and Cl − SO_4 − HCO_3 + CO_3 diagrams from samples taken in the continental saline lake of Zaragossa-Lerida and La Mancha areas in Spain (after data from Pueyo 1978/79; Marfil et al. 1975; de la Peña et al. 1982).

posits, I will review the organisms living in these settings.

Biological framework: Algal mats develop intensively on the bottom of lakes, and their position depends on the location of the surficial brine body and they are sometimes limited to small depressions. When exposed, the algal mats generate polygonal desiccation cracks. The dominant population in the algal mats is the cyanobacteria *Microcoleus,* which form 3 to 4 mm thick deposits above various detrital, wind-blown, or evaporative sediments. Underneath the surficial algal mat lies a black decaying layer resulting from intensive bacterial reduction of the upper organic layer (figure 1.27A). This is accompanied by the formation of gas which escapes to the surface and deforms the algal mat by development of gas vesicles (figure 1.27B).

A population of halophytic plants, mainly *Salicornia,* develop on the edges of the saline lakes. Other organisms such as crustaceans (mainly *Artemia salina*), ostracods (mainly *Eucipris aragonica*), and coleopterans also live in this setting. Flagellate algae, especially *Dunaliella salina,* are responsible for the red color of the very saline brines. The products of these organisms (eggs, crustacean fecal pellets, protozoan cysts) and seasonally abundant wind-blown pine pollen are carried into the lakes.

The organisms living in these continental settings are very similar to those observed in marine salt works. The extensive development of cyanobacteria in these confined environments, either marine or nonmarine, is a characteristic feature.

Surficial deposits (precipitation from surficial brines): In fall and winter, the main periods of rainfall, brines cover most of the basinal surfaces. Widespread mirabilite ($Na_2SO_4 \cdot 10H_2O$) precipitation occurs when the temperature is very low. The resulting crystals are very bright and clear, with a more or less acicular shape (figure 1.28A). At temperatures above 32°C the mirabilite crystals undergo progressive dehydration from the surface to the inside and are transformed into thenardite (Na_2SO_4). The thenardite appears as mirabilite

Fig. 1.27. Algal mats in continental saline lakes (photos courtesy of J. J. Pueyo).
A. Algal mat, up to a few millimeters thick, underlain by a black zone of decaying organic matter. Zaragossa-Lerida area. Scale bar, 5 cm.
B. Algal mat showing a deformed surface related to the development of gas vesicles that later vent to the surface. Zaragossa-Lerida area. Scale bar, 4 cm.

pseudomorphs, loose, fine-grained, and easily reworked by the wind. Another phase of comparatively very early diagenesis sometimes leads to the growth of a second generation of very fine thenardite needles (figure 1.28B).

During spring and summer, the hot arid climate and high evaporation rates cause a reduction of the surface areas occupied by surficial brines; the resulting mineralogical sequence is the more usual one with carbonates, gypsum, and halite. The gypsum deposits form an outer rim to lake areas, and are composed of millimeter-thick laminae of lenticular gypsum crystals precipitating from a 1 to 2 cm thick layer of water. The gypsum crystals are generally lenticular with surficial striae and curved faces and a size range of 100 microns up to a few millimeters (figure 1.29A). These very small crystals are easily wind-blown and commonly accumulate on the algal mats, sometimes reaching 5 mm to 1 cm thickness, usually in small depressions.

As brine concentration continues to increase, halite starts to precipitate in the somewhat deeper central parts of the basins where highly saline brines collect. Depending on the thickness of the surficial water layer, the halite deposits display different growth habits. On the edges of the lakes, just before reaching total desiccation, the residual film of brine inpregnates the gypsum deposit and precipitates tiny halite cubes with a size of 10 to 50 microns, cementing the previously formed lenticular gypsum crystals (figure 1.29B). The cementation of the gypsum deposit by halite generates a crust that follows the irregularities of the algal mat surface and shows typical desiccation phenomena such as polygonal desiccation cracks or tepee structures. Small amounts of early diagenetic glauberite ($CaSO_4 \cdot Na_2SO_4$) form at the contact between the central halitic area and the outer gypsum rim as a result of interaction between gypsum and halite-saturated brines.

Where the surficial water layer is slightly deeper (a few centimeters), halite crystallization occurs in two different ways. A film of supersaturated brines at the surface generates thin, tabular halite hoppers which coalesce rapidly and stay at the surface as long as the brine is still (figure 1.29C). As soon as the wind blows, this very delicate halite crust breaks into pieces and sinks down to the bottom of the

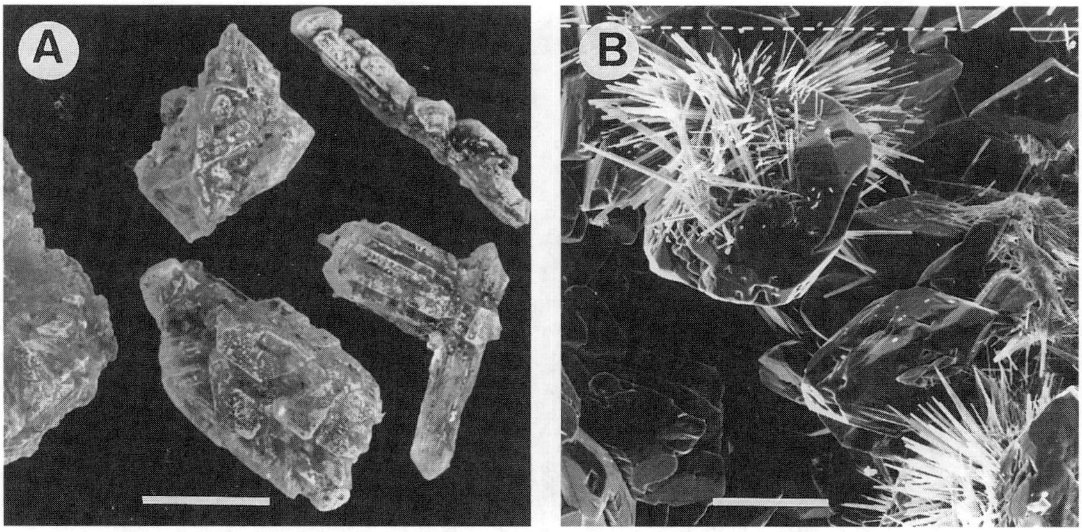

Fig. 1.28. Surficial salt deposits of continental saline lakes (photos courtesy of J. J. Pueyo).
A. Mirabilite crystals showing a more or less acicular shape. Zaragossa-Lerida area. Scale bar, 1 cm.
B. New generation of acicular thenardite crystals growing on the surface of mirabilite crystals. Zaragossa-Lerida area. Scale bar, 50 microns.

lake. On the bottom, resting on top of an algal mat or other previous deposits, halite cubes grow on the sunken crusts that act as nuclear seeds for the halite cubes. The largest halite cubes (a few millimetres up to 1 cm) grow in the center of the lake. Smaller halite cubes (0.5 mm or less) form on the edges of the lake corresponding to the very thin water layer where rapid evaporation and high supersaturation lead to very rapid crystallization.

Efflorescence crusts: Efflorescence crusts form by capillary evaporation of the interstitial brines during periods of total drying or in places that are not covered by surficial brines. The thickness of these growths is limited to a few millimeters on the edges of the lakes and reaches a few centimeters in the center where the brines with the highest salinities accumulate. Such surficial efflorescences vary from a few meters^2 to a few hundreds of meters^2 in area. The resulting crust undergoes polygonal fracture because of the forces of crystallization and thermal expansion, accompanied by the formation of well developed tepee structures (pressure ridges).

The main mineralogical species present are halite, thenardite, (Na_2SO_4) and bloedite ($Na_2SO_4 \cdot MgSO_4 \cdot 4H_2O$). The latter is only observed in these settings as an ephemeral efflorescence. During sporadic storms these efflorescence crusts are easily dissolved, generating a new brine with a new specific composition and permitting direct precipitation of bloedite in the next depositional cycle. Besides these ephemeral deposits, gypsum alone is preserved as a permanent deposit.

The surficial deposits in these saline lakes of Spain are the result of the succession in time of concentration, precipitation, and dissolution phenomena which are functions of periods of rainfall and drying.

Precipitation within the sediment: The host sediment in this area is composed of detrital particles, mainly carbonate, quartz, and clays (illite-chlorite-smectite), depending on the composition of the associated Tertiary substratum. The prevailing northwestern winds control an asymetric distribution of these detrital particles, which decrease in size and quantity from west to east.

Early diagenetic processes such as transformation of mirabilite to thenardite by dehydration or of gypsum to glauberite in a sodium-rich brine have already been described. The interstitial brines seeping through the sediment are responsible for the

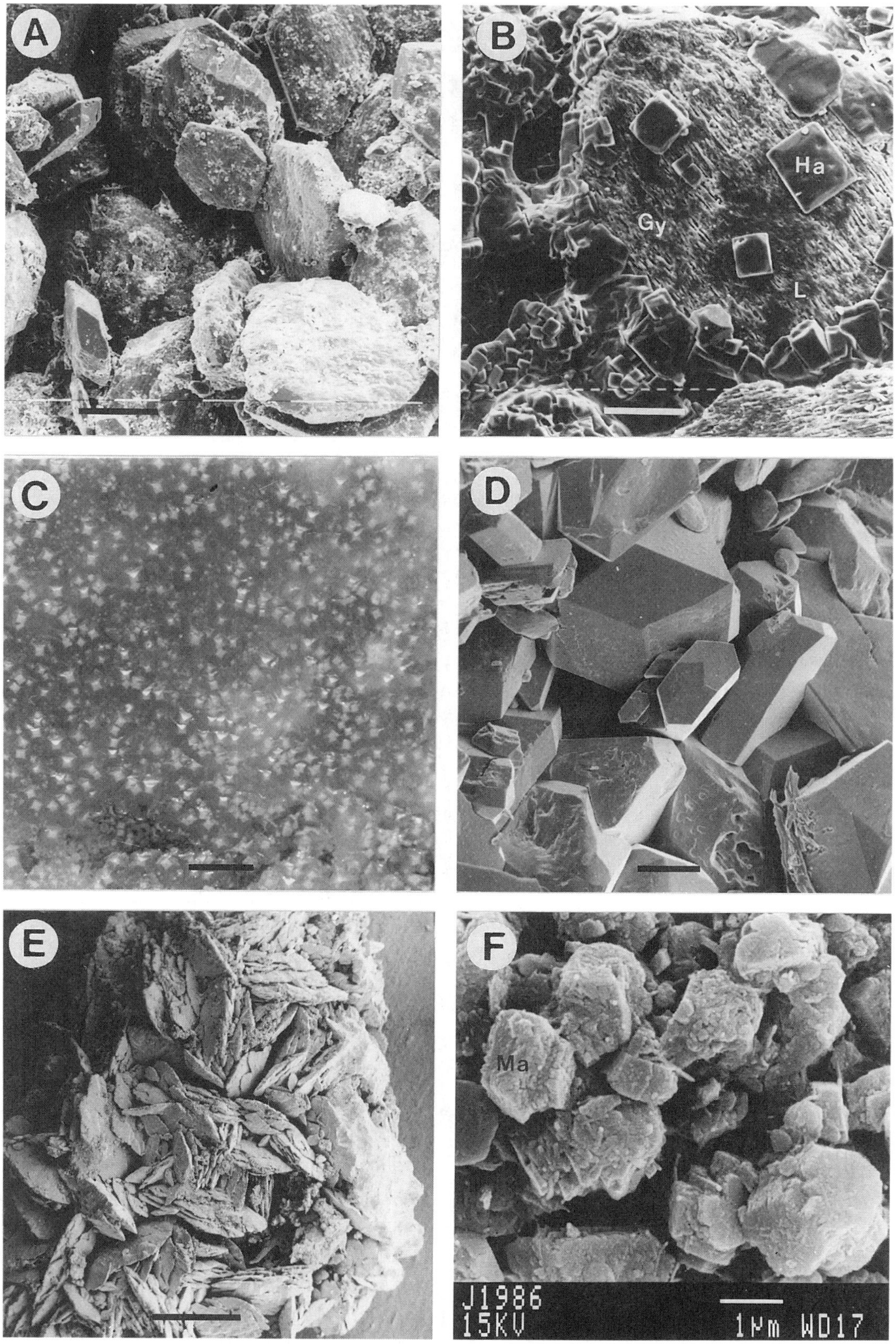

FIG. 1.29

genesis of early diagenetic evaporative minerals, such as gypsum, dolomite, magnesite, and fibrous clays.

Gypsum.

Gypsum precipitated from interstitial brines within the host sediment displays a lenticular or less commonly a nodular morphology. The lenticular gypsum crystals look very much like the surficial ones. The millimetric gypsum nodules may be subspherical or irregular in morphology. The subspherical gypsum nodules are composed of relatively loose, inframillimetric, prismatic gypsum crystals, with a sucrosic structure and a high pore brine content (figure 1.29D). In the irregular gypsum nodules, the gypsum crystals display a lenticular morphology, a brownish color, and an interlocking texture (figure 1.29E).

Dolomite.

Dolomite is limited to the edges of the lakes where brines and inflowing meteoric waters mix.

Magnesite.

There is practically no magnesite developing on the surface of the lake setting, but magnesite may form up to 50% of the total insoluble fraction of the sediment at depth, underneath the ephemeral salt-pan zone. It is likely that the mechanism responsible for this distribution is precipitation from interstitial brines below the groundwater table, because magnesite is only rarely encountered on the surface of the lakes, except where the groundwater table comes close to the surface. The magnesite rhomboedrons are well developed and grow displacively inside the host sediment (figure 1.29F). The genesis of magnesite may be related to the release of CO_2 and CO^{-2}_3 through decay of organic matter in the presence of interstitial brines with a high magnesium content, but this question is not well studied.

Fibrous clays.

Fibrous clays such as sepiolite and palygorskite are observed, commonly associated with deposits of gypsum, magnesium-bearing carbonates, and smectites. They may develop from the diagenetic transformation of smectites or from direct precipitation from magnesium-rich interstitial brines. Actual data are too limited to reach any conclusion.

As a general statement, most of the evaporative minerals formed at the surface or at depth are rapidly dissolved and the only preserved mineralogical species found in the geological record at depth is gypsum and it is not possible to find traces of many of the phenomenon occuring in these settings in ancient deposits.

Fig. 1.29. Gypsum, halite, and magnesite deposits in continental saline lakes (photographs courtesy of J. J. Pueyo).
A. Gypsum crystals displaying a lenticular habit with slightly curved faces. Zaragossa-Lerida area. Scale bar, 100 microns.
B. Lenticular gypsum crystals (Gy) cemented by halite cubes (Ha) on the edges of the saline lake. Zaragossa-Lerida area. Scale bar, 50 microns.
C. Halite crust, composed of coalescing hoppers, that develops at the surface of the highly saline brines. Zaragossa-Lerida area. Scale bar, 2 cm.
D. Highly porous prismatic gypsum crystals with sucrosic structure forming as subspherical gypsum nodules. Zaragossa-Lerida area. Scale bar, 20 microns.
E. Lenticular gypsum crystals with an interlocking texture forming irregular gypsum nodules. Zaragossa-Lerida area. Scale bar, 1 mm.
F. Magnesite crystals (Ma) displaying a rhombohedral morphology grow displacively inside the clayey host sediment. Zaragossa-Lerida area.

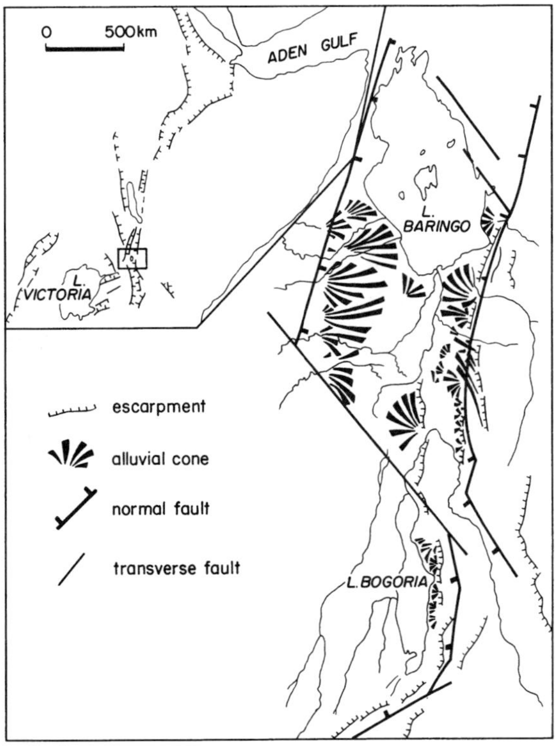

Fig. 1.30. Location map of lakes Bogoria and Baringo, Kenya, in the eastern African rift system, illustrating the structural and hydrological background (modified from Tiercelin et al. 1987).

RIFT LAKES OF EASTERN AFRICA

The great east African rift system is divided into branches: the easternmost branch, the Gregory rift, with a half graben structure, is characterized by a succession of sills and basins in which lakes may develop. Approximately in the middle of the Gregory rift are located two perennial lakes separated by a sill: Lake Bogoria to the south and Lake Baringo to the north (figure 1.30), whose hydrochemistry and sedimentology have been studied in detail (Tiercelin et al. 1987; Renaut and Tiercelin 1994).

The eastern rim of the half graben shows a succession of escarpments with foothill deposits and alluvial cones along the downdropped portion of the blocks. The water supplies mainly come from ephemeral rivers which construct alluvial fans and associated deltas. Both lakes are perennial. The setting of these lakes contains most of the subenvironments described by Hardie, Smoot, and Eugster (1978), except for the mud flats and ephemeral salt pan. This setting is characterized by large amounts of detritus. The substratum is mainly composed of volcanic rocks, basalts, and sodium-rich phonolites, so the meteoric run-off waters contain carbonate, bicarbonate, and sodium, with low K^+, Mg^{+2}, and SO_4^{-2} contents.

LAKE BOGORIA

Hydrochemistry of major elements: The water supplies in this basin have three origins: rain (600–900 mm/year), more or less ephemeral rivers that often have a very short course, and hydrothermal springs carrying $NaHCO_3 - CO_3Cl$ rich waters. The water bulk composition is the same as that of run-off waters but richer in chlorine.

The water of the lake reaches 6 to 8 m in depth and is stratified with a salinity of 55 g/l in the upper layer and 90 g/l in the bottom layer. The major elements follow the same general trends in this meromictic lake (figure 1.31). The salinity may drop down to 10 g/l close to fresh water inflows. The water stratification is a very interesting feature caused mainly by the inflow of very saline hydrothermal

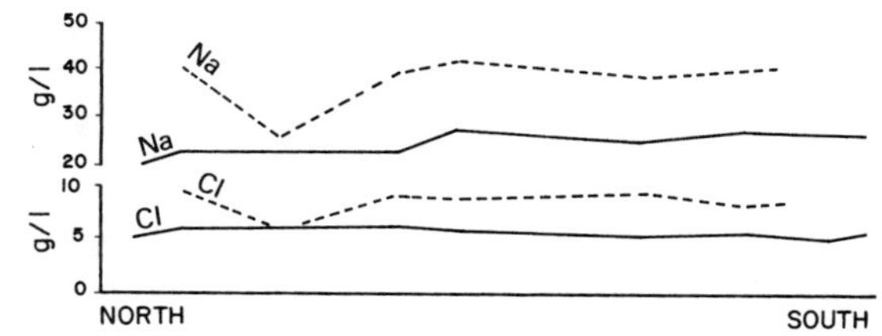

Fig. 1.31. Behavior of sodium and chlorine in the surficial (continuous line) and deeper water layers (dashed line) of Lake Bogoria, following a north-south transect (after Tiercelin et al. 1987).

brines on the bottom of the lake and some dissolution of previously formed evaporative minerals. The lack of OH^- ions in the bottom layer indicates reducing conditions allowing preservation of organic matter.

Evaporative facies: River water carries in a large amount of fine detrital material. When this fresh water mixes with the saline water of the lake some differential flocculation occurs causing rapid settling of clay particles on the bottom of the lake. Because of brine stratification, the flow is hypopycnal, and moves its detrital load in the upper layer; the flocculant is carried for some distance. The detrital supplies are distributed all around the lake, forming a peripheral rim of particles, mainly sands, silts, and clays, with decreasing size from the edges to the center of the lake. The central part of the lake is covered with a dark black or green clayey mud without internal structure or having very thin laminations. The clay mineral association is dominantly smectitic, but kaolinite and illite, or more rarely chlorite, are present. Some of the associated clays, particularly near the lake margins, are altered to zeolites by synsedimentary processes due to evaporative concentration of pore water (Renaut 1993).

The evaporative minerals develop in the deepest central parts of the lake within the black mud that forms from carbonate and sodium-rich bicarbonate brines. The minerals encountered are: gaylussite ($Na_2Ca(CO_3)_2 \cdot 5H_2O$), trona ($Na_3H(CO_3)_2 \cdot 2H_2O$), nahcolite ($NaHCO_3$), magadiite ($NaSiO_3(OH)_3 \cdot 3H_2O$), and rare calcite. Calcium carbonate (calcite, aragonite) pinnacles, bowls, chimneys, and white, millimetric, siliceous crusts form around the hydrothermal springs. Fluorite develops in fractures or as pseudomorphs of calcium carbonate.

The organisms living in the lake are made up by a very limited number of species, but constitute a significant biomass. For instance cyanobacteria, mainly filamentous *Oscillatoria,* represent 3,500 tons of fresh organic matter per day for a 35 km^2 lake. Rare diatoms are also present in the center of the lake; but they are found in large quantities in the coastal muds and algal crusts close to the hydrothermal springs. The brine stratification results in an anaerobic lower water layer, permitting the perservation of the organic matter responsible for the black color of the sediment in the central part of the lake.

LAKE BARINGO

Hydrochemistry of major elements: The water supplies originate from the same sources as for Lake Bogoria, with the major difference that there is a significant contribution by two large perennial rivers while the hydrothermal springs are limited. The bulk water composition is also sodium and bicarbonate-rich. The water depth fluctuates between 4 and 9 m and the salinity is very low, 0.5 g/l. The lake is holomictic, showing no stratification. The low salinity is probably linked to groundwater supplies, but this could not be demonstrated conclusively.

Evaporative facies: The detrital influx is similar to that in Lake Bogoria. Because of the homogenous water body, the inflowing fresh water forms a hyperpycnal flow on the bottom, where the detrital load settles directly near the area of entry. There is no occurrence of evaporative minerals in the muddy sediment in the central part of the lake.

Numerous organisms live in this setting: coccoid cyanobacteria, diatoms, sponges, ostracods, and gastropods, all of which act as sources for abundant bioclasts. The bioclasts are cemented by diffuse organic matter or thin laminated organic veils of cyanobacterial origin.

SUMMARY

These two lakes, located close together within the same rift structure, contain large variations in their composition and organization, depending on their individual water supplies, the nature of the substratum, and the structural framework. These lakes are very good examples of sedimentary differentiation between two water bodies with the same regional background: climate, vegetation, landscape, existing at the same time. The water balance and the supply of hydrothermal brines are responsible for the differences in the evaporative sedimentation and in the amount, variety, and preservation of organisms living in these lakes.

Conclusion

The purpose of this study is to give the hydrochemical and sedimentological characteristics of different evaporative environments—marine, mixed, and continental—to provide specific models that represent modern analogs to the Paleogene evaporites. The data are summarized in table 1.2 (see page 4). The modern evaporative deposits represent a wide range of environments from marine to continental through marine marginal with mixed waters, and offer a model of what could have happened on the margins of a fossil evaporative basin.

The behavior of major elements in the surficial and interstitial brines varies depending upon the environment. Of course the lithology of the basement has an influence on the chemistry of the groundwaters involved in the evaporative system. The controlled salt pond model (marine) gives the basic evolution of water bodies with increasing salinity of the major elements leading to successive precipitation of carbonates, gypsum, and halite. At the opposite end of the spectrum, the continental environments are totally dependent upon the chemical composition of the groundwater, and the evaporative minerals that form are highly variable and include mirabilite, thenardite, bloedite, gaylussite, and natron. The marine marginal model appears as an intermediate environment: the general behavior of the major elements is close to that of pure sea water; however the groundwaters introduce some differences and determine, alongside the normal evaporative sequence, the formation of additional minerals such as dolomite or polyhalite.

The evaporative deposits in the marine subaqueous environment mainly show laminated or layered structures sporadically cross-cut by nodules. The layering is also encountered in the mixed environment under the influence of both marine and meteoric waters, but the rather irregular structures such as nodules and scattered crystals are widespread and finally appear as the general feature in the continental environments. The possible presence of regular layered structures in the different environments described here does not permit clear conclusions to be drawn about the type of depositional environment in fossil evaporative deposits; however the relative abundance of each type of structure may help, if other additional arguments are available, such as trace element, fluid inclusion, and isotopic analysis.

Evaporative environments generate drastic conditions for living beings, most of which disappear rapidly as soon as salinity increases, and in fact this extreme environment produces almost no recognizable assemblage of fossil biota. The living community that adapts to these environments is largely composed of cyanobacteria that produce a tremendous quantity of organic matter which accumulates in the sediments. Upon burial the organic matter preserved in the sediment is later available for transformation into hydrocarbons.

Thanks to B. C. Schreiber, who very carefully discussed, reviewed, and improved the manuscript and also to G. Busson for his suggestions. Special thanks to J. J. Pueyo for his help in providing many of the photographic illustrations in this paper, well as to A. Cornée and J. C. Thomas.

PART 2

Paleogene Continental Rift System of Western Europe: Introduction and Stratigraphy

The next four papers cover most of the evaporative basins of the Western European rift that developed during the Paleogene. The general structural setting for the basins is given in chapter 2 and the basins themselves are presented in geographic order running from north (Alsace) to south (Basse Provence) in the chapters that follow.

All of this sedimentation was the result of a marked regional climate change with increasing aridity during the Eocene; most of the evaporative sedimentation took place and ended within the Paleogene. Alsace, however, continued to receive further evaporite deposition into the Neogene (lower Miocene). Many of the deposits that were formed seem to be partially or wholly non-marine in origin, but this question is addressed on a basin by basin basis.

Because the stratigraphy of this region is partially based upon continental sections, the nomenclature is somewhat out of phase with the usual marine designations (and boundaries). For this reason we include a stratigraphic column presenting all of the internationally agreed stage names and their more local equivalents. This table permits the reader to go back to the references cited in each section that use regional nomenclature, and to understand the non-worldwide designations. This table (table P2.1) is based on the geological column (modified) presented by Odin and Odin (1990).

TABLE P2.1.

Stratigraphic nomenclature employed in part 2.

(Adapted from Odin and Odin 1990)

ERA	SYSTEM	SERIES	STAGE	Ma
CENOZOIC	NEOGENE	PLIOCENE	PIACENZIAN	3.4
			ZANCLEAN (TABIANIAN)	5.3
		MIOCENE	MESSINIAN	6.5
			TORTONIAN	11
			SERRAVALIAN	14.5
			LANGHIAN	16
			BURDIGALIAN	20
			AQUITANIAN	23.5
	PALEOGENE	OLIGOCENE	CHATTIAN	28
			RUPELIAN (STAMPIAN)	34
		EOCENE	PRIABONIAN (LUDIAN)	37
			BARTONIAN	40
			LUTETIAN	46
			YPRESIAN	53
		PALEOCENE	THANETIAN	59
			DANIAN	

2

Paleogene Continental Rift System of Western Europe: Locations of Basins, Paleogeographic and Structural Framework, and the Distribution of Evaporites

JEAN-MARIE ROUCHY

The Western European Continental rift system is a spectacular intraplate fault system that is generally oriented N-S and cuts sharply across the Alpine foreland, extending more than 1600 km from the Mediterranean to the North Sea (figure 2.1). Physiographically, it is composed of a complex set of individual grabens with major orientation NE-SW, NNE-SSW, and locally WNW-ESE. This extensive deformational event began somewhat diachronously among the component segments during the middle-upper Eocene to Oligocene. This system is still weakly active but represents an entirely new tectonic feature intersecting a heterogeneous series of structures created by Caledonian, Hercynian, and early phases of the Alpine orogenies. The rifting was preceded by a compressive phase that is especially well marked in southern France, where it is responsible for the folding and thrusting events known as the Pyreneo-Provençal phase (Arthaud and Seguret 1981; Arthaud et al. 1980/1981).

The rifting events that created this system must be integrated not only into the kinematics of the Europe-Africa plate motion and considered within the framework of the Alpine and Pyrenean orogenies but also must be tied together with the opening of the western Mediterranean basin (Dewey et al. 1989; Helman 1989). During the period of rifting, marked compressional events were evident in the Alpine and Pyrenean areas. In the Pyrenees, this compression created the Ebro basin and Catalonian foredeep basins and is directly evidenced by intense phases of thrusting (Fischer 1984; Seguret and Daignères 1986; Vergès and Munoz, 1990), as was the main phase of shortening in the Alps (Trümpy 1980).

The different tectonic settings of continental rifts and active foredeep basins have produced physiographic conditions that, because of relatively arid climatic conditions, caused a fairly well-developed evaporative phase of sedimentation. This includes the formation of halite and potash deposits that suggest at least short phases of extreme aridity. Nevertheless, the evaporite deposits differ markedly from area to area in their thickness, composition, and regional extent. The rift settings are characterized by depocenters with evaporative accumulations of as much as 2,000 meters lying within a few extended subsiding troughs. In these troughs continental conditions of deposition prevailed in most instances (fluviatile and lacustrine) with sporadic and limited marine input. The environments evidenced rapid changes in space and time, as in comparable settings of the east African rift system or the Dead Sea. In the foredeep basins of northern

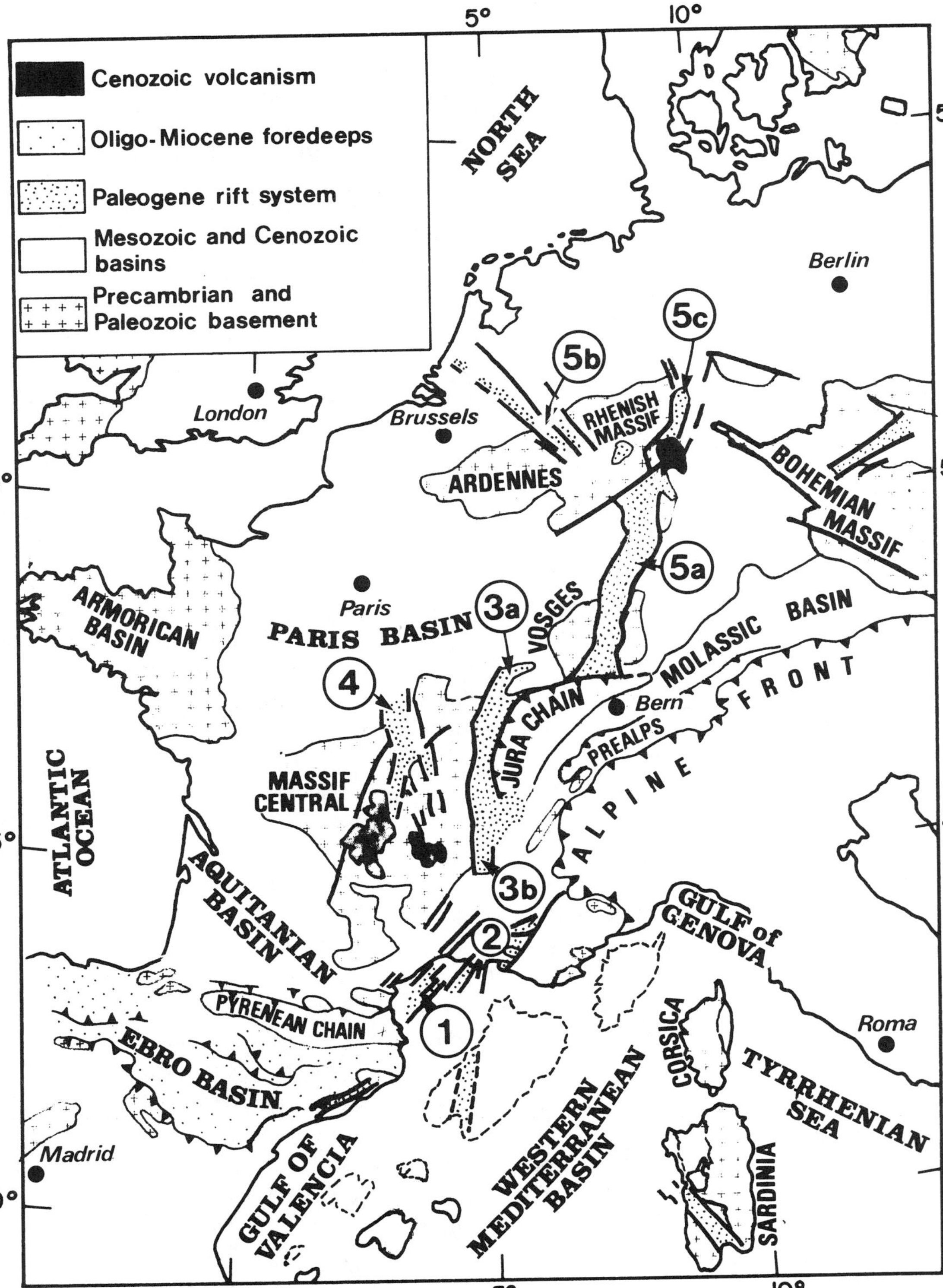
Cenozoic volcanism
Oligo-Miocene foredeeps
Paleogene rift system
Mesozoic and Cenozoic basins
Precambrian and Paleozoic basement
5°
10°
55°
50°
45°
40°
0°
NORTH SEA
Berlin
London
Brussels
5b
5c
RHENISH MASSIF
ARDENNES
BOHEMIAN MASSIF
5a
Paris
PARIS BASIN
3a
VOSGES
ARMORICAN BASIN
4
MOLASSIC BASIN
JURA CHAIN
Bern
FRONT
PREALPS
ALPINE
ATLANTIC OCEAN
MASSIF CENTRAL
AQUITANIAN BASIN
3b
2
1
GULF of GENOVA
PYRENEAN CHAIN
EBRO BASIN
CORSICA
Roma
TYRRHENIAN SEA
Madrid
WESTERN MEDITERRANEAN BASIN
SARDINIA
GULF OF VALENCIA

FIG. 2.1

Spain, the evaporative sedimentation developed over broad areas within basins fed by both marine and nonmarine waters. These conditions persisted over a long interval, from the Eocene to the Middle Miocene (Ortí and Pueyo 1977; Ortí et al. 1984).

Intensive studies based on densely spaced geophysical and well data have been carried out in some parts of these basins during exploration for hydrocarbons and potash salts, the exploitation of salt, and in studies associated with underground hydrocarbon storage. The areas of investigation include the Upper Rhine, Bresse-Valence, and Manosque segments of the Western European rift, and the southern Pyrenean foreland basins. More recently, preliminary research for nuclear waste storage sites extended our knowledge of the Bresse basin. Except for the basins of Camargue and the Gulf of Lion, where the evaporites are known in a few deep wells or inferred from seismic investigations, most of these Paleogene sedimentary areas have undergone studies that furnished a well-controlled stratigraphical, structural, and sedimentological framework.

The Continental Western European Rift: Location of Basins and Structural Trends

The continental Western European Rift does not seem to have a continuous structure but appears as a chain of disparate segments that include the northwestern Mediterranean segment (cut into Paleozoic basement), the Roussillon-Languedoc-Provence segment (superposed on the older structures of the series of Mesozoic Rhodano-Languedocian basins and the Pyrenean chain), the Bas-Dauphiné-Bresse segment (located upon Mesozoic strata), the Massif Central (located within Hercynian crystalline basement) and the Rhine graben (cutting a complex sequence including Paleozoic and Mesozoic strata). Many of the segments are controlled, at least in part, by older Hercynian structures. Although they are not dicussed in this volume, some grabens also have been cut into the Bohemian massif (northern Bohemian basin, Cheb and Sokolos basins, southern Bohemian basins of Budejovia and Trebon) (Malecha 1966; Malkovsky 1966; Vacl 1966). Each rift segment is a composite formed by individual grabens of various sizes and morphologies that have their own specific structural and sedimentary histories. Some of these segments are offset from one another along transverse directions that have been interpreted as being parts of a continental transform fault system (Illies 1972, 1974; Bergerat 1977a, 1977b, 1985). The formation of the rift did not occur synchronously along the entire length of the system; the development and subsidence started 45 million years ago (MA) (Middle Eocene) in the Upper Rhine Graben, but was active only during the Oligocene in the Marseille grabens in southern France. It continued to be discontinuously active at least into Aquitanian times and even into the Plio-Quaternary, as in the northern part of the Upper Rhine. Local interruptions, shifts in structural directions and in the rates of subsidence are also apparent.

THE NORTHWESTERN MEDITERRANEAN SEGMENT

Both applied and pure geophysical research, particularly seismic studies, carried out in the western Mediterranean have revealed the presence of tectonic depressions buried under the Mio-Pliocene sediments (figures 2.2, 2.3). Based on their morpho-

Fig. 2.1. Simplified map of the Western European rift system. The segments described in the text are listed in the order they are discussed. 1 = Gulf of Lion; 2 = Roussillon-Languedoc-Provence; 3 = Massif Central; 4a = Bresse; 4b = Bas-Dauphiné; 5a = Upper Rhine; 5b = Lower Rhine; 5c = Hesse. (The Bohemian massif, located to the east of the Rhine graben, is not discussed in the text.) Paleogene troughs and basins indicated beneath the Gulf of Lion are known from seismic studies. Note two different positions for Corsica and Sardinia: dashed lines for Paleogene and solid for modern (based on Rehault et al. 1984).

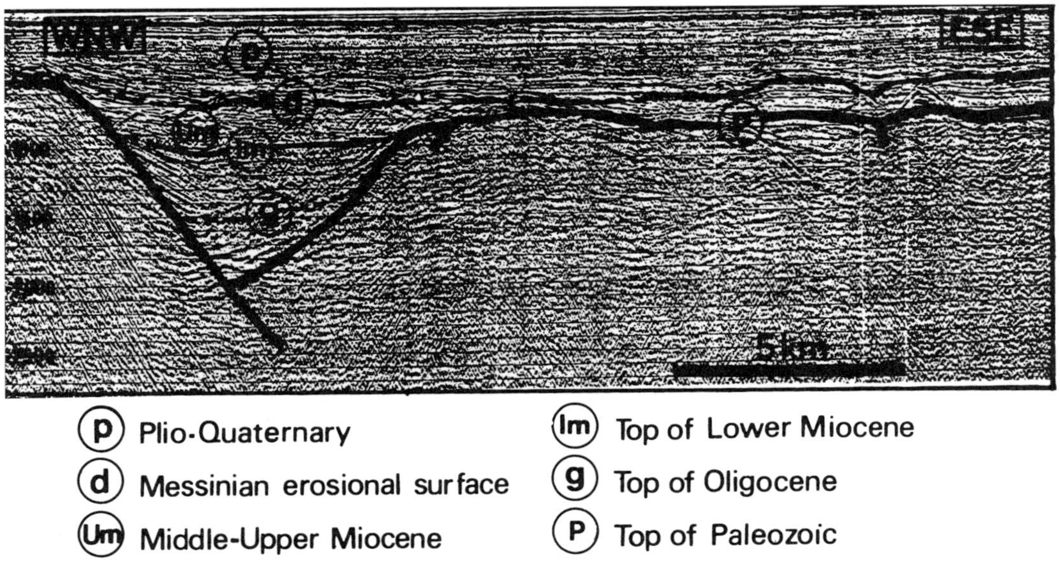

Fig. 2.2. Seismic section in the western part of the gulf of Lion crossing a half-graben which represents the off-shore extension of the Nîmes graben (Roussillon-Languedoc-Provence segment). Depth given in two-way travel time in milliseconds. Location of section given in figure 2.7. Courtesy S.N.E.A.P. and ESSO REP.

structural patterns as well as their predominant NE-SW to NNE-SSW orientation, these grabens are clearly related to the general distension that produced the Western European rift system. Indeed, for our purpose, the grabens described here have remained continental; but the initial rifting phase that evolved into the creation of the western Mediterranean oceanic basin (between 21 and 18 MA) might be of the same age (Rehault et al. 1984). None of these grabens have been penetrated by wells and the age of the initial sedimentary fill is inferred as Oligocene from projection of the ages of the overlying sediments. In the Ligurian sea, however, Oligocene conglomerates were actually dredged from the axis of one of these grabens (Belaiche et al. 1976; Rehault 1981).

Pre-rift geological and structural framework.

At the beginning of the Oligocene, before the opening of the oceanic basin, the western Mediterranean was composed of different continental blocks and subplates. In the area considered here, the Gulf of Lion was associated with Corsica and Sardinia to form a promontory of the European subplate.

In spite of large gaps in our knowledge, magnetic and gravimetric data, the results of seismic exploration, as well as that of a few deep wells (Tramontane, Autan, Mistral, and Sirocco), and some data from deep-sea dredging yield an idea of the nature of the basement in the Gulf of Lion. Offshore, based on seismic lines and boreholes, in a region lying a short distance south of the present coastline of Languedoc-Provence, the Mesozoic series is replaced by Paleozoic metamorphic schists and granitoids (Cravatte et al. 1974; Arthaud et al. 1980/1981).

Locally, the rifting has been preceded by a phase of uplift that has been documented on the periphery of the western Mediterranean, in the Maures-Esterel and southern Alps, as well as in presently submerged areas of the Gulf of Lion and the Gulf of Valencia (Aubouin and Mennessier 1963; Cravatte et al. 1974; Mauffret et al. 1981). The strong erosional processes that led to the Eocene flysch accumulations of the Grès d'Annot could have been related either to this compression or to pre-rift up-

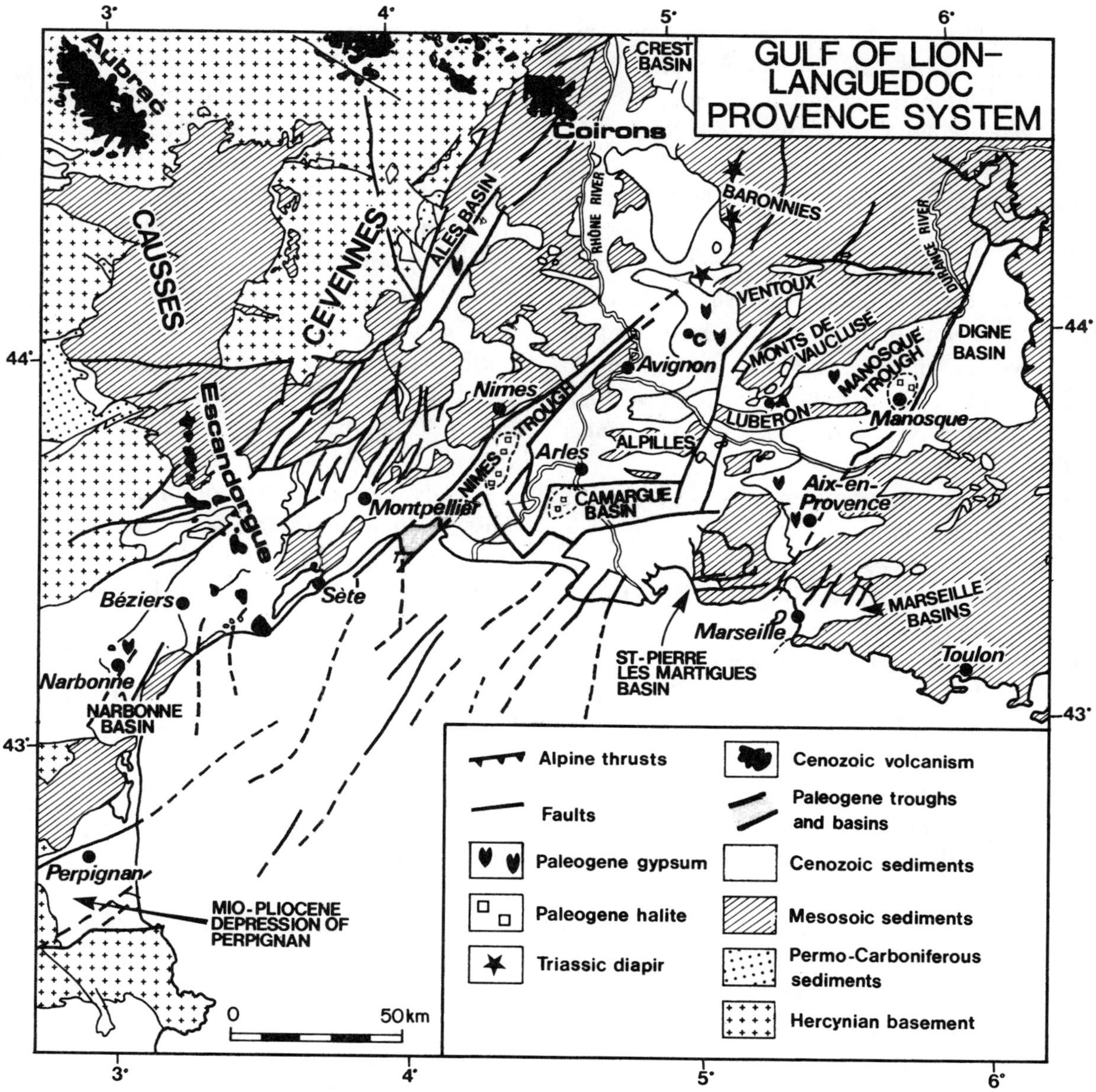

Fig. 2.3. Simplified geological map of southern France showing the location of the troughs of the Roussillon-Languedoc-Provence segment and of the Gulf of Lion (composite map based on geological map of France at 1:1,500,000 and Arthaud et al 1980/81). Two specific areas (Carpentras and Apt), mentioned in the text, are labeled C and A respectively.

lift (Stanley and Mutti 1968). The rifting phase that occurs along NE-SW fractures are more or less a prolongation of the Languedoc-Provence system. The Cévennes, Nîmes, and Durancian fault bundles are discussed in this chapter. These faults are superposed upon late Hercynian fractures and are thereafter reactivated at every compressive or distensive tectonic phase.

Location of basins and morphostructural framework.

The Gulf of Lion is cut by a system of 030° tectonic troughs in which the present depths of the base of the Tertiary floor could be 5 km below the sea level in the central graben under the continental shelf. It is noted that these troughs apparently were half-grabens filled in with thick sediments of Oligocene (including probable salt formations) and lower Miocene (Aquitanian) ages (Gennesseaux and Lefebvre 1980; Arthaud et al. 1980/1981; Rehault et al. 1981; Mauffret and Gennesseaux 1989). The age of their opening could be middle Oligocene (Mauffret and Gennesseaux 1989). These authors consider that these basins are generated in a strike-slip tectonic regime as pull-apart basins prior to the extensional regime. Locally the width of the grabens exceeds 25 km and extends over more than 100 km in length. To the south, the Gulf of Lion troughs seems to be linked to the Gulf of Valencia rift (Biju-Duval et al. 1978).

Eastward, the broad graben of Sassari-Cagliari in Sardinia, filled with Oligocene sediments, presently has a strike of 145°. If rotated into its initial position prior to the eastward motion of the Corsican and Sardinian blocks as shown in figure 2.1, the graben had a 030° trend that is approximately the same as the Marseille trough complex (Cerchi and Montadert, 1982; Nury 1987, 1988).

In the eastern part of the basin, the present canyons of Stoechades and Saint-Tropez are superimposed upon a system of narrow Oligocene grabens linked with the Gulf of Genova, which might be considered as buried NE-SW rifts as is the Valencia gulf (Rehault et al. 1984). Half-grabens filled with Oligocene to Aquitanian sediments formed within the depressed portions of tilted blocks on the margins of Corsica, Sardinia, and the Provençal basin (Mauffret et al. 1982).

Although situated in a different structural setting, the NE-SW grabens of the Menorca margin are probably related to the same Oligocene distensive phase. Thus, the lower Burdigalian section (lower Miocene) cored in the Deep Sea Drilling Project (DSDP) hole 372 overlies 800 meters of sediments that are probably Oligocene to Aquitanian age (Biju-Duval et al. 1978; Mauffret et al. 1978; Rehault et al. 1984). Biju-Duval et al. (1978) suggest that evaporative deposits possibly could be intercalated within this sedimentary series.

Rifting and post-rifting history.

As is the case in other segments of the rift, a differential subsidence of the graben floors seems to have continued into the Aquitanian, but we lack sufficient data to precisely delineate the changes in subsidence intensity. Further study of the geological history must take into account the geodynamic evolution of the entire western Mediterranean and extends beyond the scope of this book (see Dewey et al. 1989; Helman 1989). Nevertheless we note that extensional tectonics still occurred in the Sardinian grabens during the Plio-Quaternary (Thomas et al. 1988).

Calc-alkaline magmatism occurred in various regions (Sardinia, southern Spain, and north Africa) from the Oligocene until the Pliocene with a very active phase of activity that ended in the period of 24 to 17 MA. This general area (Sardinia, north Africa, and southern France) then was affected by alkaline volcanism that continued until the Quaternary (Girod and Girod 1977; Bellon 1981).

THE LANGUEDOC-ROUSSILLON-PROVENCE SYSTEM

The rift system seen in southern France is not a single, well-individualized rift but rather a set of tectonically-controlled depressions of various sizes and morphologies associated with elevated (platform-like) areas.

Detailed studies have been carried on within some of the tectonic depressions of this area, particularly in the well-exposed portions as in the Manosque-Forcalquier and Marseille areas (Destombes 1961; Nury 1987/1988), but most of these grabens remain incompletely investigated particularly with respect to their deeply buried structures. However, seismic profiles and a few deep wells made during oil exploration have yielded important data. Studies of salt formation for purposes of exploitation for hydrocarbon storage have certainly improved our knowledge but only within the immediate area of the structural limits of the salt, as in the Vauvert district of the Nîmes graben and in Manosque.

Pre-rift framework.

The tectonic depressions appear mainly but not exclusively related to three major sets of NE-SW to NNE-SSW fault systems that were active from the late Hercynian to the present (Arthaud et al. 1977; Arthaud and Seguret 1981; Arthaud et al. 1980/1981; Triat and Truc 1983). They are, from east to west, the Cévennes, Nîmes, and Aix-Durance fault bundles (figures 2.3 and 2.4). The regional structural development is widely controlled by the successive reactivation of the late Hercynian faults and the deformation associated with the Eocene Pyreneo-Provençal compresssion.

The Cévennes fault bundle initially controlled the formation of Carboniferous and Permian basins but later it linked with the Durancian fault to form the eastern limits of the rapidly subsiding part of the Mesozoic basin (Dubois and Umbach 1974; Baudrimont and Dubois 1977). Locally, significant thickening of the Jurassic strata up to 4,000 meters was apparently controlled by the Durancian fault line (Triat and Truc 1983) that also constituted the eastern border of the Paleogene graben of Manosque-Forcalquier. Although less apparent, the role of the Nîmes fault also might have been important during the Mesozoic, particularly in the Triassic where the thickest evaporites follow this same direction (Dubois and Umbach 1974; Goy and Busson 1980; Appia-Lamache 1984), and upon which the Paleogene graben of Nîmes appears more or less superimposed.

After the late Cretaceous emersion of the main part of the area, an intense period of N-S compression, generally known as the Pyrénéo-Provençal phase, began from the end of Paleocene up to the late Eocene and even continued into the early Oligocene. The Pyrénéo-Provençal tectonism results in a marked transformation of the earlier regional structure (Arthaud and Seguret 1981). The late Hercynian faults have been rejuvenated with sinistral shear displacements of as much as 20 km (Bodeur 1976; Masse 1980; Arthaud and Seguret 1981). The Mesozoic and part of the Eocene strata are affected by hectometric to kilometric folds and overturned folds, mostly oriented E-W, that are exposed in the western part of the area. Similarly, large overthrusts occurred in the Languedoc-Roussillon area (Corbières-Saint-Chinian nappe, Montpellier fold) and particularly in Provence where the thrusting was significant and has been dated as Bartonian.

Location and physiographic pattern of the rifts.

The NW-SE to E-W distension that took place shortly after the end of the Pyrénéo-Provençal structural development produced reactivation with normal displacement of the great NE-SW faults and opening of the rifts. The age of the beginning of the distension is questionable because of looseness of the stratigraphic framework. A middle Oligocene age is generally accepted but it could have begun as early as late Eocene. It reached a period of maximum activity in the Oligocene, began to slacken during the early Miocene (Aquitanian), and continued intermittently during the Miocene and Pliocene (Arthaud and Seguret 1981; Bergerat 1985; Fredet 1987). The morphostructural characteristics of the main grabens are oblique to the directions of earlier folding and thrusting (Alès, Camargue) while the orientation of some smaller depressions appears to have been partly controlled by the geometry of the overthrust fronts (as in St-Pierre-les-Martigues and Haute-Provence).

The eastern limits of the Cévennes fault bundle

FIG. 2.4. Overview of rift segments:
A. Satellite view of Languedoc-Roussillon-Provence, Bresse-Bas-Dauphiné, and Massif Central.
B. Simplified interpretative map showing the main structural trends, particularly the Cévennes, Nîmes, and Durance faults bundles. Landsat Photos.

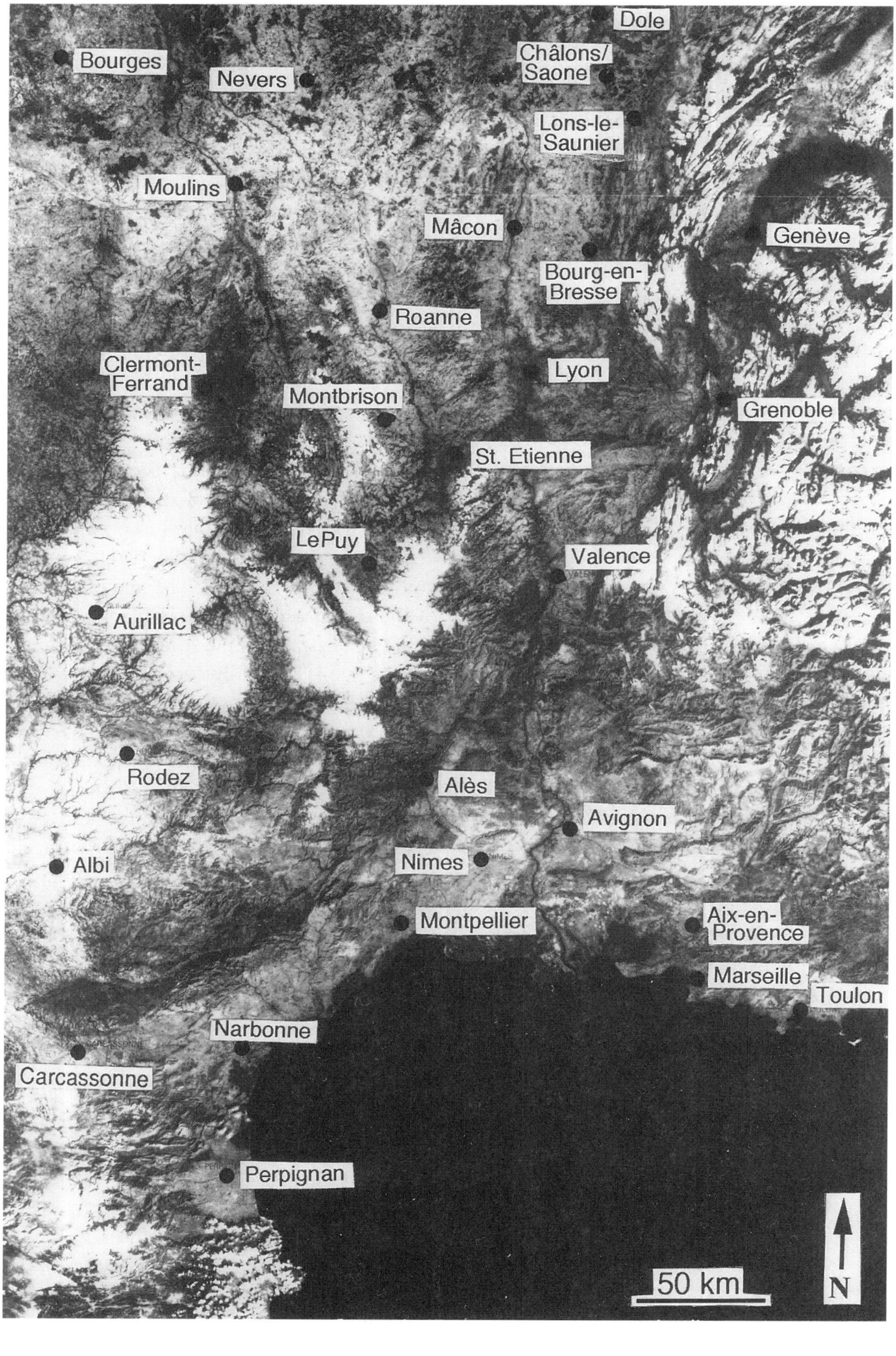
Dole
Bourges
Nevers
Châlons/
Saone
Lons-le-
Saunier
Moulins
Mâcon
Genève
Bourg-en-
Bresse
Roanne
Clermont-
Ferrand
Lyon
Montbrison
Grenoble
St. Etienne
LePuy
Valence
Aurillac
Rodez
Alès
Avignon
Albi
Nimes
Montpellier
Aix-en-
Provence
Marseille
Toulon
Narbonne
Carcassonne
Perpignan
50 km
N

FIG. 2.4A

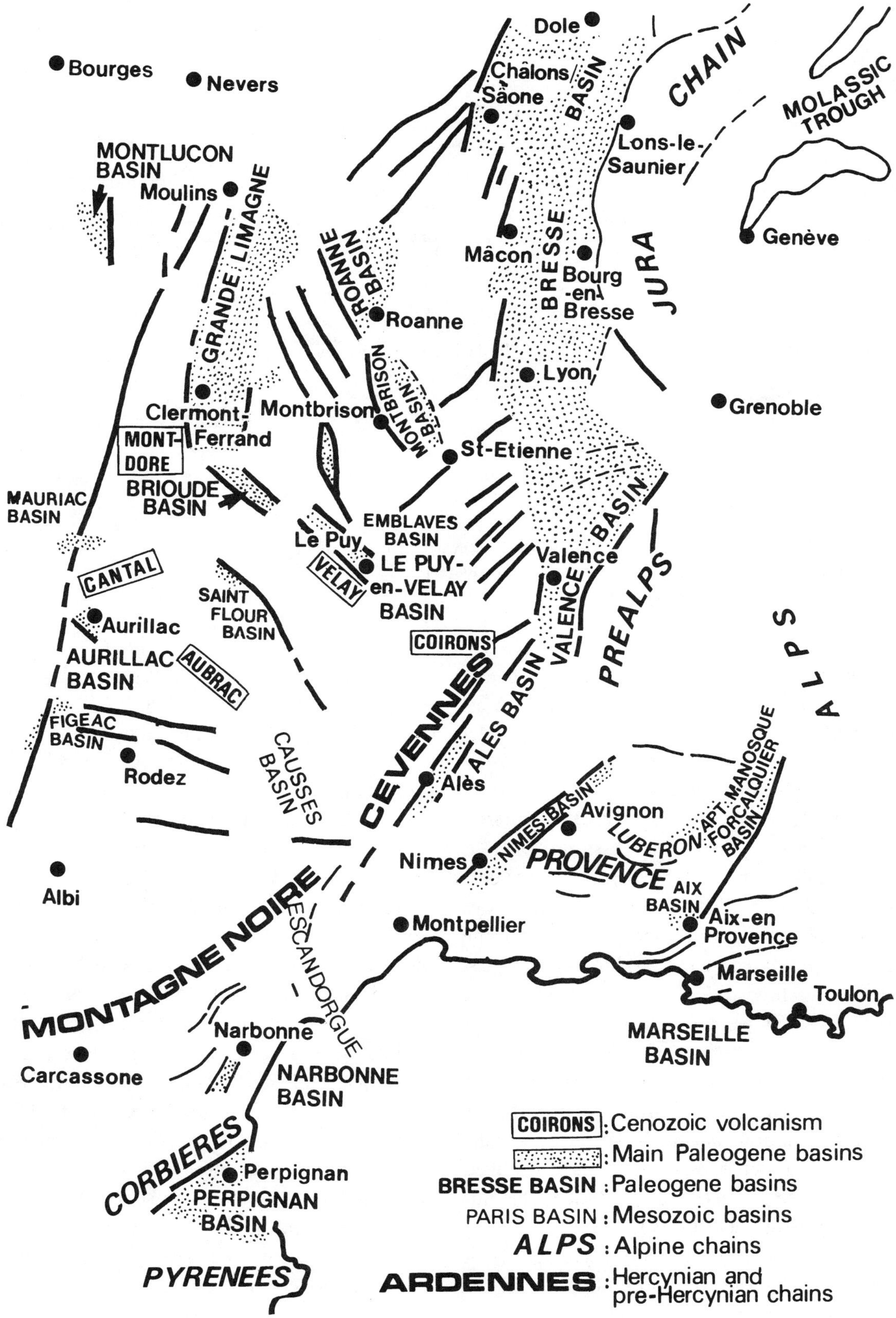
Dole
Bourges
Nevers
Châlons/Saône
BASIN
CHAIN
MOLASSIC TROUGH
Lons-le-Saunier
MONTLUCON BASIN
Moulins
GRANDE LIMAGNE
ROANNE BASIN
Mâcon
BRESSE
Bourg-en-Bresse
JURA
Genève
Roanne
MONTBRISON BASIN
Lyon
Grenoble
Clermont-Ferrand
Montbrison
MONT-DORE
St-Etienne
BRIOUDE BASIN
MAURIAC BASIN
EMBLAVES BASIN
Le Puy
BASIN
Valence
VELAY
LE PUY-en-VELAY BASIN
CANTAL
SAINT FLOUR BASIN
Aurillac
COIRONS
VALENCE BASIN
PREALPS
ALPS
AURILLAC BASIN
AUBRAC
CEVENNES
ALES BASIN
FIGEAC BASIN
CAUSSES BASIN
Rodez
Alès
APT MANOSQUE FORCALQUIER BASIN
Avignon
NIMES BASIN
LUBERON
Nimes
PROVENCE
Albi
AIX BASIN
Aix-en Provence
Montpellier
MONTAGNE NOIRE
ESCANDORGUE
Marseille
Toulon
MARSEILLE BASIN
Narbonne
Carcassone
NARBONNE BASIN
CORBIERES
Perpignan
PERPIGNAN BASIN
PYRENEES
COIRONS : Cenozoic volcanism
: Main Paleogene basins
BRESSE BASIN : Paleogene basins
PARIS BASIN : Mesozoic basins
ALPS : Alpine chains
ARDENNES : Hercynian and pre-Hercynian chains

FIG. 2.4B

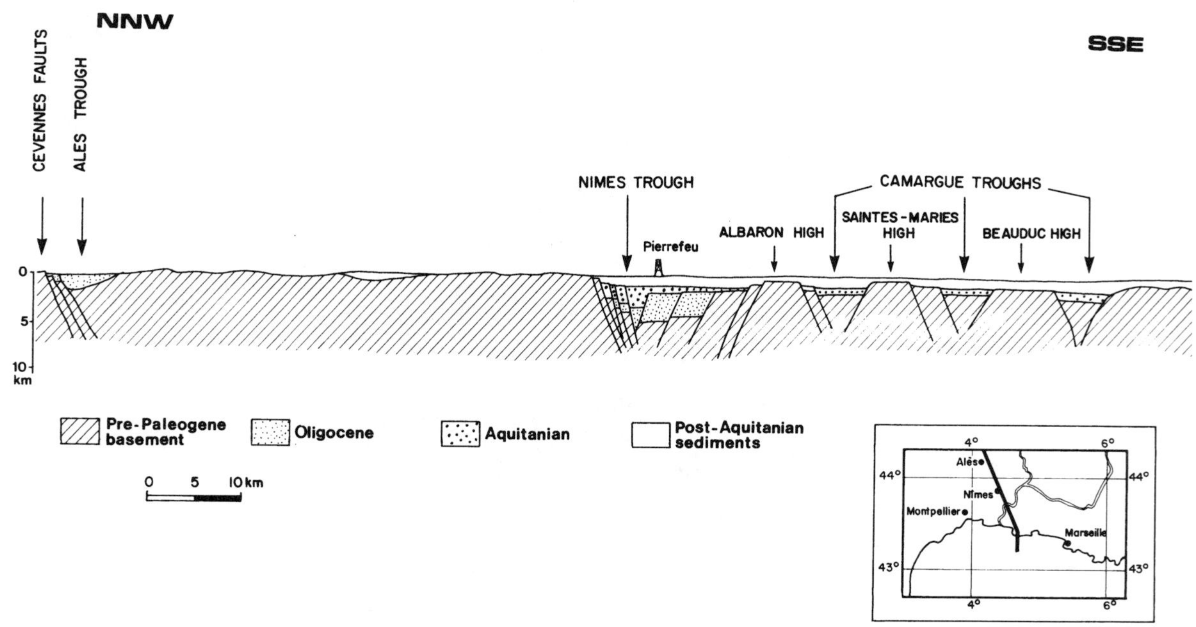

FIG. 2.5. Schematic geological cross-section in the southwest of France crossing NNW to SSE the troughs of Alès, Nîmes, and Camargue (modified from Arthaud et al. 1980/81).

lie along the Languedoc graben system. This system is composed of the asymmetrical graben of Alès, with its northern extension forming the Barjac depression (about 40 km long and 7 km wide), and a set of smaller depressions composed of half-grabens tilted westwards along NE-SW faults (Sommières, Les Matelles). Some of these depressions, like that of Saint-Martin-de-Londres, have been formed earlier during the compressive phase. In the Alès trough up to 800 meters of late Eocene-Oligocene sediments accumulated including thin gypsum beds and bituminous limestones. An additional several hundreds of meters were deposited in the deepest part of some half-grabens beginning in the Oligocene while thinner deposits occured laterally on the platform related to each half-graben as in the case of the Saint-Chaptes basin. Southward, in the Roussillon, the graben of Narbonne-Sigean is filled by about 500 meters of lacustrine sediments including gypsum (Rosset 1964; Dumas 1984; Lenguin et al. 1990; Szulc et al. 1991). The Cevennes fault changes direction to the N-S and plunges under the Gulf of Lion where the geophysical data suggest it constitutes the western limit of another local tectonic depression (Arthaud et al. 1980/1981).

The deeply buried graben system of lower Rhône-Camargue (figures 2.5, 2.6, 2.7), bounded on the west by the Nîmes fault, is the most spectacular feature developed during the Oligocene distensive phase in southern France. The NW-SE trending Nîmes graben constitutes the main trough and is about 80 km long and 10 km wide. The depression narrows and weakens northwards but then appears to continue, as represented by the small graben of Suzette, which is now occupied by a dome whose core is composed of intrusive Triassic evaporites (Casagrande et al. 1989). The Nîmes graben runs into the Gulf of Lion where it is linked to the northwestern branch of the main graben. The Nîmes graben appears in the seismic line presented in figure 2.7A as a half-graben bounded on the west by a listric fault system that flattens in the underlying Mesozoic series, probably within the Triassic evaporites. The system presents a rollover-like structure limited at the footwall of the fault by a compensation syncline along which subsidence seems to have continued during the Miocene.

In the deepest area the thickness of the sedimentary fill is estimated to be 5,000 meters, including 3,000 meters of Oligocene accumulation with evap-

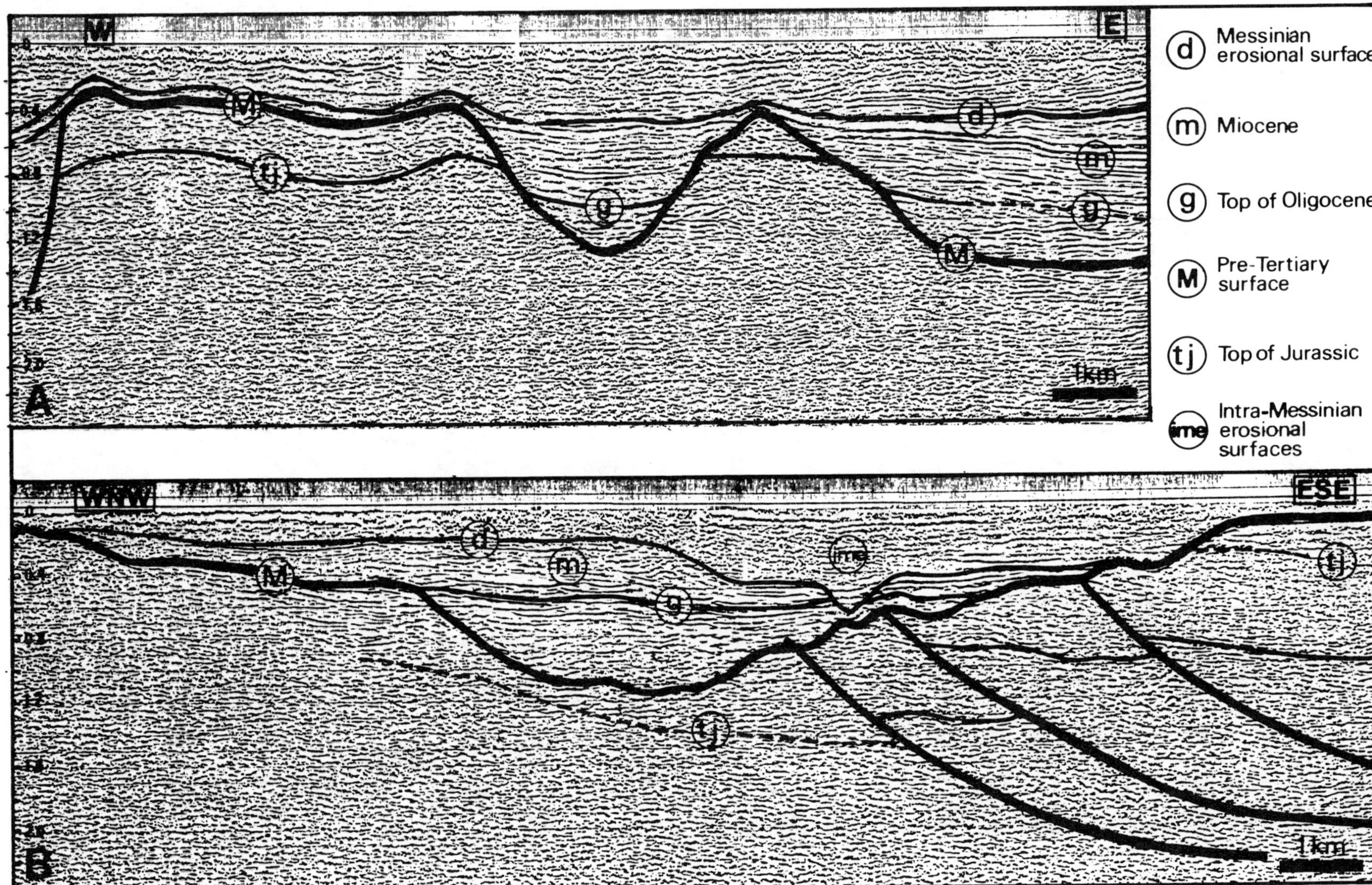

Fig. 2.6. Seismic sections of the Nîmes, Camargue, and Manosque troughs.

A. Complex system of horsts and grabens of La Crau area filled in by a thick Oligo-Miocene sedimentary succession. In the east is the Albaron high which separates the grabens of La Crau from that of Nîmes.

B. Relations between the Eocene compressive phase which produced overthrusting folds and the following distensive phase that sharply cut the compressive structures. Courtesy TOTAL.

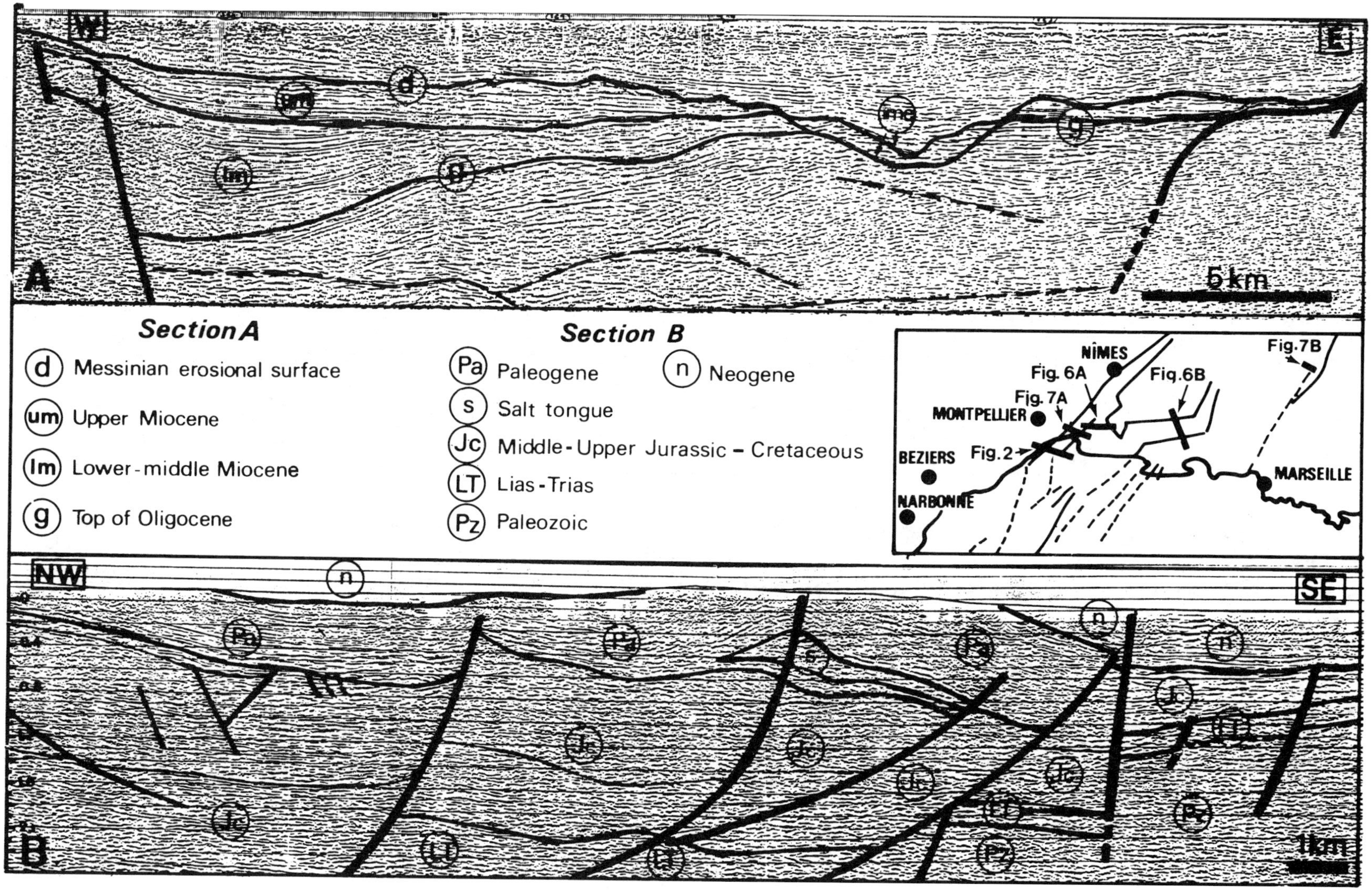

Fig. 2.7. Two seismic lines across critical basins in the Paleogene of France. The seismic lines reveal the significance of the Messinian erosion that cut the earlier sedimentary sections.

A. The Nîmes trough. The sedimentary infill in the Nîmes trough is characterized by a roll-over structure ending at the footwall of the Nîmes fault. The subsidence of this syncline remained active during the Miocene. Courtesy EURAFREP and ESSO REP.

B. The Manosque trough, showing: (1) the asymmetric morphology of the main trough; (2) the Durancian fault system that forms the eastern boundary of the trough and appears to be a listric fault system flattening within the Triassic evaporites; and (3) the lenticular shape of the salt-bearing formation and its anticlinal structure that is possibly of halokinetic origin. These structural characteristics could be related either to a post-Oligocene phase of deformation or to diapiric intrusion of Triassic salt as a detached sheet, probably during the Paleogene. Depth given in two-way travel time in seconds. Courtesy TOTAL.

orative formations that are over 760 meters thick (figure 2.6). A few deep wells that cross the entire evaporite section show that they are intensely deformed. Toward the Vaccarès and La Crau areas in the east, the graben of Nîmes is linked with a complex set of tectonic troughs oriented NW-SE, NE-SW, and E-W and are divided by well-developed horsts such as the Albaron high. Thick sedimentary deposits include a substantial salt accumulation. The most important of them, the E-W elongate trough of La Crau, is filled by 2,500 meters of sediments. This includes 1,500 meters of Oligocene with an evaporative fill of more than 400 meters (calcium sulfates and halite) that has been cored in the Sainte-Cécile #1 well. Unlike our knowledge of the Bresse/Valence/Alsace grabens, data concerning these structures is sparse and limited to a few wells plus some seismic and gravimetric information. An E-W high may have been present which might explain the lack of Oligocene deposits in some wells (Saintes-Maries-de-la-Mer, Beauduc, Peaudure). It ran along the southern edge of the grabens of Vaccarès and La Crau. This seems to indicate that no direct linkage existed with the NE-SW grabens of the Gulf of Lion in spite of their apparent inland penetration into southern Camargue (Arthaud et al. 1980/1981).

Toward the east, tectonic depressions, rather than true grabens, are aligned along the NE-SW oriented faults of Aix-Durance. These troughs mark a larger lineament running north to south from the Alps to the Mediterranean sea (Cavelier 1984). The lineament is made up of the trough of Manosque-Forcalquier with its satellites, the basin of Apt, the basin of Aix, and the basins of Marseille-Aubagne. The basin of Manosque-Forcalquier appears as a strongly asymmetric trough with its deeper eastern part abutting the complex Durancian fault; the basin is subdivided by the Monfuron high into two main troughs, the Manosque and the Forcalquier. The sedimentary fill of this basin, estimated to be about 2,000 meters, comprises evaporative intercalations and a halite accumulation of up to 700 meters of apparent thickness. The evaporites are presently located in an anticlinal structure with apparent halokinetic characteristics (Destombes 1961). Because of its structural setting and petrographic characteristics, a Paleogene age has been proposed for the salt and the possibility of intrusive Triassic salt cannot be excluded (Grappin et al. 1975). The morphologic reconstruction is difficult because of later folding, probably during the late Miocene (Destombes 1961; Triat and Truc 1974). The lenticular shape of the salt body as it presently exists together with its structural setting, shown in the seismic line of figure 2.7B, does not permit resolution of its origin as the primary morphology of an Oligocene salt formation or as an allochtonous body. As suggested by some authors (A. Maurin, oral communication), this salt body could represent a detached sheet migrated from Triassic strata in which the border faults are flattening. Thinner Paleogene deposits are present to the north, such as the gypsum outcrops of Saint-Geniès, and extend westward over a wide area surrounding the Monts de Vaucluse. These deposits evidence at least two episodes of gypsum deposition in their southern (Apt basin) and western parts (Pernes and Mormoiron, located northeast and southeast of Carpentras, respectively), while in the north sedimentation seems devoid of evaporites in the area bordering the Baronnies chain (Triat and Truc 1974).

Separated from the Apt-Manosque-Forcalquier area by the E-W high formed by the Luberon chain, the broad Aix-en-Provence basin contains a Oligocene series no more than 150 meters thick with a late Oligocene (Chattian) gypsum episode (Destombes 1964; Nury 1987/1988); the basin is limited to the east by the Aix fault, which is the southern extension of the Durancian fault.

Several basins may be recognized within the Marseille, Aubagne, Saint-Pierre-les-Martigues region, situated at the southern extension of this large lineament. The tectonic troughs of Marseille-Aubagne are represented by two different depositional units formed respectively during the lower and upper Rupelian (Stampian). The troughs opened broadly towards the Gulf of Lion in a direction similar to that of the Sassari graben before the rotation of the Corsica-Sardinia block (Nury 1970, 1977, 1984, 1987/1988). Although a main NE-SW direction predominates, except for the area of Saint-Pierre-les-Martigues which has a E-W orientation, Nury (1977, 1980/1981, 1987/1988) and Nury and Raynaud (1986) have demonstrated that the opening of the basins results from multidirectional faulting possibly influenced by rifting in the Gulf of Genova. The large volume of breccia that is present throughout the section, as well as the successive changes in the position of the subsidence axis, argues for continuous synsedimentary activity. In this area, the Oligocene accumulation is up to 1,000 meters thick, including gypsum deposits (Triat and Truc 1974; Nury 1977; 1987/1988).

In a northeastern extension of the Marseille/Aubagne basins there are several small basins with morphologies and orientations that are markedly variable in relation to the geometry of the complex Provençal thrusts (La Destrousse, Ryans, Nans-les-Pins, Sainte-Zaccharie). Other small basins were formed to the east during the same distensive phases in the Toulon area, but are not discussed here (Touraine 1971; Nury 1984, 1987/1988; Nury and Rousset 1985/1986).

Post-rifting history.

The post-Oligocene activity differed strongly among the basins. Subsidence still was strongly active in the Nîmes graben during the Aquitanian that marks the beginning of a marine transgression (Demarcq 1970). Southward the graben of Perpignan, which seems to be the prolongation of the Gulf of Lion system, contains a thick sedimentary infill that could be of Miocene and Pliocene age. The pre-Serravalian activity of NE-SW faults on the Nîmes bundle involved the formation of small subsiding basins (Triat and Truc 1983). The sea transgressed across all the earlier structures during the Miocene with a maximum extent in the early Tortonian. Although reduced, subsidence was still active in some areas. During the middle Tortonian an important paleogeographic change took place with the general retreat of the sea from all the area due to a regional uplift. The continuation of this uplift in the Messinian can be related to Alpine tectonics and also to the unloading of the Mediterranean crust during the Messinian drawdown. These events led to an important phase of erosion with the cutting of the Paleo-Rhône canyon that was later invaded by the sea in the lowermost Pliocene.

Strong compressional deformation occurred during the Mio-Pliocene interval. Three phases of post Oligocene compression have been identified in the Marseille area with different orientations (Nury and Raynaud, 1986). The N-S phase of compression described by Bergerat (1985) is dated as late Miocene to Pliocene. This period is structurally characterized by the strike-slip reactivation of the pre-existing principal fractures, a regional uplift leading to the late Tortonian emersion, or a local folding (Luberon, Manosque). In the Suzette basin, late Miocene compressive deformation reactivated the rise of Triassic evaporites that began during the Pyrenean folding and led to a present-day morphological inversion (Casagrande et al. 1989). In the Nîmes graben, the halite deposits are thickened locally by duplication (Caline, personal communication), which may be considered as an intra-salt overthrust or as a synsedimentary gravity slide (Valette, personal communication). An inversion of the subsidence pattern also is observed during the late Neogene along the Durancian fault, marked by the subsidence of the Digne basin to the east of the Manosque-Forcalquier basin. These deformational phases probably were contemporaneous with the overthrusting in other areas, such as the Jura and the sub-Alpine chains.

Some of the NE-SW faults are still active and are marked by seismically active lines. Indeed, the Durancian fault continues to be active as demonstrated by historically recorded seismicity (Barbreau et al. 1976; Triat and Truc 1983). Present-day vertical movements have been deduced from recent releveling studies and are particularly evident in the continued subsidence of some parts of the Rhône valley near Avignon (0.85 mm/year) and in the area of Aix en Provence. At the same time the external Alpine crystalline and Provence massifs are being uplifted (Fourniguet 1977, 1987).

THE TECTONIC DEPRESSIONS OF THE MASSIF CENTRAL

The Paleogene rifts of the Massif Central differ from the other segments owing to the fact that they developed entirely within the crystalline and metamorphic basement block of the Massif, which formed during the Cadomian and the Variscan orogenies (figure 2.8). The distensive phase was irregular across the area, mainly during the Oligocene, resulting in a number of tectonically controlled depressions of various sizes and morphologies, the greatest and the best known of them being the Limagne system in the northern part of the complex. Knowledge of this region has been improved by geophysical data (seismic and gravimetric) and a few deep wells drilled for petroleum exploration.

Pre-rifting history.

From the end of the Variscan orogeny in the end of the Lower Carboniferous (about 335 MA) the major part of the Massif Central remained emergent and was strongly affected by erosion and peneplanation.

In association with the tectonic activity of the late Hercynian, faults and associated small tectonic depressions were formed during the late Carboniferous and the Permian. These are filled predominantly with thick terrigenous sediment and late Carboniferous (Stephanian) coal formations. A series of coal basins are located along a major great N-S fault, the Grand Sillon Houiller, that runs across the entire Massif Central. From the middle of this fault line there is an associated fracture which curves progressively towards the north-east, crossing the Limagne system before it follows, more or less, the direction of the Le Creusot/Belfort sinistral offset that separates the Bresse segment from the Rhine. A great many of the Paleogene basins have been formed by the reactivation of these late Hercynian fractures, especially by the Grand Sillon Houiller. The Triassic and Jurassic seas only covered the margins of the Massif and invaded the Causses basin located at the southern margin. Nevertheless the presence of reworked siliceous fragments of Jurassic age in the Paleogene fill of some depressions indicates that some of the peripheral area probably was originally covered by Mesozoic sediments (Rat 1974). Most of the area affected by Paleogene rifting remain outside the limits of the Mesozoic seas.

Before the Paleogene rifting phase, several other tectonic events are suspected to have occurred, including a probable period of N-S compression considered as synchronous to the Pyrénéo-Provençal event (Burg and Etchecopar 1980; Bergerat 1985). This phase could have rejuvenated the relief and reactivated the erosion processes.

Location and morphostructure of the basins.

The largest basinal structure of the Massif Central is composed of the asymmetric grabens of the Limagne, which may be schematically divided into two sub-parallel arms diverging southward. The western arm, called the Grande Limagne (or Limagne-Bourbonnaise or Limagne of Clermont-Ferrand) extends over 100 km north to south and is up to 35 km wide and extends southward along the smaller appended structure of Brioude. The Grande Limagne arm is bordered by NE-SW faults that cut the older structural directions. The deepest part (Riom basin) is situated in the Grande Limagne and is a narrow trough filled by 2,500 meters of Tertiary sediments lying at the foot of the step-faulted western border, while the slope of the basin floor is more regular toward the east (Grolier and Tchimichkian 1963; Morange et al. 1971) (figure 3.9). Evaporites are disseminated throughout 350 meters of fill with a halite component that is about 125 meters thick and is located in the more subsident area, while gypsum layers are present in the laterally thinnest sections (Young 1946; Grolier and Tchimichkian 1963; Morange et al. 1971).

The basins of Roanne and Montbrison, separated by a high lying south of the town of Roanne, form the eastern branch of the Limagne (Forez) system. These basins are half-grabens tilted westward against the NW-SE border faults. The sedimentary fill, more than 600 meters thick in the Montbrison basin, is dominated by sands and marls and is organized into superposed sequences, a part of which is considered of Rupelian age; an accurate date for the series has not been obtained (Duclos 1967). Gypsum or other evaporites have not been described in these basins.

Many other tectonic depressions of smaller size occur in the Massif Central and are related to the reactivation of Hercynian fractures; they are the basins of Ambert, Le-Puy-en-Velay, Emblavès, Saint-Flour, Langogne, and Aurillac. The sedimentary fill is composed of sands, marls, and limestones of

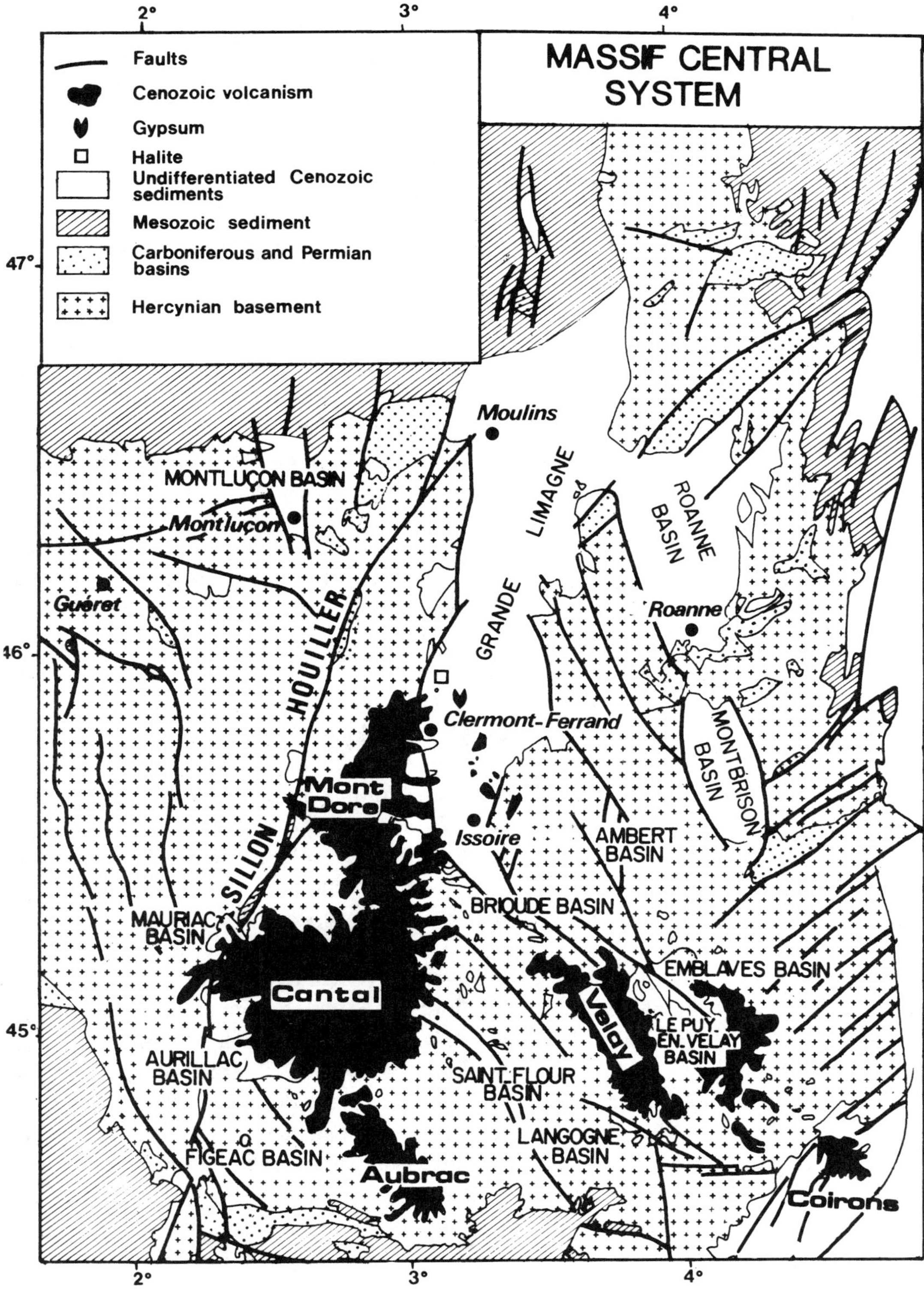

FIG. 2.8. Simplified geological map of the Massif Central segment (from the geological map of France at the scale of 1:1,500,000).

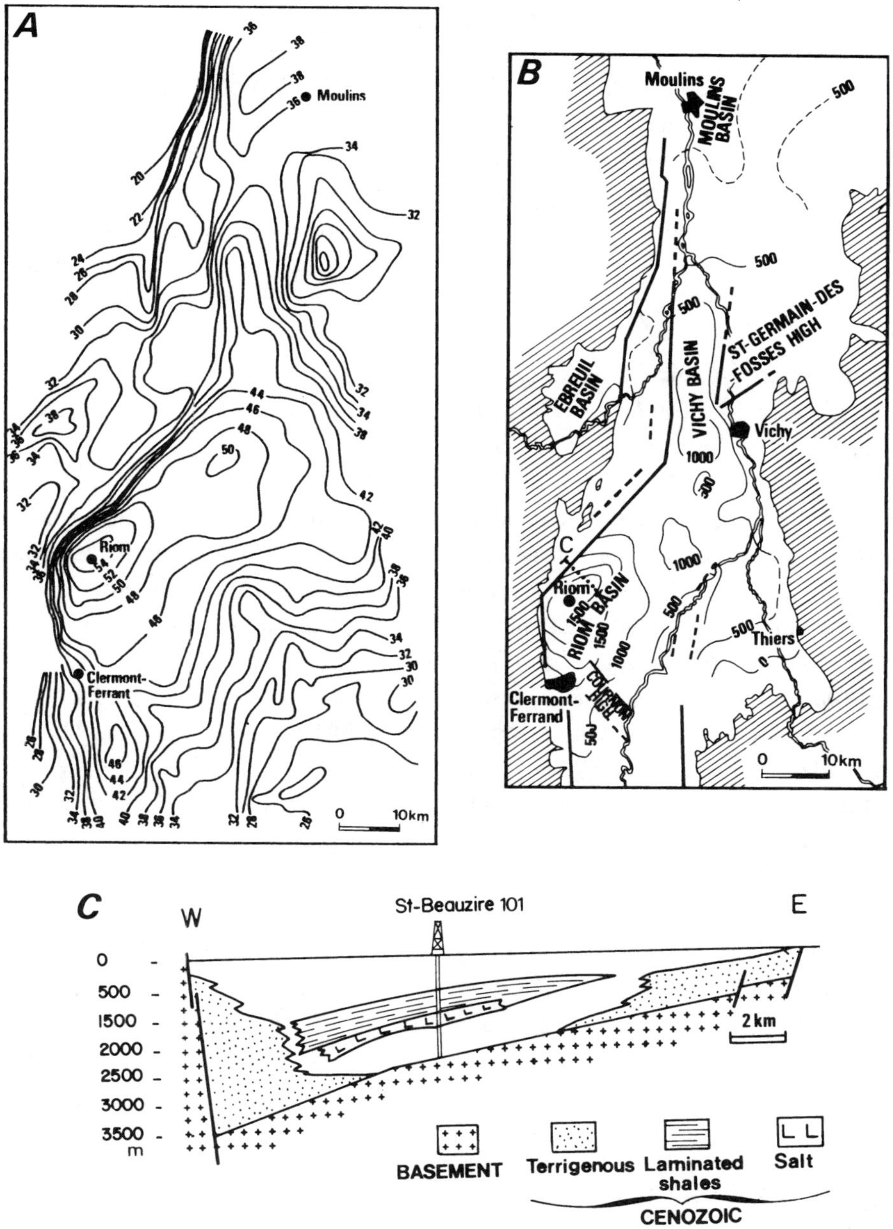

Fig. 2.9. The Grande Limagne basin in the Massif Central segment.

A. Gravimetric map (Bouguer anomaly, given in milligals): the morphostructural pattern of the basin appears clearly on the map and the trough of Riom which is the deepest part of the basin is differentiated by a strong negative anomaly (simplified from the gravimetric map of France at the scale of 1:1,000,000).

B. Isobaths of the base of the Tertiary sedimentary fill (adapted from Morange et al. 1971).

C. W-E section across the asymmetric Riom trough with the position of the halite layer observed in the well of Saint-Beauzire 101 (adapted from Morange et al. 1971).

Oligo-Miocene age. The fill reaches 300 meters in the Ambert basin but generally it is less than 100 meters and sometimes includes gypsiferous marls (Le-Puy-en-Velay). A string of smaller depressions occurs along the Grand Sillon Houiller or its satellites, such as the basins of Figeac and Mauriac, and in Mauriac some gypsum is known.

Associated volcanism.

Tertiary and Quaternary volcanic activity, one of the most prominent features of the Massif Central, are alkaline in nature. The first occurrences were manifested in the Paleogene (64 MA) in the Limagne (Vincent et al. 1977), and are an early manifestation of the extentional tectonics (Brousse and Bellon 1983). The activity, characterized by a strong differentiation of the volcanic products, was greatest from the lower Miocene to the Quaternary and has produced three main morphological types (Maury and Varet 1980): (1) local and dispersed activity (maar, neck, dikes and sills, cinder cones); (2) fissural volcanism producing large but sporadic basaltic flows (Cantal, Escandorgue, Aubrac, Velay, Coirons); and (3) large volcanoes such as the Mont-Dore and the Cantal, respectively 30 and 60 km in diameter, with associated collapse troughs (caldera) up to 500 meters deep (in the Cantal) (Vatin-Perignon and Michel 1966).

Post-rifting history.

Indications of a ENE-WSW to E-W compression have been identified (Bergerat 1985) and attributed to the Miocene, which is the period of formation of the large volcano-tectonic troughs lying underneath the Mont-Dore and the Cantal volcanoes. Sedimentation under lacustrine conditions still persisted during the Miocene in parts of the Oligocene grabens and in other little depressions, some of which are filled with fresh-water diatomites. These diatom deposits still continued to accumulate in restricted areas during the Plio-Quaternary, apparently associated with the volcanism. A recent seismic study carried out in the Ambert basin by Dorel et al. (1990) shows modern seismicity along the western border of the basin which seems to indicate a strike slip motion of the border fault.

THE BRESSE-BAS-DAUPHINÉ SYSTEM

The Bresse and Valence basins constitute a clearly developed N-S segment that runs along the Saône and Rhône valleys over a latitudinal distance of more than 350 km with an irregular width reaching 40 km in the southern Bressan area (figure 2.10). It is composed of two main tectonic depressions, that of Bresse in the north and of the Bas-Dauphiné basin in the south. The Bas-Dauphiné basin comprises the Valence trough bordered by the basins of Crest in the south and Royans in the east and the Vienne platform to the northeast. The two areas are separated by a NE-SW high, devoid of Paleogene deposits, that lies near the city of Lyon (Vienne-Chamagnieu basement high).

The general geology as well as the stratigraphy of this system has been the subject of many detailed studies (Lefavrais-Raymond 1962; Sittler 1965; Debrand-Passard 1984). The gravimetric maps (Goguel 1948), deep wells, and seismic investigations related to both petroleum exploration and more recently to the deep seismic exploration of the Etude de la Croute Continentale et Océanique par Réflexion et Réfraction Sismiques (ECORS) program (Bergerat et al. 1989), have revealed the underlying structure of the basin (figure 2.11). From these studies it appears that the late Miocene

FIG. 2.10. Schematic map of the Bresse-Bas-Dauphiné (Bresse-Valence System) segment of the graben system, synthetized from the geological map of France at the scale 1:1,500,000 and local maps given by Curial (1986), Dumas (1988), and Moretto (1986/87).

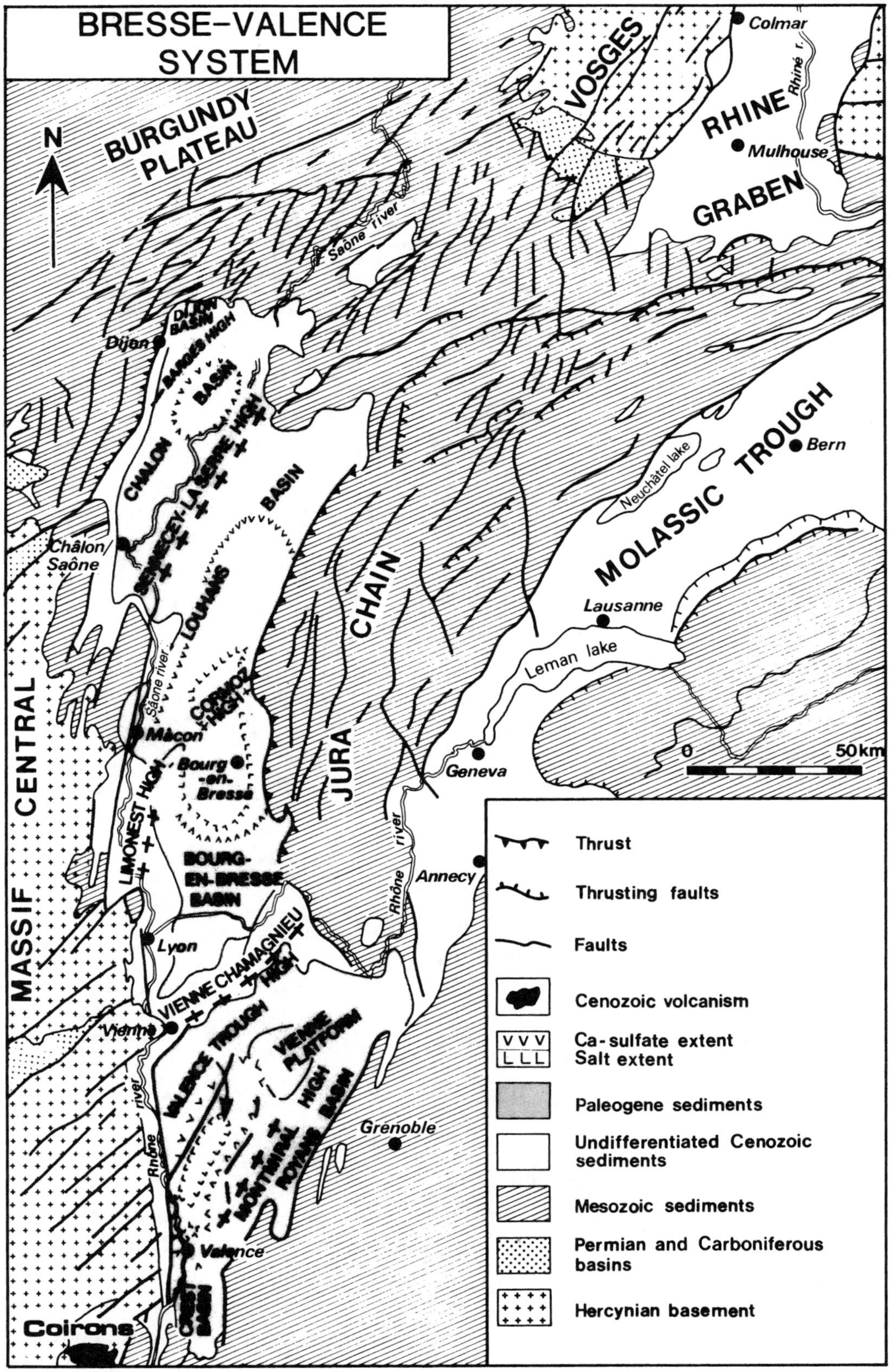
BRESSE-VALENCE
SYSTEM
N
BURGUNDY
PLATEAU
VOSGES
Colmar
Rhine r.
RHINE
Mulhouse
GRABEN
Saône river
Dijon
DIJON BASIN
CHALON
BASIN
MOLASSIC TROUGH
Bern
Neuchâtel lake
Châlon/
Saône
LOUHANS
BASIN
CHAIN
Lausanne
Leman lake
Saône river
Mâcon
Bourg
-en-
Bresse
JURA
Geneva
0
50km
MASSIF CENTRAL
LIMONEST HIGH
BOURG-
EN-BRESSE
BASIN
Annecy
Rhône river
Lyon
VIENNE CHAMAGNIEU
HIGH
Vienne
VALENCE TROUGH
VIENNE
PLATFORM
HIGH
BASIN
Grenoble
Rhône river
Valence
Coirons
Thrust
Thrusting faults
Faults
Cenozoic volcanism
Ca-sulfate extent
Salt extent
Paleogene sediments
Undifferentiated Cenozoic sediments
Mesozoic sediments
Permian and Carboniferous basins
Hercynian basement

FIG. 2.10

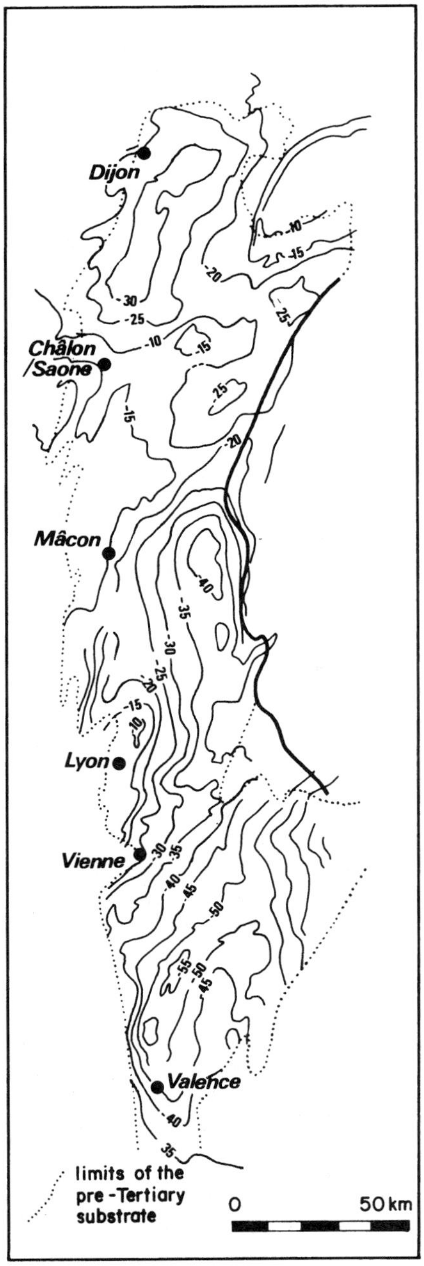

FIG. 2.11. Gravimetric map of the Bresse and Valence basins (Bouguer anomaly, given in milligals); the structural configuration of the domain into sub-basins separated by highs is clearly delineated and the two main depressions filled with the salt accumulations of Bourg-en-Bresse and Valence are differentiated by strong negative values (modified from the gravimetric map of France at the scale of 1:1,000,000 and from the map of Goguel 1948).

thrusting of the Jura and the Subalpine chain overrode the eastern parts of the Bresse and Valence basins respectively. More recent studies made for storage facilities within the halite have resulted in extremely detailed knowledge of the interior of these salt basins. Using all the data obtained thus far, three recent studies have resulted in very complete analyses of these salt basins: Curial (1986) and Moretto (1986/1987) have worked on the Bresse, and Dumas (1988) on the Valence, basin.

Evaporative deposits occur across broad areas in these basins but the halite formations themselves are comprised of at least three distinct stratigraphic layers. These layers constitute a thick but restricted depocenter limited to the more subsident parts of the graben (Goy and Busson 1980; Curial 1986;

Moretto 1986/1987; Dumas 1988). This evaporative sedimentation continued over a stratigraphic period extending from the late Eocene to the late Oligocene.

Pre-rifting history.

All of the area was emergent from the end of the Cretaceous and was affected by an intense period of erosion. Its initial structural development began with a phase of N-S compression more or less contemporaneous with the Pyrénéo-Provençal compressive phase (Bergerat 1985) although the effect of compression is less clear than in southern France. Nevertheless, in the Bresse basin these Eocene phases of deformation gave rise to transverse NE-SW to NNE-SSW structures. These structures follow pre-existing Hercynian trends that controlled the first terrigenous accumulations and the early but limited lacustrine deposits.

Morphostructural pattern.

The rifting due to the general E-W distension becomes apparent from the late Eocene onward (Lefavrais-Raymond 1962; Sittler 1965; Debrand-Passard 1984; Sittler 1965; Debrand-Passard 1984; Bergerat 1985; Dumas 1988). The geometry is controlled by the interference of the N-S trending border faults of the rift with the older transverse structural trends (NE-SW to NNE-SSW), giving an internal structural development to the sub-basins and horsts that is oblique to the orientation of the rift itself (figure 2.12). Indeed, each graben segment has a strong asymmetric step-like morphology with opposite structural polarity to those adjacent to it (Rosendahl et al. 1986). This is characterized by deepening of the rift from the NW to the SE in Bresse and SE to NW in the Valence basin.

As a result, the Bresse Basin itself is composed of three major sub-basins (Bourg-en-Bresse, Louhans, and Chalon) delimited by highs (Limonest, Cormoz, Sennecey-La Serre) (Lefavrais-Raymond 1962; Rat 1974). Several very small sub-basins, such as Dijon, are also present (separated from Chalon by the Barges high). Seismic lines show the Limonest high is superimposed upon a faulted anticline composed of Jurassic strata formed during the pre-rifting compression and against which the Paleogene clastic deposits and evaporites progressively onlap (figure 2.13A). Local Paleogene deposits are thicker than 2,000 meters. Evaporites, with massive halite interbeds, having a thickness of more than 1,400 meters are mostly restricted to the southern part (Etrez area) in the sub-basin of Bourg-en-Bresse (Goy and Busson 1980; Curial 1986; Moretto 1986/1987; Curial and Moretto 1996) (figure 2.13B). The deepest part of the trough occurs in the south-east where the Mesozoic of the Jura chain is overthrust for a distance of more than 10 km onto the late Miocene sediments of the graben (Chauve and Perriaux 1988) (figure 2.14).

The Valence trough represents a part of a more complicated sedimentary area displaying an asymmetry opposite that of Bresse. The western margin of this trough is formed by the faulted escarpment of the Hercynian Massif Central that borders the basin through a system of step-like descending blocks. The thickest sedimentary accumulations are about 2,000 meters thick and contain massive halite layers exceeding a total of 1,000 meters that characterize the narrow troughs lying at the foot of the escarpment (Lefavrais-Raymond et al. 1984; Guillevin 1981; Goy and Busson 1980; Dumas 1986, 1988). Eastward the thickness is progressively reduced and the Paleogene deposits pinch out against the high of Saint-Lattier/Montmiral, as illustrated in the seismic line of figure 2.15. The sedimentary successions on the Vienne platform and in the basins of Crest and Royans have a reduced thickness, with a lacuna extending over a large part of the early Oligocene and only traces of calcium-sulfates locally present. In the east, the westward thrusting of the sub-Alpine border (syncline of the Saint-Nazaire-en-Royans) overides the sedimentary fill of the graben and could be related to a late- to post-Miocene compressive phase.

Relationship with the other segments of the rift.

The northern end of the Bresse basin is separated from the southern part of the Rhine graben by an

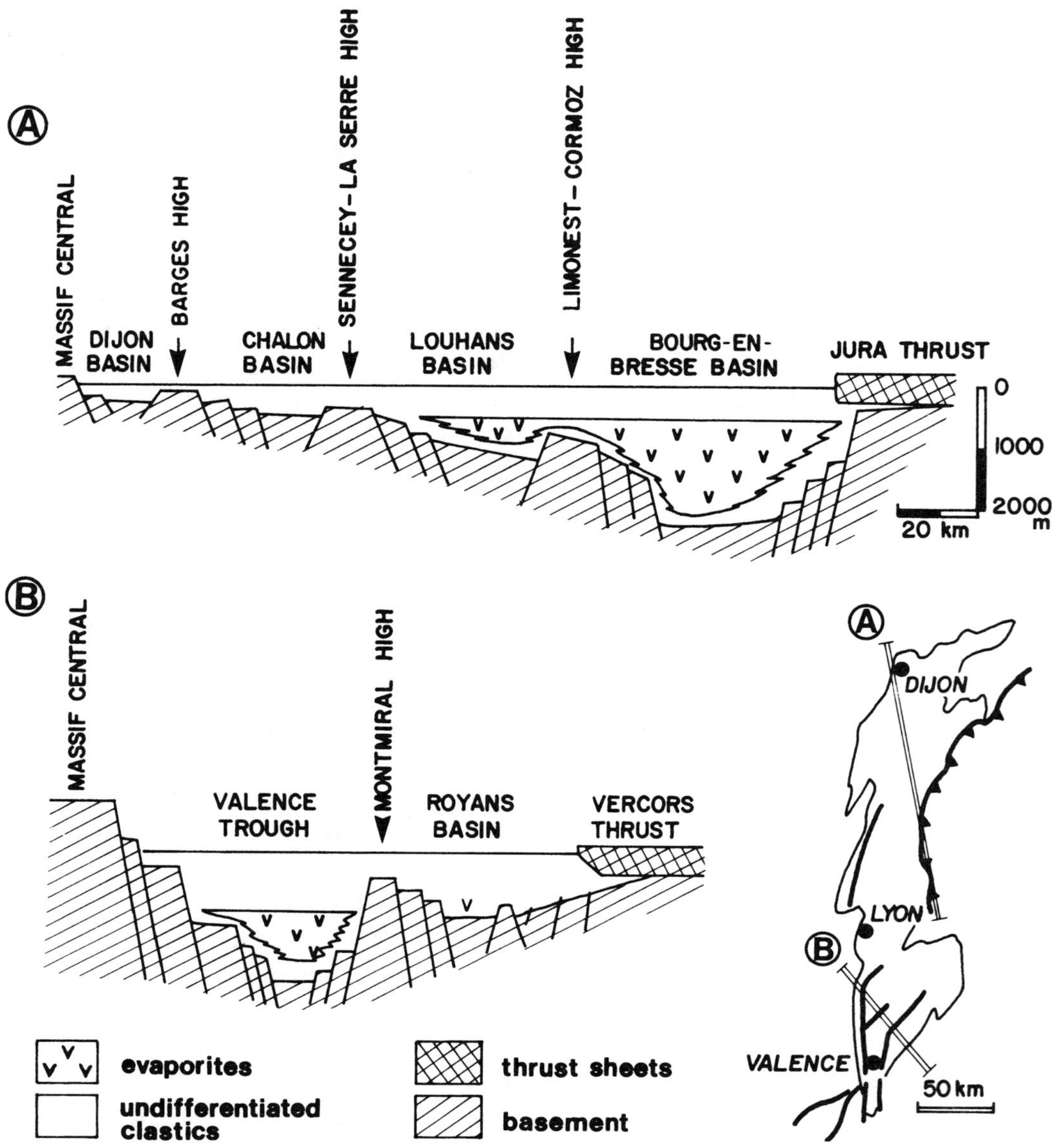

Fig. 2.12. Schematic cross-sections illustrating the reversal of the morphostructural polarity between the Bresse and the Bas-Dauphiné basins.

eastward offset of about 150 km that lies between the cities of Le Creusot and Belfort (Illies 1972, 1974; Contini and Theobald 1974). This transverse structure, oriented 060°, has been considered as a continental transform fault; the framework of 020° dominant fractures observed in surface exposures could be a Riedel shear system (Illies 1974; Bergerat 1977a, 1977b, 1985) (figure 2.16).

The southern end of the Bresse-Bas-Dauphiné rift system is marked by an important change in the main structural directions from N-S, characteristic of the northern Rhine-Rhône segments, to the NE-SW dominant directions of the Gulf of Lion—Languedoc—Provence segment. This change coincides more or less with the northeastern extension of the large lineament formed by the Cévennes fault that separates the two areas with different structural characteristics and is controlled by preexisting Hercynian structures. For Bergerat (1981/1982) a younger transform zone explains this directional

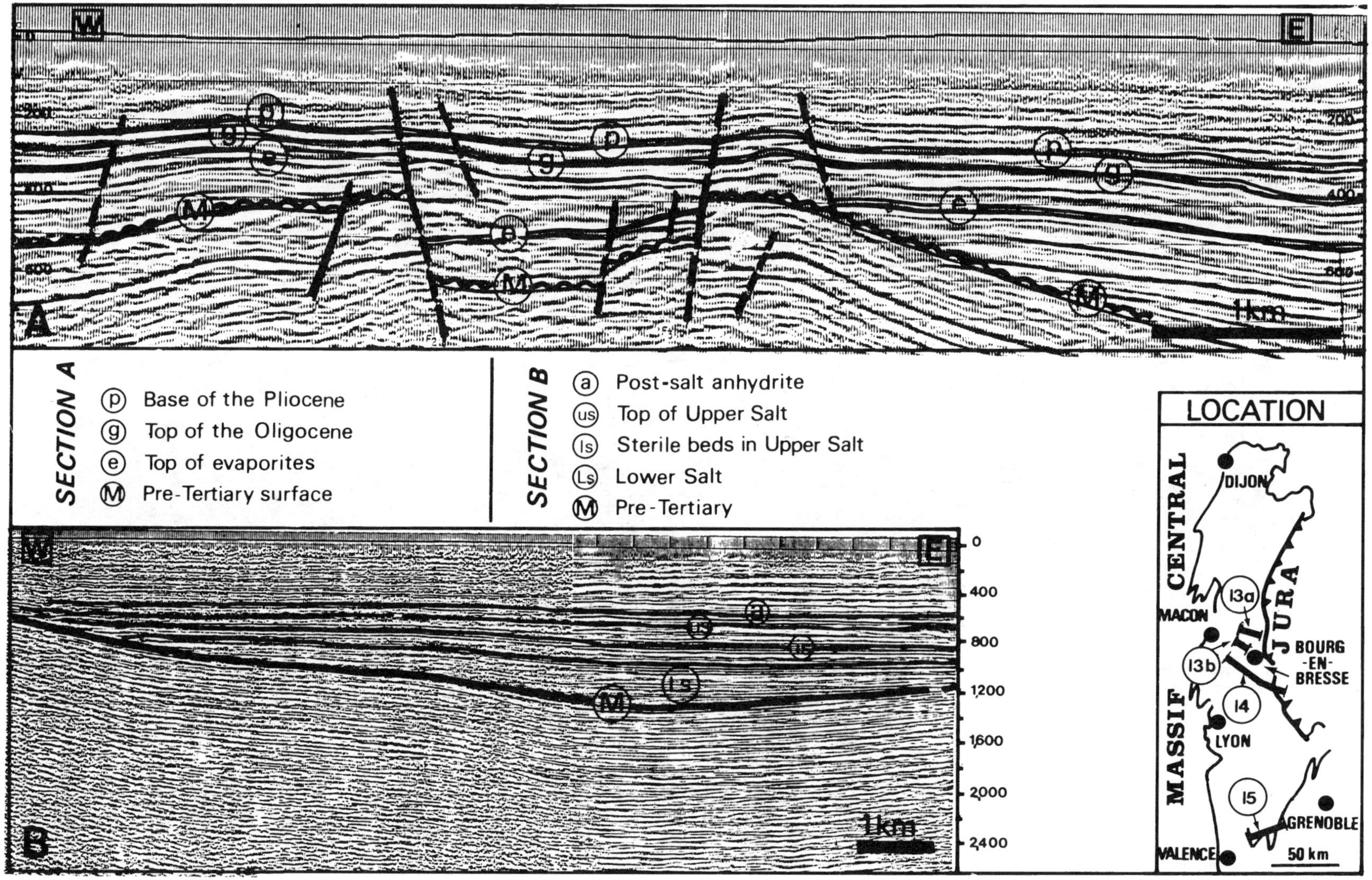

Fig. 2.13. Seismic sections across the Bresse trough:

A. The progressive onlapping of the Tertiary series onto the Limonest high which appears to be formed by a fractured anticline (composed of Jurassic formations); the halite formation is developed to the west of the section but has been bevelled in this area. Courtesy SHELL France, S.N.E.A.P., and BP.

B. Profile showing from right to left the complete sedimentary series of the basin center, including the Salt Formation and its progressive pinch-out toward the western border. Depth given in two-way travel time in milliseconds. Courtesy Gaz de France.

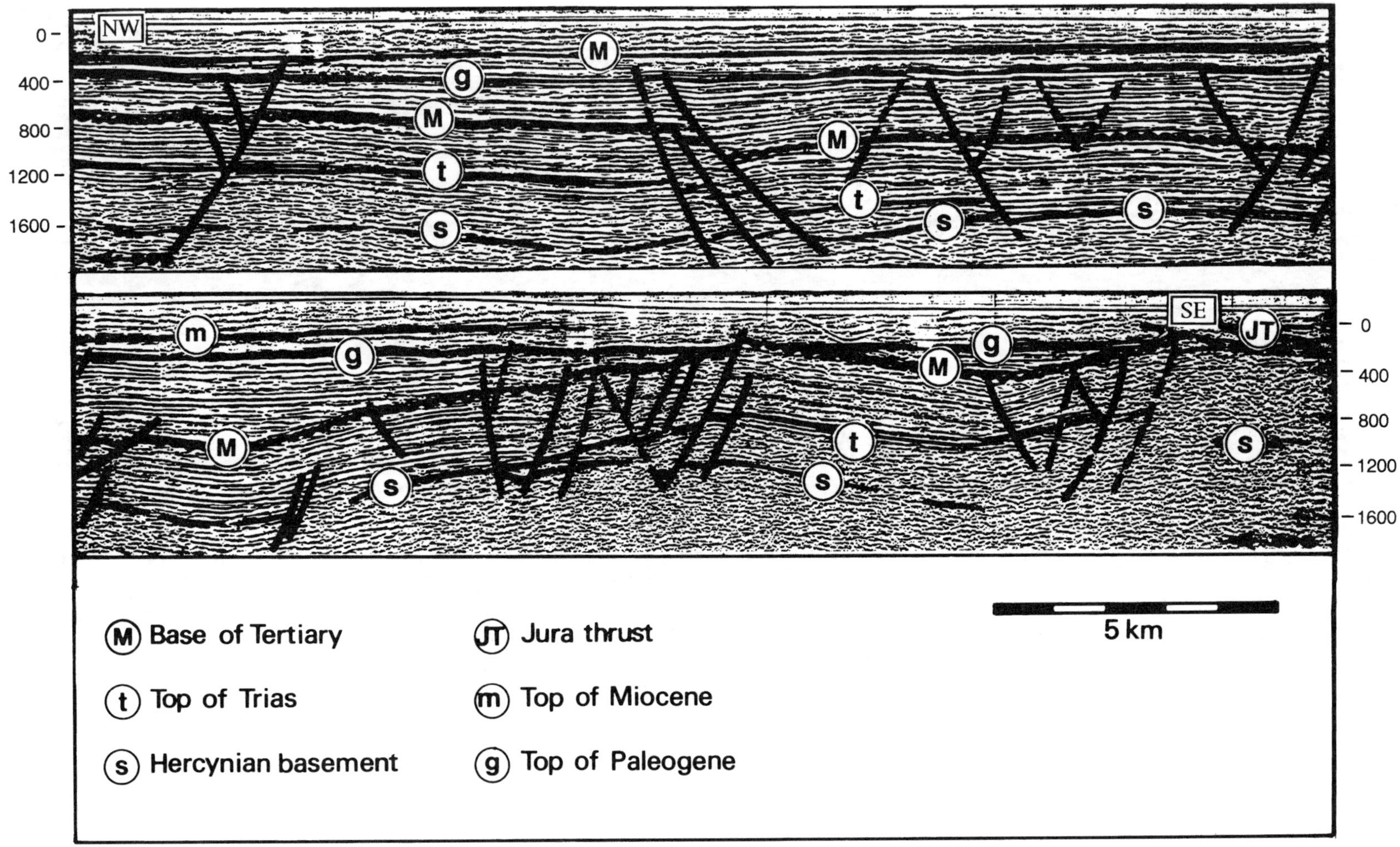

Fig. 2.14. Seismic section across the southern part of the Bresse trough south of the halite-bearing area, showing the structure of the graben with stairstep blocks (bounded by nearly vertical faults). At the east, note the overthrust of the Mesozoic formations of the Jura over the margin of the salt basin section. Depth given in two-way travel time in milliseconds. Courtesy Institut Français du Pétrole, BP, ESSO REP, S.N.E.A.P., and TOTAL.

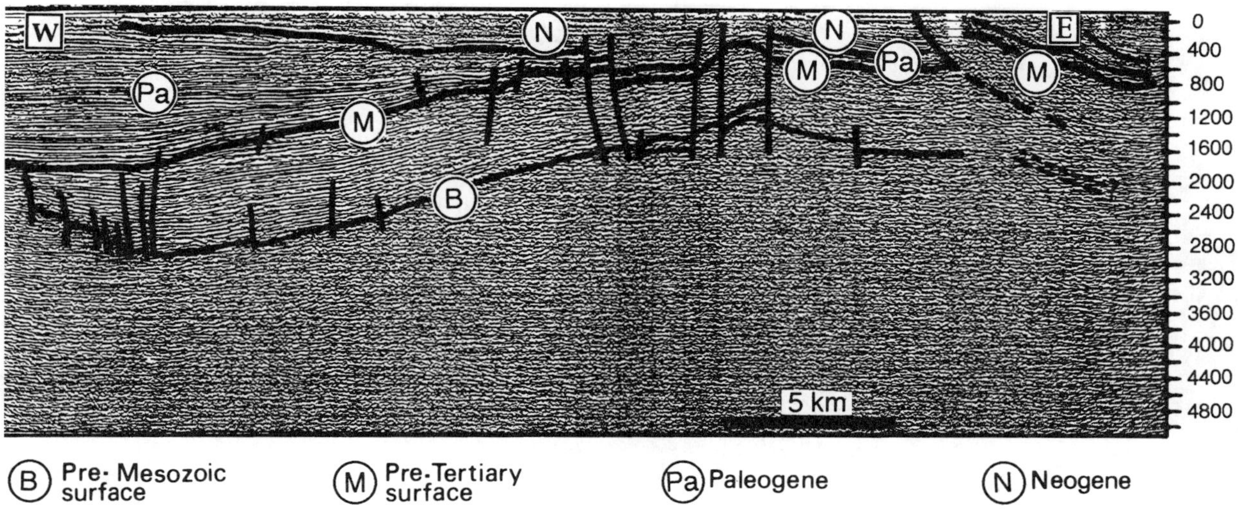

FIG. 2.15. Seismic section illustrating the morphostructural pattern of the Valence basin. Located in the southern part of the basin beyond the salt margin, this section shows: (1) the westward thickening of the Tertiary sediments and their eastward disappearance on the Saint-Lattier horst; and (2) the thrusting of the Sub-Alpine chains (at the right). See figure 2.11 for location. Depth given in two-way travel time in milliseconds. Courtesy Institut Français du Pétrole, BP, ESSO REP, S.N.E.A.P., and TOTAL.

change; it does not correspond to the Cévennes fault bundle but to a 050° oriented fault, the Saint-Remèze fault, that produced the lateral shift of the Alès and Valence grabens.

Post-Paleogene history.

Morphostructural evolution differs in the two basins. In Valence, a migration of the subsidence axis is clearly indicated by Miocene deposits that are up to 700 meters thick, occuring east of the area of maximum Paleogene accumulation (Alabouvette et al. 1984a, 1984b). However this subsidence apparently ended in the Pliocene. These deposits are limited to the Paleo-Rhône ria; this migration of the depocenter may be clearly seen on the seismic line of figure 2.15 where the thick Neogene lies upon the area of thinner Oligocene deposits, near the ridge of Saint-Lattier-Montmiral. In the Bresse basin, to the contrary, the subsidence continued in the Miocene and increased again during the Pliocene with a relatively constant subsidence superposed on the Bourg (300 m) and Louhans (200 m) subbasins respectively.

The most prominent post-Paleogene tectonic event is the westward thrusting of the Mesozoic of the Jura and of the sub-Alpine chains over the eastern parts of the Bresse and Valence depressions respectively; this is related to the phase of compression that occurs at the Miocene-Pliocene transition.

The present-day evolution of the Bresse-Jura area is characterized by an inversion of the subsidence motion resulting in the subsidence of the northern part up to the Vosgian massif and the rise of the southern area (Fourniguet, 1987).

THE RHINE SEGMENT

The rift system of the Rhine is a complex set of tectonic troughs in a Y-shape formed by the intersection of the three main grabens: the main Upper Rhine graben followed northward by the Hesse graben oriented NNE-SSW and the Lower Rhine grabens with an oblique NW-SE principal direction. A presumed triple junction is apparently located near the town of Mainz, immediately south of the great Miocene Vogelsberg volcano (figures 2.17 and 2.18). Subsidence began in the middle Eocene in the southern end of the Upper Rhine graben and con-

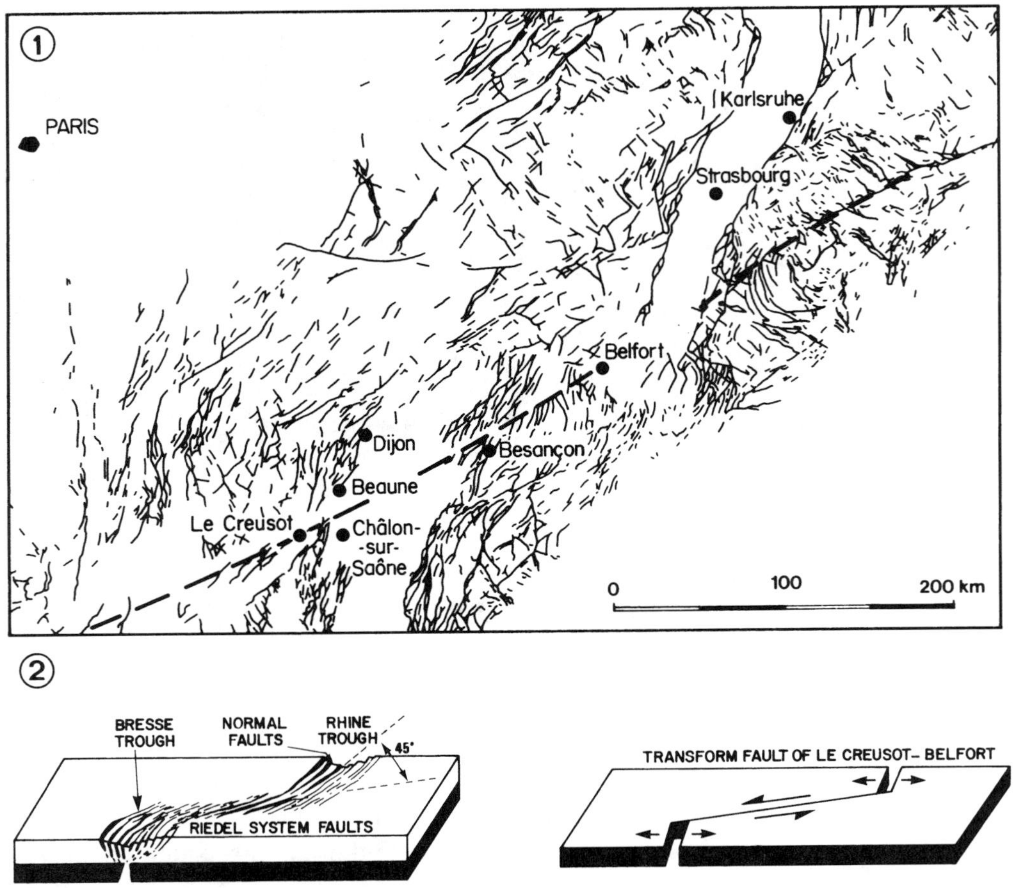

FIG. 2.16. Map showing the main structures along the Bresse-Rhine rift area (1); and interpretative diagrams (2) of the structural trends in the transitional zone between the Bresse and Upper Rhine grabens. These are considered as a continental transform fault system evidenced at the surface by the Riedel shear system (modified from Bergerat 1977).

tinued into the Miocene in its northern part. The system remains seismo-tectonically active. Evaporites are present in a large part of the Upper Rhine branch with thick halite and, in the southern part (Mulhouse trough), halite with potash deposits.

The Upper Rhine region is one of the most intensely studied continental basins owing to petroleum and potash salt exploration. Among the main works specific to petroleum and salt geology are: Schnaebele (1948), Courtot et al. (1972), Blanc-Valleron and Gannat (1985), Blanc-Valleron et al. (1989), and Blanc-Valleron (1996). Gravimetry, seismic reflection, and refraction investigations from this area include studies reported by Maïkovsky (1952), Goguel (1957), Ansorge et al. (1973), Edel et al. (1975), Hirn (1976), Edel (1978), Rousset and Bayer (1990). The stratigraphy and sedimentology are well delineated by Sittler (1965), Schuler (1988), Blanc-Valleron and Schuller (1996). For more general tectonic presentations see Illies (1972, 1974, 1975, 1977, 1981), Illies and Baumann (1982), Rulhand (1982), Bergerat (1983, 1985), Villemin (1986), and Larroque (1987). Two deep seismic profiles (Programs ECORS-DEKORP) have investigated the northern part, north of Mannheim and the middle area south of Strasbourg.

The pre-rift framework.

The Rhine segment, especially the Upper Rhine portion, sharply crosscuts an older, complex geological structure. This structure is dominated by the

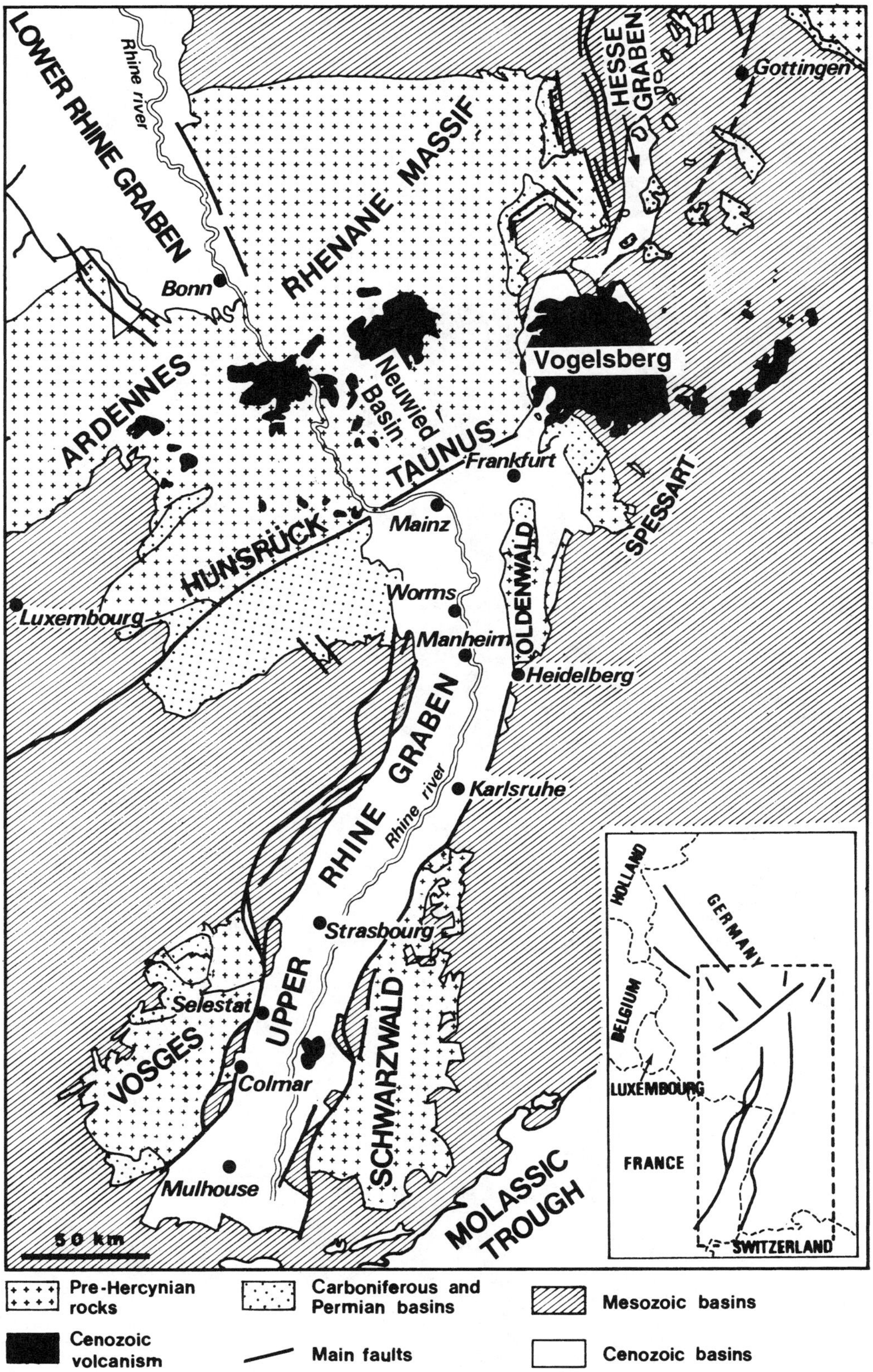

FIG. 2.17. Simplified map of the Rhine graben system.

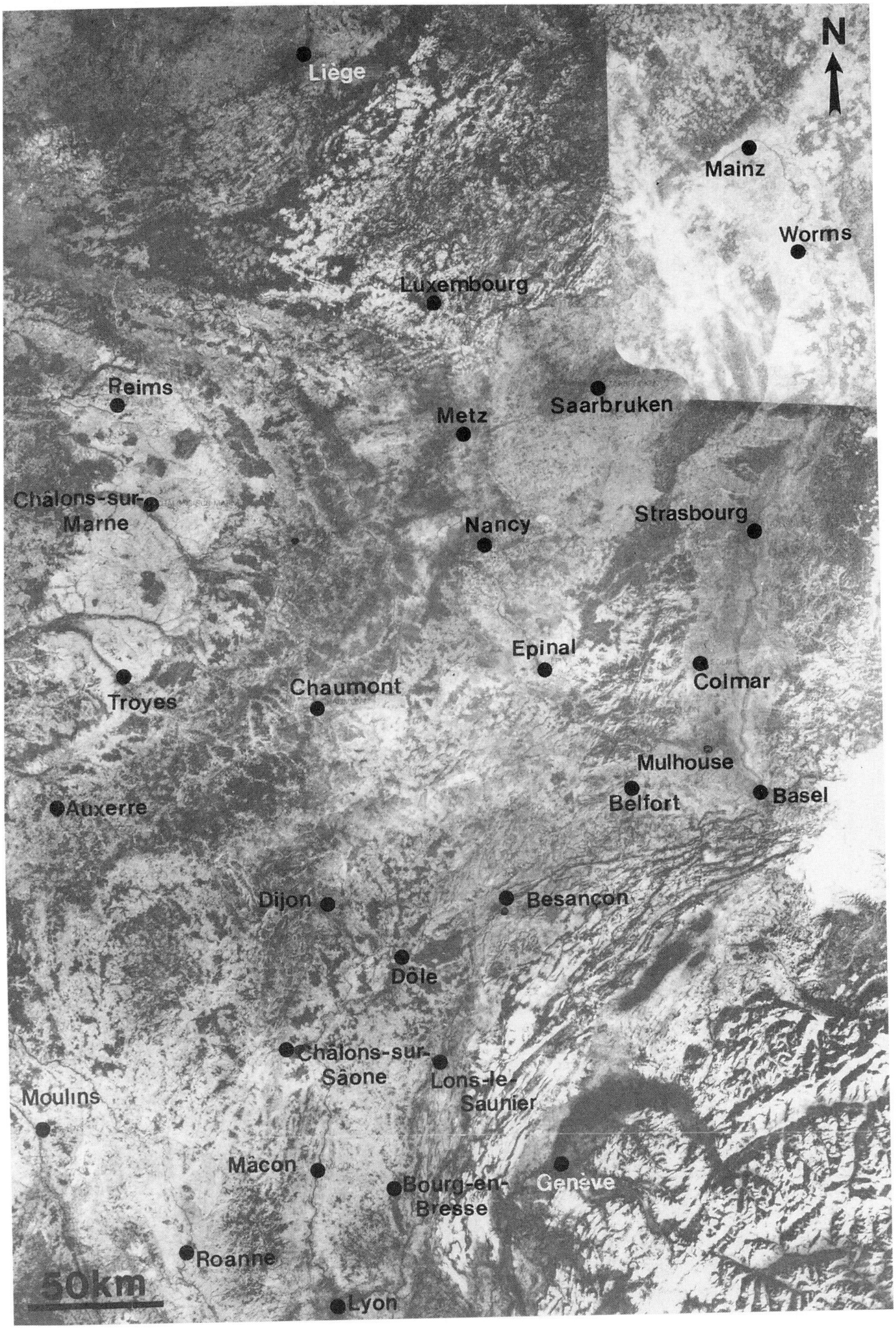
N
Liège
Mainz
Worms
Luxembourg
Reims
Saarbruken
Metz
Châlons-sur-
Marne
Nancy
Strasbourg
Epinal
Colmar
Troyes
Chaumont
Mulhouse
Belfort
Basel
Auxerre
Dijon
Besançon
Dôle
Châlons-sur-
Sâone
Lons-le-
Saunier
Moulins
Mâcon
Bourg-en-
Bresse
Genève
Roanne
50km
Lyon

FIG. 2.18A

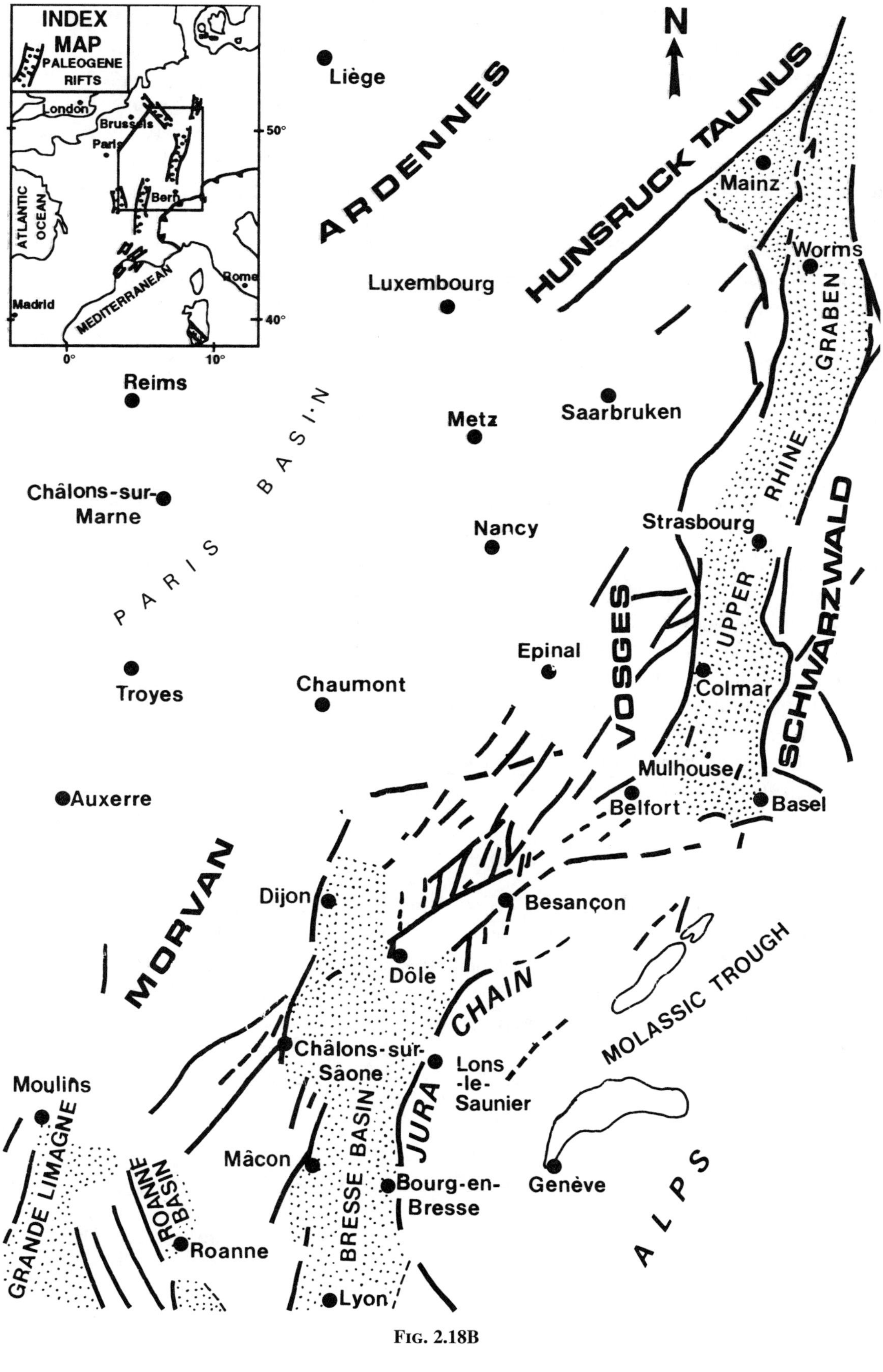
INDEX MAP
PALEOGENE RIFTS
London
Brussels
Paris
Bern
ATLANTIC OCEAN
MEDITERRANEAN
Rome
Madrid
50°
40°
0°
10°
N
Liège
ARDENNES
HUNSRUCK TAUNUS
Mainz
Worms
Luxembourg
GRABEN
Reims
Metz
Saarbruken
PARIS BASIN
RHINE
Châlons-sur-Marne
Nancy
Strasbourg
SCHWARZWALD
UPPER
VOSGES
Epinal
Troyes
Chaumont
Colmar
Mulhouse
Auxerre
Belfort
Basel
MORVAN
Dijon
Besançon
Dôle
MOLASSIC TROUGH
CHAIN
Châlons-sur-Sâone
Lons-le-Saunier
JURA
Moulins
GRANDE LIMAGNE
ROANNE BASIN
Mâcon
BRESSE BASIN
Bourg-en-Bresse
Genève
ALPS
Roanne
Lyon

FIG. 2.18B

Variscan crystalline blocks of the Vosges—Black Forest (Schwarzwald) in the south, the Odenwald-Spessart in the northeast, and the Rhenane Massif to the northwest—composed of folded Devono-Carboniferous sedimentary formations (figures 2.17, 2.18, and 2.19). Between them there is a lowland area mainly composed of Triassic sediments, the Haardt.

The structural trends inherited from the Hercynian and late Hercynian events are mainly oriented NE-SW oblique with respect to the NNE-SSW directions of the Upper Rhine-Hesse branch and NW-SE of the Lower Rhine arm. In the Upper Rhine, the Hercynian directions that control the Permian to Mesozoic regional paleogeography are clearly identified in the floor of the graben (Sittler 1965, Boigk and Schöneich 1974; Blanc-Valleron, 1996); they act as boundaries to the internal subbasins and were highs during Tertiary sedimentation. Northward, the Upper Rhine graben ends against the border of the Hunsruck-Taunus massif that coincides with a major Hercynian thrust line, while the Hesse and Lower Rhine grabens are superimposed upon older tectono-sedimentary areas active in the Permian and Mesozoic. Though it is less evident than in other segments of the rift complex, even in this area a compressive phase took place before the rifting was initiated.

Location and morphostructural characters.

The extensional events started in the middle Eocene (40–45 MA) and reached the period of greatest activity in the upper Eocene-Oligocene; Bergerat (1985) and Angelier and Bergerat (1983) consider that the main structural constraint was oriented 070° as opposed to the 020° direction of the rift itself. This induced a sinistral strike slip component of about 400 meters of post depositional transverse displacement for 1 km of normal opening. This lateral shift remains small as compared to the vertical throw in the Upper Rhine graben which is estimated to be about 5,000 meters. The vertical movements include a component of structural uplift of the adjacent blocks of about 1,500 m for the Vosges—Black Forest block and 1,250 m for the Odenwald (Illies 1972; Neugebauer and Braner 1978). The amount of differential movement is less in the northern part of the graben, although an uplift of about 200 m has been observed in Rhenane massif.

The synsedimentary tectonic activity appears to have been mainly controlled by antithetic faults that caused marked variations in sedimentary thickness (Illies 1967; Sittler 1965; Chauve and Perriaux 1988).

(a) The Upper Rhine graben, 300 km long and 30 to 40 km wide, is an asymmetrical structure whose asymmetry is reversed from the southern section, where the deeper part is located at the west, to the northern part where the deeper part is to the east. It is clearly a younger structure that sharply cuts the earlier crystalline and sedimentary Paleozoic blocks and the Triassic lowlands. The northern part of the graben widens with the development of the Mainz basin on the west. The basin then straightens into a narrow depression linked with the Hesse grabens.

The geometry of the graben is controlled by a system of 020° antithetic stairstep faults with a basinward dip of about 40 to 65°, named the Vosgian (west) and Scharzwaldian (east) faults respectively. Inside of this first faulted system, a second set of parallel faults (Rhenane faults) that dip more steeply basinward, between 75–80°, bounds the deeper Tertiary trough; close between the two sys-

Fig. 2.18. Satellite view of the northern part of the rift:
A. The Rhine segment and the transverse faulted zone of Le-Creusot-Belfort which separates it from the Bresse-Bas-Dauphine segment (Landsat photo).
B. Schematic map showing the location of the basins and the main structural trends.

Fig. 2.19. Schematic geological map of the Upper Rhine graben showing that: (1) it cuts across Paleozoic massifs as well as Mesozoic sedimentary basins; and (2) it is subdivided into a succession of subbasins and highs oriented along late Hercynian structural directions (adapted from the geological map of France at the scale of 1:1,500,000 and Sittler 1965).

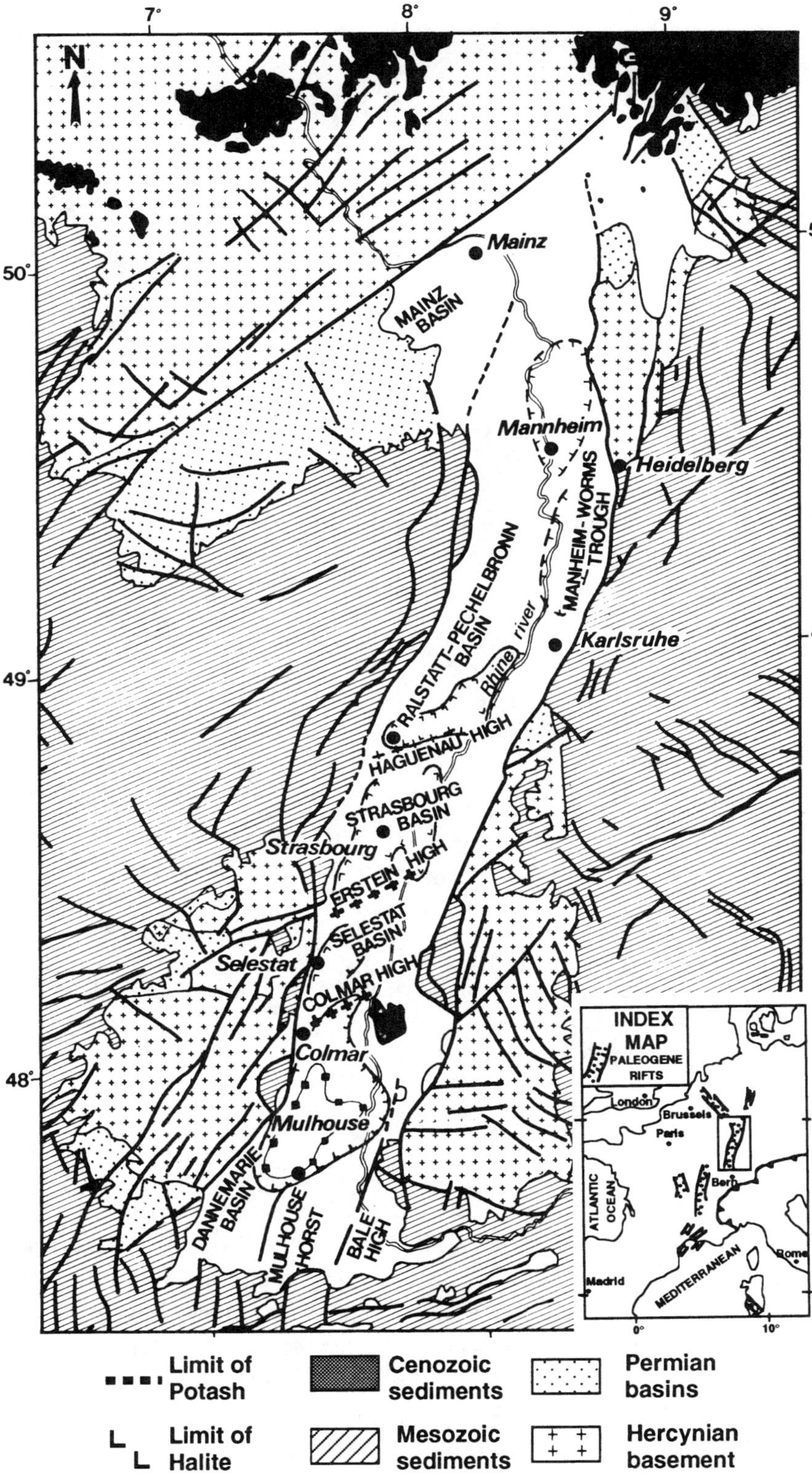

7°
8°
9°
N
50°
49°
48°
Mainz
MAINZ BASIN
Mannheim
Heidelberg
MANHEIM-WORMS TROUGH
RALSTATT-PECHELBRONN BASIN
Rhine river
Karlsruhe
HAGUENAU HIGH
STRASBOURG BASIN
Strasbourg
ERSTEIN HIGH
SELESTAT BASIN
Selestat
COLMAR HIGH
Colmar
Mulhouse
DANNEMARIE BASIN
MULHOUSE HORST
BALE HIGH
INDEX MAP
PALEOGENE RIFTS
London
Brussels
Paris
ATLANTIC OCEAN
Bern
Rome
Madrid
MEDITERRANEAN
0°
10°
40°
Limit of Potash
Limit of Halite
Cenozoic sediments
Mesozoic sediments
Permian basins
Hercynian basement

FIG. 2.19

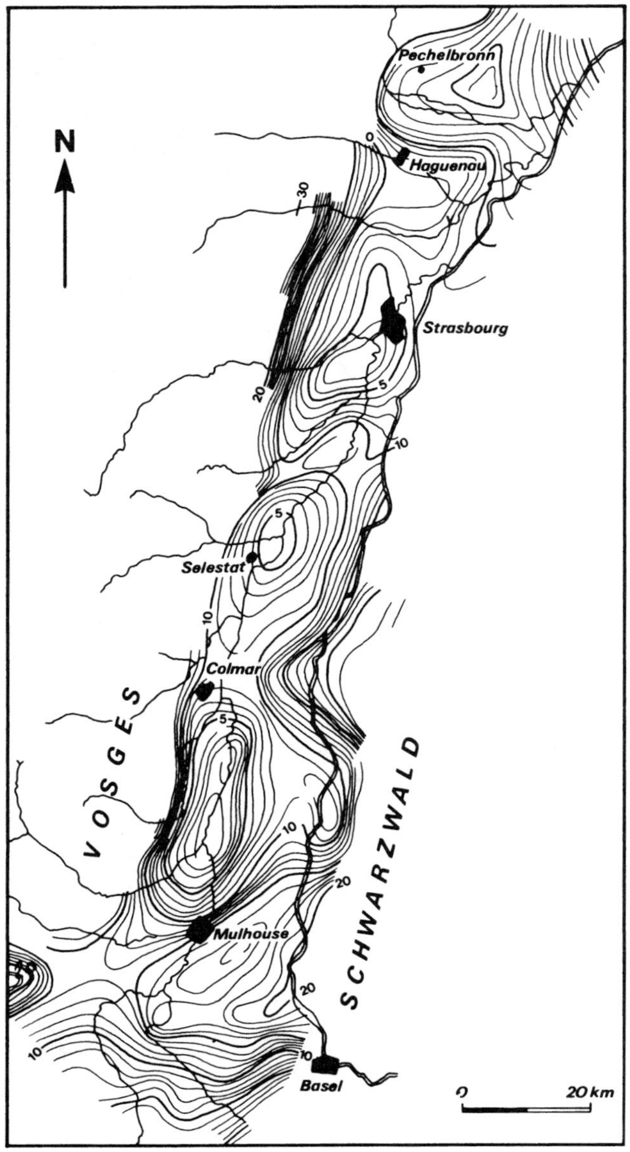

FIG. 2.20. Gravimetric map (experimental gravity anomalies) of the Upper Rhine graben showing the morphostructural subdivision into subbasins and highs, given in milligals (from Maikovsky 1952).

tems is a highly fractured strip (champs de fractures) that formed as stepped blocks of Mesozoic rocks covered by a thinner Paleogene series.

The structure of the Jurassic strata preserved on the floor of the graben shows large undulations with oblique NE-SW (late Hercynian) directions that are sharply cut along the border faults (Sittler 1965). The gravimetry map of the southern and central parts of the graben clearly reveals the subdivision of the graben into subbasins limited by highs oriented along the older NE-SW Hercynian directions (figure 2.20); successively from the south to the north, they are the basins of Mulhouse, Selestat, Strasbourg, Pechelbronn, and the trough of Mannheim.

The sedimentary filling of this rift sequence commonly reaches 2,000 meters in thickness in southern part, but the thickest sequence (up to 3,000 meters) occurs northward in the asymmetric depression of Mannheim, in an elongate sub-basin along the eastern border (Sittler 1965, 1967; Doebl 1967, 1970) (figures 2.21 and 2.22). The age of the sedimentary filling, however, is diachronous in the two areas.

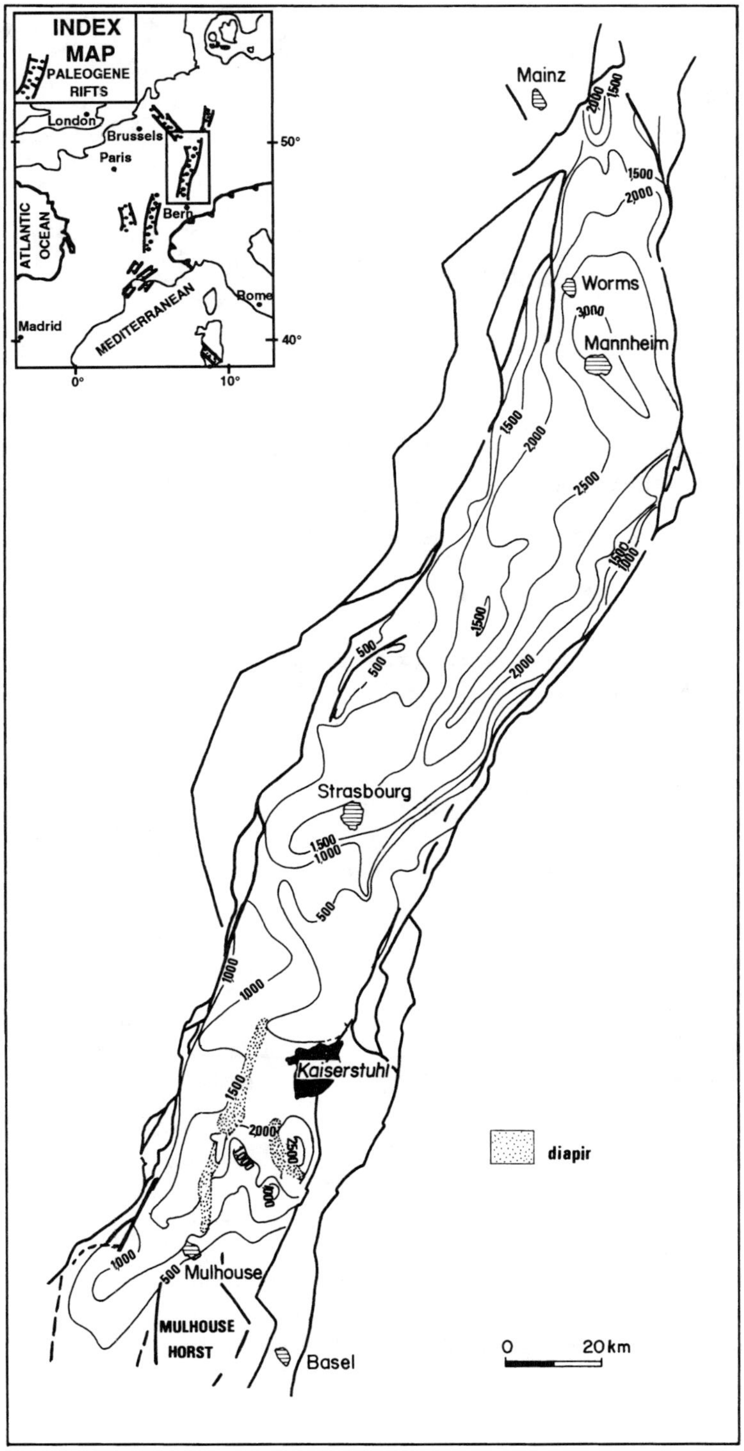

FIG. 2.21. The sedimentary fill of the Upper Rhine graben. Isobaths of the base of the Tertiary sediments given as depth in meters (modified from Doebl et al. 1972).

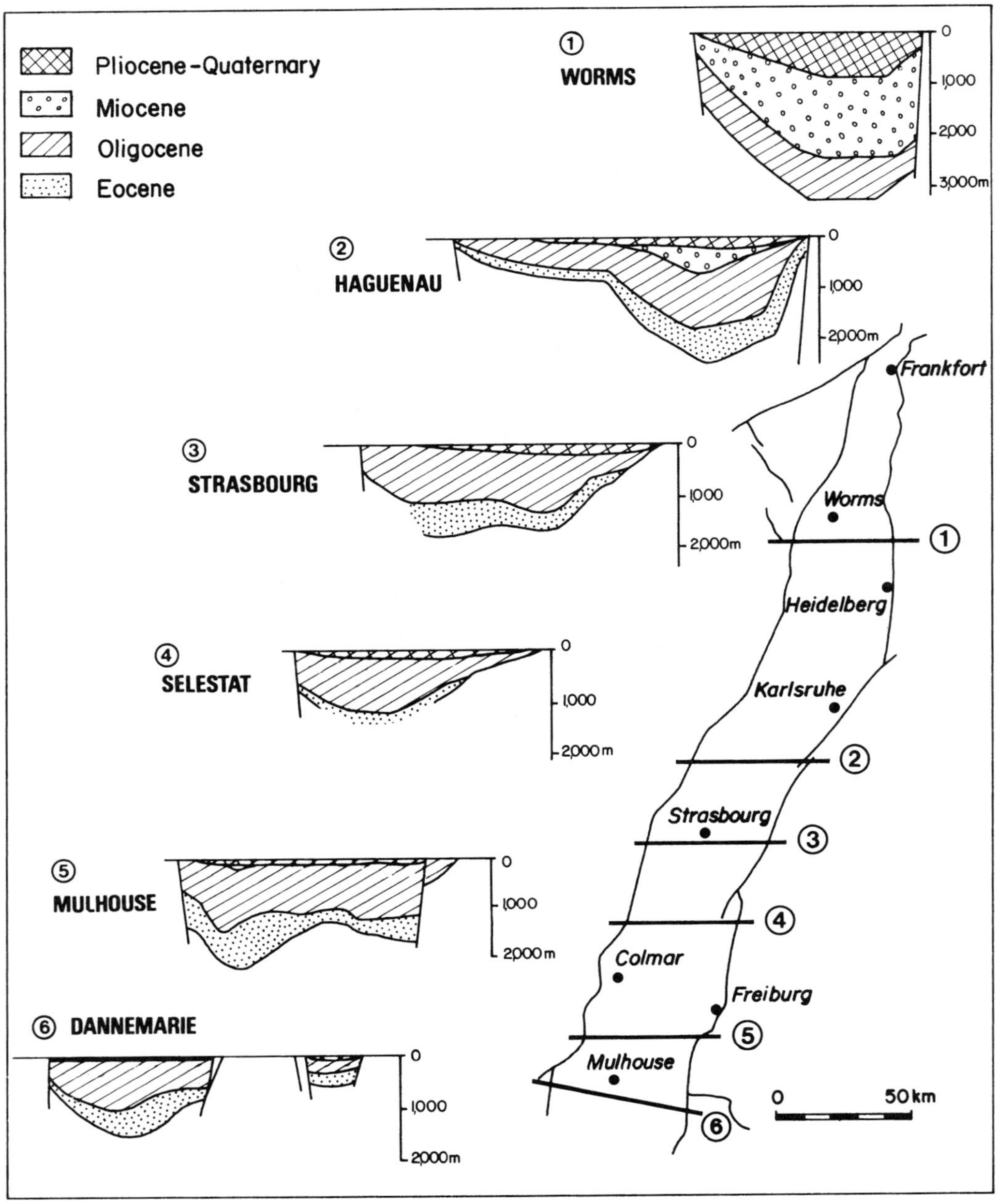

FIG. 2.22. Regional changes in thickness of the Tertiary sedimentary formations in successive parts of the Upper Rhine graben; the sections illustrate the northward tilting of the graben from the Miocene (sections made from the stratigraphical data of Doebl 1968).

Subsidence started in the middle Eocene and climaxed in late Eocene/early Oligocene in the southern and central area, but began during the late Oligocene in the north, with a maximum in the Aquitanian. This later subsidence is marked by a sedimentary thickness of 1,500 meters. The presence of more than 1,000 meters of post-Aquitanian sediments demonstrates that subsidence has continued in the north (Doebl 1967).

Thick evaporative deposits with halite beds formed during several episodes from the late Eocene to the early Oligocene and are present throughout most of the southern and central depressions. However, potash salts occur only in the southern Mulhouse area. The potash occupies the uppermost part of the formation (Courtot et al. 1972; Blanc-Valleron and Gannat, 1985; Blanc-Valleron, 1996). Northward, salt deposition develops diachronously in the latest Oligocene and even into the Aquitanian (Doebl 1970).

(b) The Hesse graben is a system about 100 km long and over 30 km at its maximum width that extends beyond the Vogelsberg volcano in the northern extension of the Upper Rhine graben. This irregular sedimentary area is limited mainly by a system of 020° trending faults, and is superimposed upon an older, narrow tectonic trough which was active in the Permian and early Triassic (Sittler 1965; Ziegler 1982). In the Tertiary basins formed by reactivation of the former fault system, the main subsidence took place between the middle Eocene and the end of the Oligocene, but remained weak as indicated by a sedimentary fill of no more than 100 meters (Sittler 1965; Illies 1974). These sediments are continental, mainly terrigenous, with intercalations of lignite.

(c) The Lower Rhine graben is a complex set of fault troughs limited by NW-SE faults linked northward with the central North Sea basin. Southward, the graben ends against the Rhenish Schiefergebirge mountains, a barrier that separates the two arms of the Rhine graben. Despite the position of the barrier, a system of NW-SE faults normal to the Hercynian directions cuts across the massif toward the northern end of the Upper Rhine graben. The small Nieuwed basin close to the massif is the last connection between the two arms, as indicated by its Eocene-Oligocene sedimentary fill. Indeed the lower Rhine system generally has a different tectonic history from the Upper Rhine: subsidence started earlier, probably in the Cretaceous, was active during the Oligocene, and continued locally in the Miocene and Pliocene, resulting in a maximum sedimentary thickness of about 1,000 meters (Teichmüller 1972). To the northwest, near the North Sea, the system progressively loses its specific pattern. Here, the tectonic evolution is marked by two main episodes of subsidence (early to middle Eocene and middle Oligocene) that are separated by a period of uplift and erosion during late Eocene/early Oligocene (Keizer and Letsch 1963; Neugebauer and Temme 1981).

Rift-associated volcanism.

An intense phase of volcanic activity that is alkaline in nature occurs in this area (figure 2.17). The earliest activity began in the late Cretaceous/early Tertiary in the Upper Rhine graben. This phase culminated in the Eocene (Horn et al. 1972; Baranyi et al. 1976; Lippolt et al. 1974; Neugebauer 1978; Ziegler 1980; Brousse and Bellon 1983). However, the main period of volcanicity was during the Miocene. Among the most important events are the formation of the Kaiserthul (18 to 16 MA) and of the largest Vogelsberg volcano, with its associated basaltic flows (Lauer et al. 1967; Wimmenauer 1967; Lippolt et al. 1974). These are located at the intersections of the NNE-SSW border faults with those faults having NE-SW directions, corresponding in the case of the Kaiserthul to the transform fault system between the Bresse and Rhine grabens and in the case of the Vogelsberg volcano to the faulted margin of the Rhenane massif. Along the NW-SE fracture field that crosses through the Rhenane massif as a prolongation of the Lower Rhine graben there are areas of volcanicity with associated lava fields that were active from the Miocene until about 10,000 years before the present (Illies 1974).

Post-rifting history.

Since the end of the Oligocene the most prominent structural feature has been the northward tilting of the Upper Rhine graben. Tilting began in the late Oligocene and progressive motion resulted in a maximum subsidence in the northern Mannheim depression during the early Miocene and in the Plio-Quaternary, expressed by more than 1,000 meters of sedimentary fill. During the same time period the subsidence ceased progressively in the southern part of the Upper Rhine graben as well in the Hesse area but remained active in the Lower Rhine. Post-Oligocene fault activity resulted in vertical throws of up to 500 meters (Courtot et al. 1972; Larroque and Ansart 1985) and is demon-

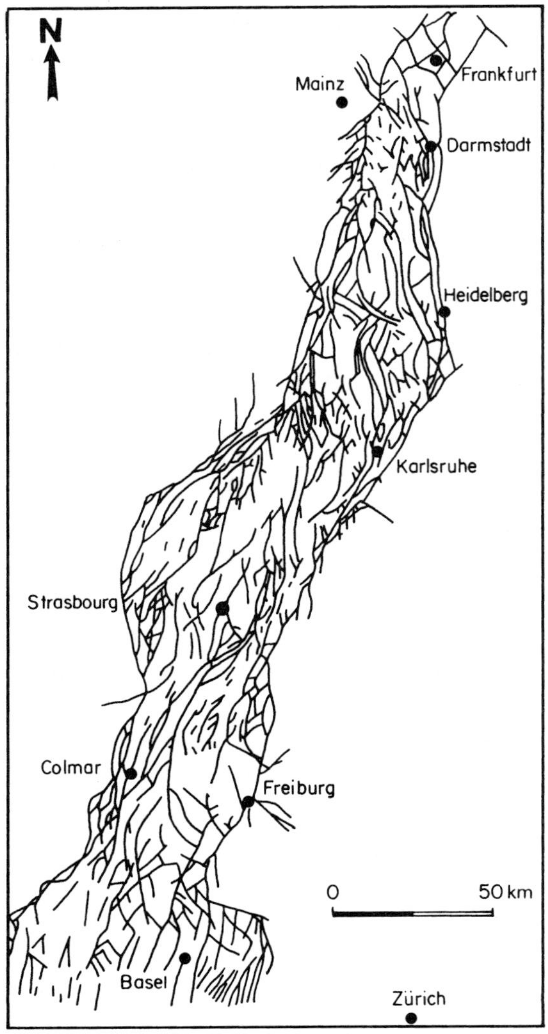

Fig. 2.23. The compressive shear system (major faults) of the Upper Rhine segment of the rift (modified from Illies 1981).

strated by the vertical migration of the salt in the Mulhouse basin.

The Upper Rhine graben system is still seismotectonically active and is affected by sinistral shear strains (figure 2.23). This deformation seems to be blocked to the north by the Rhenane massif as a result of the different mechanical properties of the basement in that area (Illies 1981; Illies and Baumann 1982).

Structural Evolution of the Rift System

The Western European continental rift is a chain of individual segments more or less related and in some cases offset one from another along shear zones that are interpreted as continental transform faults. No single segment is uniform. Variations are probably due to changes in the regional stress field through time and space in relation to the position of the segment with regard to Alpine tectonics. Variations also are probably due to differing mechanical properties of the regional blocks in which they are opened, as well as their structural framework and history.

The individual grabens that form the rift have an asymmetrical structure and most of them are typically half grabens (Gulf of Lion, Nîmes, Manosque). As shown in the regional descriptions, the Bresse-Bas-Dauphiné and Upper Rhine segments together show an inversion of the area of the maximum subsidence trough along the length of the segment.

THE PRE-RIFT REGIONAL FRAMEWORK

We have seen that the Western European rift cuts a structural complex made up by the successive orogenies such as the Cadomian, Variscan, and early Alpine phases, thus giving various structural settings for the different segments of the rift. Some segments are entirely enclosed within Hercynian blocks composed of crystalline or folded Paleozoic rocks, as in the Massif Central and the Gulf of Lion. The Roussillon-Languedoc-Provence segment occurs in an area covered by thick Mesozoic sediments strongly deformed by the late Cretaceous-Eocene compressive phase (Pyrénéo-Provençal phase), while the Bresse—Bas-Dauphiné segment runs along the border between the Hercynian block (Massif Central) and the Mesozoic basin. The Rhine segment cuts across various crystalline blocks (Vosges, Schartszwald, Odenwald-Spessart) and sedimentary areas of late Paleozoic to Mesozoic age, but is deflected along the NE-SW border of the Paleozoic Rhenane massif, probably as a consequence of differential mechanical behavior of the different segments through which it passes. To the east, the grabens of the Bohemian Massif, not discussed here, are also located in the Paleozoic massif.

The structure of the rift commonly shows the interaction of rift directions that intersect existing structures and of the rejuvenation of some older structures, mainly of Hercynian to late Hercynian origins. The general NNE-SSW direction of the rift itself appears to follow an older line of crustal weakness evidenced by a larger lineament that appears to run N-S across all of Europe from Permian Oslo graben to southern France. It is evident that local NNE-SSW trends already existed in the Rhine Graben area during the late Hercynian. If they do not give evidence of any relationship to the border faults of the upper Rhine graben that are part of a new structural system, they did control other features, such as the superposition of the Permian and Tertiary grabens of Hesse.

The fault systems formed during the late Hercynian were fundamental to the formation of many Western European rift grabens by having controlled their general orientation or their internal structure. One example of such structural control is clearly demonstrated by the NE-SW oriented fault bundles of Cévennes, Nîmes, and Durance which were reactivated through time with either normal or reverse motion. These faults have successively controlled the geometry of the Mesozoic basin and then governed the orientation of the Tertiary Alès and Nîmes grabens and the Manosque-Aix-Marseille tectonic troughs. Similarly this is the case for the Grand Sillon Houiller of the Massif Central with its string of small Carboniferous and Tertiary sedimentary depressions. The Bresse basin is another example, limited by N-S to NNE-SSW faults cutting obliquely across the dominant NE-SW Variscan directions of the Massif Central. Nevertheless these Variscan trends control the orientation of the Paleogene highs (Limonest, Cormoz, Sennecey—La Serre) and subdivide the graben into sub-basins that progressively deepen from NW to SE. In the Upper Rhine graben, the internal subdivision is imposed by older transverse NE-SW to NNE-SSW structures which are conformable with Variscan features. These trends were rejuvenated by the compressive pre-rift folding.

The lateral shift between the Bresse and the Rhine grabens evidenced on the surface by its Riedel system could be the result of shear reactivation during the N-S pre-rifting compression of an important basement fault that earlier controlled the formation of the Permo-Carboniferous basin of Blanzy—Le Creusot (Gelard 1978; Bergerat 1986). The Lower Rhine graben is also clearly associated with older structures.

TIMING AND CHARACTER OF THE MAJOR PHASES OF DEFORMATION

The N-S pre-rift compression.

We have seen that the Paleogene rifting was preceeded by a general compression perceptible in all the rift segments, although with strong differences in regional intensity. The reconstruction of the paleostress directions shows that the N-S compression was a general phase of shortening, the effects of which are apparent in all areas of peri-Alpine platform. Here the shortening direction is relatively constant, nearly N-S, with the s1 (principal) stress orientation ranging between 170° and 010° (Letouzey and Tremolières 1980; Bergerat 1985, 1987). If the deformation is weakly expressed in some of the segments, as in the Massif Central where it produced only local rejuvenation of faults, this phase also produced different fault bundles as well as the sinistral shear rejuvenation of most of the older faults in other areas as in southern France (Bodeur 1976; Arthaud et al 1977; Masse 1980; Arthaud and Seguret 1981; Philip et al. 1987; Tempier 1987; Viallard 1987).

The deformation is most strongly expressed by: (1) the sinistral shear reactivation of the late Hercynian fault bundles, such as that of Cévennes, Nîmes, and Durance where the lateral throw may be greater than ten or several tens of kilometers (Bodeur 1976; Masse 1980); (2) the folding of the Mesozoic series in northern Languedoc with E-W fold axes; and (3) the reverse folds associated with backthrusting (Montpellier fold) and the thrusting of segments such as the Corbières and Saint-Chinian nappes and the South-Provençal thrust in different directions during the Alpine period around the border of the Mediterranean. The main orogenic phase occurred at the same time in the Pyrenean chain where thrusting was active up to the Oligocene (Fischer 1984; Seguret and Daignières 1988; Vergès and Munoz 1990). Several basins formed during this compression event, such as Saint-Martin-de-Londres and Alès, although in Alès maximum subsidence occurred later (Fredet 1987). A similar succession of tectonic events has been proposed by Mauffret and Gennesseaux (1989) for the Gulf of Lion grabens with the opening of pull-apart basins by strike-slip activity of the fault system prior to the distension regime.

All the deformation features attributed to the pre-rifting compression could not have been exactly synchronous and locally several episodes apparently occurred anywhere from the late Cretaceous to late Eocene, but generally the paroxysmal phase was Bartonian (Arthaud and Seguret 1981).

The rifting

a—Chronology of the subsidence.

In spite of common characteristics, the chronology of the local structures is not identical in all the segments. Except for an area of the southern Bohemian Massif where the subsidence has been continuous from the late Cretaceous to the Miocene (Malkovsky 1966), the beginning of the subsidence had its earliest expression (early Eocene) in the Lower Rhine graben. In the Upper Rhine graben, the beginning of the subsidence is dated as middle Eocene (45 MA) while in the Marseille trough it is of early Oligocene. In the Alès graben, the main subsidence began in the late Eocene (Fredet 1987). Maximum subsidence began in the Priabonian (Ludian) and is marked by the deposition of the Lower Massive Halite Formation in deeper areas of the Upper Rhine, Bresse, and Valence basins as well as sulfate deposits in numerous platform areas (e.g. the Carpentras Basin) and grabens (e.g. Le-Puy-en-Velay, Alès). To the extent that there is sufficient statigraphic control, this period of intense subsidence in the northern basins seems to correspond to the period of folding and thrusting in most of the southern area. Thus the opening of most of the southern grabens is usually considered to be Oligocene.

In all the grabens, the Oligocene was a period of intense subsidence characterized by the deposition of several thousand meters of sediments including: thick halite formations of Rupelian (Stampian) age in the main troughs of Rhine, Bresse, Valence, and perhaps in the Camargue; the salt layer in the Grande Limagne; and dispersed gypsum layers in some of the platform area (Carpentras Basin, Marseille, and Saint-Pierre-les-Martigues). While synsedimentary tectonics appear to have been active throughout this period, the rate of subsidence was not continuous and successive changes in structural directions occurred. In the Valence basin the detailed works of Dumas (1988) show 020° oriented faults controling the deposition of the First Salt Member while 110° faults governed the morphology of the trough in which the Second Salt Member was deposited. In Marseille the successive changes in paleogeography were controlled during all of the Oligocene by variations in the main structural trends (Dumas 1987/1988). Migration of the subsidence axis has been revealed by the detailed sedimentological studies in most of these grabens, particularly in the Bresse basin (Curial 1986; Moretto 1986/1987) and in Manosque-Forcalquier.

While subsidence ceased in some grabens at the end of the Oligocene (e.g. southern part of the Upper Rhine graben, Hesse and Lower Rhine, Massif Central), it remained strongly active in other areas. Thus a progressive northward tilting in the Upper Rhine graben produced the migration and the reversal of the direction of maximum subsidence in the northern Mannheim depression where halite was deposited at the end of the Oligocene (Chattian) and in the Miocene (Aquitanian). Thick terrigenous deposits (up to 1,700 m) accumulated during the Aquitanian in the Nîmes graben. Although imprecise stratigraphy does not allow accurate dating of the subsidence, it seems that the Roussillon depression sank considerably during the Miocene (perhaps partly in the Oligocene) and was filled by more than 1,000 meters of sediments. This subsidence and sedimentation was then followed (after a stratigraphic gap) by 700 meters of Pliocene. In the grabens of the Gulf of Lion, the main subsidence probably was Oligocene-Aquitanian age, as in the Sardinia basin.

b—Kinematics of the deformation.

The rifting processes appear as a relatively independent tectonic development related to the kinematics of the Alpine foldbelt, interfering with the Pyrenean orogeny and with the mechanisms of Mediterranean opening.

The opening of the Western European rift was controlled by extension in which the direction of the least compressive stress (s3) is E-W (075° to 105°), governing the dominant N-S to NNE-SSW directions of the Rhine, Great Limagne, Bresse, and Valence grabens. However, the dominant NW-SE direction of the rift in southern France and the Gulf of Lion reflects an earlier structural framework formed by a system of older lineaments (late-Hercynian for the most part) that were then activated during the Pyrénéo-Provençal compression. The role of the late Hercynian lineaments and of the Eocene compression is clearly evident in the control of the Oligocene Massif Central grabens and in the internal morphology of most of the grabens. These structures are made up of juxtaposed sub-basins separated one from another by highs that follow Variscan structural trends. The deformation was not synchronous in all the basins, and the stress direction did not remain stable during their formation, as mentioned above. Some of the more important changes included a shift in direction within the Valence basin, Manosque-Forcalquier, and tilting in the Rhine and Bresse grabens; the southern part of the Provence region also tilted southward near the end of the Oligocene (Nury 1987/1988).

Through the deformation style in the graben area is clearly extentional, most authors note that in the setting just outside of the grabens but within their direct borders, the extensional constraints apparently change into compressional constraints (Bergerat 1985; Villemin 1986). At the time of the extension the general tectonic environment in the central and northern Pyrénées was compressive where the thrusts developed from the Eocene to the Oligocene and in the Alps, which were affected by the main phase of shortening (Trümpy, 1980). We have seen that compressional events occurred from the late Cretaceous to late Eocene and even in the early Oligocene in southern France and that the Pyrenean thrusting involved Oligocene strata as well. Thus the stratigraphic data show the opening of some grabens started at the end of the middle Eocene while the beginning of the maximum subsidence is late Eocene. These events actually would have occurred within a period of compression. Some authors consider that the first stages of basin opening may have developed in a period of compression along transverse wrench faults that were crustal transform systems (Masse 1983).

Triassic evaporites whose role is evident in the deformation due to the Pyreneo-Provençal compression also appear to have contibuted actively to the extensional faulting, at least in southern France (Maurin 1984; Appia-Lamache 1984; Rouchy 1987; Roure et al. 1988). Roure at al. (1988) have shown that the structure of some grabens are characterized by listric faults that flatten within the Triassic beds, roll-over anticlines, and compensation synclines along the footwall of the faults. These structures are similar to those illustrated in the seismic line of the Nîmes and Manosque grabens (figure 2.7). Lenguin et al. (1990) has pointed out the halokinetic processes associated with the Narbonne graben formation. Moreover, there is some evidence that the salt body of Manosque was also an allochtonous sheet migrated from the Triassic.

In southern France and the Gulf of Lion just to the south of the Nîmes graben there is another complication due to the tectonic intersections of the opening of the Western European rift, which remained intracontinental, and the western Mediterranean, which formed an oceanic basin. In this latter case, the initial rifting is of the same age and has been followed by an oceanic accretion phase during the Aquitanian-Burdigalian stages (21 to 17 MA). Nury (1987/1988) and Nury and Raynaud (1986) consider that the multidirectional opening of the Marseille and the St-Pierre-les-Martigues depressions resulted from the interference of the opening directions of the Western European rift and that of the Gulf of Genova (a segment of the Mediterranean rift complex that remained continental).

c—The post-rifting history.

The main distensive phase ceased in the late Oligocene Aquitanian interval, but subsidence did not stop everywhere at the end of the Aquitanian. The post-Aquitanian history is marked by a succession of compressive and distensive phases throughout the area until the Pliocene and some of the associated faults still remain seismo-tectonically active.

In the depression of Perpignan (southern France), which is more or less the continuation of the Gulf of Lion grabens, subsidence occurred mainly during an undated Miocene-Pliocene interval. The northern part of the Upper Rhine graben is the best example of more or less continuous subsidence, where the cumulative thickness of the Mio-Plio-Quaternary sediment locally reached 2,500 m and is largely controlled by the S-N tilting of the basin. In some basins, the axis of post-Aquitanian maximum subsidence coincides with that of pre-Aquitanian maximum subsidence, although there are some differences. Several examples illustrate these paleogeographical changes:

1. the Valence area, where the thickest Miocene deposits (up to 700 m) are not superposed directly upon the Paleogene trough, but overlie the earlier laterally eastern platform;
2. the Provence region, where the main area of subsidence was transferred to the east of the Durancian fault in the Digne basin, resulting in some 2,000 m of Miocene terrigenous filling;
3. the Carpentras graben, formed to the east of the Nîmes faults after the end of the Burdigalian;
4. the successive motion of the main area of subsidence is the post-depositional backtilting of the segments of the Rhine graben as well as the activation of subsidence in the Bohemian segment during the Miocene. In contrast, the differential subsidence of the former deepest trough continued in Bourg-en-Bresse as well as in the Nîmes graben.

So the area of the Western European rift has remained tectonically active with a succession of variably intensive tectonic phases depending on the area considered. A NE-SW compression observed throughout the entire system occurred at the end of the Aquitanian and produced the reactivation of older faults and local unconformities (Bergerat 1985). Distentional deformation also occurred in southern France during the Langhian-Serravalian along the widespread 050° faults (Triat and Truc 1983).

The main tectonic episode affecting this area is the late Miocene-Pliocene compressive event, dominated by a NW-SE direction (Platt and Vissers 1989). It produced the folding of the Jura and its thrust over the margin of the Bresse basin, as well as overthrust folding of the sub-Alpine chains on the eastern part of the Dauphiné basin. These main changes in basin morphologies resulted in newly formed late Miocene to recent basins and are also responsible for the reactivation of faults and folds in some of the southern France area. For example, Casagrande et al. (1989) showed that this compressional event reinitiated the rise of the Triassic diapir of Suzette leading to the morpho-tectonic inversion of a small Oligocene graben. As indicated by Casagrande (1989), these phases of deformation do not result from a single tectonic phase but are rather the sum of several events occurring from the Serravalian to the Pliocene. In part of the Alpine system these events are related to the neo-alpine phases that could be responsible for the thrust of the Helvetic nappes onto the sub-Alpine molasse (Trümpy 1980). This structural event is correlative with the marine regression that affects the Languedoc-Roussillon area at the end the middle Tortonian, long before the Messinian lowering of the Mediterranean level. Probably the deep incision of the Rhône and other valleys, a spectacular regional feature, is the result of these two interrelated events (Rouchy 1983; Casagrande 1989).

The most recent phase of tectonics is illustrated by the reactivation of the subsidence of the northern part of the Upper Rhine graben, characterized by the deposition of more than 1,000 meters of fluviatile sediments from the late Pliocene as well as neotectonic deformation affecting the terraces of the Rhine river. Presently there is noticable seismotectonic activity; an active stress field indicates a sinistral shear parallel to the graben axis (Illies and Greiner 1978; Illies and Baumann 1982). In southern France too, some older fractures such as the Durancian faults seem to have a historical record of seismic activity, and subsidence of about 0.8 mm/y has been measured in the Rhône valley near the northern end of the Nîmes graben, while the Bresse graben has been affected by recent S-N tilting (Barbreau et al. 1976; Fourniguet, 1977, 1987; Fourniguet et al. 1981). Seismic activity is also recorded along the western border of the Ambert basin (Dorel et al. 1990).

The Sedimentary Position of the Evaporites Related to the Structural Development of the Rift System

Although the volume of these rift deposits is small compared to giant evaporite basins such as the Zechstein of northern Europe or the Messinian of the Mediterranean, numerous economically important evaporite deposits are present in parts of this rift system, containing gypsum/anhydrite, halite, and potash. The genesis of these evaporative deposits has been the subject of considerable controversy: did they form from marine derived sources, continental waters enriched in salts by leaching of older evaporites, or from hydrothermal input from rift-related springs? These have all been suggested as possible ionic sources. Although episodic penetration of marine waters into some parts of the basins is known (e.g., Upper Rhine graben, Marseille grabens), it is clear that the paleogeographical history of the rift is dominated by lacustrine conditions. Thus, most of the evaporative deposits would have been formed in hypersaline lakes, although episodic seawater input or hydrothermal water may have contributed to the ionic budget.

EVAPORITE DISTRIBUTION IN THE PALEOGENE RIFT

We have seen that evaporative sedimentation took place mainly from the late Eocene (Priabonian/Ludian) up to the early Oligocene (Rupelian/Stampian), while in some areas it occurred in the late Oligocene and continued into the lower Miocene, as in the northern part of the Upper Rhine graben (Doebl 1967).

Evaporative deposits are separated into two types:

1. The first type is exclusively composed of gypsum (locally anhydrite) intercalated within carbonates or terrigenous deposits either as massive beds (several decimeters to several tens of meters thick) or as thin layers and/or dispersed sulfates. These formations are encountered irregularly in most of the shallower platform areas or in tectonic depressions where the subsidence was less active than in the main troughs: Marseille, St-Pierre-les-Martigues, Aix-en-Provence, Alès, Narbonne, Apt, and Carpentras (southern France segment), Le Puy-en-Velay and Mauriac (Massif Central segment) (figure 2.24).

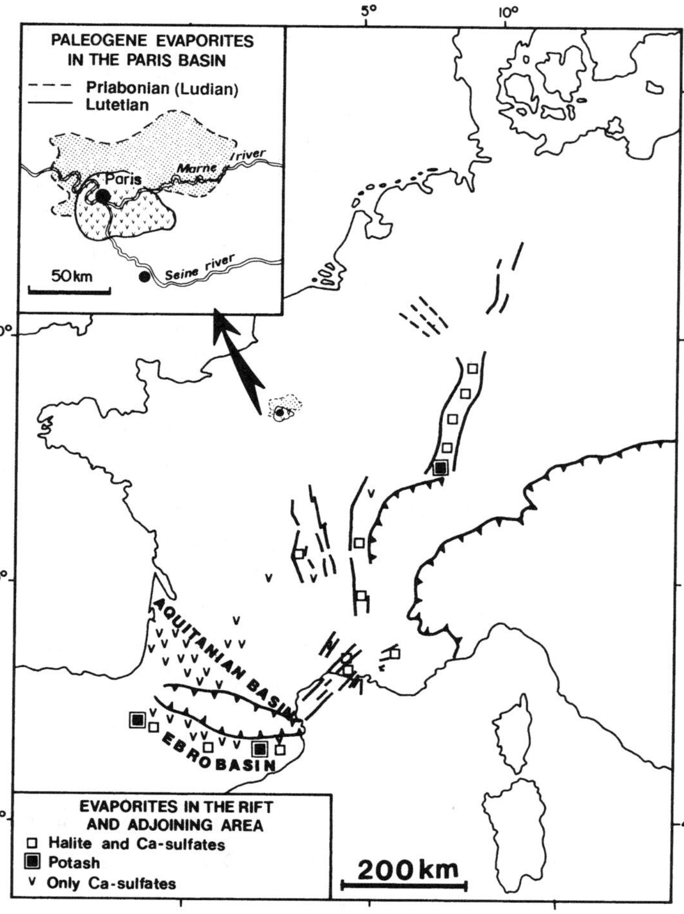

FIG. 2.24. Location of the Paleogene evaporites in the Western European rift system and the adjoining region showing the Paris basin, the Aquitaine basin, and the Ebro foredeep (from Winnock 1974; Ortí et al. 1984). The insert at upper left maps the extent of the Lutetian and Priabonian evaporites in the Paris basin (modified from Toulemont 1987).

2. The second type consists of thick salt formations in the most subsident parts of the rift: Upper Rhine, Bresse, Valence, Grande Limagne, Nîmes, and La Crau and, probably, in some of the Gulf of Lion depressions, as suggested by seismic data. Generally but not systematically, the halite is correlated with those areas having the maximum thickness of Tertiary sediments and is associated with variably-developed calcium sulfate intercalations that are extensive in areas peripheral to the halite and in the shallower parts of the basins, as in the first type of deposit. Tectonic control of such a typical depocenter for halite is well illustrated in all these basins. Thus, in the Upper Rhine graben the period of halite deposition changes from south to north through time, in response to the tilting of the northern part of the graben. The main evaporative deposits are Priabonian (Ludian) to Rupelian (Stampian) in the basins of Mulhouse, Selestat, and Strasbourg while thinner evaporative deposits occur in the Aquitanian of the Worms-Mannheim depression.

The area covered by the halite sedimentation within the deepest parts of a basin is a small fraction of the total surface of the basin, at least as it can be deduced from the present limits of the salt

TABLE 2.1.
Characteristics of the evaporative deposits in the different segments of the western European Rift.

LOCATION		STRATIGRAPHIC POSITION				SURFACE AREAS COVERED
		PRIABONIAN (LUDIAN)	RUPELIAN (STAMPIAN)	CHATTIAN	AQUITANIAN	
UPPER RHINE	MULHOUSE	Lower/Middle Salt E > 500 m	Upper Salt E = 400 m	Ca -sulfates		B =7,000 km² H =2,200 km² P =700 km² (except for Mannheim)
	SELESTAT	Lower Salt E ?	Upper Salt E = 345 m (*) H = 60 m	Ca -sulfates		
	STRASBOURG	Lower Salt E = 600 (*) H = 260 m	Upper Salt E = 350 m (*) H = 30 m	Ca -sulfates		
	PECHELBRON-RALSTATT	Lower/Middle Salt	Upper Salt	Ca -sulfates		
	MANNHEIM-WORMS			E = 300 m H = 12 m		
BRESSE	CHALON		Ca -sulfates			B = 8,000 km² E = 2,000 km² H = 800 km²
	LOUHANS	Lower/Intermediate Salt E = 400 m H = 260 m	Upper Salt E = 300 m H = 240 m	Ca -sulfates A = 20 m		
	BOURG	Lower/Intermediate Salt E = 200 m H = 700 m	Upper Salt E = 500 m H = 450 m	Ca -sulfates A = 20 m		
DAUPH.	VALENCE	Lower Salt (T6) E = 200 m (*) H = 150 m	Lower Salt (T5-T4) E = 500 m (*) H = 300 m			B = 1,800 km² H = 200 km²
	ROYANS		Gypsum (traces)	Upper Salt E = 450 m H =150 m		
	CREST		Gypsum (traces)			
SOUTHERN FRANCE	ALES	Gypsum				
	NIMES	Evaporites E = 1100 m (*) (+) H = 300 m				
	LA CRAU	Evaporites E = 400 m (*) H = 90 m (*) A = 400 m (*)				
	APT-CARPENTRAS-MANOSQUE	Gypsum G = 200 m(**); Salt H = 600 m (+) A = 200 m	Gypsum	Gypsum G = 10 m		
	AIX-EN-PROVENCE +MARSEILLE		Gypsum G = 3 m			
	NARBONNE			Gypsum G = 10 m		
M. CENT.	GRANDE LIMANGE	Evaporites E = 360 m (*) H = 128 m				
	LE PUY-EN-VELAY	Gypsum				
	MAURIAC		Gypsum (traces)			

- Thicknesses and surfaces of the evaporites are indicated by E (evaporites), H (halite), P (potash salts), A (anhydrite), G (gypsum); B indicates the surface of the basin. The values are followed by (*) when evaporites are not massive but dispersed through the indicated thickness; (**) when it is the maximum thickness observed in the area; (+) when tectonic or halokinetic deformation has probably modified the sedimentary thickness.
- For the Nîmes basin, the values are only representative of the Vauvert district, and represents only a part of the basin.
- For the La Crau and the Grande Limagne basins, the values are taken from only one borehole each, respectively Sainte-Cécile 1 and Saint-Beauzire 101.
This data have been obtained from Curial 1986 and 1987 (Bresse), Dumas 1988 (Valence), Blanc-Valleron 1990 (Rhine), Triat and Truc 1974 + Nury 1987 (Southern France).

(without any consideration of the possible effects of dissolution), but the thickness of the salt accumulation is considerable in some locales. The differences between the total basin area and the zones restricted to calcium sulfate, halite, and potash are given in table 2.1. From this data we see that a basin in a generally evaporative area with a large total area or thickness of sedimentary fill does not guarantee a thick halite accumulation, although in most cases the most subsident portions also have the thickest salts preserved.

There are some basins that are problematic. The

evaporites in the Manosque depression are somewhat ambiguous because of the possibility that part of the formation, if not all, was a diapir of Triassic evaporites. Recent data, however, seems to indicate that the salt is Paleogene. In contrast, some depressions appear devoid of any evaporative deposits, e.g., the Bohemian massif, Hesse, Lower Rhine, and Roanne-Montbrison. The sedimentation exhibits great variations in these narrow and rapidly subsiding depressions, and the fill could be either evaporative or terrigenous.

These summary data illustrate several aspects of evaporite sedimentation. The thickest deposits are correlative with the deposition of the more soluble salts and are generally correlated with the deeper and more subsident areas of the basins; the migration of the concentrated brines in the deepest part is correlated with a significant narrowing of the sedimentary area. In such continental basins characterized by their narrowness, and with highly-differentiated morphology and synsedimentary tectonic activity, changes in sedimentation occurred rapidly and the value of chemical sedimentation appears to be inverse to the volume of the terrigenous input that entered the main depot-centers.

PALEOGENE MARINE EXTENSION AND SALINE SEDIMENTATION IN WESTERN EUROPE

During the late Eocene-Oligocene, evaporites were also deposited in subsident areas adjacent to the rift, such as the Paris and Aquitaine basins, where they are thinner and only composed of calcium sulfate. The Priabonian (Ludian) evaporative episode in the basins of Paris and probably in Aquitaine is contemporaneous with the rift segment where evaporite deposition generally constitutes the first episode of sedimentation. The main depositional control is probably climatic while the thickness and the nature of the deposits is influenced by the physiographic characteristics of the basins and their drainage pattern.

Evaporative deposition occurred in the areas adjacent to the major fault-controlled areas somewhat earlier in the Paleogene, especially in the subsiding Paris basin (figure 2.24; figure 2.25; table 2.2). The first evaporative sedimentation was Mass 4, formed in the late Bartonian. However, the greatest part of the evaporative deposit in the Paris Basin is of Priabonian age (Ludian) (Masses 1, 2, and 3) and was more or less synchronous with the first episode of evaporites in the rift (Toulemont 1987). The marine waters that covered the Paris Basin up to the middle part of the middle Eocene (middle Lutetian) progressively withdrew because of the successive uplift of highs (Artois high, then the Bray anticline) or more generally because of regional uplift or eustatic sealevel drop. From the end of the middle Eocene (Lutetian), marine influence became episodic and later this area was dominated by lagoonal to lacustrine conditions. The detailed lithology and stratigraphy of the Paleogene evaporites of the Paris basin are given in table 2.2.

Elsewhere in western Europe, Tertiary evaporites are mainly developed in the South-Pyrenean basin. In the North Pyrenean area gypsum occurs locally in the lower and upper Eocene but the deposits appear to be of limited thickness and volume (Winnock 1974; Kieken 1980; Plaziat 1984).

Gypsiferous marls occur in the lower fresh-water molasse of the late Oligocene in the Molassic basin; in this case the development of playa type environments was due to the retreat of the sea because of the Oligocene tectonism of the Alpine chain but it is also associated with climatic fluctuations (Sittler 1965; Homewood et al. 1989).

EVAPORATIVE DEPOSITION IN THE RIFT: LACUSTRINE VERSUS MARINE ORIGIN

The relationship between the rift system and the open sea during this period remains ambiguous. In the upper Eocene, the embayments of the Paris and Aquitaine basins that represent the inland penetration of the Atlantic domain are closed towards the east and no connections existed with the North Sea (Rat 1974; Ziegler 1982; Cavelier 1984). The Alpine sea was present as a narrow peri-Alpine depression, but the sedimentary record of its westward margins, now modified from its former disposition by Alpine deformation, ends more than one hundred kilome-

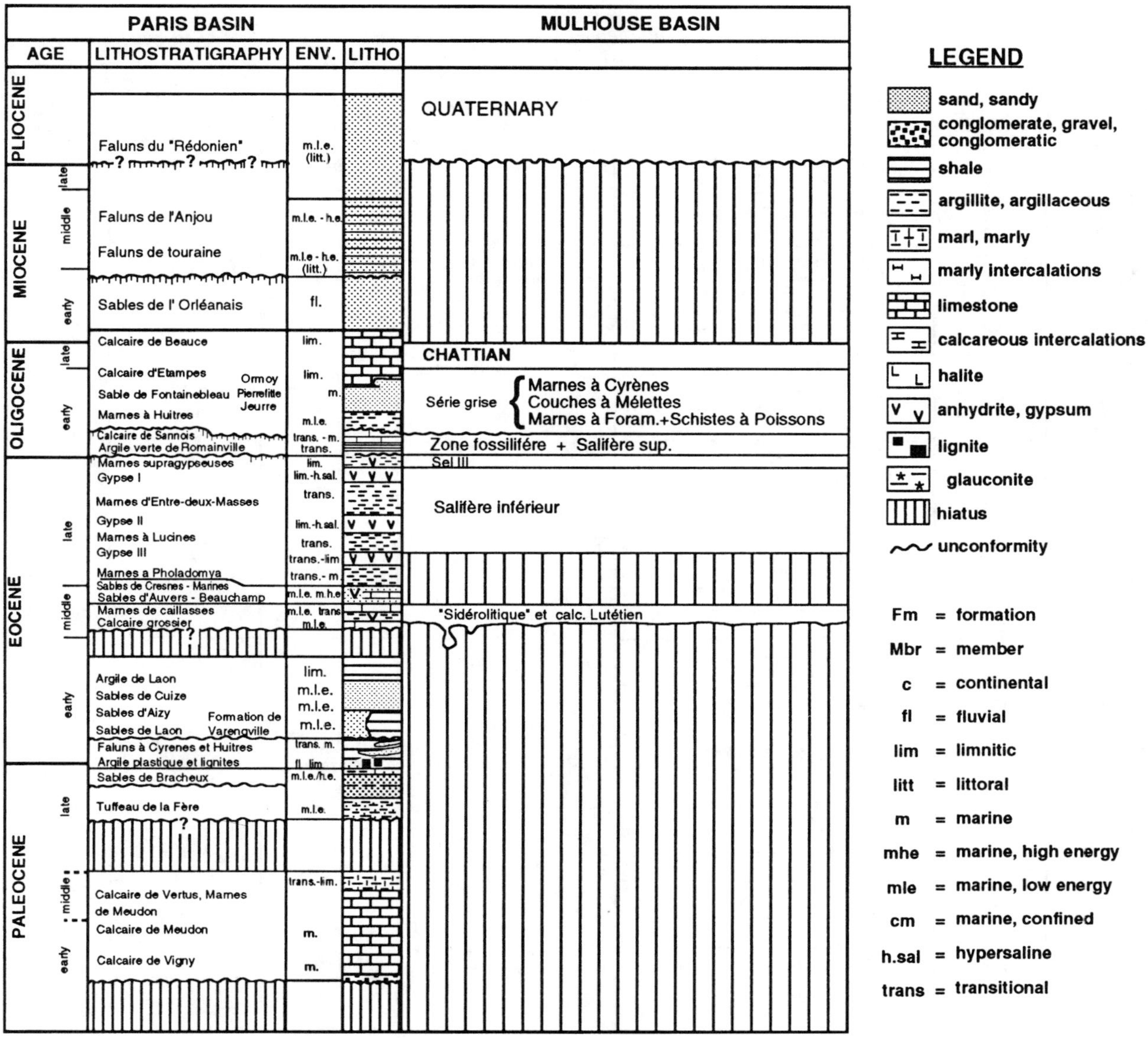

FIG. 2.25. Stratigraphic sections comparing the sedimentary record of the Paris Basin with that of the nearby Mulhouse Basin (Paleocene through Pliocene).

ters from the Bresse—Valence—southern France segments of the rift.

In Oligocene deposits, phases of distinct marine influence are identified in parts of the rift. In the upper Rhine graben, an influx of sea water has been suggested for the "Zone Fossilifère" (base of the Stampian) on the basis of paleontological evidence but the origin of the seawater influx is still uncertain (either peri-Alpine sea from the SE or North Sea from the NW). Marine conditions could have existed in this area from the time of the foraminiferal marls and fish beds (the "Rupelian transgression"), with this domain being connected with the North Sea via the Hesse graben. In this way, the North Sea could have been linked with the peri-Alpine sea (Sittler 1964). Southward, the peri-Alpine seaway probably extended locally nearer to the Bresse-Valence and southern France basins, but direct connections are a matter of conjecture (Lefavrais-Raymond 1962; Rat 1974; Triat and Truc 1974;

TABLE 2.2.
Stratigraphic distribution of the gypsum deposits in the middle-upper Eocene of the Paris Basin.
(Soyer 1960 and Toulemont 1987)

EPOCH	STAGE	FORMATIONS	DEPOSITIONAL ENVIRONMENTS	GYPSUM LAYERS	THICKNESS	AREA COVERED
EOCENE	PRIABONIAN	**Marnes blanches de Pantin** (White marls of Pantin)		(Marabet)	1.5 m	
		Marnes bleues d'Argenteuil (Blue marls of Argenteuil)				
		Première masse (First masse)	Brackish to hypersaline	Massive Gypsum	20 m	
		Marnes d'entre-deux masses (Marls between the two masses)				4,000 km^2
		Seconde masse (Second mass)	Paralic or continental	Massive Gypsum	4.5 m	
		Marnes à Lucines (Lucina marls)		Gypsum beds	1 m	
		Troisième masse (Third mass)		Massive Gypsum	3.3 m	
		Marnes à Pholadomyes (Pholadomya marls)	Marine to paralic conditions			
	BARTONIAN	**Quatrième masse** (Fourth mass)		Massive Gypsum	2 m	
		Calcaire de Noisy-le-Sec (Noisy-le-Sec limestone)				
		Sables de Monceau (Monceau Sands)		Gypsum layers	1 m*	
		Sables de Cresnes (Cresnes Sands)	Shallow marine			
		Calcaire de Saint-Ouen (Saint-Ouen Limestone)		Gypsum layers	4.3 m*	
		Horizon de Mortefontaine (Mortefontaine Horizon)				
		Calcaire de Ducy (Ducy Limestone)	with paralic or	Gypsum layers	1.80 m*	
		Horizon d'Ezanville (Ezanville Horizon)				
		Sables de Beauchamp (Beauchamp Sands)	continental intervals	Gypsum layers	1.3 m*	
		Horizon d'Auvers (Auvers Horizon)				
		Horizon de Mont Saint-Michel (Mont Saint-Martin Horizon)				
	LUTETIAN	**Marnes et Caillasses** (marls and conglomerates)	Restricted to hypersaline environments (probable saline lake episodes)	Gypsum layers	15 m	1,500 km^2
		Calcaire grossier (Coarse limestone)	Restricted marine		sporadic nodules	

Gaudant 1981; Moretto 1986/1987; Nury 1987/1988; Dumas 1988; Dromart and Dumas, this volume). The penetration of an arm of the sea from the Paris Basin into the Massif Central depressions has also been suspected (Rey 1971). Although marine influx took place episodically in some parts of the rift, the dominant conditions through the history of the rift were undoubtedly lacustrine; an evaporative deposition within hypersaline lakes with only local and occasional contibutions of sea water seems the best documented hypothesis.

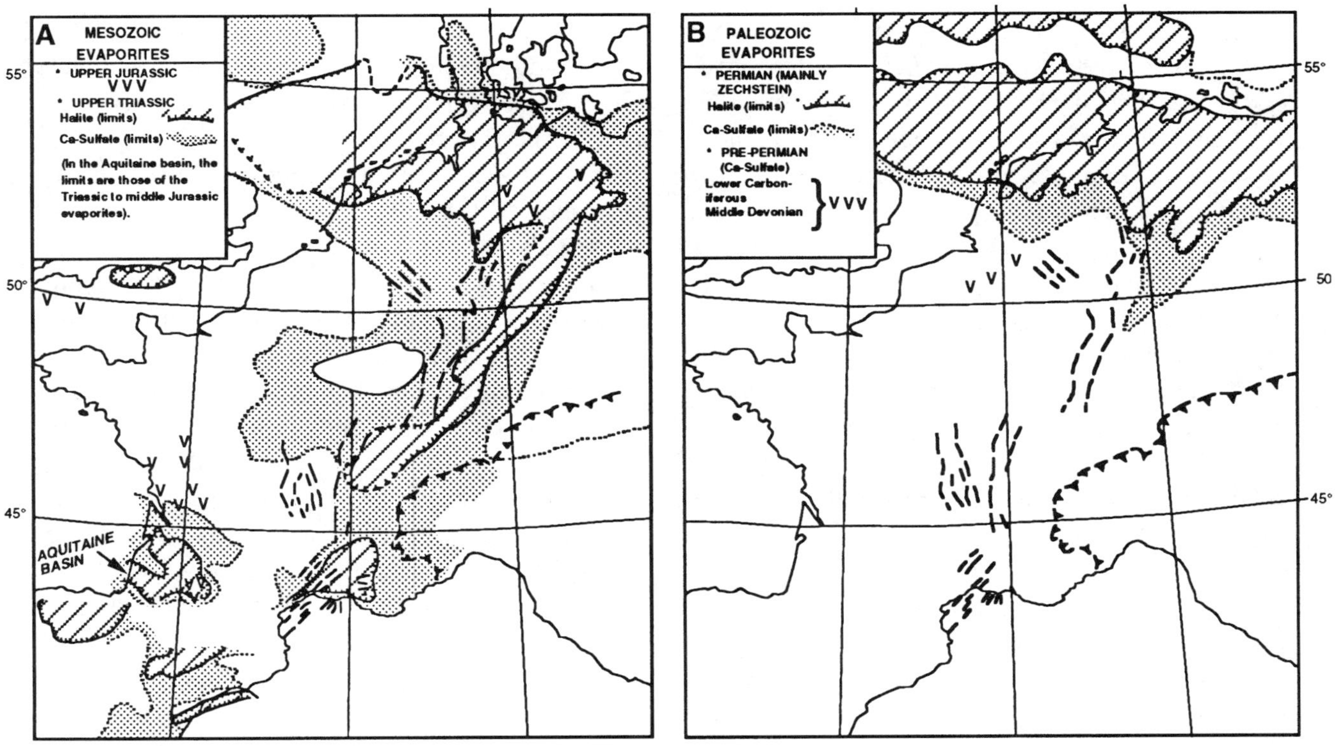

Fig. 2.26. Maps showing the extent of the main pre-Paleogene evaporites:
A. Mesozoic evaporites (from Zarkhov 1981; Ziegler 1982; Rebelle 1986; Rouchy et al. 1986).
B. Paleozoic evaporites (from Dubois and Umbach 1974; Castillo-Herrador 1974; Ziegler 1982; Curnelle 1983; Marchal 1983).

PRE-PALEOGENE EVAPORITES: EXTENT AND STRUCTURAL AND PALEOGEOGRAPHICAL ROLES IN TERTIARY BASINS

Thus far in this review, only those evaporative formations having a direct relationship to the Paleogene rift area have been considered. Other evaporative deposits are present in the same general area, but are of different ages (figure 2.26).

(a) *Pre-Permian evaporites.* The oldest evaporites in western Europe (figure 2.26A) are known to be as old as the late Precambrian (Riphean) in Norway (Siedlecka 1976) and the early Cambrian in the Montagne Noire (southern France) and Sardinia (F. Boyer and J. M. Rouchy, unpub. data; Courjault-Rade, 1982). Most of the evaporites in these older deposits occur as dispersed carbonate or silica pseudomorphs after gypsum or anhydrite.

Thicker evaporative occurrences of Devono-Dinantian age are located around the Brabant Massif; they have also been observed in petroleum or stratigraphic wells in northern France, Belgium, the Netherlands, and Great Britain or as scattered carbonate/silica pseudomorphs after gypsum/anhydrite in outcrops (Llewellyn et al. 1968; West et al. 1968; Bless et al. 1980; Rouchy et al. 1986). Several hundreds of meters of alternating carbonates and anhydrites have been found in the Annappes 1, Tournai, and Vieux-Leuze wells (Northern France and Belgium) located at the southern border of the Brabant Massif. Localized evaporative conditions can be recognized in the upper Devonian in Belgium (Goemaere et al. 1985; Rouchy et al. 1986). In the lower Carboniferous (Dinantian) evaporative deposits probably formed a large but irregular belt around the Brabant Massif, especially at the northern edge of the Paris Basin. In that area the thickest deposits are composed of alternating carbonates and anhydrite without known halite, and have been observed in two wells, Saint-Ghislain and Epinoy 1, where their cumulative thickness is respectively 760 m and 900 m. These discontinuous formations, strongly deformed by the Hercynian compressions, especially at Epinoy 1, probably represent the remnants of a more widely spread evaporative region extending southward. They have been fragmented by the cumulative effects of earlier paleogeographic differentiation, tectonics, and later dissolution. Widespread areas of breccia, considered as a solution-collapse feature, delineates the considerable volume of evaporites that were removed, probably after the Permian. The solution process was enhanced by the intense tectonic faulting and disruption of the beds (Rouchy et al. 1986, De Magnée et al. 1986). Locally the composition of water from Devonian and Carboniferous aquifers could indicate the contribution of dissolved evaporites, perhaps including rock salt deposits that have never actually been identified in that area (Glasbergen 1985).

(b) *Permian.* At the regional scale, one of the most important evaporative deposits of northern Europe and the North Sea was undoubtedly the Zechstein series (Ziegler 1982; Zarkhov 1981, 1984; Rebelle 1986). To the west of the Rhenane Massif, an extension of this calcium sulphate belt reaches and underlies the lowermost part of the Rhine Graben area. To the east of the same massif, the southern tongue of the Zechstein evaporite, including salt deposits (southern part of the Werra and Fulda basins), extends far to the south within the Permian Hesse depression (Richter-Bernburg 1955). The Permian depression was far larger than that of the Oligocene, and the Zechstein deposits are present in the eastern border of the Paleogene graben, extending from Hersfeld to Fulda.

(c) *Triassic.* The Triassic evaporites (figure 2.26B) cover nearly all of western Europe and most of the Paleogene rift segments (Upper Rhine Graben, Bresse, and Languedoc-Roussillon-Provence) are more or less directly superposed upon them. From the north the evaporative basin forms a straight tongue that first parallels the Hesse—Upper Rhine segment and then intersects it in its southern parts. From the Muschelkalk to the Keuper the salt remained localized in the eastern part of the Paris Basin (Dubois and Umbach 1975; Maget and Maiaux 1980; Ziegler 1982; Marchal 1983; Geisler-Cussey 1986) and did not reach the area of the Eocene evaporative basins. Westward, the calcium sulfate belt extended underneath the easternmost part of the upper Paleogene deposits of the (Ludian) basin and to the south it reached the northern end of the Grande Limagne. The salt extended into the Burgundy trough which corresponds in the Tertiary to the Bresse and Jura domains. The Upper Rhine graben and the Bresse systems were cut inside the depositional area of Triassic evaporites and, for a part of the Bresse, within that of Triassic halite. The deepest Paleogene trough, Bourg-en-Bresse, is superposed upon a 2 to 300 m thick Triassic halite formation whose thickness increases eastward toward the Jura (Lienhardt et al. 1984).

Southward, a large evaporative basin covered most of the Languedoc-Roussillon-Provence area from the border of the Massif Central in the west to the Mediterranean coast in the south. Northward the basin communicated with the Burgundy trough through the Bas-Dauphiné area in the eastern part of the Paleogene trough of Valence, while it was open toward the Alpine area in the east. A gulf penetrated between the southern Massif Central (Montagne Noire) and the Gulf of Lion and probably communicated with the large Aquitaine basin (Dubois and Umbach 1974; Baudrimont and Dubois 1977; Goy and Busson 1980; Appia-Lamache 1984; Lienhardt et al. 1984). Because of strong halokinetic or compressive deformation the original thicknesses are uncertain, but it seems that the more subsident areas were also controlled by the NE-SW fault bundles that later created the Paleogene grabens (Cévennes, Nîmes, and Durance).

All of the southern Paleogene depressions lie close to or within the geographic limits of the Triassic evaporative basins. The Valence, Marseille, Alès, and Narbonne basins are in the belt of calcium sulfates, while the basins of Nîmes, La Crau-Aix-en-Provence, and Manosque are in the halite zone. Indeed, Triassic saliferous deposits with a maximum thickness of 900 meters lie in a narrow NE-SW area directly underlying the Paleogene Nîmes graben.

(d) *Post-Triassic.* The evaporite deposition continued locally in the early Jurassic as in the Aquitaine basin where halite deposits are present. However they lie outside of the area where they could have been recycled into the Paleogene rift basins. Upper Jurassic (Purbeckian) calcium sulfates occur in the Aquitaine basin, in the western part of the Massif Central, and in some parts of the London—Paris basin but the deposits are too restricted to be considered as a significant source for recycled evaporites. In Aquitaine there was an evaporative tendency during the Jurassic and the early Cretaceous with local accumulation of calcium sulfates or residual evaporative breccia in the Purbeckian and Neocomian stages (Curnelle et al. 1982).

STRUCTURAL AND PALEOGEOGRAPHICAL IMPORTANCE OF TRIASSIC EVAPORITES

Some diapiric intrusions of Triassic evaporites occur in the Vocontian area of southern France in the vicinity of some Paleogene evaporative basins, such as those of the Carpentras area (e.g., the diapirs of Suzette, Propiac, Condorcet) or the northern end of the Manosque depression. Althought not accurately dated, the main salt intrusions could be related to the fracture development during the first stage of the Paleogene rifting as indicated both by the structural data and the K/Ar age dates from associated neoformed feldspars (Emre and Truc 1978; Perthuisot and Guilhaumou 1983). Casagrande et al. (1989) have defined the successive stages of the rise of the Suzette diapir. They show, after a first phase of intrusion in the late Eocene, that the ascension of the diapir ceased during the Oligocene, characterized by the development of a small superposed graben. The diapir then was reactivated during late Miocene compression.

It is also known that the Triassic basin underlying the Languedoc-Roussillon-Provence segment has been subjected to significant halokinetic deformation (Maurin 1984). These evaporites were also involved in the Pyrénéo-Provençal phases of deformation and served as the décollement surface for many of the overthrusts. As previously discussed, the border faults of most of the grabens in southern France appear as listric faults that flatten in the Triassic evaporites (as in Nîmes, Manosque, Narbonne). This may be similar to the structure of the mobility of the Manosque salt body and the role of similar halokinetic processes in the Narbonne basin (Lenguin et al. 1990). Thus we must consider the importance of the role of Triassic evaporites in the structural evolution of the rift system, at least in southern France.

If one postulates that the major source of the Paleogene evaporites was the recycling of older deposits into saline lakes environments, the Triassic evaporites represent the most available ionic source. The Triassic evaporites more or less directly underly most of the Paleogene segments except for the Rhine Graben, where the area of the thick Triassic evaporative deposits borders the Paleogene graben. In addition to the influx of sea water, dissolution by groundwaters circulating along the border faults of the rift may have provided the salt content of the Paleogene saline lakes. In the southern part of France the Triassic evaporites were probably present along the main tectonic lineaments (e.g. faults and thrusts) and they may have even outcropped in many localities. Leaching of the Triassic evaporites that migrated because of salt deformation along the border faults up to the surface is possibly one of the ionic sources for the Paleogene salt lakes. Several periods of migration may have occurred: Pyrénéo-Provençal compression, Eocene-Oligocene rifting, and late Miocene deformation. Examples of solution-collapse features in areas of active faulting during the Paleogene are well documented (Appia-Lamache 1984; Rouchy 1987).

In the Rhine segment, the Paleogene evaporative deposits are mainly concentrated in the Upper Rhine arm near the area where Triassic evaporites are located. Paleogene evaporites however are absent in the Lower Rhine and Hesse arms, where the basins lie in close proximity to extensive, available Permian evaporites (Zechstein). Thus there is a significant problem in modeling: why weren't the Permian evaporites near the lower Rhine area also recycled to become the parent of other Paleogene evaporites? In all probability there are other sources for the ions that make up the Paleogene evaporite deposits, and both additional marine influx and hydrothermal contributions were part of the overall source for these evaporites.

This paper was submitted for publication in 1992.

Jean-Marie Rouchy is grateful to the following companies and their directors for permission to use the seismic illustrations provided in this chapter: Société Nationale Elf Aquitaine Production (SNEAP), TOTAL Compagnie Française des Pétroles, EURAFREP, SHELL France, BP France, ESSO REP, Institut Français du Pétrole (IFP), and Gaz de France (GDF). He thanks particularly: Roger Ricateau, exploration manager of the France branch of SNEAP, Jean Perrot (assistant exploration manager), and Robert Curnelle; Philippe Picard and Pascal Lalouel, successively exploration managers of the France/Spain branch of TOTAL-CFP, Michel Morizet, assistant exploration manager, together with Alain Laumondais and Dominique Conte; Jean Demians D'Archimbaud, exploration manager (EURAFREP), and Bernard Mervoyer, assistant exploration manager; Gérard Olivier, exploration manager of SHELL France; Jean Taborin, exploration manager of BP France; Henri Navail, exploration manager of ESSO-REP; Lucien Montadert, exploration manager of IFP; and François Verdier, exploration manager of GDF.

Some of the profiles presented is this chapter come from exploration areas and contain only simplified data concerning actual location and associated geology (figures 2.2, 2.6, 2.7, 2.13, 2.14, 2.15).

J. M. Rouchy acknowledges helpful discussions with Marc L. Helman (Oxford University), André Maurin (TOTAL), Bertrand George (ATOCHEM), and Maurice Gennesseaux (University P. and M. Curie, Paris VI) and valuable comments on the manuscript by Michel Séguret (University of Montpellier) and Michel Valette (ATOCHEM). Agate Cambreleng assisted with the figure preparation.

3

The Salt Basins of Alsace (Southern Rhine Graben)

MARIE-MADELEINE BLANC-VALLERON AND
MONIQUE SCHULER

Paleogene Evaporitic Basins of Alsace: Geological Setting

Alsace belongs to the area of the Upper Rhine Graben, part of the Western European Continental Rift system: details of its position with respect to that system have been reviewed by Jean-Marie Rouchy in chapter 2, which outlines the pre-rifting framework, its morphostructural characteristics, associated volcanism, and post-Paleogene evolution.

The thick Paleogene series of evaporitic deposits in the southern Rhine graben (over 2,500 m) is particularly well documented due to extensive potash and petroleum prospecting. Numerous studies have been made and published, including large numbers of mineralogical, geochemical, petrographical, paleontological, stratigraphical, and cartographical analyses (for a review, see Blanc-Valleron 1990/1991). The density of subsurface data available (almost 500 boreholes) is indicated in figure 3.1. Two regions have been particularly well studied: the oil fields in the northern part of the basin, and the potash basin in the south.

In the northern part (Pechelbronn area), a great many boreholes (5000+) were drilled because of extensive exploration for oil (Sittler 1985), but most of them are too old to yield information useful to this study. Some boreholes will be discussed, mainly those for which: (a) depth or nature of the Mesozoic substrate is known, and (b) data (well logs) permit the firm establishment of characteristics of at least one of the Paleogene formations studied (mostly the recent ones).

The data set available in the southern part of Alsace (near Mulhouse) has been largely drawn from data collected due to the discovery of potash in 1904 and its exploitation (the first mining shaft was completed in 1909). Our present knowledge is based on data from more than 300 deep boreholes (about 150 in the mine concession area alone), 23 shafts reaching the potash seams, and from hundreds of kilometers of associated seismic sections.

From south to north the four areas marked by low gravimetry anomalies were first illustrated by Maikovsky (1952) (figure 3.20) and are interpreted as four discrete sedimentary basins (near Mulhouse, Sélestat, Strasbourg, and Pechelbronn-Rastatt). The basins are separated by the Colmar, Erstein, and Brumath sills, which have high gravimetric values. The map of the depth of the Jurassic substrate (figure 3.2) (which has been drawn from the results of deep boreholes) confirms and refines this structural image by illustrating the thickness of the rift filling by the Tertiary and Quaternary sediments within the basins as compared with that on the intervening sills.

The Tertiary sediments were laid down on a Jur-

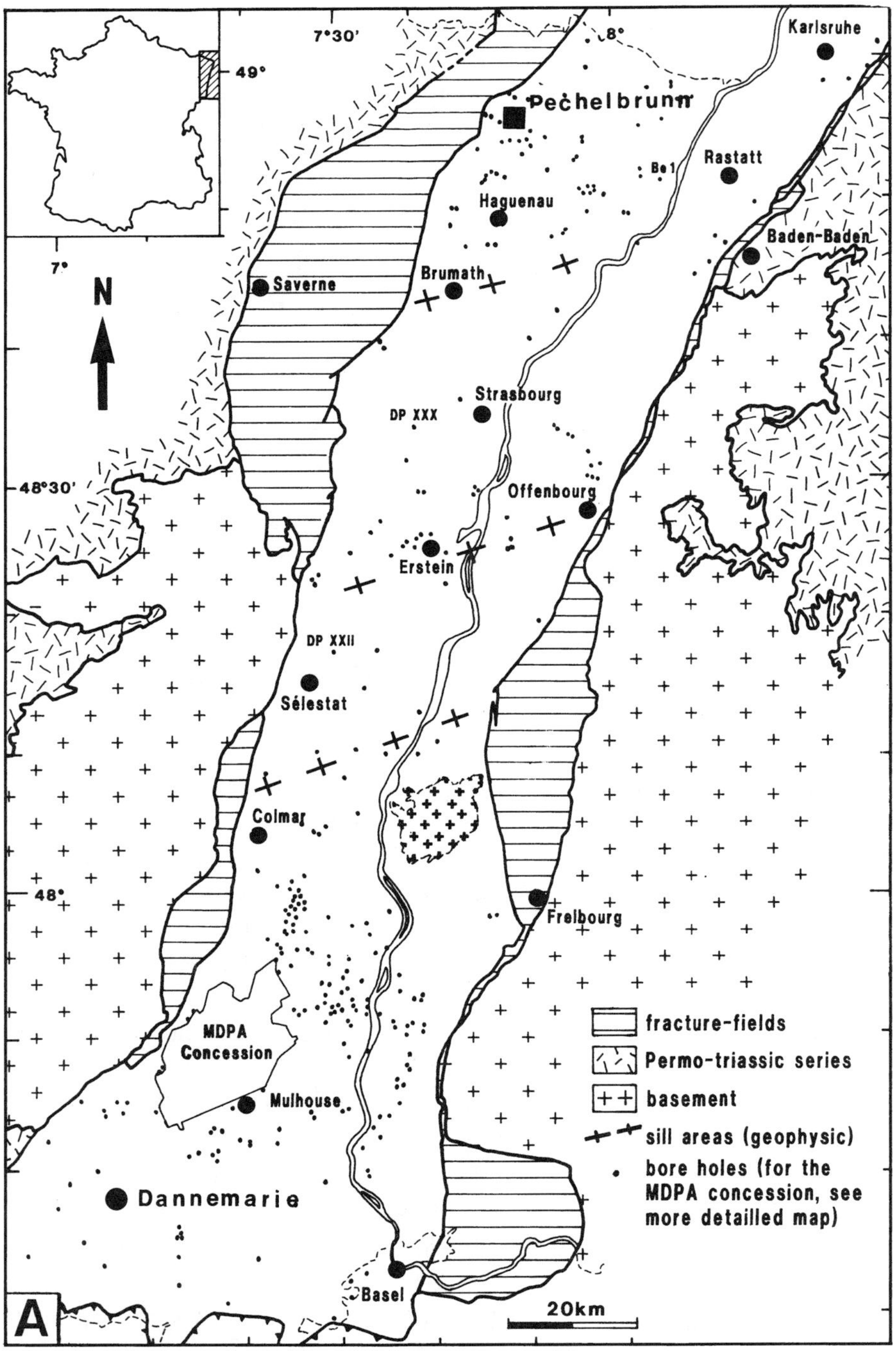

FIG. 3.1. Geographic maps of the Rhine Graben and its component sections.
A. The deep boreholes (black dots) used in this study of the Paleogene basins of the Alsatian Rhine Graben (Blanc-Valleron 1990/91, modified). Two boreholes (DP XXII and DP XXX) are named: they are presented in figures 3.21 and 3.22, respectively.
B. Enlargement showing boreholes within the MDPA concession and the outline of diapiric zones in the Mulhouse Basin (Blanc-Valleron 1990/91, modified).

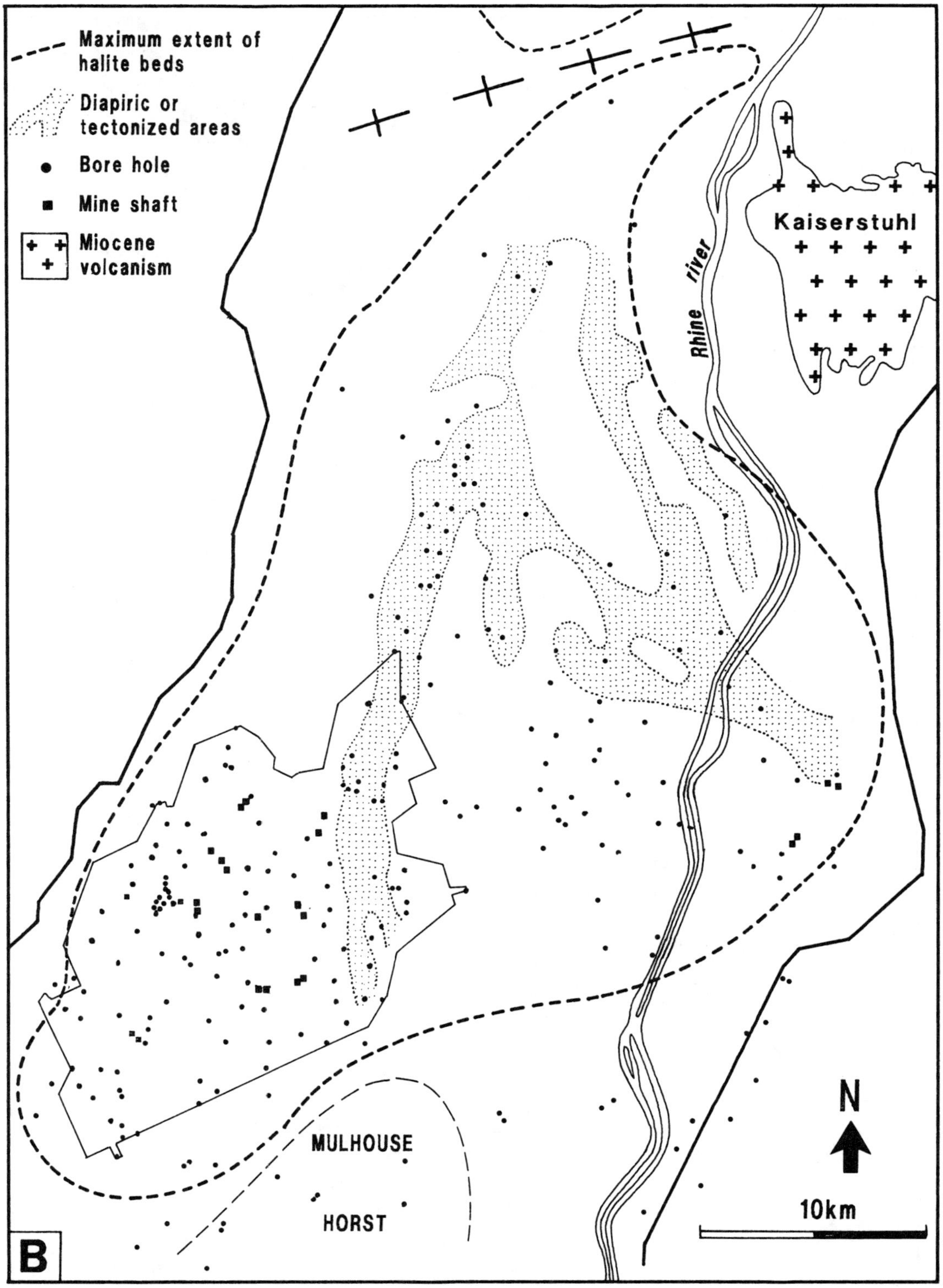

Maximum extent of halite beds
Diapiric or tectonized areas
Bore hole
Mine shaft
Miocene volcanism
Kaiserstuhl
Rhine river
MULHOUSE
HORST
N
10km
B

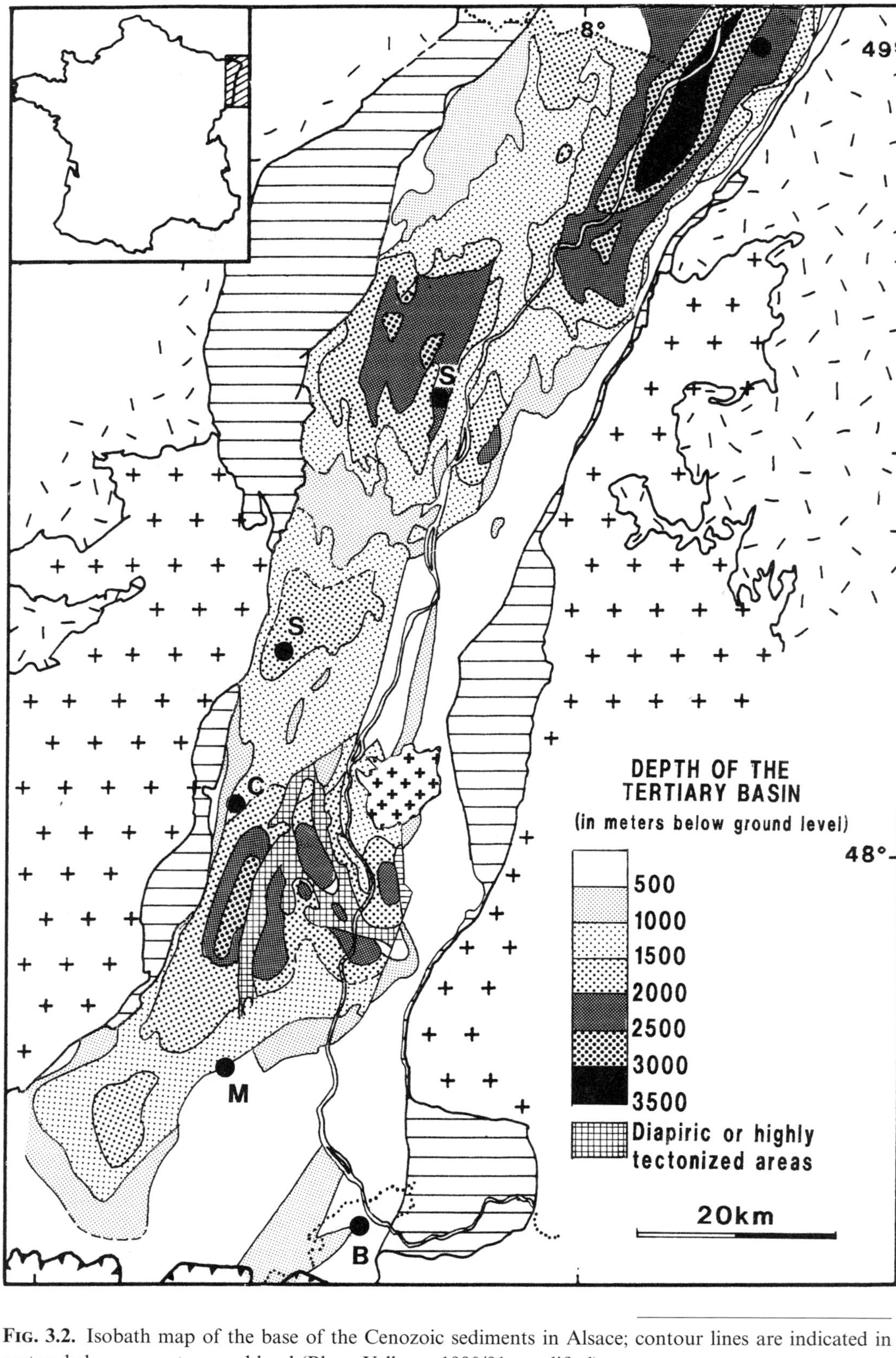

FIG. 3.2. Isobath map of the base of the Cenozoic sediments in Alsace; contour lines are indicated in meters below present ground level (Blanc-Valleron 1990/91, modified).

assic substratum (Pliensbachian to Oxfordian) composed of synclinal and anticlinal ridges (figure 3.3) which were eroded before the middle Eocene. The anticlinal structures were more resistant to subsidence and formed sills as in the Colmar-Erstein areas. On extensions of these highs near the borders of the rift, the substratum may have been totally flooded only during the middle part of the Saliferous Series (i.e., throughout the entire period of salt deposition).

The comparison between the rift filling thickness and the present limits of the potash, halite, and gypsum/anhydrite realms illustrates the coincidence between the general basin depocenters and halite depositional areas (figure 3.4). For each basin the maximum cumulative thickness of halite within the Paleogene evaporitic series is given together with the approximate surface area in order to give a rough evaluation of NaCl volumes. The Mulhouse area is the only basin containing potash; it is the major salt deposit, with the volume of halite being about two times greater than in all the other basins of Alsace together. In this book it will be the basis for the general description of the salt basins of Alsace.

The Mulhouse Basin displays a thick Tertiary and Quaternary succession (more than 2,500 meters) that is partly evaporitic in nature (Maikovsky 1940; Sittler 1965; Courtot et al. 1972; Blanc-Valleron 1990/1991). The main evaporitic sequence—including halite, anhydrite (and gypsum), and sylvite with marly interbeds—is late Eocene to early Oligocene in age (Schuler 1988/90; Blanc-Valleron 1990/1991) and has been subdivided into three formations (i.e., Lower, Middle, and Upper Salt formations) (figure 3.5). The two sylvite seams, whose average thickness is 2 and 4 meters, currently worked by the MDPA (Mines de Potasse d'Alsace) are located at the base of the Upper Salt (Salt IV member).

The upper Eocene to lower Oligocene evaporitic series is about 1,600 meters thick. It lies upon continental sediments (including the Lutetian *Planorbis pseudoammonius* limestone) and is covered by Rupelian marine sediments. The "Foraminifera Marls" marking the so-called Rupelian transgression have been dated as the extreme base of the NP 23 biozone (Martini 1973; Doebl et al. 1976; Cavelier 1979; Schuler 1988–1990).

Within the Paleogene evaporitic series of Alsace, well defined stratigraphic dates are rare. Mammals are present in the "Calcaire à Mélanies" of the Mulhouse horst (time equivalent of Salt II and Salt III) and in the lower Pechelbronn strata of Rot-Malsch (near Heidelberg) that correspond to the Montmartre zone (Franzen 1968; Tobien 1968; Cavelier 1979). Mammals present in the "Calcaire asphaltique de Lobsann" (top of the upper Pechelbronn strata) belong to the Ronzon zone (Cavelier 1979); nannoplankton of the Fossiliferous Zone corresponds to the NP 22 biozone (Martini 1973).

Palynological data (Schuler 1988–1990) show that:

1. the Lower Salt Formation and the base of the Middle Salt Formation (Salt III and time-equivalent formations) correspond to known Priabonian sediments of the Paris basin ("Marnes à Lucines" with "Marnes d'entre-deux masses" and "Marnes bleues d'Argenteuil" with "Marnes blanches de Pantin");
2. the top of the Middle Salt Formation (Fossiliferous Zone) and the Upper Salt Formation correspond to the Rupelian sediments of the Paris basin (Stampian: Sannoisian facies) ("Argile verte de Romainville" and "Caillasses d'Orgemont" + "Calcaire de Sannois").

The relationship of the tectonic style to the patterns of sedimentation in the Mulhouse Basin shows that rifting did not produce high relief at the borders at the beginning of the Paleogene sedimentation—these topographic expressions developed later in the evolution of the rift area. This has been demonstrated by paleogeographical reconstruction of the hinterland, deduced from palynological studies (Schuler 1988/90).

Conglomerates formed only during limited periods of time. They became a significant part of sedimentation, but were restricted to narrow foredeeps, limited to river mouths and did not spread far out into the basin. In the central part of the basin, saliferous sediments appear to be very well stratified, except near the diapiric zones, as illustrated by seismic profiles (figures 3.6 A, B, and C).

North of Alsace sedimentary formations are

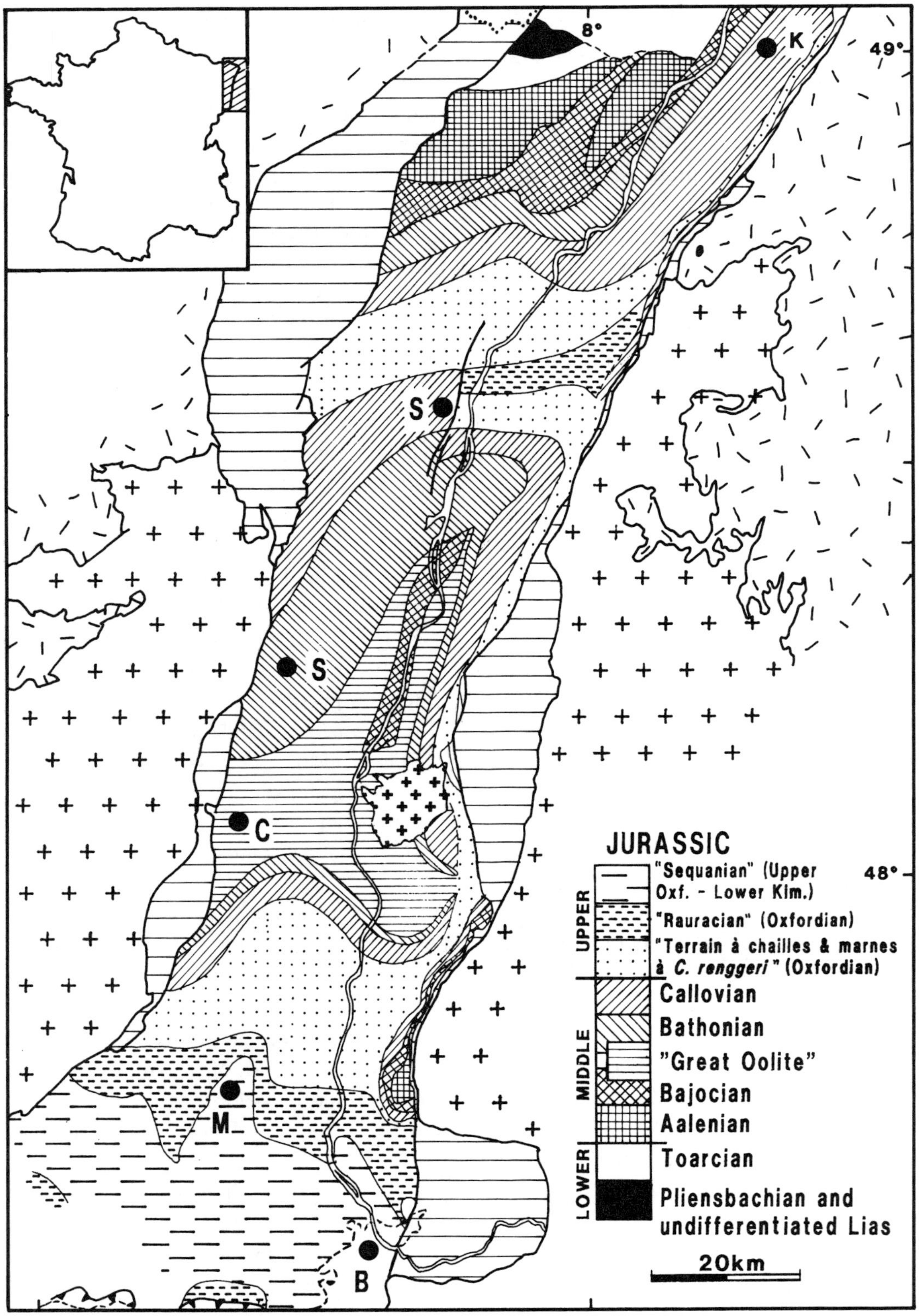

Fig. 3.3. Map illustrating the nature of the Jurassic substratum in Alsace (Blanc-Valleron 1990/91, modified).

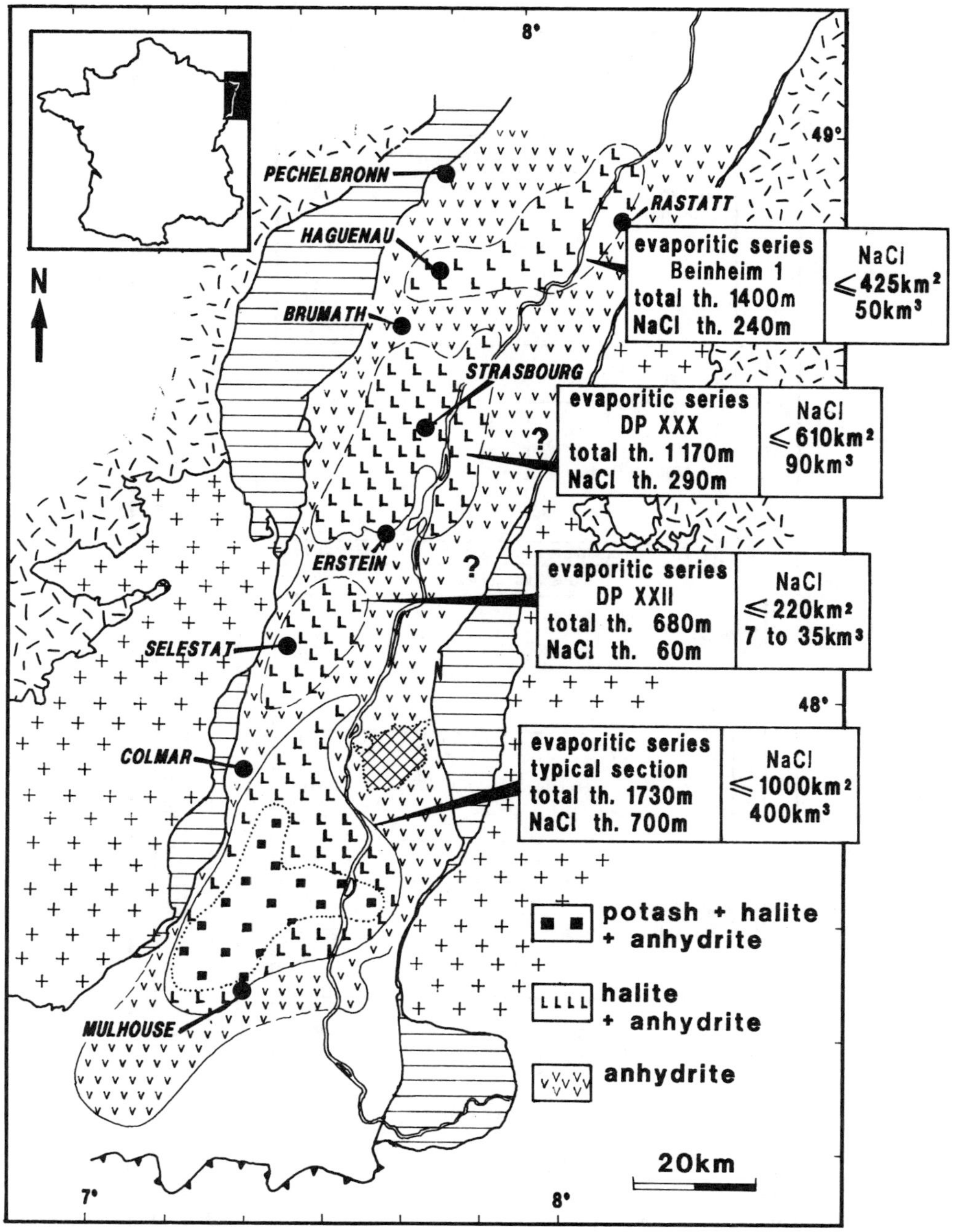

Fig. 3.4. Location map of Paleogene evaporites in Alsace. For each basin the thickness of the evaporitic series, halite sediments, maximum halite surface, and an approximate volume of NaCl are also given (Blanc-Valleron 1990/91, modified).

known in the Mainz and Hesse basins which can be correlated with those of the Mulhouse Basin (see figure 3.7 for the interval, corresponding to the main Paleogene evaporitic episode of the Mulhouse Basin, that falls between the pre-evaporitic formation and the Foraminifera Marls). $CaSO_4$ is sporadically present in the Chattian series and some halite has even been noted in the Chattian-Aquitanian series of Worms, in the northern part of the Rhine Graben, but these deposits are not discussed here.

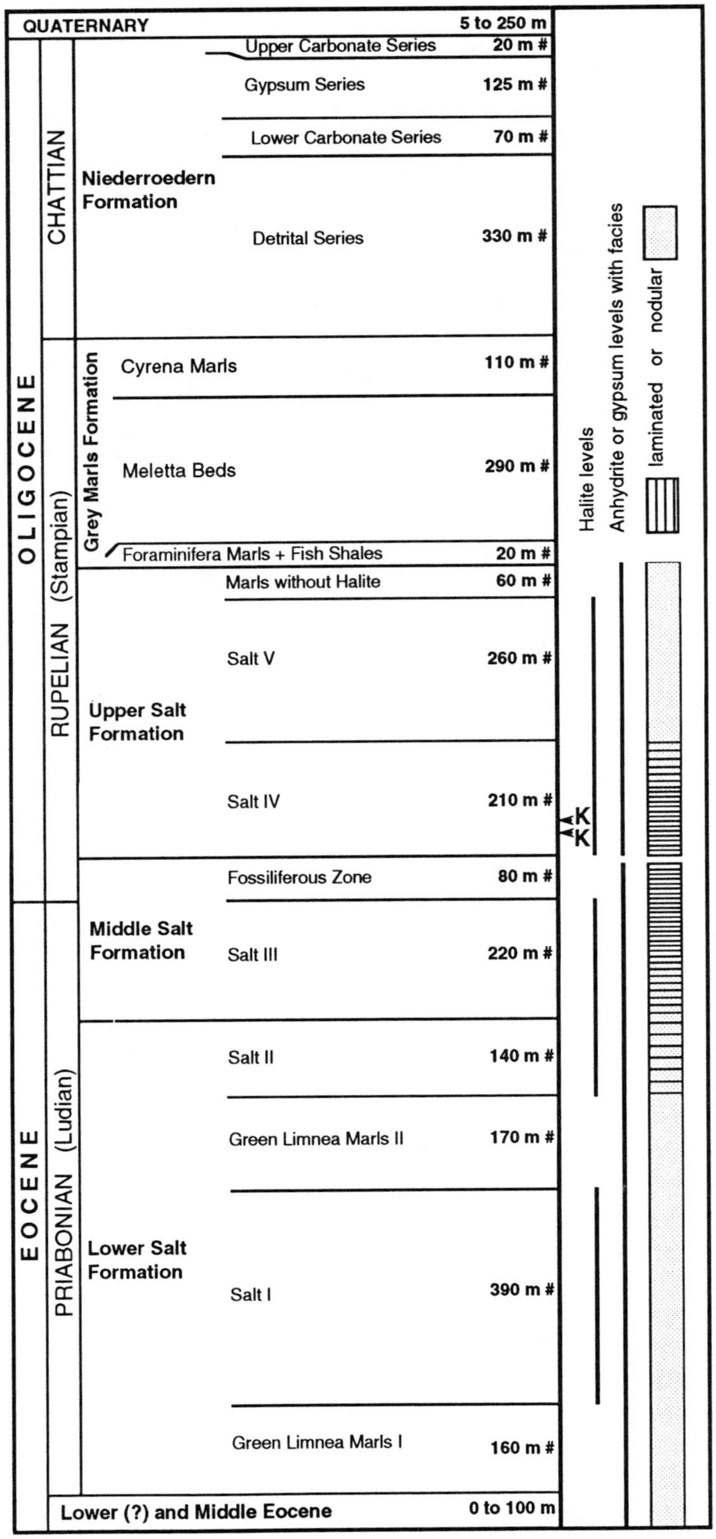

Fig. 3.5. General stratigraphic section of the Mulhouse Paleogene series with main evaporitic facies indicated (K = the two potash seams) (Blanc-Valleron 1990/91, modified). # symbolizes average section thickness.

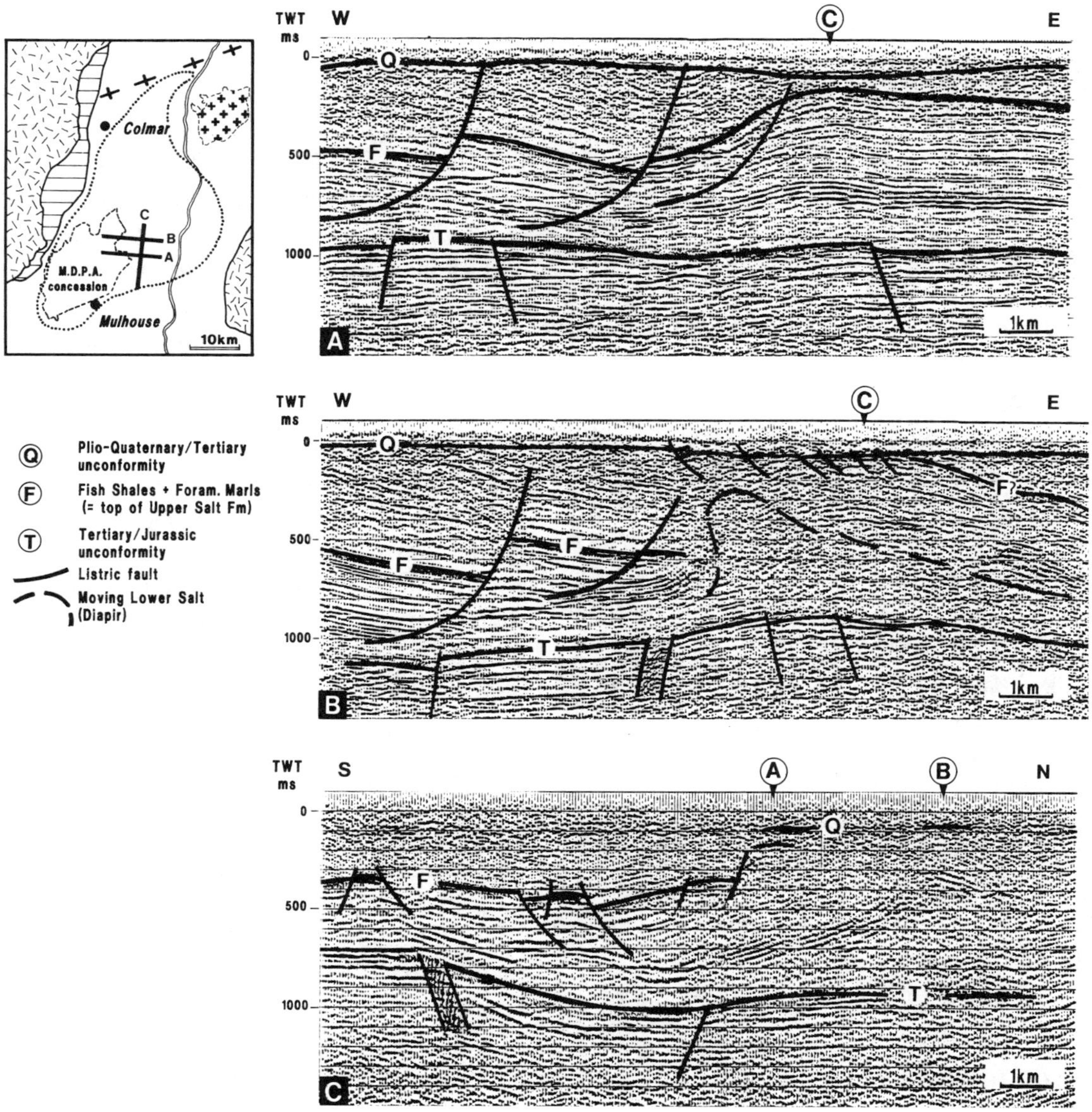

FIG. 3.6. Seismic sections through the Mulhouse Basin on the eastern flank of the diapiric area (Blanc-Valleron 1990/91, modified). See inset for positions of sections.

Epoch	Stage	MULHOUSE BASIN	PECHELBRONN - KARLSRUHE BASIN	MAINZ BASIN	HESSE BASIN (BORKEN - KASSEL)
OLIGOCENE	RUPELIAN (STAMPIAN)	Foraminifera Marls	Foraminifera Marls	Lower Rupelton	Lower Rupelton
		Upper Salt Formation$_M$ - M. without Salt - Salt V - Salt IV	Upper Pechelbronn Beds Fm	?	Melanienton (E D C B A)
		Middle Salt Formation$_M$ - Fossil. Zone	Middle Pechelbronn Beds Fm	Pechelbronn Beds	
EOCENE	PRIABONIAN (LUDIAN)	- Salt III			
		Lower Salt Formation$_M$ - Salt II - Green Marls II - Salt I - Green Marls I	Dolomitic Group Fm - Red Beds - Dolomitic Zone	?	
	MIDDLE AND LOWER	Preevaporitic Fm	Transition Zone Fm	Basal Clay Fm	Weissblauer Ton Lignites of Borken

FIG. 3.7. Lithostratigraphic correlations for the main Paleogene series of the Rhine Graben.

Geological History of Paleogene Evaporitic Basins of Alsace and Their Deposits

MINERALOGY

The marly units throughout the section have the same mineral paragenesis whether they occur in the anhydritic clays or in the halite-bearing beds. The grouping consists generally of clay minerals (dominantly illite, with kaolinite and chlorite), quartz, K-feldspar, calcium sulfate (dominantly anhydrite with traces of gypsum), calcite (sometimes magnesian), some dolomite (commonly calcium-rich), and halite. Pyrite and celestite may also be present but only as traces. Near the potassic strata, sylvite (occasionally carnallite) is present, but otherwise the mineralogy of the marls does not vary as a function of the type of interstratified evaporites.

Illite is the dominant clay mineral throughout the whole series. It is generally associated with some kaolinite or chlorite (table 3.1). Sporadically near the potash seams, some palygorskite and sepiolite may be present, but only in the central part of the basin (Sittler 1964).

GEOCHEMISTRY

The bromine content of halite samples is well known in the Middle and Upper Salt formations of the Mulhouse basin. The profile from the first halite above the Fossiliferous Zone up to the base of the lower potash seam is classically interpreted as the product of progressive concentration of sea water in the basin (Kühn and Roth 1979). Other geochemical studies of brines associated with the potash seams (seepages in the mine drifts, fluid inclusions in halite samples) characteristically fall on a curve obtained from studies of the evaporation pathway of marine waters (Poutoukis 1991).

The isotopic composition of sulfates and carbonates from the evaporite layers suggests that the deposits are the result of secondary precipitation of redissolved sulfates of Permo-Triassic origin within a closed hydrological system. Some authors (Fontes et al. 1991) argue against a marine origin for all of the salts. However, distinct marine influences are clearly identified (Schuler 1988–1990; Blanc-Valleron 1990/1991), especially in the upper part of the middle salt-bearing unit (Fossiliferous Zone), and suggest an associated marine influx for the Middle and Upper Salt formations.

Organic material forms an integral part of the studied sediments. A systematic study of evaporitic formations in the Mulhouse Basin (Blanc-Valleron 1986a, 1986b) has shown that some clayey anhydrite-bearing horizons are rich in organic matter, particularly the Salt III (basal part of the Middle Salt Formation) and the Salt IV members (basal part of the Upper Salt Formation). Furthermore, a significant variation in total organic carbon has been detected in the core material of the Max borehole recovered from the base of Salt IV (Blanc-Valleron et al. 1991). Total organic carbon (TOC) analyses from 450 samples show that marly layers interbedded within the evaporites have TOC that fluctuates in a rhythmic manner. Geochemical (elemental analysis of kerogen, gas chromatography of bitumen) and palynological studies (Blanc-Valleron et al. 1991; Schuler et al. 1991) of 26 samples show that the organic matter is mainly of algal origin. Highest TOC values are situated towards the tops of anhydritic shaly beds and correspond to a dominant contribution of phytoplanktonic biomass (group A of the groups of organic matter (OM)) characterized by an abundance of amorphous organic matter and algal tissue remains. During deposition of the halitic facies, the TOC decreases to a minimum at the tops of halitic beds; organic matter is likely to have originated from phytoplanktonic and bacterial components (group B) or may have been derived from a mixture of continental supply and in situ bacterial productivity (group C).

PETROGRAPHY

Marginal regions.

On the Mulhouse horst and on the border zones terrigenous, carbonate, and gypsum deposits exist which are time equivalent to at least part of the

TABLE 3.1.
Nature of clay minerals (K = kaolinite, I = illite, C = chlorite, P = palygorskite, S = sepiolite) within the central part of the Mulhouse Basin.
(Blanc-Valleron 1990/1991, modified)

Formation	Lithostratigraphic unit	mean composition of < 2 μm particles	other clay minerals	illite crystallinity (width h/2)
Upper Salt	Marls without Halite	25 % K + 45 % I + 30 % C		7
	Salt V	25 % K + 60 % I + 15 % C		8
	Salt IV	25 % K + 75 % I	chlorite, P, S	9
Middle Salt	Fossiliferous Zone	40 % K + 60 % I		10
	Salt III	15 % K + 85 % I	chlorite	10
Lower Salt	Salt V	80 % I + 20 C	kaolinite	7
	Green Limnea Marls II	no sample studied		
	Salt V	80 % I + 10 C		8
	Green Limnea Marls I	5 % K + 80 % I + 15 C		8

evaporitic sediments of the Mulhouse Basin. In these regions conglomerates have been studied by Kessler (1909), Schirardin (1954) and Duringer (1988). They were first considered as marine coastal conglomerates ("Küstenbildungen des Oligocänmeerses" cf. Andreae 1884, map 2 and pages 291–313) but further studies have shown that they are fluviatile deposits. Duringer (1988) argued that there is no possible connection between the existing alpine seaway and the Rhine Graben, even for the Fossiliferous Zone. This interpretation is weak because none of these conglomerates has been dated specifically from the Middle Salt Formation in the southern part of the graben: thus the existence of a continuous belt of fluviatile conglomerates has not been proven. Sandstones are associated with most conglomerates (figure 3.8). They are also well represented in the Haustein time series (equivalent to the Upper Salt Formation) on the Mulhouse horst.

Carbonate deposits are well represented in the area of the Mulhouse horst within the "Calcaire à Melanies" (time equivalent of part of the Lower and Middle Salt Formation) and the Fossiliferous Zone, and are present as mechanical laminites, algal laminites, mottled carbonates (palustrine), and other palustrine facies with vadose cementation. Gypsum is present as displacive crystals (in lenses) and is known in the Middle Salt Formation on the Mulhouse horst.

Basin center.

Within the central part of the basins (halitic realm), the main depositional facies are grouped as argillaceous and carbonate sediments (figure 3.9), sulfate deposits (figure 3.10), halite deposits (figures 3.11, 3.12, and 3.13), and sylvite (figures 3.13, and 3.14).

Argillaceous and carbonate sediments. Marls are of two types: homogeneous sediments in which detrital and bioturbation features are common (sometimes they include algal laminar sediments displaying desiccation and bioturbation structures), or marls present as very fine scale laminae. Formation and preservation of these laminae testifies to the variations in particle influx, lack of grazing or burrowing organisms, absence of strong bottom currents, and protection from coarse clastic input. The clays are oriented parallel to the bedding plane (as seen under the petrographic microscope by their uniform and maximum illumination within clay lamina in 45° position under crossed polars); this shows that the detrital influx (fluviatile input) did not usually move along as bottom currents but probably spread out along the interface between the brine and fresher surface water, where clay particles remained suspended for some time before slowly sinking and reaching the bottom (Sonnenfeld and Hudec 1985).

Minor quantities of dolomite are present within

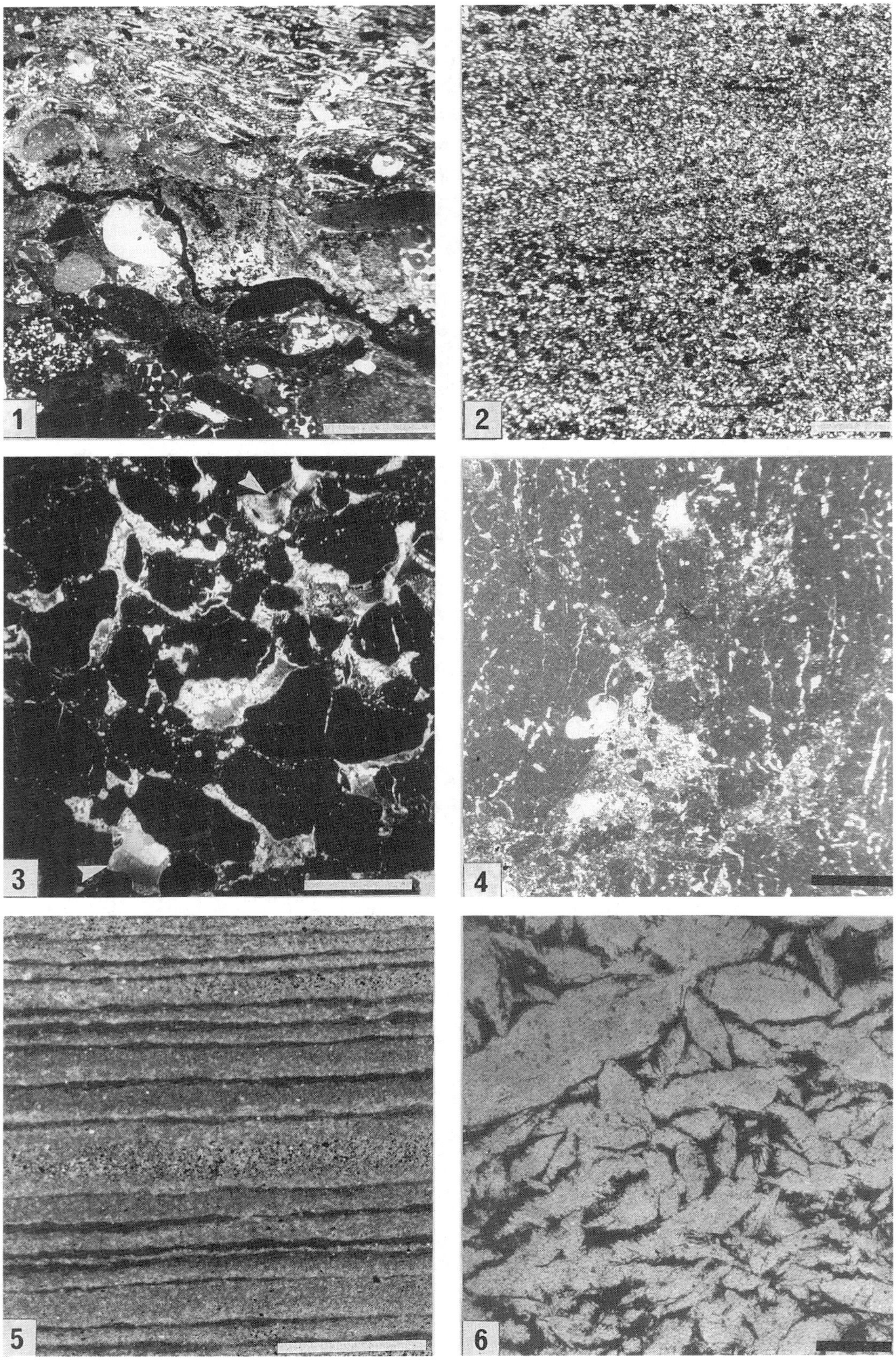

FIG. 3.8

crystal pseudomorphs after halite, at the contacts between mudstone and halite, and along the boundaries between halite crystals. This dolomite is more or less ferroan and appears to have been the product of the latest diagenetic phase. As in the Permian San Andres halite rocks of Texas (Gao et al. 1990), it is suggested that the dolomite was precipitated during burial compaction of mudstones from a halite-saturated brine, which was derived from the mixing of mudstone-released water (from the clays) with evaporated brine from halite.

Sulfate deposits. Sulfate deposits appear as interlayered levels within halite or marl deposits. They formed displacively from saturated interstitial brines within existing sediments or by precipitation from the surface brine. Within the marls, the presence of lenticular crystals is typical of gypsum that formed in the presence of trace amounts of organic matter (Cody and Cody 1988). When these crystals are ubiquitous, the sediment can appear as laminar gypsum or anhydrite.

Nodular sulfates are common, especially in the Lower Salt Formation. Because anhydrite nodules show the relics of gypsum lens-shaped forms on their borders and are commonly associated with small pseudomorphs (of halite, anhydrite, or carbonate) after gypsum, they are not interpreted as being the result of a classic sabkha environment as described in supratidal sediments from the Trucial Coast. These nodular sulfates probably correspond to nucleation of gypsum from interstitial brines within the first centimeters of sediment burial (Purser 1973).

Massive laminar sulfates are present in the Middle Salt Formation and at the bottom of the Upper Salt Formation. As seen with the microscope, crystals are lath-shaped and randomly oriented (with a felted texture). Within these anhydrites, lamination corresponds to more or less clay-rich bands or is marked by thin films of insoluble material where, occasionally, foraminifera tests have been noted. Laminar anhydrite can be more or less rich in dolomite corresponding to scattered replacive crystals or to pseudomorphs after displacive halite crystals. Tiny carbonate crystals have also been noted near the boundaries between anhydrite and halite, forming a sub-planar layer that looks like the upper limit of a former gypsum crust, such as those known from modern and ancient evaporites. Additionally, sulfate appears as:

- thin films of small anhydrite crystals truncating the top of halite crystals, corresponding to synsedimentary dissolution horizons due to new water influx;
- small anhydrite crystals scattered within halite or potash;
- large anhydrite laths in halite or potash, sometimes present as radial clusters.

Halite deposits are commonly present as milky halite showing remarkable preservation of depositional textures or as clear halite with, on the con-

---

FIG. 3.8. Sedimentary microfacies noted on the Mulhouse horst and in the border zones, time-equivalent to the evaporitic sediments of the Mulhouse Basin. (In thin section seen in cross-polarized light except as noted.)

(1) Border detritus. Conglomeratic and calcareous sandstone (bottom) overlain by marl with *Mytilus.* Scale bar, 5 mm. STR 20 sample.

(2) Horst detrital influx, Haustein. Sandstone-like limestone with oblique lamination. Scale bar, 2 mm. ALT 80 sample.

(3) Palustrine facies, pseudobreccia limestone, showing vadose structures (pendant and silty vadose cements). Scale bar, 5 mm. TAG 23a sample.

(4) Palustrine facies: mottled carbonates. The sediment is intensely bioturbated (bottom); subvertical fenestrae may have been produced by burrowing organisms, plant roots, or the decay of organic matter and gas migration within an organic-rich mudstone. Scale bar, 2 mm. TAG 27 sample.

(5) Laminar limestone. Scale bar, 1 mm. ALT 42a sample.

(6) Detail of laminar gypsum (seen in plane light). Gypsum has crystallized early as displacive lenticular forms within a calcareous mud, been transformed in anhydrite and back into gypsum (lenticular forms are now polycrystalline gypsum with some anhydrite remains). Scale bar, 1 mm. ALT 0 sample.

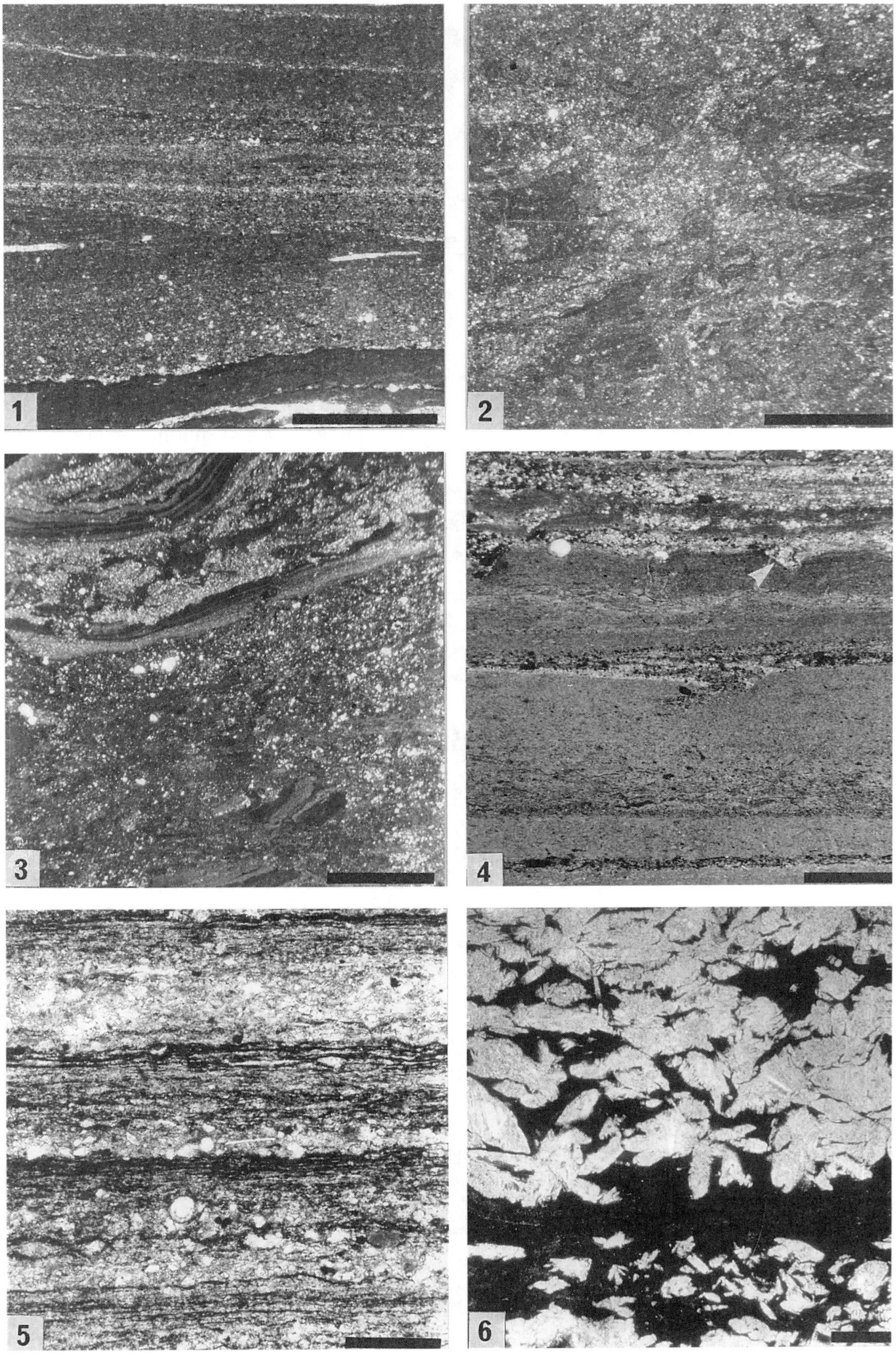

FIG. 3.9

trary, a strong diagenetic overprint. Milky halite has been most particularly studied adjacent to the potassic strata. Here, it commonly shows vertically oriented chevrons and cornets indicating bottom growth of halite crusts. Tooth-like textures have been found within centimeter-thick levels of anhydrite-rich halite and could have been formed by the exaggeration of chevron texture within the halite. Because these tooth-like structures are closely associated with anhydrite, this facies may record the time when halite and gypsum are coprecipitates, at the very begining of the halitic domain. Other halite layers are made up of well-sorted millimeter-sized tabular plates and cubes indicating nucleation of crystals at the air-brine interface or within the brine, and accumulation at the bottom. Unsorted halite crystals, often of larger sizes and randomly oriented, are also present with the rather unusual forms of skeletal and pagoda shapes.

In some layers, the uppermost apices of halite

Fig. 3.9. Argillaceous and carbonate sediments within the halitic realms. Microfacies illustrated in thin section, seen in crossed-polarized light.

(1) Siltstone with oblique lamination. Scale bar, 1 mm. DP XXV, 1578.8 m, Green Limnea Marls I sample.

(2) Bioturbated sediment. Scale bar, 5 mm. DP XXV, 1697 m, Green Marls I sample.

(3) Algal laminar sediments showing desiccatioon and bioturbation structures. Scale bar, 5 mm. DP XXV, 481.9 m, Salt V, U45 sample.

(4) Siltstone with erosion features such as impact or gullying by current structures, surmounting a laminar claystone. Scale bar, 2 mm. DP XXV, 259.3 m, "Marls without salt" sample.

(5) Laminar claystone, organic-rich (1.7% TOC) with foraminifera and ostracoda. Scale bar, 200 microns. VA 34, Salt IV, Mi sample.

(6) Laminar gypsum, highly altered. Scale bar, 1 mm. DP XXV, 232.5 m, Salt V sample.

Fig. 3.10. Dolomite and anhydrite microfacies illustrated in thin section, seen in plane light (except as noted).

(1) Perfect dolomite rombs within halite. Ghost of lenticular gypsum crystals; small carbonate crystals on the borders possibly caused by sulfate reduction. Scale bar, 1 mm. DP XXV, 1414.8 m, Salt I sample.

(2) Composite aggregate (seen in cross-polarized light) of quartz, fibrous silica, anhydrite, and more or less ferroan dolomite within halite. Scale bar, 500 microns. DP XXV, 1513.5 m, Salt I sample.

(3) Small nodules of anhydrite within halite. Scale bar, 5 mm. DP XXV, 1414.8 m, Salt I sample.

(4) Laminar anhydrite. Dark bands correspond to more or less clay-rich seams where, at higher magnification, small foraminifera tests are visible (tests are about 50 microns in diameter: possible dwarfism due to unfavorable environmental surroundings?). Scale bar, 5 mm. F1, 23.1 m, Salt IV, S stratum sample.

(5) Laminar anhydrite. Near the anhydrite-halite limit, the dark layer at top (arrow) may correspond to the preserved outline of the top of the crystal crust. Scale bar, 1 mm. VA 2a, Salt IV, Ci basis (contact between "faux mur de marnes" and "faux mur de sel gemme") sample.

(6) Detail of the dark layer of (5) showing round structures outlined by small carbonate crystals. These may have been caused by algal calcification which thus preserved some original structure (halite or gypsum crust). Scale bar, 500 microns.

Fig. 3.11. Halite microfacies illustrated in thin section, seen in plane light (except as noted).

(1) Intrasediment halite cube (seen in crossed-polarized light). Scale bar, 2 mm. VL 103, Salt IV, Ci, sample "a."

(2) Intrasediment halite cube replaced by dolomite (and some small laths of anhydrite) seen in crossed-polarized light. Scale bar, 5 mm. VA 14, Salt IV, Ci, sample "a."

(3) Intrasediment halite cube replaced by anhydrite (and some dolomite crystals). Scale bar, 500 microns. VL 107, Salt IV, Ci, sample "c."

(4) Recrystallized halite with curved crystal boundaries meeting at triple junctions with angles approaching 120°. Scale bar, 1 mm. DP XXV, 1551.1 m, Salt I sample.

(5) Vertically oriented chevrons. Scale bar, 1 mm. VA 17a, Salt IV, Ci, sample "b."

(6) Cube with depressed faces. Scale bar, 1 mm. VA 3a, Salt IV, Ci, "faux mur de sel gemme" sample.

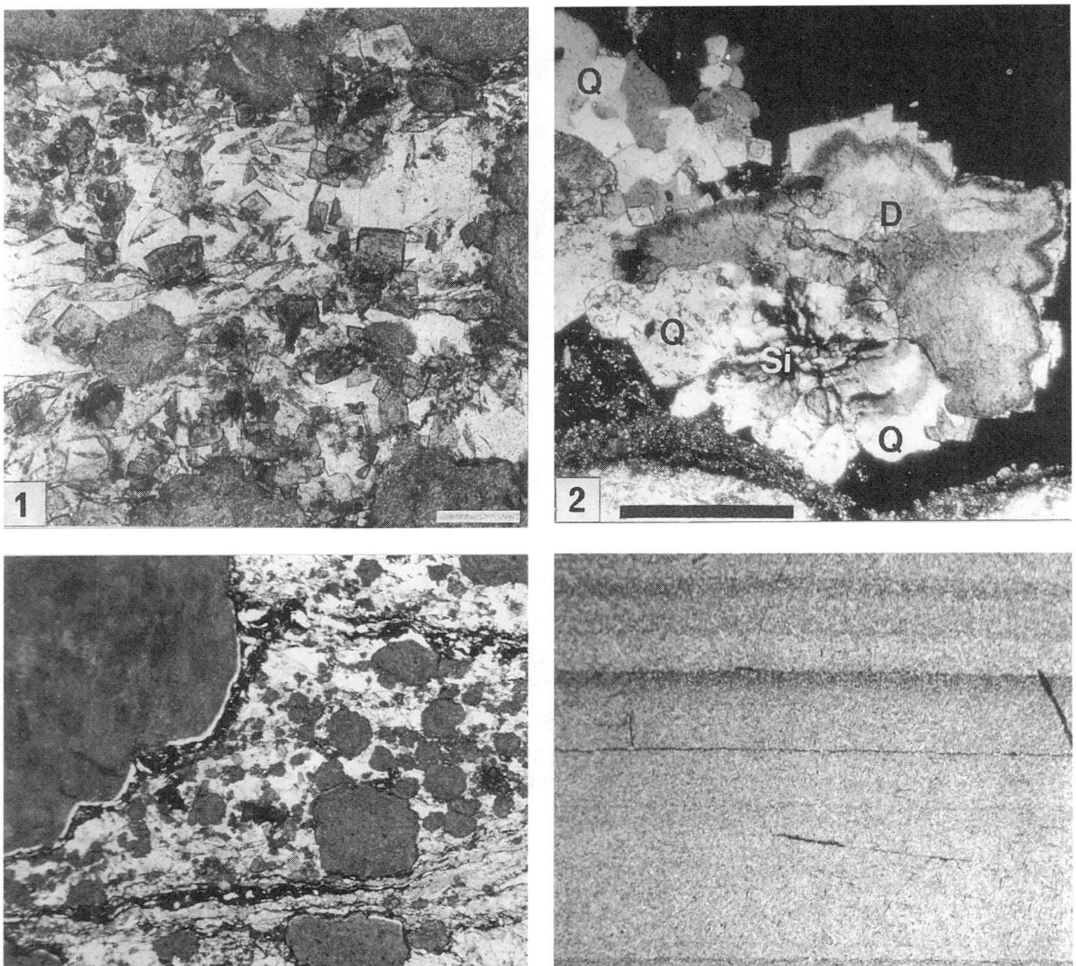

Fig. 3.10

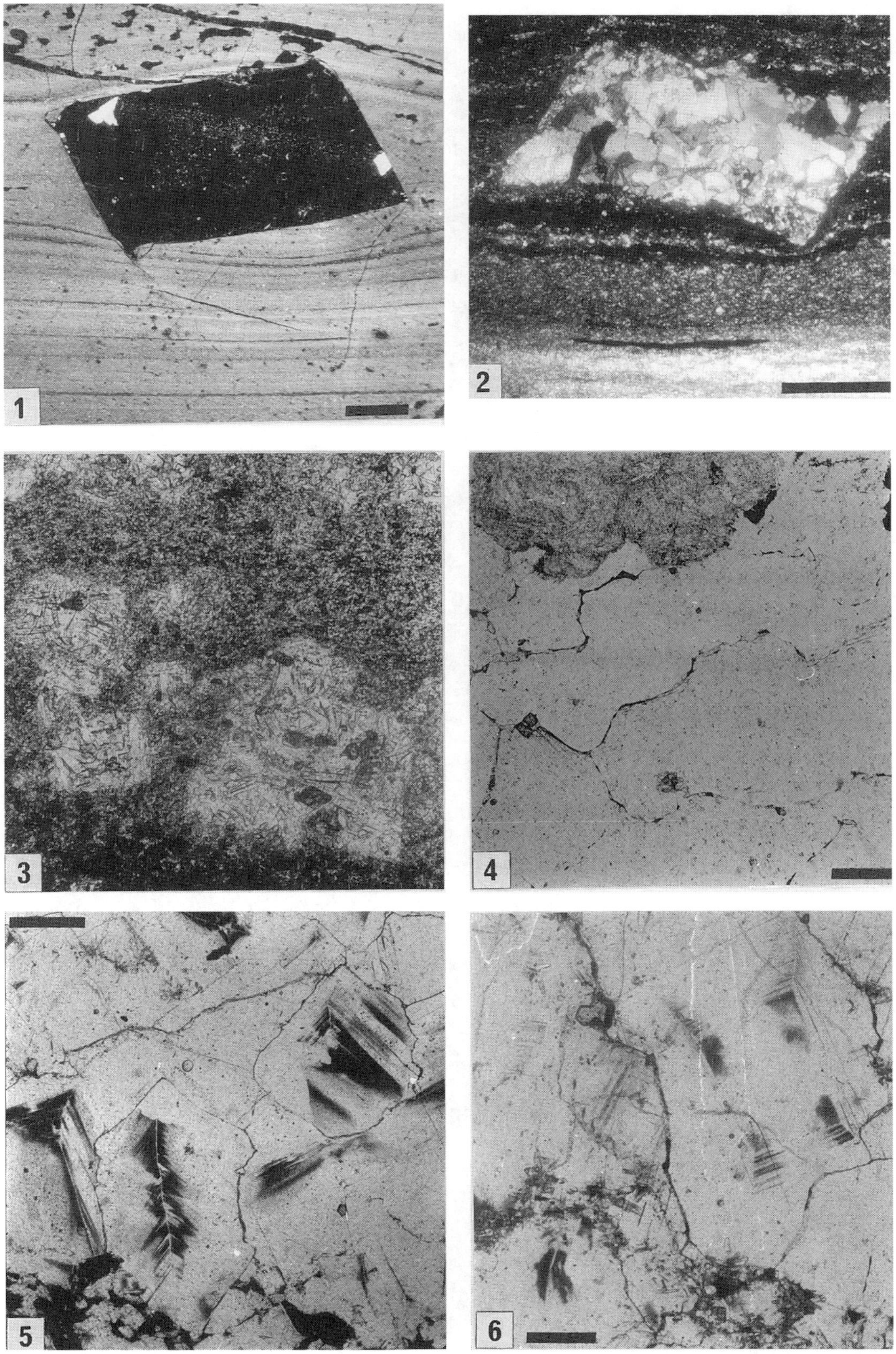

FIG. 3.11

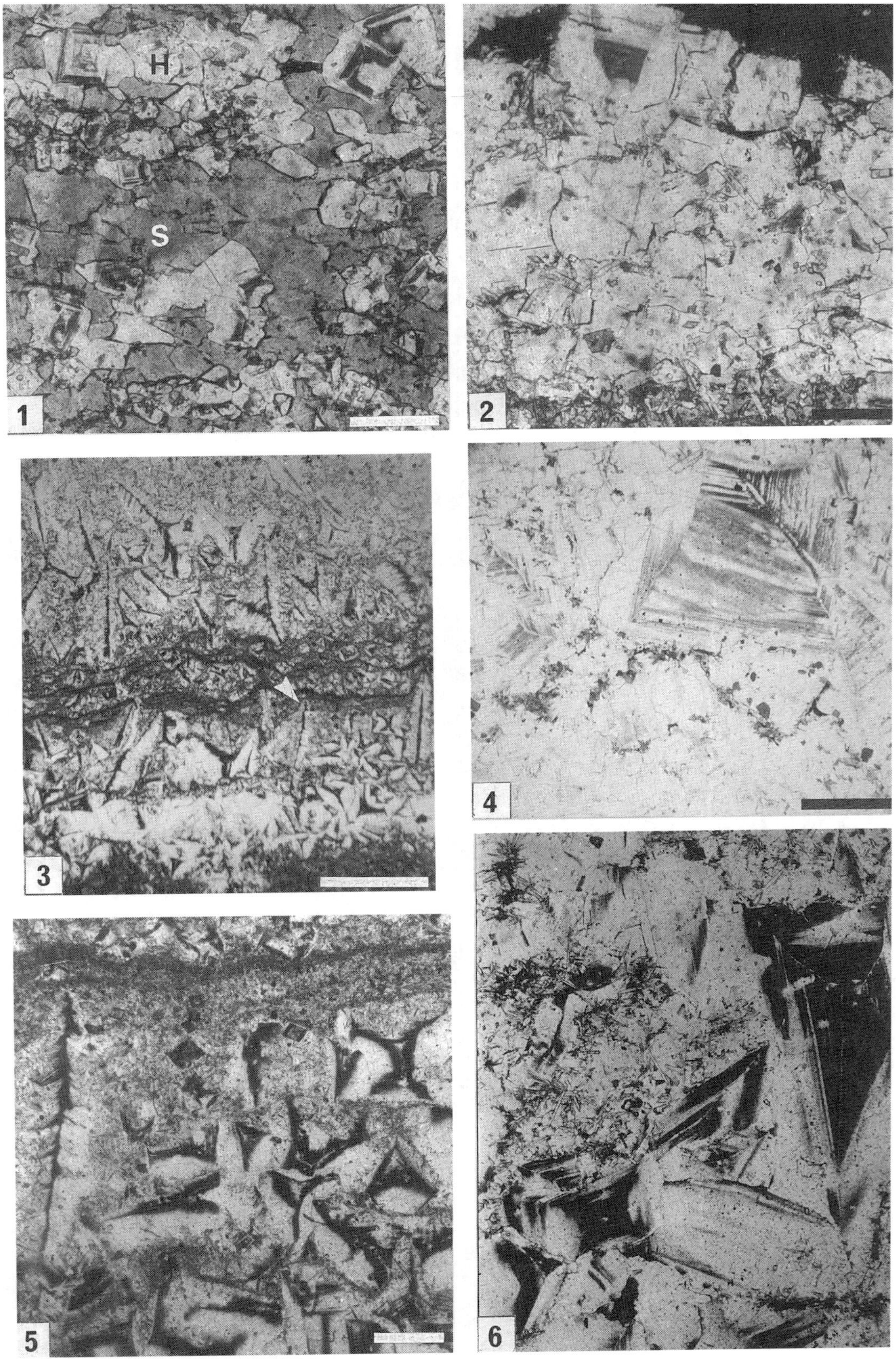

FIG. 3.12

crystals are partly dissolved, apparently by new influxes of undersaturated water into the basin. They can also be truncated during early diagenesis by the downward growth of dolomite crystals proceeding from marly layers.

Displacive halite (early to late diagenetic) is present as intrasediment halite cubes that can be replaced by dolomite or anhydrite.

Halite recrystallization is well delineated by curved crystal boundaries meeting at triple junctions with angles approaching 120°. Carbonate is sometimes present between halite crystals and small nodules of anhydrite; carbonate spherulites have also been observed within halite.

Sylvite deposits. The two potash seams worked by Mines de Potasse d'Alsace (MDPA) in the Mulhouse basin are well bedded and consist of laminar mudstone, anhydrite, halite, and sylvite forming centimeter- to decimeter-scale mineralogical sequences which are present in lateral continuity throughout virtually the entire basin. Because diagenetically altered deposits rarely show such lateral continuity, the potash and associated sediments are classically interpreted as primary.

The macroscopic appearance of sylvite, however, is not homogeneous; variations are mainly related to the color (it varies from pale orange to deep red, due to tiny hematite inclusions) and grain size. In general, color becomes deeper as grain size decreases. Usually, the sylvite layers of the lower pot-

---

FIG. 3.12. Halite structures illustrated in thin section seen in plane light.
(1) Equant halite cubes (hoppers) within sylvite: cumulative structure. Scale bar, 1 mm. VA 152a, Salt IV, Ci, "C2" KCl stratum, sample 4.
(2) Cornets. Scale bar, 1 mm. VA 152a, Salt IV, Ci, "C2" NaCl>KCl strata, sample 4.
(3) Anhydrite-rich halite with tooth-like crystals. Anhydrite (dark) forms a drape above the teeth (on the right). Scale bar, 5 mm. VL 107, Ci, sample "c."
(4) Unusual halite morphology. Scale bar, 5 mm. VL 97, Salt IV, Mi sample.
(5) Halite teeth: detail of (3) above. Scale bar, 1 mm.
(6) Pagoda structure? Scale bar, 1 mm. VL 105b, Salt IV, Ci, "b plaquettes de sel gemme" sample.

FIG. 3.13. Halite and sylvite microfacies illustrated in thin section seen with plane light (except as noted).
(1) Synsedimentary solution boundary, truncation of a bed of halite. Scale bar, 1 mm. VT 140 sup., Salt IV, Ci, sample "b."
(2) Diagenetic truncation of a bed of halite: dolomite crystals apparently growing downward from a thin marly layer, destroying primary structures of halite cubes. Scale bar, 1 mm. VA 145, Salt IV, Ci, sample "C2."
(3) Large laths of anhydrite (needle-like crystals) in halite associated with marly remains. Scale bar, 1 mm. VL 100, Salt IV, Ci, "mur-couche" sample.
(4) Polycrystalline carbonate formed between halite crystals (seen in cross-polarized light). Scale bar, 200 microns. VA 29b, Salt IV, CS, KCl stratum, sample 26.
(5) Small nodules of anhydrite and carbonate spherulites within halite (seen in cross-polarized light). Scale bar, 200 microns. VA 29c, Salt IV, CS, KCl strata, sample 26.
(6) Normal sequence: marls → anhydrite + dolomite → halite → sylvite. Scale bar, 2 mm. VA 17b, Salt IV, Ci, sample "b."
(7) Sylvite filling a halite cube. Scale bar, 1 mm. DP 96, Salt IV, Ci, "A," stratum 6 basis sample.

FIG. 3.14. Sylvite microfacies illustrated in thin section seen in plane light.
(1) Sylvite capping halite crusts. Scale bar, 5 mm. VA 5, Salt IV, Ci, "A," KCl strata, sample 3.
(2) Detail of (1): small equant sylvite and halite crystals capping large halite cornets. Scale bar, 1 mm.
(3) Detail of (1): halite hoppers within anhydrite-rich potash. Scale bar, 1 mm.
(4) Alternating halite and sylvite layers, a product of rapid salinity fluctuation. Scale bar, 5 mm. VA 152a, Salt IV, Ci, "C2," KCl strata, sample 4.
(5) Large laths of anhydrite in potash. Scale bar, 1 mm.
(6) Sylvite replacement of halite. Scale bar, 1 mm. VA 148, Salt IV, Ci, "E" basis sample.

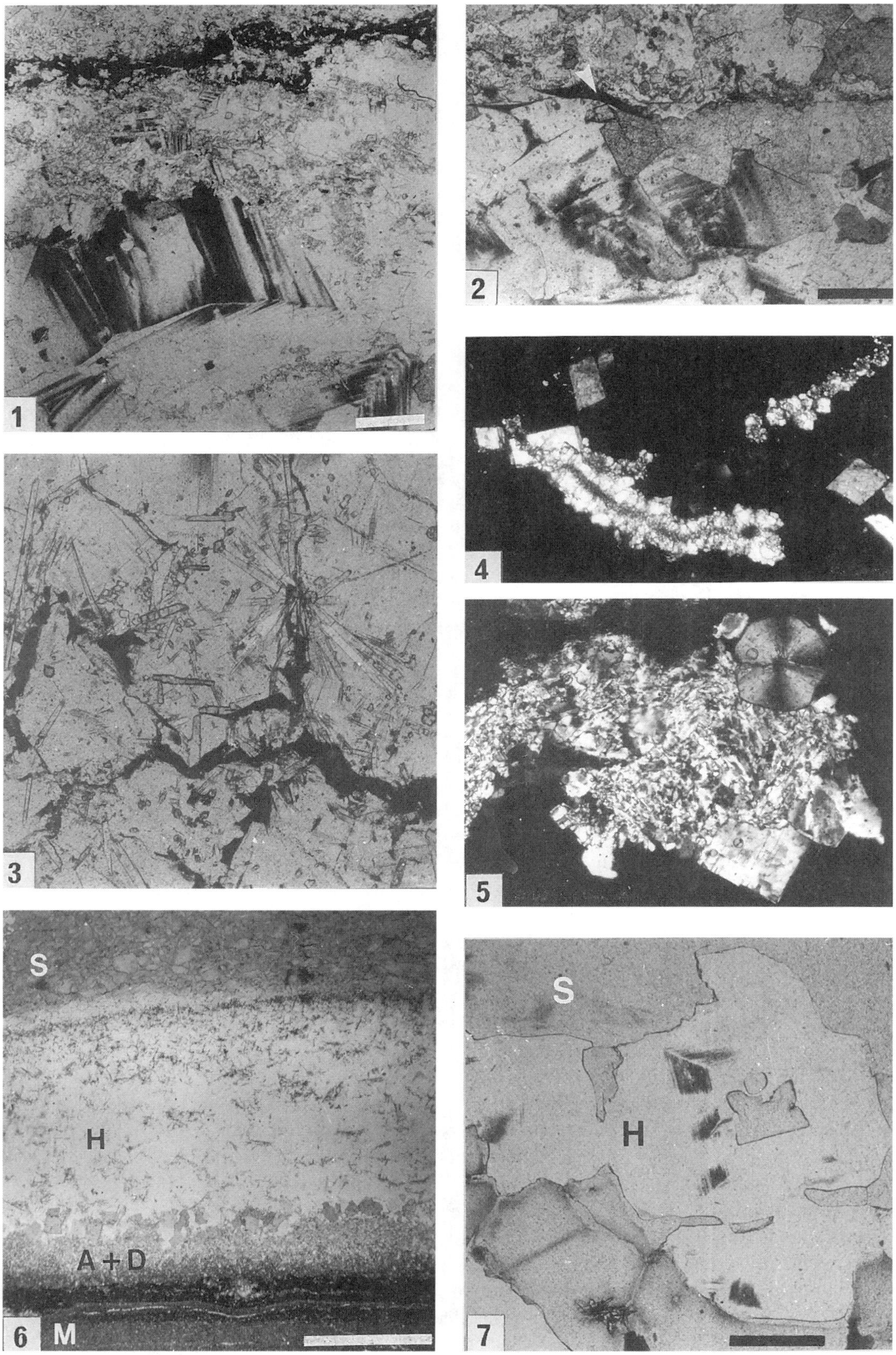

FIG. 3.13

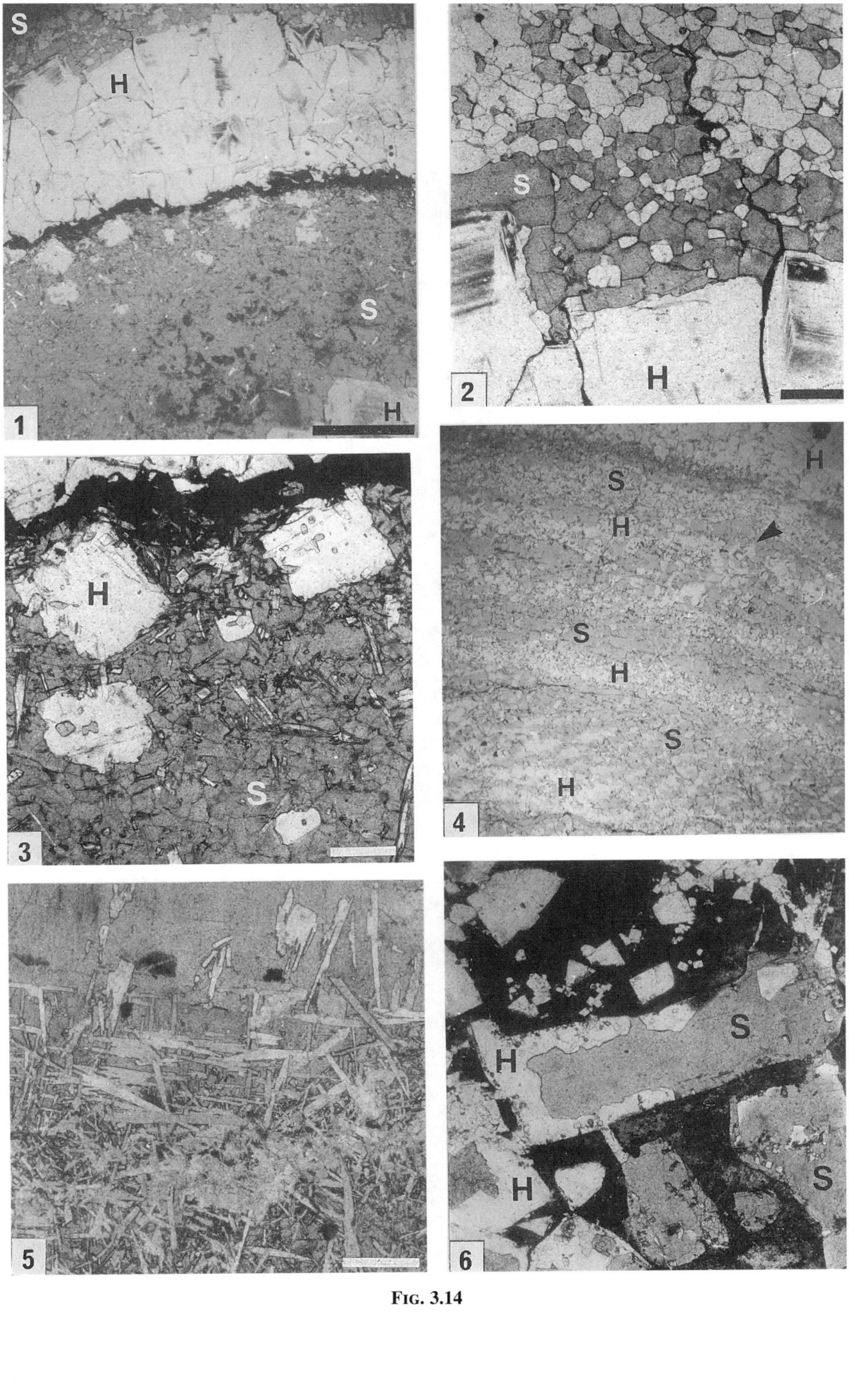

Fig. 3.14

ash seam drape the irregular surfaces of underlying chevron or cornet halite so they have an irregular lower surface whereas the upper surface is commonly horizontal. Sylvite also infills skeletal halite cubes within the underlying layer. Some potash layers are composed of alternating bands of halite and sylvite crystals, suggesting that conditions of crystallization fluctuated rapidly. Because the halite and sylvite crystals are almost equigranular (same size), with a clear cumulate structure in the halite, the alternate sylvite also can be interpreted as a primary cumulate that precipitated at the brine surface or in the water column.

Preservation of primary structures within halite associated with sylvite, together with the lack of dissolutional pipes and intercrystalline cavities typical of salt pan deposits (Lowenstein and Hardie 1985), indicate that these chemical deposits accumulated in a shallow perennial surface brine pool. Sylvite also may have formed syndepositionally during or just after deposition by processes controlled by the existing environment, because it does not form only as a subaqueous deposit. This idea has been based upon the presence of diagenetic (early) structures which are present within the upper part of the lower potash seam.

SALT BASIN GEOMETRY, LITHOSTRATIGRAPHY, BIOSTRATIGRAPHY, AND SEDIMENTOLOGY

Mulhouse Basin.

Middle Eocene (Lutetian). During Cretaceous times, the Jurassic substratum was deeply incised and a karstic network and residual products of continental weathering were generated (laterite phase). Later, during the Eocene, a flexure of the sedimentary cover developed with brecciation in the areas located along major faults. A pre-evaporitic series was deposited in the basinal areas which were only slightly differentiated in comparison with the more meridional domain of continental sedimentation. Products of erosion and alteration of the Mesozoic substratum accumulated during this time and are about one hundred meters thick (the "Siderolithic"). Siderolithic refers to the residual ironstones that developed earlier as alteration products and were then reworked, as first defined in Switzerland by Fleury (1909). Some carbonate layers containing continental fauna are interlayered within the Siderolithic zone (lacustrine sediments such as the Bouxwiller Limestone, further north). Anhydrite nodules occurred locally in the upper part of the series, perhaps as the result of later diagenesis (downward percolation of brines later associated to the Green Limnea Marls I in some parts of the basin). So the bituminous shales known in the central part of the basin (in the Blodelsheim oil well) may record the first restricted Paleogene facies.

The presence of these sediments can be interpreted as indicative of a first subsidence of the Mulhouse basin, with continental detritus (reworking of older laterites) and lacustrine sediments deposited and preserved within small depressions. Palynological data indicate a warm, humid, tropical to subtropical climate with a non-accentuated topography (Schuler 1988/90).

Lower Salt Formation (Green Limnea Marls I, Salt I, Green Limnea Marls II, and Salt II members)

No Bartonian sediment has been recorded, so it seems that the period of true and well-defined subsidence and filling of the graben only began during the upper Eocene (figure 3.15). From the basal part of the Lower Salt layers (~900 m thick), relative subsidence became an important factor. A large SSW-NNE subsidence zone evolved in the southern part of the Rhine graben, extending from the Dannemarie to the Mulhouse sub-basin. South-east of this area, a more resistant zone existed (particularly the Mulhouse horst). This structure was produced by active border faults marked by the erosion of previously folded Jurassic strata, as indicated by the presence of small rounded clasts of Jurassic rocks in the center of the basin. Alteration products were reworked and deposited together with evaporitic deposits which are well represented by the Salt I.

The Green Limnea Marls I (~160 m thick) are composed of marls with nodular anhydrite. The marls are usually compact, with a conchoidal or irregular breakage surface. Their color is generally green or gray. There is almost no fossil content except for scarce plant remains and limnic mollusca (*Limnaea*). The facies are relatively homogeneous on horizontal and vertical scale. The depth of the basin was not trivial; but no other characteristic enables us to define it. In very limited parts of the basin (between Bollwiller and Feldkirch) some halite beds are present and indicate that hypersaline conditions were present. In the same area, development of anhydrite nodules within the siderolithic facies may be related to downwards percolation of brines generated by later compaction of the halite beds.

Salt I (total thickness about 400 m). The lower part of Salt I ("grand banc de sel gemme," about 200 m thick) is mainly halitic with almost no marl intervals. It is the main source for diapiric salt. In the upper part of Salt I (about 200 m thick), halite beds alternate with green marls. Nodular anhydrite is present within both the halite and marls. There is again almost no fossil content, except for some plant remains and mollusca (scarce *Limnaea* and *Hydrobia* remains). Within Salt I, no primary structures are known within the halite facies. Sulfates and carbonates are apparently the product of later diagenetic processes.

Palynological data for the Green Limnea Marls I and Salt I intervals indicate a subtropical climate with a short arid season; as before, the topographic relief does not seem very important (very low percentages of taxa characteristic of higher elevations are present).

Green Limnea Marls II (about 170 m thick). The lithologic facies are the same as for the first Green Limnea Marls and again some halite beds (between Bollwiller and Feldkirch) indicate that hypersaline conditions were present. Macrofossils seem to be more abundant, indicating a lacustrine or brackish environment (*Limnaea marginata, L. crassula, Melania albigensis, Nystia polita*). Microfossils (charophyte oogones and some ostracods) are also noted. The first appearance of a significant development of Pinaceae (pines) indicates that areas of high elevation were present on the basinal borders.

Salt II (~140 m thick). The Salt II is composed of halite, anhydrite (nodular and laminar), and marly facies. The marls have the same appearance as those of the Limnaea Marls. The green color is less evident and gray tints are more common. In the marls interbedded within evaporite facies, there are lacustrine and brackish fauna (*Limnaea marginata, L. crassula, Melania albigensis, Hydrobia, Nystia polita*). Forest populations indicative of high altitudes are still dominant.

Within the upper part of the Lower Salt Formation, in the Green Limnea Marls II + Salt II (as the Mulhouse Basin is structured), fossils become slightly more abundant but are still, for the most part, characteristic of fresh waters. Palynological data for the Green Limnea Marls II and Salt II intervals indicate a subtropical climate with a short arid season. During the Salt II, a regression of the swampy low areas, which were well developed around the basin, has been noted: acceleration of basin subsidence may have permitted the development of more marked highlands on the borders, which are colonized by Pinacea.

Within the Lower Salt Formation, data are too scarce to precisely characterize the sedimentary environment. Only some of the deeper bore holes go through this formation, which has been affected by strong halokinetic deformation.

Middle Salt Formation (Salt III and Fossiliferous Zone).

The Middle Salt Formation comprises the Salt III and the Fossiliferous Zone, both of which can be followed across the basin (figure 3.16). As compared to the Lower Salt Formation, the structural scheme is more complex: two depocenters exist, one on the western side of the basin (about 300 m of sediments), the other on the eastern side (about 260 m); whereas halite isopercentage lines (up to 40%) are located in a more central position (figure 3.17).

The Salt III (upper Eocene) is composed of about 220 m of halite, anhydrite, and gray marls that are variably bituminous and contain thin dolomitic beds. Anhydrite beds are present within all the intervals that are associated with nodular anhydrite in the lower part of Salt III. This member is the first organic-rich interval with about 1.8% TOC in the central part of the basin.

Macro- and microfauna are well developed with

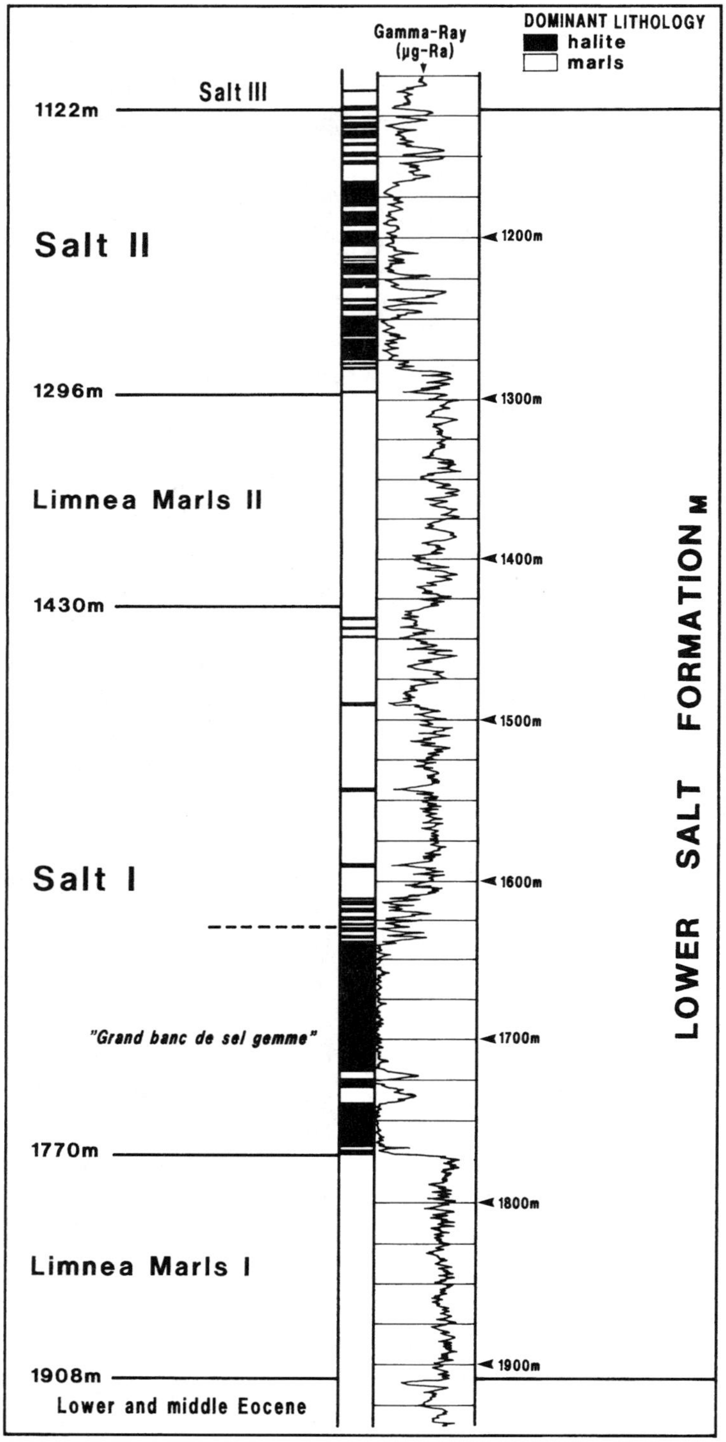

Fig. 3.15A. Lower Salt Formation: (A) gamma-ray and lithologic section.

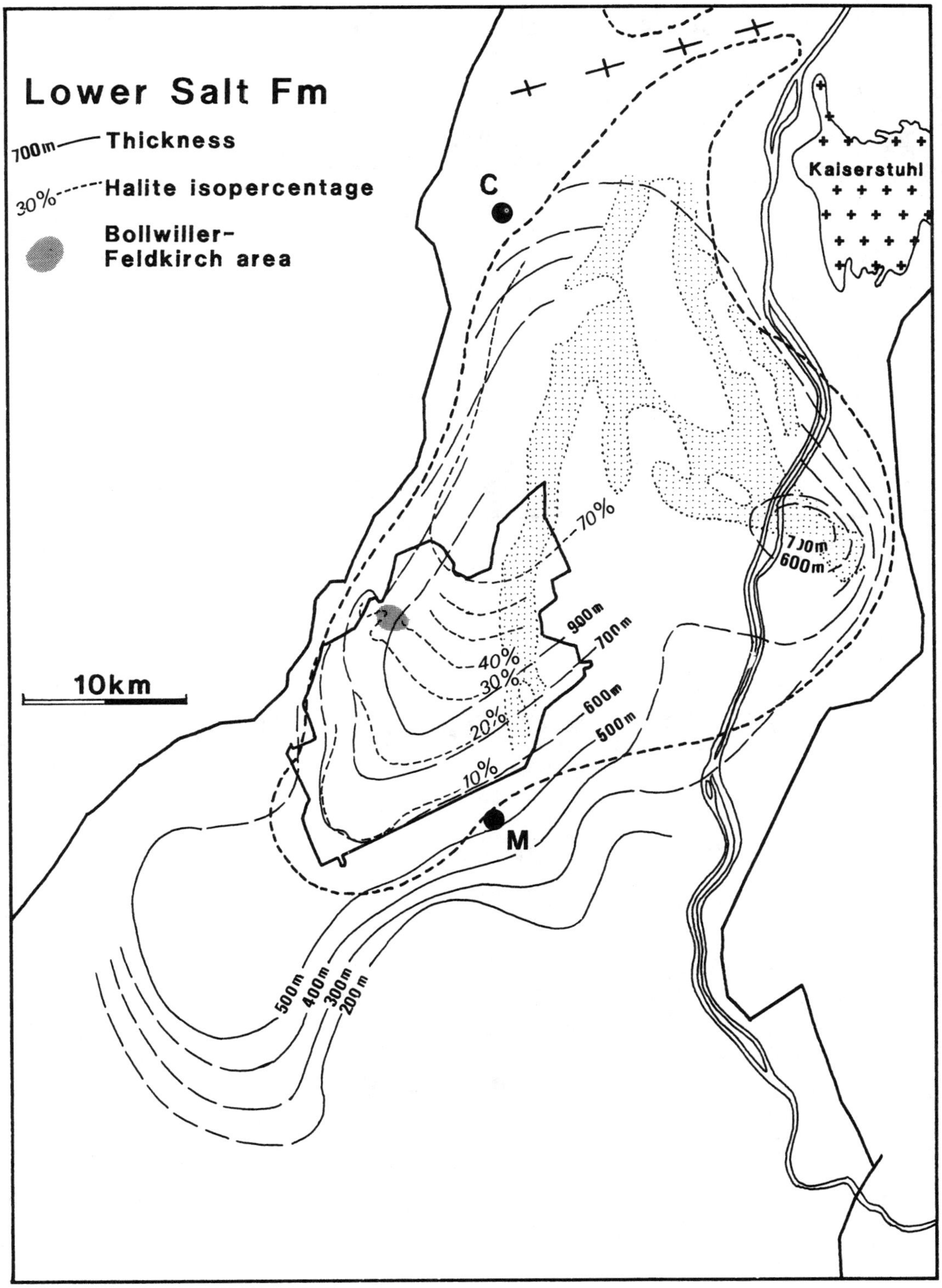

FIG. 3.15B. Lower Salt Formation: (B) isopach and halite isopercentage map of the Mulhouse Basin (Blanc-Valleron 1990/91, modified).

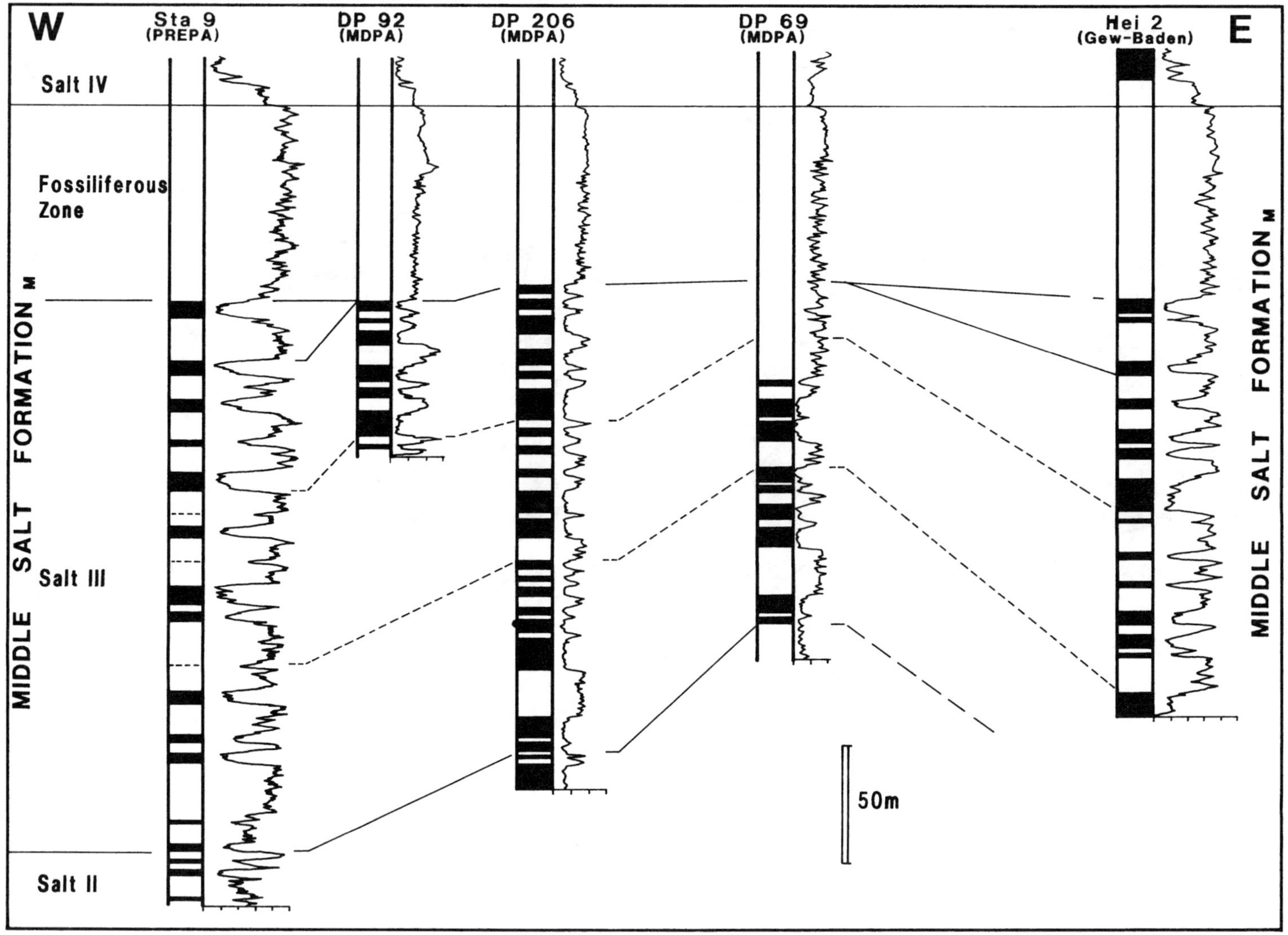

Fig. 3.16. Correlation, based on borehole logs, between the sediments of the Middle Salt Formation of the Mulhouse Basin (Blanc-Valleron 1990/91, modified).

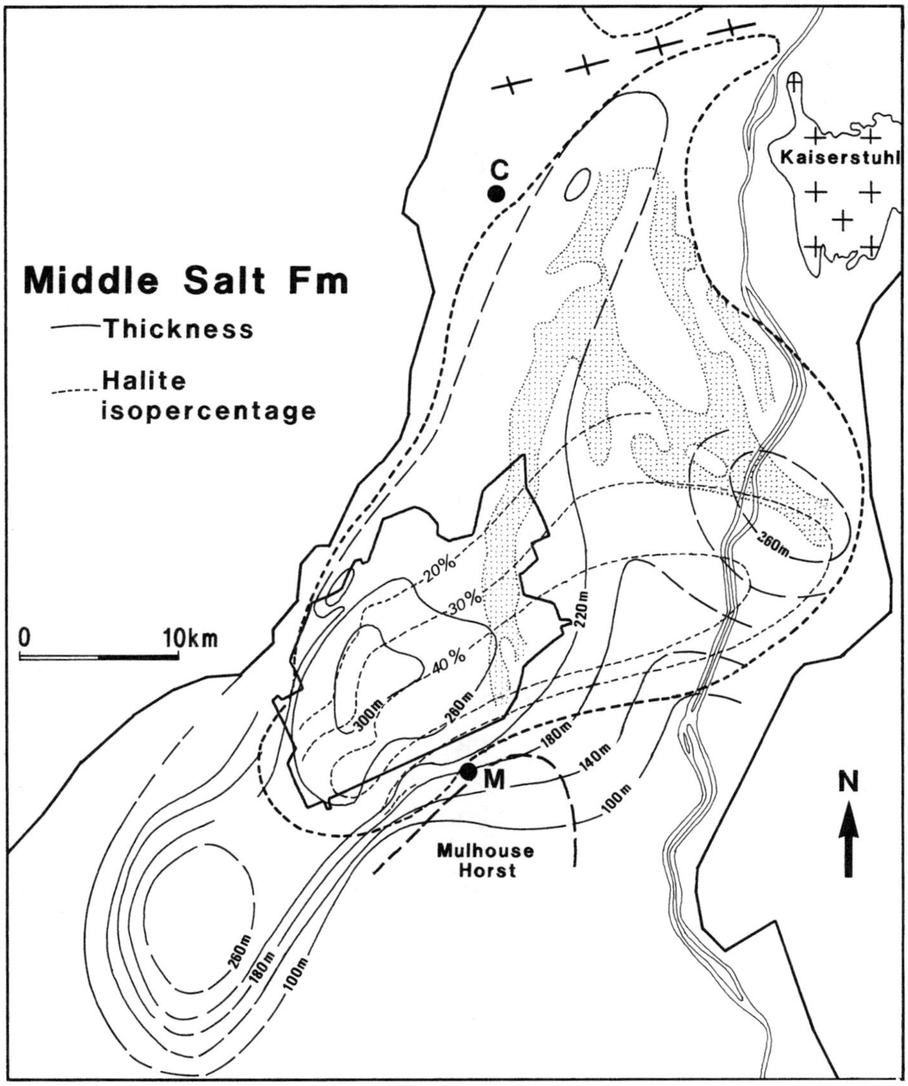

Fig. 3.17. Isopach and halite isopercentage map for the Middle Salt Formation of the Mulhouse Basin (Blanc-Valleron 1990/91, modified).

many mollusca (*Cyrena, Hydrobia, Mytilus*), fish, ostracoda, and amphipoda (Gammaridae). Other mollusca, some isopoda, foraminifera, insects, macro- and microfloral remains (higher plant remains, algae, charophyta) are also present. The sporadic presence of Tertiary nannoplankton and dinokystes permits consideration of a possible connection with nearby marine waters. As compared with the Lower Salt Formation, the fossil assemblage testifies to the disappearance of lacustrine influence in favor of brackish and marine environments (*Cyrena, Hydrobia, Mytilus, Tympanotonos,* foraminifera, some fish, nannoplankton, and dinokystes).

Palynological studies in Alsace (Schuler 1988–1990) show that the climate was intermediate between a subtropical and a mediterranean climate, with a weak regression of the dense, humid forest and expansion of the drier forests.

Fossiliferous Zone (lower Oligocene) (~80 m thick). In the lower part ("Marnes en Plaquettes," ~60 m thick), the gray marls contain many anhydrite beds with the same fauna as the Salt III layers. In the upper part ("Marnes à Hydrobies," ~20 m

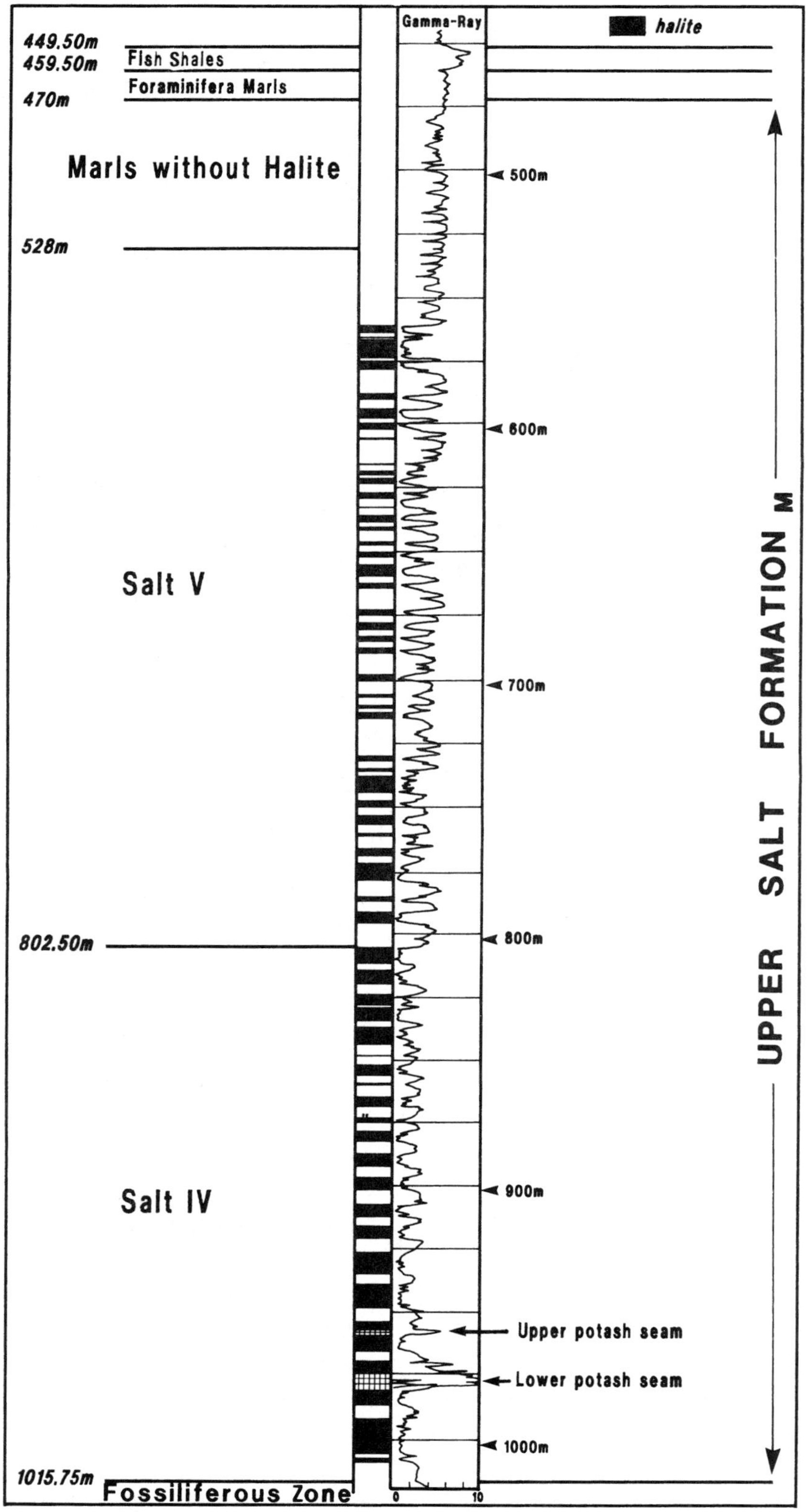

Fig. 3.18A. Upper Salt Formation: (A) gamma-ray and lithologic section.

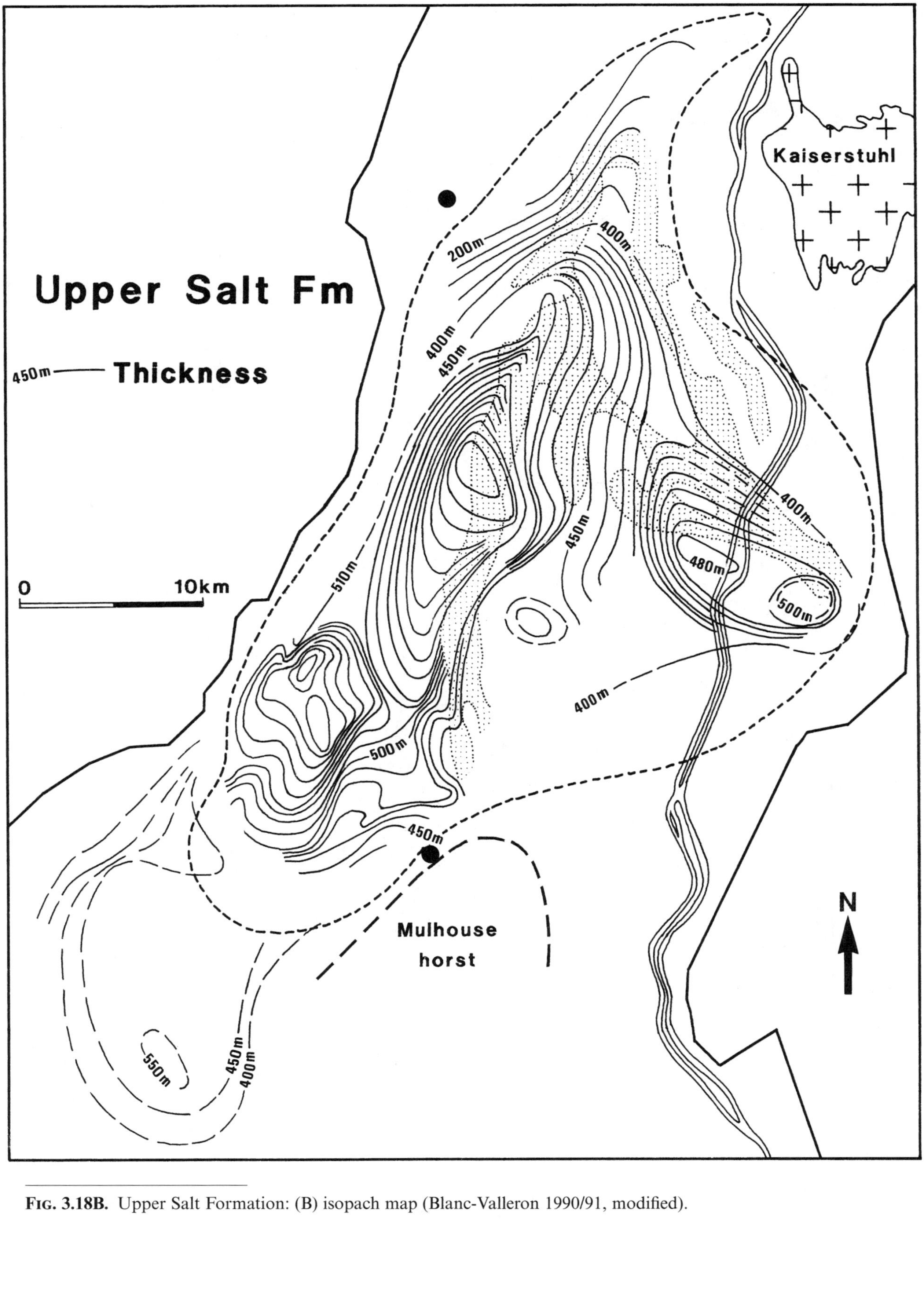

Fig. 3.18B. Upper Salt Formation: (B) isopach map (Blanc-Valleron 1990/91, modified).

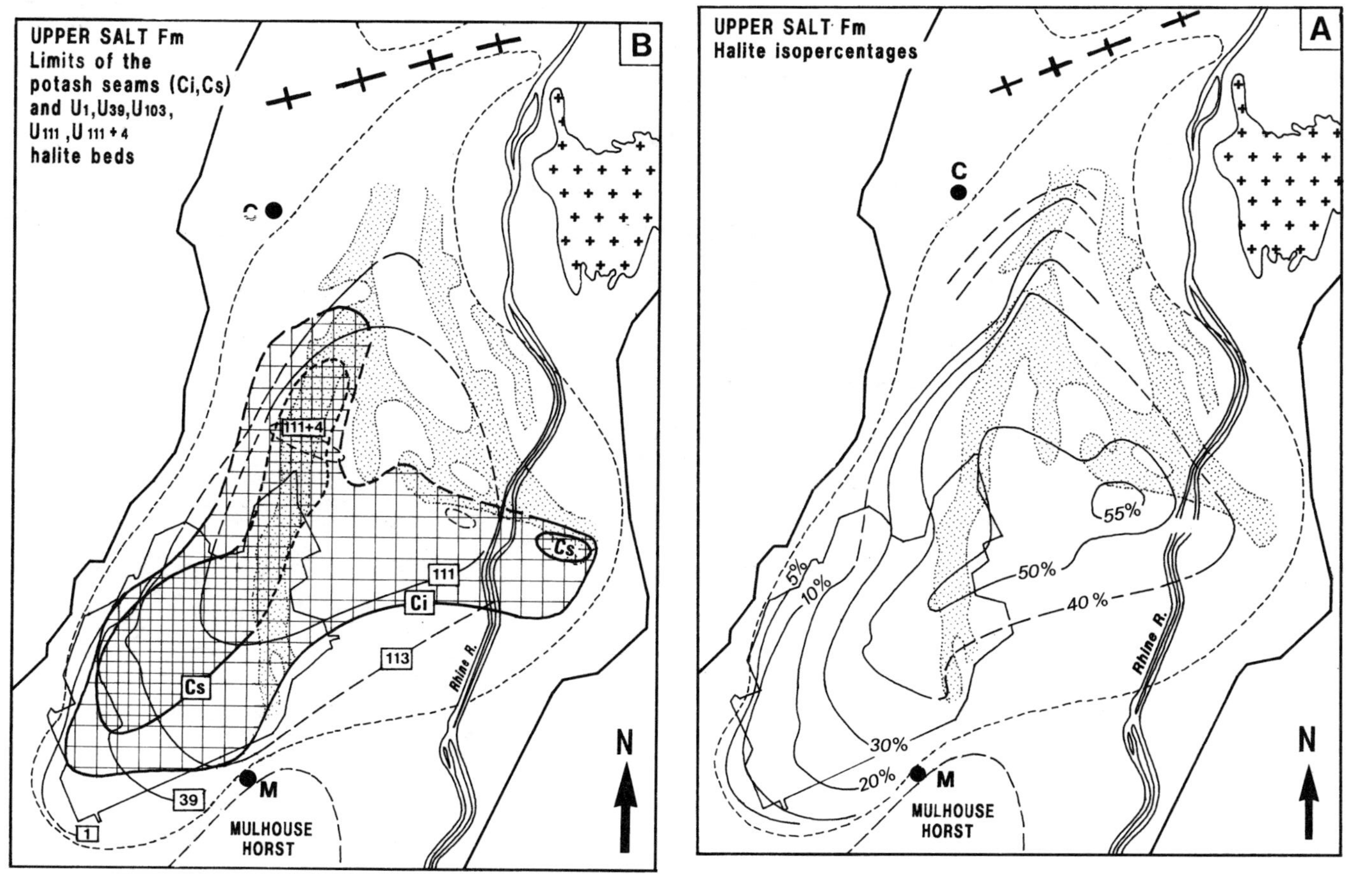

Fig. 3.19. Maps for parts of the Upper Salt: (A) Isopercentage map for the Upper Salt Formation of the Mulhouse Basin; (B) Restriction of the halitic realm during the Upper Salt Formation as shown by the maximum area covered by beds U1, U39, U103, U111, and U111+4. The depositional limits of the potash strata (Ci, Cs) are also given (Blanc-Valleron 1990/91, modified).

thick), in contrast, there is almost no anhydrite. TOC values are less important than in the Salt III member (about 1.3%) but fossils are far more numerous. Within the fauna, Hydrobia and ostracodae are quite abundant. Other mollusca, some isopodae, amphipodae (*Melita palmata, Gammarus retzi*), fish (*Prolebias rhenanus, Pomatoschistus bleicheri*), and insects are also present together with macro- and microflora remains.

Marine influences are identified for the Fossiliferous Zone on the basis of the presence of marine nannoplankton (*Reticulofenestra sp., Markalius inversus, Toweius sp., Ismolithus recurvus, Braarudosphaera sp.*), shallow water benthic foraminifera (*Bolivina, Eponides, Bulimina, Discorbis*), and well diversified dinokyste assemblages (Rauscher and Schuler 1988; Schuler 1988–1990). But connection to the ocean is still a matter of discussion (peri-Alpine Sea from the SE?, North Sea from the NW?).

Palynological studies in Alsace (Schuler 1988–1990) show the development of forests on the higher topographic areas and small hills during the same period that vegetation from humid and swampy palustrine lowlands disappears. This period is characterized by pronounced tectonic activity with formation of high relief and substantial detrital input on the borders.

Upper Salt Formation (± 630 m thick, Salt IV, Salt V, and "Marls without Halite").

The lithological succession of potassic (Ci and CS), halitic (S1, Mi, Ti, Ms, Ts, U1, U3, U5, U7, U9, etc.), and marl (S2, S, T, U, U2, U4, U6, U8, U10, etc.) strata is illustrated in figure 3.18. The sediment infill of the basin shows that maximum halite isopercentages (figure 3.19A) do not correspond to the area of maximum sediment thickness.

Salt IV (~210 m thick). In the lower part of the series (S2 → U5), anhydrite is layered and potassic salts (sylvite and carnallite) are present at the center of the basin. The lower potash seam (Ci, 0 to 6 m thick) is localized about 40 meters above the top of the Fossiliferous Zone; the upper potash seam (CS, 0 to 2 m thick) about 20 meters above the lower seam. In the upper part of the series (U6 → U33), anhydrite occurs as bedded and nodular facies. Figure 3.19B outlines the limits of the lower and upper potash seams and the gradual restriction of the halite beds to the center of the basin. The sediment infill of the basin for the lower part of the Salt IV member is shown in figure 3.20. The lower and upper potash seam limits correspond to the isopach limits but not to the isopercentage lines.

Remains of higher plants and macro- and micro fauna are less abundant than in the Fossiliferous Zone, but some mollusca (*Limnaea marginata,* in the upper part of the interval), isopoda, ostracoda, insects, fishes, and high plant remains have been cited in the bore hole descriptions. Rare levels containing dinokystes or nannoplankton have been observed, but the associations are poor and do not prove the existence of true marine communication.

Palynological studies (Schuler 1988–1990) show that the climate was drier with a prevalent Mediterranean climate having a long dry season (more than half the year).

Salt V (~260 m thick). During Salt V deposition, synsedimentary faults were more active and detrital deposition apparently increased. Salt V is composed of alternating halite and marl intervals, with nodular anhydrite. The marl beds are of two types: (1) gray and layered; and (2) brown or green-red with a more homogeneous appearance.

High plant remains, macro-, and microfauna are indicated in borehole descriptions; some mollusca (*Limnaea marginata*), ostracoda, insect, fish, and charophyta remains also have been noted.

Marls without Halite (~60 m thick). As evaporitic conditions progressively diminish the composition of this interval corresponds to red-brown and gray marls with gypsum or anhydrite nodules. Higher plant remains, charophyta, foraminifera, and mollusca (*Limnaea* sp.) have been cited from the borehole descriptions. Palynological studies have shown the presence of varied dinokyste associations which may presage the "Rupelian transgression"; but many other lacustrine or brackish levels have been also noted.

Palynological reconstruction for the Salt V and "Marls without Halite" in Alsace (Schuler 1988–1990) indicates that the climate was a mixed one

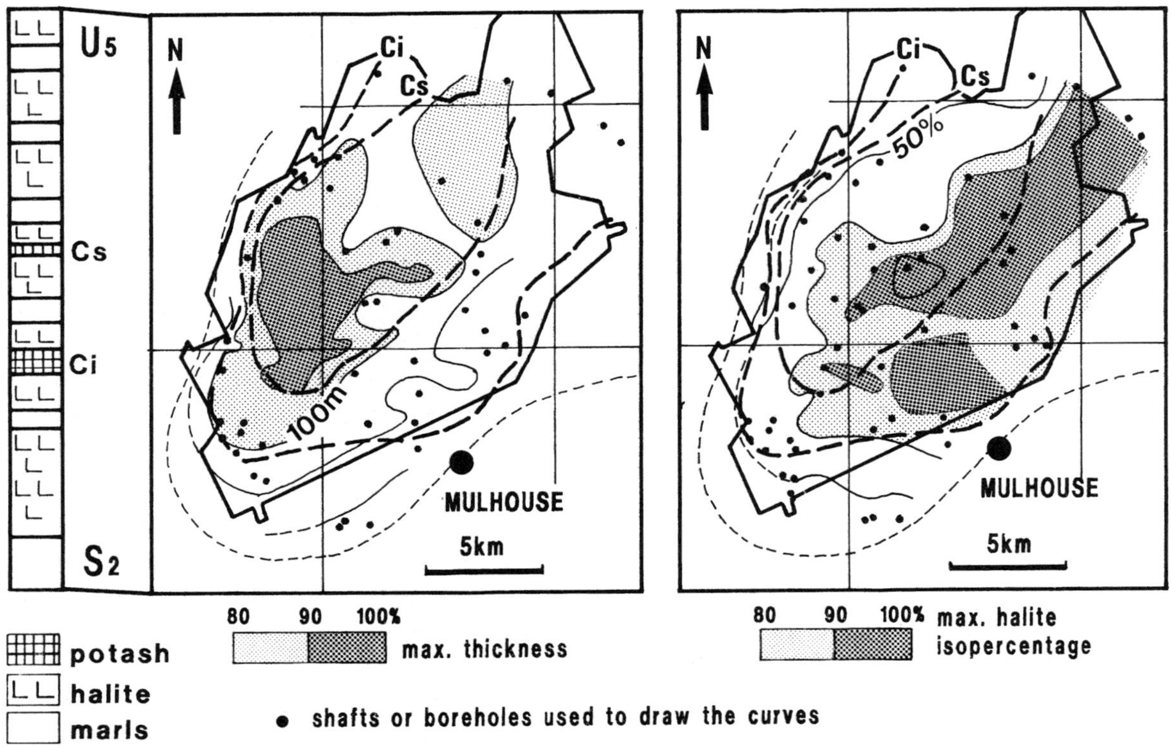

Fig. 3.20. [S2-U5] isopachs and isopercentages for lower Salt IV (Blanc-Valleron 1990/91, modified), drawn every 10 units. On each map, areas with maximum values have been shadowed to permit comparison.

with subtropical, Mediterranean, and temperate influences.

Foraminifera Marls and Fish Shales (~20 m thick). The Foraminifera Marls and Fish Shales cover everything (transgressive), over-stepping some new parts of the substratum—there are no evaporites in these mostly marine strata (dated NP 23). This is the sedimentary product of the so-called "Rupelian transgression" that came from the North Sea through the Hesse graben. However, the Upper Rhine Basin also may have been connected with the peri-Alpine sea to the south.

Other Basins.

Two general logs are given to illustrate the lithostratigraphy of basins situated north of the Colmar sill: Sélestat (DP XXII) and Strasbourg series (DP XXX) (figures 3.21 and 3.22). Detailed descriptions of these areas can be found in Blanc-Valleron (1990/91), including the detailed correlations with the Mulhouse Basin, isopach and halite isopercentage maps, and other analytical data. The main results (sedimentological, paleontological, and geochemical data, location and evolution of the evaporitic

Fig. 3.21. DP XXII (Sélestat Basin): lithostratigraphic section (Blanc-Valleron 1990/91, modified). Key: 1 = laminar anhydrite (or gypsum); 2 = nodular anhydrite (or gypsum); 3 = dolomites and dolomitic marls; 4 = sandy or conglomeritic levels; 5 = pyrite; 6 = bituminous impregnations; 7 = lignite fragments; 8 = organic (macro- and micro-) remains.

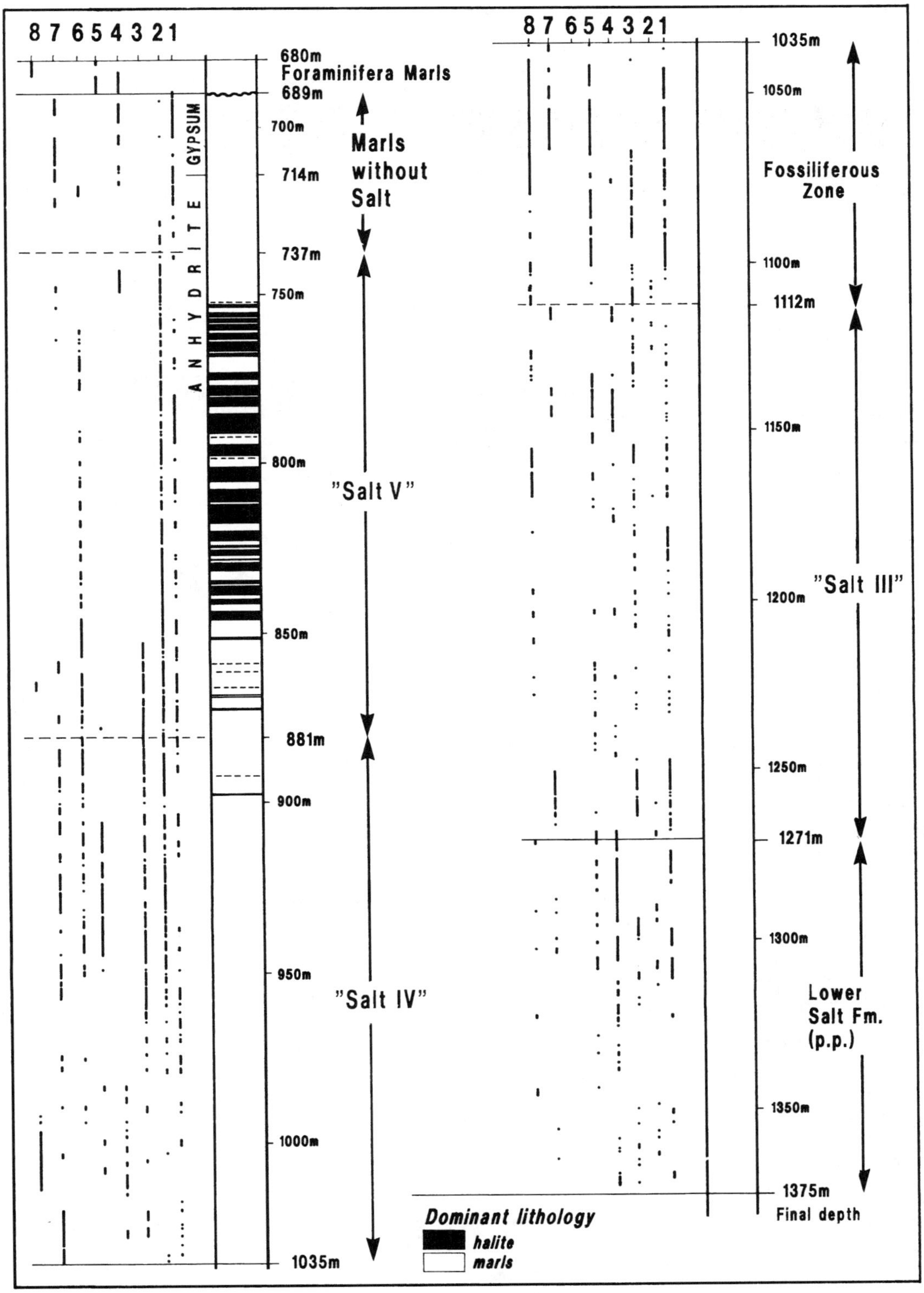

8 7 6 5 4 3 2 1
680m
Foraminifera Marls
689m
700m
714m
737m
750m
800m
850m
881m
900m
950m
1000m
1035m
GYPSUM
ANHYDRITE
Marls without Salt
"Salt V"
"Salt IV"
1035m
1050m
1100m
1112m
1150m
1200m
1250m
1271m
1300m
1350m
1375m
Final depth
Fossiliferous Zone
"Salt III"
Lower Salt Fm. (p.p.)
Dominant lithology
halite
marls

FIG. 3.21

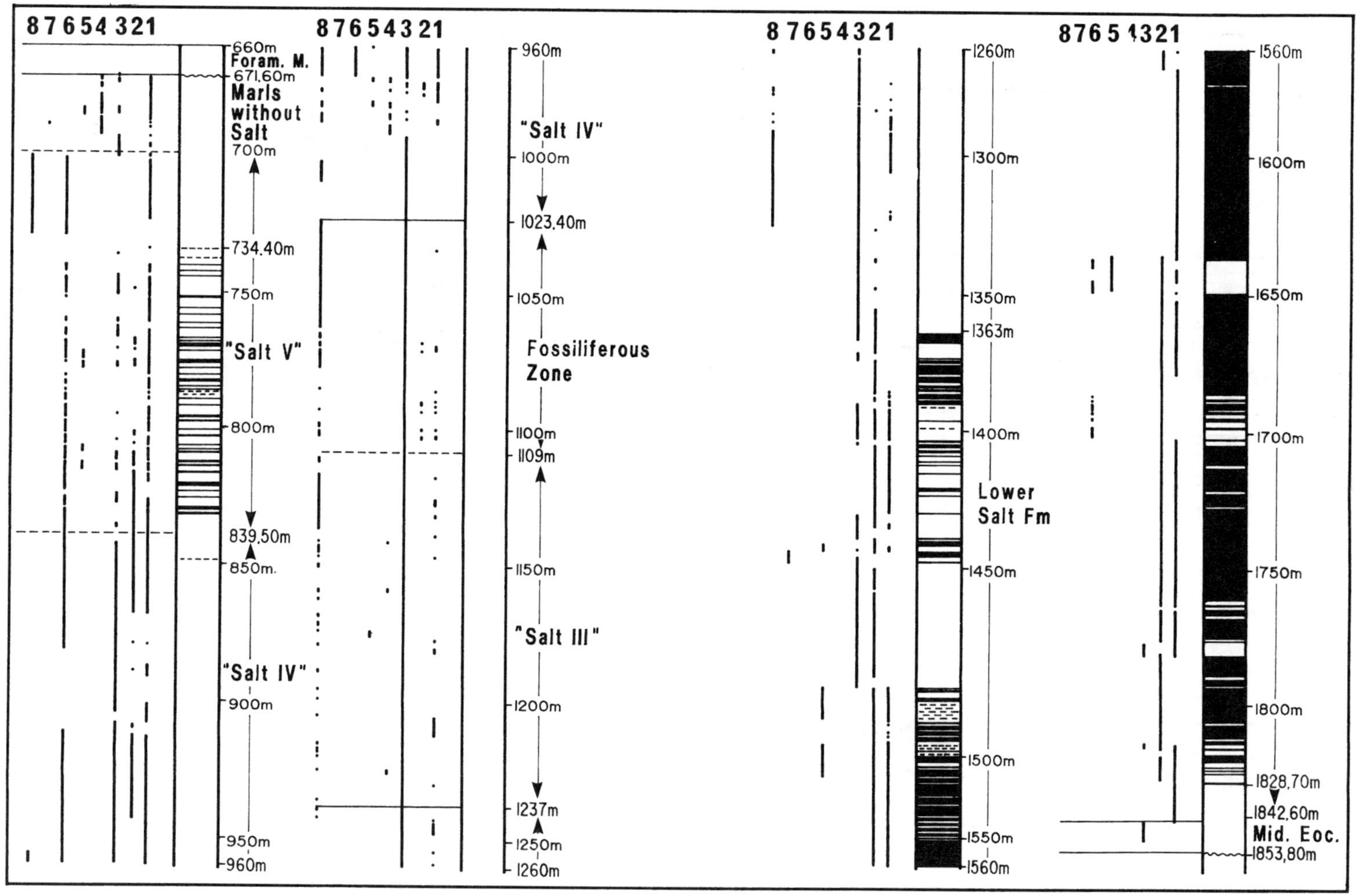

Fig. 3.22. DP XXX (Strasbourg Basin): lithostratigraphic section (Blanc-Valleron 1990/91, modified). Key: 1 = laminar anhydrite (or gypsum); 2 = nodular anhydrite (or gypsum); 3 = dolomites and dolomitic marls; 4 = sandy or conglomeritic levels; 5 = pyrite; 6 = bituminous staining; 7 = lignite fragments; 8 = organic (macro- and micro-) remains.

realms through time) concerning basins situated north of the Colmar sill are reviewed below in order to discuss the evolution of environments within the Rhine Graben and the problem of the origin of salt deposits.

General Synthesis and Discussion

MULHOUSE: RECONSTRUCTION OF PALEOENVIRONMENTS

The evolution of subsidence and confinment within the Mulhouse basin from the Salt I interval to the Salt V interval is illustrated in figure 3.23. For each halitic interval, areas with maximum thickness and maximum halite isopercentages have been outlined. For the Salt I and Salt II intervals, halite isopercentages are greatest wherever the thicknesses are greatest. As the contour lines outline the salt deposit, conditions of halite deposition seem to depend strongly on the basin morphology. The pan (basin center) scheme, as defined by Busson (1978, 1980), can be applied to the Mulhouse Basin, with the more soluble salts accumulating in the more central and subsident part of the basin. But this scheme may be a little too simplified, because the data used to map these intervals are less abundant than for the other intervals.

Two depocenters near the border faults were differentiated during the deposition of Salt III (Middle Salt Formation), Salt IV, and Salt V (Upper Salt Formation). These two sub-basins were separated by a shelf area that corresponds to a northeastern extension of the Mulhouse horst. Maximum halite isopercentages are located in a more central position, between the two depocenters. Thus the pan (basin center) scheme defined by Busson cannot be strictly applied to the Mulhouse Basin. The small scale of this basin, together with active tectonism which generated coarse detritus that was deposited and rapidly trapped on the borders, may explain the pattern observed.

RECONSTRUCTION OF PALEOENVIRONMENTS

An overall picture of climate, landscape, and morphology of the Alsace sub-basins is necessary in order to study the evolution of subsidence and confinment, and the problem of the origin of salt.

Lower Salt and time equivalent formations (figure 3.24). Study of the different basins show that the Mulhouse area was the most subsident and contains the most halite of the four related basins, with a total of about 800 m of sediments and a 70% halite isopercentage at the depocenter. In the Strasbourg basin the depocenter accumulation is 500 m thick with 40% halite; in the Rastatt Basin, 700 m with 20% halite. Halite isopercentage values decrease going north because the upper part of the formation is halitic only in the Mulhouse Basin (Salt II), whereas true lacustrine associations are known in time equivalent levels of the Sélestat, Strasbourg, and Rastatt basins. As marine organisms are absent and only lacustrine or brackish ones have been found, connection with a marine realm is unlikely in these northernmost three basins. Chlorides and sulfates, where present, were probably reworked from older sources (Permo-Triassic) as groundwaters migrated and rose to the surface along faults generated by the graben formation.

Middle Salt and time equivalent formations (figure 3.25). As in the Lower Salt Formation, the Rastatt (250 m) and Mulhouse (300 m) basins are the more subsident areas, but halite is only present in the Mulhouse area (40% halite isopercentage). At the end of the Lower Salt Formation the marginal borders were already differentiated. During the Middle Salt Formation, structural evolution continued, and the Mulhouse horst and the sills between the basins were less subsident. This internal structural differentiation of the graben continued and caused significant variations in sedimentation through time:

1. Fossils are no longer only lacustrine but alternatively lacustrine and brackish (Salt III within the Strasbourg and Rastatt basins) or always brackish (Salt III within the Sélestat and Mulhouse basins, Fossiliferous Zone). The presence in the Fossiliferous Zone and

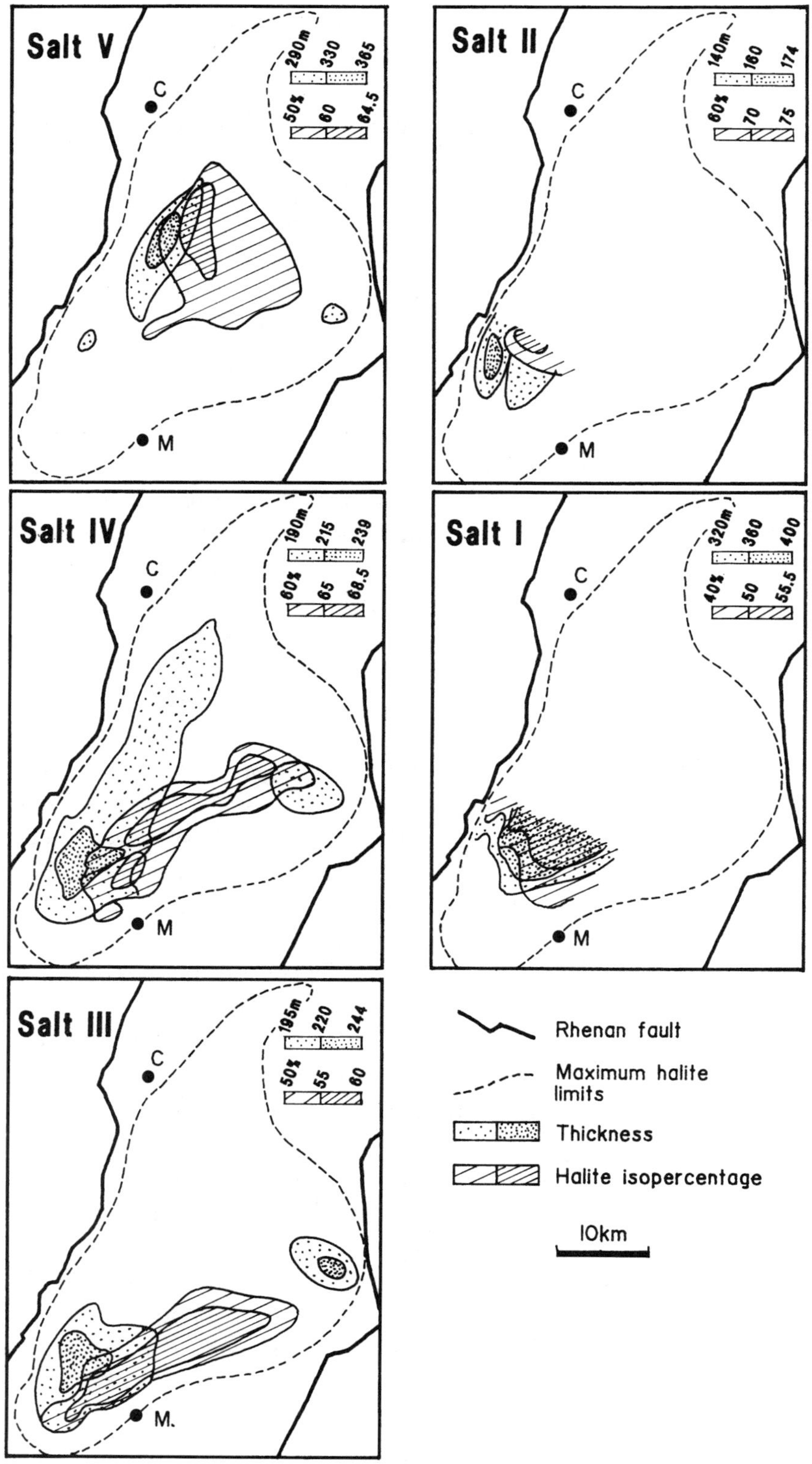

Fig. 3.23. Evolution of subsidence and confinement in the Mulhouse Basin: location and evolution of evaporitic realms through time (Blanc-Valleron 1990/91, modified). For each halite member, curves corresponding to about 80% and 90%, respectively, of maximum thickness and halite isopercentages have been drawn.

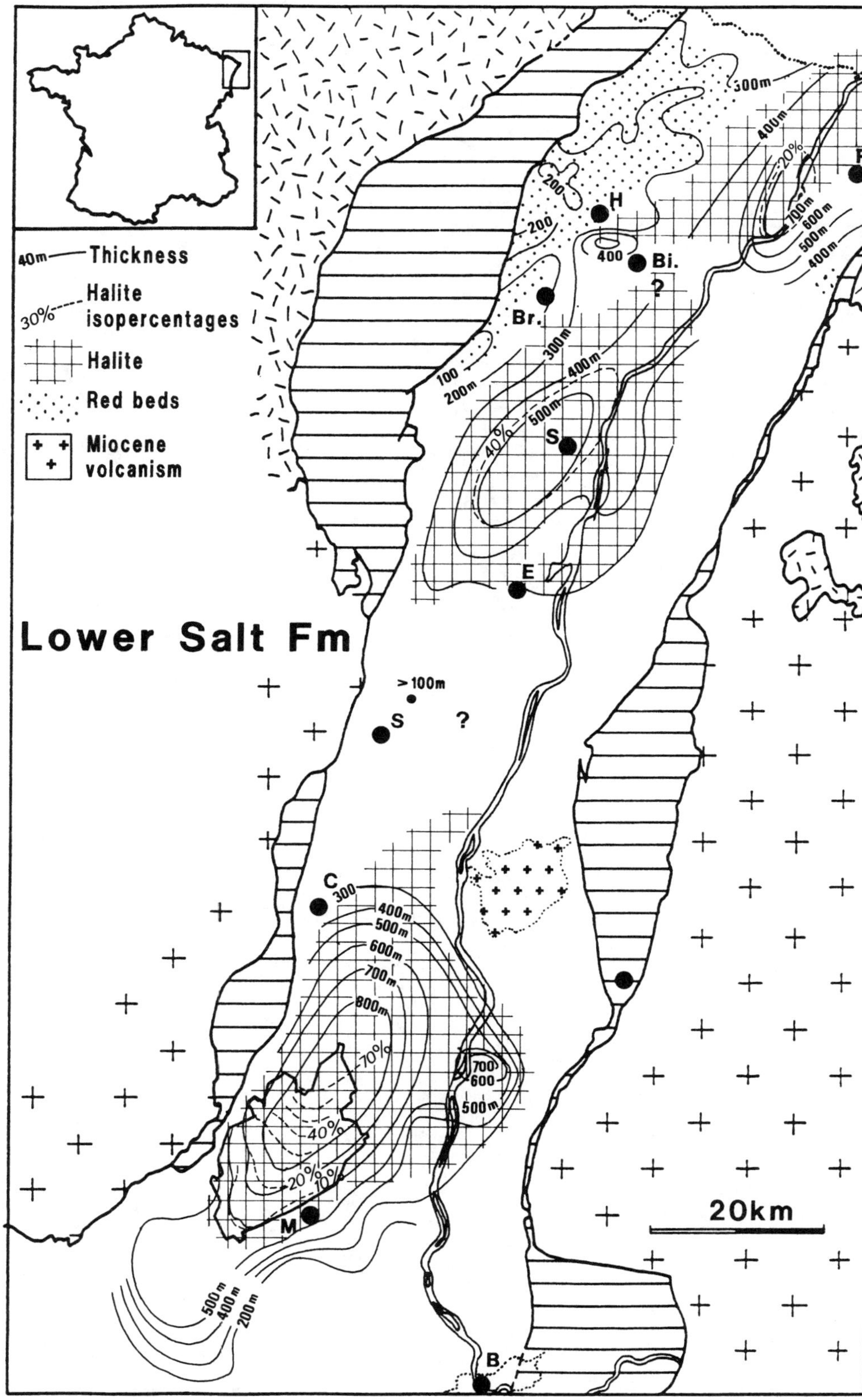

FIG. 3.24. Isopach and halite isopercentage map for the Lower Salt Formation and time-equivalent formations of Alsace (Blanc-Valleron 1990/91, modified).

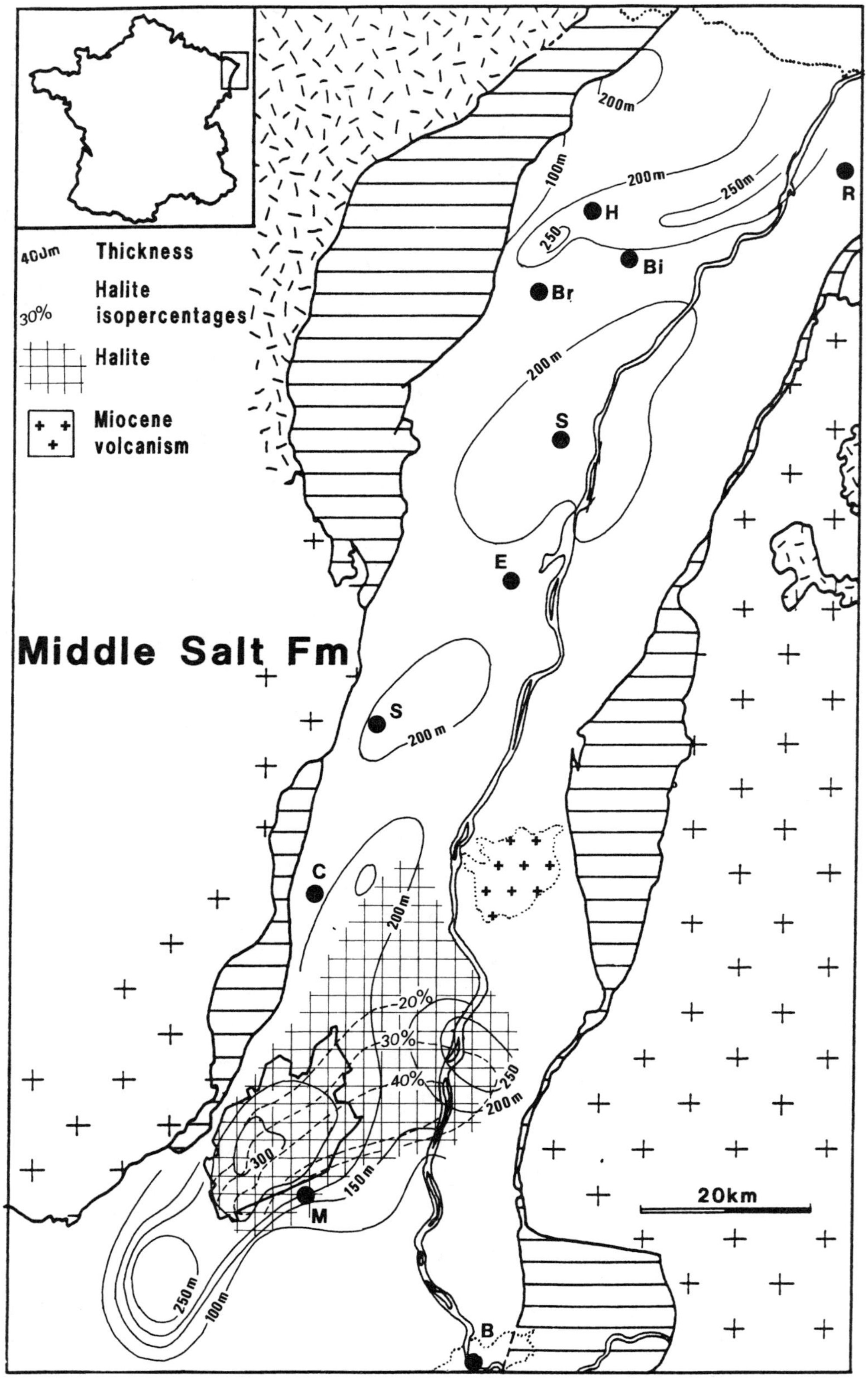

FIG. 3.25. Isopach and halite isopercentage map for the Middle Salt and time-equivalent formations of Alsace (Blanc-Valleron 1990/91, modified).

Salt III, of certain marine remains (nannoplankton, dinokystes, foraminifera) calls for strong consideration of a possible influx of nearby marine waters.

2. The appearance of the marly facies changes: bioturbation trails are fewer, finely laminar facies are abundant; this modification goes with an increase of mean TOC values.

   Within Salt III, halite is only present in the Mulhouse Basin, which is the most subsident (about 230 m of sediments instead of 150 m further north of the Colmar sill) and the more organic-rich (1.8% instead of about 1 to 1.1% in the center of the Sélestat and Strasbourg basins).

The Fossiliferous Zone that overlies almost all of the older sediments is rather homogeneous within the Rhine Graben; while the Mulhouse Basin is somewhat particular because of the greater quantities of anhydritic facies (10 to 20% instead of about 3 to 4% in the center of the Sélestat and Strasbourg basins). The amounts of organic matter is also much higher in the Mulhouse basin (1.3% instead of about 0.8 to 0.9% in the center of the Sélestat and Strasbourg basins).

Upper Salt and time equivalent formations (figure 3.26). Halite is present in all the basins, with decreasing maximum halite isopercentages going north from the Mulhouse (50%), Sélestat (20%), and Strasbourg (10%) to the Rastatt (5%) basins. As in the Lower and Middle Salt formations, the Rastatt (400 m) and Mulhouse (600 m) basins are the more subsident.

Whereas the Fossiliferous Zone is fairly constant across the entire Rhine Graben, evaporitic conditions at the base of the Upper Salt Formation in the Mulhouse Basin return again (as in the Salt III) with broad variations of the sedimentary environment, specially on each side of the Colmar sill. The Salt IV and its time equivalent formations show:

there was a more marked subsidence within the Mulhouse Basin (220 m as opposed to only 155 m and 185 m for the depocenters of the Sélestat and Strasbourg basins);

halitic and potassic facies, interlayered with laminar marls, were only present in the Mulhouse Basin;

detrital input was limited to the marginal areas near the border zones within the Mulhouse Basin, whereas they extend as far as the depocenters north of the Colmar sill.

For the Salt IV deposits of the Mulhouse Basin bromine values of halite, together with the presence of some marine nannoplankton, phytoplankton, and some foraminifera indicate the possibility of a connection with a marine realm. The connection with the marine seaway would have been from the south (peri-Alpine sea) as further to the north the marls are homogeneous and contain only lacustrine and terrestrial fossils.

In the Salt V no marine evidence has been observed, whereas lacustrine remains are present. For the "Marls without Halite," well marked by detrital and lacustrine influences, the renewed presence of dinokystes and foraminifera suggest that a marine influx took place near the top of the Upper Salt Formation.

Conclusion

The Southern Rhine graben allowed us to study a thick Paleogene evaporitic series within several basins as copious subsurface data were available; but, as in many other segments of the Western European Continental Rift system, the origin of salts is still a much debated question: calcium sulfate has been interpreted (from isotopic analysis of O and S) as being of non-marine origin, (chemically) reworked from Permo-Triassic sources; and paleontological assemblages and bromine values within some levels of the Middle and Upper Salt formations suggest an associated marine influx.

A rough estimate of NaCl volume (about 550 km^3) within the Paleogene series has been made (figure 3.4); it is compatible with the estimate made of the Permian and Triassic halite volumes (about 800 km^3) that are likely to have been dissolved during that time period (Blanc-Valleron 1990/1991). So, at least for the Lower Salt Formation where nothing argues for a marine origin of the evaporites,

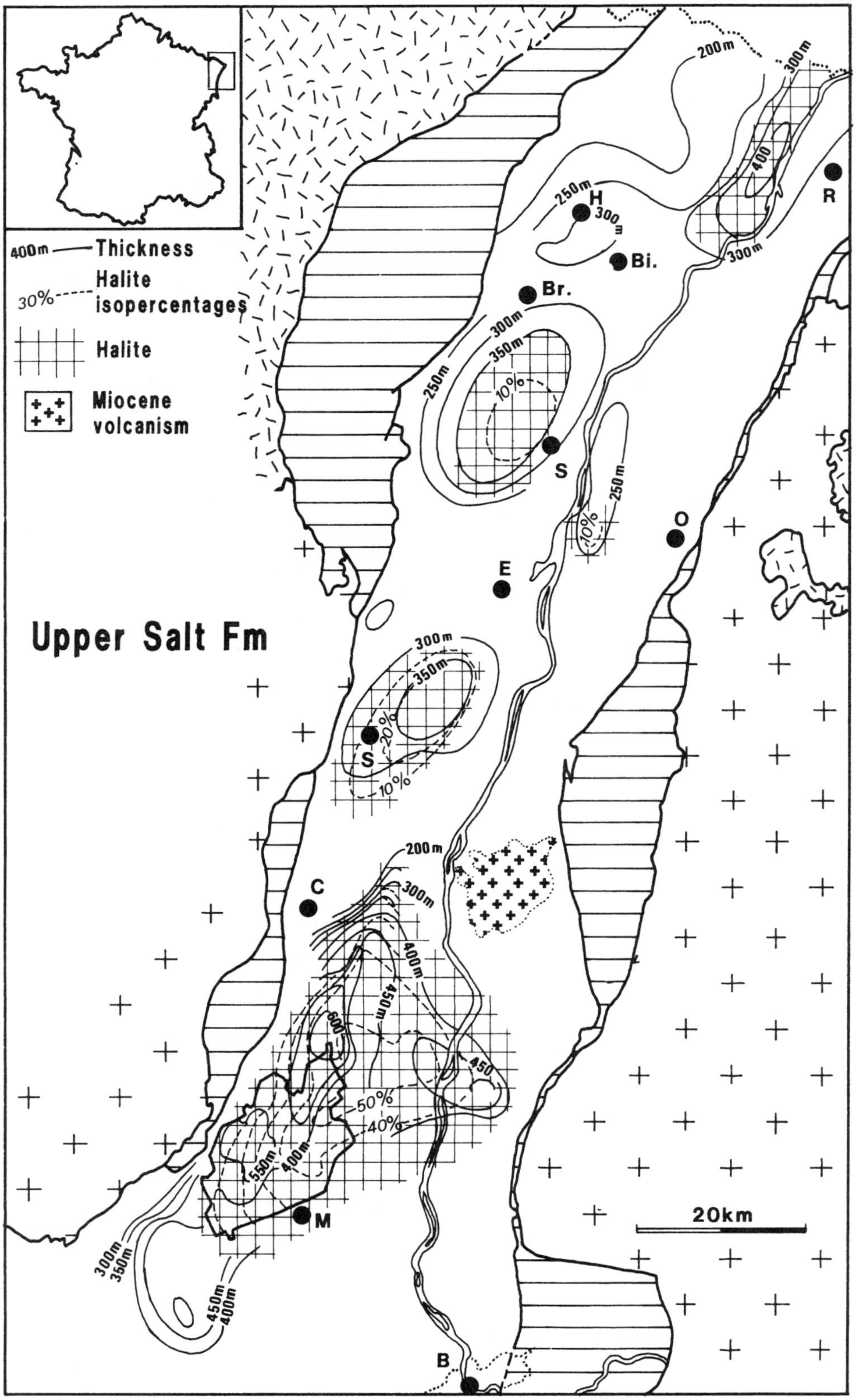

Fig. 3.26. Isopach and halite isopercentage map for the Upper Salt and time-equivalent formations of Alsace (Blanc-Valleron 1990/91, modified).

the continental salt lake model is consistent with the data.

For the Middle and Upper Salt formations, marine ingressions may have come from the north of the graben (more probably from Hesse, to the northwest) or from the southeast (see Blanc-Valleron 1990/1991, p. 304, for a review). So the problem of the origin of evaporites is not resolved. It is a difficult one, because the time span of evaporitic sedimentation covered about 3 to 4 MA and many parameters may have varied and strongly influenced the sedimentary record.

4

The Salt Basin of Bresse: Southern Saône Graben

ALAIN CURIAL AND ROBERT MORETTO

Geological Setting

LOCATION AND EXTENT OF PALEOGENE EVAPORITIC DEPOSITS

The Saône Graben forms a tectonic depression filled by Cenozoic sediments, elongated approximately in a 015° to 020° direction. It is 200 km long and its width ranges from 30 to 60 km (figure 4.1). This graben is composed of 3 main sub-basins separated by basement highs: the Chalon Basin, the Louhans Basin, and the Bourg Basin (this volume). The northern area of the Saône Graben (Chalon Basin) is limited to the northwest by the Bourgogne high (Rat et al. 1975). Eastward, the Chalon Basin passes into a tectonized area that separates the Rhine Graben and the Saône Graben (Gélard 1978). The structure of this tectonized sector apparently results from the sinistral movements of a 060° transform fault (Bergerat 1977; Illies 1981). To the south, the Saône Graben is separated from the Valence Basin by the Vienne-Chamagnieu basement high, oriented 050° (this high and its northern slope form the "Seuil Lyonnais"). The Saône Graben is bounded to the east by the Jura chain (Mesozoic formations) and to the west by the Massif Central (crystalline basement and Mesozoic formations).

The areas of the Saône Graben in which subsidence was greatest during the Cenozoic are located in the southern part of the Louhans Basin and in the Bourg Basin. In these areas, usually termed the "salt basin of Bresse," thick salt formations of Paleogene age developed. In the peripheral areas of the salt basin (e.g., Louhans Basin), thin evaporite deposits have accumulated (mostly as carbonates and sulfates). Studies based on these deposits (Lefavrais-Raymond 1962; Choignard 1965) deal primarily with stratigraphy and do not provide other data concerning these peripheral evaporites. The actual limit of the main salt deposits has been deduced from seismic data (Leroy 1984). The area of halite deposition was apparently controlled by two main structural trends: 020° (both the Limonest horst and the axis of subsidence in the Bourg Basin and the Louhans extension), and 050° (the Cormoz basement high). Most of the terrigenous supply entered the salt basin from the south and competed with halite deposition in the progressive filling of the depressions induced by synsedimentary subsidence. The abundance of halite is closely related to the thickness of the salt sequence, as shown by the superposition of isopachs of Paleogene salt series and corresponding isopercentages of halite (figure 4.2).

The Bourg Basin shows an E-W asymmetry with deepening and thickening of the salt toward the east (figure 4.3). Between the Massif Central margin and the Limonest basement high, a small subbasin developed (the Western sub-basin: BR2 and MaMaCD1 wells), filled by terrigenous deposits

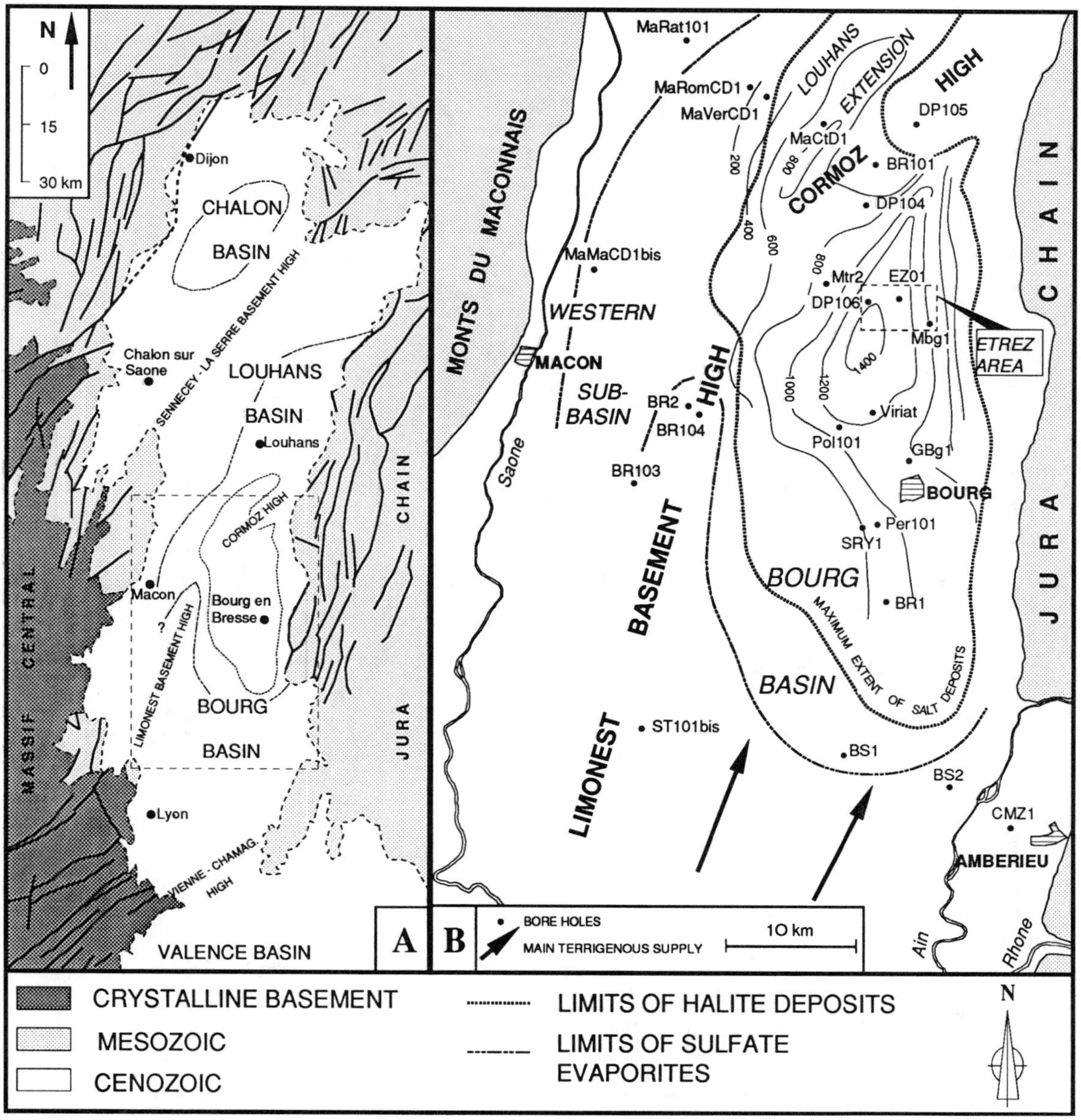

FIG. 4.1. General geological setting of the Bresse Graben:
A. Geological map with the main structural elements of the entire basin and its environs.
B. Detail of Bourg Basin (the main depocenter), showing an isopachous representation of the halite sequence and its lateral equivalents together with present limits of halite and other evaporites (deduced from well logs and seismic data); the important bore holes employed in this study are also indicated. The Etrez study area is located near the depocenter area of the salt basin.

(conglomerates, sands, and shales) but also including some evaporite horizons (gypsum and anhydrite). The main salt basin is confined between the Limonest horst to the west and the fringing N-S faults which abruptly rise eastward to the pre-Tertiary basement.

LITHOSTRATIGRAPHY OF THE BRESSE SALT BASIN

The best known Paleogene halite sequences are located in the Etrez region (Bourg Basin), where the salt is thickest, due to the number of boreholes

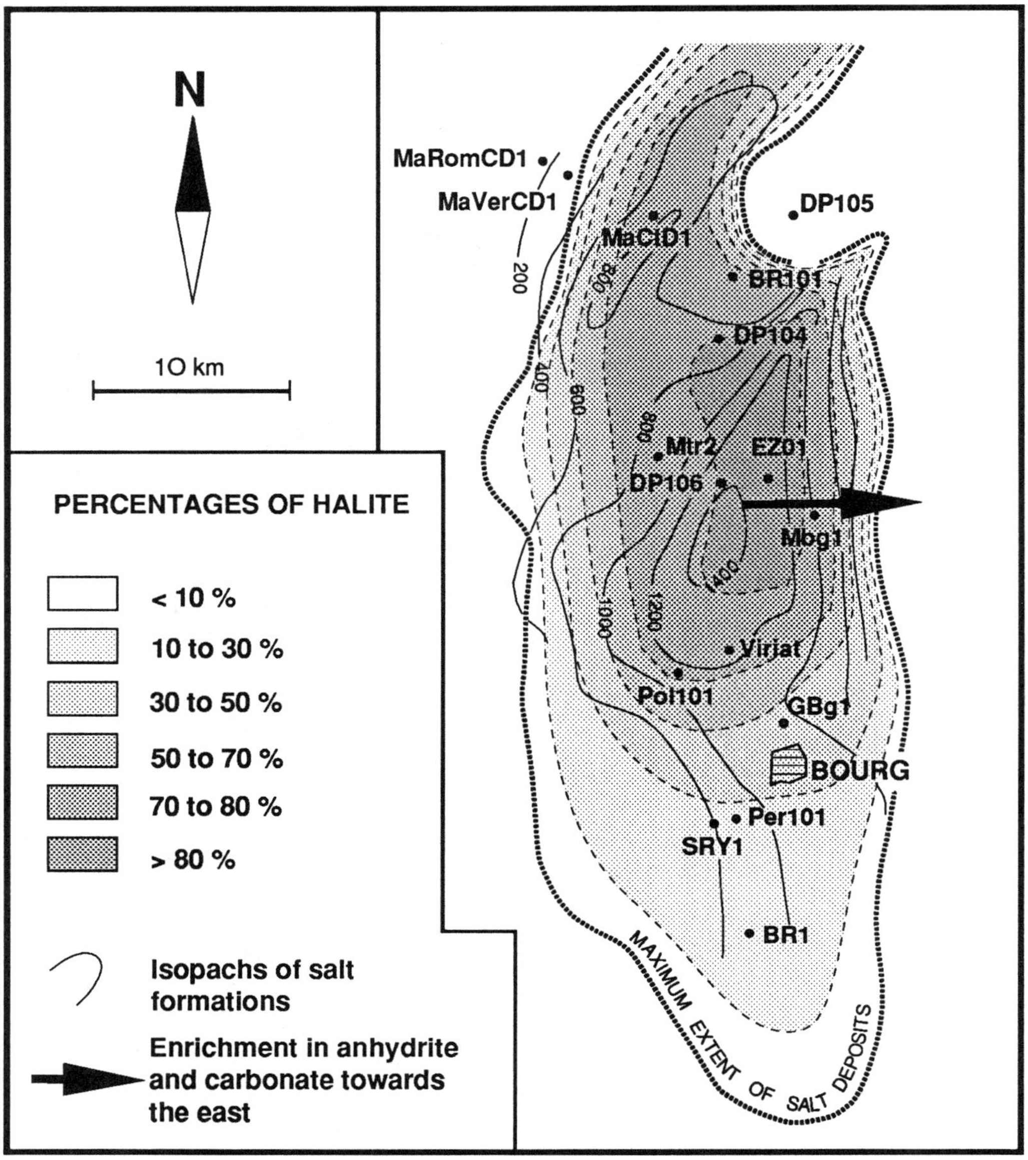

FIG. 4.2. Map showing thickness of the Bresse halite sequence (isopachs deduced from well logs and seismic data) plotted together with the isopercentages of halite (estimated from well logs). Note the enrichment in anhydrite and carbonate to the east evidenced from well logs in the Etrez area. (After Curial 1986/87; seismic data from Leroy 1984; estimation of halite isopercentages partially from Goy and Busson 1980.)

drilled there by Gaz de France for storing gas in salt cavities. The lithostratigraphy of the Bresse salt basin has been defined in this area based on marker horizons present within salt and within overlying deposits. These horizons are clearly recognized on well logs and in seismic profiles (reflectors H1 to H7; figure 4.4). Based on these correlations, the salt deposits have been subdivided into five stratigraphic units referred to as E1 to E5 and together form three main depositional sequences: the Lower Salt Formation$_B$ (Unit E5), the Intermediate Salt Formation$_B$ (Unit E4), and the Upper Salt Formation$_B$ (units E3–E2–E1) (nomenclature from Gaz de France).

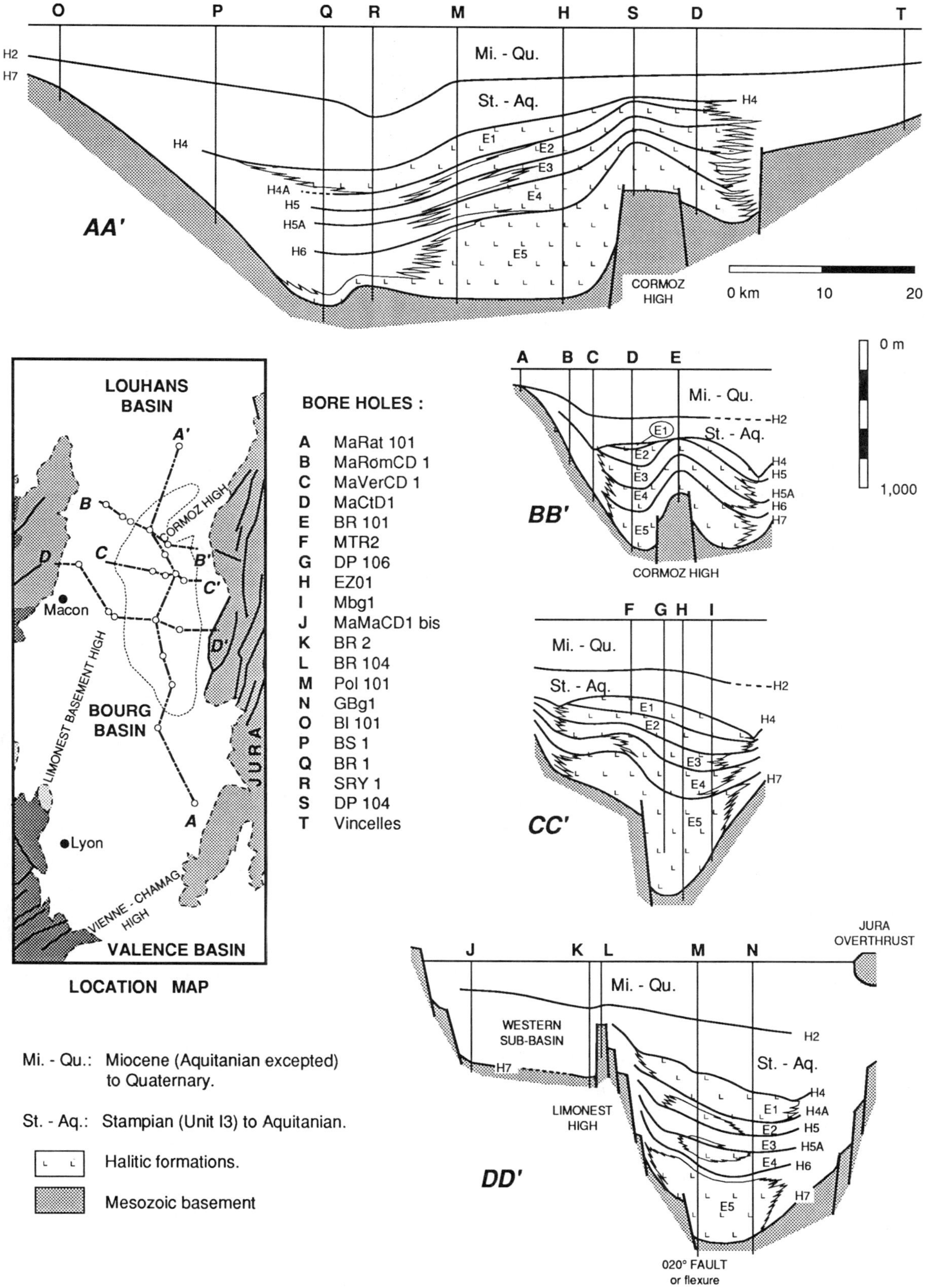

FIG. 4.3. Cross-sections of the Cenozoic sequence in the Bresse area including sections of the Bourg and Louhans basins plus a longitudinal profile with location map (see figure 4.4 for stratigraphic explanation) (adapted from Curial 1986/87).

In the Etrez field, Paleogene deposits lie directly upon a basement of probable early Cretaceous age (Mbg1 well). Immediately above the basement lies the Subsalt Formation$_B$, formed by glauconitic sands and shales enclosing chert, and in turn overlain by the E6 Unit (defined by Goy and Busson 1980), mainly composed of anhydrite and shales.

The Lower Salt Formation$_B$ (Unit E5) is primarily made up of halite affected by strong diagenetic recrystallization. This halite contains variable amounts of nodular anhydrite and clay-carbonate sediments in which interstitial halite growth has strongly disrupted the depositional structures (80% halite). Non-halitic layers are never more than 1 meter thick, except in the lowermost part of Unit E5. According to Triat and Truc (1974) and Trauth et al. (1984), it is highly probable that the Lower Salt Formation$_B$ and the evaporite sediments forming Unit E6 may be Priabonian (Ludian) in age.

The Intermediate Salt Formation$_B$ (Unit E4; 50 to 60% halite) displays two distinct lithologic intervals. The lower section, so-called "Argilites de Base," consists of primary salt, whose structures have been only slightly altered since deposition, and greenish claystone interbeds. The thicknesses of major beds of both salt and claystone range from 1 to 10 meters. The upper part of Unit E4 is composed of alternating milky halite and laminated carbonates, and the bed thickness is similar to that of the "Argilites de Base." Nodular anhydrite recurs throughout the formation in the form of minor intercalations. The occurrence of the pollen species *Boehlensipollis hohli,* which is a reliable indicator of Rupelian age (Sittler and Schuler 1975), has been found in the middle part of Unit E3 and is present continuously until the upper part of Unit E1 (determined by M. Schuler in Moretto 1986–87). Based on this age assignment, the Eocene-Oligocene boundary is believed to correspond approximately to the top of the Intermediate Salt Formation$_B$ (marker horizon H5A).

The Upper Salt Formation$_B$ (units E3–E2–E1) is mainly composed of halite with sporadic claystone horizons, several meters thick, corresponding to seismic reflectors H4A and H5. In Unit E3, halite is mainly of primary origin (halite ~70%), rhythmically interbedded with numerous laminated carbonate beds commonly several tens of centimeters thick. E2 and E1 units are made of secondary salt (halite ~80%), similar to that of Unit E5. Unit E2 contains the lowest amount of porous sediments (carbonates and clays) but it displays a notable increase in nodular anhydrite. Unit E1 includes carbonate and argillaceous horizons of several decimeters and rarely up to 1 meter thick.

The Paleogene cover of the salt formations has been divided into three units (I3 to I1; Curial 1986/87). The first sediments of the Intermediate Marl Formation$_B$ (Unit I3) form a breccia which has been related to salt dissolution. Overlying sediments are constituted of porous and clayey limestones, including sporadic anhydrite beds. In certain areas of the Etrez field, the Upper Evaporite Formation$_B$ (units I2 and I1) displays abundant calcium-sulfate deposits, concentrated more particularly in the "Barre Sulfatée" zone. From correlations made between Bresse and Valence graben, and based on the similarity of the major sedimentary events (Curial et al. 1988), it is thought that the Upper Evaporite Formation$_B$ is of Chattian age.

The "Aquitanian" Limestone Formation$_B$ is formed by chalky limestones enclosing chert and marls rich in sepiolite (Sittler 1965). This formation, which was deposited in a lacustrine environment, closes the Paleogene sedimentary cycle (Lefavrais-Raymond 1962). Above the "Aquitanian" lie marine sands and gravels enclosing argillaceous and limestone beds. These marine deposits have been dated from late Serravallian to early Tortonian (Demarcq et al. 1984). They are, in turn, overlain by lacustrine argillaceous deposits interfingering with sand and gravel horizons (late Tortonian; Demarcq 1970). With a few exceptions, the uppermost part of Miocene (Messinian) is unknown anywhere in the

FIG. 4.4. Cenozoic lithostratigraphy in the Etrez area (true thicknesses not given because of broad variation).

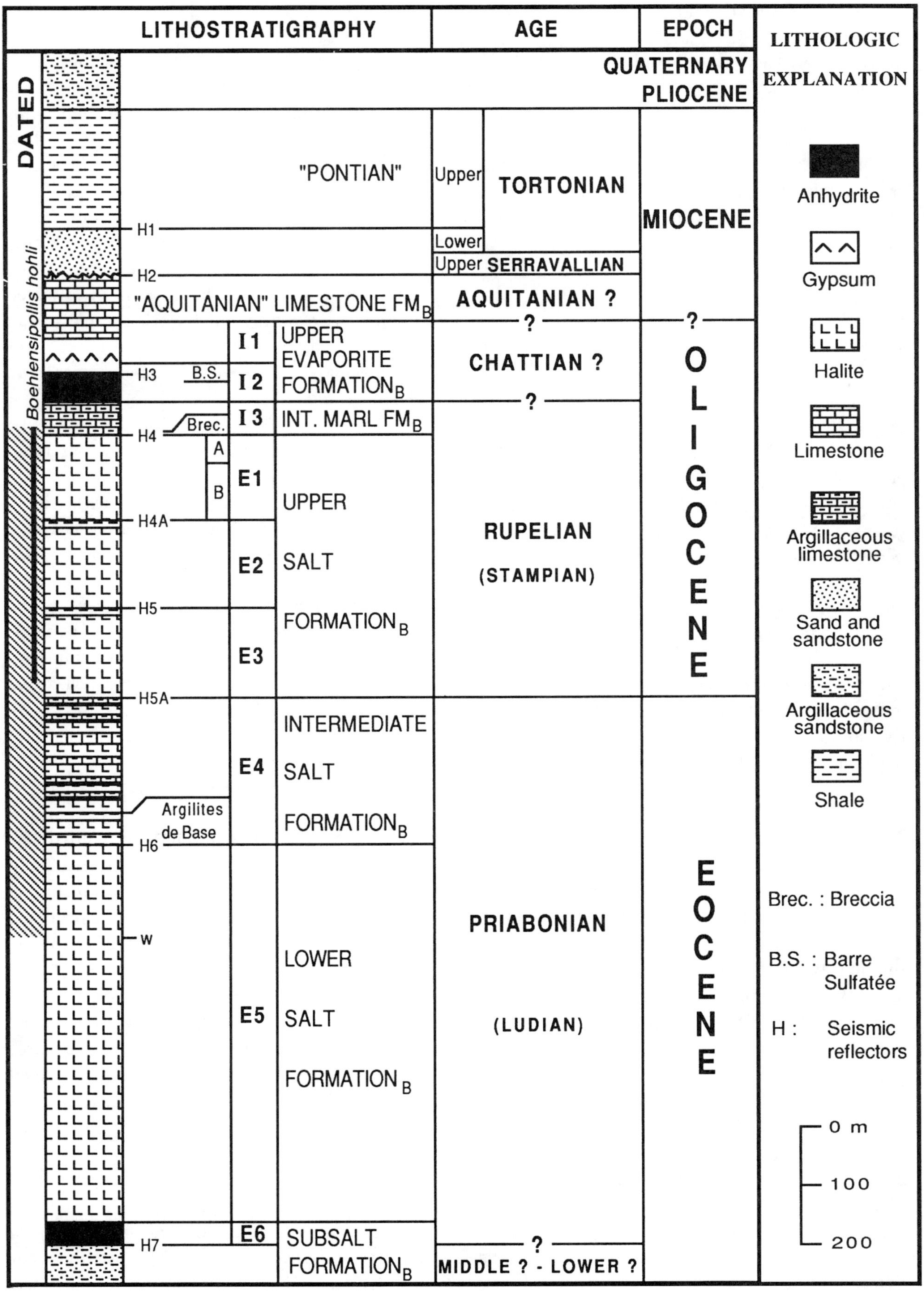
LITHOSTRATIGRAPHY
AGE
EPOCH
LITHOLOGIC EXPLANATION
DATED
Boehlensipollis hohli
QUATERNARY
PLIOCENE
"PONTIAN"
Upper
TORTONIAN
MIOCENE
H1
Lower
H2
Upper SERRAVALLIAN
"AQUITANIAN" LIMESTONE FM B
AQUITANIAN ?
?
?
I 1
UPPER EVAPORITE FORMATION B
CHATTIAN ?
OLIGOCENE
H3
B.S.
I 2
?
I 3
INT. MARL FM B
Brec.
H4
A
B
E1
UPPER SALT FORMATION B
RUPELIAN
(STAMPIAN)
H4A
E2
H5
E3
H5A
E4
INTERMEDIATE SALT FORMATION B
Argilites de Base
H6
E5
LOWER SALT FORMATION B
PRIABONIAN
(LUDIAN)
EOCENE
w
E6
SUBSALT FORMATION B
H7
?
MIDDLE ? - LOWER ?
Anhydrite
Gypsum
Halite
Limestone
Argillaceous limestone
Sand and sandstone
Argillaceous sandstone
Shale
Brec. : Breccia
B.S. : Barre Sulfatée
H : Seismic reflectors
0 m
100
200

FIG. 4.4

Bresse Graben (Lienhardt and Rat 1974). At the end of the Miocene, the Mesozoic section of the Jura chain was overthrust upon the eastern part of the Saône Graben (Michel et al. 1951). The extent of the Jura chain overthrust is estimated to be approximately 10 km. The last Cenozoic sediments are represented by Pliocene and Quaternary clays and sands.

Lithology, Mineralogy, and Well Log Expression

A general presentation of sediment types is given in this section. Most of the observations reported come from boreholes made in the Etrez area (which acted as the main depocenter through a large period of the overall salt sedimentation) and are represented by numerous logs and cores (on the order of 1,000 meters of core). The more marginal areas are only poorly sampled. Many of these boreholes have no cores, or only selective coring, and cuttings of salt are rarely useful in general lithologic studies. Each facies group is first outlined in tabular form and is followed by a discussion of the sediment. For more details about the petrography of the evaporites and associated sediments of the Bresse Basin, see Moretto (1986/87).

TERRIGENOUS DEPOSITS

Terrigenous deposits mainly develop on the basin margins and occur in the depocenter area of the basin (Etrez field) only as claystones, carbonate-sulfate claystones, and rare silty claystones. A listing of the mineralogy of sandstones of the southern part of the Bresse Basin taken from Tchimichkian et al. (1958), is included in table 4.1 and is briefly cited here.

1—*Sands and sandstones.* The clastic component enters the Bresse Basin primarily from the south end of the basin and most of the deposits are confined to the depocenter in its southern portion. These clastic components seem to be derived from the Mesozoic exposures of the surrounding region and are apparently deposited within largely nonmarine conditions.

2—*Claystones.* The best studies of this facies type have been carried out in the deposits of the central area of the basin (Etrez) where there is dense sampling. These claystones often bear plant debris and molluscan shells (gastropods and pelecypods). Commonly the deposits enclose cm-sized skeletal crystals of displacive halite crystals coated with iron oxides, and are mixed together with anhydrite nodules and finely disseminated pyrite.

The clay components have been investigated through X-ray diffractometry of samples from shales as well as the carbonate-sulfate facies (over 80 X-ray analyses). Three clay assemblages have been recognized:

illite/kaolinite/chlorite in the most argillaceous facies (argillite);
illite/mixed layers of illite-smectite/kaolinite/chlorite in the intermediate facies (argillaceous carbonate);
smectite/illite/(kaolinite, chlorite) in the carbonate facies.

Illite is the dominant clay mineral, except for the carbonate layers that contain a substantial percentage of smectite. The kaolinite and chlorite minerals are not abundant, and are associated with layers containing silica and feldspar.

CARBONATE DEPOSITS

In the southern part of the basin there are fossiliferous limestones (with gastropods), limestones rich in continental organic matter, limestones including characea, and sandy limestones. In the central area of the basin the carbonates are composed of calcite, and dolomite occurs in trace amounts only. Dolomite may be more abundant within facies of marginal areas, but the amounts cannot be estimated because there are so few samples available (table 4.2). In general, the carbonates are fine-grained with a grumeleuse texture, and the micrite consists of minute rhomboids arranged in a rosette-like fabric.

The carbonate deposits may be divided into three general sedimentary groupings:

1—*Peloidal coprolitic muds and carbonates with gypsum crystal pseudomorphs.* These usually occur in massive beds. They enclose ghosts of millimeter-

TABLE 4.1.
Main characteristics of detrital deposits found in the Paleogene series of the Bourg Basin.

SEDIMENT TYPE	SEDIMENT COLOR MINERAL COMPONENTS	BEDDING STRUCTURES	SPECIAL FEATURES COMMENTS
Sands and sandstones	Quartz, feldspar, micas, glauconite, rock fragments, vegetal remains	Undetermined	Probably reworked from Mesozoic deposits
Argillites and shales	Dark greenish or brown Illite, kaolinite, smectite, chlorite, mixed layer clays, micrite, halite, vegetal remains	Thin bedded to laminar. Laminar variation may be due to alternations of amount of $CaCO_3$ and organic material	- May have grumeleuse texture with plant debris, mollusca (gastropods, pelecypods) - May contain displacive halite (skeletal)

TABLE 4.2.
Main characteristics of carbonate deposit found in the Paleogene series of the Bourg Basin.

SEDIMENT TYPE	SEDIMENT COLOR MINERAL COMPONENTS	BEDDING STRUCTURES	SPECIAL FEATURES COMMENTS
Carbonate	White, tan, grey to bluish white Calcite with traces of dolomite, halite, anhydrite nodules, disseminated anhydrite, pyrite	1- Peloidal 2- Algal laminae 3- Depositional laminae	1- Grumeleuse structure with coprolites and gypsum crystal ghosts 2- Micritic with small tube-like filaments, variable porosity, layer by layer 3- Very regular millimetric or finer crystals may have alternances of clay and sulfates: with displacive halite.

size gypsum crystals that lie parallel to the bedding. Along with the gypsum pseudomorphs there are particulate elements composed of carbonate muds having a cylindrical shape that sometimes are found as accumulations in small lenses and thin interlayers (figure 4.5A). These cylindrical elements are composed of micrite, are up to 1 millimeter long and 100 microns in diameter, and do not show any internal structures. They have been referred to as the coprolites of insect larvae such as Chironomes (Triat, Truc, and Ortí, pers. com. 1986); supporting evidence comes from the presence of *Culicidae* (mosquito) remains in the salt sequence. Such good preservation of the pellets suggests a low-energy depositional environment with no significant bioturbation.

2—*Rhythmical algal laminae* (centimeter- to decimeter-thick) are made of alternating compact micrite and micrite with a higher porosity, and exhibit slightly corrugated to convoluted structures (figure 4.5B). The high porosity of some of these laminae could be the result of the solution of sulfates or of the alteration of organic matter (figure 4.5C). Such structures resemble those of deposits related to benthic "cyanophycea" layers that exist in modern salt pans (Geisler-Cussey 1985 and 1986). In these layers there are tubular structures, a few micrometers in diameter that may be relics of other organically produced components.

3—*Depositional laminae* result from the alternation of very thin laminae (microns to millimeters thick), flat or corrugated, composed of carbonate, sulfate, and clayey-carbonate enriched in organic matter (figure 4.5D). This alternation produces a mechanical layer-cake appearance in the rock. While the carbonate laminae are almost pure, the sulfate laminae result from the coalescence of very small nodules wrapped in a filamentous network of

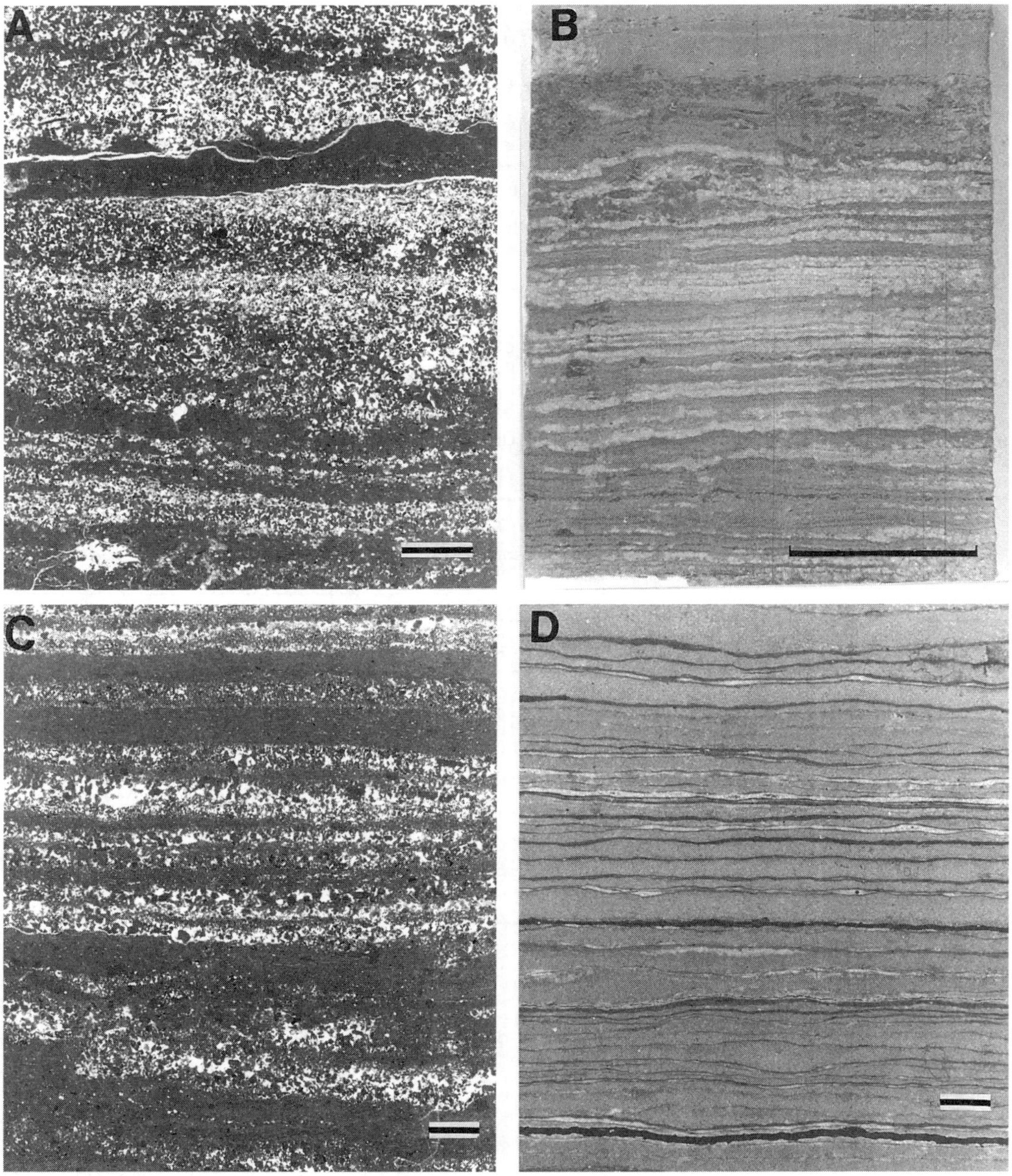

Fig. 4.5. Representative laminar, micritic limestone samples taken from various levels in the section.
A. Photomicrograph showing alternation of micritic laminae with coprolitic-pelletal laminae. The coprolites are nearly uniform, apparently uncompacted, and very porous (in white). Coprolitic laminites are usually restricted to Units E2 and E3. Scale bar, 1 mm (seen in cross-polarized light).
B. Photomicrograph of slabbed core showing rhythmical micritic laminae similar to those of algal mats. Scale bar, 2 cm (seen in cross-polarized light).
C. Photomicrograph showing a detail of laminae with algal structures observed in (A). The laminated aspect is caused by the alternation of compact micritic laminae and laminae with high porosity (in white) probably resulting from sulfate or halite solution. Scale bar, 1 mm (cross-polarized light).
D. Photomicrograph showing the alternation of thin, flat, or corrugated micritic layers and very thin laminae rich in organic matter (in black). This microfacies is common in Units E3 and 4. Scale bar, 1 mm (plane light).

FIG. 4.6. Photomicrograph showing the alternation of laminae composed of carbonate and sulfate. The sulfate-rich layers are made of anhydrite pseudomorphs of lenticular gypsum crystals. These crystals originally grew displacively within algal mats. Note the growth of halite crystals within a few laminae, some of which are in the form of "pagoda" structures with extended, skeletal corners (arrows). Scale bar, 1 mm (plane light).

organic matter. The presence of lenticular crystal morphs suggests that these nodules derive from the alteration of gypsum crystals which developed displacively, presumably within algal mats (figure 4.6).

On the whole, these facies are poor in biological traces and coprolites, except for the presence of many Culicidae remains. The regular rhythmic pattern of the laminar sequences, the absence of bioturbation traces, and the preservation of organic matter (Moretto 1986–87; Curial and Moretto 1988) suggest that deposition took place beneath stratified water bodies under somewhat anoxic conditions (Curial et al. 1990). Fundamental aspects of deposition beneath a system of stratified water layers have been extensively discussed by Busson and Noël (1972) and Busson (1978, 1988).

SULFATE DEPOSITS

Throughout the part of the salt sequence sampled in this study, sulfate facies are rarely pure or massive (table 4.3). The individual layers are no thicker than 20 cm and are made up of beds and lenses of juxtaposed and coalescent nodules. Other parts of the basin, and other parts of the section, apparently have massive gypsum beds; however, these are recognized only through well logs and are not discussed here.

1—*Gypsum.* Evidence of original, primary gypsum has been recognized within the halite sequence, but only through remains of its primary crystal forms as pseudomorphs. This mineral no longer exists but its epigenetic forms are common. In several instances, layers of small halite crystals, packed together and set perpendicularly or slightly obliquely to the bedding plane, resemble the grass-like growth habit of gypsum (figure 4.7). Such layers of crystals suggest an initial growth of gypsum crusts on the depositional floor similar to those encountered in modern salt pans (Geisler-Cussey 1985, 1986) with later halite replacement (Schreiber and Walker 1992; Horovka 1992).

The original presence of gypsum is commonly indicated by the ghosts of lenticular crystals. These crystal pseudomorphs are mm- to cm-sized, and in the case of the largest ones sometimes occur as swallow-tail twin crystals. When these forms are en-

TABLE 4.3.
Main characteristics of sulfate deposits encountered in the Paleogene series of the Bourg Basin.

SEDIMENT TYPE	SEDIMENT COLOR MINERAL COMPONENTS	BEDDING STRUCTURES	SPECIAL FEATURES COMMENTS
Sulfate	Lenticular **gypsum** crystal ghosts	-Disseminated within carbonate - Small beds or lenses - Gypsum crusts	Replacement by halite, dolomite, anhydrite, silica . . .
	Anhydrite		
	1- Nodules, white to bluish white	Disseminated or with mozaic structures, sometimes entherolitic	Replacement of gypsum crystals or direct crystallization Partially replacement of carbonated matrix
	2- Acicular crystals	In nodules and as cement of other sediments	Associated with dolomite, silica, celestite, clear halite . . .
	3- Euhedral crystals and elongate prisms	Disseminated or in patches	
	Celestite	Disseminated or in patches	May result from replacement of calcium sulfate

closed within carbonate sediments, the crystal phantoms are found as lenticular cavities in which epigenesis developed a mixture of anhydrite, dolomite, euhedral quartz crystals, quartz spherulites, and patches of calcite (figure 4.8). The rest of the cavity is infilled by clear halite. The replacement also can be made by a single mineral form, such as anhydrite, halite, or silica as lutecite and/or quartz crystals having a mosaic fabric. When these crystal ghosts are enclosed within clear halite they are detectable only owing to the presence of a film of micritic calcite which outlines the initial crystal. This carbonate film indicates the complete replacement of gypsum (pseudomorphs). The gypsum ghosts are halite matrix supported and do not exhibit any deformation, suggesting replacement before compaction. A number of mineral forms occur as infilling within these structures.

The crystal ghosts are assumed to be the remains of gypsum that formed as synsedimentary displacive crystals in the sediments (precocious formation) according to their shape and distribution; the lenticular habit of the original gypsum is caused by crystal modification due to the presence of traces of organic matter (Cody 1979; Rouchy et al. 1984; Cody and Cody 1988). The replacement of the gypsum is believed to have occurred not long after deposition, thus accounting for the preservation of the primary crystalline forms.

2—*Anhydrite* occurs as white to bluish nodules ranging in size from a few millimeters to centimeters, composed of random acicular crystals grouped in clusters, or as felted needles in fine corrugations. When scattered in the matrix, they can deform it, becoming coalescent and developing a nodular mosaic fabric (figure 4.9A) which obliterates any depositional bedding (Maiklem et al. 1969). In some cases, the interpenetration or coalescence of nodules results in enterolithic structures. For some nodules, the geometric forms follow the original forms of the gypsum precursor (figure 4.9B). For others, the nodular shape plus the presence of ma-

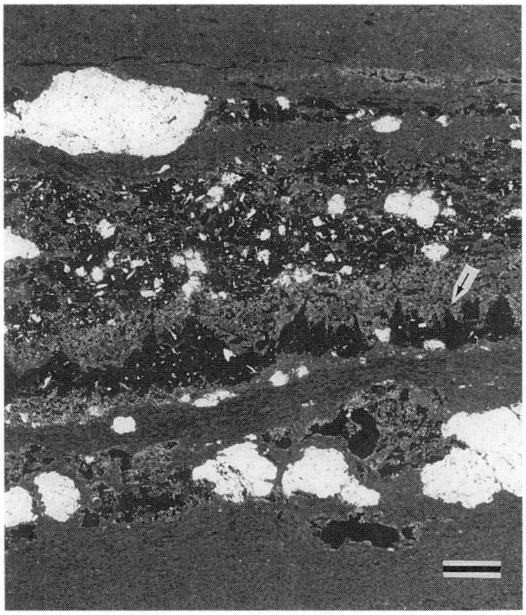

Fig. 4.7. Photomicrograph showing a framework of small halite crystals arranged in palisade fashion and pointing up (arrow). These probably result from a replacement of primary gypsum crust. The under-layering laminae shows nodules of anhydrite (in white) whose growth has distorted the primary structures of the deposit. Scale bar, 1.5 mm (plane light).

trix irregularly incorporated within the crystals may be related to a direct crystallization as anhydrite within the deposits. Massive anhydrite beds are only known from log studies: no cores have been taken, and their texture can only be inferred.

The presence of both a sulfate component in carbonates and also argillaceous carbonates causes increasing lithification. Anhydrite cement occurs as short or acicular crystals (tens of microns), and most are elongate conformably to the bedding plane. Crystals are commonly scattered within the sediment, with a flow-induced appearance (flame-like fabrics) making corrugated structures which possibly resulted from physical processes such as uneven compaction. Crystals that are packed together tightly may form anhydrite and carbonate patches having a nodular appearance and may also obliterate the depositional laminae, resulting in a chicken-wire fabric (nodular mosaic), outlined by the presence of thin streaks of organic matter. Some of the structures resemble those formed by benthic cyanophycea in salt pans (Geisler-Cussey 1985/86). Sulfate diagenesis is variably developed, possibly related to the replacement of carbonates by sulfates. Whether anhydritization operated directly or passed through a gypsum phase is not known. Yet anhydritization seems to have been initiated early, because in some samples it is affected by compaction-related deformation.

Anhydrite as euhedral crystals (up to 1 mm long) and anhydrite as elongate prisms (of a few microns in length) are not abundant. These forms together with minerals such as dolomite, silica, and celestite are encountered within pseudomorphs of gypsum. These replacement minerals are always associated with secondary halite. Sometimes they are included among microcrystals of displacive halite or are inserted between crystals of the clear halite closely associated with milky halite.

3—*Celestite.* Euhedral crystals of celestite, commonly up to 2 mm in size, are present either scattered or clustered in the sediments, making up aggregates with indistinct or thin layers. Celestite is often enclosed within clear halite, together with dolomite, silica (figure 4.10A), or associated with patches of acicular anhydrite crystals (figure 4.10B). In this last case, development of celestite caused the deformation or the fracture of the anhydrite laths. Celestite occasionally is found as crystals elongated parallel to the bedding plane where it pushed aside the matrix within carbonate-sulfate layers. The presence of carbonate, anhydrite, silica, and pyrite inclusions within the celestite crystals indicates a late growth phase for the celestite. This

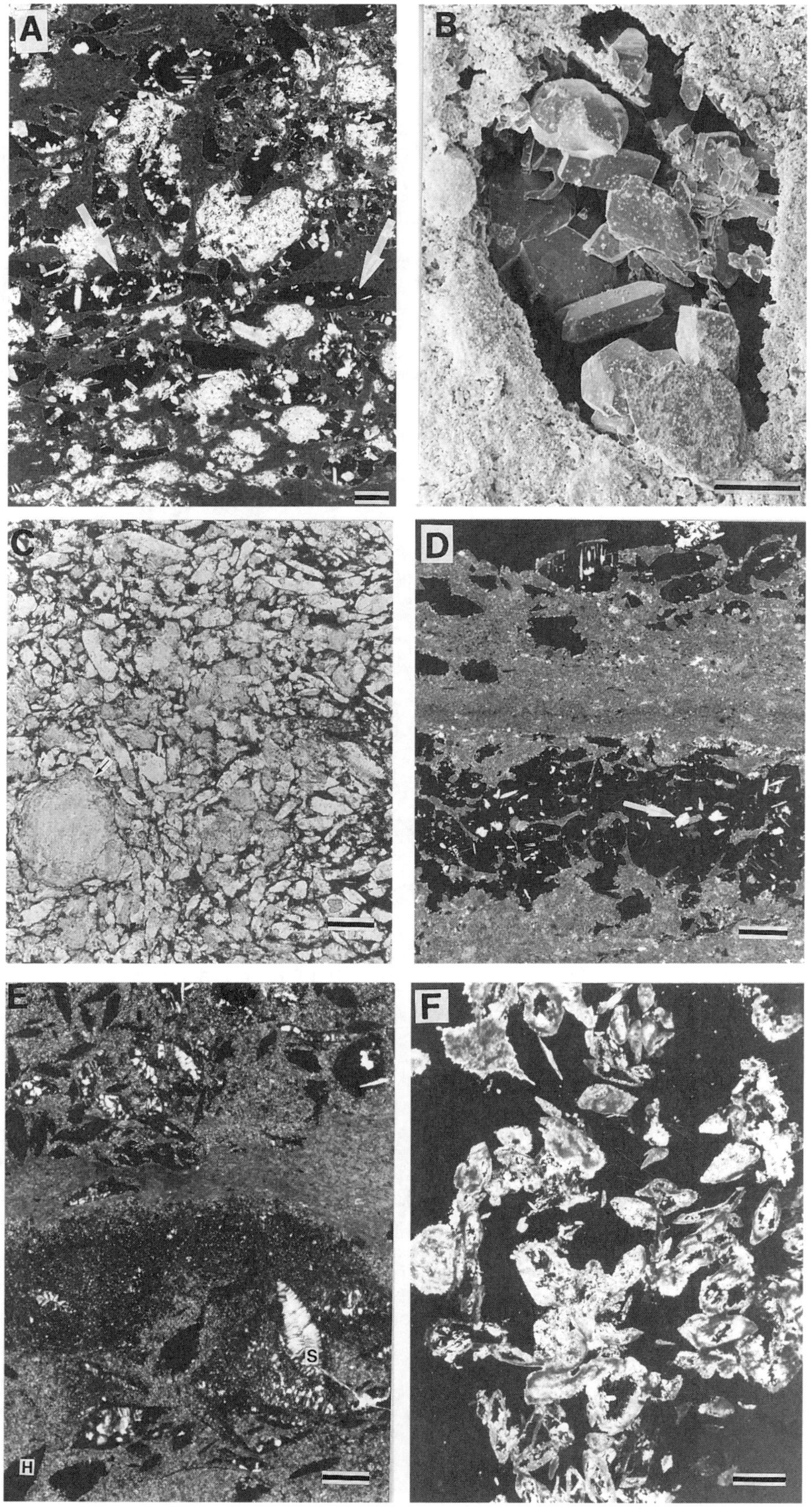

FIG. 4.8

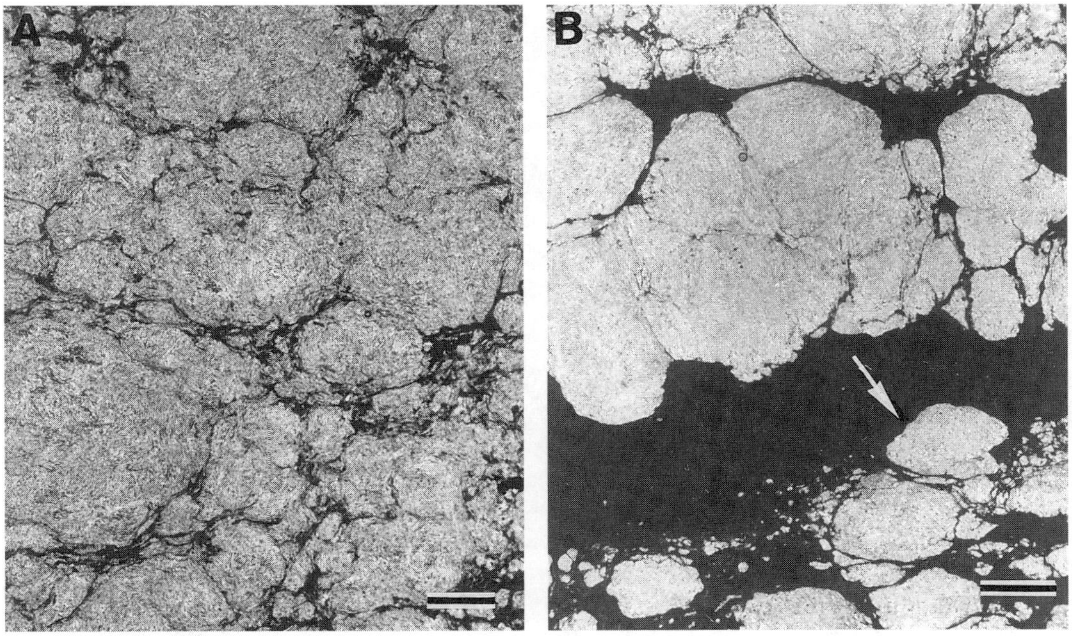

FIG. 4.9. Photomicrographs of anhydrite textures.

A. Numerous nodules of anhydrite of variable size, developed within a mixed carbonate-argillaceous layer, and resulting in a mosaic structure. Nodules are murky due to the incorporation of matrix during their growth and show no geometric shapes that correspond to a gypsum precursor. Scale bar, 2 mm (cross polars).

B. Enterolithic sulfate layers developed within a bedded micritic carbonate zone. The enterolithic layers result from coalescence or interpenetration of anhydrite nodules, but here are themselves the pseudomorphs of lenticular gypsum, whose shapes are still clearly recognizable (arrow). Scale bar, 2 mm (cross polars).

FIG. 4.8. Pseudomorphs of lenticular crystals of gypsum within micritic carbonate:

A. Photomicrograph of gypsum pseudomorphs with original crystal shapes preserved (arrow); the replacement has taken place by development of crystals and nodules of anhydrite, crystals of dolomite, and spherulites of silica. The remaining porosity was infilled with clear halite cement (in black). Scale bar, 400 microns (cross-polarized light).

B. SEM photograph of a "ghost" of a lenticular crystal of gypsum. The cavity is partially occupied by crystals of dolomite and anhydrite. Remaining porosity was infilled with clear halite which was dissolved during sample preparation. Scale bar, 100 microns.

C. Photomicrograph showing pseudomorphic anhydrite replacement of lenticular gypsum that originally grew within a micritic carbonate matrix (in black). Original crystal shape of the gypsum is very well preserved, except in the lower left where an anhydrite nodule has grown (arrow), destroying the earlier morphology. Scale bar, 650 microns (plane light).

D. Photomicrograph of displacive lenticular gypsum formed within micritic carbonate, later replaced by halite (in black). The crystal shapes of gypsum are still obvious in spite of the halite growth. A few laths of anhydrite are included in the halite (arrow). Scale bar, 500 microns (cross polars).

E. Photomicrograph showing replacement of lenticular crystals of gypsum by silica (S) and halite (H). The lutecite fibres (quartz) are developed perpendicular to the outlines of crystals. Scale bar, 600 microns (cross polars).

F. Photomicrograph of nearly perfect pseudomorphs of lenticular gypsum crystals by calcite. The replacement has developed in a centripetal manner. The crystals are included within halitic patches in a carbonate layer. Scale bar, 800 microns (cross polars).

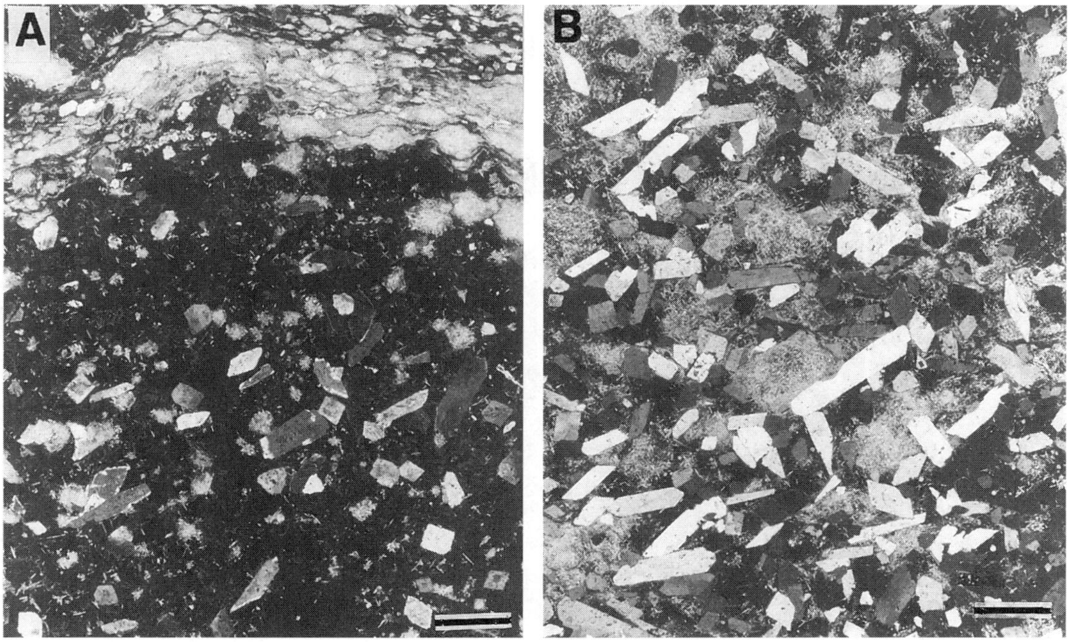

Fig. 4.10. Photomicrographs showing celestite morphology.
A. Crystals of celestite, associated laths of anhydrite, and dolomite crystals randomly located within a halitic zone. In the upper part, within a carbonate-sulfate layer, crystals of celestite are smaller and layered parallel to the stratification, but in the lower portion they are random. Scale bar, 2 mm (plane light).
B. Photomicrograph showing euhedral celestite crystals randomly scattered within an anhydrite-rich area. The growth of the celestite took place later than anhydrite because bundles of acicular crystals of anhydrite are deformed by the celestite crystals. Scale bar, 2 mm (seen in cross-polarized light).

appearance of celestite is possibly related to the replacement of calcium sulfate by strontium sulfate. Such epigenesis has been documented by West et al. (1968). In addition, dolomitization and the bacterially induced reduction of sulfates may also have favored its development (Kinsman 1969; Butler 1970).

HALITE DEPOSITS

1—*Halite.* Two distinct types of halite are found throughout the salt deposits of Bresse: milky and clear (table 4.4). They are differentiated based on the presence or absence of abundant fluid inclusions within the halite crystal network and also based on their general morphology.

a—*Milky halite.* Because of the great abundance of fluid inclusions, the halite rock has a white, cloudy, milky appearance. Inclusions commonly range in size from a few micrometers to about 30 micrometers (up to 100 micrometers). The inclusions are in the form of negative cubes, and they occur in layers parallel to crystal sides, alternating with clear zones devoid of inclusions. Such zonation owes its origin to changes in growth rates related to changes in evaporation rate or in temperature (Shearman 1978; Holser 1979; Roedder 1984). When growing slowly, crystals remain free of any inclusion, whereas when the crystallization rate is high, they develop a milky appearance by enclosing fluid inclusions (Arthurton 1973).

Several types of fabrics have been distinguished throughout the milky halite microfacies. A clearly aligned distribution of the fluid inclusions outlines fabrics of the hopper and chevron type of crystals, already mentioned in numerous studies (Dellwig 1955; Wardlaw and Schwerdtner 1966; Shearman

TABLE 4.4.
Main characteristics of halite deposits found in the Paleogene series of the Bourg Basin.

MINERAL TYPES	SEDIMENT TYPES	SEDIMENT COLOR(S) MINERAL COMPONENTS	BEDDING STRUCTURES	SPECIAL FEATURES COMMENTS
Cloudy halite and clear halite	Cloudy halite salt	White colored, milky	Hopper, raft, and chevron structures. Small bedded layers <15 cm alterning with other sediments	Primary crystallization from a free water body (interface air-water for hoppers and rafts, on the bottom for chevrons). Syntaxial growth and cementation by clear halite. Dissolution surfaces at the top of every sequence.
		Cloudy and clear halite Xenomorphic crystals		
	Displacive halite salt	Greenish to tan	Not very well bedded. Crystalline growths more or less important within diverse sediments	Primary depositional laminae disturbed, deformed by halitic growths. Early diagenetic growths within a soft mud.
		Clear displacive or incorporative halite - euhedral to subhedral crystals - skeletal crystals (depressed side, dendritic and pagodalike growths) Carbonate, sulfate and clays		
	Clear and macrocrystalline salt	Large euhedral clear crystals (up to 10 cm in size)	Intersect the primary halitic sequences	Filling of dissolution cavities.

1970; Ortí 1977a, 1977b; Holdoway 1978; Geisler-Cussey 1985/86).

Hoppers. Hopper-shaped fabrics are of a small size (millimeters to 1.5 cm). They mostly occur as the core of crystals that are enclosed in clear halite (figure 4.11A). In thin section, the hoppers appear as small plates having a square or rectangular base. The occurrence of such small plates is considered by Llewellyn (1968) to have been the result of multistage growth within a film of oversaturated brine. The core of the hopper consists of a clear zone which possibly corresponds either to a crystallization nucleus or to an infill of a hollow body. According to Dellwig (1955) and to Arthurton (1973), hoppers originate at the surface of the brine pool within a film of saturated brine. Initially these nuclei consist of small cubes held at the brine-air interface by surface tension. The hoppers are composed of hollow inverted pyramids whose apexes point downward. When coalescing, such hoppers can form rafts. The halite in the rafts thickens due to continued crystal growth until the crystals become grounded by their own weight.

Rafts made of grouped hoppers recur throughout the salt deposits of Bresse (Moretto 1986/87) (figure 4.11B). The rafts are found as elongated plates about 5 cm long. The deposition of the hoppers on the bottom of the water body does not appear to have caused any disruption to the raft fabrics. The laminae, whether they are rich in fluid inclusions or clear, are laterally constant through plates of adjacent hoppers, showing that the individual hoppers grew simultaneously and at the same rate.

Chevrons. Chevron-shaped fabrics are assumed to have developed on the floor of the brine pool.

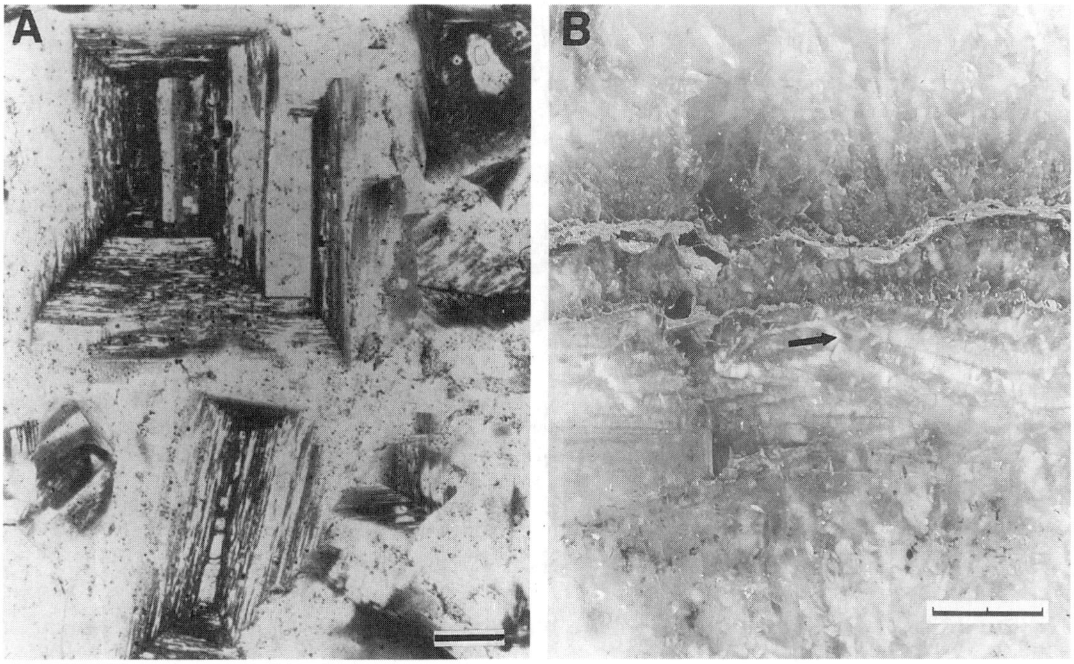

Fig. 4.11. Primary halite textures:
A. Photomicrograph of hoppered halite crystals seen in a "cloudy" salt matrix. Some crystals are approximately cubic, others have more rectangular shapes. Fluid inclusions are very abundant, usually small, and are located along the planes of crystal growth. Scale bar, 2 mm (plane light).
B. Photograph of cut core face showing several halite rafts made up of the coalescence of hoppers. The rafts are piled up and approximately arranged parallel to the stratification. Note the chevron structures (arrow) of later halite overgrowths that are initiated upon rafts. Scale bar, 2 cm (plane light).

Arthurton (1973) mentions that chevron halite, first formed at the brine-air interface, may sink and continue to grow on the floor. Preferential growth develops on crystals in which one corner or one plane (110) points upward (Wardlaw and Schwerdtner 1966; Shearman 1970). When growth develops from a cube resting upon one of its sides it may result in either a hollow pyramid or a cornet (Arthurton 1973; Lowenstein and Hardie 1984; Moretto 1986/87).

Chevron fabrics are well developed throughout the salt sequence of Bresse. They are variable in size (5 mm to 3 cm), and are mostly subvertically or vertically oriented with a corner pointing upward. As in the case of hoppers, fluid-rich chevrons make up the core of the crystals and are commonly encased in clear halite (figure 4.12). Some of the samples show chevrons initiated upon hoppers and rafts, creating mixed lithologies. For both hoppers and chevrons, most authors agree that such fabrics have a primary origin.

Cubes. Large halite crystals (centimeter-size), with a hazy appearance, show fluid-rich planes parallel to the plane referred to as (100). Such crystals have been encountered within solution cavities infilled by clear halite of diagenetic origin.

Similarly, small cubic crystals (0.5 to 1 cm) can develop in a non-halite matrix. Crystal cores are usually milky in appearance and one plane (110) of the cube is often oriented upward. Because of its relation to the matrix and because the deposition of insoluble material occurs simultaneously with the formation of hoppers, it is very unlikely that this form of milky halite is the product of diagenesis.

b—*Clear halite.* This type of halite owes its clearness to the very low number of fluid inclusions

FIG. 4.12. Photomicrograph of chevron growth structures seen in a "cloudy" salt. These stuctures are developed during rapid halite growth by the incorporation of numerous, tiny, fluid inclusions along growth planes. The crystals are oriented with their corners (coigns) upward. Scale bar, 1 mm (plane light).

within it. When inclusions exist, they are larger (mm-sized) than they are in milky halite. Clear halite occurs in several distinct forms.

Cementation halite. Clear halite commonly occurs as cement surrounding hopper- and chevron-shaped fabrics of milky halite (figure 4.13). The passage between the milky core and the overgrowth is usually gradual. Sharp contacts occur only when solution-related structures develop. Indeed, the same clear halite fills solution cavities and at the same time makes up the crystal overgrowth that develops in crystallographic continuity to that of the early milky halite. The crystal continuity is evidenced by the continuance of the same cleavage planes between the two species of halite. Such a pattern indicates syntaxial growth of clear halite upon milky halite. According to Dellwig (1955), these two species are from the same stage of generation, in spite of developing under somewhat different conditions. Shearman (1970) thinks that clear halite is an early internal cement whereas Kühn (1979) thinks that it originates while milky, fluid-rich crystals are still in touch with their mother brines.

A number of clear, xenomorphic, cm-sized halite crystals are associated with milky crystals and with their overgrowth. They appear to be a cement that infilled voids. Is this porosity primary or is it related to solution processes? This clear halite is in contact with milky crystals without any traces of solution, suggesting that the clear halite participated as cementation or void fill and not replacement. According to Wardlaw and Schwerdtner (1966) and Roedder (1984), this clear halite is possibly the result of the recrytallization of milky halite. Burial-related increases in temperature and pressure might be supposed to cause replacement by clear halite, but Roedder (1984) and Hardie et al. (1985) think that all milky halite would necessarily disappear in such a case. Because the milky halite is still present, it is likely that the clear halite is an early first cycle cement and not a recrystallization product.

FIG. 4.13. Photomicrograph of syntaxic overgrowths and cementation of halite hopper stuctures by clear halite. This is the typical appearance of cloudy, inclusion-rich, salt. Scale bar, 3 mm (plane light).

Displacive clear halite is well documented from deposits of various ages and areas (Smith 1971; Holdoway 1978; Handford 1981) and is common throughout the salt sequences of Bresse. It occurs as crystals referred to as "phenoblasts" (or phenotopes) that are millimeters to centimeters in size. The phenoblasts develop within a non-halite matrix (clayey to carbonate) as well as within halite ones (between the crystals of milky halite).

Euhedral to subhedral crystals are the most frequent forms of displacive halite in the Bresse Basin. They sometimes have a corner or a plane (110) pointing upward and commonly trap impurities during crystallization. The relative amount of halite in the matrix is variable. In some cases there is very little matrix present and halite crystals are xenomorphic, but vertical elongation is noted and may be the result of primary competitive precipitation.

In some cases, skeletal crystals of clear halite (edges up to 4 or 5 cm long) have been encountered within dominantly claystone lithologies. They present an overgrowth on the edges of crystals due to differential growth. Such a mode of growth can also result in syntaxial accretion upon milky halite-bearing levels. In other cases, skeletal crystals occur mostly within a matrix of carbonate-dominant lithology. They range from 1 to 3 cm in size (figure 4.14A). Soft sediments were incorporated along the crytallographic planes, parallel to the faces of the cubes. Such developments have been described by Gornitz and Schreiber (1981) from the deposits of the Dead Sea. Skeletal crystals can also present dendritic and pagoda-like growths formed from a cube, millimeter sized, that acted as a nucleus, and whose one plane (110) pointed upward (figure 4.14B) (Southgate 1982).

Clear halite as macrocrystals. In the Bresse Basin, halite also occurs as very large, euhedral, clear, clean crystals, up to 10 cm in size. It infills solution cavities of variable sizes and contains few, if any, fluid inclusions.

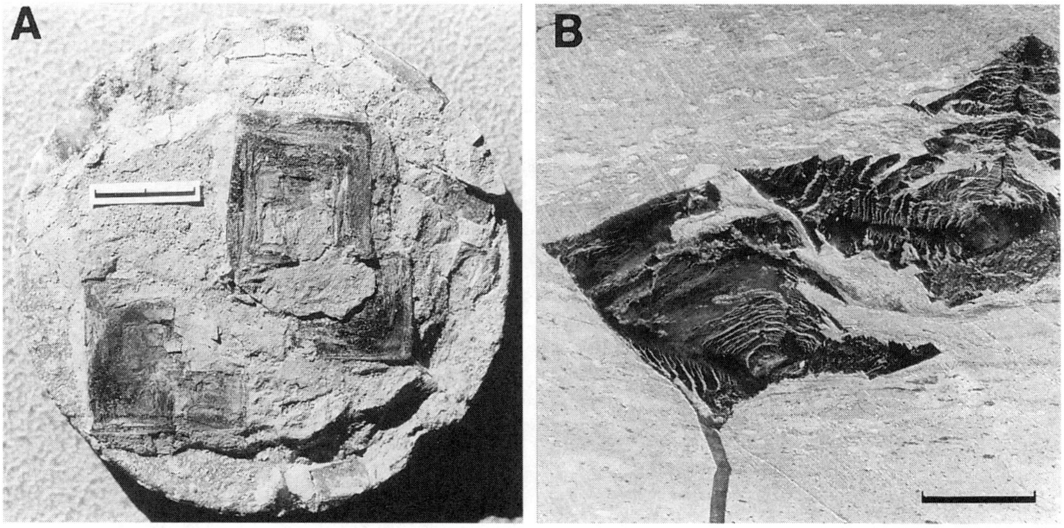

Fig. 4.14. Halite textures in core slabs.
A. Bedding surface showing skeletal crystals of halite with an incomplete growth of crystalline faces with matrix inclusion between the halitic planes of growth. Scale bar, 2 cm (plane light).
B. Vertical section showing pagoda halitic growth within a mixed carbonate-sulfate sediment. The pagoda growth is particularly well developed in the crystal at the right. Scale bar, 2 cm (plane light).

2 — SALT FACIES

a—*Facies of initial precipitation: milky halite deposits* (table 4.4). Milky halite deposits are composed of hopper- and chevron-shaped (associated or exclusive) crystals, together with a variable amount of clear halite (figure 4.15A). The milky halite has an incipient cloudy to white appearance according to the relative volume of fluid inclusions. Such halite rock consists of a mosaic of xenomorphic crystals arranged in an interlocking fabric, and whose outlines are clearer than they are for the chevron-shaped crystals. Whereas chevron-shaped crystals that grow on the depositional floor are mostly vertically oriented, hopper-shaped crystals which apparently sink from the air-brine interface are randomly oriented ("cumulates" of Hardie et al. 1985). The syntaxial, clear halite encloses any fluid inclusion bearing fabric, but is best developed around the chevron-shaped crystals. The surviving porosity has also been filled by clear halite. Milky halite deposits are found as individual, well-bedded levels less than 15 cm thick. Some layers consist of a single horizon made of chevron-shaped crystals whose size coincides with the thickness of the horizons. Hopper-shaped crystals are sometimes size-graded, usually coarsening upward, which may reflect either an increasing rate of evaporation or increasingly effective concentration of the brine. Within a single bed, a hopper-rich zone can grade upwards into a chevron-rich zone. In contrast, sharp contacts may exist between the two types of fabrics, suggesting a hiatus or marked change in deposition. A sharp contact between a chevron-rich zone and a hopper-rich zone is usually outlined by solution traces (figure 4.15B).

All the layers made up of milky halite deposits in this study were affected, more or less deeply, by solution processes which altered primary structures. The voids so produced were filled by non-halitic sediments and by clear halite.

b—*Facies of secondary precipitation: displacive halite within carbonates, sulfates, and clays.* This salt facies consists of an association of subhedral to eu-

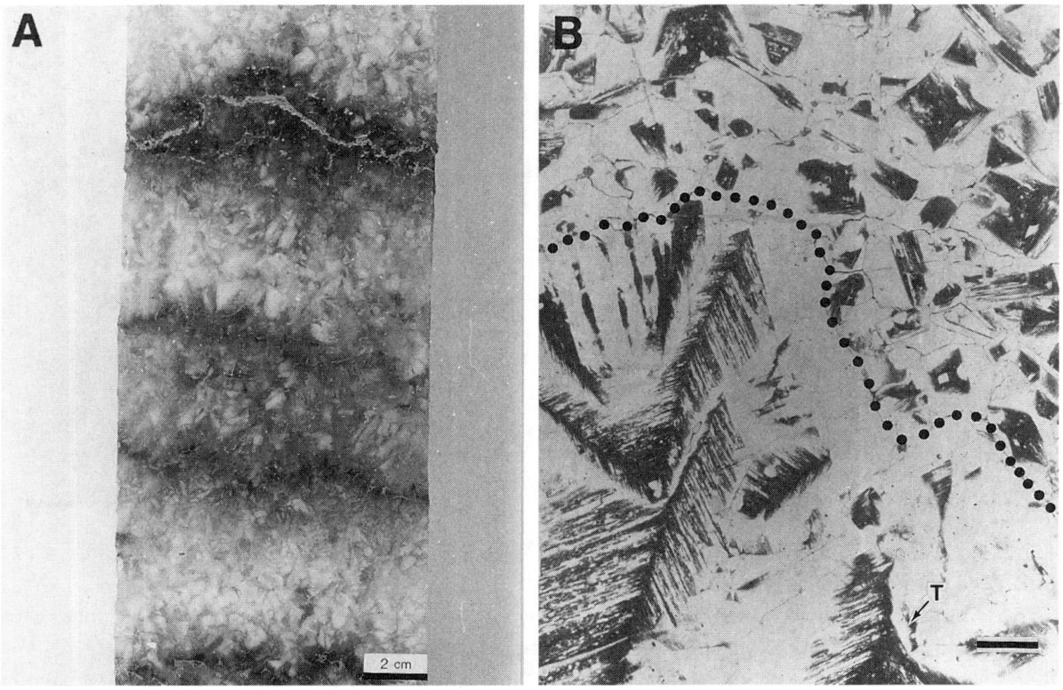

Fig. 4.15. Primary structures in halite.
A. Slabbed core of halite with hopper and chevron structures within primary cloudy salt. The salt layers are separated by small carbonate-rich beds (in black). Scale bar, 2 cm.
B. Photomicrograph of a salt layer with chevron structures overlain by a salt layer with hopper structures. The contact between these two beds is distinct (black circles) and corresponds to a dissolution surface made visible by the truncation of the top of the chevron structures (T). The dissolution results from a very small variation in salinity. Scale bar, 1 mm.

hedral crystals of clear halite, 60% to 90% of the bulk, mixed with sediments of varied natures (figure 4.16). In contrast to the milky halite, this facies does not show clear bedding but rather forms horizons, up to 1 m thick, containing fine claystone, carbonate, and sulfate lithologies.

During their growth, some crystals apparently pushed aside the surrounding host sediment and the resulting crystals are clear. Others incorporated these sediments within the crystal network, and these crystals became brownish to greenish. Depending on the volume of crystal growth and the number of crystals, the matrix is variable deformed and occurs as patches and narrow plates in an intercrystalline position. When coalescent, several single crystals make an interlocking fabric, vertically elongate, which may be the result of competitive growth (Holdoway 1978). Within most patches, clearly visible primary depositional laminae remain, permitting the recognition of early deformation due to the halite accretion. Bits of non-halitic beds show fractures, slices, and displacement of the matrix. Intercrystal patches sometimes no longer have any clear primary texture. In such cases, the crystal outlines are hazy, producing a texture referred to as "chaotic, mottled" by Holdoway (1978), who thinks it is the result of multistage solution. Smith (1971) uses the term "haloturbation" for this feature.

Levels bearing salt with phenoblastic halite sometimes show a gradual upward increase in the total halite volume, concomitant with a decrease of the intercrystal patches. Such salt may occur generally in this basin as clear crystals, vertically elongate, in an interlocking puzzle-like fabric.

FIG. 4.16. Photograph of cut core showing subhedral to euhedral crystals of halite and interstitial patches of carbonate-sulfate matrix in a phenoblastic salt. Some phenoblasts show growth edges oriented up (arrow). Scale bar, 2 cm.

PARTICULAR DIAGENETIC OCCURRENCES

1—*Dolomite* is not common in the central part of the basin. When present, crystals are variable in size, up to 700 microns (figure 4.17A). They have perfect crystal faces and their brilliance is similar to that of quartz. Dolomite is commonly associated with clear halite, as infill of small solution cavities (in the uppermost part of the levels of milky halite), as well as in the layers containing pseudomorphs of gypsum crystals. Its pervasive association with diagenetic halite permits the assumption of a comparable origin (figure 4.17B). The appearance and the distribution of these dolomite crystals resemble those described by Naiman et al. (1983), who suggest a diagenetic origin.

2—*Silica and silica epigenesis.* Silicification primarily affected sulfates, and more rarely carbonates, and apparently developed independently from any specific mineral phase. Silicification is most frequent in the evaporitic deposits (Chowns and Elkins 1974; Tucker 1976; Milliken 1979; Arbey 1980; Rouchy et al. 1984).

Silicification affected both anhydrite and gypsum, but to different degrees. Silicification of anhydrite nodules (figure 4.18A) developed as spherulites of microquartz, 400 to 500 microns in size, and whose core may have been made of opal. The replacement transformation sometimes began with a mosaic of isometric quartz crystals (around 400 microns) composing the core, whereas the cortex was made up of either smaller crystals (around 50 microns) or of lutecite fibers whose growth is marked by films of impurities. When silicification developed at the boundary of the anhydrite nodules, it incorporated a part of the matrix. In this case, when micrite is present, a carbonate rim is preserved from which quartz crystals developed, with a palisade arrangement. Epigenesis also occurred as isometric, petal-like quartz and as euhedral crystals. The transformations are partial or total and commonly centripetally developed.

Replacements rarely affected gypsum crystal

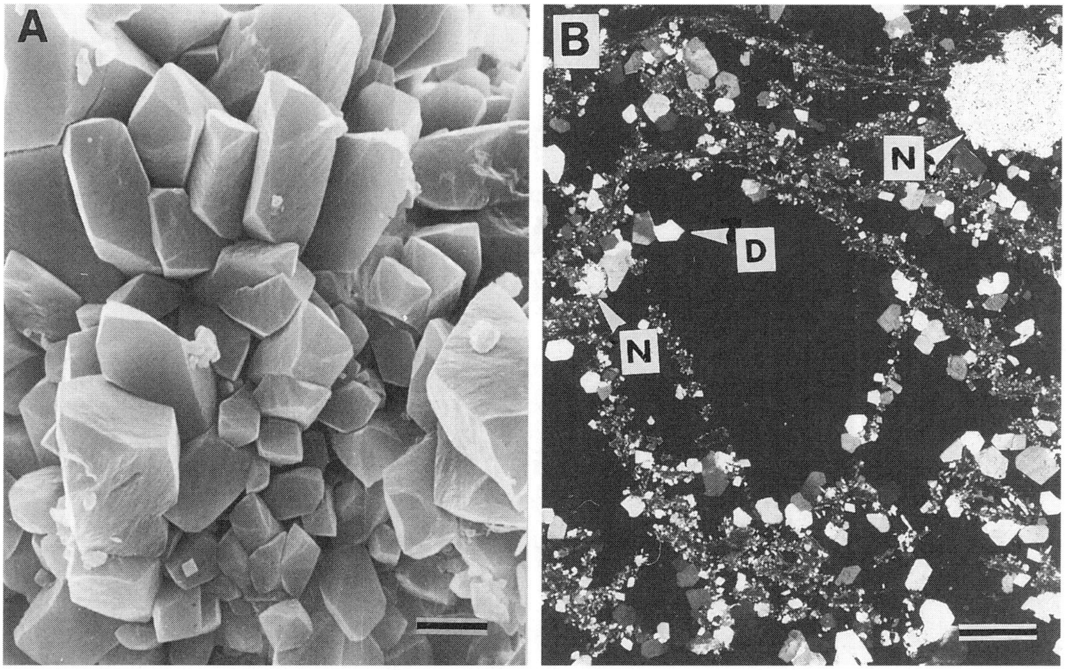

Fig. 4.17. Photomicrographs of various crystal morphologies of dolomite.
A. Dolomite observed with SEM at the contact with a micritic lamina. The crystals float within a clear halite zone. Scale bar, 4 microns.
B. Crystals of dolomite (D) within clear halite (in black), between laminae composed of carbonate and organic matter. Nodules of anhydrite (N) and a few crystals of celestite and anhydrite are associated with them. Scale bar, 700 microns (plane light).

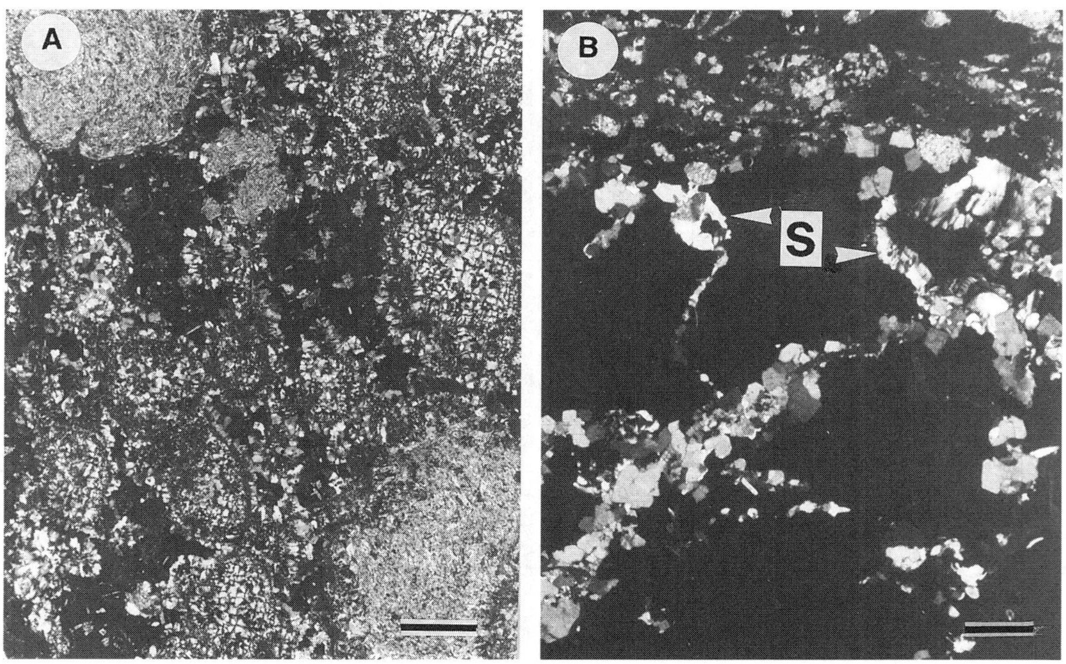

Fig. 4.18. Silicification (seen with crossed polars): A. Photomicrograph of anhydrite nodules showing different stages of silicification. Sample is located within a halite layer in a carbonate matrix containing numerous "ghosts" of lenticular gypsum crystals. Scale bar, 900 microns. B. Photomicrograph showing spherulites of silica (S) in an intercrystalline position at the top of a cloudy salt layer. The salt layer is overlain by a carbonate bed with pseudomorphs of lenticular crystals of gypsum. Scale bar, 500 microns.

pseudomorphs in the deposits of the Bresse basin. When present, the epigenesis developed through the formation of lutecite bundles or of mosaic quartz that crystallized perpendicularly to the outlines of the crystal. It is apparent that epigenetic processes operated directly on gypsum, prior to any anhydritic transformation or any bacterially induced reduction. Silica also is found as spherulites and as patches of mosaic quartz, together with clear halite, in an intercrystal position or touching the phenoblasts (figure 4.18B).

Silica possibly derived from the solution of tests of associated siliceous microorganisms such as diatoms (Gouleau 1982; Gouleau and Noël 1983) or from quartz and clay minerals (Holdoway 1978). But it is most likely that the bacterially induced reduction of sulfates was a major factor in the precipitation of silica. Indeed, the bacterial reduction caused the release of organic acids H_2S and CO_2, which triggered the precipitation of silica via the subsequent changes in pH (Arbey 1980).

EVAPORITE RECOGNITION USING WELL LOGS

Well log data has proved helpful in the recognition of the stratigraphy and mineralogy of the evaporite formations of the Bresse basin. The detailed geometry of the deposits has been reconstructed based on both seismic lines and well log correlations, permitting observation of the results of particular phenomena such as widespread salt dissolution and lateral changes in lithology. Well logs are the single continuously recorded source of information and are perfectly related to depth: cores are taken only through some short intervals and cuttings do not have sufficient resolution to provide reliable recognition of the details of the vertical lithology. So it has been fundamental to correlate existing cores and well logs in order to recognize characteristic patterns and then extrapolate this knowledge to uncored intervals (Curial 1986/87).

The well logs used and their main points of interest are given in table 4.5. More detailed information concerning the function of the tools and the application of well log data has been already published (Serra 1979, 1985; Doveton 1986; Rider 1986). The application of various types of well logs in the interpretation of evaporite sequences has been presented by Nurmi (1978, 1988), Serra (1980, 1986), Serra and Curial (1985), and Curial (1986/87, 1988). Our short discussion concerns the recognition of halite facies and structures through dipmeter curves, the mineralogical determination using various well logs, and particularly Pef (photoelectric factor) measurements. Other applications of well logs for recognition of the evaporite formations of the Bresse basin are exemplified in case studies in other sections.

1—*Recognition of halite facies and structures through dipmeter curves.* Within the salt deposits of the Bresse area, two basic salt types can be distinguished according to the development of diagenetic alteration of the original depositional structures. These are: the syngenetic (primary) salt, slightly to moderately altered since its deposition on the bottom of the brine pool; and the diagenetic (secondary) salt in which the depositional structures have been strongly disrupted or destroyed by intrasedimentary halite-crystal growth. These two fundamental macrofacies can also be distinguished by dipmeter curves, mainly because of the relative distribution of porous material and the great electrical contrast between compact, resistive components (halite, anhydrite, gypsum) and conducive constituents (clays). Conductivity is mainly related to the water present within the rock pores. The variably conducive rocks are claystones and porous matrix lithologies (carbonates, sands, and sandstones).

Preservation of original bedding is a reliable indicator of primary halite deposits. Owing to their high vertical resolution, the dipmeter curves reflect the bedded appearance of primary salt horizons and variations in the depositional rhythmicity with great clarity (figure 4.19A). On the other hand, diagenetic salt deposits are recognized by their mixed or variable resistivity signals resulting from their heterogeneity and their rubbly character (figure 4.19B). They differ strongly from syngenetic salt in that resistivity contrasts cannot be correlated between all curves. Different types of secondary halite can be distinguished by the examination of the resistivity curves, depending on the abundance of po-

TABLE 4.5.

Main well logs used in this study with principal applications to the recognition of the evaporite series.

LOGS	MEASUREMENTS	MAIN APPLICATIONS	REFERENCES
Gamma ray	Total natural radioactivity.	Recognition of radioactive lithologies such as shales, potassium minerals, etc.	In review: see references in text
Spectrometry of the natural radioactivity	Evaluation of U (ppm), Th (ppm), and K (%) amount by analysis of the energy spectrum of the natural gamma ray.	Specific recognition of radioactive minerals. K : potassium feldspar, potassium evaporites . .. Th : various minerals mainly related to lithologies of detrital origin. U : may be associated with organic matter, phosphate deposits... Shales are recognized by a simultaneous increase in Th, K and U amounts.	Serra *et al.* (1980). Quirein *et al.* (1982). See Hassan & Hossin (1975) and Quinif *et al.* (1982) for the Th-K-(U) amounts related to clay species.
Neutron	Hydrogen index (commonly, logs are scaled in porosity units).	Evaluation of the matrix porosity and lithology recognition : hydrated minerals such as gypsum, polyhalite ...and the shales rich in water show high neutron porosity values. Very compact, non-porous minerals (anhydrite, halite) may have a negative HI.	CNL: Schlumberger (1987), in review.
Sonic	Sonic transit time (µsec/foot).	This measurement is influenced by the porosity, the lithology and the structure of the formations.	In review: see references en text
Density	Bulk density of the formations (g/cm3).	Evaluation of the matrix porosity and lithology recognition (e.g.: halite, 2.04 g/cm3; anhydrite, 2.98 g/cm3; celestite, 3.79 g/cm3).	FDC: Wahl *et al.* (1964). In review: see references in text.
Photoelectric factor (Pef)	Photoelectric absorption cross-section (barns/electron).	Pef measurement is barely affected by porosity and it provides additional helpful information for the recognition of the lithology. Minerals containing heavy atoms (Sr, Ba...)may be easily detected (celestite, barite...).	Gardner & Dumanoir (1981).
Dipmeter curves	Microresistivity	Because they are recorded with a high vertical frequency (every 2.5 to 5 mm), dipmeter curves give precise information about rock fabrics, texture and structure. These data are later processed to give dip angles and azimuths.	In review: see references in text. HDT, SHDT : review in Schlumberger (1987).

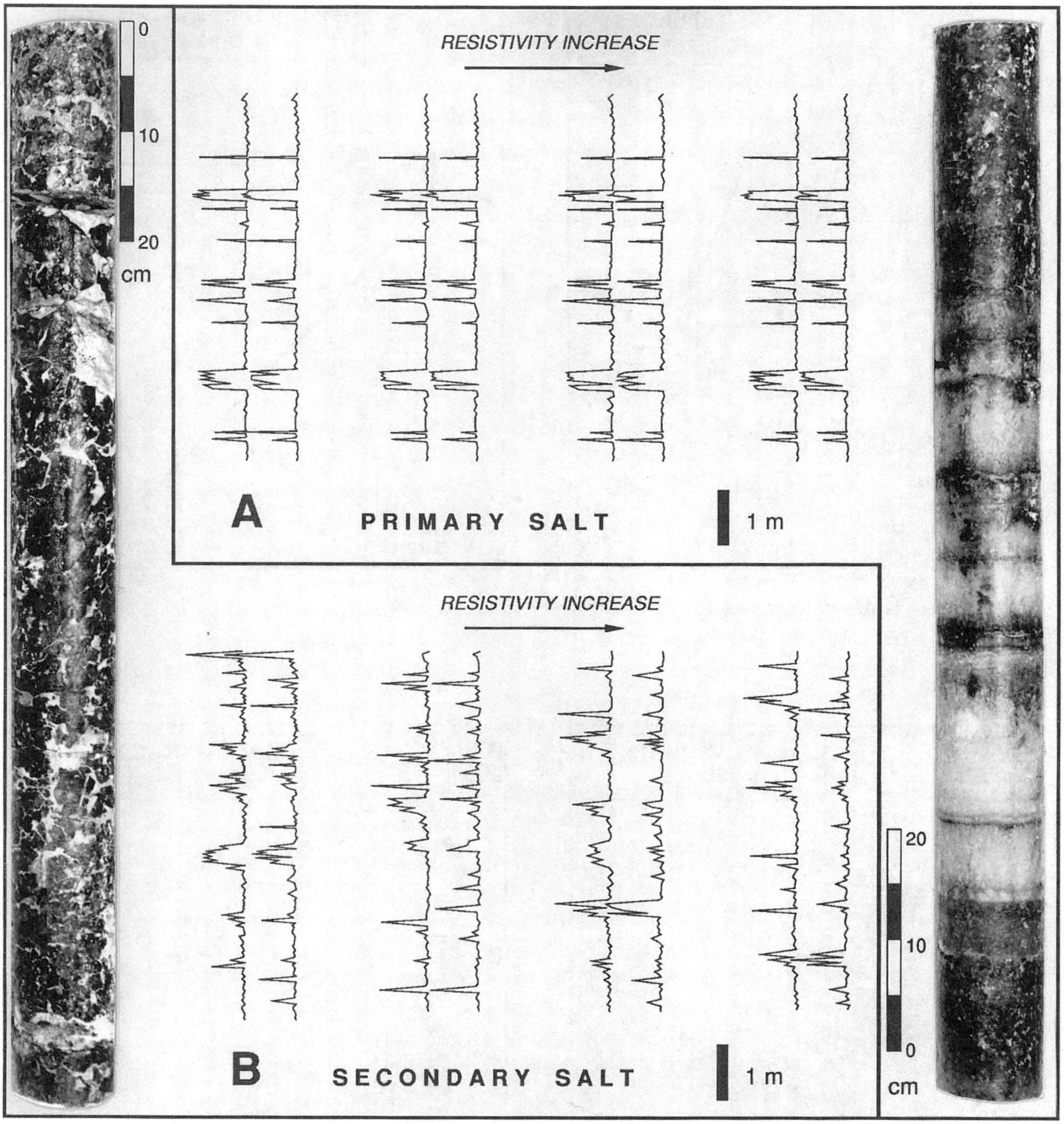

Fig. 4.19. Examples of microresistivity curves recorded by dipmeter (SHDT of Schlumberger) within a primary halite sequence (A) and within secondary halite (B).

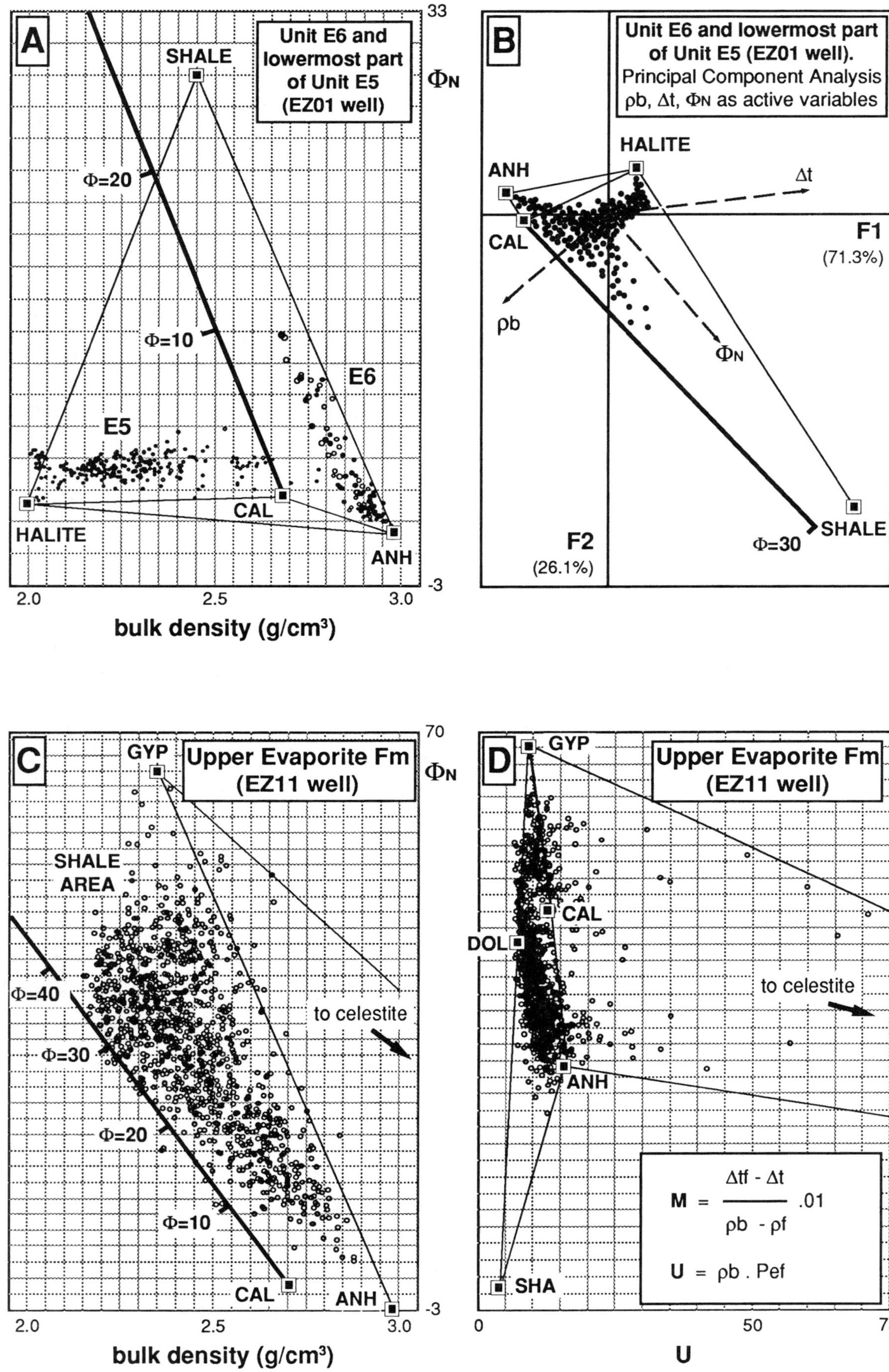
A
Unit E6 and lowermost part of Unit E5 (EZ01 well)
SHALE
Φ=20
Φ=10
E6
E5
HALITE
CAL
ANH
33
ΦN
-3
2.0
2.5
3.0
bulk density (g/cm³)
B
Unit E6 and lowermost part of Unit E5 (EZ01 well).
Principal Component Analysis
ρb, Δt, ΦN as active variables
HALITE
ANH
Δt
CAL
F1
(71.3%)
ρb
ΦN
F2
(26.1%)
Φ=30
SHALE
C
Upper Evaporite Fm
(EZ11 well)
GYP
70
ΦN
SHALE
AREA
Φ=40
Φ=30
Φ=20
Φ=10
to celestite
CAL
ANH
-3
2.0
2.5
3.0
bulk density (g/cm³)
D
Upper Evaporite Fm
(EZ11 well)
GYP
M
1
CAL
DOL
to celestite
ANH
M = (Δtf - Δt)/(ρb - ρf) .01
U = ρb . Pef
.5
SHA
0
50
75
U

Fig. 4.20

rous impurities localized along intercrystal planes (Curial 1988). The primary or secondary origin of salt deposits thus may be deduced from dipmeter curves through the uncored intervals.

2—*Recognition of lithology using the cross-plot method.* In Unit E6 and the lower part of Unit E5 (EZ 01 well, Etrez area) (figure 4.20) the neutron porosity vs. density cross-plot demonstrates the co-occurrence of anhydrite (which is dominant), low porosity limestones, and shales (the shale point is defined by comparison with other plots) forming Unit E6. It also indicates the association of halite with low porosity limestones, probable anhydrite, and small amounts of shales in the lower part of Unit E5. A principal component analysis follows a synthetic approach using the neutron porosity, density, and sonic logs as active variables (figure 4.20B). The second example concerns the Upper Evaporite Formation$_B$ (EZ11 well, Etrez area), through a neutron porosity vs. density cross-plot (figure 4.20C) and a U factor vs. M cross-plot (figure 4.20D). The co-occurrence of anhydrite, gypsum, porous limestone, celestite, and small amounts of clays is clearly evidenced. The Pef log (through the U factor) is useful for the detection of enrichment in minerals that bear heavy atoms such as celestite (Sr) (figure 4.20D).

Some techniques of lithologic recognition using cross-plot methods have been reviewed in Augier (1980) and Fertl (1981).

Geochemistry

BROMINE GEOCHEMISTRY

In a geochemical profile made of the upper 250 meters of Unit E5 to the top of Unit E1, the bromine distribution shows a large range of variability (figure 4.21) (Moretto 1986/87). In the upper 250 meters of Unit E5 bromine varies between 30 and 125 ppm bromine with a very small but general increase from the base to the top of the section. The values obtained in the beds of diagenetic halite are not very different from those obtained in primary halite.

Bromine concentration progressively increases from the base of Unit E4 to the base of Unit E3. The values range from 100 to 180 ppm, with some significant fluctuations, and are characterized by facies made up of primary salt, rich in fluid inclusions, and an early diagenetic salt, with limpid and macrocrystalline halite.

In the lower part of Unit E3, bromine concentrations are about 80 ppm and continue to decrease progressively until the values reach a low in the teens ppm at the top of Unit E1. At the bottom of Unit E1, the bromine content in halite rich in fluid inclusions is generally greater than that obtained in clear halite. The results obtained from a few samples after elimination of brines trapped in the fluid inclusions show that the bromine content in cloudy halite is comparable to that obtained in clear ha-

FIG. 4.20. Examples of graphic visualisation of well log data.

A. Cross-plot density versus neutron-porosity for Unit E6 (anhydrite dominant with thin levels of shales) and the lowermost part of Unit E5 (halite with beds of anhydrite, low porous limestone; small amounts of clay) crossed by EZ01 well (Etrez area).

B. Principal Component Analysis for same interval using density, velocity, and neutron-porosity logs.

C. Cross-plot density versus neutron-porosity for the Upper Evaporite Formation$_B$ (EZ11 well, Etrez area). Note the high porosity of limestone on the line drawn from the pure, compact calcite point and the co-occurrence of anhydrite and gypsum ("Barre Sulfatée" member).

D. Cross-plot U factor (computed from density and Pef logs) versus M parameter (computed from density and interval transit time logs) for interval in (C). Presence of celestite is clearly evidenced by a drift of dots toward high U factor values.

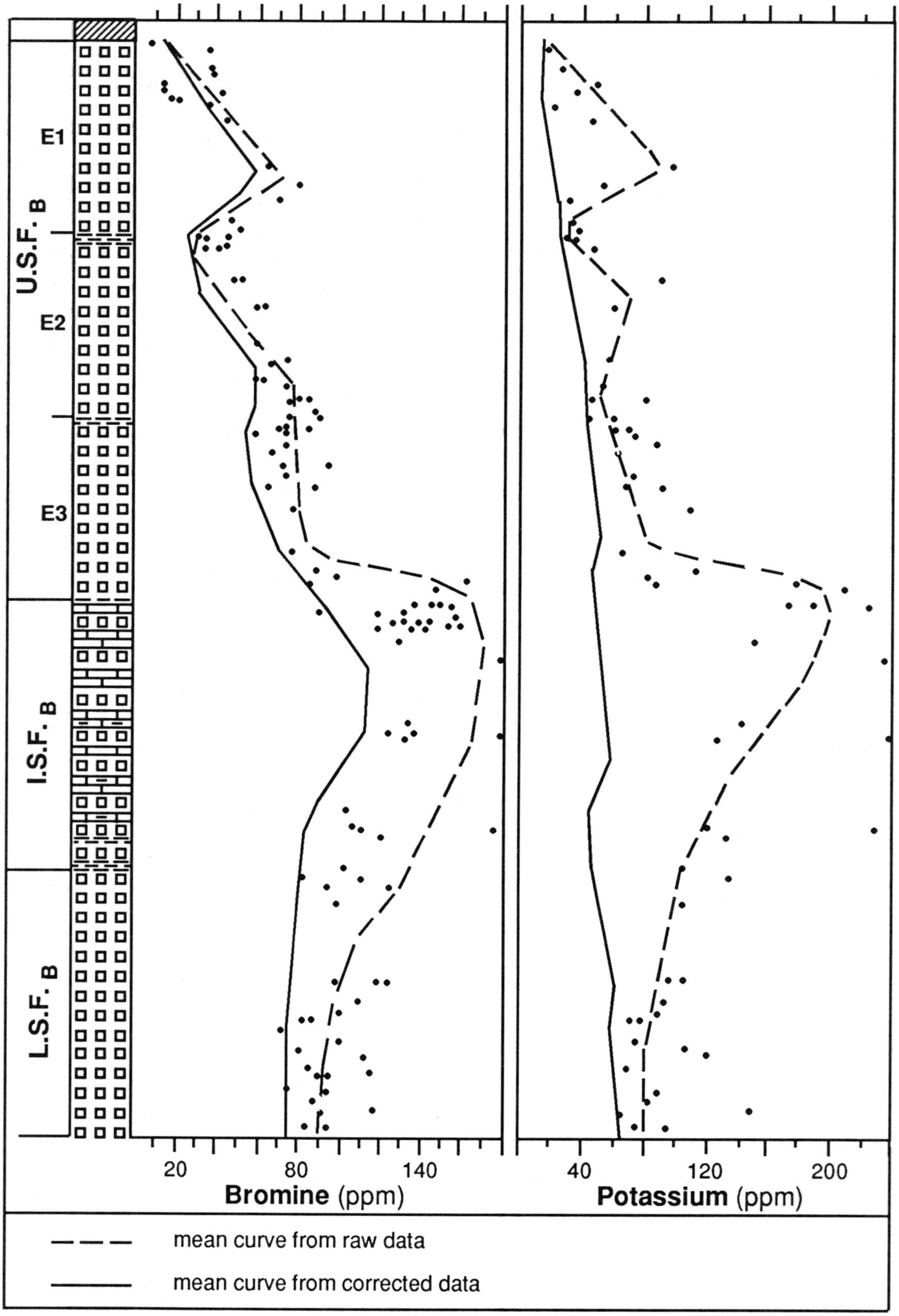
E1
E2
E3
U.S.F. B
I.S.F. B
L.S.F. B
20
80
140
Bromine (ppm)
40
120
200
Potassium (ppm)
mean curve from raw data
mean curve from corrected data

FIG. 4.21

lite (see Moretto 1986/87, 1988). These observations suggest that there was a similar evolution of the initial brines (from which cloudy halite crystallizes) and the interstitial brines (from which clear displacive halite grows).

The overall profile shows from Unit E5 to the base of Unit E3 a progressive increase in the bromine content followed by a decrease until the top of Unit E1. The potassium content discloses a quite similar profile. Interpretation of the lower part of the profile (Unit E5 $\rightarrow$ base of Unit E3), where the values are greater than 75 ppm with many fluctuations, suggests that the salt was deposited under restricted marine conditions (Holser 1966; Raup 1966). This also suggests a progressive increase in salinity due to an increase of the evaporation rate or a decrease of the diluted water influx. The maximal values recorded in the upper part of Unit E4 are doubtless due to a warmer and/or more arid climate and more rapid evaporation during this period. In the same way, the apparent deepening of the basin during this time permitted the protection afforded by a larger water body less affected by any influx of diluted water, hence fewer fluctuations.

From Unit E3, the bromine content of the salt progressively decreases, going below the theoretical 75 ppm considered appropriate for true marine halite, until finally the values fall to 8 ppm, going up section. Having suggested a marine origin for the underlying deposits, it is somewhat difficult to admit another origin for Unit E3 to El salt. Possibly the lower bromine content is the result of repeated recycling of initial marine salt. The lowered bromine may, on the other hand, be the result of an increasing contribution of nonmarine waters, or a combination of both sources. The climatic change with a more humid character indicated at the base of Unit E3, and an associated southern detrital input, corresponding to a diluted water influx, support this hypothesis. From the top of Unit E2, the southern diluted influx decreases and a later small increase in bromine probably corresponds to the beginning of brine reconcentration. But this one period of reconcentration was short, because a later influx of diluted water (coming from the north) caused intense phases of dissolution.

CLAY GEOCHEMISTRY

Chemical analysis of the major elements in clay minerals reveals a linear distribution of aluminum and magnesium (figure 4.22). The increase in magnesium content is parallel to an increase of total $CaCO_3$ amount, and indicates that the magnesian character of clays is more marked when the sedimentation becomes less detrital. The positive correlation between magnesium and $CaCO_3$ content may be inherent to an early diagenesis, which favored the aggradation of clays to form smectites from interstitial waters rich in magnesium (Geisler-Cussey 1985, 1986). More precise information is documented in Moretto (1986/87).

FIG. 4.21. Vertical profiles showing bromine and potassium content in the halite of the Etrez area (adapted from Moretto 1986/87).

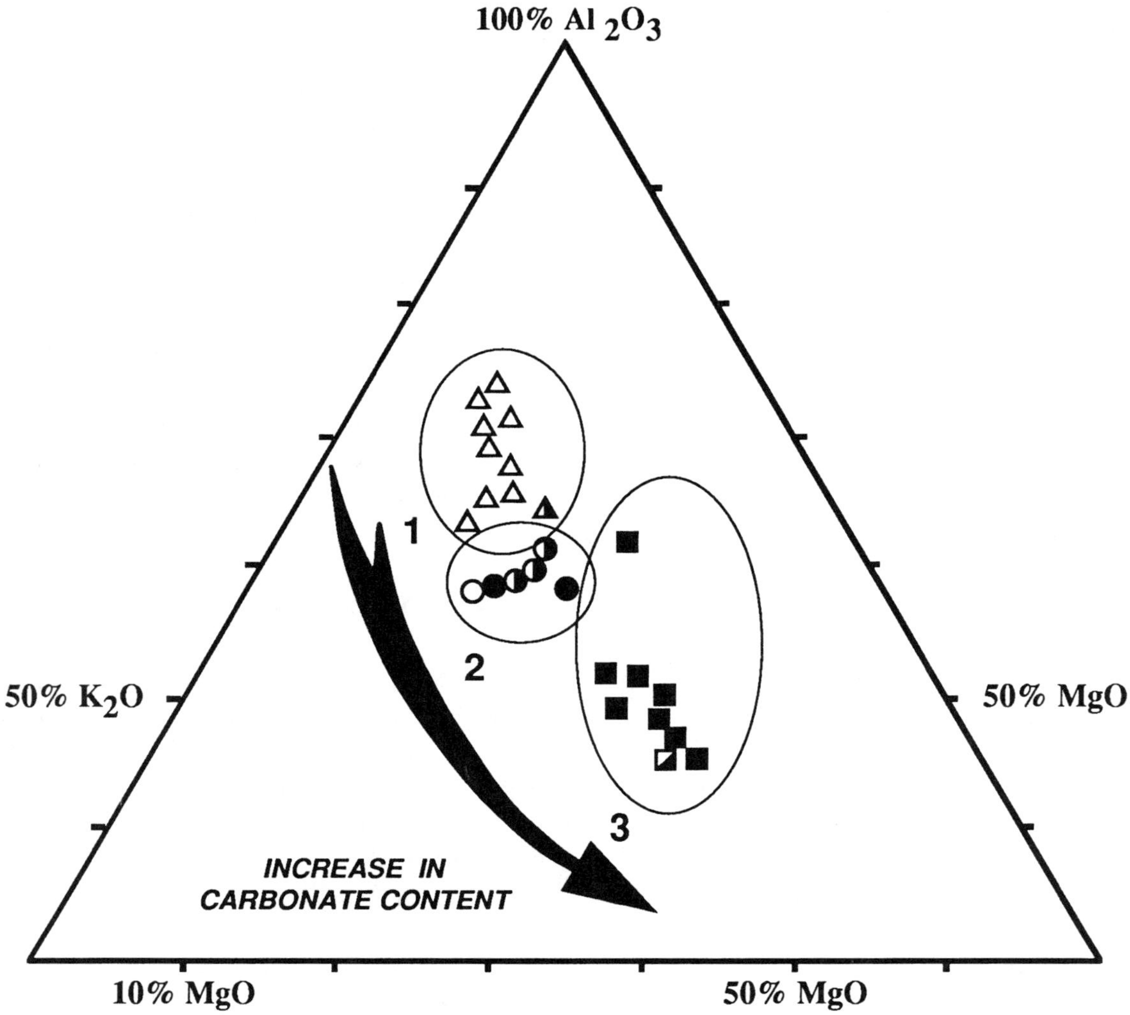

0 to 30	30 to 50	> 50	% carbonate / Clay assemblages
△	◭		1 : Illite - Kaolinite - Chlorite
○	◑	●	2 : Illite - Kaolinite - Chlorite - Mixed layer
	◪	■	3 : Illite - Smectite - (Kaolinite, Chlorite)

FIG. 4.22. Geochemical characteristics of the silico-aluminous phases present within non-halitic horizons of the salt sequence (after Moretto 1986/87).

Salt Basin Geometry, Facies Distribution, and Sedimentary Dynamics

The most striking feature of the evaporitic deposits of Bresse is their layered aspect, very often described in other salt formations (Richter-Bernburg 1955; Wardlaw and Schwerdtner 1966; Ortí and Pueyo 1977a, 1977b; Geisler-Cussey 1985, 1986). Any given bed (usally representing one macrofacies) is rarely thicker than tens of centimeters without the presence of the interlayering of another facies. The rapid succession of interlayers, sometimes on a millimetric scale, results in a laminated facies. Of course, growths of diagenetic materials (halite, sulfates) disturb the primary order but, considering only the least altered facies, it is possible to propose, through sequential analysis, the initial modes of deposition.

Micrite and clay minerals (sometimes silts) are the principal primary constituents of non-halitic beds in the Bresse salt series. Allochthonous continental organic matter (vegetal fragments, spores, and pollen) or autochthonous organic matter, from algal mats, is commonly associated with them. The finely laminated aspect is due to the rapidly changing sequence of different microfacies. These associations correspond to primary variations in deposition. An association, most common within Unit E4, is composed of an argillaceous or clayey carbonate basal layer, more or less rich in organic matter, immediately overlain by a micritic layer, followed by a thin lamina of organic matter. Sulfate laminae with ghosts of lenticular gypsum crystals can coincide with them and, because such gypsum crystal growth is early, gypsum can be integrated into the primary depositional sequence. Sulfates correspond to an appreciable increase of salinity.

When films or thin beds (centimeters thick) of non-halitic sediments are interlayered with cloudy salt, they form the base of primary sequences of halite deposits. In these salt deposits, the diastem which separates two successive sequences usually corresponds to a minor dissolution surface. This dissolution is evidence for a temporary decrease in the salinity of the formative water. These primary sequences are the result of the repetition of dilution/concentration episodes. They may be part of plurimetric larger order sequences (most clearly observable through Unit E4) and represent a progressive drift towards facies that correspond to a general environment of increasing salinity.

Facies associations and their distribution permit differentiation of the major component units of the Bresse series and correspond to environmetal changes. One of the factors to be considered as a control of the sequential distribution of facies is the climate. Elementary or primary alternations apparently result from minor climatic fluctuations, while general climatic evolution occurs over much longer time periods, bringing about larger order (long term) sequence formation.

THE SUBSALT FORMATION$_B$ (E6) AND THE LOWER SALT FORMATION$_B$ (E5)

During the early part of the Paleogene, the Bresse region was exposed and affected by an active period of erosion. The erosion of the pre-Cenozoic basement was more effective along the Limonest basement high (020°) and on the ancient Vienne-Chamagnieu horst (050°) and its northern slope ("Seuil Lyonnais") than it was in the depressed areas of the future salt basin. The Cormoz region, however, did not act as a basement high through this period but seems to have become active later, after the deposition of the Subsalt Formation$_B$.

Non-evaporitic deposits (several tens of meters thick) that form the lower part of the Subsalt Formation$_B$ are made up of shales and sandy shales with only a few exceptions (figure 4.23): lacustrine limestones developed in the area of the future Louhans extension (MaCtD1 well) and Mesozoic rocks underwent weathering and *Microcodium* alteration on exposed surfaces (DP 104 well; Allard et al. 1959). Weathered surfaces are overlain by intrabasin conglomerates in the Cormoz area. The presence of intrabasinal conglomerates suggests the initiation of the tectonic activity of the Cormoz

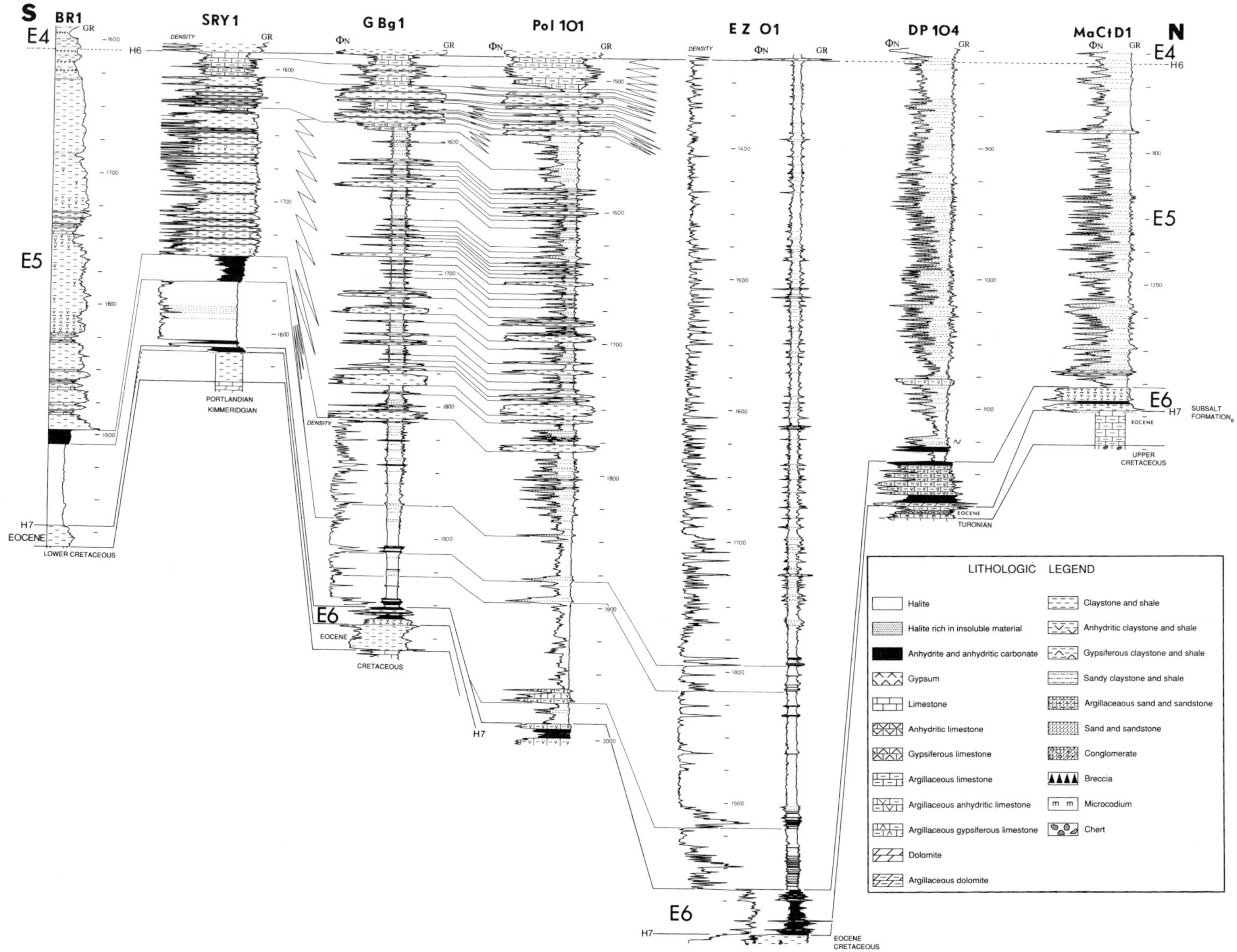

FIG. 4.23. N-S cross-section of the Bourg Basin showing the Lower Salt Formation$_B$ based on well log correlation (after Curial 1986/87).

basement high. Therefore in this basin the 020° fault trend seems to have been active earlier than the 050° structural trend.

Unit E6 (from 0 to 60 m thick), forming the upper part of the Subsalt Formation$_B$, is composed of anhydrite and includes argillaceous and carbonate layers. It thins towards the south and disappears (BR1 well). The boundary of the Subsalt Formation$_B$ to Lower Salt Formation$_B$ seems to be diachronous based on well log correlations, and is thought to have formed earlier in the Etrez area than in the southern regions. Indeed, calcium-sulfate deposition appears to have commenced in the depocenter area and then spread gradually over the basin. About the time the calcium-sulfate deposition overran the Bourg area (GBgl and SRY1 wells), halite began to precipitate in the depocenter area.

Following a pattern similar to the sulfate deposition of Unit E6 (onlap of evaporites with sealing of the potential marginal reservoirs), the salt sedimentation of Unit E5 began in the depocenter area (Etrez) and extended southward through time (figure 4.23). In the southern regions (BR1 and SRY1 wells) sulfate deposition resumed after this period of salt sedimentation. The evaporitic domain then again shifted towards the north in response to a major influx of water bearing the terrigenous material coming from south.

The detrital input first accumulated in southern sectors and then extended toward the north, interrupting salt deposition that meanwhile developed in the Bourg region (GBgl and Pol 101 wells). The detrital material in the southern sectors of the basin consists of sandy claystones and glauconitic sands and sandstones, including conglomerates (BR1 well; Tchimichkian et al. 1958). Farther northward, both the number and the thickness of beds of terrigenous material decrease. Even further northward, siliciclastic deposits become finer and grade into claystones and silty claystones. In the Etrez field, terrigenous deposits are not abundant. They are commonly associated with carbonate-sulfate sediments in which the structure was largely disrupted or destroyed by early intrasedimentary halite-crystal growth. Some solution-related breccia deposits are intercalated within these salts. These periods of dissolution suggest influx of continental water runoff during which terrigenous deposits accumulated in the southern margin.

In the middle part of the basin, spatial distribution of terrigenous input and salt sedimentation was controlled by ≈020° structural trends (figure 5.24A). The thicker deposits are localized along a 020° axis corresponding to a synsedimentary fault. To the west, salt deposits are bounded by the Limonest basement high (020°), and to the east by a bundle of normal N-S faults. The relative uplift movement of the Cormoz basement high, which initiated during the deposition of the Subsalt Formation$_B$, became more intensive. This horst controlled salt deposition along a 050° structural trend in the Louhans extension.

THE INTERMEDIATE SALT FORMATION$_B$ (E4)

1—*The "Argilites de Base"*. The lower part of the Intermediate Salt Formation$_B$, called Argilites de Base in the Etrez field, is composed of thick claystone layers alternating with primary salt horizons (figure 4.25). This interval evidently reflects a period of increased continental water inflow.

Southward, salt layers pass into thin anhydrite-rich beds and the claystone horizons become thicker (Pol 101, GBgl wells). In the more southern regions of the basin terrigenous deposits are evaporite-free and are mainly composed of sandstones of various grain sizes that include angular quartz, micas, feldspar, and glauconite (Tchimichkian et al. 1958). These siliciclastic sediments have a mixed composition throughout deposition of the units E4-E3. Most of the Paleogene terrigenous material was reworked from Mesozoic formations (particularly Aptian and Albian marine glauconitic sands and sandy shales). This has been pointed out by various studies of detrital deposits of the southern Bresse (Tchimichkian et al. 1958; Moretto 1985, 1986/87, Curial and Moretto 1985), of other areas of the Saône Graben (Duplaix et al. 1965), and also of the Jura chain (Duplaix and Guillaume 1962, 1963).

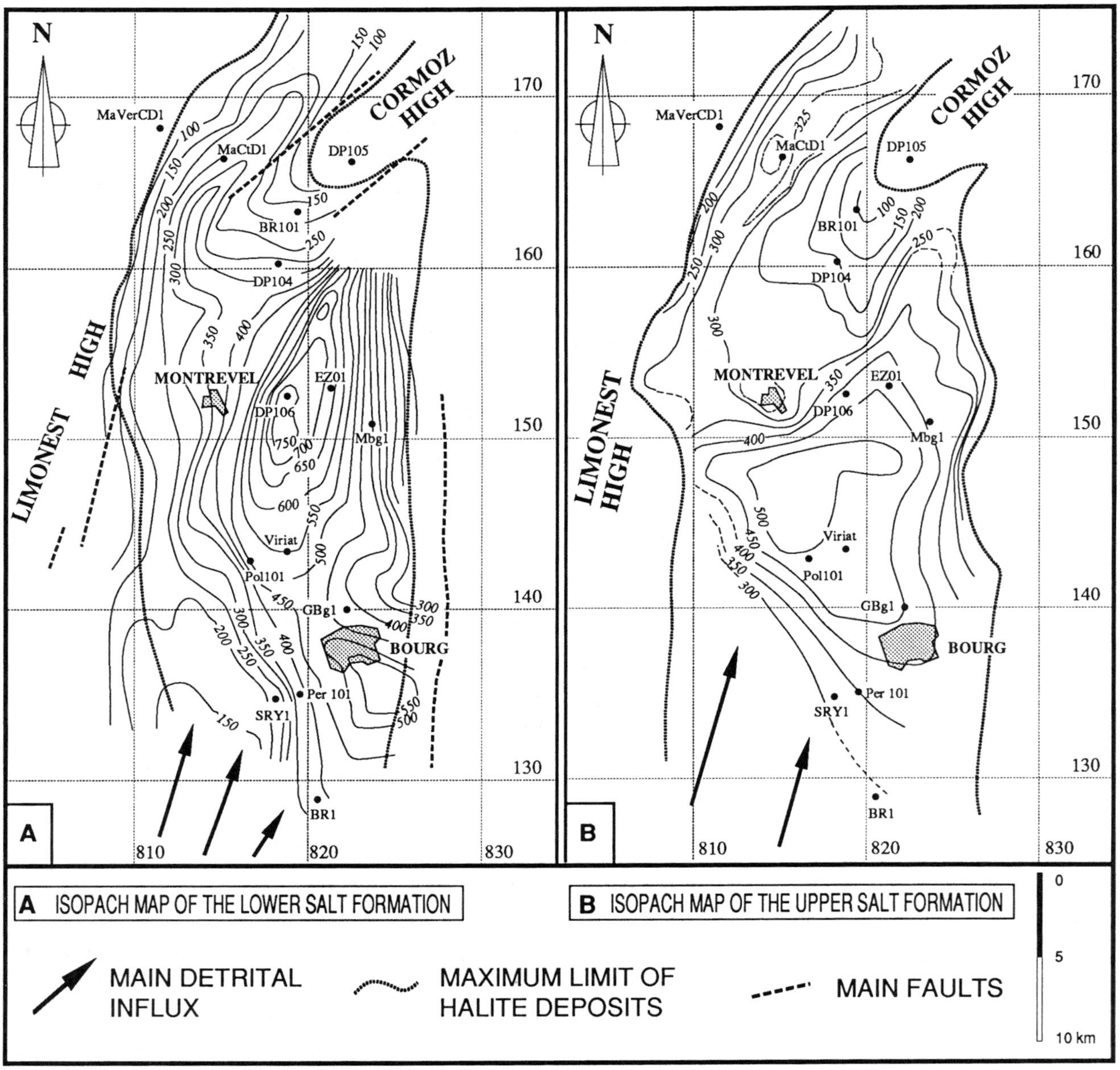

FIG. 4.24. Maps of the Bourg Basin showing isopachs of (A) the Lower Salt Formation$_B$ and (B) the Upper Salt Formation$_B$ (from well log and seismic data; after Leroy 1984, modified). Present limit of halite deposits is indicated and the south to north sense of detrital influx is given by arrows.

Across the Etrez area, the total thickness of claystone beds decreases progressively from south to north following the 020° direction (see figure 4.26), showing the influence of this structural trend on the distribution of terrigenous input (Curial 1987).

2—*The interval of laminar carbonate-salt alternation.* The laminar carbonates are interbedded with primary salt throughout the upper part of the Intermediate Salt Formation$_B$ and have a wide distribution over all of the basin. Laminitic horizons have clearly been recognized from the Louhans extension (MaCtD1 well) to the Bourg area (GBg1 well), whereas intercalated halite beds commonly pass laterally into carbonate, variably rich in anhydrite, but do not exist everywhere. From the base to

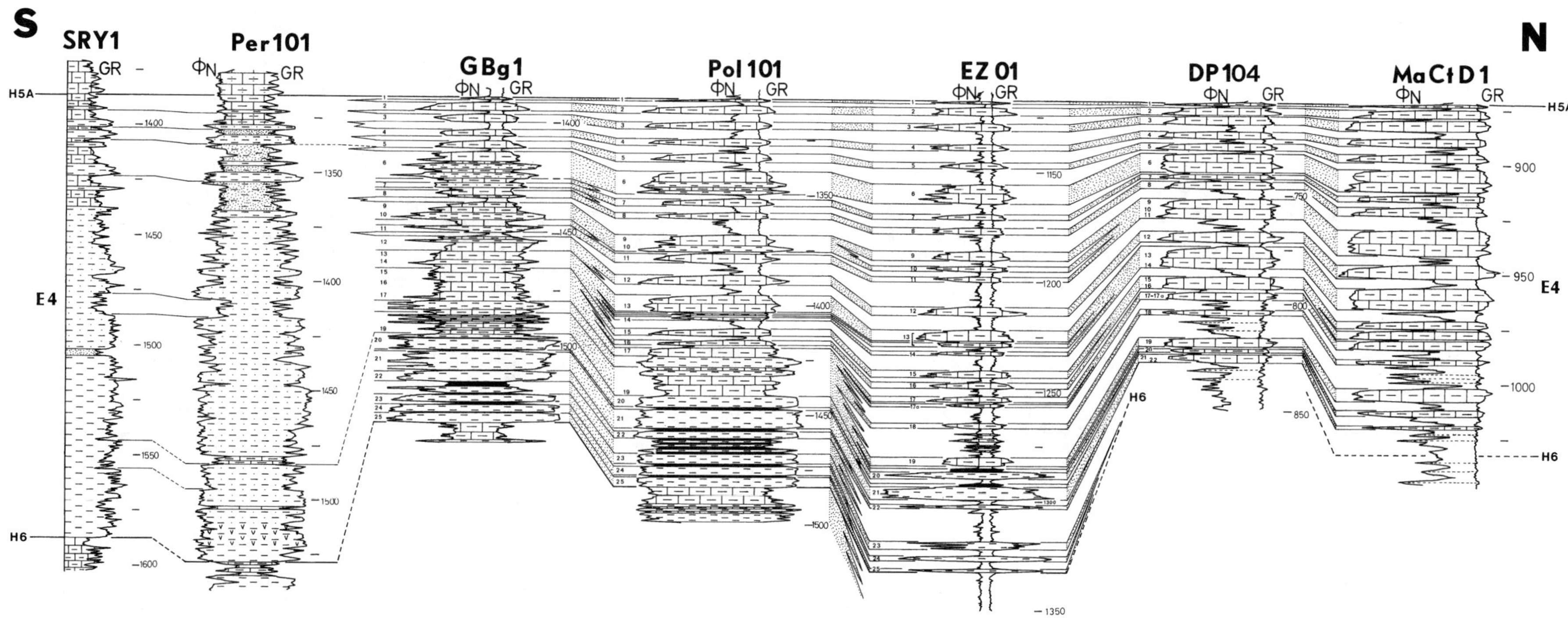

Fig. 4.25. N-S cross-section of the Bourg Basin showing the Intermediate Salt Formation$_B$ based on well log correlation (after Curial 1986/87). See figure 4.23 for lithologic legend and figure 4.1 for borehole locations.

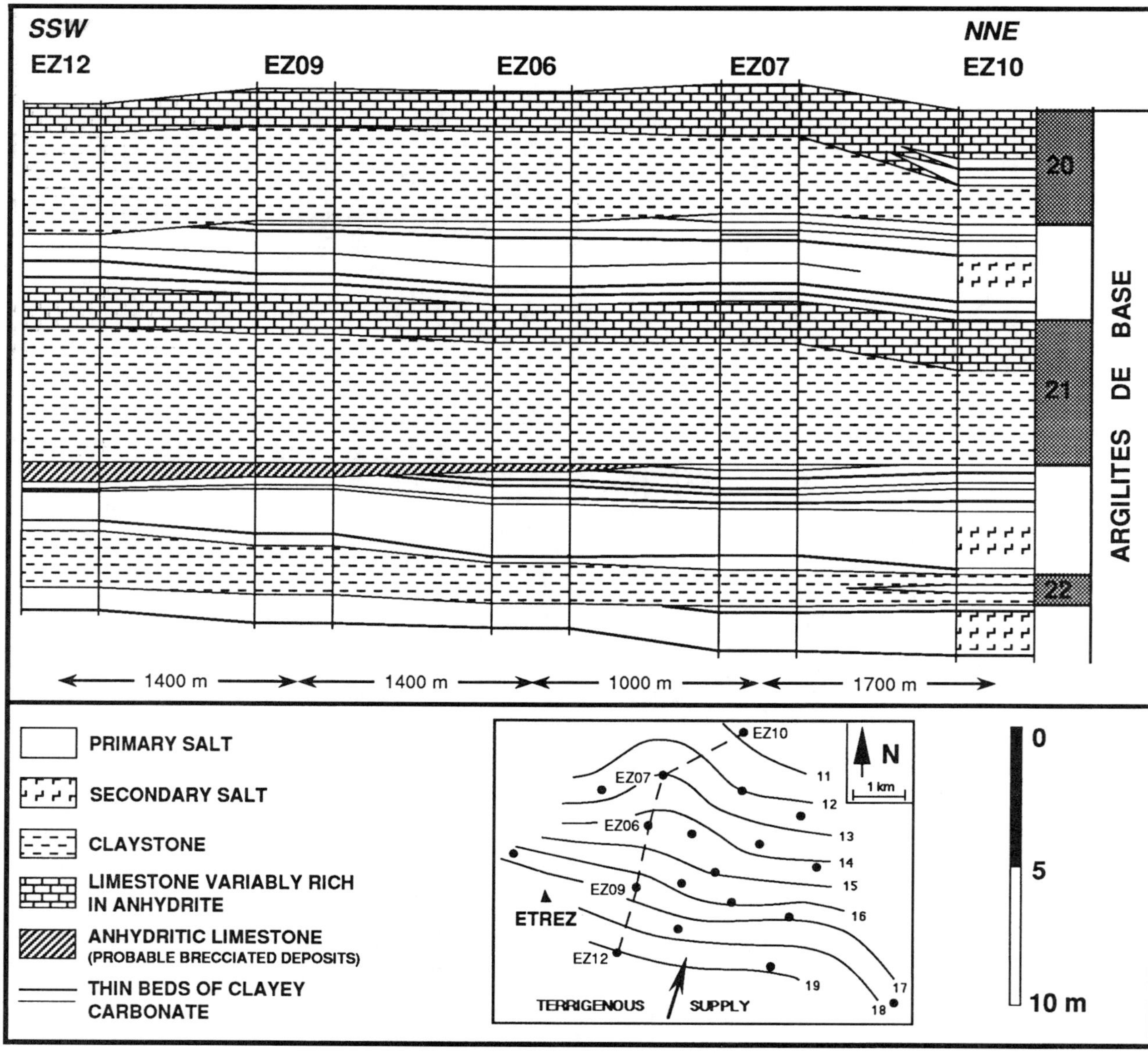

Fig. 4.26. N-S section across five bore holes of Etrez area for the upper part of "Argilites de base" (Intermediate Salt Formation$_B$) showing dissolution surfaces at the bottom of claystone beds (beds 20 and 21). Lithology is inferred from well logs and dipmeter curves (after Curial 1988).

the top of Unit E4, halite deposits extend into sectors further to the south. Because horizons of cryptalgal carbonates showing desiccation- and brecciation-related structures exist within detrital facies of southern regions (Per 101 well; Moretto 1986/87), it seems likely that the halite had extended further but has undergone dissolution.

THE UPPER SALT FORMATION$_B$ (E3–E2–E1)

In the Etrez area, Unit E3 is mainly composed of primary salt enclosing argillaceous limestone and claystone beds several tens of centimeters thick. In the lower part of Unit E3 limestone facies resemble those of Unit E4 (laminated limestones containing organic matter with a probable depositional origin under stratified water bodies). In the middle part of Unit E3 limestones are poor in clays and display a massive structure. In the upper part limestones are often laminated, but in this case laminations are produced by distinct algal fabrics and aligned laminae of insect coprolites or of gypsum-crystal ghosts. In the uppermost part of Unit E3 two thick horizons of greenish to brownish claystones enclosing plant debris and separated by primary salt reflect the increase of the siliciclastic supply. These claystones disappear northward in the vicinity of the Cormoz basement high.

Salt deposits and associated sediments forming the E3 Unit in the Etrez area pass southwards to claystone and carbonate interbeds devoid of halite (Pol 101, GBg1 wells). Farther towards the south the grain size of siliciclastic deposits become coarser and the limestone horizons disappear.

Throughout Unit E2 halite is mainly of a diagenetic origin (i.e., is displacive) in contrast to the primary salt of underlying units E3 and E4. The detrital input from the south, widespread within the deposition of Unit E3, was more restricted in extent for the early deposition of Unit E2. Correspondingly, salt facies developed in the more southern areas (Pol 101, GBg1 wells) and resulted from growth of diagenetic halite crystals within shales and carbonates. Later, major terrigenous influx resumed in these sectors alternating with periods of salt deposition. This detrital influx never reached as far north as the Etrez region. In the Etrez area, nonhalitic facies contain few clays: halite crystals grew within a sulfate-carbonate matrix and also seem to have preferentially developed in the northern part of the basin (MaCtD1 and DP 104 wells). The limited northward extention of the terrigenous input is related to the tectonic shift of the depocenter sector towards the south. This tectonic drift was initiated at the beginning of Unit E2 deposition and lasted throughout deposition of Unit E1, in response to the tilting of the Bresse graben and the relative uplift of the Cormoz basement high and adjacent areas.

The migration of the depocenter towards the south had considerable influence on the deposition of Unit E1. Intensive synsedimentary halite dissolution developed in the northern sectors of the basin, in alternation with periods of salt deposition. Halite occurs as very coarse crystals wrapped in carbonate-sulfate sediments. Numerous salt-solution cavities have been recognized in the Etrez area (up to several meters in vertical extent), pointing to the intensity of the halite dissolution processes. Salt deposits spread widely towards the south (BR1 well) and thicker sedimentary accumulation took place in the Bourg region (GBg1, Pol 101 wells: figure 4.27). The terrigenous supply seems to have decreased during this period and its accumulation was confined to the most southern regions (e.g., BRS1 well).

The isopach map of the Upper Salt Formation$_B$ is different from that of the Lower Salt Formation$_B$, partly because of salt dissolution in the northern part of the basin. The Louhans extension appears as a trough orientated in a 030° to 040° direction (influence of both 050° and 020° structural trends). In the Bourg basin the salt-sedimented area seems to have spread eastward and westward because of the less intensive activity of the 020° faults: the deposits are transgressive on the eastern slope of the Limonest high and the axis of subsidence, clearly expressed for the Lower Salt Formation$_B$, is less marked for the Upper Salt Formation$_B$.

THE TERMINAL EVAPORITE FORMATION$_B$

1—*Overview*

a—Lithologic description of the Etrez area.

The Paleogene sediments overlying the salt formations have rarely been cored (except for the lowermost part, for which a precise description is given below), but their continued presence is known owing to the various well logs which have been recorded in the boreholes of the Etrez field.

The tens of meters of sediments which commonly form the immediate cover of the salt series (breccia, figure 4.28) have been tied to salt dissolution. The lower part (1.5 meters) is represented by a brecciated carbonate formation showing an heterogeneous structure. More or less angular,

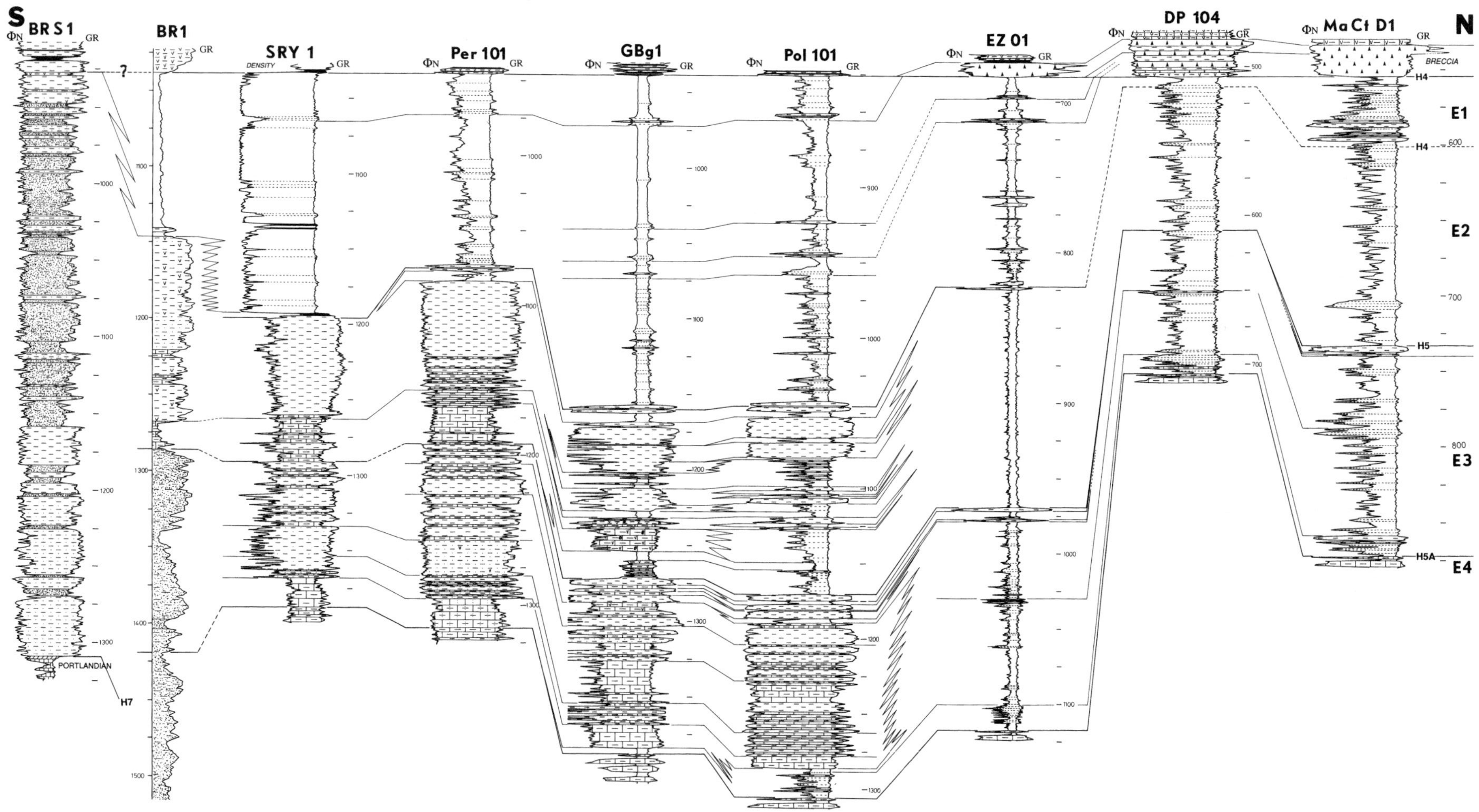

Fig. 4.27. N-S cross-section of the Bourg basin in the Upper Salt Formation$_B$ based on well log correlation (after Curial 1986/87). See figure 4.23 for lithologic legend and figure 4.1 for bore hole location.

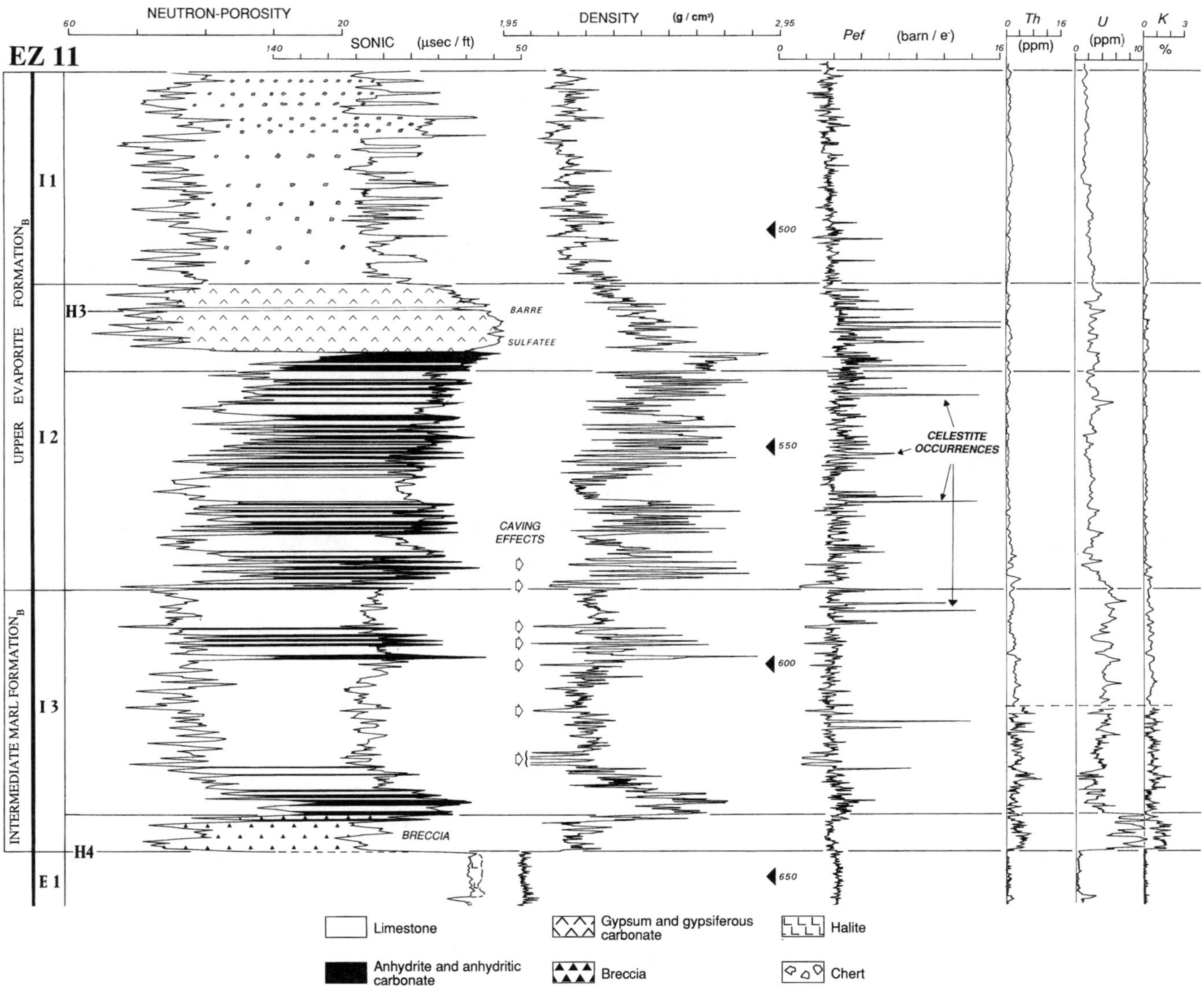

FIG. 4.28. Lithostratigraphy of the Paleogene (extending to bottom of "Aquitanian") cover of salt formations based on various types of well logs recorded in the EZ 11 well (Etrez area) (after Curial 1986/87). Anhydrite is detected by low neutron-porosity values and high density, gypsum is marked by high neutron-porosity and medium density, and celestite occurrences are clearly detected by Pef measurements (fine peaks to high values).

centimeter-size calcareous fragments are present and correspond to pieces of broken, dislocated, collapsed, or reorganized in situ carbonates. An argillaceous or sulfate phase (anhydritic nodules with regular geometrical shapes and also as deformed masses) is associated with them. At the base of the solution-related breccia there are still some residual salt enclaves. Over this profondly disrupted and brecciated zone lie alternating beds of: (1) laminar carbonate facies with fluidal and wavy structures (sometimes with microslumps), (2) more argillaceous facies with grumeleuse structure (showing many fractures and slide planes), and (3) brecciated facies in which primary aspects of deposits are better preserved than in the previously mentioned breccia facies. The solution-related breccia is covered by anhydrite beds interlayered with fine laminar argillaceous mudstones and calcareous clays. The breccia is recognized through well logs owing to elevated radioactivity levels mainly reflecting an

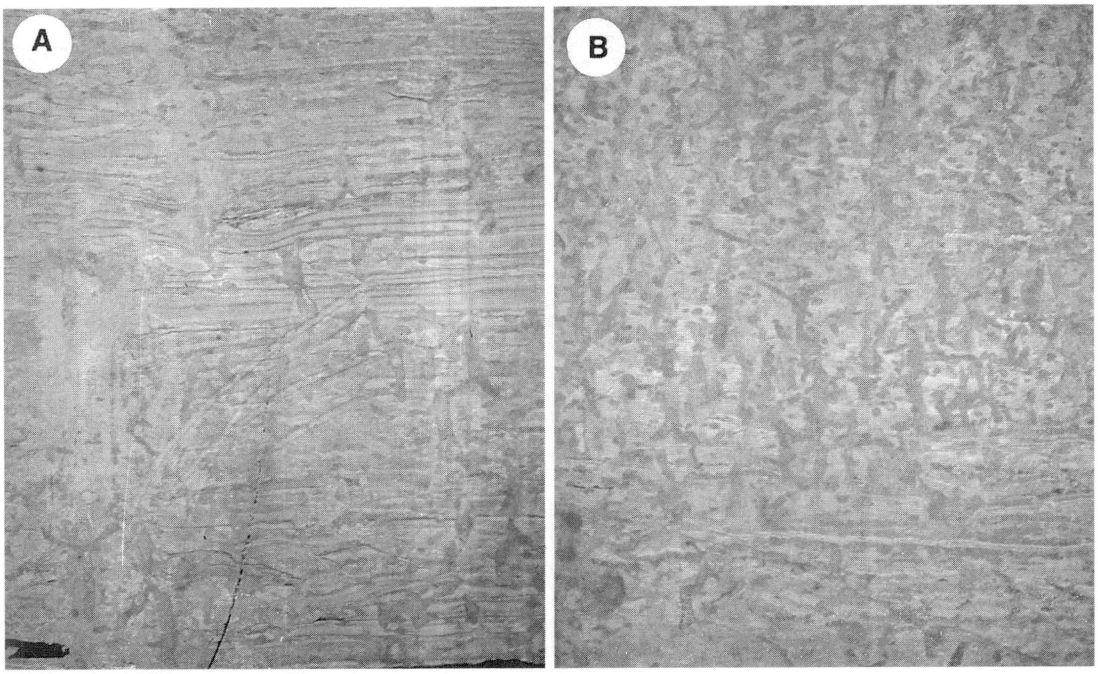

Fig. 4.29. Photographs of cut core sections of laminated limestone crosscut by small animal burrows. A. An early stage of burrowing with discrete, clearly developed structures cross-cutting the laminae, plus one larger root structure (?) at left, also subsequently burrowed. Photo 10.5 cm left to right. B. Profusely burrowed section in which laminar structure is almost destroyed. Photo 10.5 cm left to right.

increased amount of uranium (up to 10 ppm on the natural gamma-ray spectrometry logs, figure 4.28). In this horizon, the uranium abundance has been tied to enrichment in organic matter (Curial 1986/87; Curial and Moretto 1988).

Above the breccia, the Intermediate Marl Formation$_B$ (Unit I3) is made up of clayey carbonate including two small intervals rich in anhydrite. Our core study has shown the presence of laminated limestone that is variably bioturbated (figure 4.29). Some levels disclose paleosoils with well developed root structures (figure 4.30). This indicates a lacustrine/palustrine environment of deposition (shallow water). In certain sectors of the Etrez field, the lower part of Unit I2 is formed by an alternation of beds of carbonate and anhydrite, and the upper part ("Barre Sulfatée") is mainly made up of sulfates (anhydrite and/or gypsum). Gypsum, a compact and hydrated mineral, is recognized on logs by the co-occurence of low sonic transit times and high neutron-porosity values. Anhydrite is recognized by low sonic transit times with very low values of neutron-porosity. Above the "Barre Sulfatée" occur limestones, weakly argillaceous (low K and Th contents), enclosing cherts, and rare thin layers of sulfate (Unit I1). Together, units I2 and I1 make up the Upper Evaporite Formation$_B$.

Numerous thin beds of limestone rich in celestite exist throughout the three units, clearly evidenced by the high values of the photoelectric factor (Pef > 5 barns/electron; figure 4.28).

b—Overview of the Bresse Basin.

Whereas precise correlations using well logs have been established through the Paleogene cover of the salt formations between the boreholes of the Etrez area, the geometry of these deposits is less known at the basinal scale (figure 4.31). The distinct formations and sedimentary units defined in the Etrez field have not been distinguished clearly in the other areas of the Bresse Basin. The Paleogene cover above the salt is thinner in the northern half of the basin (Etrez area, DP 104 well, Louhans extension) than in its southern part. This is basically due to synsedimentary subsidence which seems to have

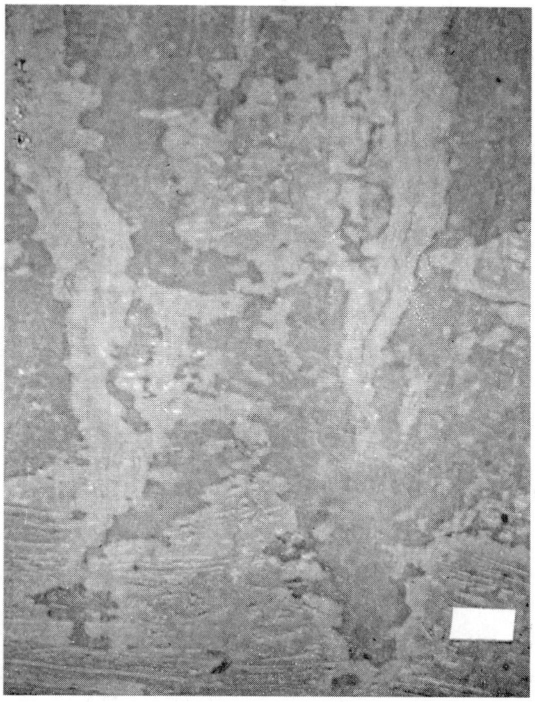

FIG. 4.30. Photograph of slabbed section of laminated limestone crosscut by large vertical desiccation structures. Zoned areas around cracks are marked by a mottled limestone texture. Original unaltered laminar structure is preserved at the bottom of the slab. Scale bar = 1 cm.

been more effective and relatively homogenous over the central (Viriat, Pol 101, and GBg1 wells) and more southern areas of the basin (Per 101 and SRY1 wells).

The Upper Evaporite Formation$_B$ corresponds to the second major evaporitic period in Bresse, the first period being represented by the salt formations (units E5 to E1). In contrast to the Valence Basin, in which the second evaporitic period corresponds to halite deposits containing glauberite and anhydrite (Dumas 1988; Curial et al. 1988; Dromart and Dumas, this volume), no halite and no evaporite typically of continental origin (such as glauberite) has been detected thus far in the Bresse Basin. The evaporite deposits are mainly made up of calcium-sulfates (gypsum and/or anhydrite) enclosing variable amounts of celestite.

2—Lateral changes of lithology and incidental gypsification in Etrez area

a—Deposit geometry and well log interpretations.

Major lateral changes of the lithology (anhydrite → gypsum → carbonate) occur in the Etrez area over short distances (0.5 to 2 km). In particular, they affect units I2 (especially the "Barre Sulfatée") and I1. This interval has never been cored. So the problem is to determine the origin of the lateral changes (depositional, diagenetic) and the mechanisms causing these changes, using the well log data. We illustrate these phenomena with a SW-NE section across EZ 09, EZ 53, and EZ 14 wells, showing sonic and neutron-porosity logs (figure 4.32).

In the EZ 09 well the "Barre Sulfatée" is made up of massive anhydrite with thin intercalated clay-carbonate beds. Gypsum exists in Unit I1 only, and is limited to rare and thin horizons. In the EZ 53 well the "Barre Sulfatée" is mainly made of anhydrite (lower part) and gypsum (upper part). Several gypsum-rich beds exist in Unit I1. The two clay-carbonate horizons (time marker H3)—which are in fact two bundles of thin layers—are present in a gypsum host facies (EZ 53) as well as in equivalent anhydrite lithology (EZ 09).

In the EZ 14 well units I2 and I1 are almost devoid of sulfate deposits. This is also true for the wells located in the so-called "Carbonate Area" (northeastern part of the Etrez sector), in which sediments are mainly composed of porous, slighty argillaceous, limestone (low K and Th amounts identified on GR spectrometry logs). This lateral transition between sulfates and limestones does not

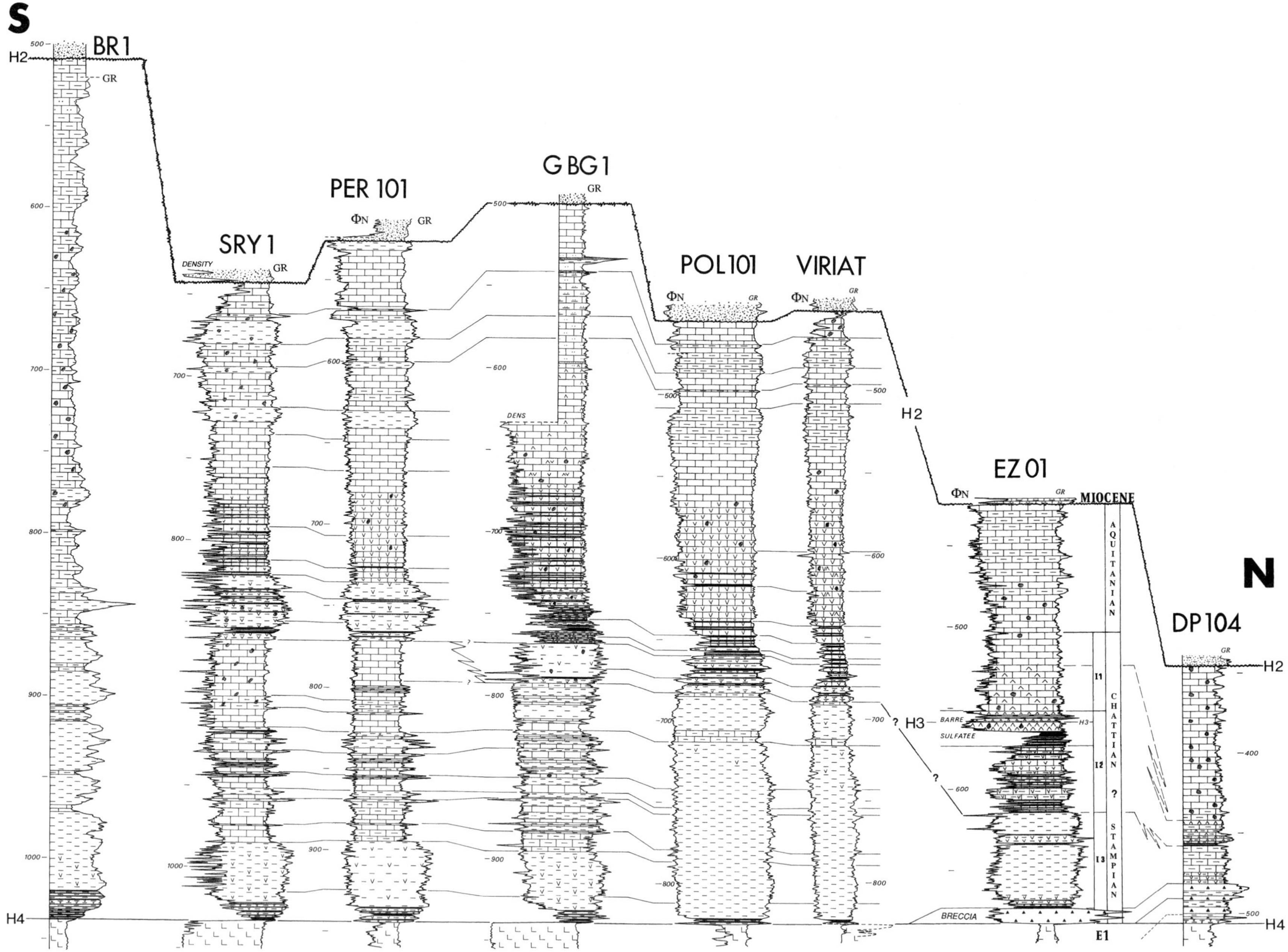

FIG. 4.31. N-S cross-section of the Bourg Basin in the Paleogene cover of the salt formations based on well log correlation (after Curial 1986/87). See figure 4.23 for lithologic legend and figure 4.1 for bore hole location.

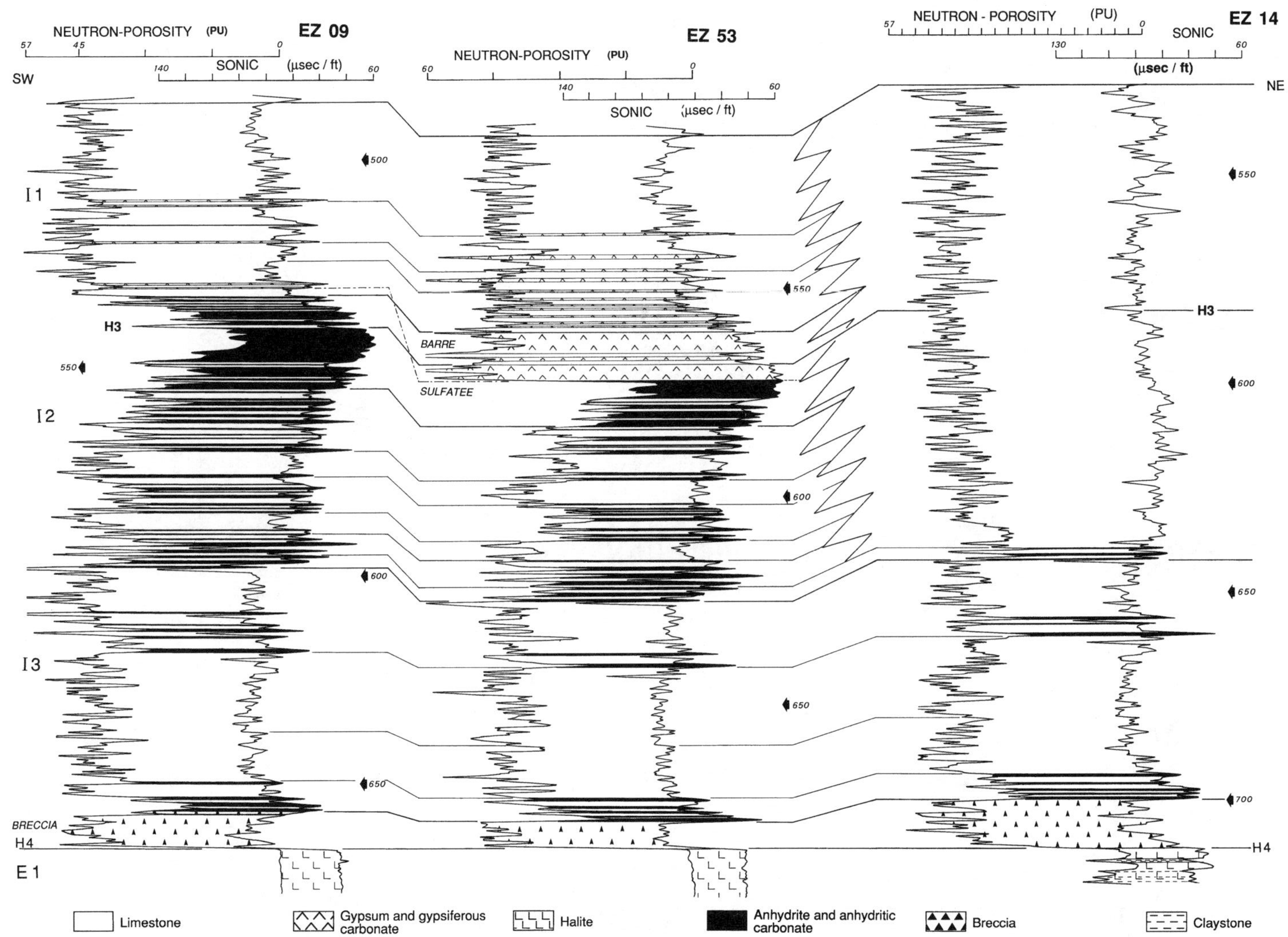

Fig. 4.32. Detailed SW-NE section across three wells of Etrez area for the Paleogene cover (until the bottom of "Aquitanian") (after Curial 1986/87). Lithology is inferred from well-logs. The lateral change of lithologies from anhydrite (EZ 09 well) into gypsum (EZ 53 well) mainly affects the "Barre Sulfatée." Also note the passage from sulfates into carbonates (EZ 14 well).

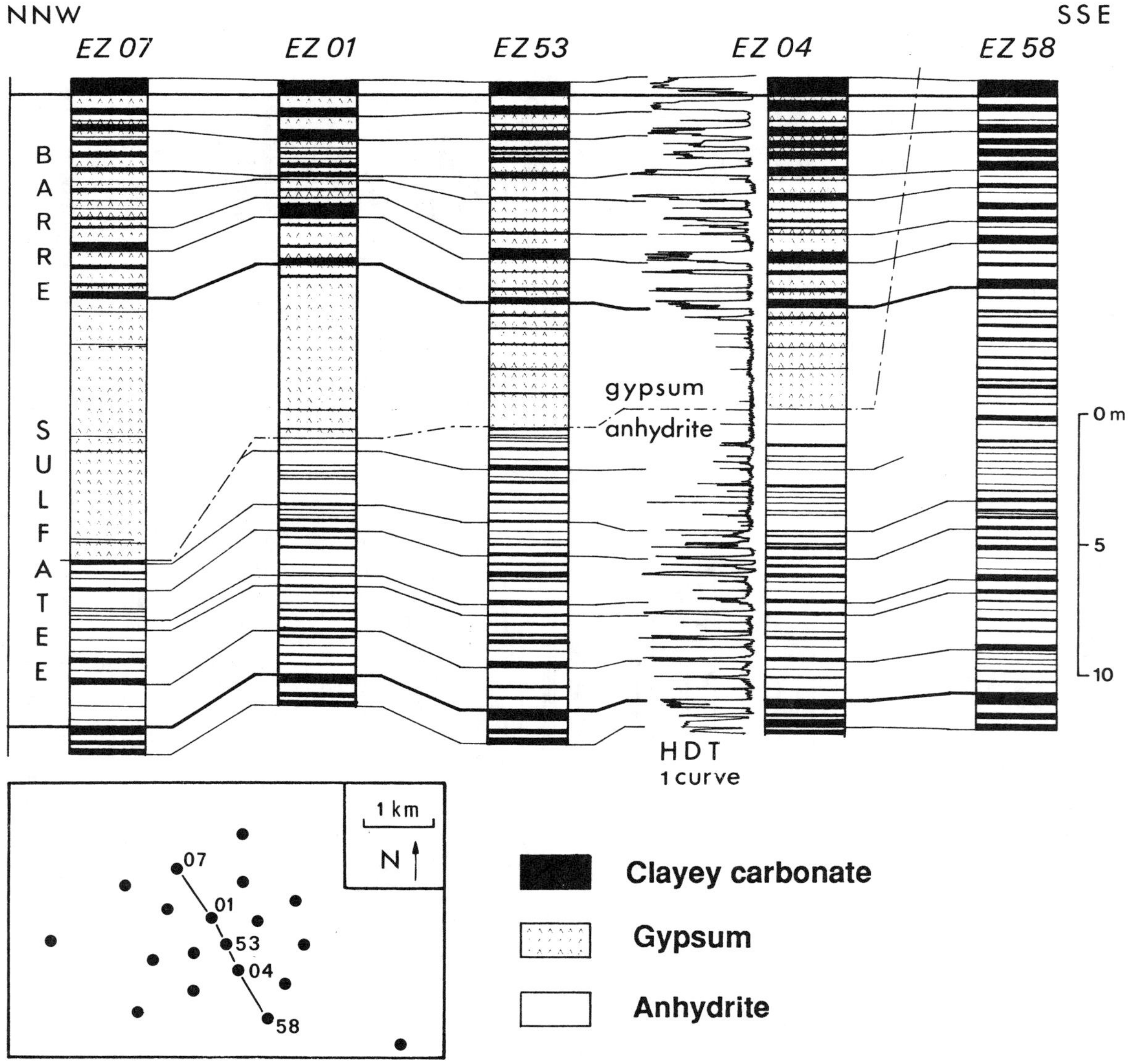

FIG. 4.33. Detailed section across five bore holes of Etrez area for the "Barre Sulfatée" (Unit I2) showing lateral passage from gypsum into anhydrite (after Curial 1986/87). Lithology is interpreted from well logs and dipmeter curves. The thin clay-carbonate beds disappear when lithology becomes gypsiferous, suggesting that the last sulfate transformation is an anhydrite hydration rather than a gypsum dehydration.

involve any appreciable change in thickness. Where the "Barre Sulfatée" develops, gypsum always occurs overlying anhydrite. A NNW-SSE section across the "Barre Sulfatée," constructed from interpretation of combined logs and dipmeter curves, displays the lithological changes in great detail (figure 4.33). The anhydrite-gypsum contact is sharp and oblique to the general bedding (it regularly dips down from E to W; this contact lies below a sedimentary cover that varies in thickness from 500 to 580 meters). The thickest of the interbedded clay-carbonate layers (at least 10 centimeters) are laterally constant, whatever the host sulfate lithology. In the dominantly anhydritic lithology a number of thin clay-carbonate beds (less than a few centimeters) have been detected and may be correlated between wells. Most of the clay-carbonate beds disappear on dipmeter curves when the lithology changes to gypsum. This observation provides supporting evidence for a late anhydrite hydration and the change in volume that caused the destruction of the thinnest and weakest beds.

In the Etrez area, distinct maps have been constructed from well log data so as to disclose the spatial distribution of the main lithofacies (Curial 1986/87). We present here only the isolith maps (total thicknesses of a given lithology) covering the whole of the section from units I3 to I1. The thickness of the calcium sulfate deposits (both anhydrite and gypsum) decreases rapidly towards the NE, along with the lateral change from sulfate to carbonate (figure 4.34A). The area with the highest sulfate contents surrounds the northeastern part of the Etrez sector. Towards the S and SW, the broad pattern of isolith curves reflects a gradual decrease in the amount of total sulfate. The spatial distribution of the gypsum facies (figure 4.34B) follows the same general trends as those of sulfates. Gypsum increases progressively in volume from the southern sector (no gypsum has been detected in EZ 12 well), towards the center of Etrez field where the greatest thicknesses are localized along a NW-SE curved, elongated zone. This zone corresponds to the one where the thickness in total sulfate is the greatest. Farther towards the NE, gypsum accumulation decreases rapidly and then disappears.

Three discrete facies areas can be distinguished from the data of the Etrez area (figure 4.34C):

- The Carbonate Area (CA) is located in the northeastern part of the Etrez field, with some calcium-sulfate deposits (anhydrite) occurring in Unit I1 and in the lowest part of Unit I2 only (see lithologic column in EZ 14 well, figure 4.33).
- The Mixed Ca-Sulfate Area (MA): calcium-sulfates are composed of both anhydrite and gypsum in variable amounts. The gypsum/total sulfate ratio increases progressively towards the CA. Near the limit between CA and MA, gypsum appears in the "Barre Sulfatée" (see figure 4.8). Towards SW and S, the contact of gypsum/anhydrite moves up-section regularly through the general bedding, until calcium-sulfate, forming the "Barre Sulfatée," is represented by anhydrite only (for instance, see EZ 09 well, figure 4.32 and EZ 58 well, figure 4.33).
- The Anhydrite Area (AA) (EZ 12 well), where no significant gypsum deposits have been detected.

Before discussing the origin of these lateral changes in lithology, it is of interest to examine the spatial distribution of celestite, clearly detected by Pef (photoelectric factor) measurements. Pef curves have been recorded in three recent boreholes of the Etrez field (EZ 11, 12, and 15) which are fortunately located in each of the three lithological areas defined above. Celestite-rich deposits are more abundant in the Mixed Ca-Sulfate Area and in the Anhydrite Area than in the Carbonate Area. Within the "Barre Sulfatée" celestite exists as well when the dominant lithology is anhydrite as when it becomes gypsiferous laterally (see the example of EZ 11 well, figure 4.28). However, no trace of celestite has been detected using Pef curves within the limestones in the CA which pass laterally to the "Barre Sulfatée" and the overlying Unit I1 (MA). According to Holliday (1967, 1970) and West (1973), strontium may be mobilized by subsurface water circulation during gypsum-anhydrite-gypsum transformations (dehydration or hydration). Because celestite is less soluble in water than calcium-sulfate, it readily reprecipitates and accumulates in the vicinity of gypsum or anhydrite horizons affected by these transformations.

b—Synthesis and discussion

In the I2 and I1 units, sulfate deposits are restricted to the western and southern sectors of the Etrez field (AA and MA). The lateral change of sulfate to carbonate which occurs between these sectors and the CA is thought to reflect an original sedimentary process. The hypothesis of a sulfate dissolution or calcitization can be rejected, because such a phenomenon would cause some collapse structures and a decrease in thickness, but this is not the case here: the interval I3 to I1 is relatively uniform in thickness across the entire field. In addition, celestite does not exist in limestone deposits which are equivalent to the I2 and I1 units in the CA. In the hypothesis of dissolution or calcitization of sulfates, and because of the low mobility of Sr, it is presumed that the celestite would not been entirely removed. For these reasons the lateral change of sulfate to carbonate is believed to result from a

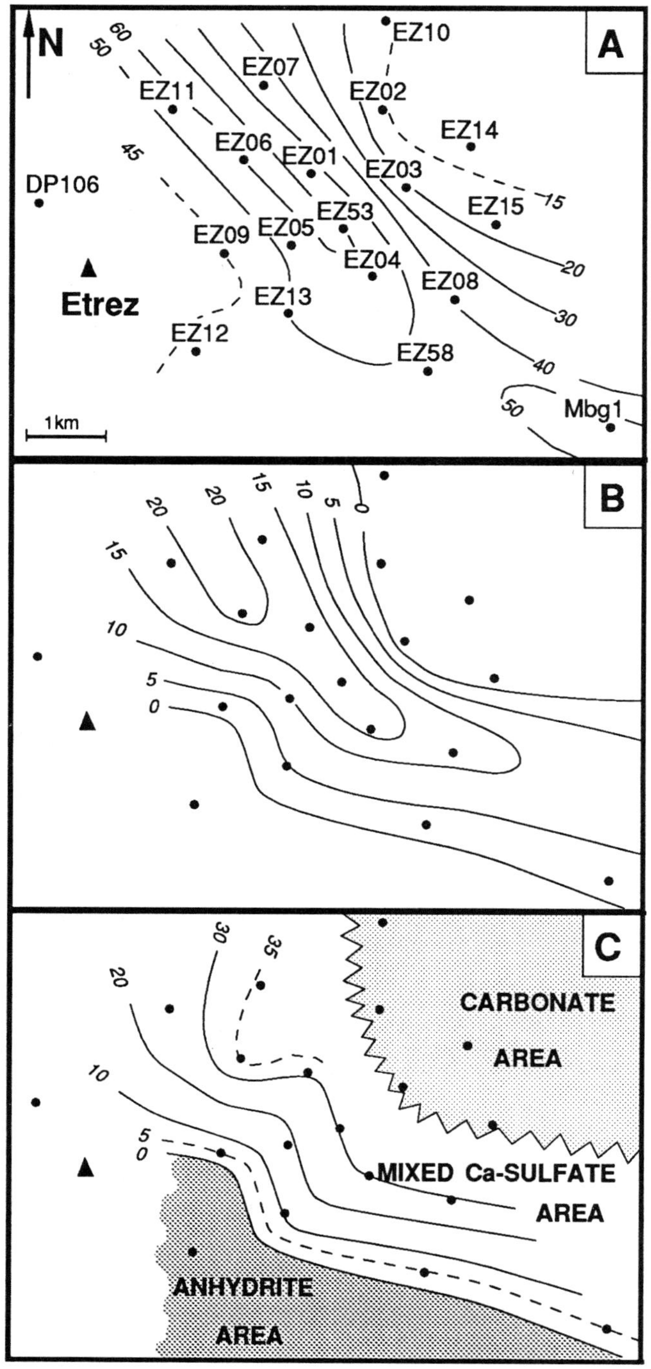

FIG. 4.34. Maps of the Intermediate Marl Formation$_B$ (Unit I3) and the Upper Evaporite Formation$_B$ (I1 and I2 units) in the Etrez area (after Curial 1986/87):
A. Map showing isoliths of sulfates deduced from logs.
B. Map showing isoliths of gypsum deduced from logs. Note the aureole pattern of gypsum occurrence.
C. Map showing isopercentages of gypsum/total sulfates. The boundary between the mixed Ca-sulfate area and the carbonate area marks the disappearance of the "Barre Sulfatée" towards the NE. No gypsum has been detected fom logs in the anhydrite area.

lateral segregation of deposits related to the salinity gradient of the brines during sedimentation and early diagenesis. So the CA characterizes a sedimentary depositional environment located near the margin of the basin where brines had lowest salinities (or at least had greater availability of $CaCO_3$).

On the contrary, the lateral anhydrite-gypsum change has a diagenetic origin. What is the manner of this transformation (hydration or dehydration)? Jauzein (1974), based on McDonald (1953) and Zen (1965), suggests that anhydrite after gypsum may have formed at various burial depths according to the geothermal gradient and the chlorinity of the underground brines: with the presence of chloride-rich water, gypsum dehydration can occur under lower pressure and temperature than with the presence of fresh water. On the other hand, former anhydrite can be rehydrated under exposure or in near-surface conditions or abnormal pressures. In contrast, under the stability conditions of anhydrite, the less concentrated in chloride the underground waters are, the deeper the anhydrite to gypsum transformations may develop.

In the Etrez area, the supporting evidence is available for a formation of gypsum as result of hydration of former anhydrite (possibly itself derived from a gypsum precursor):

1. The initial evidence comes from the correlation between wells using dipmeter curves. The thinnest clay-carbonate beds are numerous and continuous when the dominant lithology is anhydrite. They disappear laterally when anhydrite passes to gypsum. Only a hydration process can explain this destruction of the thinnest, weakest beds.
2. The most abundant gypsum concentration develops in sectors where the thickness of calcium-sulfates is the greatest. This fact suggests an increase in volume due to an anhydrite hydration (a 50 percent increase in volume resulting from the transformation). According to the concepts of Jauzein (1974), this fact suggests anhydrite hydration.
3. The spatial distribution of calcium-sulfates in concentric sectors surrounding the CA (centrifugal decrease in gypsum) is consistent with anhydrite hydration. This configuration may have resulted from a circulation of underground waters of low salinity coming from the eastern margin of the basin through the CA. As these carbonates have never been cored we do not have precise information concerning their hydrological features such as permeability or fractures. Well logs indicate only that they have high porosity with values ranging from 30 to 50 porosity units).

Syn- and post-depositional histories (especially for the I2 and I1 units) can be summarized in the following scenario. The depositional environment was subjected to cyclic increases/decreases in salinity which gave rise in certain places (AA and MA) to gypsum layers in alternation with limestone beds. Closer to the edge of the basin (CA) waters were not concentrated enough to produce gypsum. During early or later burial diagenesis, with the presence of saline ground waters, gypsum was dehydrated and transformed into anhydrite. Later, circulation of undersaturated water from the margin induced a partial re-hydration of the "Barre Sulfatée" and overlying anhydrite-rich layers. Even though the hydrological properties of the deposits in the CA are still unknown, it is assumed that they had a significant influence on the movements of this water.

DISSOLUTION PHENOMENA OF HALITE

1—Dissolution within the salt sequences.

Salt deposits are frequently affected by variably intense dissolution evidenced by traces left on the remaining primary structures. These occur on several different scales ranging from very fine phases of solution (crystal scale) to the basinal scale.

a—Crystal scale.

Hopper-shaped crystals commonly occur as fragments bearing a number of dissolution pits and in turn are enclosed within clear halite. Dissolution can produce residues consisting of mere skeletons or broken fragments that retain primary structures (figure 4.35A). Both Dellwig (1955) and Llewellyn (1968) have suggested that the solution-precipitation processes were very early and took place within the formative waters. This is supported by the observations of Geisler-Cussey et al. (1985).

b—Bedding scale.

Evidence of dissolution occurs most strongly along the upper surface of salt layers as erosional structures penetrate unevenly into a given level (each sequence of salt deposition is no more than 15 centimeters thick). In core, the beds of chevron-shaped crystals commonly have truncated surfaces and show corrosion pits (figure 4.35B). Finger-like dissolution cavities, vertically-oriented, penetrate the sediment following the discontinuities in the crystals (see Shearman 1970; Casas and Lowenstein 1989). Subsequent cavities are filled up by clear

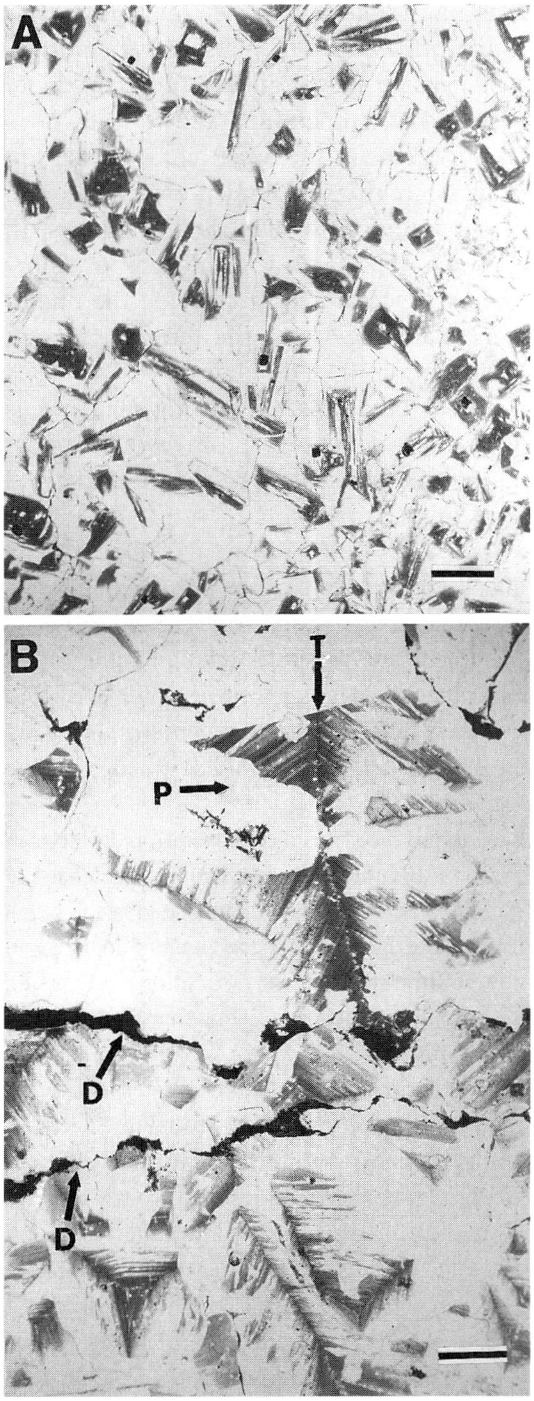

Fig. 4.35. Dissolution features as actually observed in cores:
A. Photomicrograph showing skeletal halite hoppers in cloudy salt, with numerous dissolution cavities that result from partial solution during the settling of the hoppers to the bottom through the water column. Scale bar, 2 mm.
B. Photomicrograph showing dissolution at the top of small salt layers (D). Chevron structures are truncated and associated dissolution cavities are infilled by clear halite. The large chevron in the upper part of the photo shows a well developed truncation (T) and many dissolution pits (P). Scale bar, 2 mm.

halite together with a small amount of non-halite sediment that seeped into the pores (figure 4.36A). When halite developed syntaxially, clear halite occurred as overgrowths around the milky halite and filled the penetrative voids. Various other mineral particles (dolomite, silica, anhydrite) commonly accumulate in the solution voids and on surfaces and are incorporated into the beds by the clear halite overgrowths. Dissolution also affected several salt levels of milky halite interbedded with other sediments as cavities centimeters to decimeters in size (figure 4.36B) which are plugged by euhedral crystals of clear, macrocrystalline halite. Particles of non-halite sediments and of anhydrite nodules, derived from the collapse of interbeds, have been enclosed within the infilling halite.

Phases of dissolution that are somewhat more profound and affect several beds at a time resulted in formation of an intermediate salt type (mixed salt) composed of milky halite, mostly as chevron-shaped crystals, together with phenoblastic (phenotopic) clear halite. Such a mixed salt occurs as well-layered beds, decimeter-thick, in which zones of milky halite represent remains of primary salt beds. Subsequent to dissolution, insoluble sediments partly derived from the disruption of interbeds were deposited in the cavities.

Intense dissolution affecting both primary salt and secondary halite produced breccia-like facies. These clayey to carbonate facies are thin, of a rubbly appearance, and are composed of subangular fragments similar to the surrounding matrix. These intraclasts are sometimes aligned along the bedding planes and appear to be the product of the breakup of thin layers. The morphology and the arrangement of the fragments suggest that they were moved only a short distance and that they most likely result from an in situ collapse due to the dissolution of soluble levels.

Breccia-like facies also occur on the marginal areas of the basin. Algal laminar textures have been fragmented into randomly oriented clasts that can result from the desiccation of algal mats during periods of subaerial exposition or from the dissolution of evaporitic minerals intercalated into the laminae. Whether they are expressed as truncated surfaces, finger-like structures, variably deep cavities, or breccia-like facies, the dissolution phases are early, and may be penecontemporaneous with deposition (Geisler-Cussey et al. 1985).

c—Large scale dissolution surfaces.

Widespread salt dissolution surfaces can be detected quite clearly using dipmeter curves, providing that the boreholes are sufficiently numerous and in proximity (Curial 1988). In the Etrez field such dissolution surfaces are documented in the "Argilites de Base" located in the lower part of Intermediate Salt Formation$_B$ (E4). A reconstructed NNE-SSW cross-section is presented for the upper half of the "Argilites de Base" (figure 4.26). For each well the dominant lithology is inferred from standard well logs. The evidence for thin clay-carbonate beds and the salt facies is deduced from the dipmeter curves. The thickness of the claystone layers decreases progressively from south to north, opposite to the change in thickness of salt horizons. The thin clay-carbonate beds intercalated within halite are useful lithological markers because of their broad continuity. The truncation of the uppermost part of salt layers by dissolution is clearly evidenced when considering the correlations based on these beds. Halite solution became more and more effective towards the south. Claystone horizon 21 is underlain by some anhydrite-rich carbonate deposits whose thickness decreases from the south to the north (between EZ 12 and EZ 06 wells) and which disappear farther north. These deposits are thought to represent an in situ accumulation of insoluble material after dissolution of the halite matrix.

This particular geometry emphasizes the importance of the phenomenon of chemical remobilization of halite that seems to have been very active towards the basin margins because it occurred even in the main depocenter area during certain periods.

This geometry of the beds also discloses the influence of basin physiography on salt deposition/dissolution. Successive phases of major detrital influx coming abruptly from the south toward the Etrez area formed a S-N sediment wedge. This new topography is distinct from that more slowly created by synsedimentary subsidence. The depressed areas (induced by a differential lack of sediment and not by increasing subsidence) shifted to the north independently of tectonism. During a new period of evaporite sedimentation, salt deposition resumed in the northern low, filled it, and rapidly extended and overstepped, covering all the basin. When a new major freshwater inlet occurred (together with associated terrigenous input), preexisting salt was unaffected in the lowest areas (in the starved low formed in the northern part of the basin). On the contrary, it seems that salt solution was more effective towards the south, because the salt deposits were localized in higher topographic position.

Finally, the progressive decrease in thickness of the terrigenous deposits towards the north of the basin (figure 4.26), as well as the lithology of the

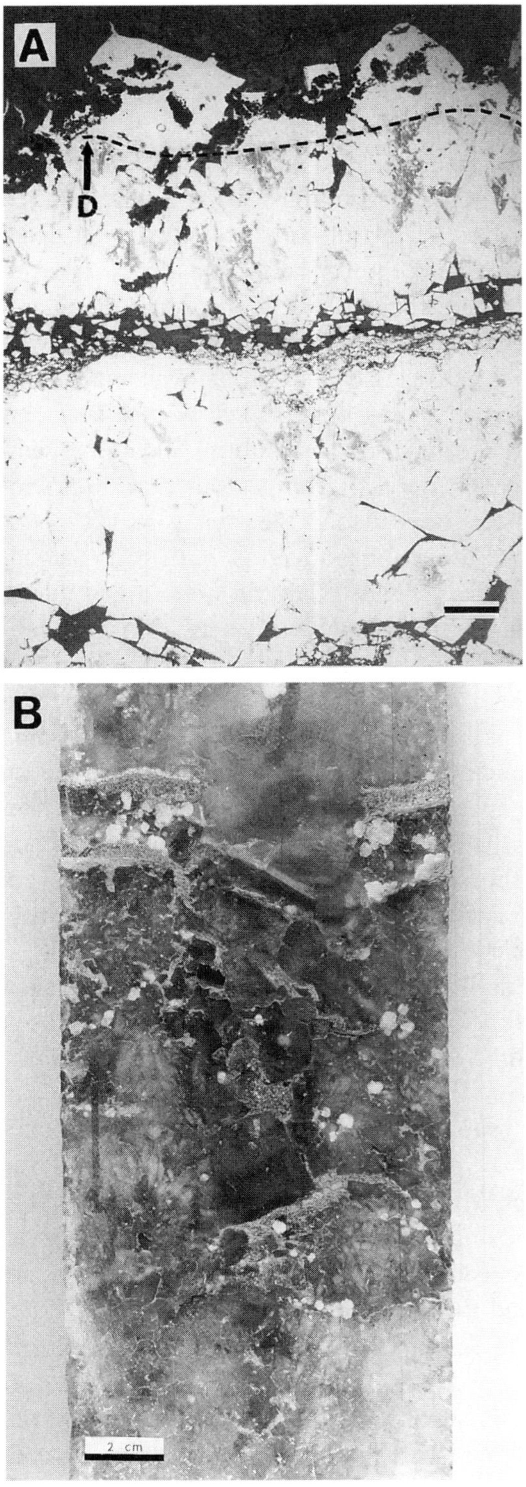

FIG. 4.36. Dissolution features in halite:
A. Photomicrograph showing dissolution along the upper part of a cloudy salt layer. The overlaying non-halitic sediment infiltrated down through the salt along the dissolution pits. Note the syntaxial clear halite overgrowth above the dissolution surface (D) (along dashed line) and under the carbonate layer. Scale bar, 2 mm.
B. Photograph of cut core section showing a dissolution cavity passing through several depositional sequences marked by horizontal carbonate layers. The carbonate layers collapsed as a result of dissolution and the pieces have fallen into the cavity and are now included within a clear halite infill. Scale bar, 2 cm.

halite, suggests that the southern Bresse graben was a relatively flat area covered by a water body of variable depth and extent, but which was never very deep. The same spatial distribution of the detrital material for the "Argilites de Base" is also observed for the thick claystone layers located between the E3 and E2 units and at the top of Unit E2. Indeed, if the graben had steep slopes and a deep water body, fine terrigenous material would have easily spread all over the basin, and the thickness of these deposits would have been more constant in the central area.

2—Salt dissolution in the upper part of the Upper Salt Formation$_B$.

Evidence for salt dissolution in the Upper Salt Formation is provided by the presence of dissolution breccia covering the Upper Salt Formation$_B$ in the Etrez area (Kiener 1984). Later successive studies based on core examination and well log correlations were undertaken to determine the spatial extent, the mechanisms, and the chronology of this salt dissolution feature (Moretto 1985; Curial and Moretto 1985; Curial 1986).

Three cross-sectional reconstructions of Unit E1 and overlying breccia deposits have been made from well logs (figure 4.37; the lines of the cross sections are traced in figure 4.38B). The thickness of sub-unit E1A decreases progressively towards the NE (section A), and towards the N (sections B and C) (see also E1A isopach map, figure 4.38A). The clay-carbonate layers seem to be truncated (horizons a, a′, and b). The isopach map of the solution-related breccia (figure 4.38B) exhibits a similar trend to that indicated by the E1A isopach map: the breccia becomes thicker and thicker as salt deposits are more and more reduced.

It is unquestionable that such a geometry for the Upper Salt Formation$_B$ in the northern part of the basin is caused by salt dissolution due to circulation of undersaturated water, the genesis of the breccia being closely related to this solutional phenomenon. The relation between breccia thickness and salt dissolution intensity indicates that a subtantial part of the breccia results from an in situ accumulation of residual insoluble material. The relative thickness of salt removed by solution between the DP 104 and CUET 1 wells is estimated to be around 140 meters. This estimate of thickness is virtual and permits an approximation of the volume of reworked salt on the order of 0.07 km^3/km^2.

Salt dissolution may have occurred: (1) in alternation with salt deposition periods (synsedimentary and multistage dissolution); (2) immediately after the end of salt deposition (early monostage dissolution); or (3) in the later history of the basin, as dissolution beneath sedimentary cover. The last of these three hyptheses must be ruled out because of the sedimentological character of the breccia described above. In addition, isopach and isobath maps for the overlying formations in Etrez area do not reveal any collapse features. Two main facts provide supporting evidence for synsedimentary and multistage salt dissolution. Some horizons of brecciated deposits that lie intercalated within salt have been examined in cores in the Unit E1 of the VIRIAT well and of the Etrez wells. These horizons show the existence of several distinct dissolution phases. Within the solution-related breccia, typical brecciated facies only occur at certain levels, intercalated with greenish clay-carbonates that do not display any particular internal structure (e.g., cores from the DP 104 well). This indicates that the deposition of the breccia did not develop as a continuous process; rather the breccia originated in response to successive periods of salt dissolution, not from a single major solutional event.

From previous information and from that deduced from isopach maps, it is most likely that the breccia in the northern areas of the basin is the time-equivalent of the upper part of salt Unit E1 preserved in more southern regions, as suggested on the cross-sections (figures 4.24 and 4.37). In this case, the top of Salt (H4 limit) cannot be considered as a time marker. On the other hand, the horizons of clay-carbonate intercalated within halite (e.g., horizons a and a′ in figure 4.37) are not truncated towards the N but pass into the breccia. Thus it seems that the breccia was formed during successive salt sedimentation-dissolution cycles by complete leaching of pre-deposited halite and by in situ accumulation of insoluble residues.

Through time (i.e., up section) the dissolution

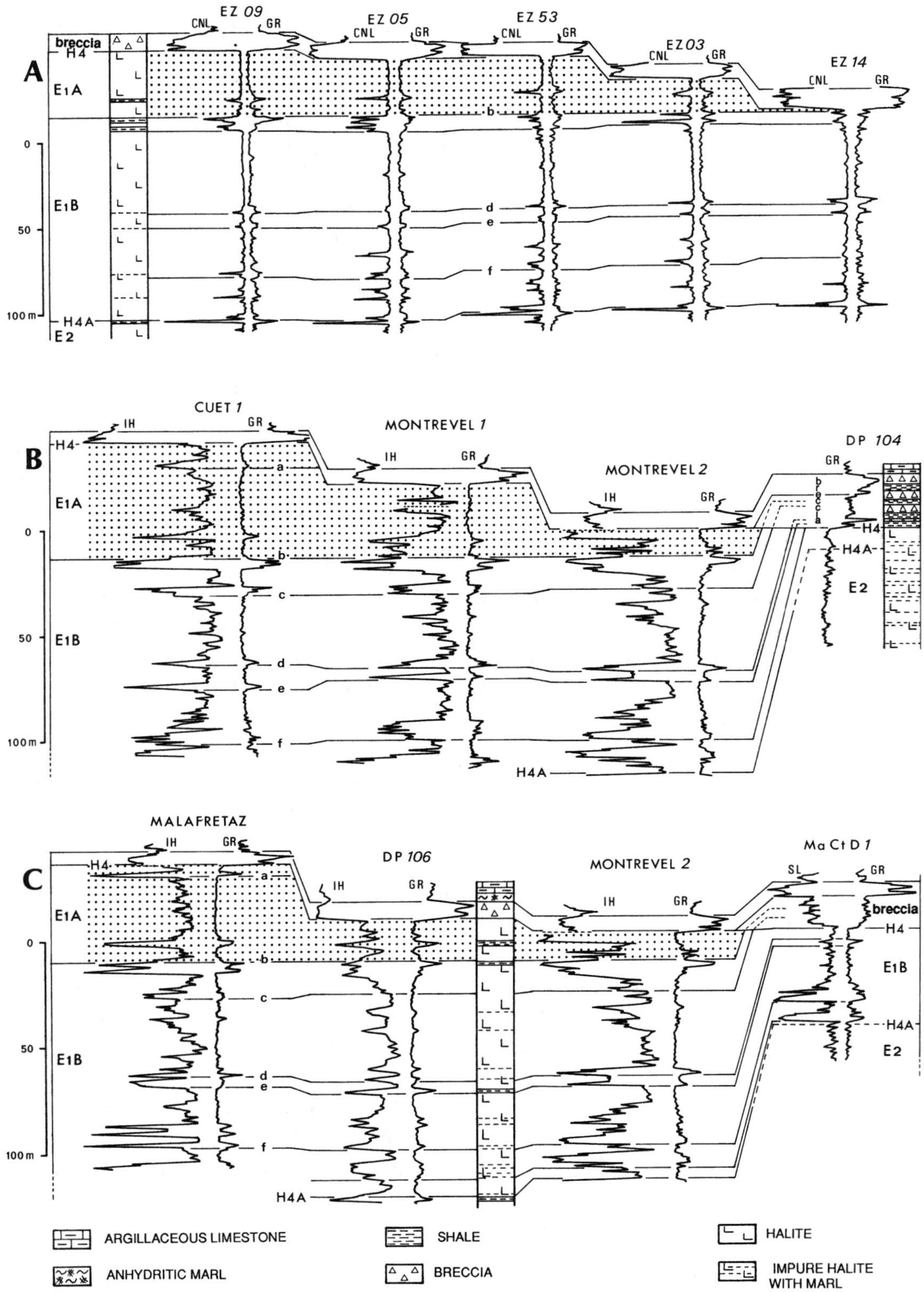

FIG. 4.37. Well log correlation in the northern part of the Bourg salt basin (after Curial 1986). See figure 4.38 for location of sections. Stippled interval represent the E1A sub-unit most affected by halite dissolution. IH, CNL = neutron-porosity; GR = gamma ray; SL = sonic log.

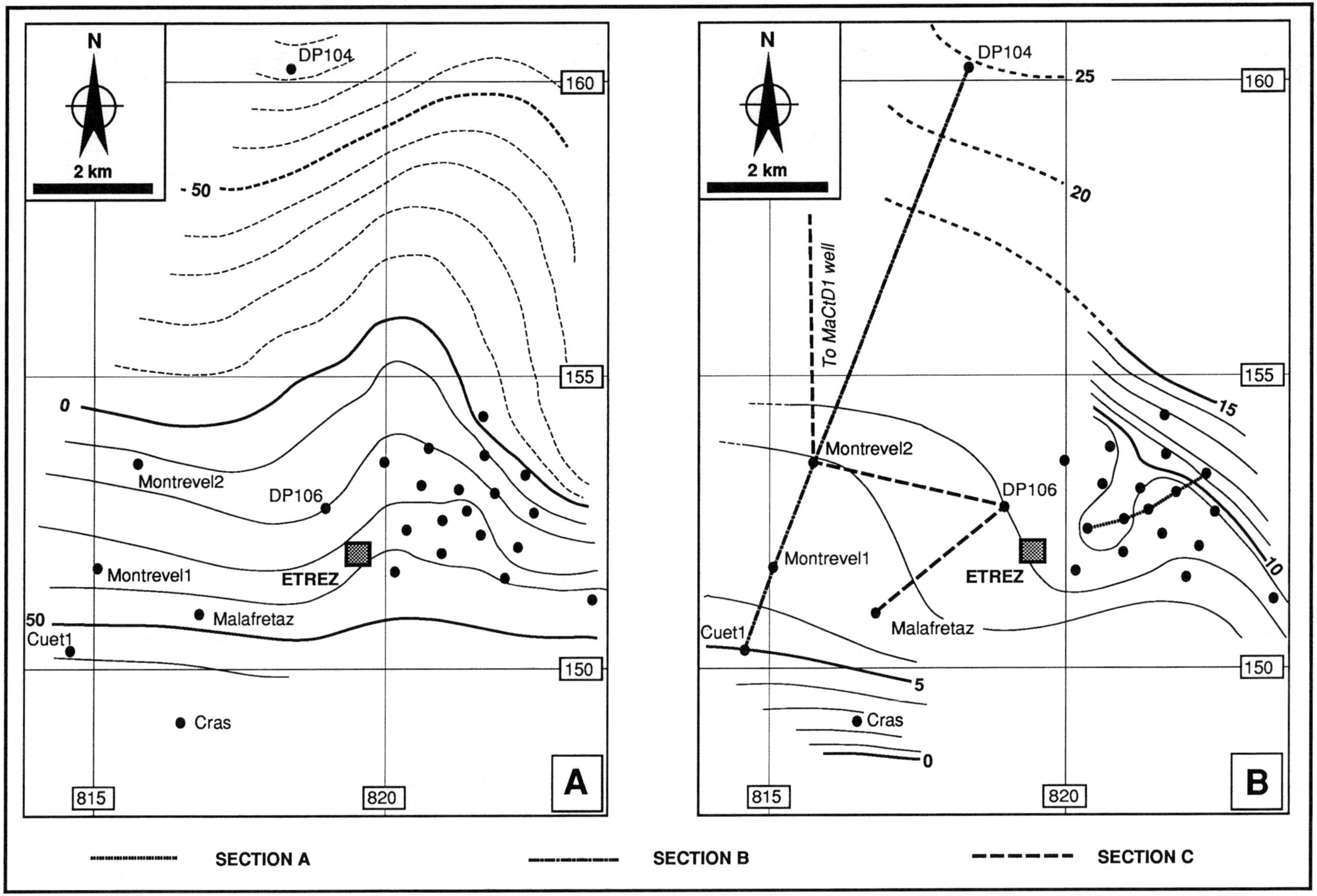

Fig. 4.38. Isopach maps:
A. Map showing E1A sub-unit isopachs (positive values in meters) in the northern part of the Bourg salt basin (after Curial 1986). Dotted line extrapolated from Montrevel and Etrez areas. Curve zero represents the disappearance of the E1A sub-unit, and negative values give an estimation of dissolved and wedging of salt thicknesses for the E1B sub-unit. B. Map showing breccia thickness as isopachs in meters (after Curial 1986).

phases became more and more effective towards the southern regions. This is related to the drift of the depocenter towards the south induced by the tilting of the Bourg Graben initiated at the beginning of deposition of Unit E2 (relative uplift of northern sectors). The salt sub-unit E1B is not greater than a few meters thick near the Cormoz area (DP 104 well) but its thickness increases to 100 meters and shows little variability across the middle sector of the basin (Etrez and Montrevel regions). In contrast, the salt sub-unit E1A is absent in northern areas (DP 104 and MaCtD1 wells) and it is greatly affected by dissolution across the middle sector (especially in the NE area of the Etrez field, located closer to the edges of the basin), whereas it seems to be well preserved in more southern regions (Pol 101 and GBg1 wells; see figure 4.27).

Tectonic activity seems to be intense throughout this period with block adjustments and a southward shift in subsidence, and is believed to be the main factor responsible for this dissolution phenomenon. During deposition phases, topographic irregularities produced by differential subsidence were compensated by the accumulation of salt and associated sediments. When the sedimentation rate became less than the subsidence rate (deficit in salt water influx and/or less efficient evaporatic conditions due to minor climatic changes) all new salt deposits became localized on the highest sectors, and such areas as the southern slope of the Cormoz basement high and eastern edge of the basin became vulnerable. On the contrary, in the low areas of greater subsidence, salt deposits were preserved. In fringing sectors, salt deposits could be either progressively exposed or covered by low salinity waters (as the upper part of a stratified water body). Therefore, salt could have been leached and removed to the southern central depression. The low bromine contents in halite in these areas are consistent with such salt dissolution/reprecipitation processes (figure 4.21).

Finally, an additional type of salt dissolution other than those discussed above also is noted in the Bresse Basin (Curial 1986) based on seismic data which have been synthetized in the form of isobath maps (Leroy 1984 and re-interpreted by Curial 1986/87). The isobath curves of the top of Salt (H4 marker) show two deep lateral depressions located along the western and eastern edges of the basin (figure 4.3) related to marginal salt dissolution (this hypothesis was suggested by Busson, oral communication, 1984). These troughs are filled by upper Oligocene and "Aquitanian" formations as illustrated by the BB′ cross-section for the western margin of the basin (figure 4.3). These facts point out that this second kind of major salt dissolution occurred later, beneath a sedimentary cover, probably because of the relative uplift of the northern part of the salt basin, and that such events were presumably active till the end of the "Aquitanian." However in the absence of more precise information about these marginal areas, we cannot be certain that dissolution processes are presently absolutely inactive.

Discussion and Conclusion

We will first discuss the facies observed in the Bresse salt basin in order to reconstruct the depositional environment (according to the geometry of the deposits), particularly for the main depocenter where the data are the most abundant. Then we review the origin of the mother brines which supplied the salt basin and gave birth to a volume of halite presently estimated to be about 160 km^3. Finally, we conclude with a synthetic history of the basin during the Paleogene, giving the main tectonic and sedimentary events.

FACIES REVIEW AND DEPOSITIONAL ENVIRONMENTS

The terrigenous material is composed of conglomerates, sands, and sandstones with variable amounts of clays, silts, and argillites, that are mostly reworked from the Mesozoic formations. The coarser clastic components are commonly restricted to the southern part of the basin, in the vicinity of the source-area. Conglomerates are also known in the periphery of the basin, near the basement highs and in the vicinity of the major faults. The southwestern and southern parts of the basin seem to correspond to an alluvial fan which was reactivated episodically during wetter periods. During these wetter periods deposition of the finest material—silts, clays, and vegetal remains—extended far to the north and reached the main depocenter where they sporadically accumulated as thick greenish to brownish claystone beds (e.g., Argilites de Base). The passage between these claystones (Etrez area) into carbonates (Louhans extension), as evidenced from well log correlation, suggests a longitudinal (S to N) gradient of increasing salinity during sedimentation.

The carbonates primarily occur as micritic calcite. In the southern part of the basin they sporadically contain clearly recognizable algal deposits, and some of them also exhibit desiccation and brecciated structures that indicate very shallow water conditions to probable subaerial exposure. In the central area of the basin carbonate rocks are commonly associated with calcium-sulfate (anhydrite) and sometimes with celestite. Anhydrite occurs mostly as nodules but evidence for formation of gypsum precursors is given by the presence of lenticular gypsum pseudomorphs replaced by anhydrite or halite during early diagenesis.

Primary halite facies mainly occur within units E4 and E3. They are composed of hopper-shaped crystals, sometimes grouped together in rafts, and chevron-shaped crystals. These fabrics formed within a free water body, and initiated at the air-brine interface (hoppers and rafts) or grew on the bottom (chevrons). The abundance of fluid inclusions in these types of primary halite resulted from a rapid crystallization due to a high evaporation rate. Numerous dissolution structures developed in the form of solution pits affecting the crystals, vertical solution cavities of several tens of centimeters, and solution surfaces more or less conformable to the general bedding. Each elementary sequence or cycle of salt accumulation is never more than 15 centimeters thick, and the top of each sequence corresponds commonly to a solution surface that truncates the crystals. This indicates that the brine pool was affected by rapid variations in salinity, these variations also being suggested by the vertical changes in bromine content within the individual beds. Such rapid oscillations in salinity may only affect a shallow water body which is easily influenced by changes in external conditions. As previously discussed, the spatial distribution of the terrigenous input across the depocenter area (e.g., Argilites de Base) also indicates a depositional environment covered by a shallow water depth. Primary salt facies are associated with horizons of laminar carbonates, commonly of depositional origin (horizons decimeters to several meters thick), which suggest long and stable periods during which the water body was stratified.

The secondary salt is dominant through units E5, E2, and E1. It is mainly made up of clear halite crystals that grew displacively within pre-existing sediment and variably altered the original bedding. These intrasedimentary halite precipitates are penecontemporeneous with the sedimentation and originated within a soft mud from interstitial saturated brines. Generally it is supposed that deposition of this type of salt usually takes place within a near-surface sediment exposed in subaerial conditions (salt flats). From analogous modern deposits from the Dead Sea, Gornitz and Schreiber (1981) also suggest that displacive halite may grow in a subaqueous depositional environment (no more than a few meters in water depth) by a per descensum circulation of the brines through the sediment. According to this concept, and because no trace of subaerial exposure has been clearly evidenced in the depocenter area of the Bresse basin, this type of halite is thought to be mainly formed within a mud covered by a shallow water depth. However, periods of subaerial exposure are not excluded. Sometimes the primary structures of the non-halite matrix are completely destroyed; this may be related to successive halite solution/reprecipitation processes (the haloturbation of Smith 1971). Such phenomena may only take place beneath shallow water and in subaerial conditions. The absence of fluid inclusions in these halite deposits indicates a slow growth of halite crystals and may be related to wetter periods during which evaporation rate was relatively low. The nucleation of pyramidal hoppers at the surface of the brine body may be less effective than during drier periods. In the absence of a sufficient number of nuclei sinking to the bottom of the pool, on-floor crystallization cannot easily

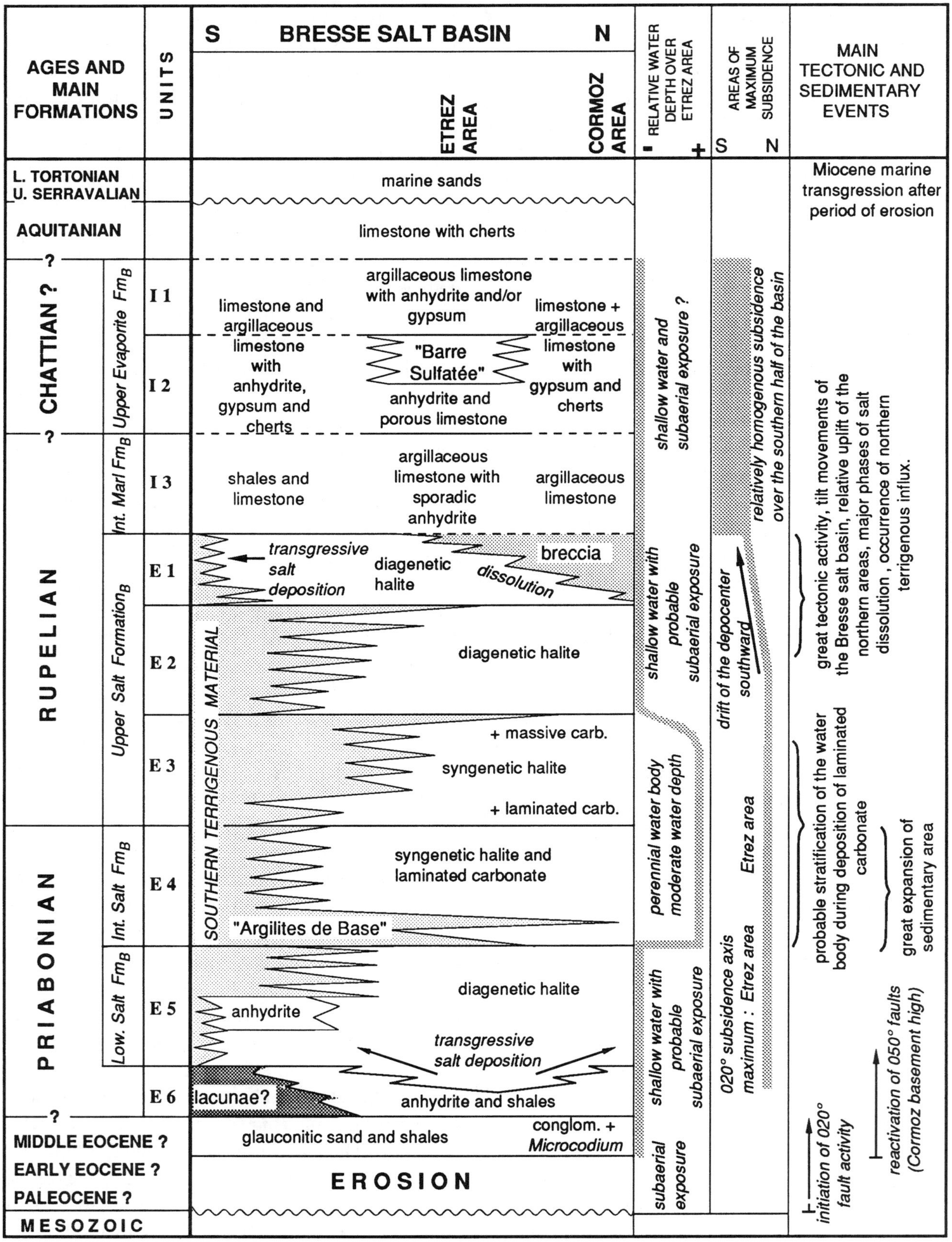
AGES AND MAIN FORMATIONS
UNITS
S BRESSE SALT BASIN N
ETREZ AREA
CORMOZ AREA
RELATIVE WATER DEPTH OVER ETREZ AREA
- +
AREAS OF MAXIMUM SUBSIDENCE
S N
MAIN TECTONIC AND SEDIMENTARY EVENTS
L. TORTONIAN
U. SERRAVALIAN
marine sands
Miocene marine transgression after period of erosion
AQUITANIAN
limestone with cherts
CHATTIAN ?
Upper Evaporite Fm
I 1
limestone and argillaceous
argillaceous limestone with anhydrite and/or gypsum
limestone + argillaceous
I 2
limestone with anhydrite, gypsum and cherts
"Barre Sulfatée"
anhydrite and porous limestone
limestone with gypsum and cherts
Int. Marl Fm
I 3
shales and limestone
argillaceous limestone with sporadic anhydrite
argillaceous limestone
RUPELIAN
Upper Salt Formation
E 1
transgressive salt deposition
diagenetic halite
dissolution
breccia
E 2
diagenetic halite
E 3
+ massive carb.
syngenetic halite
+ laminated carb.
PRIABONIAN
Int. Salt Fm
E 4
syngenetic halite and laminated carbonate
SOUTHERN TERRIGENOUS MATERIAL
"Argilites de Base"
Low. Salt Fm
E 5
diagenetic halite
anhydrite
transgressive salt deposition
E 6
lacunae?
anhydrite and shales
MIDDLE EOCENE ?
glauconitic sand and shales
conglom. + Microcodium
EARLY EOCENE ?
PALEOCENE ?
EROSION
MESOZOIC
shallow water and subaerial exposure ?
shallow water with probable subaerial exposure
perennial water body moderate water depth
shallow water with probable subaerial exposure
subaerial exposure
relatively homogenous subsidence over the southern half of the basin
drift of the depocenter southward
Etrez area
020° subsidence axis maximum : Etrez area
great tectonic activity, tilt movements of the Bresse salt basin, relative uplift of the northern areas, major phases of salt dissolution, occurrence of northern terrigenous influx.
probable stratification of the water body during deposition of laminated carbonate
great expansion of sedimentary area
reactivation of 050° faults (Cormoz basement high)
initiation of 020° fault activity

FIG. 4.39

develop and the precipitation of halite occurs preferentially within the sediment.

In summary, the salt facies observed in the Etrez area within units E5, E2, and E1 suggest a shallow water depositional environment with sporadic subaerial exposures (figure 4.39). During the deposition of units E4 and E3 the water depth was shallow but the brine body was substantially deeper than for the deposition of the other units, and presumably was never completely evaporated. At the basinal scale, the Bresse basin may be considered a wide flat area in which salt deposits and terrigenous supply competed in progressive fill of the depressions induced by synsedimentary subsidence.

SOURCES OF MOTHER BRINES

The origin of mother brines (continental or marine) is presently uncertain because certain facts seems to be inconsistent:

1. The depositional assemblage of the evaporites of the Bresse basin (halite and calcium-sulfate) suggests formation from marine water. None of the particular evaporitic minerals (such as natron, trona, mirabilite) that are common within evaporite sequences of continental origin (Eugster 1984) have ever been recognized in the Bresse deposits.
2. The bromine content in halite through units E5 to E3 (more than the theoretical 70 ppm appropriate to marine water) also suggests a marine source of the mother brines.
3. It is unlikely that the Paleogene evaporites of the Bresse basin (the volume of salt deposits is estimated to be 160 km^3) may be completely reworked from more ancient evaporites. The evaporite sequences of the Trias, which are the single previous salt formations in this region, are precisely correlated over a large surface area from the Bresse basin (Herry 1983) to the Alsace and Lorraine regions (Marchal 1983). Locally these sequences are truncated by dissolution surfaces, but the volume of reworked evaporites is not sufficient to explain the whole of the Bresse evaporites.

In spite of these facts a completly marine source, or influence of marine water at certain periods only, cannot be firmly demonstrated: no un-reworked biological remains typical of marine origin (such as foraminifera) have been detected through the Paleogene formations of the Bresse basin (Moretto 1986/87). In the absence of intermediate deposits between these confined facies and the nearest Paleogene marine deposits located in the Subalpine Ranges (about 100 km apart toward the ESE), no entry for potential marine water inlet has been clearly recognized.

In summary, none of the proofs for either a continental or a marine origin (or mixed) for the mother brines are presently evidenced. But the previous facts and the existence of unquestionable marine deposits in the Valence Salt basin (Dumas 1986, 1988; Dromart and Dumas, this volume) suggest the possibility that some if not all of the salt of the Bresse basin may be formed from marine water: inlets of marine water from the Alpine sea via a channel system might have come across the Jura chain, which had a low relief during the Paleogene, or across the southeastern end of the Bresse basin.

SYNTHESIS AND CONCLUSIONS

Two major evaporitic periods occurred during the Paleogene that gave rise successively in the Bresse basin to halite (and sulfate) deposits which mainly form units E6 to E1 (first period), and to sulfate and carbonate deposits of units I2 and I1 (Upper Evaporite $Formation_B$ = second period). From palynology the first evaporitic period has been dated as partly of the Priabonian and also Rupelian ages, owing to the presence of the pollen *Boehlensipollis*

FIG. 4.39. Table contrasting time lithologic units and relative tectonic and depositional environments.

hohli, which suggests that the Eocene-Oligocene boundary falls approximately at the top of the Intermediate Salt Formation$_B$. However, more recent palynological studies concerning the same sequences in the entire Rhine-Rhône complex slightly contradict early works (Schuler 1988).

The first evaporitic period begins with sulfate-bearing deposits (Unit E6) located at the beginning in the center of the basin. While this type of sedimentation rapidly spreads centrifugally across the basin, the salinity of the brines increases and brings about halite precipitation. The maximum extension of the depositional area of halite is reached during the deposition of the lower part of Unit E5 (figure 4.39). Then detrital input coming from the south restricted the deposition of halite within the central and northern part of the basin. The terrigenous supply became more and more significant and spread toward the north with an associated decrease in grain size. In the lower part of Unit E4 (Argilites de Base) the detrital material reached the depocenter area and alternated with brief periods of halite deposition. These oscillations in the surface area of halite deposition result from a competition between the evaporite sedimentation and the southern detrital input in the progressive filling of the depressions induced by synsedimentary subsidence (this depositional style is evidenced throughout the entire period of salt sedimentation).

The deposition of Unit E4 and the lower half of Unit E3 took place under distinctly different conditions than prevailed during the sedimentation of the other units: an increase of the water depth probably occurred, which would have been favorable to the establishment of a stratified water body and which may have persisted for a long time during certain periods. Such a perennial stratified brine pool would have allowed: (1) a wide distribution of fine clay particles and the development of laminated undisturbed facies all over the basin; (2) the development of planktonic organisms in the upper water layer having lower salinity and the preservation of the organic matter on the bottom due to low oxygenation of the lowest water body (see Curial et al. 1990); (3) preservation of the primary halite facies without intense dissolution/recrystallization processes in the deepest areas of the basin.

During the deposition of the upper part of Unit E3, salt deposits are restricted to the northern region of the basin, but they spread again to the south during Unit E2 and reached their greatest southward extension during Unit E1. The general southward transgression of the halite deposits results from the tectonic uplift of the northern part of the salt basin (Cormoz high and adjacent areas). This tilting seems to have been initiated during deposition of Unit E2 and became more effective during Unit E1 with major phases of deposition/dissolution of halite deposits which affected all of the northern part of the salt basin. This phenomenon induced the formation of a widespread breccia which occurs at the top of the Upper Salt Formation$_B$.

Preceded by a small episode of sulfate deposition and then a prolonged period of argillaceous carbonate sedimentation (Intermediate Marl Formation$_B$), the second evaporitic period then began (Upper Evaporite Formation$_B$). These restricted conditions gave rise to widespread sulfate deposits. Then the sedimentation evolved in a presumably lacustrine context and closed the Paleogene sedimentary cycle.

The distribution in space and time of the evaporitic deposits of the Bresse salt basin results from two main factors which governed the sedimentation independent of the salt water supply and the question of its origin (marine or nonmarine). The first factor is climate: climatic oscillations determined a succession of drier periods favorable to evaporite formation and wetter episodes during which the influx of continental water carried the terrigenous material into the basin. These two opposite types of sediment competed in the progressive filling of the depressions. The second factor is tectonism: subsidence determined the physiography of the basin, created the necessary space for sedimentary filling, and variously permitted the deposition and preservation of the halite facies.

5

The Salt Basin of Valence (France)

Gilles Dromart and Daniel Dumas

Geological Setting

The Paleogene salt basin of Valence is situated in southeastern France, in the Rhône valley, just north of the city of Valence (figure 5.1). In plan view it is shaped like a boomerang about 50 km long, 10 km wide, and trends roughly N-S. It lies at the eastern edge of the Paleozoic Massif Central and just west of the subAlpine ranges of Vercors.

Paleogene salt deposits are not exposed at the surface and the data presented in this paper come entirely from subsurface investigations: seismic profiles, borehole cuttings, cores, and well logs. Study of the area was initiated in 1948 with mapping of gravimetric anomalies. The earliest work focused on oil exploration, but was abandoned in 1963. Later work focused on the salt formations for hydrocarbon storage repositories, salt exploitation by in situ induced dissolution, and exploration for potash. A total of 30 boreholes have been drilled in the salt basin, providing substantial and valuable information (figure 5.2).

The graben of Valence is structurally linked to the major Paleogene rift system of western Europe (Rouchy, 1996) and is filled by a thick Paleogene section that exceeds 2,000 m at the depocenters (figure 5.3). Cross-sections drawn across the basin (figure 5.4) have been reconstructed from seismic lines together with associated well-logs. They illustrate the structurally controlled double asymmetry of the graben: an E-W asymmetry, in which the steeply dipping, sharply defined, western margin contrasts with the gradual slope east of a transverse basement high (020°-oriented Montmiral horst) and the marginal Royans subbasin; and a N-S asymmetry, in which normal faults converge irregularly towards the north and pinch out against a 050°-oriented uplift, in contrast to the regular updip rise of the basement to the South. Two main fault directions controlled basinal development: 020° and 160°, together with rare 120°-oriented faults. The Mesozoic portion of the Vercors ranges is overthrust upon basinal sediments at the eastern margin of the graben (up to several tens of km, according to Doudoux et al. 1982).

Stratigraphy

LITHOSTRATIGRAPHY

The Paleogene sequence ranges from 1,000 to 2,250 m in thickness and rests unconformably on a basement that is variably Hercynian to Cretaceous in age. The lower limit is the seismic reflector referred to as marker H4, clearly recognizable throughout the basin.

The lowermost part of the sequence, referred to as the Subsalt Formation, is poorly known because it was only penetrated at two sites (Cl1 and GVA1 wells, figure 5.2). It consists mainly of greenish-gray shales with various amounts of silt. Close to the

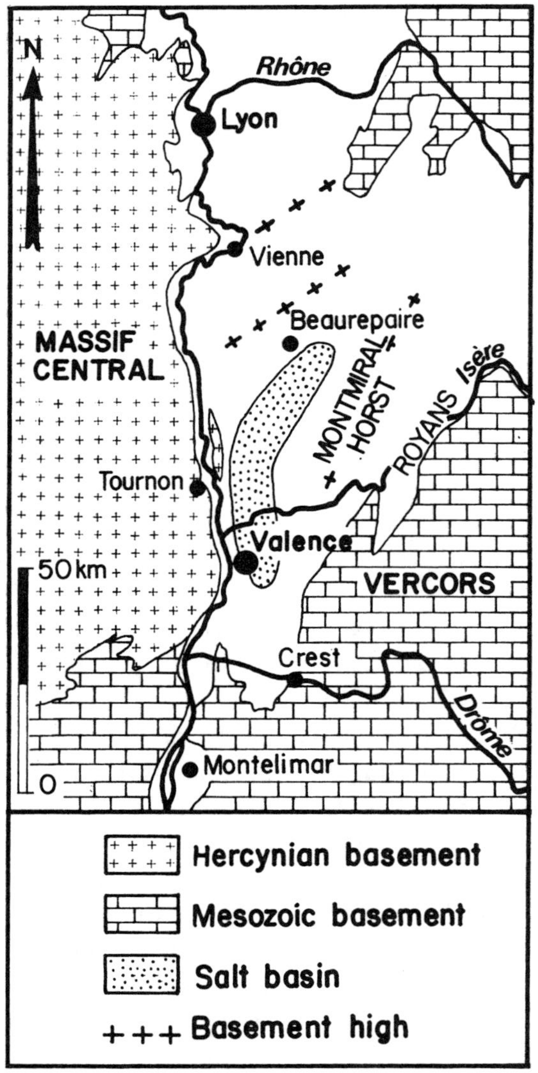

FIG. 5.1. Geological setting of the salt basin of Valence, southeastern France (adapted from Dumas 1986).

top, these lithologies grade into the presalt deposits of finely laminated limestones enriched in organic matter, massive fine-grained limestones, and sulfates in the form of massive anhydrite beds.

The overlying Lower Salt Formation is better known because of intensive coring. Here the dominant lithology is halite developed as several different lithofacies, such as bedded, milky, and argillaceous salts. The formation has been subdivided into three units (figure 5.5) described in stratigraphically ascending order:

The lowermost unit, T6, is made of halite enclosing thin, rhythmic shale layers frequently interrupted by calcareous shale beds, meters thick, that bear anhydrite. The thickness of Unit T6 ranges from 200 m at the depocenter to 100 m on the margins of the basin.

Unit T5, 100 to 250 m thick, subdivided into T5–3, T5–2, and T5–1, consists of intercalations of halite or limestones (massive or laminated) and calcareous shales. Anhydrite is common throughout the unit.

The uppermost unit, T4, is fairly uniform in thickness (ca. 250 m). Unit T4 has been split into six horizons (A to F from the top to the bottom, figure 5.6). The dominant lithology is halite with shaley intercalations, meters thick, with the shale mostly restricted to the lower part. The top part is composed of interbeds of laminated carbonates and anhydrite (Horizon B),

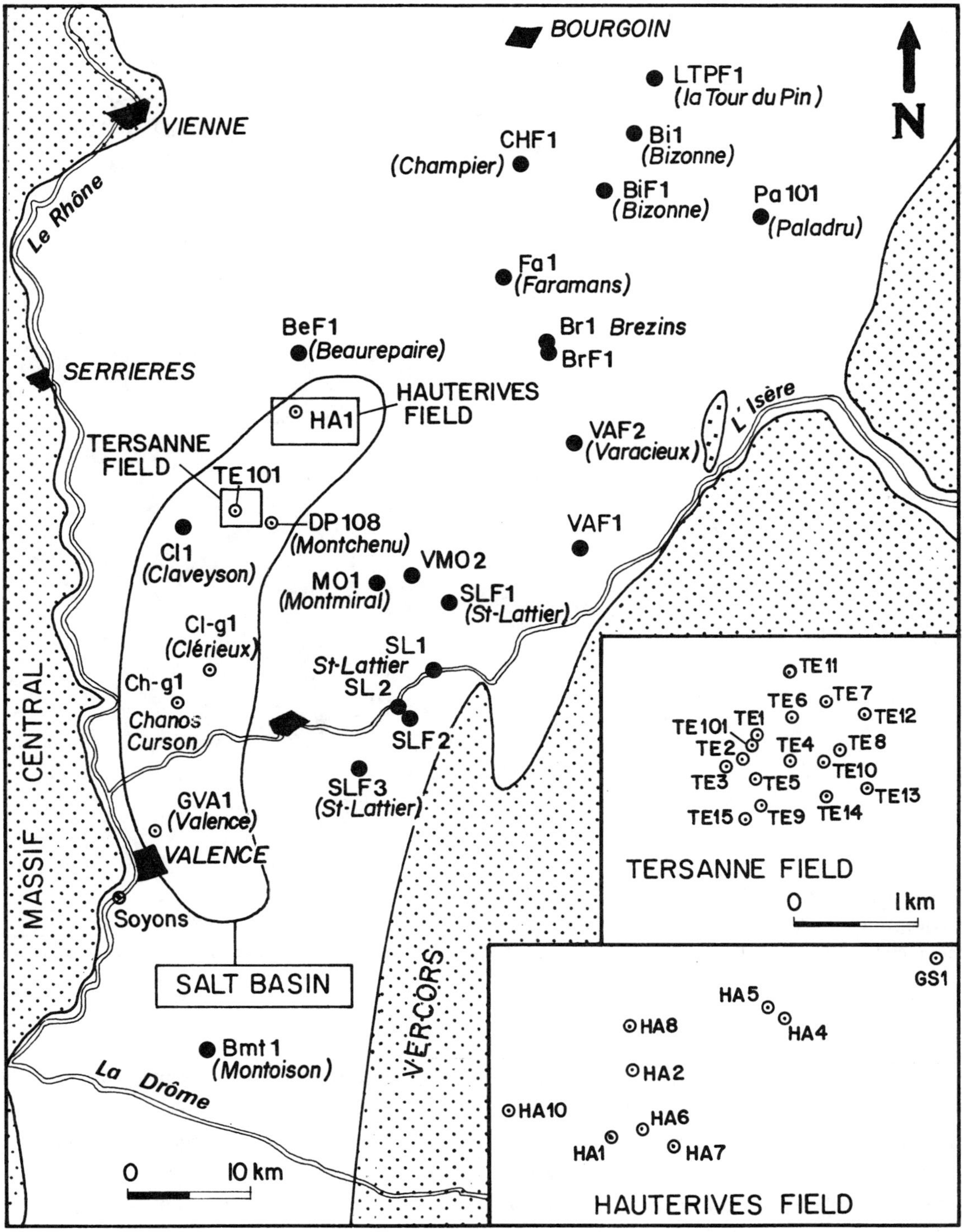

FIG. 5.2. Location map of wells through the Cenozoic basin of Valence (adapted from Dumas 1988).

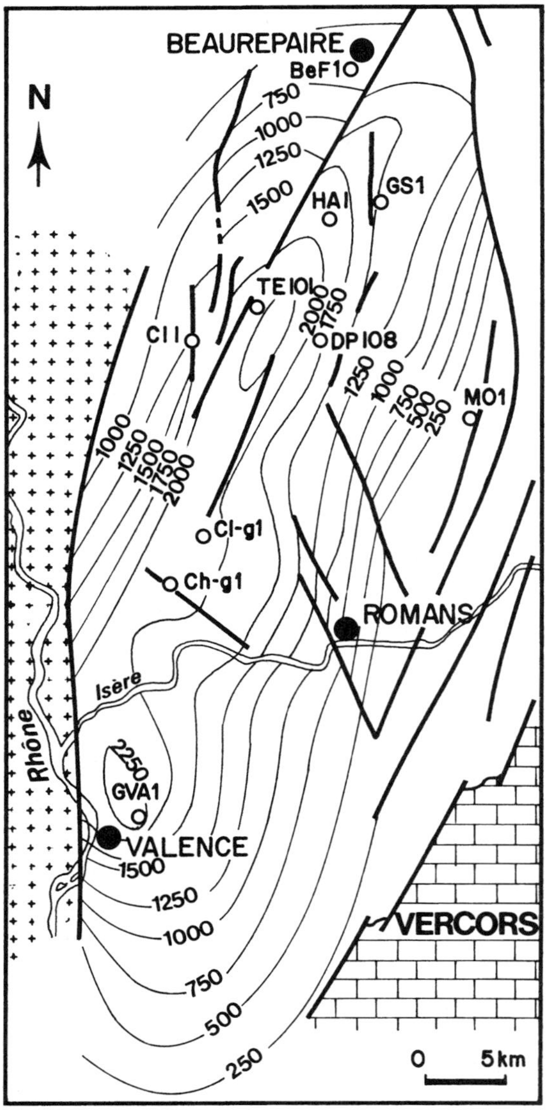

FIG. 5.3. Restored isopach and tectonic map for the Paleogene sequence of the Valence Basin. Thicknesses are in meters (adapted from Dumas 1988).

overlain by a zone, about ten meters thick, of pure halite (Horizon A). The upper limit of the Lower Salt Formation coincides with the seismic reflector, H2, that is clearly present and well developed throughout the basin. This reflector is formed by a sharp lithologic change from deposition of salt to shales.

The Intermediate Marl Formation is largely composed of siliciclastic material. The total thickness is fairly constant (ca. 500 m) and the deposits show only minor lateral changes through the central graben. Four successive lithologic types have been defined in ascending order:

gray calcareous shales with silty interbeds, carbonate layers enriched in organic matter, and anhydrite beds;

silty shales interbedded with argillaceous sands and bearing mollusk shells and sporadic foraminifer tests together with calcareous nannoplankton;

variegated shales enclosing beds of black shales and carbonate-cemented siltstones;

variegated shales with carbonate beds containing variable amounts of anhydrite. Close to the top, the carbonates contain oolites and have undergone early dolomitization and locally severe diagenetic anhydritization.

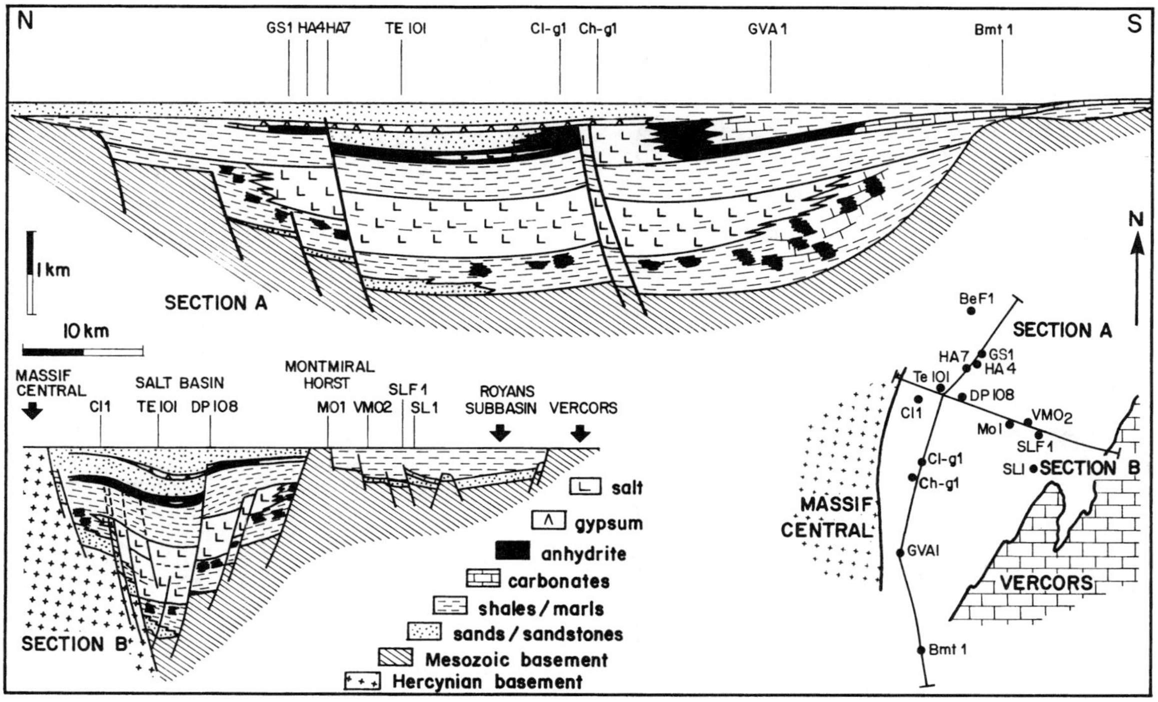

FIG. 5.4. Longitudinal and transverse diagramatic cross-sections of the Paleogene Valence Basin. The section lines are indicated on the map (lower right) (adapted from Dumas 1986).

The Upper Evaporite Formation is less uniform and shows major lateral changes in both nature and thickness of deposits (figure 5.7). The lithologic section is made up of halite, sulfates, and carbonates with a minor amount of shale. Where present, the halite deposits are up to 500 m thick and pass transitionally to the north and south into carbonate-sulfate facies having a decrease of one order of magnitude in thickness.

The Terminal Paleogene Formation is largely composed of terrigenous material. Where no halite facies are developed, the lowermost deposits of the Terminal Formation are either carbonate/shale interbeds (southern area, GVA1 well), or "Radioactive" sands, bearing both feldspar and heavy mineral components (northern area, Tersanne field). "Radioactive" refers to the characteristic response signature on the gamma-ray log. Above these sands are gray marls (which contain gypsum) and then a clearly defined well-log marker, a glauconite sand bed. Overlying the glauconite are thick deposits (300 m) of silty variegated shales, in turn locally covered by coarse sands that terminate the Paleogene lithologic sequence.

BIOSTRATIGRAPHY

Biostratigraphic markers are scarce throughout both the terrigenous and the evaporitic Paleogene accumulation (figure 5.8). Calcareous nannofossils have such a low diversity that they are of no stratigraphic value. Only spores and pollen provide reliable age assignments despite the commonly poor conditions of these remains. The first studies in the Valence Basin were made by Chateauneuf (in Alabouvette et al. 1984) and additional investigation is consistent with them (Schüler, personal communication in Dumas 1988).

On the basis of these studies the uppermost part of the Lower Salt Formation (Unit T4 and top of Unit T5) has been assigned to Early Rupelian because of the co-occurrence of *Boehlensipollis hohli* and *Slowakipollis hippophaeoides.* The first appear-

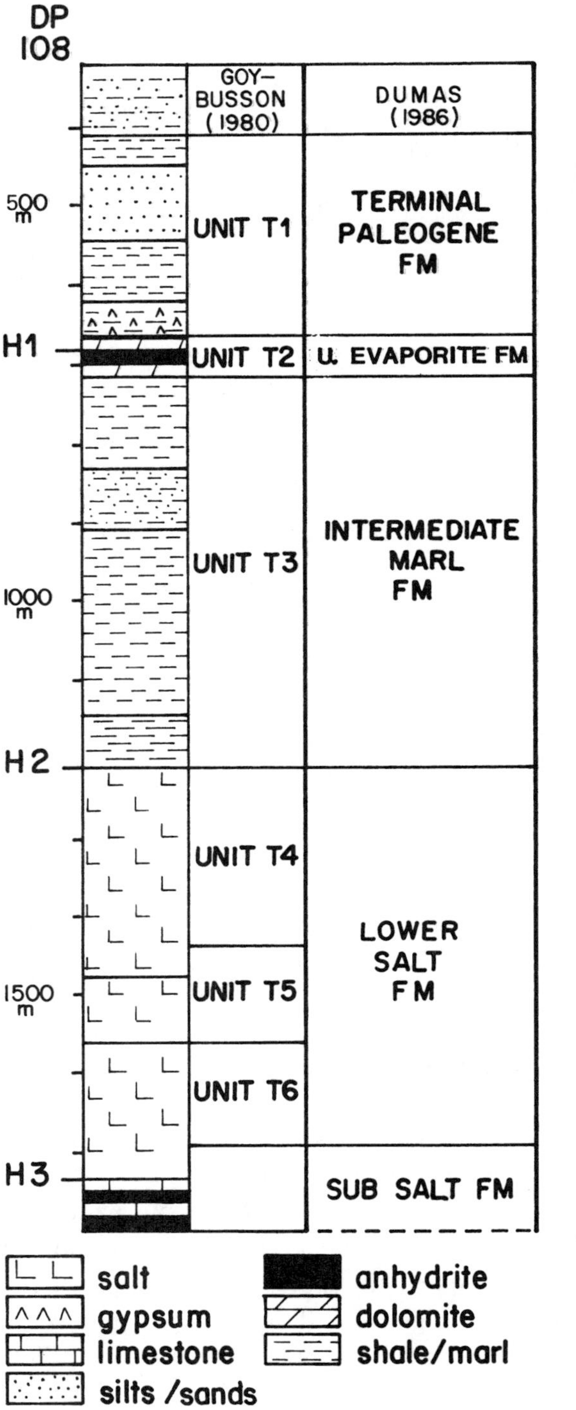

Fig. 5.5. Lithostratigraphic column and terminology of the Paleogene Salt sequence as developed from the DP 108 well (core, well-log, and seismic data; location in figure 5.2) (adapted from Dumas 1988).

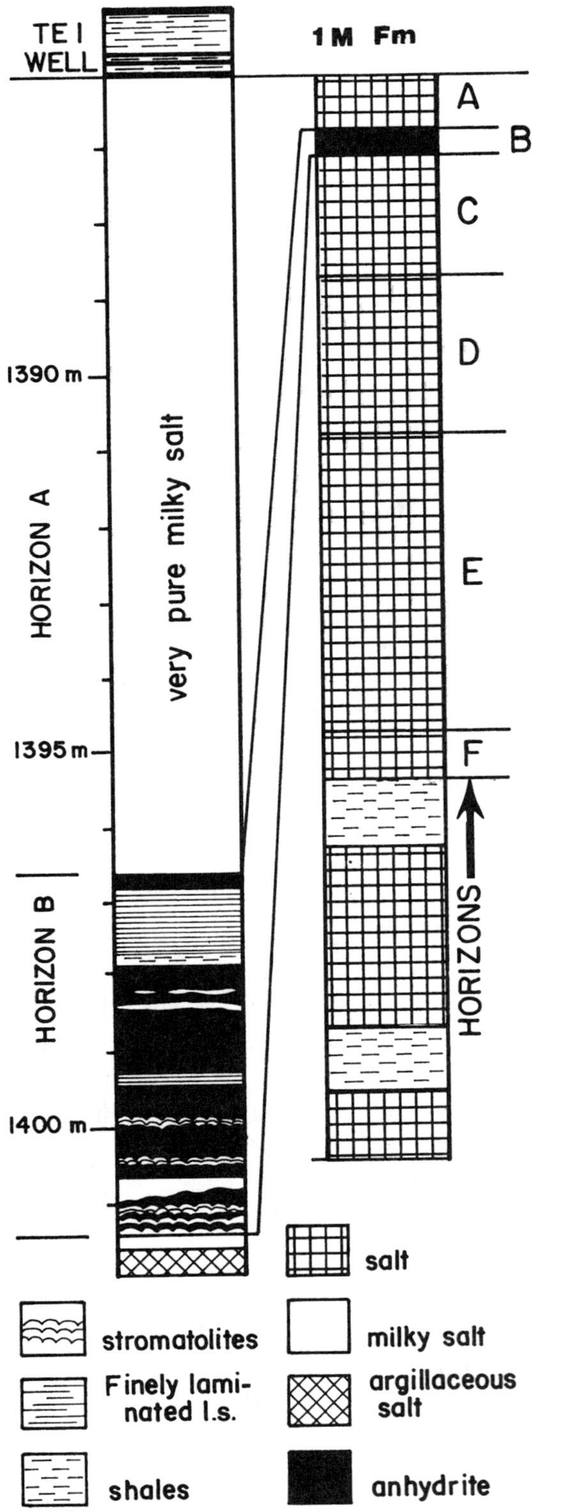

Fig. 5.6. Lithostratigraphic terminology for Unit T4 of the Lower Salt Formation (Tersanne field, Valence basin) and detailed lithologic column of Horizons A and B (TE1 well) (adapted from Dumas 1988).

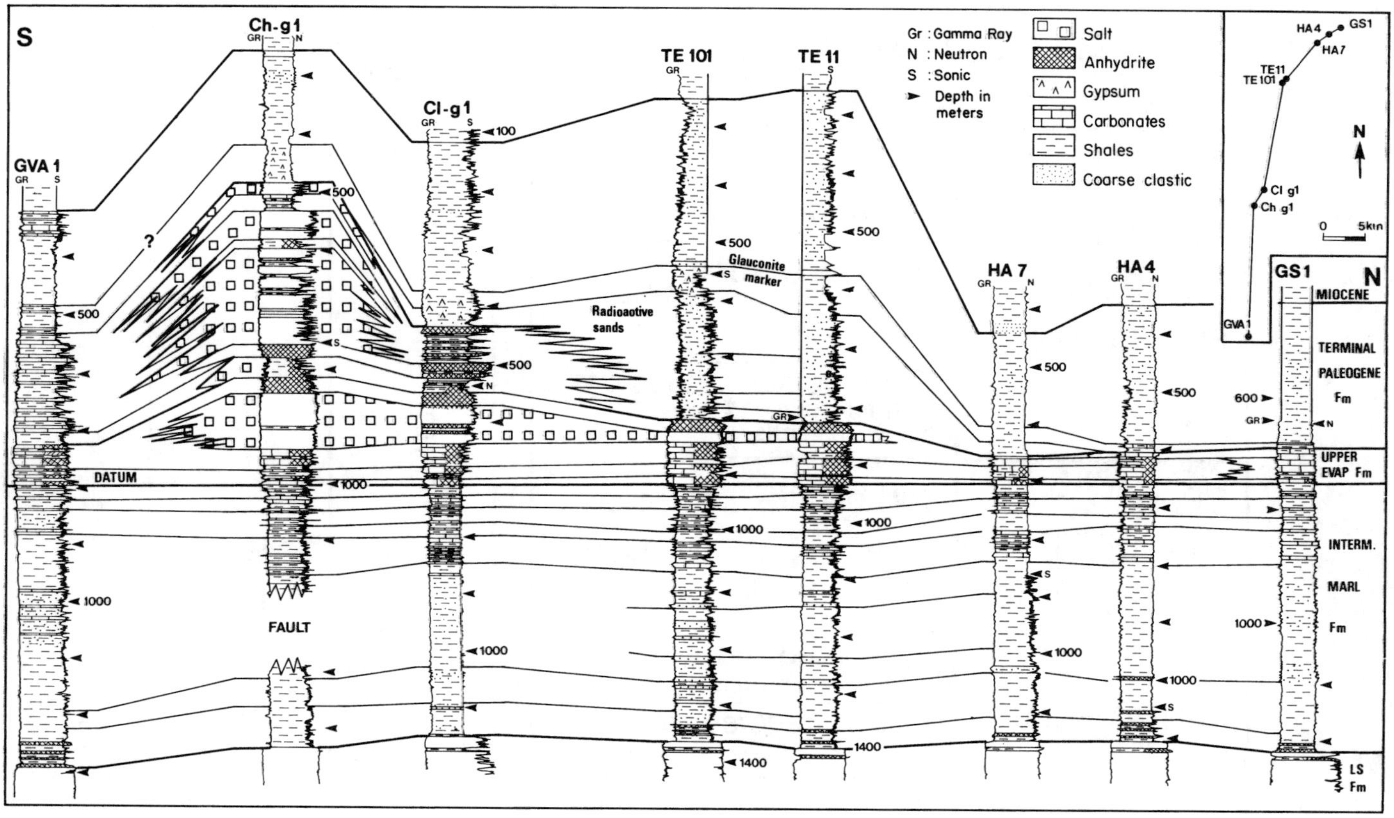

FIG. 5.7. Tentative well-log correlation for the Intermediate Marl, Upper Evaporitic, and Terminal Paleogene Formations across a sublongitudinal section (see inset, upper right, for locations) of the Valence basin (adapted from Dumas 1986).

ance of the pollen genus *Boehlensipollis* above the bottom of the sequence suggests that the Eocene-Oligocene boundary may be placed approximately at the passage from Unit T6 to Unit T5. It may be below this level but no more accurate assignment is possible because of the very poor preservation of the flora within Unit T6 (Schüler 1988). Finally, the palynological assemblage present within the mollusk-bearing sandy shales of the Intermediate Marl Formation indicates a late Rupelian age.

LITHOLOGIC NATURE AND SEDIMENTARY STRUCTURES OF DEPOSITS

The Paleogene evaporative section of the Valence Basin is made up of many different lithologies, almost all of which seem to have formed under somewhat restricted conditions. Based on mineralogy, there are at least five basic sediment types—terrigenous, carbonate, sulfate, halite, and mixed sulfate-halite—with numerous subdivisions.

Terrigenous deposits.

Two basic terrigenous lithologies are developed throughout the Paleogene sequence of Valence: sandstone and shale. The salient lithologic features are presented in table 5.1.

The coarse clastics include rounded quartz grains that were reworked from the Mesozoic section and lower Eocene sands that lie on the margin of the basin. In addition, the large number of quite fresh feldspar grains along with many angular quartz particles suggest that part of the coarse clastic component was probably derived, together with

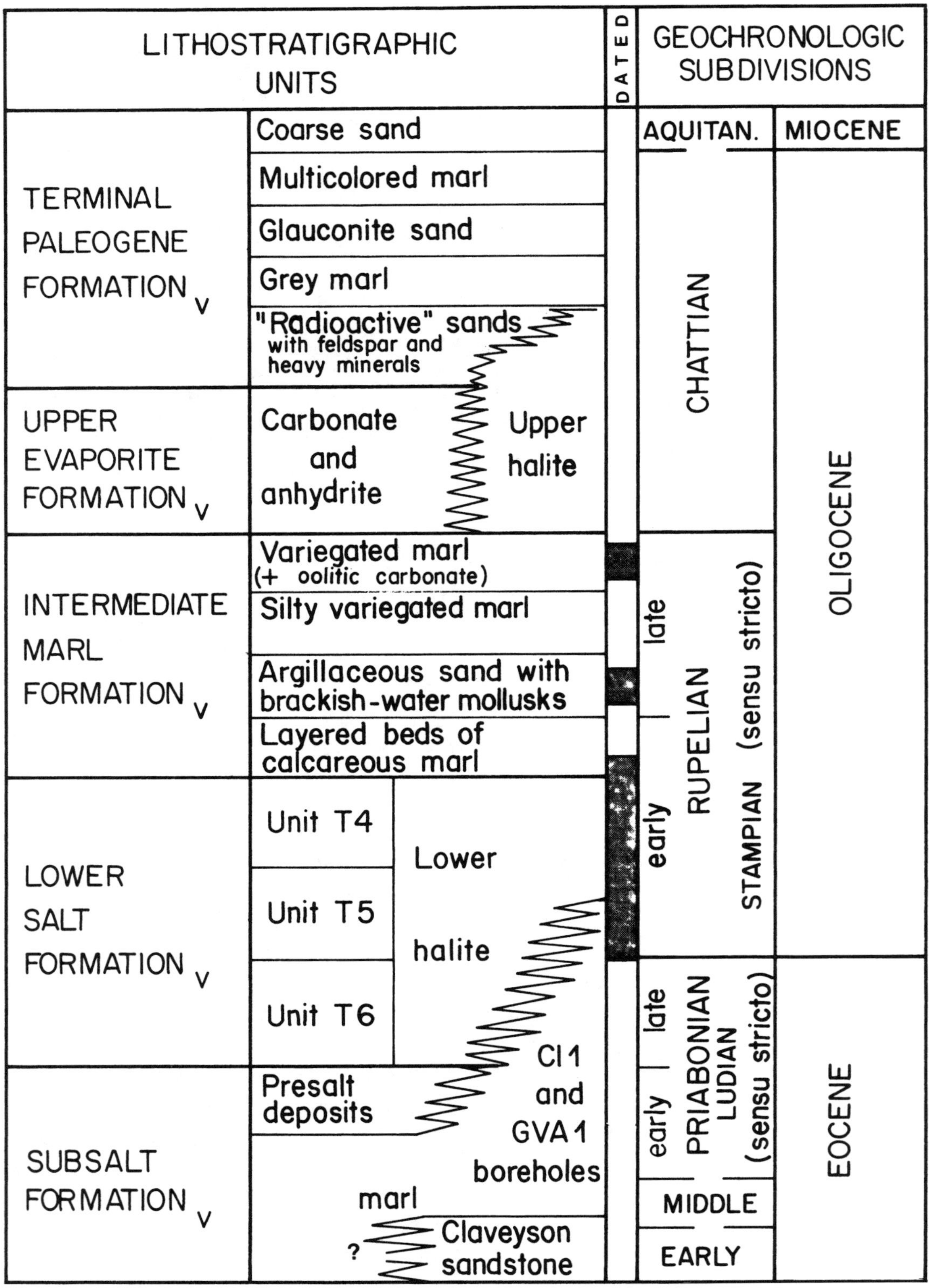

FIG. 5.8. Stratigraphic diagram for the lithologic units of the Valence Graben. Age assignments are based upon palynological data (adapted from Dumas 1986).

TABLE 5.1.
Sedimentological characteristics of the terrigenous facies in the Paleogene sequence of the Valence Basin.

LITHOLOGY	TYPE	STRATIGRAPHIC POSITION	MACROSCOPIC FEATURES	COMPOSITION	DEPOSITIONAL STRUCTURES	DIAGENETIC FEATURES	OTHER FEATURES
Sandstones		I M Fm S S Fm	Gray, matrix-, to grain-supported	Silt- to sand-sized grains: Quartz, Feldspars, (Biotite, Muscovite, and Chlorite) Clayey-carbonate matrix	Cross-bedding (low-angle) ? flute,groove, and ripple casts burrows		
Shales (Marls)	Laminated Silty Shales	Throughout	Thin, parallel lamination of dusky reddish shales and light-colored siltstones	Two clay assemblages: 1) Illite - (Smectites - Kaolinite - Chlorite) 2) Illite - Smectites - (Kaolinite-Chlorite) Origin Illite: mixed Smectites: tied with carbonate-rich facies Kaolinite: detrital Chlorite: detrital	Rhythmical bedding Microsequence: shales silts Sharp basal contacts Load casts Desiccation cracks	Illite (partly diagenetic)	Organic Matter Content: SEEN IN: IFm; TE & Well * Type III (plant debris) * TOC: 0.88% * Petroleum Potential : medium, S2 = 2.88
	Homogenous Shales		Gray or variegated (blue, green/red)		Regular, horizontal bedding	Smectite possibly diagenetic	

kaolinite and chlorite, directly from nearby igneous and metamorphic rocks of the Hercynian basement.

Ripples, commonly noted in the Paleogene sand lithologies, are a signature of current movement (figure 5.9A). The ripples are thought to have formed in shallow water because it is unlikely that repeated and marked bottom-currents can have developed in a deep water basin with so limited a volume. Handford and Fredericks (1980) describe similar depositional structures from the Red Cave formation (Permian, Texas) and characterize such deposits as the distal zone of wadi plain systems.

The shale occurrences, apart from the "laminated silty" facies (figure 5.9B), suggest deposition under low-energy conditions, and do not provide any significant information about the depositional depth.

Carbonate deposits.

The features of carbonates found in the Paleogene evaporitic sequence are summarized in tables 5.2 and 5.3 under three main groupings: fine-grained carbonates, cryptalgal limestones, and coarse-grained limestones. The classification is somewhat arbitrary because some components may appear within more than one general grouping.

1—*Fine-grained carbonates* from the Paleogene deposits may be conveniently divided into three categories, based on bedform and composition:

(a) Massive bedded, fine-grained limestones, composed of somewhat burrowed, micritic carbonate, appear to have accumulated primarily under low-energy and low-salinity conditions.

(b) Finely-laminated limestones (figure 5.10 A and B and table 5.3) that are almost entirely lacking in traces of bioturbation were apparently deposited

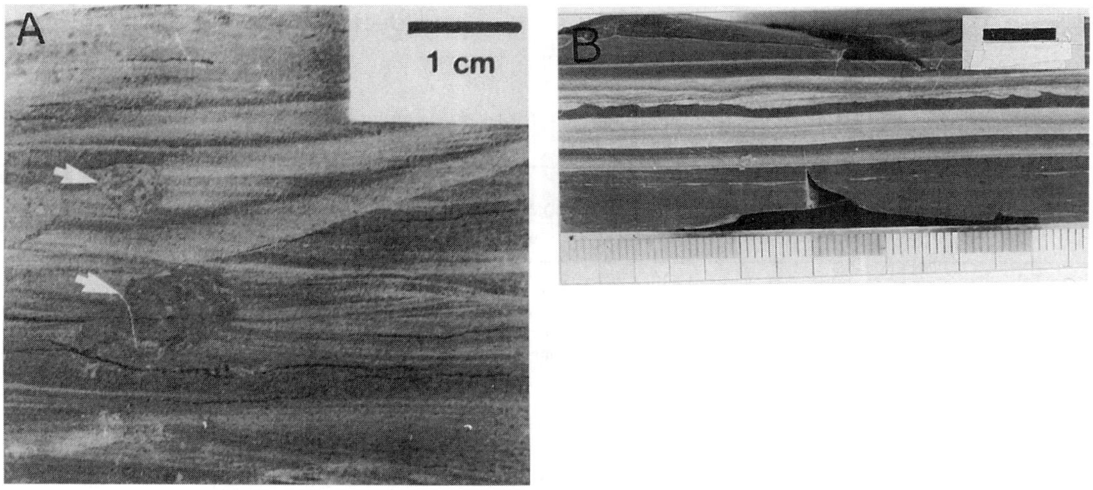

FIG. 5.9. Siliciclastic lithologies in the Intermediate Marl Formation seen in polished core slabs.
A. Sandstone lithology with low-angle cross-bedding in a shaley sandstone. Burrowing traces are distinguishable (arrows). DP 108 well; 1048 m.
B. Laminated silty-shale lithology with fine interbedding of silts (light-colored) and shales (dark-colored). Passages from silt layers to overlying shale layers may be either gradational or abrupt. Soles of silt layers are very sharp with occasional minute load casts (b). TE1 well; 1343.3 m. Scale 1 cm.

TABLE 5.2.
Sedimentological characteristics of the carbonate facies in the Paleogene sequence of the Valence Basin: (A) Described by grain size and composition; (B) Described by bedding structure.

A LITHOLOGY	TYPE	STRATIGRAPHIC POSITION	DEVELOPMENT AND MACROSCOPIC FEATURES	B COMPOSITION	DEPOSITIONAL STRUCTURES	DIAGENETIC EVIDENCE	OTHER FEATURES
Fine-grained carbonates (mudstones)	Massive bedded limestone	I M Fm L S Fm (Ut 5)	Up to 1 m thick beds Beige to light brown colored	Micrite	Minute, vertical burrows	Nil	Organic Matter content : TOC up to 1.87% Free-hydrocarbon content up to 14.31 mg HC/g Feeble to good source rock OM of Type I Nanno (coccolith) -bearing laminae (loosely to closely packed)
	Finely-laminated limestones	Lower I M Fm L S Fm (Ut 4; H B) SS Fm	Dm's to m thick beds Very thin, parallel laminae of light-colored micrite and dusky shales	Micrite with clays + (Organic Matter)	Common microslumps	Compaction-related with slight deformation of the thin layers (flaser - like)	
	Dolomicrite	U E Fm	Most of carbonates U E Fm Mostly chalky, white- to beige- colored With dusky, horizontal, cm's thick bands	Dolomicrite : minute, euhedral crystals (precipitates)	Nil	Nil	
Cryptalgal limestones (boundstones)	Domal stromatolites Planar stromatolites	L S Fm Ut 4 (H B) Ut 5	light gray- to beige-colored Isolated and coalesced (LLH type) hemispheroids 2-10 cm in diameter cm's in height Crust - like appearance cm's thick	Micrite	Internal dome-shaped lamination Truncation Planar, wavy, corrugated lamination Desiccation cracks Ghosts of lenticular gypsum	Early, clear halite crystals, pushing aside the laminae Full replacement by anhydrite, leaving original structures	Nil
Coarse-grained limestones	Pelletal limestones	L S Fm Ut 5 (Hauterives Field)	Beige- to light brown-colored	Grains: pellets cylinder-shaped 100-200 um long 50-100 um wide presumed to be pellets of chironomidae larvae Cement Micrite (little) + Clear halite / anhydrite +Fibrous calcite overgrowth around pellets? after aragonite	Cryptalgal, micritic laminae Grain-supported Pellets elongated to the bedding plane	Clear halite/ anhydrite pore fills	Nil
	Oolitic limestones	Uppermost I M Fm	Dm to m thick beds Light - colored Variably consolidated	Grains: oolites - mm-sized - nucleus: ? - cortex: dolomicrosparite; equigranular to fibrous rhomboids Cement Micrite (little) + Dolomite/ Anhydrte Porosity up to 40%	Inverse grain-size grading Grain-supported	-Dolomite-replaced cortex - Dissolved nuclei - Anhydrite as intra/ intergrain pore fill - Dolomite as inter-grain pore fill	Nil

TABLE 5.3.

Sedimentological characteristics of the depositional environments in the Paleogene sequence of the Valence Basin: (A) for terrigenous deposits; (B) for carbonates.

A	**DEPOSITIONAL ENVIRONMENT**				**INFORMATION SOURCES**					
Facies	**Salinity**	**Energy**	**Depth**	**Type**	**Structures**	**Content**	**Associated lith./min.**	**Recent reference**	**Fossil reference**	**Related bibliography concerning depositional environments**
Sandstones	Variable: very low to medium	Medium - high	? low	Distal Zone of Wadi Plain System	*	*			*	Handford and Fredericks 1980
Laminated silty shales		High to low	very low		*					
Homogenous shales	Variable: very low to medium	very low			*	*				

B	**DEPOSITIONAL ENVIRONMENT**				**INFORMATION SOURCES**					
Facies	**Salinity**	**Energy**	**Depth**	**Type**	**Structures**	**Content**	**Associated lith./min.**	**Recent reference**	**Fossil reference**	**Related bibliography concerning depositional environments**
Fine-grained limestones	Low	Low	Low		*					
Dolo-micrite	Medium - high	Low	Low	? sabkha			*	*		Folk and Land 1975 Hardie 1987
Finely-laminated limestones	Stratified waters	Very low	High to low	Isolated basin	*	*	*	*	*	Busson+Noel 1972 (1) Anderson et al. 1985 (2) Sonnenfeld and Hudec 1980 (3) Hudec and Sonnenfeld 1980 (4) Eugster 1984 (5)
Cryptalgal limestones	(Low) - medium	Medium	Medium to very low		*		*	*		Perthuisot 1975 Thomas+Geisler 1982 Thomas 1984 Ortí et al. 1984
Pelletal limestones	Low	Low	Low					*		Ortí et al. 1984
Oolitic limestones	Medium	Medium - (high)	Low		*		*	*		Freeman 1962 Kahle 1974 Loreau 1982

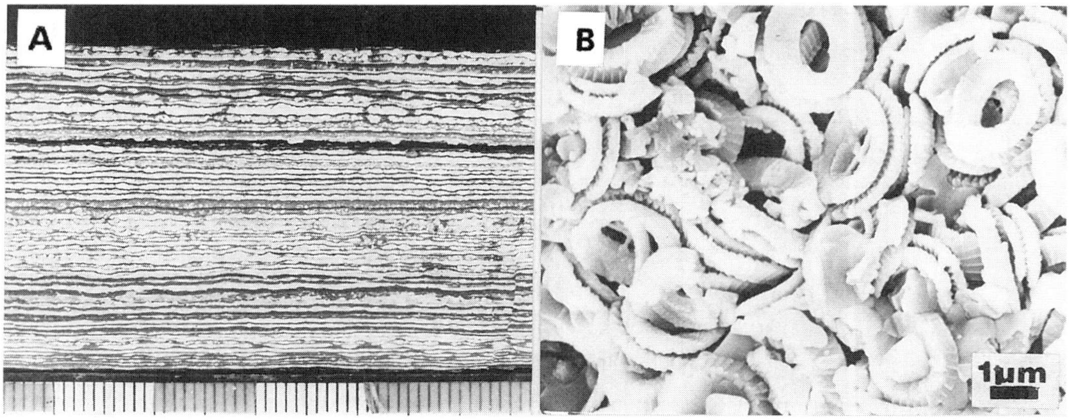

Fig. 5.10. Finely-laminated limestone lithology in the Lower Salt Formation:
A. Polished core slab with thin-layered, regular alternations of light-colored limestones and dark-colored, organic-rich shales. TE4 well; 1464.1 m (Unit T4; Horizon B).
B. SEM photomicrograph of a single, light-colored lamina within a sample of finely laminated limestone lithology. The lamina is made up of coccolith debris (plates) derived from a single species (*Cruciplacolithus tarquinius*). TE1 well; 1396.7 m (Unit T4; Horizon B).

Fig. 5.11. SEM photomicrograph of dolomicrite exclusively composed of minute rhombs of dolomite with well-formed crystalline surfaces. Cl-g1 well; 646 m; Upper Evaporite Formation.

in a poorly-oxygenated environment beneath a stratified water-body (Busson and Noël 1972).

(c) Dolomicrites (figure 5.11 and table 5.3), which may have diverse origins, are found in association with evaporite minerals (sulfates, halite). This suggests that the dolomicrite was formed in a hypersaline environment, probably in a shallow subaqueous setting and having a sabkha-type source.

2—*Cryptalgal limestones* occur as planar to domal stromatolites (figure 5.12 and table 5.3). They

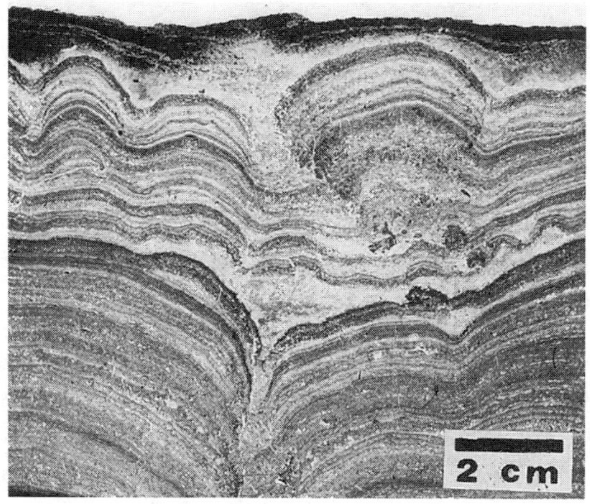

FIG. 5.12. Cryptalgal limestone lithology: Planar stromatolite crust passing upward into a domal form ("laterally linked hemispheroids" of Logan et al. 1964). Polished core slab; HA7 well; 1494 m. Scale in mm. Lower Salt Formation (Unit T5). Scale bar, 2 cm.

are found in association with accumulations of evaporite minerals and erosional surfaces, and display sheet-crack development. This association suggests deposition in very shallow-water to exposed conditions (peritidal environment). Similar deposits have been met with in sabkhas (Perthuisot 1975), and in pre-concentration areas of modern salt pans (Thomas and Geisler 1982; Thomas 1984).

3—*Coarse-grained limestones* are made up of either pelletal or oolitic components. The pelletal elements mostly occur in close association with stromatolitic laminae (figure 5.13A) whereas the oolitic limestone composes a distinct carbonate lithology (figure 5.13B).

(a) Pelletal limestones are usually made up of micritic calcite formed into cylindrical pellets about 100 microns long and are variously cemented by micrite, halite, anhydrite, and fibrous carbonate (figures 5.13 and 5.14). Based on studies of modern evaporitic environments, the pelletal elements found throughout limestones of this basin closely resemble the fecal pellets of chironomidae larvae (Diptera insects) that live in the lime muds of pre-concentration areas of modern salinas (Santa Pola, Spain) at a salinity range of 40 to 60 g/l (Ortí et al. 1984). Such deposits form in low energy, shallow waters of moderate salinity. However, such pellets may have been resedimented into depositional zones of somewhat higher salinity, as evidenced in some beds by the presence of aragonite as overgrowths around the pellets (Trichet 1967), and by the development of halite as primary cement.

(b) The oolitic limestones encountered in this basin have undergone severe alteration (figures 5.14 and 5.15, and table 5.2). Similarly altered sediments have been described from shallow depositional depths in modern environments such as the Laguna Madre in Texas (Freeman 1962), the Salt Lake in Utah (Kahle 1974), and the upper tidal zone in the Persian Gulf (Loreau 1982). The oolitic deposits found here are thus believed to reflect a shallow environment of substantial salinity.

SULFATE DEPOSITS

The sulfate deposits in the Valence Basin occur almost entirely in the form of anhydrite, glauberite, and celestite (table 5.4). Many of the rocks contain clear relics of their original mineralogy and morphology, which was gypsum. Some anhydrite morphologies, however, show either primary anhydrite textures or syndepositional alteration of other sediment types, which are also common (Ortí et al. 1984; Geisler-Cussey 1985).

1—Distinct *primary gypsum* relics are shown in figure 5.16 and are summarized in table 5.4. Gypsum developed in the form of tiny lenticular gypsum crystals (figure 5.16A), as are commonly met with in modern salt pans, mostly as syndiagenetic crystals in the uppermost sediments within a few meters of the pan floor, and more rarely on the depositional floor (Ortí et al. 1984; Geisler-Cussey 1985). In contrast, the larger lenticular shapes of

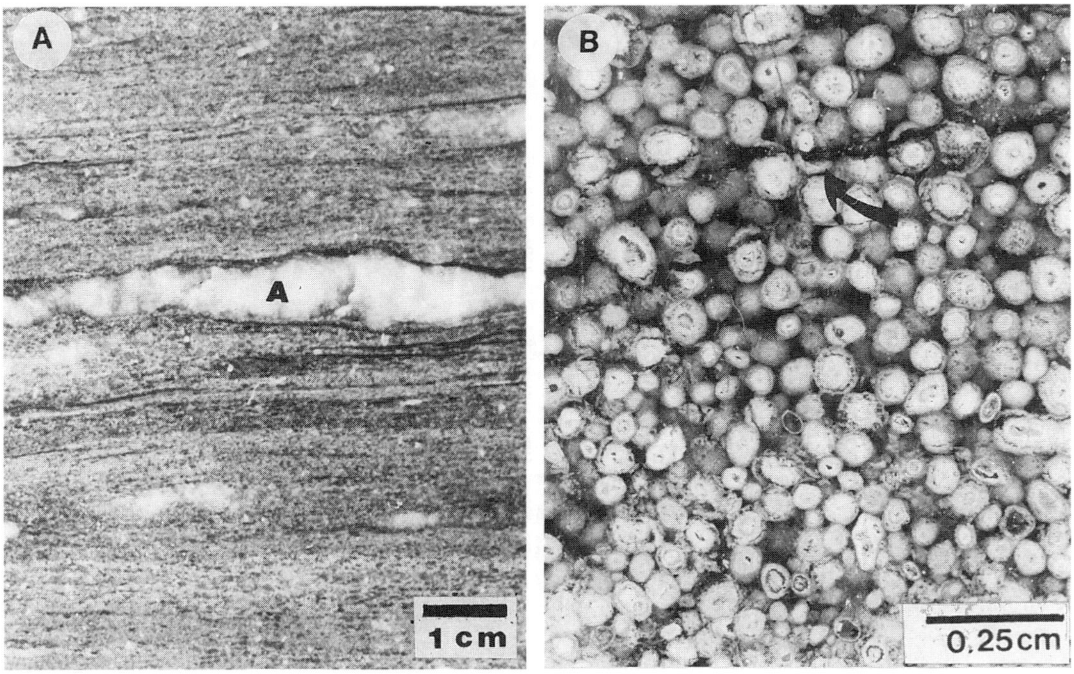

FIG. 5.13. Coarse-grained limestone lithology seen in polished core slabs:
A. Pelletal limestone bearing subhorizontal, dark-colored laminae that were algally originated. An anhydrite, lens-shaped layer (A) is conformably intercalated. HA7 well; 1519.8 m. Lower Salt Formation (Unit T5).
B. Substantially porous oolitic limestone with a grain-supported texture. The grain-size coarsens upwards. The minute, subhorizontal fracture (arrow) is healed by anhydrite. DP 108 well; 762.1 m; Intermediate Marl Formation.

gypsum have been encountered as displacive crystals in sabkhas, and possibly may grow in somewhat lower salinity (Shearman 1963; Perthuisot 1975; Butler et al. 1983). Gypsum also occurs as grass-like crusts (figure 5.16B) that initiate and develop at the sediment-water interface and commonly continue to grow, incorporating other sediments that may enter the basin. The present day analog of such gypsum is the hummocky crusts of selenite gypsum found in salt pans (Ortí et al. 1984; Geisler-Cussey 1985).

Detailed petrographic observations (Dumas 1988) have disclosed that the initial gypsum has been altered to anhydrite by transformation and replacement related to dehydration. Replacement by halite commonly implies a phase of dissolution prior to the replacement but the clear preservation of the original lenticular shape indicates that replacement probably occurred within indurated sediments. The dissolution of sulfate and replacement by halite may occur very early during burial as a synsedimentary feature (Hovorka 1988, 1992; Schreiber and Walker 1988, 1992). The dissolution may also occur due to the later migration of chloride-rich brines from salt horizons in the sedimentary pile (Ortí and Rosell-Ortíz 1980b; Hardie et al. 1985). Subsequent porosity was obliterated by anhydrite or halite that precipitated separately, concurrently, or successively.

2—*The anhydrite facies* appears in a nodular form and as two types of anhydrite veins. The vein forms are matrix-dependent while the nodular form is more universal in occurrence.

(a) Nodular anhydrite facies are commonly found within the shale and carbonate lithologies (table 5.5; figure 5.17). Nodules have variably developed contacts with the host sediment, ranging from hazy to sharp, and are found as isolated forms or,

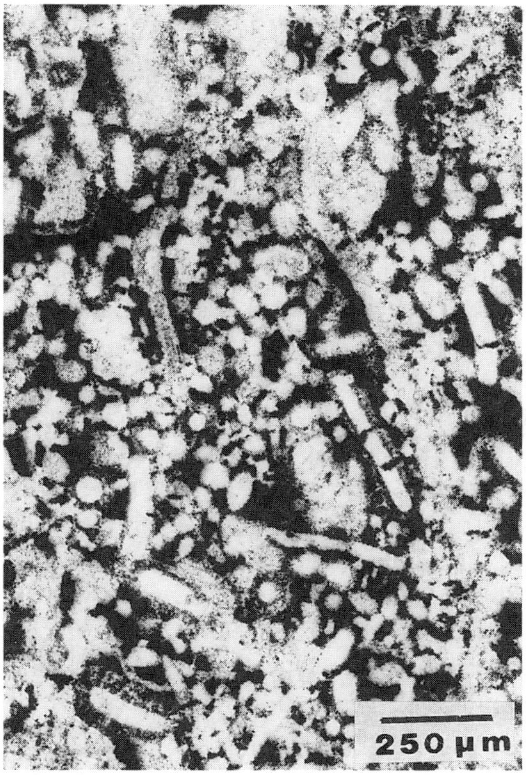

FIG. 5.14. Thin section photomicrograph of a pelletal limestone showing a grain-supported texture. Fecal pellet lengths range from 200 to 1000 microns with diameters ranging from 50 to 100 microns. The light-colored cement consists of anhydrite. Some of the fecal pellets display a rim of fibrous radial calcite. Plane light. HA7 well; 1468.8 m; Lower Salt Formation (Unit T5).

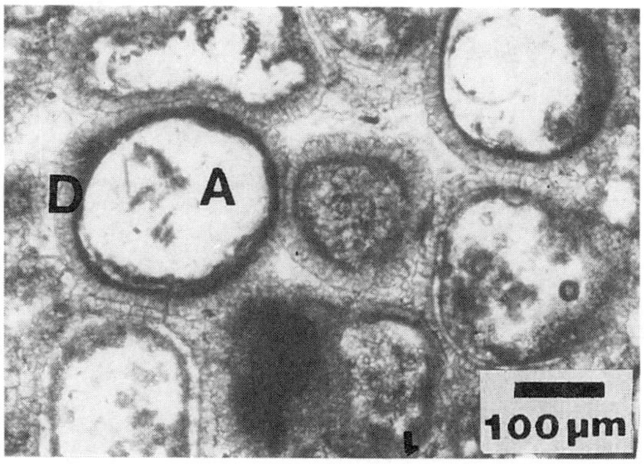

FIG. 5.15. Thin section photomicrograph of an altered oolitic limestone. Oolite rims are composed of dolomite crystals (D) with fibrous and equant shapes. Oolite cores and intergranular spaces are plugged with anhydrite (A). Plane light. DP 108 well; 751.8 m; Intermediate Marl Formation.

where they coalesce, with a mosaic or chicken-wire appearance. The nodules are made up of microgranular, microprismatic, and acicular crystals. Most of them are arranged in a more or less aligned pattern suggesting fluidal or compactional deformation.

The depositional laminae of the host sediment are commonly deformed due to the internal growth of the anhydrite nodule(s) together with later, differential compaction. It is thus suggested that the nodules originated early in the sediment history, in response to subsurface migration of oversaturated waters during periods of subaerial exposure. Whether the nodules derive from a gypsum precursor or from initial anhydrite is unknown.

(b) Veins of subvertical anhydrite are developed throughout the argillaceous salt lithology (figures 5.18 and 5.19). These veins have a contorted appearance and may exceed one meter in length. Width and internal folding of each vein decrease upward, and the veins end as a bundle of minute subvertical veinlets (mm-wide) that penetrate salt beds. The anhydtrite in these veins consists of crystals having an internal fluidal fabric arrangement (figure 5.19A). The anhydrite crystals include microgranular forms, along with superposed, phenoblastic, rectangular forms.

These vertical veins are believed to have originated subsequent to the release of trapped fluids (Shearman 1985) that moved up along tectonically related or hydraulically forced fractures. Decrease in fluid pressure resulted in the precipitation of anhydrite.

(c) Anhydrite veins, cm-wide, crosscut carbonate deposits and have lateral extensions that are conformable with bedding (figure 5.19B). The veins do not have straight borders but zigzag through the sediment, first following one bedding plane and then jumping to another. Within the veins the component crystals change centripetally from a microgranular to a lath form. The small laths have a lineation parallel to the vein direction.

TABLE 5.4.
Sedimentological characteristics of the primary gypsum (now seen as pseudomorphs) in the Paleogene sequence of the Valence Basin.

FORMS	SIZE	ENCLOSING LITHOLOGIES	PRESENT MINERAL COMPOSITION	CRYSTAL ARRANGEMENT
LENTICULAR SHAPED	mm- to cm-sized	shales/marls beneath halite carbonate beds of massive anhydrite levels	anhydrite - halite	scattered, elongated parallel to the bedding plane "rosette"-like fabric
	cm's- to dm's-sized	dark-colored shales	anhydrite	sparse mostly grouped as "rosette"-like fabric
GRASS-LIKE CRUSTS	as layers cm-thick	anhydrite horizons cryptalgal carbonates	anhydrite	laterally juxtaposed crystals

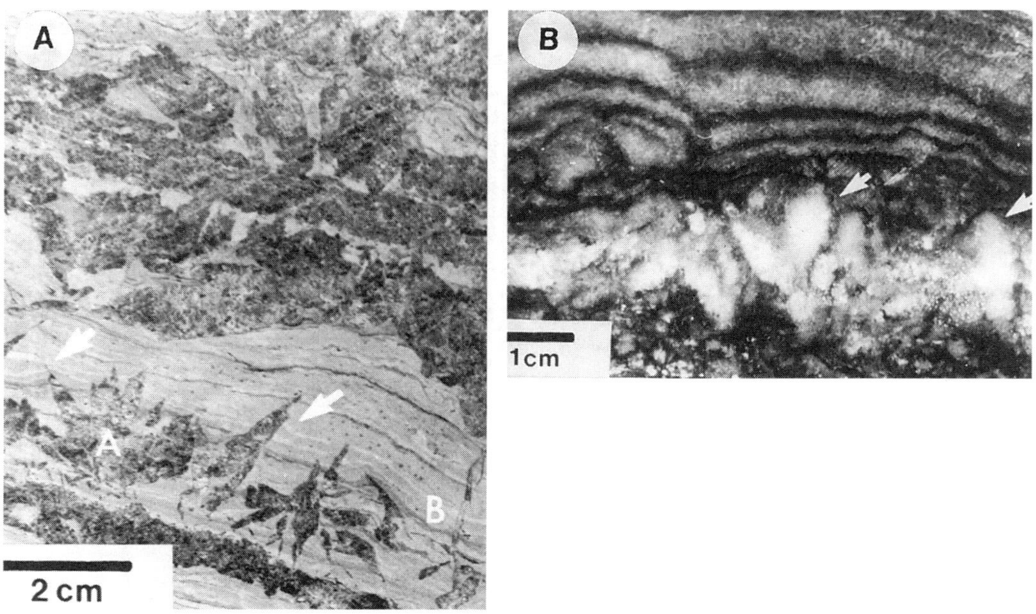

FIG. 5.16. Pseudomorphs (ghosts) of gypsum crystals seen on polished core slabs.
A. Lenticular gypsum, pseudomorphed by anhydrite. Initial forms of gypsum, having a rosette-like fabric (A), are layered within carbonate host sediments which also display algal-derived lamination. The laminae are cut and deformed by the sulfate lenses (arrows). This deformation suggests an early and displacive development of the original lenses. Soyons well; uncertain position.
B. Lenticular gypsum altered to anhydrite. Crystals are vertically elongated and juxtaposed (arrow), forming a grass-like crust. The crust is overlain by layered anhydrite. The dome-shaped layering may reflect the algal origin of the rock. TE 1 well; 1361.8 m; Lowermost Intermediate Marl Formation.

TABLE 5.5.
Sedimentological characteristics of the depositional environments in the Paleogene sequence of the Valence Basin: (A) for sulfate lithologies; (B) for salt lithologies.

A

SEDIMENT TYPE	DEPOSITIONAL ENVIRONMENT				INFORMATION SOURCES					
Facies	Salinity	Energy	Depth	Type	Structures	Content	Associated lith./min.	Recent reference	Fossil reference	Related bibliography concerning depositional environments
Gypsum small lenses grass-like crusts	High (150-320 ppt)	Nil (internal)	Very shallow to exposed					*		Geisler-Cussey 1985 Ortí et al. 1984
Gypsum large lenses	? medium			? sabkha				*		Perthuisot 1975
Nodular anhydrite		Nil (internal)	Very shallow to exposed	Mud-flat			*	*		Ali and West 1983 Dronkert 1977
Glauberite		Nil (internal)	Very shallow to exposed	Continental mud-flat Sabkha			*	*	*	Conrad 1969 + 1975 Hardie et al. 1968 Hunt et al. 1966 Ortí et al. 1979 + 1983 Ortí and Pueyo 1980 Ortí 1982

B

	DEPOSITIONAL ENVIRONMENT				INFORMATION SOURCES					
Facies	Salinity	Energy	Depth	Type	Structures	Content	Associated lith./min.	Recent reference	Fossil reference	Related bibliography concerning depositional environments
Reticulate Salt	High	Very low	Very low		*				*	Southgate 1982
Bedded Salt	High (just>320 ppt)	Low	? low		*	*	*			
Argillaceous Salt		Nil (internal)	Very shallow to exposed	Mud-flat	*					Handford 1988 Gornitz+Schreiber 1981 Curial+Moretto (this volume)
Milky Salt	Very high (up to 370 ppt)	Very low	Low		*			*		Geisler-Cussey 1985 Shearman 1970 Lowenstein+Hardie 1983 Ortí et al. 1984

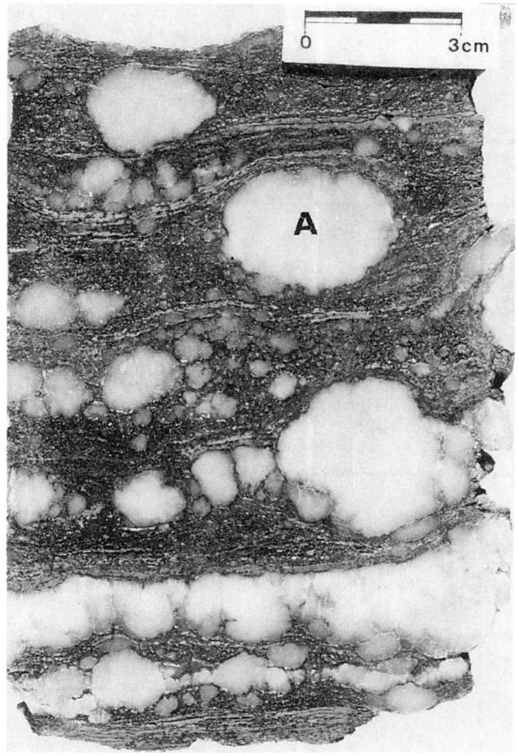

Fig. 5.17. Nodular anhydrite lithofacies seen on slabbed core sample. Depositional laminae of the host carbonate sediment are deformed in response to early growth of displacive sulfate nodules (A) and later, differential compaction. Differential compaction must have been minor because the carbonate matrix is halite cemented. HA 7 well; 1505.6 m; Lower Salt Formation.

There are also oblique veins of anhydrite that consist of non-deformed nodules closely associated with halite (figure 5.19C). Such features possibly document a late crystallization of anhydrite (and then halite cements) from brines moving along planes of high permeability, such as fractures or minor diastems.

3—*Secondary gypsum* has been encountered in the DP 108 well within the Upper Evaporite Formation at a depth of about 700 m (figure 5.20). The facies consists of pale nodules of gypsum mostly formed of crystals that are small but heterogenous in size. This gypsum also occurs as megacrystals or porphyroblasts that enclose anhydrite remains in the form of patches of microgranular anhydrite. Primary lenticular morphs of gypsum have locally been preserved. A two step process of replacement is believed to have occurred in which primary gypsum (lenticular form) is converted to secondary anhydrite which in turn is later gypsified.

4—*Glauberite-bearing facies* occur in beds that are stratigraphically restricted to the Upper Evaporite Formation (Cl-g1 and Ch-g1 wells, figure 5.21A). Glauberite-bearing lithologies are usually beige but may display a honey-color or chalky appearance because of the degree of hydration of the mineral (Ortí et al. 1984). The glauberite is developed as two crystal habits:

(a) *small plate-like crystals* (figure 5.21B). These cm-sized crystals are present within a dolomicritic matrix. They may be isolated or closely packed and thus appear to form incipient beds;

(b) *prismatic crystals* (figure 5.21C). These crystals also develop within a dolomicritic matrix, are less than one cm long, and are homogenous in size. They are grouped as beds and nodules. Crystals are closely packed in the core of the nodules and become more diffuse toward the edges. The nodules are thought to have originated within soft sediment during early diagenesis and to have deformed the bedding of the host lithology. In addition, slightly larger crystals that compose massive beds are somewhat associated with halite. These crystals display a fluidal arrangement within the beds, thus causing the structure to resemble an incipient enterolithic contortion.

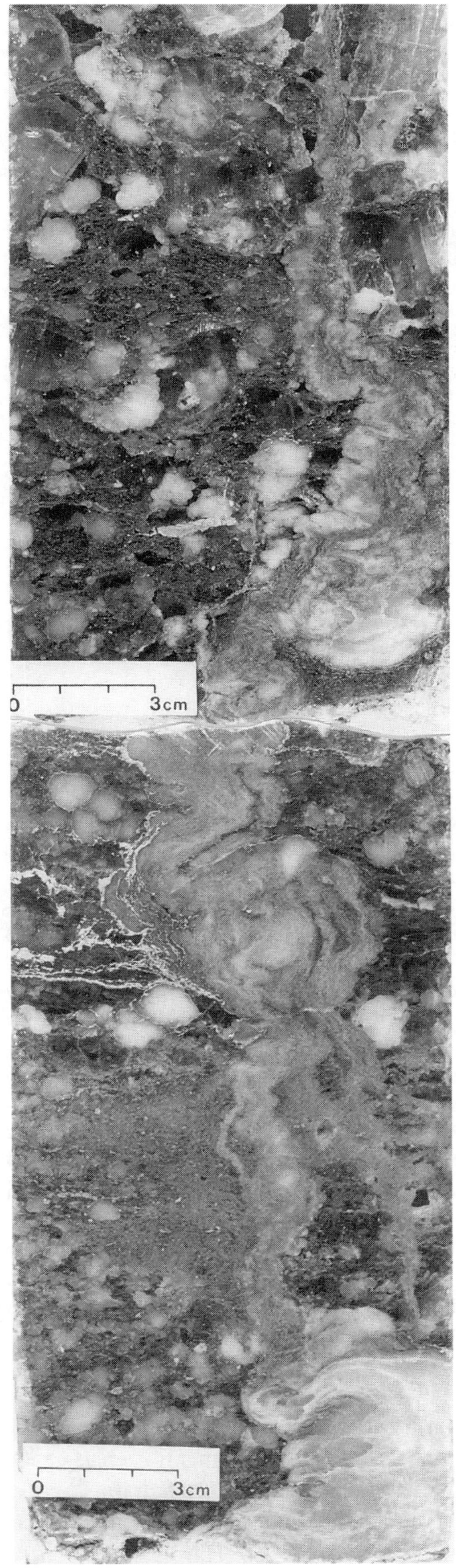

FIG. 5.18. Slabbed core sample showing vertical vein of anhydrite developed in argillaceous salt lithology. The vein is contorted. The width of the vein and the internal folding decrease in the uppermost section of the vein. TE 1 well; 1480 m (top of the section); Lower Salt Formation (Unit T4; Horizon E).

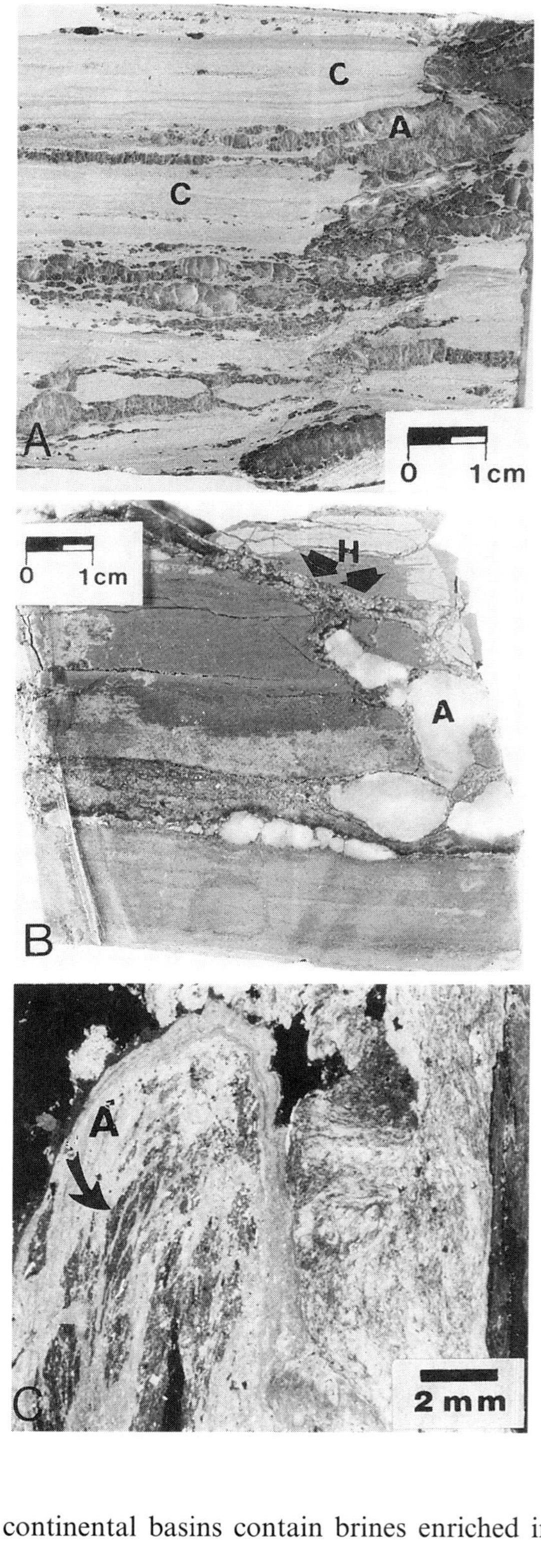

Fig. 5.19. Anhydrite as veins:
A. Thin section photomicrograph of a vertical folded anhydrite vein. The vein is made up of anhydrite microcrystals (C) arranged with a fluidal fabric. In the fold core, anhydrite is closely associated with halite which forms dark, spindle-shaped patches (arrows). Crossed-polars. TE 1 well; 1481.3 m; Lower Salt Formation (Unit T4; Horizon E).
B. Slabbed core showing anhydrite veins (A) crosscutting carbonate layers with some horizontal extensions inserted within the beds. Anhydrite originated from brines migrating in conduits of higher permeability such as fractures (oblique development) and minor diastems (horizontal development). DP 108 well; 1761m; Subsalt Formation.
C. Slabbed core showing light-colored nodules of anhydrite (A) present as veins both along a depositional layer and in a fracture oblique to the bedding. The anhydrite nodules together with halite are not deformed, suggesting that they are not products of any tectonically forced injection of ductile material, but rather derive from brine migration. DP 108 well; 1630 m; Lower Salt Formation (Unit T6).

Modern glauberite-bearing deposits are fairly common in continental environments such as continental sabkhas and playas (western Sahara: Conrad 1969, 1975) and in more permanent salt lakes (western U.S.A.: Hunt et al. 1966, Hardie 1968). These continental basins contain brines enriched in sulfate and sodium. In the Saline Valley (California), from the basin margins through the mud flat to the central halite brine pond, belt-like zones successively contain calcium-sulfate, calcium-sulfate with

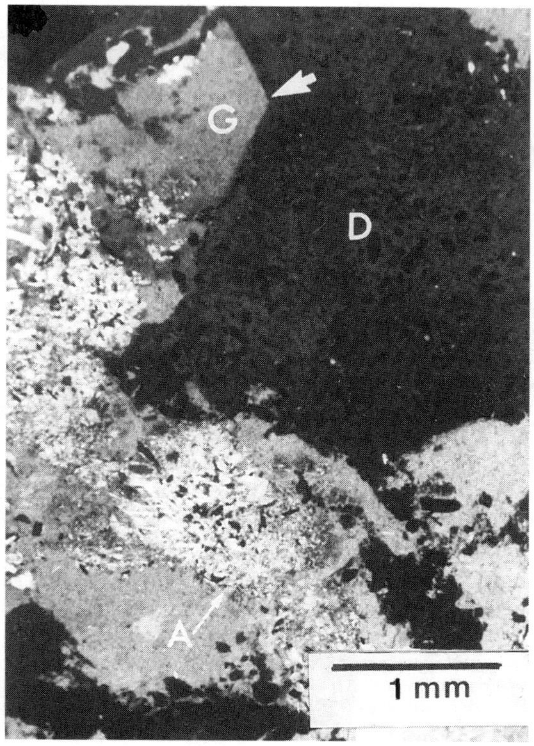

Fig. 5.20. Thin section photomicrograph of the secondary gypsum lithofacies. Large crystal of secondary gypsum (G) with well-defined surfaces locally (arrow). The matrix is a dolomicrite (D). Gypsum crystal encloses remains of microgranular anhydrite (A) that composed the nodule prior to rehydration. Crossed-polars. DP 108 well; 677.5 m. Upper Evaporite Formation.

glauberite, and glauberite. According to Hardie, the glauberite of the proximal mixed facies zone could have resulted from the transformation of gypsum in contact with subsurface hypersodic brine. Alternatively, the glauberite of the distal pure zone could have crytallized directly from the subsurface brine.

Glauberite occurrences are rare in the rock record. In the continental deposits of Tertiary basins in Spain (Ortí et al. 1979; Ortí and Pueyo 1980; Ortí 1982; Ordóñez and García del Cura 1994; Salvany and Ortí 1994), most glauberite appears to have derived from calcium-sulfates during early diagenesis, in playa lake and sabkha-type environments for the bedded and nodular facies, respectively (Ortí et al. 1985). Glauberite in the Valence Basin is therefore thought to have developed by saline mud flat deposition from shallow waters in continental basins. Such ponds were subjected to depositional rhythms of flooding/concentration/desiccation of brines enriched in sulfates and impoverished in calcium (type Na-SO_4-Cl of Hardie et al. 1978).

5—*Celestite-bearing facies* contain celestite as tiny, euhedral prisms together with calcite within clear halite and anhydrite. Lithologies bearing celestite can derive from the replacement of anhydrite (West 1973). Dolomitization and bacterially induced reduction of sulfates would have a substantial influence on enrichment in strontium within sulfates (Kinsman 1969a, 1969b; Butler 1970).

Celestite can also precipitate directly from sea water but the salinity range for the crystallization is still poorly known (Müller 1962, 1964; Usdowski 1973). Precipitation is strongly governed by the Eh-pH relationships of the parent water body as well as ionic concentration. The critical (Eh-pH) boundary between celestite and strontianite (for example) is directly in the middle of the conditions available within hypersaline water bodies (Brookins 1989) so that the development of celestite helps to constrain the formational conditions.

6—*Sulfate-related silcates.* Silicates, which are commonly associated with sulfates, belong to two disinct groups: silica (quartz) and analcime. Silica-replaced patches are pervasive within the sulfate lithologies. Several crystal habits of quartz have been encountered (figure 5.21C): quartzine spherulites, mostly as coalescent zones (200 to 600 microns);

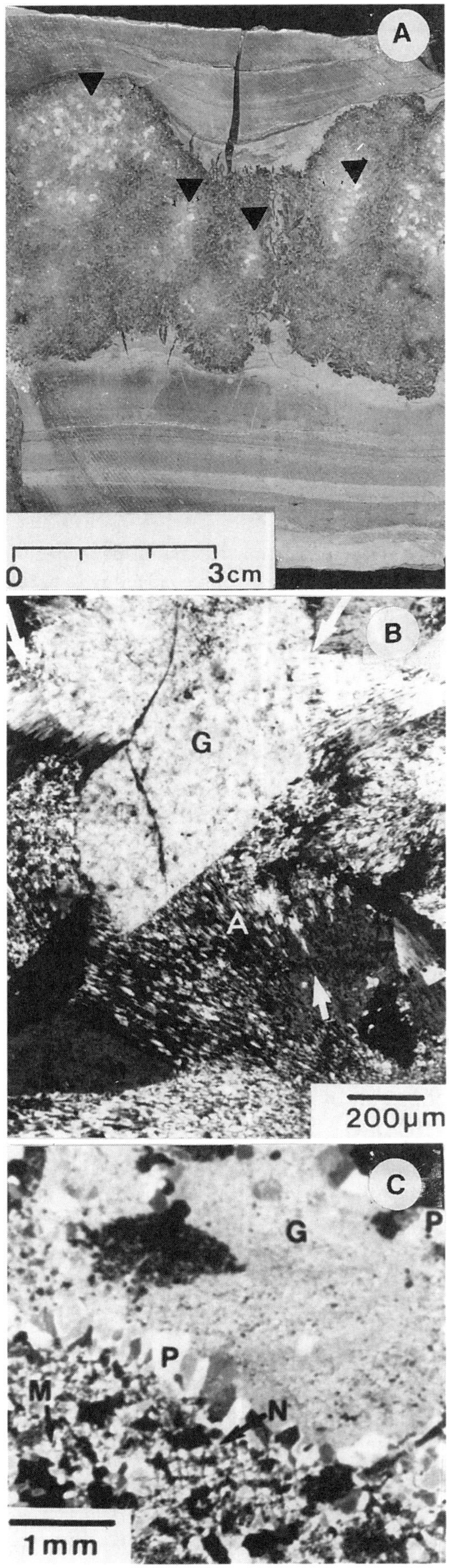

Fig. 5.21. Glauberite as seen in hand sample and thin section from the Cl-g1 well, Upper Evaporite Formation.

A. Slabbed core showing nodular glauberite with deformed depositional laminae in the host dolomicrite. Deformation was probably in response to the early displacive growth of the glauberite nodules. In this sample several nodules are coalescent. The light-colored patches enclosed within the nodules (arrows) are epigenetic siliceous products. Sample from 643.5 m.

B. Thin section photomicrograph of glauberite; crystals developed as prismatic forms (G). Anhydrite (A) is locally intercalated between the glauberite crystals. Anhydrite occurs as fibrous forms which originated along the cleavage planes of the glauberite crystals (arrows). Crossed-polars. Sample from 639 m.

C. Thin section photomicrograph of glauberite; crystals developed as small plates (G). The glauberite has been extensively silicified. Several siliceous habits can be distinguished: bladed quartz (P), microquartz (M), and quartzine spherulites (N). Crossed-polars. Sample from 649.5 m.

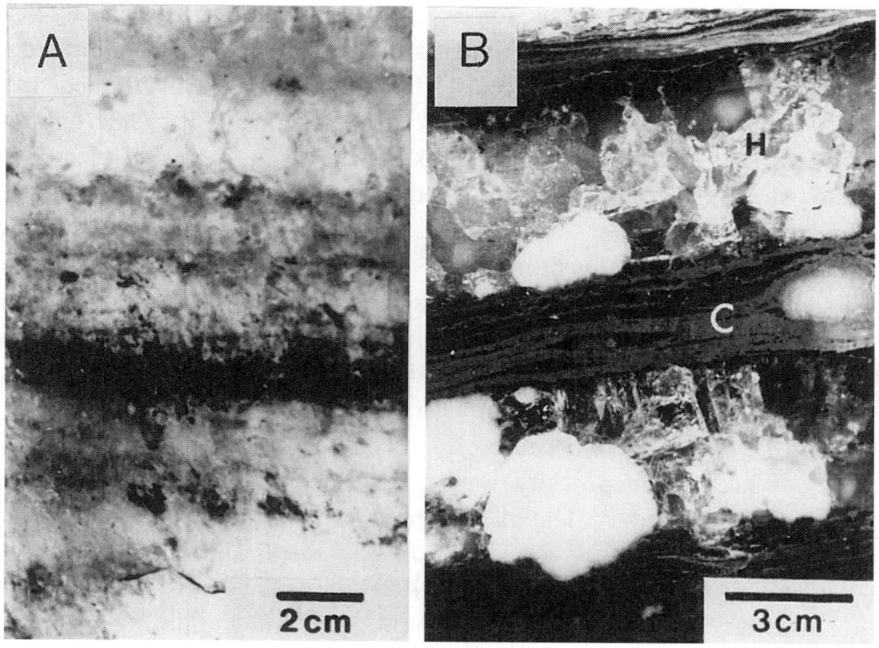

Fig. 5.22. Bedded salt lithologies; Lower Salt Formation.
A. Thin bedded alternation of clear halite and clayey/calcareous material. This rhythmic pattern is evidence supporting the primary origin of this type of halite. Core slab from TE 1 well; 1421.8 m (Unit T4, Horizon C);
B. Regular interlayering of clear halite (H) and poorly-laminated, fine-grained carbonates (C). Additional, light-colored, cloud-shaped nodules of anhydrite are well developed at the contact between the salt and the carbonate layers. Core slab; DP 108 well (1677.5 m in Unit T6).

fibrous lutecite; vertical bladed quartz; and as isometric or euhedral crystals. In the carbonate beds, the silica beds are made up of minute, aligned chalcedony spherulites whereas epigenetic products after anhydrite may be composed of quartzine spherulites and of large-sized quartz crystals within the lath-bearing anhydrite. The crystal habits of the silica seem to have been induced by the initial shape of the minerals that underwent epigenesis.

Analcime has been detected through X-ray analysis and is not readily recognized visually. It is present at the top of the Lower Salt Formation. As no volcaniclastic material, which is analcimes most common association, has been found, an origin from the interaction between highly alkaline brines and silica in solution or direct reaction with clay minerals in the sediments is likely (Renaut and Tiercelin 1994).

HALITE FACIES

The salt deposits in the Valence Basin occur in a large variety of primary and secondary facies (table 5.5B). The basic lithologies are bedded salt, argillaceous salt, and milky salt. Subordinately, salt develops as veins. All salt lithologies show some evidence of extensive dissolution.

1—The *bedded salt* lithology is composed of carbonate/sulfate and clear halite interbeds (figure 5.22A). The individual halite beds are relatively thin (5 to 20 cm), and the bedding can be very fine (up to 20 couplets per meter). Some of the clear halite beds contain clayey/calcareous films lying conformably with the bedding. The clear halite is commonly yellow because of the presence of sulfate and carbonate particles along cleavage planes and between the crystals.

This halite is made up of anhedral crystals, some of which are vertically elongated, resembling cornet-type crystals. Some of the beds have crystals clustered as spindle-shaped patches that are composed of clear halite (with the exception of the core of the crystal). These patches consist of cubes, mm-sized, rich in fluid inclusions.

Anhydrite nodules are common in this lithology. They are mostly localized at the contact between the salt beds and carbonate interbeds (figure 5.22B). These nodules may be isolated or coalescent with an enterolithic appearance. They apparently grew selectively within the halite layers because of their higher porosity. Whether the porosity was original (syndepositional) or secondary (late dissolution) is not resolved.

This clear halite is present as thin beds, regularly interbedded with anhydrite/carbonate lithologies. The halite probably precipitated from a barely saturated brine as indicated by the presence of sulfates and carbonates as intra- and intercrystal associations. The absence of trapped fluid inclusions indicates a slow rate of crytallization; moreover, it reflects a low degree of brine saturation.

2—The *argillaceous salt* lithology is common throughout the Lower Salt Formation and occurs as horizons one to several meters thick. It is generally composed of porphyroblasts of clear halite, within a greenish, shaley matrix, in the form of mm-sized isolated to closely packed crystals. These crystals have a cubic habit, either with irregular outlines or as depressed-face (hoppered) cubes. Some crystals are fully clear, the matrix having been pushed aside during slow crystal growth. This form is referred to as displacive halite (Handford 1988) (figure 5.23A). Other crystals enclose impurities aligned parallel to the faces of the cubes, as a result of a faster rate of crystallization. Such halite is here considered as incorporative, i.e., a type of halite cement.

Apart from clear halite, the matrix also contains common pseudomorphic ghosts of lenticular gypsum and nodules of anhydrite. The original depositional structures have mostly been obliterated. The host sediment is now either incorporated within halite crystals or squeezed and deformed between the evaporite patches. In addition, calcareous fragments are common, apparently derived from an in situ brecciation of thin, indurated carbonate layers of intercalated deposits that collapsed into solution cavities (figure 5.23B). The overall disruption of the shale-carbonate deposits, referred to as "haloturbation" (Smith 1971), was caused by early internal development of a series of evaporitic crystals (gypsum, anhydrite, and halite). The disruption of these sediments was probably caused by solution/concentration events related to the periodic influx of undersaturated waters that also carried in the terrigenous material.

The argillaceous salt is considered to have formed in a subaerial environment such as a saline mud flat or in an intrasedimentary position beneath an ephemeral shallow body of water (Handford 1988).

3—The *milky salt* lithology is composed of halite having a white, cloudy appearance because of the presence of fluid inclusions (figure 5.24). The horizons of this lithology range from a few centimeters to several meters in thickness and are relatively pure. They offer a rhythmic lithologic pattern related to:

- change in the crystal form of milky halite: depressed-face (hoppered) cubes—chevron-shaped crystals—cornet-shaped crystals
- occurrence of non-salt material as weakly distinguishable thin carbonate-shale layers or as carbonate-sulfate impurities or beds enclosed in the cornet-type lithology
- variable development of syntaxial clear halite as overgrowths around milky cubes and chevrons
- development of thin layers (mm-thick) of clear halite without any evident hiatus.

Hoppered cubes appear in white horizons, decimeters thick, in which they are scattered or grouped as thin layers. The chevrons are one to several centimeters, and are closely packed along thin beds. Chevron-bearing beds are rare and usually are developed upon crusts composed of simple cubic crystals. Cornets occur as vertically elongated, cm-sized coalesced crystals. Fluid inclusions are less abundant here and are arranged in horizontal bands. The cornet-bearing beds range from 5 to 10 cm

FIG. 5.23. Argillaceous salt of the Lower Salt Formation:
A. Slabbed sample with well-shaped, large crystals of clear displacive halite (phenoblasts) which displaced the host shales as they grew, causing contortion in laminae (arrows). Core slab, TE 1 well, 1413.5 m (Unit T4; Horizon C).
B. Argillaceous salt seen on slabbed surface with displacive phenoblasts of halite (H) developed in a shaley matrix. Wrapped in the matrix are a number of carbonate pebbles (arrows) suggesting an in situ brecciation of the initial deposits. Core slab; TE 1 well; 1413 m (Unit T4; Horizon C).

in thickness and are intercalated between the carbonate-sulfate beds. The cornets seem to be an intermediate form between clear salt and milky salt composed of cubes and chevrons.

The milky salt is primary and is normally formed at the sediment-water interface. The distinct crystal morphologies described here have been met with in modern salt pans and are related to differences in mother-brine concentration (Geisler-Cussey 1985) or water depth (Logan 1987). The cubic crystals develop in a high salinity environment (370 g/l in Geissler-Cussey 1982). The chevron-bearing salt is very pure, suggesting a high rate of crystallization and little time for influx of detritus. Supporting evidence is available from the presence of copious fluid inclusions. The chevrons developed, as did the cubes, from very concentrated brines. In contrast, the cornets may have formed from a brine only slightly oversaturated, as suggested by the lesser amounts of fluid inclusions. An alternative proposal is that chevron-shaped and columnar (cornet-type) halites are initiated from cuboids formed on the floor of the basin, in very shallow (a few cm) and in slightly deeper (up to 1 m) ponds, respectively (Logan 1987). The crystal fabric formed on the basin floor continues upward growth, being relatively unaffected by changes in brine concentration.

The regular interruption of milky salt layers by very thin, clear halite layers, without any clear evidence of a dissolution-related hiatus, reflects periodic decreases in salinity. These could correspond to seasonal fluctuations in evaporation and influx without large scale flooding and dissolution.

The fine rhythmic depositional pattern of this primary salt reflects rapid changes in the brine pa-

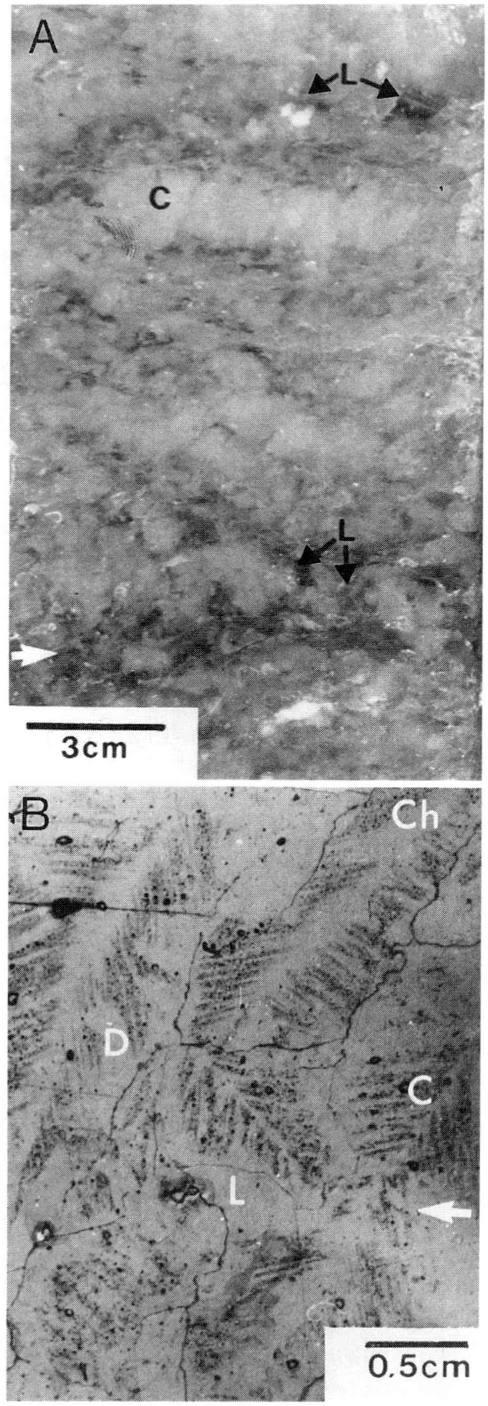

Fig. 5.24. Milky salt lithology of the Lower Salt Formation: A. Core slab sample made up of depressed-face cubes of milky halite (C) surrounded by clear halite (L) that appears as dark patches (photograph taken on a black background). TE 1 well; 1394.5 m; (Unit T4; Horizon A).
B. Thin section photomicrograph showing halite with depressed faces (hoppered) and chevrons (Ch). At center of the hopper, the initial cube is still distinguishable (arrow). The chevrons are outlined by minor dissolution pits (D) and are surrounded by clear halite (L). Plane light. TE 1 well; 1413 m; (Unit T4; Horizon C).

rameters induced by influx of undersaturated waters. Ideally, an increasing input of undersaturated water would cause the halite crystallization to pass successively from cubes, to chevrons, to cornets, to clear halite, then to the development by internal solution-precipitation of clear halite cement, and finally the deposition of non-salt material.

4—*Salt veins and fibrous salt* (figure 5.25) are mostly subvertical but locally horizontal, and are developed within carbonate-shale lithologies. This

FIG. 5.25. Slabbed core from the Lower Salt Formation showing salt vein of fibrous, light-colored halite (F) interrupting shale/carbonate layered deposits. The bending of the salt fibers close to the wall of the host lithology indicates a slight differential movement which in turn suggests that the dilational fracture was re-activated. TE 1 well; 1546 m (Unit T4; Horizon F).

halite morphology is similar to that of satin spar in gypsum. The halite occurs here as an infill of dilational fractures, and consists of fibers up to several centimeters long and several millimeters in diameter. The direction of crystal elongation is perpendicular to the wall of the veins as in the case of satin spar. Most of the veins are orange/pink because of the presence of small amounts of hematite. The veins of fibrous halite link salt horizons that lie above and beneath the shaley beds, but never penetrate them. They differ from the vertical fracture fills, meters deep, of the Permo-Triassic series in England that appear to have formed by repetitive dilation/shrinkage produced by temperature changes at the surface of the basin during deposition (Dellwig 1963; Tucker 1981). They are similar to fibrous veins documented in shale beds of the Permian rocks in Texas (Hovorka 1983a, 1983b), and are interpreted as having developed in the upper ten meters of the sedimentary pile.

5—*Dissolution features* occur at all scales, from minute millimetric solution pits to meter-high solution pipes. Larger solution structures have not been seen in this study.

(a) Crystal-scale effects, seen as minute solution pits plugged by clear halite, affect the chevron-shaped crystal of milky halite and are referred to as fishbone-like features (Lowenstein and Hardie 1985). Slightly more evident solution traces are marked by the infiltration of shaley material through the halite of milky interbeds (figure 5.26). Such effects were certainly produced during the periods of terrigenous input and caused the initial intercrystalline porosity of the salt deposit to increase. The milky salt encloses pocket-shaped, decimeter-sized cavities infilled by clear halite together with carbonate-sulfate fragments derived from the collapse of the immediately overlying bed. These internal solution cavities occur in the uppermost sediments and have been defined as an intraformational solution (Geisler-Cussey et al. 1985).

The most evident crystal-scale solution features are horizontal surfaces, truncating layers of clear and milky halite crystals (figure 5.27), which are overlain

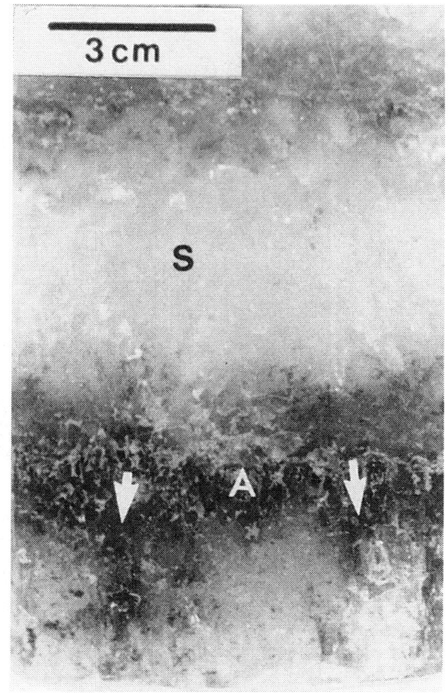

FIG. 5.26. Slabbed core showing solution in salt lithologies from the Lower Salt Formation. Interbeds of milky salt (S) and shales (A); minor, purse-shaped cavities (arrows) filled with detrital material outline the upper surface of each salt bed. TE 1 well; 1425.8 m (Unit T4; Horizon C).

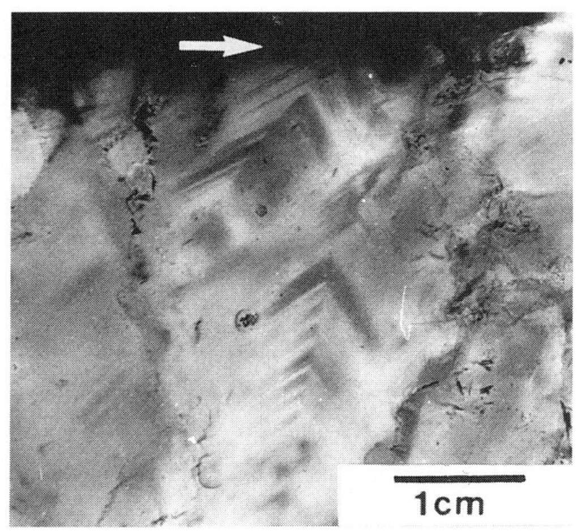

FIG. 5.27. Core slab from the Lower Salt Formation showing evidence for solution in salt lithologies. Chevron-shaped crystals of halite are truncated and overlain by an anhydrite/carbonate bed. This sharp change documents a solution event. TE 1 well; 1413.5 m (Unit T4; Horizon C).

by shale-carbonate-sulfate deposits. These are clearly related to the influx of the less concentrated water from which the shale, carbonate, and sulfate are deposited.

(b) Core-scale dissolution effects (figure 5.28), meters in size, are common throughout Unit T4 (Lower Salt Formation) in the Tersanne field, as sharp, subvertical contacts between primary salt, mostly milky, and argillaceous salt. Solution pipes are largely plugged by argillaceous salt, enclose carbonate-shale fragments, and have a rubbly appearance. The outline of the pipe is commonly marked by a band of clear halite that appears as having developed as a rim cement upon the wall of the cavity.

A distinct infill type of halite in some solution pipes, present in horizons D and E of Unit T4, Lower Salt Formation, consists of milky halite having anhedral shapes arranged randomly.

Similar solution pipes, with depths up to tens of meters, have been reported from modern salt depositional areas such as California (Death Valley;

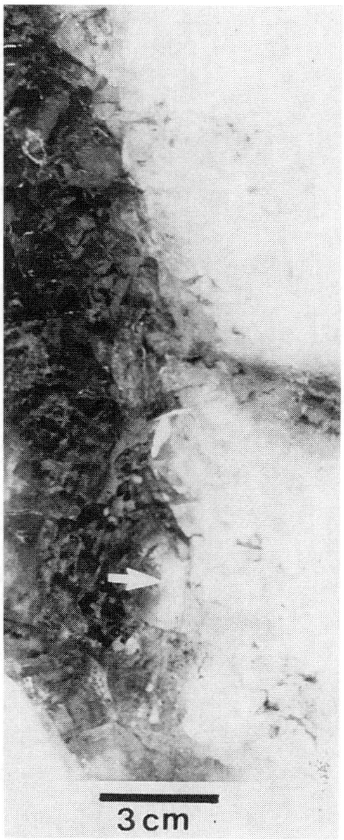

Fig. 5.28. Core slab from the Lower Salt Formation showing solution in salt lithologies. Large-sized, subvertical, solution cavity (left-hand side) in a milky salt lithofacies (right-hand side). The cavity is plugged by argillaceous salt. A narrow clear zone (at arrow) marks the contact between the two distinct salt lithologies. This zone is made up of clear halite which consists of syntaxial crystals that formed from the chevrons of the milky salt prior to the infill of the cavity. TE 1 well; 1424.8 m (Unit T4; Horizon C).

Hunt et al. 1966), Chile (salars; Stoertz and Ericksen 1974), and Tunisia (Sabkha el Melah; Perthuisot 1975; Busson and Perthuisot 1977). The pipes are thought to originate from underground flows of undersaturated waters along fault planes or through sedimentary conduits such as distal alluvial fans. Similar structures have also been explained as resulting from a per descendum seepage of undersaturated waters from the desiccated surfaces down to the underlying brine sheet (Hovorka 1983a, 1983b). It is uncertain which of these processes account for the present case.

MIXED SULFATE-HALITE FACIES

The halite often is closely associated with sulfates, resulting in a large number of mixed, singular facies including reticulate salt, anhydrite-bearing salt, and macrocrystalline halite.

1—*Reticulate salt* (figure 5.29) is a mixed facies, developed as cm-thick beds, that exhibits a close interrelation between anhydrite and halite. The sulfate crystals are elongate plates, about 5 mm long, that enclose the boundaries of similar sized halite crystals. They form a regular layered network developed in three perpendicular directions, corresponding to the cube faces of the halite (Dumas 1988). Such a crystal pattern probably is the result of a cocrystallization of sulfates and halite, possibly from a very satured brine. It is reminiscent of the spinifex structures seen in komatiites.

A somewhat similar halite facies has been documented by Southgate (1982) in modern highly oversaturated lagoons of the Australian coast. According to Southgate, such facies can only be produced within very shallow, hypersaline brines.

2—*Anhydrite-bearing halite* (figure 5.30A) is found within any primary halite facies where a number of anhydrite laths are developed in an intercrystal position or within the cleavages of the halite. However these forms never make up a regular network such as in the reticulate salt. Enclosed anhydrite laths are sometimes grouped as a rosette-like

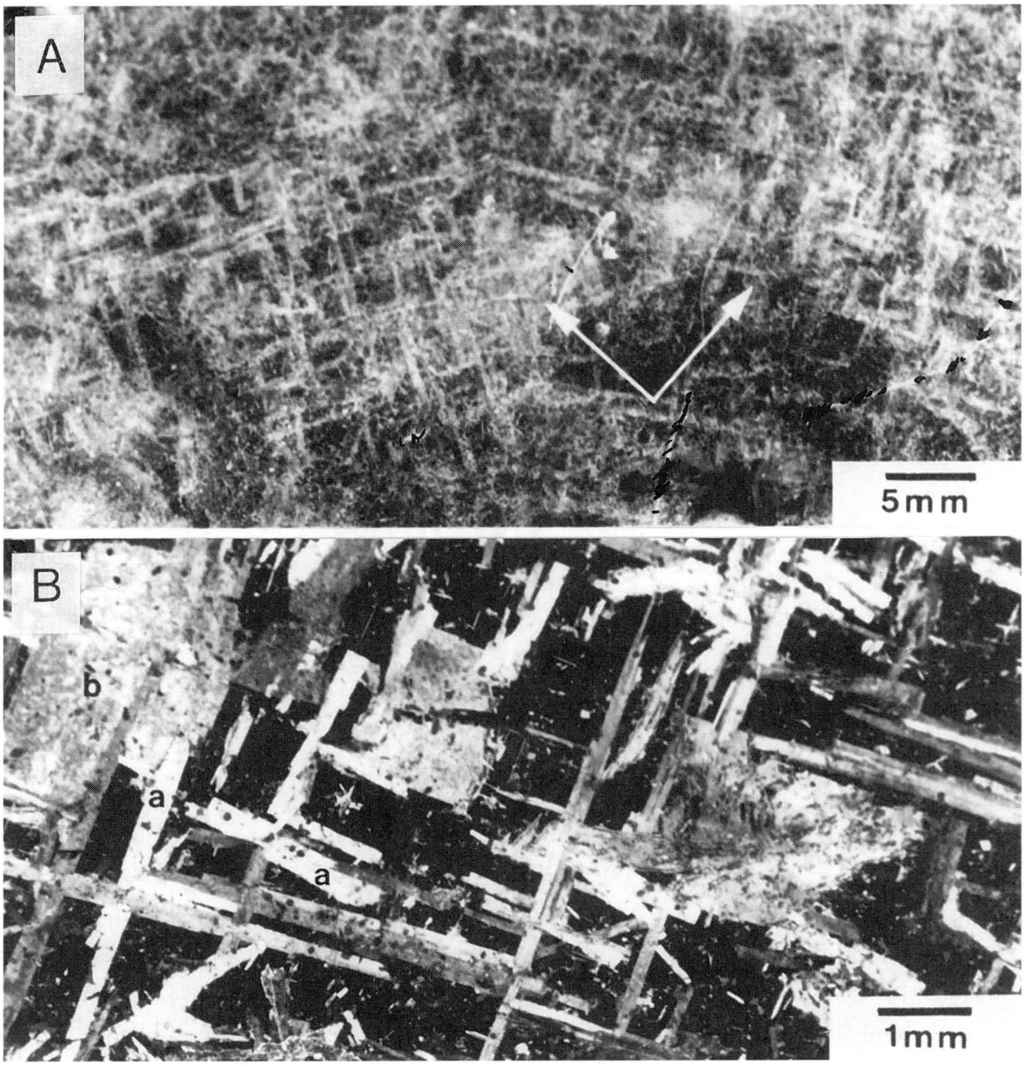

FIG. 5.29. Thin section photomicrographs of mixed sulfate-halite facies in the Lower Salt Formation. A. Sample composed of reticulate salt showing the regular network of elongated plates of anhydrite within halite. Crossed-polars. DP 108 well; 1494 m (Unit T5).
B. Mixed sulfate-halite facies showing details of the reticulate fabric. Anhydrite plates are set in the cleavage planes of the halite crystals in three perpendicular directions. Crossed-polars. DP 108 well; 1494 m (Unit T5).

fabric but in most instances they are arranged randomly. The laths occur together with small nodules composed of microgranular anhydrite at the core with radial laths at the edges. This crystal assemblage is probably the result of an early crystallization of anhydrite in the initial pores of halite deposits before early cements or crystal overgrowths have closed them.

3—*Macrocrystalline halite* (figure 5.30B) is dominantly constituted of halite in the form of clear (inclusion-free), clean (no impurities), cm- to dm-sized crystals, usually anhedral but sometimes euhedral in form, developed within large voids in the Argillaceous Salt facies. Anhydrite occurs as connected veins, mm-wide, that enclose the halite crystals. Some of the veins, composed of microgranular

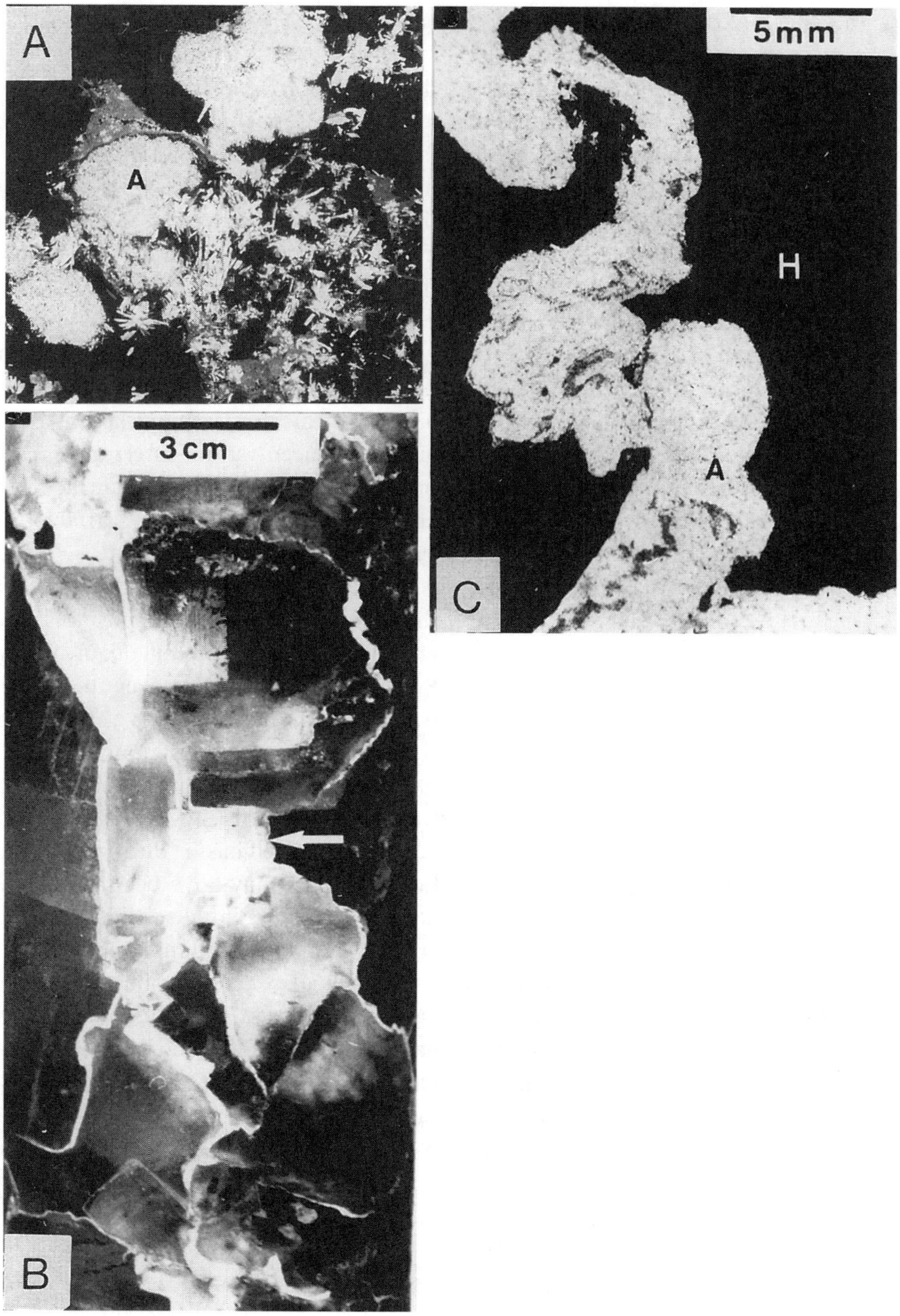

Fig. 5.30. Mixed sulfate-halite facies in the Lower Salt Formation: anhydrite-bearing halite.
A. Thin section photomicrograph of a nodule of microgranular anhydrite (A) within halite (H). In addition to the nodules there are anhydrite plates inserted within the cleavage planes of the halite (see arrows). Crossed-polars. TE 1 well; 1425.8 m; (Unit T4; Horizon C.B).
B. Slab sample of macrocrystalline halite. Light-colored, anhydrite veins (arrows) outline casts of large crystals of clear halite. Anhydrite in these veins is very fine grained. Core, TE 1 well; 1437.5 m (Unit T4; Horizon D).
C. Macrocrystalline halite in a thin section photomicrograph showing the detail of the internal stucture of an anhydrite vein (A) within halite macrocrystals (H). Vein shows tight internal folding (contorted appearance) due to development in a very restricted space. Crossed-polars. TE 1 well; 1463.2 m (Unit T4; Horizon E).

crystals, have an enterolithic form (figure 5.30C), suggesting that the formation of the anhydrite took place in a restricted space, probably after the halite was deposited.

This facies appears as large patches, decimeter- to meter-sized, that lie oblique to the bedding of the surrounding lithology, which is typically milky and argillaceous salt. In agreement with Moretto (1986), this facies is believed to reflect infill of a solution cavity. Development of pocket-shaped cavities and sporadic presence of associated breccias suggest a dissolution phase that occurred much later than the usual haloturbation that affects the argillaceous salt. Such late development could explain the clearness of the halite crystals protected from contamination by clays during their slow growth.

Distribution of Deposits, Sedimentary Dynamics, and Syndepositional Tectonics

THE SUBSALT AND THE LOWER SALT FORMATIONS

The Subsalt Formation is mainly composed of terrigenous sediments such as sandstones and reddish shales, that are dominant in the southern area of the basin (GVA 1 well). The siliclastic material came from the south and was deposited there by a distal fluvial system. Bedding structures that would help to define the exact environment are not available for study due to lack of coring, and interpretation is based purely on provenance analysis of the sands. Higher up in the sequence, sandstones pass gradationally into marls and carbonates that probably reflect slack periods in the terrigenous input, even though they still contain silty and sandy material. Anhydrite within the siliclastic deposits may have formed during periods of drying and subaerial exposure. Rare anhydrite-bearing carbonate beds possibly originated in response to expansion of the central brine body onto this southern margin.

In the subcentral part of the basin (DP 108 well), the upper portion of the Subsalt Formation shows an "ideal" depositional sequence as presented in figure 5.31 (at left). The finely laminated limestones, intercalated between the massive carbonates and the sulfate deposits, seem to have been deposited beneath a stratified water body, possibly caused by the sporadic influx of undersaturated waters. The progressive concentration of the uppermost waters by evaporation forced mixing and resulted in the deposition of sulfate and then salt. The sedimentary history of this area was marked by periods of progressive concentration, interrupted by episodes of undersaturation caused by surface water runoff, and was accompanied by a variable influx of siliclastic material.

The terrigenous deposits of the southern part of the basin and the presalt deposits of the subcentral area are apparent lithologic equivalents within the Subsalt Formation. However the correlation illustrated in figure 5.31 shows that, at the sequence scale, the marginal terrigenous material was probably deposited prior to the accumulation of the central evaporites. This suggests that the Valence Basin acted as a starved basin (Busson 1980). This depositional pattern may have been maintained and enhanced by a differential subsidence of the basin relative to its margin. Sediment buildup on the margin itself may also play an important part (termed "auto-aggravation" by Busson 1974). The dominant laminated-limestone lithology, very rich in organic matter, that is commonly found underlying salt deposits, corresponds to the presalt units recorded in a number of other evaporite basins (Busson 1979).

LOWER SALT FORMATION

Unit T6 of the Lower Salt Formation is largely composed of halite (figures 5.32 and 5.33). Bedded salt ranges from 70 to 150 meters in cumulative thickness and the halite represents 70% to 80% of the total lithology in an elongate 020°-trending area from the Ci-g1 well to the Hauterives field at the basin center. Differential sedimentation, however, developed across the basin. Bedded salt accumu-

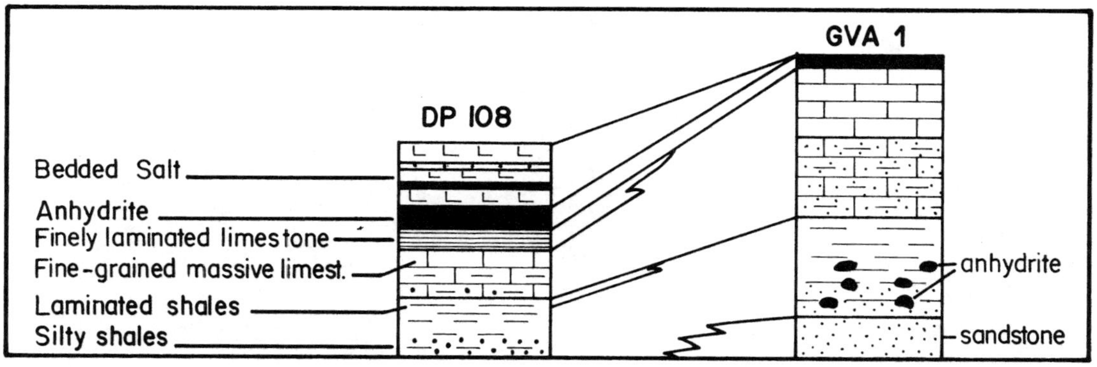

FIG. 5.31. Tentative correlations of lithologies in the Subsalt Formation between the southern and central portions of the Valence Basin. Basic lithologic sequences employed are from the GVA1 well (southern part) and the DP108 well (subcentral part) (adapted from Dumas 1988).

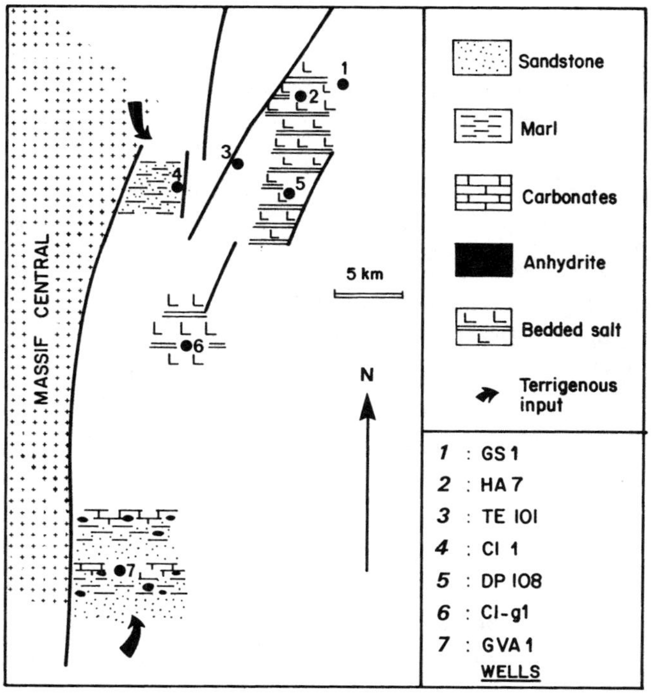

FIG. 5.32. Schematic depositional facies zones of Unit T6, Lower Salt Formation, Valence Basin. The salt zone is restricted to a 020°-trending elongated area and is rimmed by a clastic zone. A carbonate belt is suspected to lie between these two areas but has not been penetrated by a well bore. Well designations at lower right. Arrows represent terrigenous input (adapted from Dumas 1988).

lated in the central and northeastern parts whereas terrigenous products were deposited peripherally in the western and southern portions.

The typical depositional sequence of the central evaporitic area is tens of centimeters thick and is composed, in ascending order, of laminated marls, carbonates, anhydrite, and bedded salt. This basic sequence of sedimentation (in an ideal depositional format) probably reflects a gradual concentration of dilute water that is interrupted during halite deposition by repeated and abrupt dilution of the shallow water body. In the marginal zones, these same run-off waters carrying the clastics from the land caused the dissolution of salt beds previously deposited during short-term extensions of the central brine body onto these areas.

Unit T5 contains less salt than does Unit T6 (figure 5.33). The total thickness ranges from 20 to

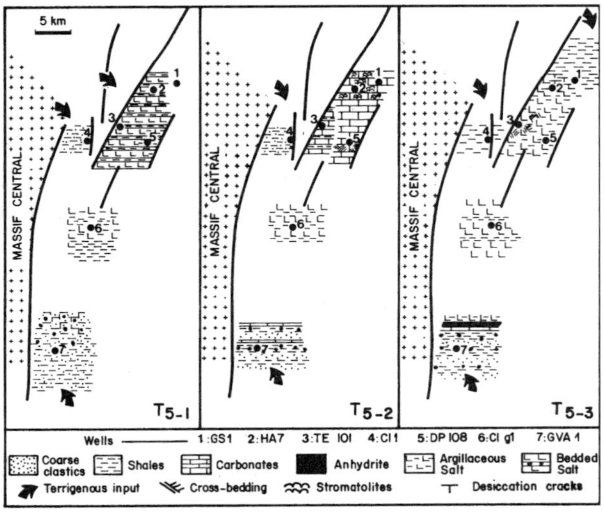

Fig. 5.33. Schematic depositional facies zones for subunits T5–1 (base), T5–2, and T5–3 (top) of Unit T5 of the Lower Salt Formation, Valence Basin (adapted from Dumas 1988).

15 meters and halite makes up 30% to 60% of the total lithology. The greatest amounts of salt are restricted to the Cl-g1 well area. The depositional components that make up Unit T5 are highly variable from one area to another and generally are composed of six discrete lithologies: sandstones, shales, limestones (fine-grained and stromatolitic), anhydrite (nodular and massive), reticulate salt (as thin occurrences), and primary salt (clear or milky).

The distribution of depositional facies through the subunits of T5 (figure 5.33) shows several aspects of interest in comparison with under- and overlying units:

the area of salt accumulation is significantly reduced with an expansion of carbonate and terrigenous deposits in many of the former salt areas;

thick carbonate beds develop, which are generally rare in the Lower Salt Formation: these carbonates bear traces of bioturbation and contain mollusk or foraminifer shells, which suggests a relatively lower salinity, formative water body;

a belt-like centripetal distribution of the deposits formed of halite, carbonate, and terrigenous material (with a modified concentric pattern). The carbonate rim is sporadic and is only present in the eastern marginal area where it is probably developed due to the low terrigenous input. Carbonate deposits exist in the Tersanne area but are mixed with clastics.

The restriction of salt deposition to the depocenter areas is probably a result of such combined processes as (a) decrease of the total brine volume; (b) drainage of brines to the basin center because of differential subsidence and accumulation of clastic and carbonate sediments on the margins; and (c) repetitive dissolution of salt deposits on the margins by localized inflow of less-saline water and re-concentration in the depocenter area.

Even though the dominantly terrigenous/carbonate sediments are the lithologic equivalent of the central salt deposits, it should be stressed that salt beds pass at the margins into dissolution surfaces within argillaceous beds, sometimes marked by accompanying anhydrite layers that were not affected by significant dissolution.

Unit T4 displays a relative high and constant salt content, 70% to 80% halite, with a cumulative thickness ranging from 160 to 210 meters. The isopach and isolite (relative content) lines both follow an apparent 020°-trend from the area of the Cl-g1 well to the Hauterives field.

The lowermost section of Unit T4 is composed of argillaceous and milky salt interbedded with thick shale layers containing nodular anhydrite, whereas in horizons F to C halite predominates. The rhythmic depositional successions are basically made up of argillaceous salt, enclosing carbonate and massive anhydrite beds, overlain by crusts of cubic halite. These in turn are capped by salt with chevron-shaped crystals, most of which are truncated by dissolution surfaces. In addition, evidence of internal dissolution, such as pocket- and pipe-shaped cavities, is present. The vertical succession of facies reveals that the sedimentation pattern consisted of the repeated influx of undersaturated waters succeeded by evaporitic periods.

The uppermost section of Unit T4 is composed of two subunits present in both the Tersanne and Hauterives fields. Horizon B (about 5 meters thick) exhibits a loose interbedding of laminated carbonate (up to 1 meter thick), massive anhydrite, and cryptalgal carbonate (Guillevin 1981). Horizon A (about 10 meters thick) is composed of pure halite (cubes and chevrons) intermixed only with flakes of clay. It is overlain by the basal laminated limestones of the Intermediate Marl Formation. Horizons B and A, taken together, are quite singular in their makeup. Horizon B contains finely laminated limestones, rich in organic matter, coccoliths (Goy and Busson 1980; Dumas 1988), and fish remains (Gaudant and Guillevin 1979). Each of the different lithologies is present as pure beds, one meter to meters thick, and not as co-occurrences or in tight successions. Horizons A and B contain very little terrigenous material and the salt is pure and devoid of any trace of dissolution. Halite of Horizon A has a high bromine content that ranges from 70 to 120 ppm (Dumas 1986, 1988) whereas it is low (0 to 30 ppm), with an irregular pattern of concentration, in the underlying units of the Lower Salt Formation (Alabouvette et al. 1984; Dumas 1988). The depocenter of Horizon A is shifted toward the northwest (north of the Tersanne field); and Horizons A and B (at least in part) are absent in the nearby southern area (DP 108 and Cl-g1 wells).

The disappearance of Horizons A and B in the southern portion of the basin is probably related to dissolution and/or erosion processes that took place in areas of lesser subsidence (Dumas 1986). The deposits preserved in the Hauterives-Tersanne area may have been protected from dissolution by an oversaturated brine layer developed at the base of a stratified water body. This model is supported by the presence of finely laminated limestones that lie immediately above Horizon A.

Throughout these pre- and true evaporitic periods, the structures generated by 160° and the 020°-oriented distensive faults exerted a substantial influence upon the accumulation of sediments, as documented by:

- the overall pattern of the isopach lines for the Subsalt Formation and the Lower Salt Formation (figure 5.34)
- the distribution of depositional facies of the units
- the isolite (cumulate thickness) and isopercentage (relative content) lines for the salt deposits (Dumas 1988).

The seismic profile of the Valence Basin discloses a progressive thinning of the Lower Salt Formation toward the Montmiral basement high, probably accompanied by a change of facies from the halite to carbonate and anhydrite lithologies. The absence of the Subsalt Formation and the Lower Salt Formation upon this horst is due either to non-deposition or early re-solution and/or erosion of a thin sequence.

Both of the two basal formations show a systematic depositional evolution from one that is dominantly terrigenous/carbonate (Subsalt Formation and Unit T5) to one that is dominantly saliferous (Units T6 and T4). The first type is marked by restriction of salt accumulation to a confined central area. In contrast, when salt deposits dominate (Units T6 and T4) they become widespread over large areas. In the Valence Basin, these uppermost phases of widespread salt formation are marked by (1) a low-differential net subsidence, as suggested by the constancy of the thickness of the deposits, and (2) the progressive spreading and overstepping of the salt deposits across the siliclastic fans, a process regularly noted in many evaporitic basins.

INTERMEDIATE MARL FORMATION

Basal deposits of this unit are typical of restricted evaporitic environments (laminated limestones, anhydrite beds). Higher in the section, the lithology passes to dominantly silty marls, argillaceous sands, and sandstones, all containing cross-bedded current structures. Rare anhydrite beds and laminated facies are intercalated within this section. The molluscan assemblage, including genera *Hydrobia,*

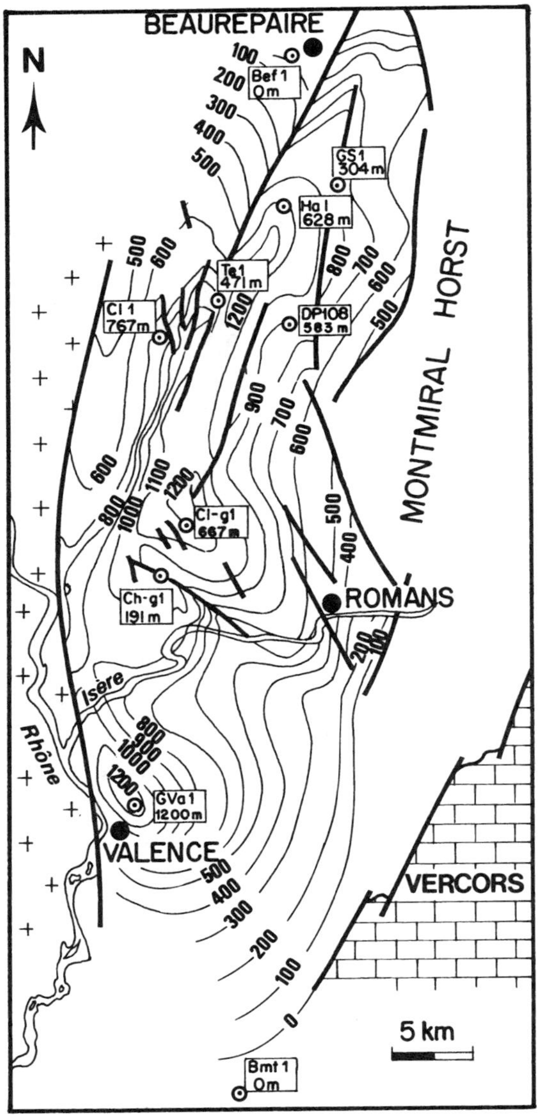

Fig. 5.34. Restored isopach and tectonic map for the Subsalt Formation together with the Lower Salt Formation, Paleogene sequence, Valence Basin (faults indicated by strong black lines) (adapted from Dumas 1986).

Cyrena, and *Potamides,* suggests environments of low to moderate salinity. The calcareous nannofossil assemblage indicates that marine Cretaceous and lower Paleogene sections were reworked into the formation. Definite marine deposits only existed to the east of the Valence Basin (Charollais et al. 1980; Fabre et al. 1986). For these reasons it is suggested that most of the clastic material of the Intermediate Marl Formation was sourced from the nearby incipient Alpine uplands.

Moving stratigraphically higher in the section, the increasing frequency of anhydrite-carbonate beds and the presence of dolomitized oolitic limestones suggest that evaporitic conditions resumed. This change in styles of deposition, from shallow water (cross-bedding in clastic deposits and sparse evaporite accumulation) to evaporite deposition, suggests an alteration of the water balance in response to climatic change and/or progressive closure of the basin outlet.

The lithologies of the Intermediate Marl Formation are uniform across the basin, including the area of the Montmiral horst (with the exception of the basal deposits). No coarser deposits have been recorded in the western marginal area (Cl-1 well). Isopach lines also indicate only a minor effect of differential subsidence throughout this period (figure 5.35).

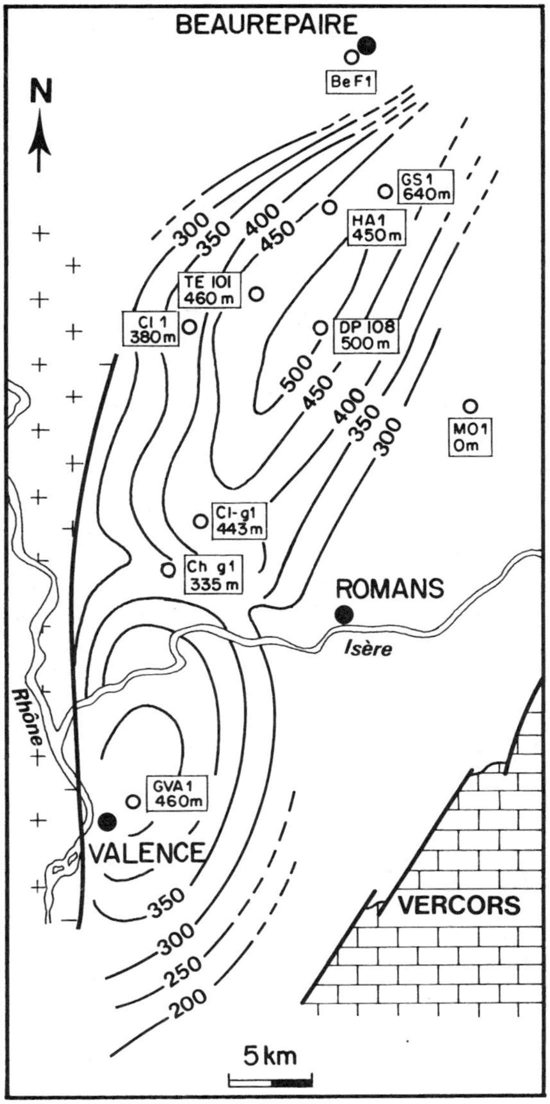

FIG. 5.35. Restored isopach map for the Intermediate Marl Formation, Paleogene sequence, Valence Basin (thicknesses are in meters) (adapted from Dumas 1986).

THE UPPER EVAPORITE AND THE TERMINAL PALEOGENE FORMATIONS

Salt deposits of the Upper Evaporite Formation are restricted in area. Only three wells (TE 101, Cl-g1, and Ch-g1 have penetrated them (figure 5.36A). Their total thickness is no more than 150 m (depocenter at Ch-g1), but the composition ranges up to only 30% halite. The total formation thickness is 500 m at the depocenter. It is composed of carbonates (dolomicrite), sulfates (anhydrite and glauberite), halite (pure, milky, and chevron microfacies), and rare greenish marls. The glauberite, a sulfate mineral, is found admixed with halite in the depocenter area in the lower part of the section (in wells Cl-g1 and Ch-g1).

The distribution pattern of the deposits also changes towards the basin margins, as seen from the extent of siliciclastic and carbonate/sulfate sediments in the lower and upper parts of the Upper Evaporite Formation illustrated in figure 5.11. The depositional zones have a roughly concentric distribution parallel to the isopach lines. The initial marginal carbonate and sulfate zones (figure 5.37A) were replaced by a siliciclastic zone (figure 5.37B) whose progradation into the basin caused the progressive shrinkage of the central evaporitic area.

It is apparent that salt deposition was controlled

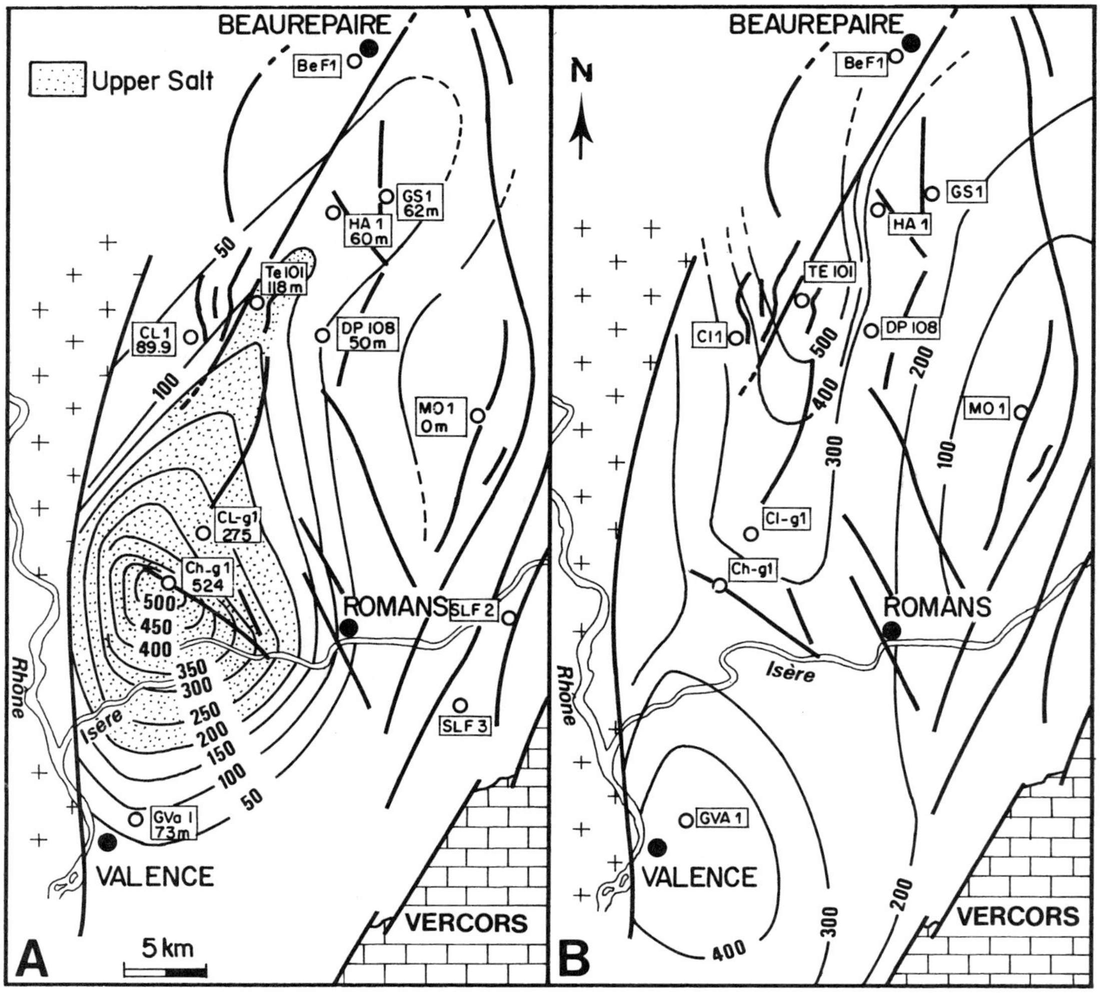

FIG. 5.36. Restored isopachs and tectonic map for the Upper Evaporitic Formation (A) and the Terminal Paleogene Formation (B), Valence Basin (thicknesses are given in meters) (adapted from Dumas 1986).

first by the combined activity of 020°- and 110°-oriented normal faults and later by the activity of a 110°-oriented fault alone (see reconstructed cross-section in figure 5.31). The salt also may have been deposited on the margins, but due to differential subsidence, may have been dissolved and then recollected in the central area (erosion, re-solution/re-precipitation).

Above the Upper Evaporite lithologies and the "Radioactive" sands lie (secondary) gypsum-bearing marls capped by glauconitic sandstones. Higher up in the section are thick, extensive, terrigenous deposits that cover a much broader area than the underlying glauconites. These deposits are coarse-grained in the north (Tersanne field) and much finer in the south (Valence area). The distribution of the sediments of the Terminal Paleogene is marked by the inflective pattern of the isopach lines together with the disappearance of the "Radioactive" sands between the Tersanne field and the Hauterives field, which shared a similar depositional evolution until this point. It is suggested that syndepositional activity of a 110°-oriented (and possibly also of a 160°-oriented) normal fault began late in the development of the basin. Higher up, the Paleogene section of the Valence Basin is covered by the thick and extensive marine deposits of the Miocene Molasse.

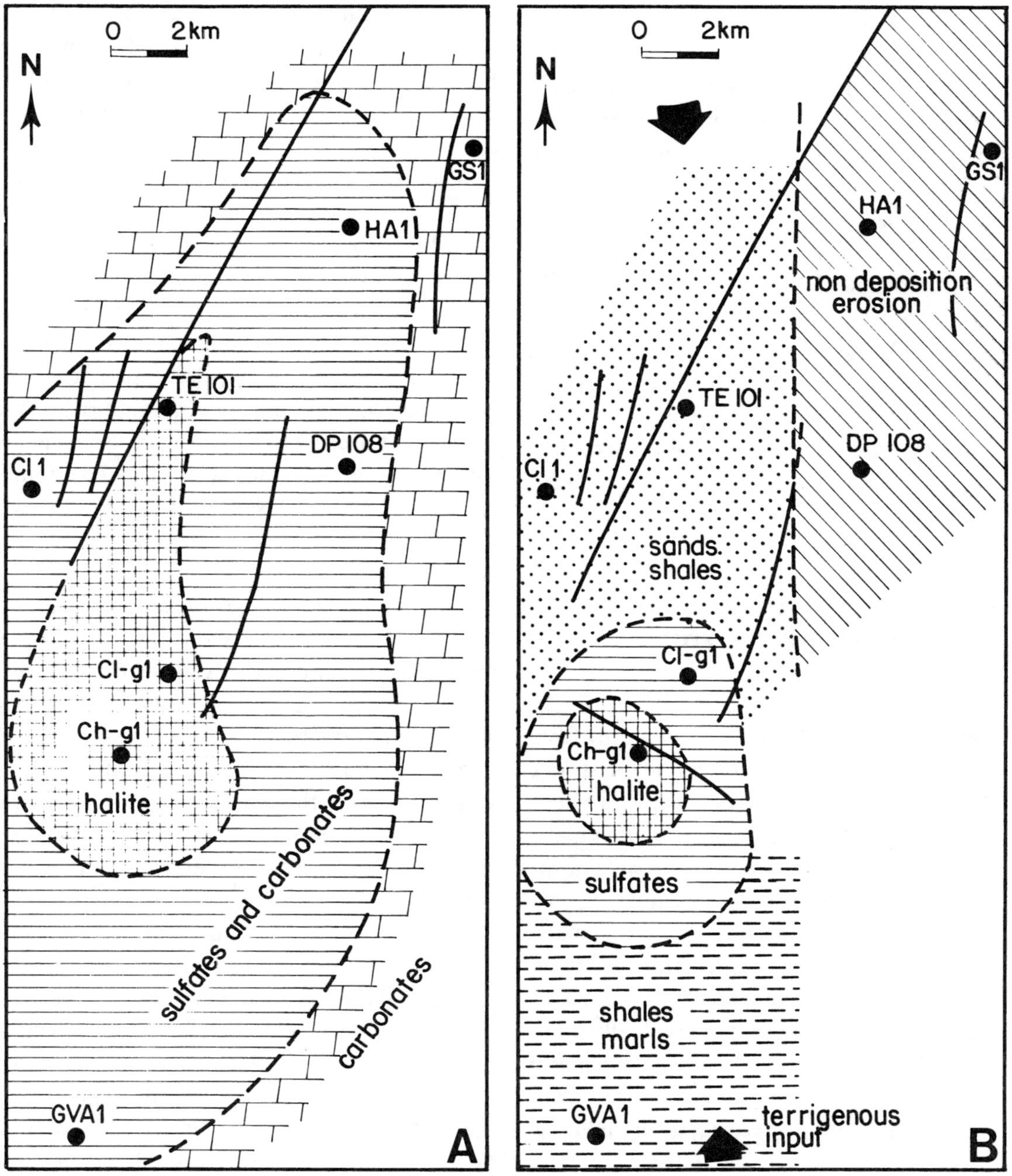

FIG. 5.37. Schematic representation of the depositional facies zones for the lower part (A) and the upper part (B) of the Upper Evaporitic Formation, Paleogene sequence, Valence Basin. Note general concentricity of facies during this depositional phase. The halite zone in "A" contains some admixed glauberite (Ch-g1; Cl-g1) (adapted from Dumas 1988).

Discussion and Conclusions

We have shown that during the Paleogene the basin of Valence evolved as a physiographic entity filled, as it developed, with a series of sediments most of which were deposited within very restricted environments. These sediments may be considered from four separate but related points of inquiry: (1) the overall environments of deposition, including climate; (2) the general modes of diagenesis of the various sediments; (3) the sources of water from which these sediments were deposited and altered early in their burial history; and (4) the general tectonic evolution of the basin and its control on sedimentation.

DEPOSITION: FACIES REVIEW AND THE ENVIRONMENTS OF DEPOSITION

A number of sedimentary facies developed in the Valence Basin, some of which are known only from rare core samples and from identification on seismic lines and well logs (i.e. the detrital facies). Others, with more plentiful sampling, are better known and their development can be traced across parts of the basin (carbonates, halite, and glauberite).

(a) *Terrigenous clastics.* Most of the coarse material appears to be restricted to the basin margins. The finer materials (silts and clays) were transported into the basin proper and are typically intermixed with other sediments. The environments of sedimentation cannot be clearly defined, but their limited peripheral position suggest a wadi plain system (table 5.3A).

(b) *Carbonates.* There are several significant facies developed in the Valence Basin (table 5.3B).

Microbially-originated limestones, whether with well-developed stromatolitic forms or not, may develop in a wide range of marine and nonmarine environments. Since the early Paleozoic these algae and bacteria have been the victims of extensive predation. When they do accumulate (as in parts of units T4 and T5 of the Lower Salt Formation) and have clear stromatolitic form, they commonly do so under conditions that are too restricted for their predators. Usually the restricted environment is hypersaline (either marine or nonmarine), as reflected here by the common co-occurrence of anhydrite.

The oolitic carbonates usually form in shallow environments, often somewhat hypersaline as suggested here by the traces of early alteration (dolomitization and anhydritization).

The pelletal limestones, in which the pellets are well-formed and uniform in size, are otherwise unfossiliferous and have well-developed fibroradial cements surrounding and separating most of the pellets. These are likely to be facies of re-deposition into very shallow water with elevated salinities as evidenced bythe presence of aragonite as overgrowths around pellets and halite as primary cement.

Finely laminated limestones are also helpful in facies interpretation. They are not bioturbated or mechanically reworked. Even though it seems that finely laminated limestones were deposited under stratified waters, there is some question about the depositional depth, together with the causes of the stratification of the water. A density difference of 15 g/l is large enough to induce a water body to have stable stratification even in a shallow environment. The stratification may owe its origin to an abrupt influx of low-salinity waters, as documented in continental salt lakes.

Because the rate of sedimentation is very low for such facies (Degens and Ross 1974), it can be inferred that the meter-thick horizons detected in the Subsalt Formation (the presalt deposits) and in the Lower Salt Formation (Unit 4; Horizon B) reflect periods of water stratification that lasted several thousands of years at least. Such a period of long-term water stability is not consistent with an extremely shallow environment. Thus it is probable that the slow accumulation, together with a continued differential subsidence, may have caused the progressive deepening of the central part of the depression, in a starved basin setting.

Some laminae in these limestones are rich in coccoliths, sometimes preserved as entire coccospheres. The laminae consist either of regular-sized,

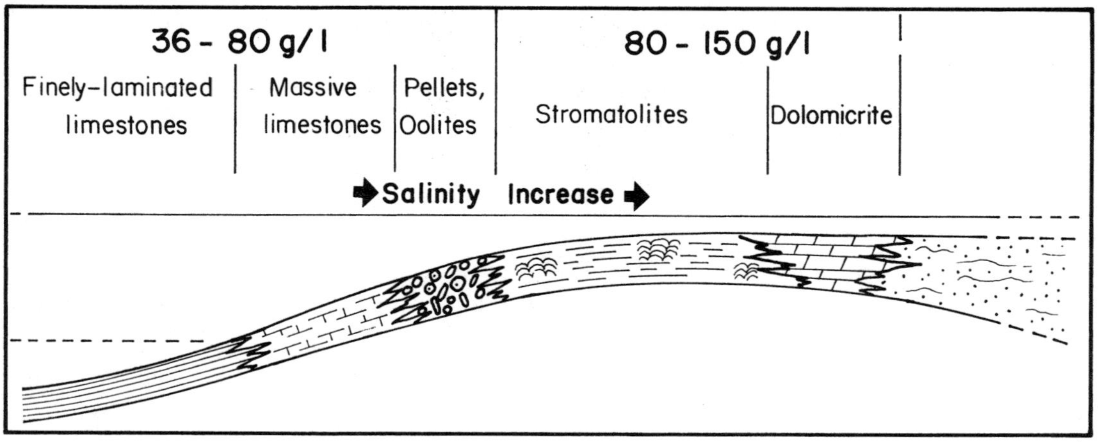

FIG. 5.38. Correlation of salinity levels with depositional environment. Tentative depositional model versus salinity for carbonates of the Paleogene salt sequence in the Valence Basin (adapted from Dumas 1988).

multispecific forms or of diminutive single species. Such a nannofossil content suggests that the uppermost water was of variable salinity but basically marine derived. It has been shown that certain coccolith species can be abundant in marginal marine waters of comparatively low salinity due to contamination by fresh nonmarine waters (Noël 1982). Therefore, it is suggested that the bloom and deposition of the single-species forms in the Valence deposits was induced by sporadic salinity drops within the uppermost waters. This opinion is fully consistent with the presence of organic matter of Type I, which is thought to characterize lake (Pelet 1985) or confined environments (Espitalié et al. 1985 and 1986). Such a freshening of shallow waters can also explain the absence of any other planktonic marine forms within the deposits.

Thus the carbonate facies in general demonstrate two formational milieus: shallow water with elevated salinities for the cryptalgal, oolitic, and pelletal facies but with no definite water source (i.e., marine or nonmarine); and somewhat deeper, stable, stratified waters that are probably marine-sourced, with upper water of more-or-less reduced salinity.

A depositional zonation of the carbonate facies versus salinity has been traced (figure 5.38), based upon this work and observations of modern salt pans in Spain by Ortí et al. (1984). Their zonation model seems applicable and valid because the entire Valence sedimentary sequence is evaporitic.

(c) *Sulfates.* Calcium sulfates (gypsum and anhydrite), glauberite, and celestite are present in this basin (table 5.5A).

Those sulfate facies that yield definitive information concerning environments of primary deposition are not well developed. Grass-like gypsum crusts (shallow water deposition) are not common in this basin while the other sulfate forms that are present are variously early to late diagenetic, and their environmental interpretation is not at all clear. The one certain form, grass-like gypsum, associated with stromatolites, thus seems to be the only reliable depositional indicator. Both the gypsum and the stromatolites are considered to be of shallow water origin.

Glauberite, on the other hand, gives a definite suggestion of a continental water source. While the actual settings for this mineral are somewhat uncertain, most of the known glauberite deposits seem to originate in continental playas and mud flats.

(d) *Halite.* There are several types of halite that yield significant information concerning depositional settings (table 5.5B). The white, milky salt, filled with fluid inclusions, develops during rapid precipitation from a very saturated brine during intense evaporation. According to a number of authors (Ortí et al. 1984), the intracrystalline alternation of milky and clear bands is the result of the daily variation in evaporation intensity, which also implies a very shallow brine-body. Such milky halite deposits are easily affected by early diagenetic tranformation. Original porous crusts suffer early phases of dissolution, cm to dm in magnitude, followed by cementation and syntaxial overgrowths of clear halite. Such occurrences have been observed in shallow, high salinity lagoons and appear to characterize cycles of flooding—concentration—drying with subaerial exposure—reflooding (Shearman 1970; Lowenstein and Hardie 1985).

Some of the halite facies, such as the clear salt and the cornet-shaped crystals, cannot reliably be referred to a precise water depth. In contrast, others such as the milky and reticulate salts have an apparent signature of shallow environments.

The common development of diverse dissolution traces overlain by unaltered layers within the milky salt deposits, reveals that drops in salinity affected the depositional areas and were not caused by later dissolution. Environments within a shallow water body are more prone to record such events than are deep water bodies (Busson 1982). The fine rhythmical pattern of some primary salt (bedded salt, milky salt) results from the rapid influence of variably effective dilution of the brine upon the sediment-water interface. This also indicates shallow depositional depths.

REVIEW AND DISCUSSION OF DIAGENESIS

The dolomitization of calcium carbonate deposits.

Subhedral rhomboids (isogranular to fiber-shaped) of dolomicrosparite are very likely the heritage of initial crystals of calcium carbonate, although the crystal-size may have increased slightly during the transformation. Indeed, the primary fiber-shaped dolomite is uncommon (Weijermars et al. 1986). Supporting evidence for such a style of replacement comes directly from the evident preservation of depositional textures.

Dolomicrosparite-bearing oolitic beds display a high porosity. This dolomite habit does not show any variation at the edges of the intergranular voids, which testifies to a dolomitization prior to the void generation. The formation of the dolomicrosparite originated early in the oolite-supporting muds, subsequent to the migration of hypersaline solutions.

The sulfate welter.

The diagenetic history of the various sulfates throughout the section appears to have been a complex and multistage evolution.

Early in the deposition, various alteration events have been documented:

- the precipitation of anhydrite in clear halite within bedded salt (syndepositional or due to early dissolution)
- the development of gypsum in different forms (lens-shaped patches, grass-like crusts, also formation of gypsum as possible cement of cryptalgal carbonates)
- the accretion of anhydrite and glauberite as nodules within shale/carbonate deposits directly from a subsurface brine-sheet or as a replacement of initial gypsum.

Burial-related transformation and neoformation features are also numerous:

- the common change of gypsum into anhydrite by simple dehydration
- the decrease of the anhydrite crystal-size with concomitant change from granular to microgranular texture
- the variably effective obliteration by sulfates of later-formed, intergranular porosity of the coarse-grained carbonates
- the replacement of anhydrite by celestite
- the dissolution of gypsum, followed by the internal crystallization of anhydrite or halite as lenticular pseudomorphs (the initial dissolution was presumably related to the release of chloride-rich brine)
- the development of oblique cross-cutting veins of anhydrite, reflecting the internal and late migration of sulfate-bearing fluids.

These processes illustrate how the diversification of the sulfates occurred throughout the burial history. Transformation and recrystallization are inherent because of the temperature and pressure increases due to burial. The intimate association of sulfate-chloride deposits results in alteration because halite alters the solubility and stability of gypsum and anhydrite, and makes up horizons impervious to brines. The development of fractures can trigger and favor the deep migration of brines trapped between layers of salt. Such solutions also migrate due to the presence of sedimentary conduits such as coarse-grained carbonate and terrigenous deposits.

Finally, late fresh-water circulation through the sedimentary section is shown by the re-hydration of anhydrite, producing secondary gypsum with variably-preserved original structures (Curial and Moretto, this volume).

Halite variants.

The milky primary salt was subjected to an early evolution in morphology due to syndiagenetic alteration. Salt crusts are initially of high porosity (up to 40%, Perthuisot 1975; Busson and Perthuisot 1977; Hardie et al. 1985), and are affected by fine-scale dissolution succeeded by crystallization of clear halite as cement and as syntaxial growth.

The argillaceous salt is of early diagenetic origin. The clear-halite porphyroblasts grew through and within a shale matrix, subsequent to substantial dissolution by runoff waters that deposited the argillites (terrigenous clastics), followed by progressive concentration due to halite re-solution and/or evaporation. Supporting evidence for such brine dilution is provided by the close connection existing between this particular salt facies and solution cavities.

The fibrous halite within veins, the macrocrystalline clear halite within large solution cavities, and the meter-sized dissolution pipes apparently formed after the early diagenetic phase. No evidence permits clear identification of their time of generation. However, these features disclose that the salt sequences supported active fluid migration through part or all of the burial period.

A correlation of core-scale dissolution features was attempted for the Tersanne field to estimate the extent of the depositional surfaces that marked the dilution of brines and salt dissolution. Such correlation was not possible because only the section at the TE-1 well was cored in continuity; available well-log data from other sites are not suitable for reliably distinguishing primary and argillaceous salt (Curial 1988); and the description of the cuttings of the uncored intervals was too rough for accurate stratigraphic correspondence.

SOURCE OF MOTHER BRINES (BY FORMATION)

Evaporites may have originated from an ion stock derived from either marine waters or continental waters, or a mixture of these sources.

Lower Salt Formation (late Priabonian—early Rupelian).

Mineralogic assemblages similar to the one encountered here, consisting of halite, calcium-sulfate, calcite, and subordinate dolomite, are fairly common in marine-related deposits. In addition, the lateral constancy of this simple mineralogy throughout the West European Rift (Curial et al. 1988; and this volume) appears to contradict the hypothesis of purely continental deposition because continental origins are commonly much more complex and variable from one basin to another. in addition the presence of indigenous coccoliths in Horizon B (Unit T4) of the Lower Salt reflects a marine source during that interval (Noël 1982). As mentioned earlier, this group of microorganisms is apparently tolerant of short-term lower salinities induced by water runoffs, which produce single-species blooms.

The high bromine content detected in Horizon A of Unit T4 (70 to 120 ppm) is usually ascribed to a marine origin of the parent brine. The leaching of potassic salt or volcano-clastics can be excluded in this case, because no evidence for such deposits has been encountered either in the basin or in its vicinity during or prior to this time period (the activity of the nearby "Coirons" volcano initiated only at the end of the Miocene). The low bromine content of the underlying units T6 and T5, together with their variable bromine concentration pattern, may reflect nonmarine water sources but also may merely be related to the common solution/re-

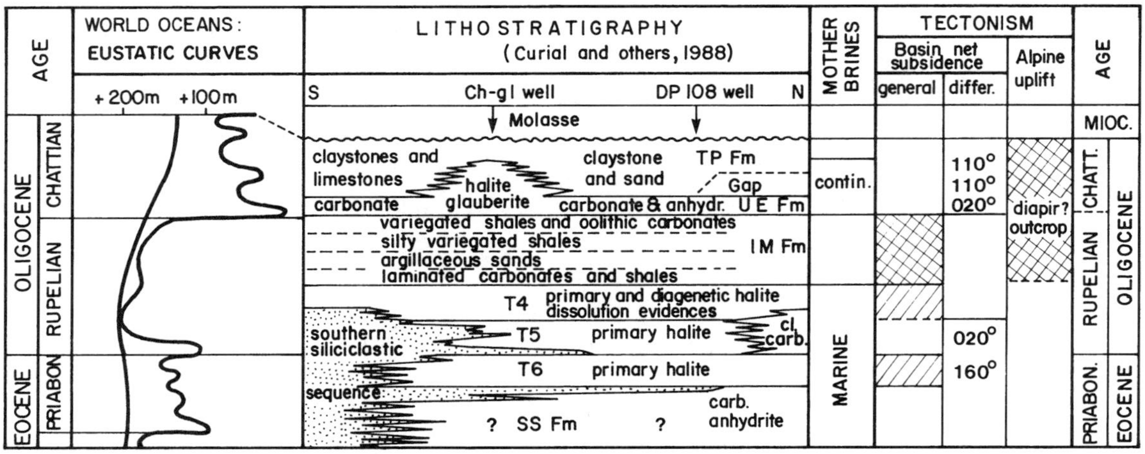

FIG. 5.39. Synthetic diagram depicting the tectonic-sedimentary history of the salt basin of Valence during the Paleogene.

precipitation processes that affected these salt deposits.

The regional paleogeography suggested by Pairis et al. (1984) shows that marine deposition from the direction of the northern Alps migrated toward the west during the early Oligocene. The sedimentary trace of that shoreline lies only 50 km distant from the depositional area of the Valence Basin. In fact, it actually may have lain farther to the east because of the overthrust of the Alpine sections upon the graben (shortening the distances), or been closer because of the fact that some Oligocene marine deposits possibly have been covered, overthrust, or removed by erosion (Paleogene material is known to be reworked into the Intermediate Marl Formation). In summary, it is clear that the early Oligocene sea was not farther than a hundred kilometers from the Valence Basin, making marine input possible along 020°-oriented, fault-related channels.

Finally, the likelihood of possible marine ingression is also suggested by the eustatic curves of Haq et al. (1987) that disclose a worldwide highstand for the early Rupelian seas. It should also be stressed that at the Priabonian-Rupelian boundary the successive overlaps of the salt deposits upon the southern clastics appear to coincide with these worldwide sea-level rises.

Separately, none of the data permit unmistakable recognition of marine mother brines for the Lower Salt Formation. However, the many lines of supporting evidence and consistent data make a predominantly marine origin almost certain, at least for the uppermost section of the formation (Horizons A and B), and likely for parts of the underlying units.

Upper Evaporite Formation (? Chattian).

The presence of glauberite and dolomicrite indicates evaporitic deposition from a calcium-impoverished residual brine such as commonly characterizes continental-sourced waters. Support for the idea of a lacustrine setting arises from the likelihood of non-marine saline deposits in the Bresse Basin (Curial et al. 1988; Curial and Moretto, this volume), as well as a paleogeographic restoration (Pairis et al. 1984) that shows continental facies in the northern Alps from the late Rupelian onward, and a major sea-level fall that marks the Rupelian-Chattian boundary (Haq et al. 1987).

Even though it is quite clear that evaporites of the Upper Evaporite Formation were deposited in a continental environment, the actual source of the ion stock is far from resolved. In southeast France, older, evaporite-bearing lithologies are present and the potential ion sources are from Triassic rocks. Triassic sections of the eastern margin of the Massif Central, as well as of the border of the Valence Basin (seen in the St. Lattier well), are only slightly evaporitic and are composed solely of sulfates. However, Triassic salt is believed to exist in the Vercors area (Busson et al. 1984). Such a source would suggest a salt diapir similar to that inferred just south of the Valence Basin, in the Nyons area, during the Oligocene (Perthuisot and Guilhaumou 1983). An extrusive salt body may be concealed along the Vercors overthrust plane and be too deformed to be resolved on seismic sections. A localized salt source such as from an emerging diapir is fully consistent with both the smaller volume of the Upper Halite in comparison with the Lower Halite and the restriction of the late salt occurrence to the Valence Basin (there is no salt at that time in the nearby Bresse Basin). All the interpretation is still very speculative, but cannot be ruled out a priori.

THE TECTONIC-SEDIMENTARY SCENARIO

The Paleogene history of the Valence Basin is marked by two evaporitic periods separated by a widespread and uniform clastic episode (figure 5.39). The first salt deposition, of Priabonian-Rupelian age, is apparently general through the West European rift, and the mother brines are at least partially marine. The second salt occurrence, presumably of a Chattian age (but not well constrained by fossil evidence), is local with deposits that are much reduced in volume, and seems to have originated in a continental environment. Furthermore, it is evident that the dominant direction of the syndepositional faults also changed through time, reflecting the modification of the overall structural stress pattern, which could readily have provided the control for the changes in water sources.

We propose two interpretations for the evolution of the Paleogene deposits. The first is based upon the variation in subsidence. Units T6, T4, and continuing into the Intermediate Marl Formation mark a phase of general, weakly differentiated basinal subsidence whereas units T5 and the Upper Evaporite Formation illustrate a higher rate of differential subsidence. The change in the character of the sedimentation between unit T4 and the Intermediate Marl Formation, and the renewal of evaporitic conditions in the Upper Evaporite Formation, could be the result of mere climatic alterations.

The second interpretation is based upon the variation of the quantity and the nature of water and sediment provided to the basin, the differential subsidence being constant through time. For a given time-span, high ionic or clastic influx results in a poor record of differential subsidence. In contrast a low supply favors preservation of such a record, because of exposure time, the reworking of the clastics, and solution/re-precipitation of the evaporites, and thus permits a much thicker sediment accumulation in areas of higher subsidence. Thus, units T6 and T4 may have developed in response to sea-level rises that favored a large ionic supply and associated sedimentation rate, and the apparent lateral overstepping of salt deposits. On the other hand, it seems that clastic deposition dominated during the time of Intermediate Marl deposition, due to the formation of the nearby Alpine uplift that interrupted marine communication and furnished a large quantity of detrital material. This supply apparently diminished temporarily during deposition of the Upper Evaporite Formation. Whether this apparent drop in supply was due to a regional climatic change, the development of a marked salt source due to formation and exposure of a salt diapir, or a slackening in the Alpine uplift is not resolved.

6

The Paleogene Basins of Southern Provence

DENISE NURY AND B. CHARLOTTE SCHREIBER

General Background

GEOGRAPHIC LOCATIONS

The Paleogene basins of Provence (Basse Provence) extend from the Fos-sur-mer on the west, the Luberon massif on the north, and eastward to Toulon (figures 6.1 and 6.2). These basins evolved and became separated by the evolution of the Provençal massifs and chains that developed during the tectonic activity from the late Eocene and onwards. The main structural outlines of these basins are illustrated in figure 6.2A. It is important to note that in the region of southern Provence, the Paleogene evaporites formed only in the Oligocene basins of Marseille, St Pierre-les-Martigues, and Aix-en-Provence. Before the Oligocene, only detrital clastics and limestones accumulated in the Aix basin and no record of sedimentation of any kind exists before the Oligocene in either Marseille or St. Pierre-les-Martigues.

CRUSTAL MORPHOLOGY

The crustal morphology of the Basse Provence has been pieced together based principally on the work by Gourvernet et al. (1979); Guieu and Rousset (1980); and Tempier (1987). In review, during the Eocene the thrusting evidently was from the south and moved toward the north, displacing basement together with the overlying Mesozoic section. This tectonized zone constitutes the allochthonous portion of Provence (figure 6.2B). The thrust sheets may be subdivided into a number of overthrust units, and the location and behavior of the associated fault ramps are apparently a consequence of preexistent heterogeneity, the variability in thickness of the overlying sections, and the effects of fault reactivation. As a consequence the units are sheared and uneven, with twisted axes, and are not displaced in a homogenous manner. The division of the thrust sheet into only three important units in earlier literature does not seem entirely justified (Tempier 1987).

After the Variscan orogeny, the basement underwent considerable erosion and truncation followed by a Mesozoic rise in sea level. The sea covered the Variscan basement complex from the Triassic through the early Cretaceous, forming an extensive marine sequence. The first structural episode to affect these Mesozoic sediments took place during the Albo-Cenomanian. The deformed region became emergent and was then exposed to erosion and profound weathering under a humid climate. Considerable sedimentary reworking of the weathering residues resulted in the formation of bauxite. During the late Cretaceous the sea again covered the region, but sedimentation was very much controlled by the earlier structural and erosional irregularities. The sedimentary environment slowly changed over time and by the Campanian it became

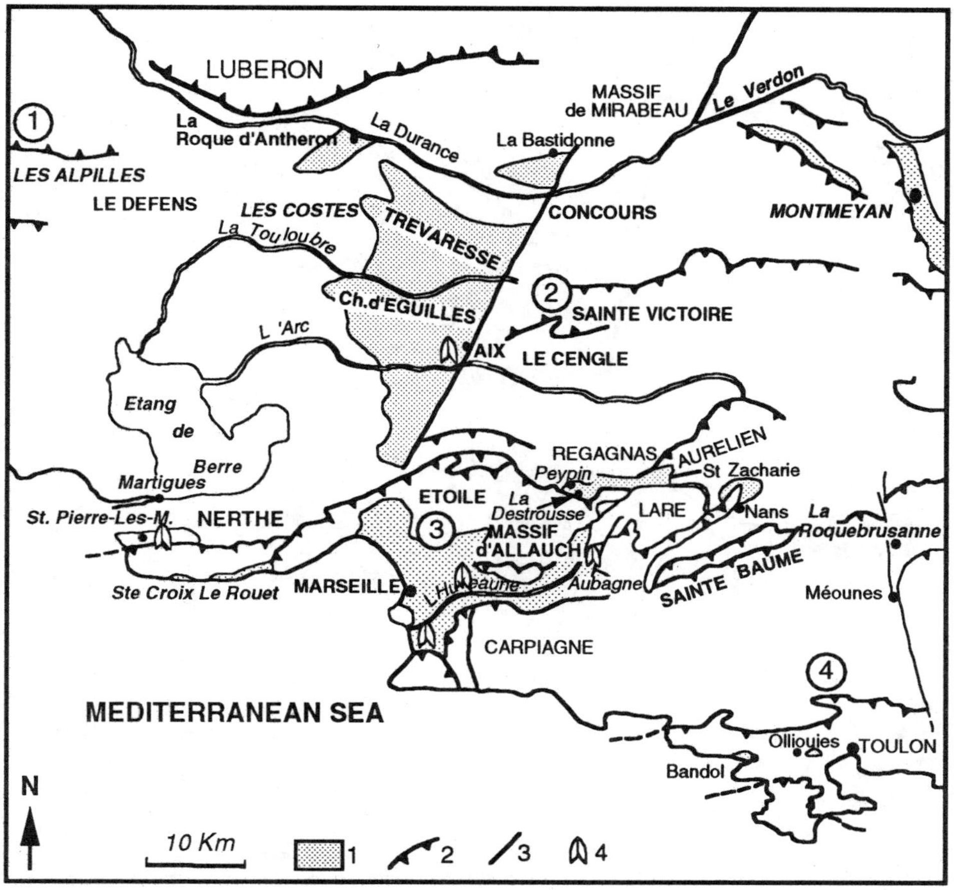

Fig. 6.1. Geographic location map of the major outcrop areas of the Oligocene deposits in Southern (Basse) Provence. Key: 1 = Oligocene outcrop; 2 = limits of overthrust; 3 = faults; 4 = gypsum.

a fluvio-lacustrine environment. By the latest Cretaceous (Maastrichtian), due to further tectonics and uplift, the exposed area increased rapidly and erosion became profound, cutting down through the Jurassic and into the underlying Triassic. This relief, developed during the late Cretaceous, controlled the fundamental outlines of the ensuing Paleogene basins of Provence that then became the sites of continental deposition until the middle Lutetien. The underlying Paleozoic basement, which lay exposed in areas to the southwest (the Pyrenees and Sardinia) and to the southeast (Corsica), became the sources for the sedimentary infill of these Paleogene basins (Durand 1984; Guieu et al. 1987).

By the late Lutetian, or more generally by the end of the middle Eocene (Cavelier and Pomerol 1986), the major phase of tectonism in Provence created major thrusts of the crystalline massif with associated décollement of the overlying Triassic section (Guieu and Rousset 1990). These thrust sheets are made up of the thick basinal sequences that later overrode the very thin sedimentary sequences that formed on the highs, apparently without the development of inverse fold flanks common in such structures. The over-thrust sheets were stretched to the point of breaking and thus created a series of arcuate belts or bands of displaced Triassic segments. It seemed that in this area there was no existing record of sedimentation before the Oligocene in southern Provence (de Brun and Châtelet 1923):

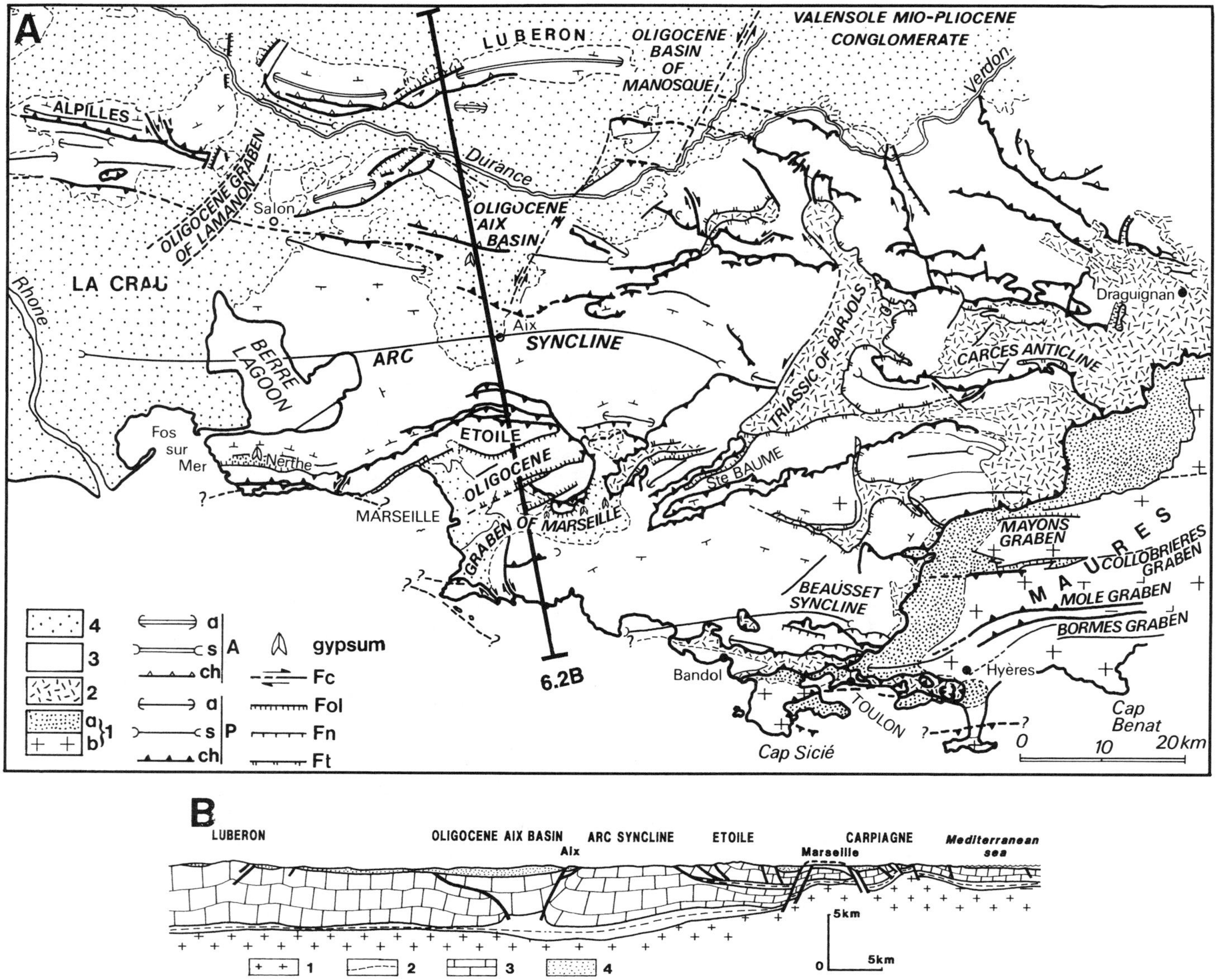

FIG. 6.2. A schematic structural map of southern Provence with main structural components shown. Associated section crosses Marseille, Aix-en-Provence, and north to Luberon.

A. Schematic structural map. Key: 1a = Variscan basement; 1b = upper surface of Variscan basement (Permian to lower Triassic); 2 = sedimentary layers below décollement surface (middle and upper Triassic); 3 = cover; 4 = Eocene to Quaternary (post-tectonic) formations; A = Alpine structures (late Miocene); a = anticlines; s = synclines; ch = thrusts; P = Provençal structures (late Eocene age); Fc = strike slip faults; Ft = normal listric faults at border of Triassic units; Fn = normal faults (late Eocene or older); Fo = normal faults (Oligocene distensive phase) (Tempier 1987, modified).

B. Balanced cross-section of Provence. Key: 1a = basement; 1b = upper surface of Variscan basement (Permian to lower Triassic); 2 = décollement level (sole thrust); 3 = cover; 4 = post tectonic formations (location of section given in figure 6.2A) (courtesy of Tempier, unpublished).

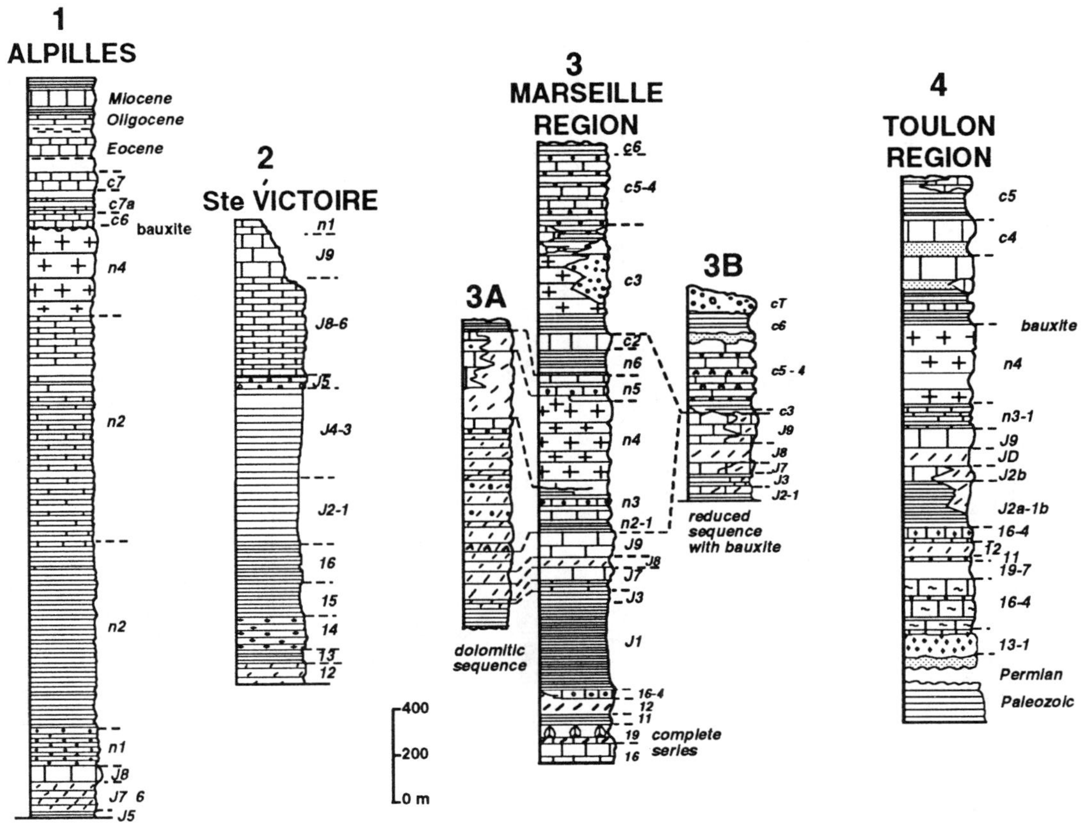

FIG. 6.3. Representative stratigraphic columns for the folded mountain blocks of southern Provence. See individual basins for matching lithologies. Key: t3–1 = Bundsandstein; t6–4 = Muschelkalk; t9–7 = Keuper; l1 = Rhetian; l2 = Hettangian; l3 = Sinemurian; l4 = Pliensbachian (Domerian); l5 = Toarcian; l6 = Aalenian; j1 = Bajocian; j2 = Bathonian; j3 = Callovian; j4 = Oxfordian; j8 = Kimmeridian; j9 = Portlandian; jD = undetermined Jurassic (dolomitic); n1 = Berriasian; n2 = Valanginian; n3 = Hautervian; n4 = Barremian; n6–5 = Aptian; c2 = Cenomanian; c3 = Turonian; c4 = Coniacian; c5 = Campanian; C7 = Maastrichtian. For complete stratigraphic discussion see notations for Geologic Map of France (Gouvernet et al. 1979).

however Rey and Grambast demonstrated that Priabonian (Ludian) sediments were indeed deposited soon after the period of tectonism in some areas (based on charophyte biostratigraphy; in Rey 1966).

GENERAL SEDIMENTATION AND TECTONIC CONTROL

The depositional sequences developed during the Mesozoic are primarily composed of carbonates, dolomitic carbonates, evaporites, and calcareous marls, but sections made up of conglomerates and sands are present in the Triassic of Toulon and Alpilles and the Upper Cretaceous in Marseille (figure 6.3). The period of relative sea-level drop during the Albian-Cenomanian, mentioned earlier, resulted in a break in sedimentation with profound weathering and attendant erosion, and also the formation of a residual bauxitic horizon (les Baux) in a number of sections (see summary in table 6.1 which deals only with the Paleogene formations).

TABLE 6.1.

Stratigraphic columns for the Paleogene sequences (including basal Miocene) present in the various subbasins of Southern (Basse) Provence. (Nury 1988; Nury 1994; Durand 1984 modified)

EPOCH	STAGE	AGE	BASINS AND FORMATIONS: MARSEILLE	PEYPIN LA DESTROUSSE St ZACHARIE	NANS-LES PINS	REGION TOULONNAISE	ST PIERRE-LES-MARTIGUES	LE ROUET	AIX-EN-PROVENCE AND NORTHERN AREAS	NORD VAROIS	*
MIOCENE	AQUITANIAN	early Aquitanian						Sausset	Rougiers		12
OLIGOCENE	CHATTIAN	late Chattian, latest	Cap Janet				St Julien	Carry Rousset / Cap du Nautes	Aix-en-Provence		11
		late Chattian, middle	Piédautry Collet-Redon					Poudingues du Rouet	Poudingues de Galice		10
		late Chattian, early	Poudingues de Marseille		Conglomerates	Méounes					9
		middle Chattian	Arnavaux-Ste Marthe								8
		early Chattian	St Henri-St André						Les Milles		7
	RUPELIAN (*sensu strictu* STAMPIAN)	late Rupelian, latest	L'Estaque								6
		late Rupelian, early	Pointe-rouge	La Destrousse		Bandol					5
		middle Rupelian, late	Les Camoins	Peypin (moyen)	Laminites and red argillites	Ollioules	St Pierre-les-Martigues		St Pons	Montmeyan	4
		middle Rupelian, early	St Marcel supérieur	Peypin (inférieur)							3
		early Rupelian, late	St Marcel inférieur	Laminites à plantes			Ponteau		Luynes		2
		early Rupelian, earliest	Calcaires à striatelles	Calcaires à striatelles and lignites	Calcaires à striatelles		Calcaires à striatelles			Les Bourdas	1
EOCENE	PRIABONIAN	late middle Ludian							La Daubergue		I
	BARTONIAN	Marinesian Auversian							Plan d' Orgon (Alpilles)		H
	LUTETIAN								Cuques		G
	YPRESIAN	Ilerdian							Montaiguet		F
									Palette	Rians	E
PALEOCENE	THANETIAN	latest late Thanetian							Langesse	Calcaires à Bithynies	D
		late Thanetian							St Marc		C
		early Thanetian							Meyreuil		B
	DANIAN	Vitrollian ?							Vitrolles		A

Stratigraphy, Lithology, and General Paleontology of the Paleogene Sedimentation of Southern Provence

PALEOGENE STRATIGRAPHY

The stratigraphic nomenclature for the sedimentary accumulations throughout southern Provence during the Paleogene is shown in table 6.1 and figure 6.4. The formations themselves are described below, basin by basin, with specific lithologies, fauna and flora. The environments and sedimentary development of each area of southern Provence are illustrated in figure 6.4. Due to the various phases of tectonic activity detailed in the discussions of each basin, the depositional record contains a considerable number of breaks that are not synchronous throughout Provence.

Because the general purpose of this paper is the understanding of sedimentation in the Paleogene of this region, the following detailed discussion is

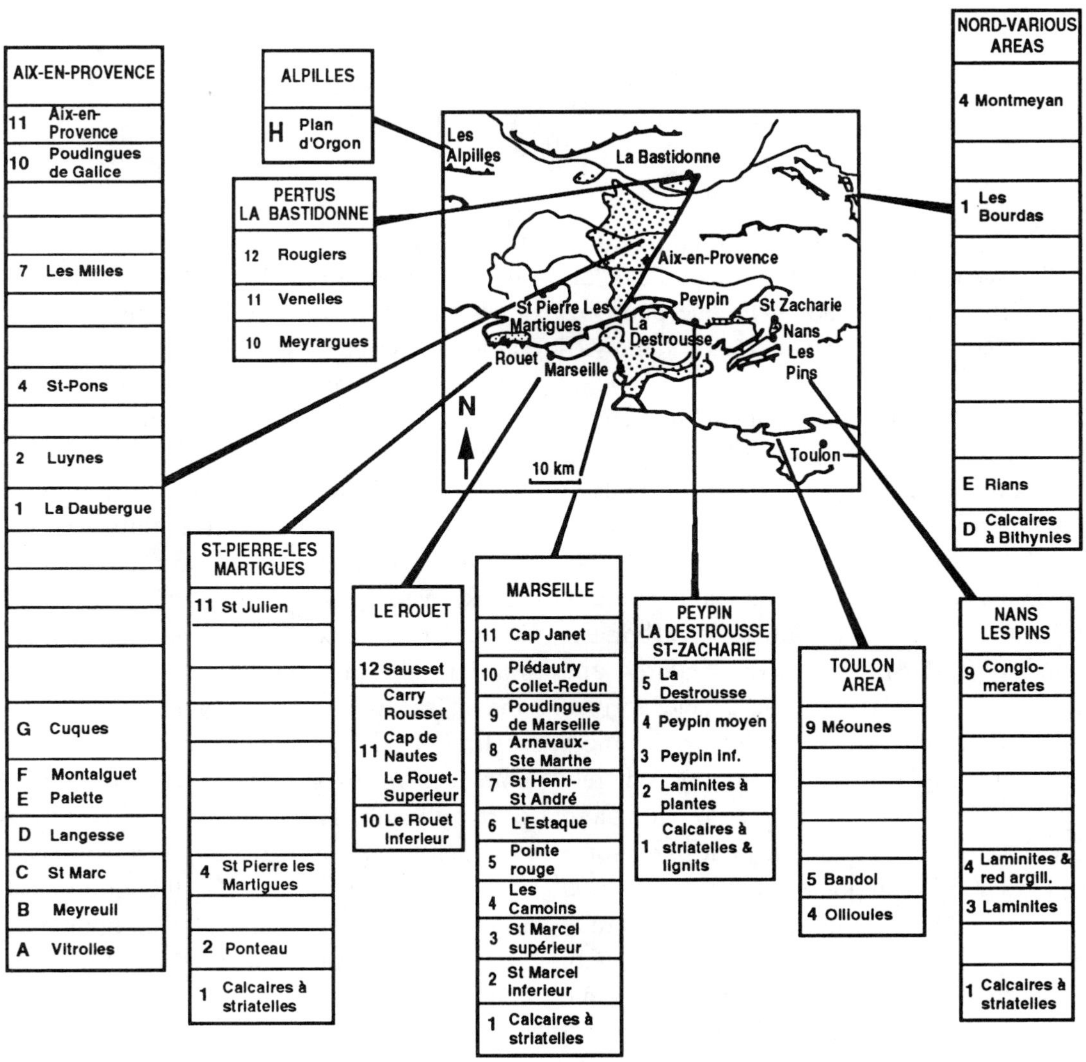

FIG. 6.4. Representative stratigraphic sections for the Paleogene basins of southern Provence. Stage designations and formation names are given. See table 6.1 for individual basins. Matching lithologies indicated by numbers 1 to 12 for the Oligocene series and letters A to I represent Paleocene and Eocene.

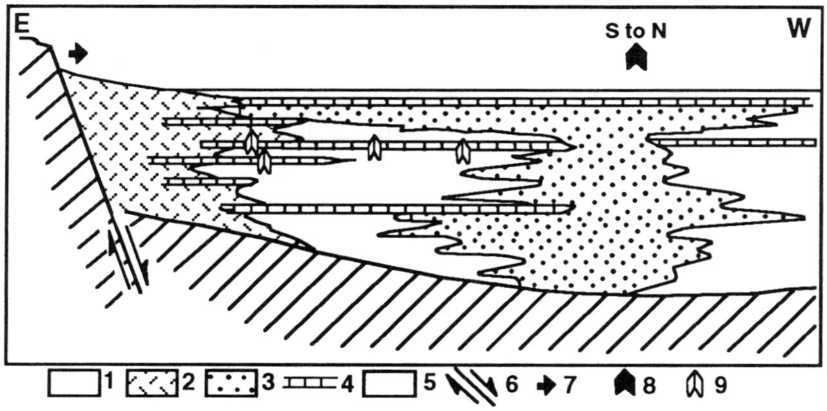

Fig. 6.5. Generalized lithologic model for Oligocene deposits and structure in the area of Southern (Basse) Provence. Key: 1 = substrate; 2 = breccias and megabreccias (olistoliths); 3 = conglomerates (puddingstones); 4 = bedded and/or laminated limestones, with or without gypsum; 5 = argillites, sands, and sandstones; 6 = normal faulting; 7 = transport direction of local detritus; 8 = transport direction of exotic detritus (external sources); 9 = gypsum.

given. Later in the presentations of the individual basins, reference will be made to these lithologic descriptions, rather than repeating the fine points of the depositional and diagenetic regimes individually.

Sedimentologic studies have shown that a general facies association is shared by all of the basins in southern Provence. The character and origins of the various lithologies can be described based on six basic lithologic groupings. These are (1) olistolithic and brecciated accumulations (fault related); (2) fluvial conglomerates and sands; (3) carbonates (five distinct types); (4) gypsum; (5) argillites; and (6) lignite (sporadic). Not all of these facies formed at the same time, nor are they similar distances from the basin margins. This presentation is merely a description of the general sedimentological and diagenetic features of each of the facies and some of their physical interrelationships. The apposition of the various sedimentologic facies can generally be represented by the model shown in figure 6.5. The various facies are not all formed at the same time, and it must be noted that within each basin the regional relief, climate fluctuations, water levels, sources, and chemistry also varied with time.

1—*Olistoliths and Breccias.* Accumulations of beds composed of olistolithic and brecciated components are primarily made up of reworked carbonates and dolomites (Mesozoic and also Oligocene). They are commonly observed at or near the normal faults at the basin margins and adjacent to horst blocks within the basins (figure 6.6A). These deposits usually accumulate on the downdropped sides of the faults, and the breccias are sometimes noted within the faults themselves (as a fault gouge). In general the olistoliths, megabreccias, and breccias are quite angular and vary in size. The largest blocks, usually of Mesozoic age, may be up to 40 m long. These breccia beds tend to thin rapidly toward the basin centers, and generally the blocks and clasts are without apparent stratification or sorting. This morphology suggests formation and emplacement by gravity movement, without transport by water. However some breccia beds show channelized reworking over short distances, and it is possible to determine the direction of flow from the orientation of sedimentary components as well as from tool and scour marks in the base of the channels.

The brecciated and olistolithic deposits are typically within a heterogeneous carbonate matrix (figures 6.6B and 6.7). While the blocks of Mesozoic carbonates represent the largest fragments, Oligocene blocks may be up to 20 m in length (oc-

FIG. 6.6. Outcrop photographs of Oligocene basin marginal megabreccia accumulations.

A. A roadcut with Mesozoic carbonate bounded by a fault at left (at arrow) and flanked by olistolithic chaotic megabreccia at the right. Blocks are Mesozoic carbonates in a heterogenous matrix. Outcrop is 13.5 m, top to bottom.

B. Natural exposure of breccia showing large chaotic blocks of Mesozoic carbonates. Photos of l'Estaque fault zone (Marseille area). Outcrop is 14 m, top to bottom.

Fig. 6.7. Outcrop exposure (natural) of gray dolomitic Mesozoic blocks in an Oligocene matrix composed of white limestone (le Marinier, Marseille area). The hammer visible in the photo is 35 cm long.

casionally much longer). The thickest accumulations of this facies grouping formed by renewed tectonic activity after each phase of relative quiescence. It seems that the prolonged exposure of the limestone surfaces during periods of tectonic inactivity prepared the blocks for movement by the development of open joints and comminution in the form of pseudobreccia development due to groundwater and soil processes. With a new phase of tectonism, the altered limestone was well prepared for ready gravitational reworking.

2—*Fluviatile Conglomerates.* The detrital deposits formed by rivers are composed of repetitive sequences of conglomerates (puddingstones), gravels, sands, shales and, rarely, intercalated carbonates. A representative fluvial sequence begins at the base with an erosion-cut surface into the underlying sediments, followed by a fining upward sequence (figure 6.8A). The upper portions of a sequence may include argillaceous or carbonate laminites. The onset of the next overlying cycle commonly truncates the previous bed, commonly developing as well formed channels (figure 6.8B). Synsedimentary tectonics seems to have controlled the development of many deposits. Faulting affected some beds during deposition and the offsets were infilled and then covered by the next sedimentary sequences, the whole sequence being again faulted, infilled, and then covered (figure 6.9).

The mineralogy of the fluvial sequences throughout Provence (in all three major basins) is quite similar, including siliceous elements such as clasts of vein quartz, quartzites, quartz-phyllites, radiolarites, and carbonates including marbles. The basic components vary only slightly from place to place and change little through time. The Variscan source areas that were then to the south (now separated some distance by tectonics) are known to have provided ample supplies of all of these components. On the basis of a series of studies of general grain mineralogy, heavy mineral groupings, grain size, and Krumbein rounding analyses, Nury (1988) shows both the sources and the transport distances are appropriate for sediment transport of 250 to 350 km from Pyreenean-Corsican-Sardinian sources.

3—*Carbonates.* The various carbonates of the Oligocene in southern Provence are here organized into somewhat arbitrary groupings, so that they are easier to present in this compendium of facies. It should be noted that carbonate microfacies 3a and

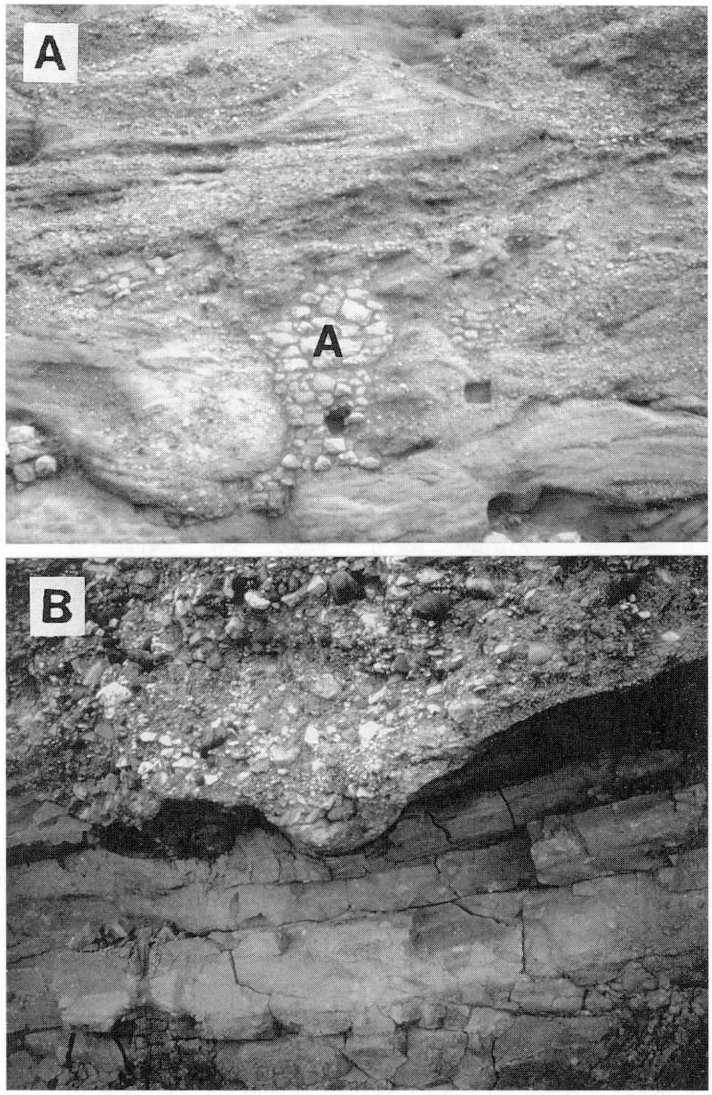

FIG. 6.8. Oligocene fluvial deposits: A. imbricated channels filled with coarse conglomerates (Marseille) (A = artifact, bricks) (Fort Saint Jean); B. channel cut into lacustrine clay and filled with coarse conglomerates (Le Rouet, Nerthe chain).

FIG. 6.9. Oligocene fluvial deposits with synsedimentary faulting and repeated infilling of fault steps (Le Rouet, Nerthe chain).

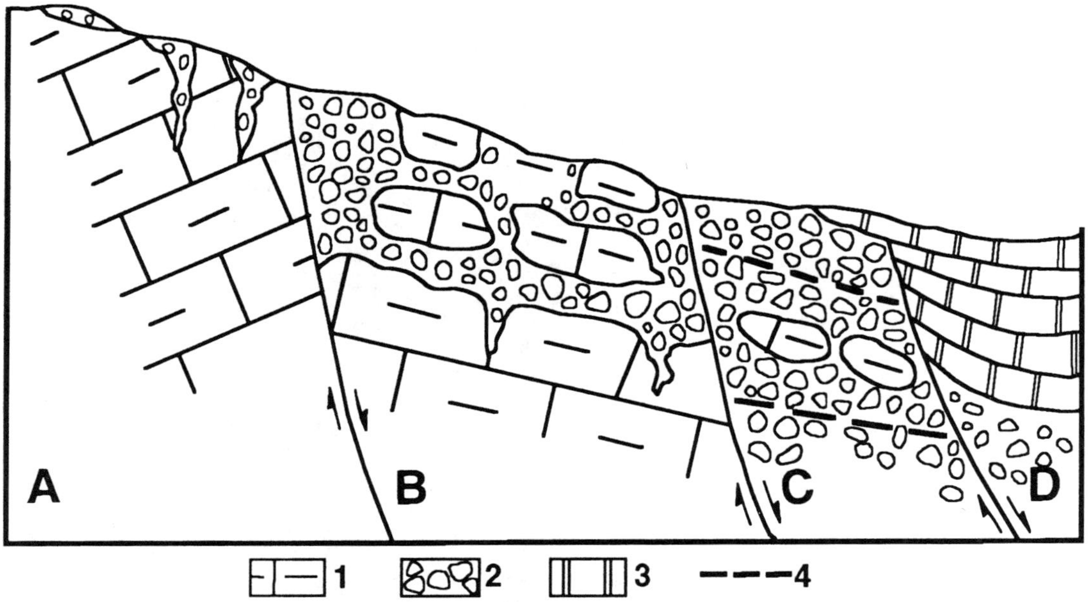

FIG. 6.10. Schematic diagram illustrating the development of pseudo-breccias from original dolomite beds (Jurassic). Note progressive alteration from left to right (A, B, C, and D). Key: 1 = original Jurassic dolomite; 2 = pseudo-breccia; 3 = Oligocene limestones; 4 = trace of original bedding.

b (pseudobreccias and calcretes) are related diagenetic constructs upon existing carbonate deposits. Microfacies 3c, d, and e are primary sedimentary deposits and are not created by the reworking or alteration of previous materials.

(a) *Pseudobreccias.* The development of pseudobreccias by the alteration of earlier dolomitic carbonate deposits provides both an unusual microfacies ("calcaires désorganisés") and a mechanism for the formation of large volumes of carbonate breccia (or pseudobreccia). These broken, disorganized zones of carbonate rock originally began as coherent Jurassic dolomites (figure 6.10A) into which groundwater gradually infiltrated along joints and faults. This resulted in both dissolution along the fractures and progressive dedolomitization. As the process continued, the section became more and more fragmented (figure 6.10B). Finally almost all of the original section was fragmented but the original bedding generally remained preserved (figure 6.10C). In detail the development of the fragmented carbonate may be seen to proceed by formation and widening of fissures, dedolomitization, and fissure filling by fine-grained limestone cements which are first gray or tan and then increasingly red (figure 6.11, stages 1 and 2; figure 6.12, A–C). Partial solution of the dedolomites and of some of the cement took place, and the whole rock became cut by a new generation of fissures that also became filled with a red-stained carbonate cement (figure 6.11, stage 4).

(b) *Calcrete and soils.* These carbonates, formed basically as soil profiles, appear as indistinct surficial veneers in some instances while in others (as in the Aix basin) the calcrete development is very clear. In Aix the calcretes begin just upon the upper surface of a lagoonal argillite as layers containing nodular micritic carbonate. These nodules developed into massive nodular beds or formed as beds with columnar groupings of nodules (figure 6.13). As the calcretization process progressed, zones of attapulgite clays developed and a true soil sequence was formed. Localized dolomitization of the carbonate occurred in some areas.

(c) *Palustrine and travertine limestones.* Palustrine (lake-marginal) and travertine limestones appear as spongy, porous beds commonly containing remains of terrestrial gastropods. In the Paleogene

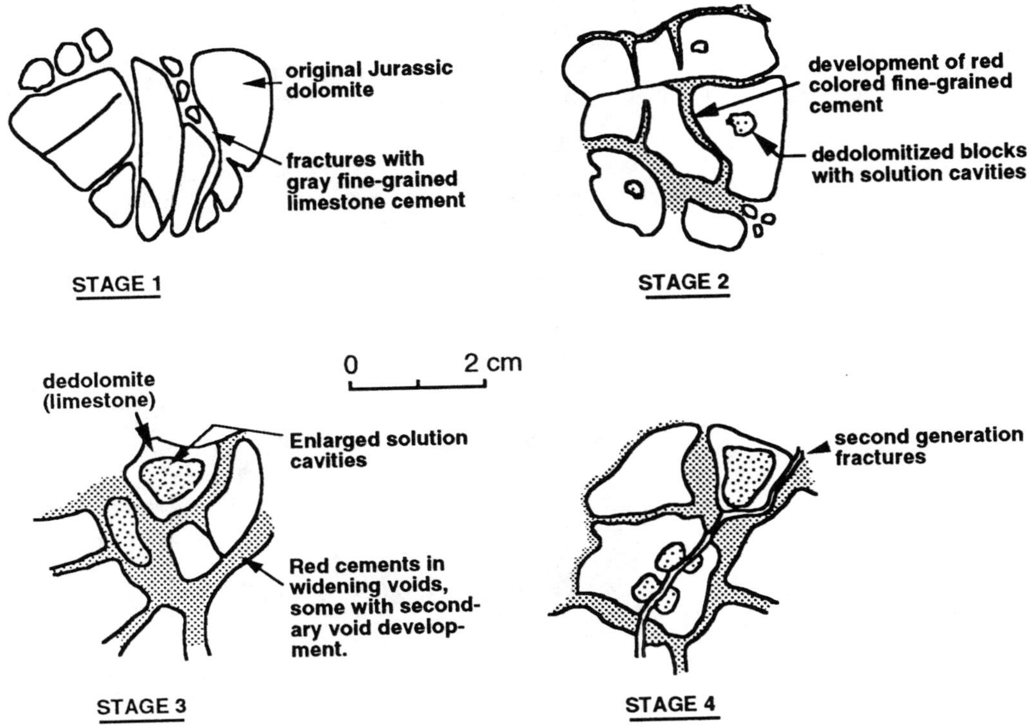

Fig. 6.11. Successive stages of alteration and resulting internal disorganization in the development of pseudo-breccias: Stage 1: original dolomitic carbonate crossed by fractures now filled with gray limestone cement; Stage 2: dedolomitization of fragments and early void development, beginning of formation of red limestone cement; Stage 3: widening of fractures, infilling with a red limestone cement and enlargement of voids; Stage 4: re-fracture of breccia and beginning of second stage of fissure filling.

deposits of southern Provence, these accumulations display uneven bedding and other poorly developed internal structures (figure 6.14). Such accumulations appear to have developed near fault zones.

(d) *Stromatolites and algal limestones.* The stromatolitic and algal accumulations form deposits that may be divided into two morphological groupings (Casanova and Nury 1989). The first is in the form of biohermal accumulations composed of metric-sized masses of branching cylindrical carbonate growths with a fairly high organic content. The second group consists of nodular, concentric masses, decimeters in size, commonly formed as accumulations in small depressions within other carbonate sediments and having lower organic content (figure 6.15A and B). In thin section the first type of stromatolite is seen to be made up of small pustules or cylinders in which filamentous micritic carbonate is mixed together in laminae containing quartz or other clastic particles. The nodular form is composed of very thin layers, which are so fine that it is almost impossible to differentiate the individual laminae even with a hand lens, and which contain no discrete internal structures.

(e) *Laminar and thin-bedded carbonates.* Laminar and thin-bedded carbonates are found in moderately thick, repetitive accumulations and are largely in the form of limestone, although some dolomite may be present. The beds may be composed of very evenly layered, millimetric laminae, or the layers may show gentle undulations having decimetric wavelengths (figure 6.16A and B). Usually those two types of carbonate do not appear in the same portions of the stratigraphic section. In some deposits interbeds of carbonate breccias are present (figure 6.16B).

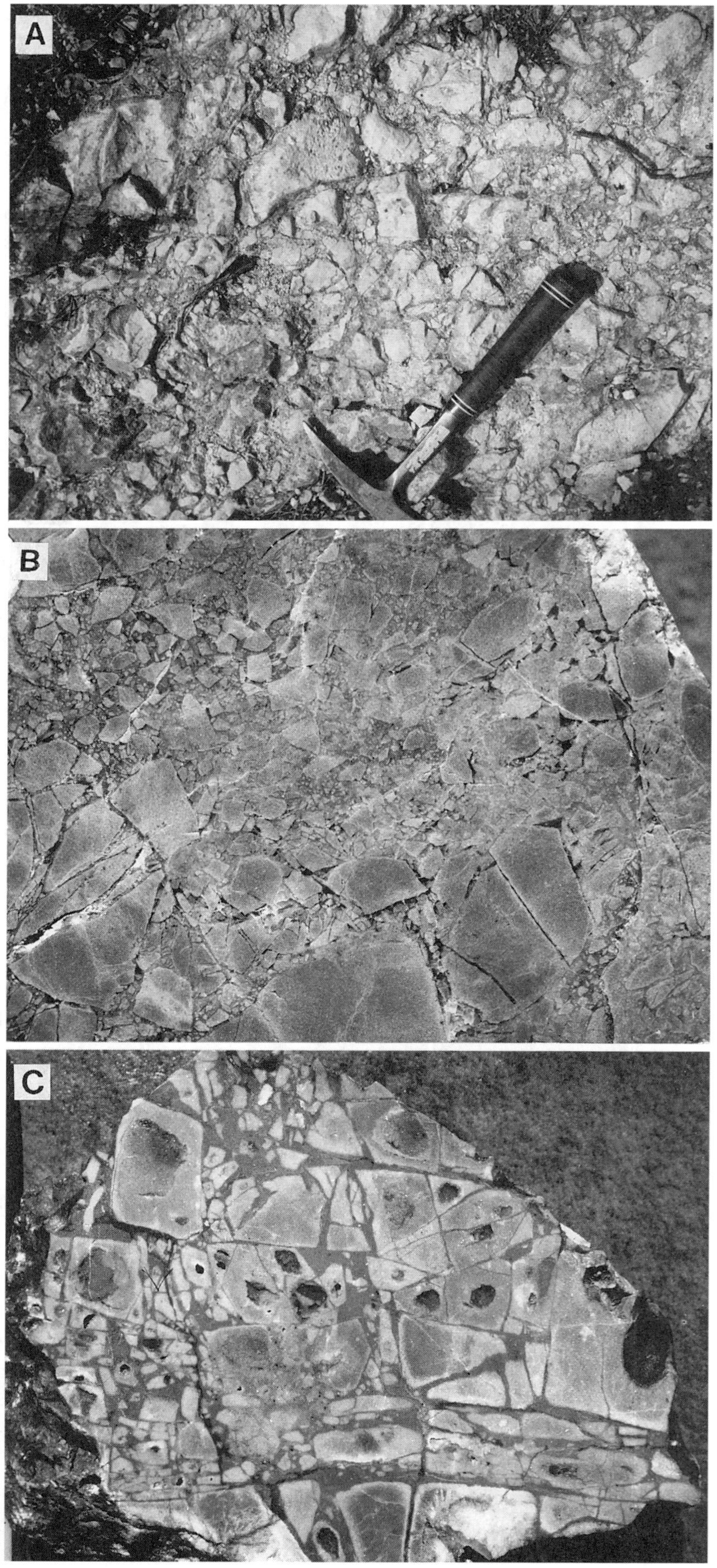

Fig. 6.12. Photographs illustrating the development of pseudo-breccia.

A. Outcrop photograph of pseudobreccia illustrating the apparent disassociation of dolomite fragments (Marseille Basin).

B. Hand sample of an early stage of brecciation in which the carbonate is still largely dolomite.

C. Hand sample of late stage pseudo-breccia in which the fragments are calcite (dedolomite) and the fractures are filled with red-colored limestone (dark in photo). N.B. secondary voids in both the dedolomite fragments and in the red cements.

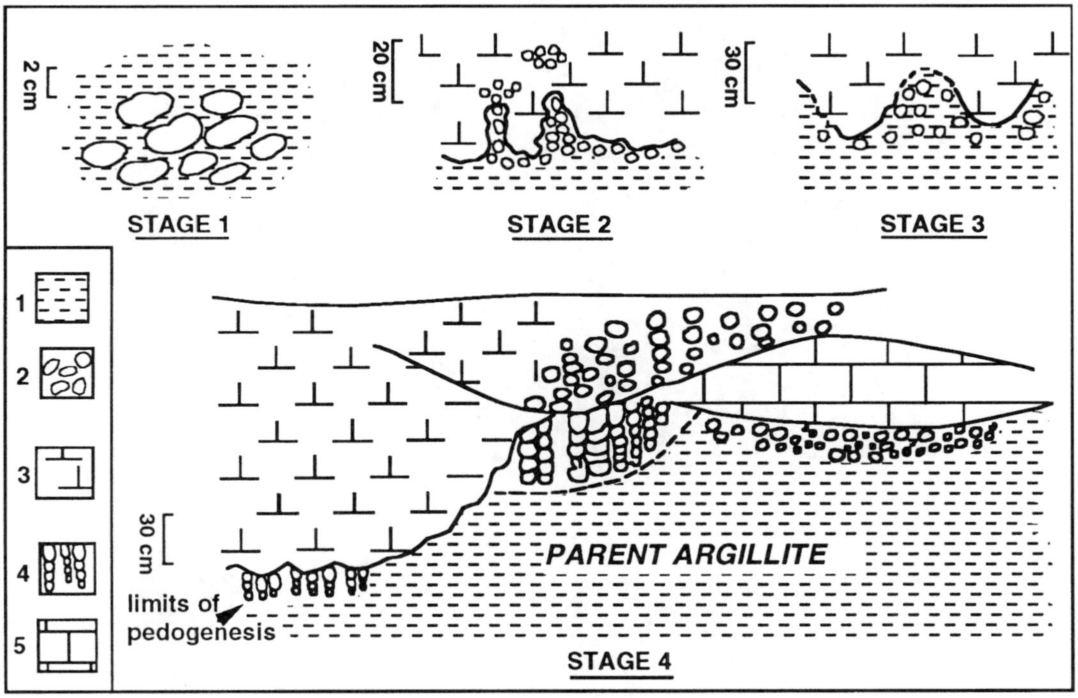

FIG. 6.13. Development of calcrete crusts, by stages, observed in the Aix basin (area of Venelles). Key: 1 = continental clays; 2 = calcitized nodules; 3 = massive calcitization; 4 = columnar nodule buildups; 5 = pervasive micritic carbonate.

4—*Gypsum.* The sulfate deposits in southern Provence are usually composed of fine-grained lenticular gypsum crystals, commonly mixed with micritic carbonate. They are found in laminar accumulations and also as more massive beds (figure 6.17A and B). Intercalations of thin argillaceous layers are common, as are interbeds of micritic carbonate. Various sedimentary structures are sporadically observed showing weak ripples, synsedimentary slumps (figure 6.18), and beautifully developed

FIG. 6.14. View of natural exposure showing irregular morphology due to vertical columnar buildups of calcite nodules in association with root structures forming a massive palludial limestone, observed in the Aix basin (Venelles).

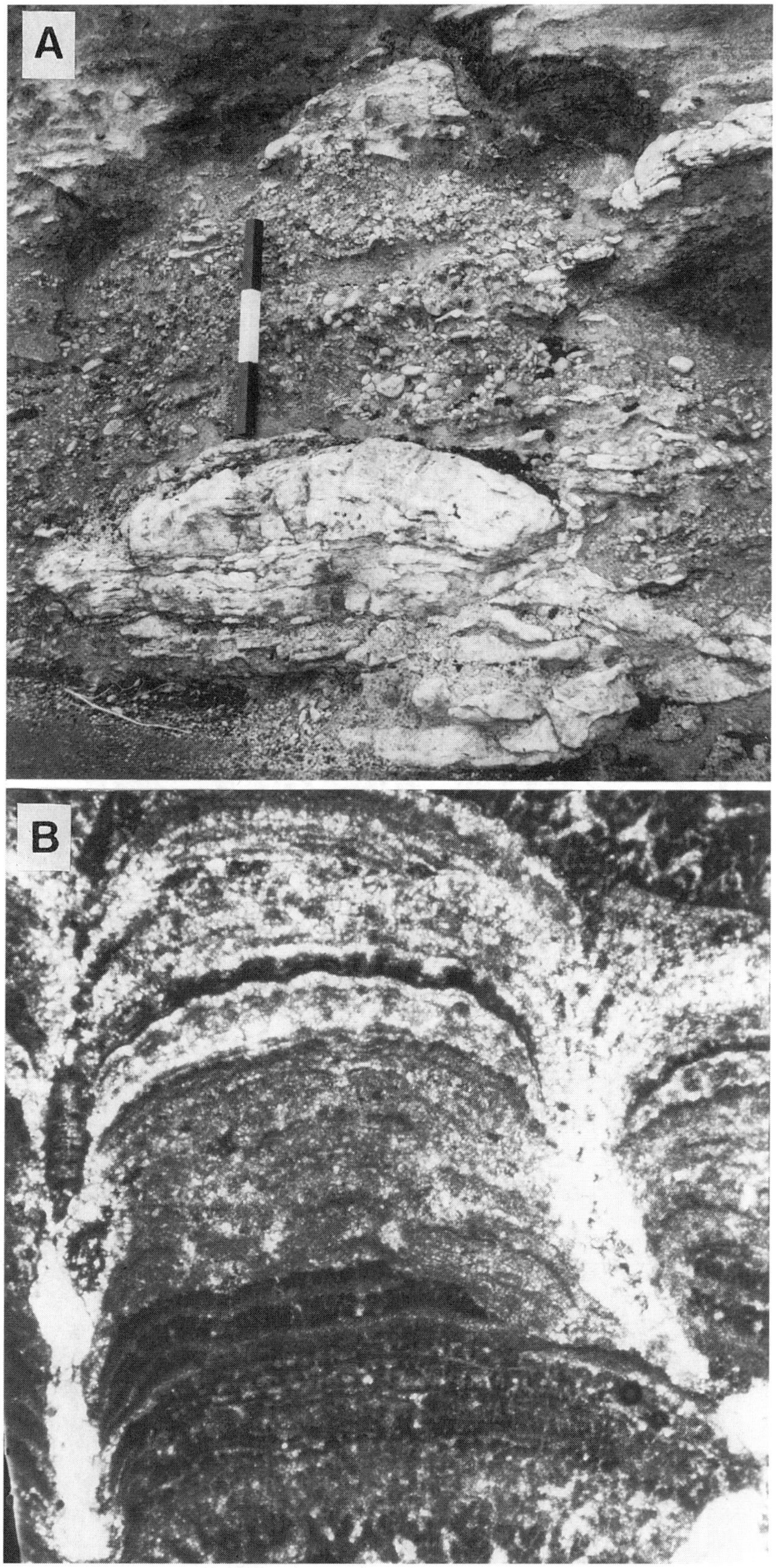

FIG. 6.15

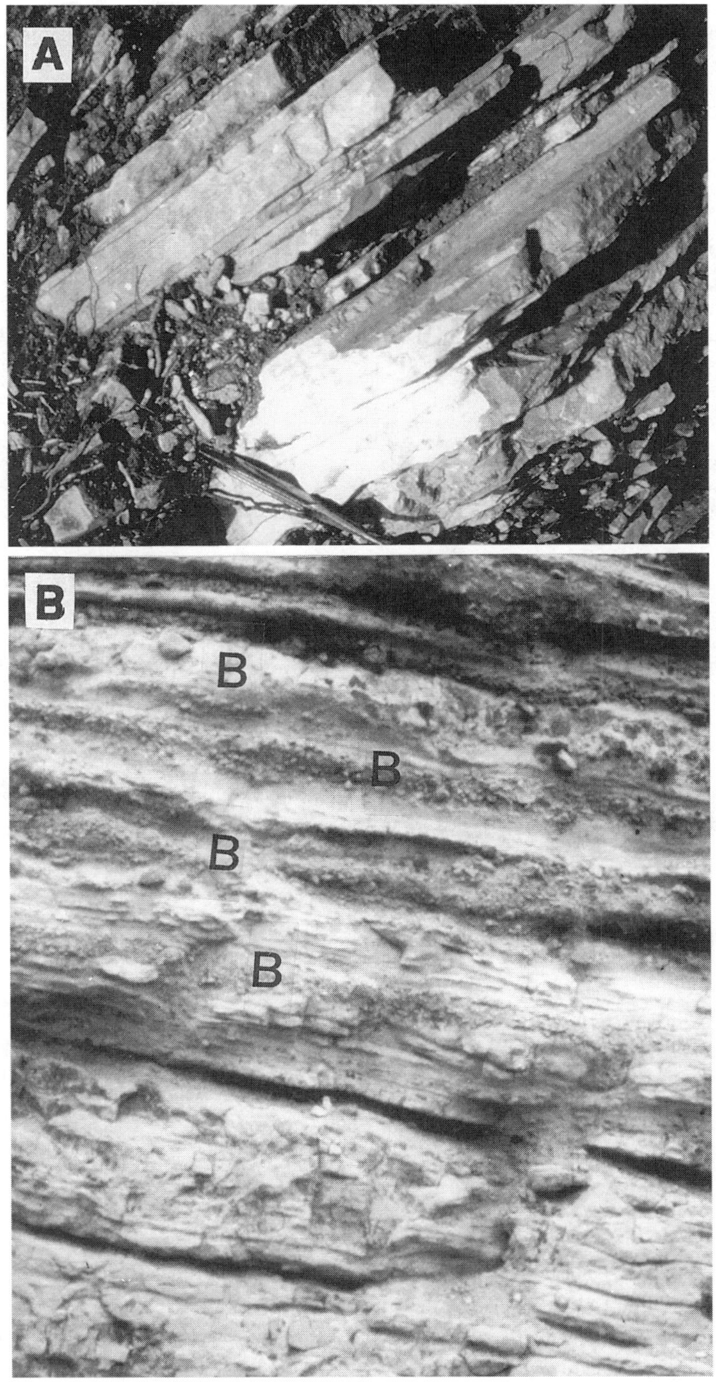

Fig. 6.16. Thin to laminar carbonate bedding, lacustrine deposits (fresh to mesohaline water), Marseille Basin. A. Fresh-water calcareous laminites (les Maurins). B. Thin-bedded, mesohaline sequence interstratified with reworked breccias (B) (l'Estaque).

Fig. 6.15. Stromatolitic buildups in the Marseille Basin, seen as macro- and micro-views. A. Stromatolite dome seen on outcrop. B. photomicrograph of columnar stromatolites (seen in plane light). Photograph 3 mm top to bottom. (photo courtesy J. Casanova).

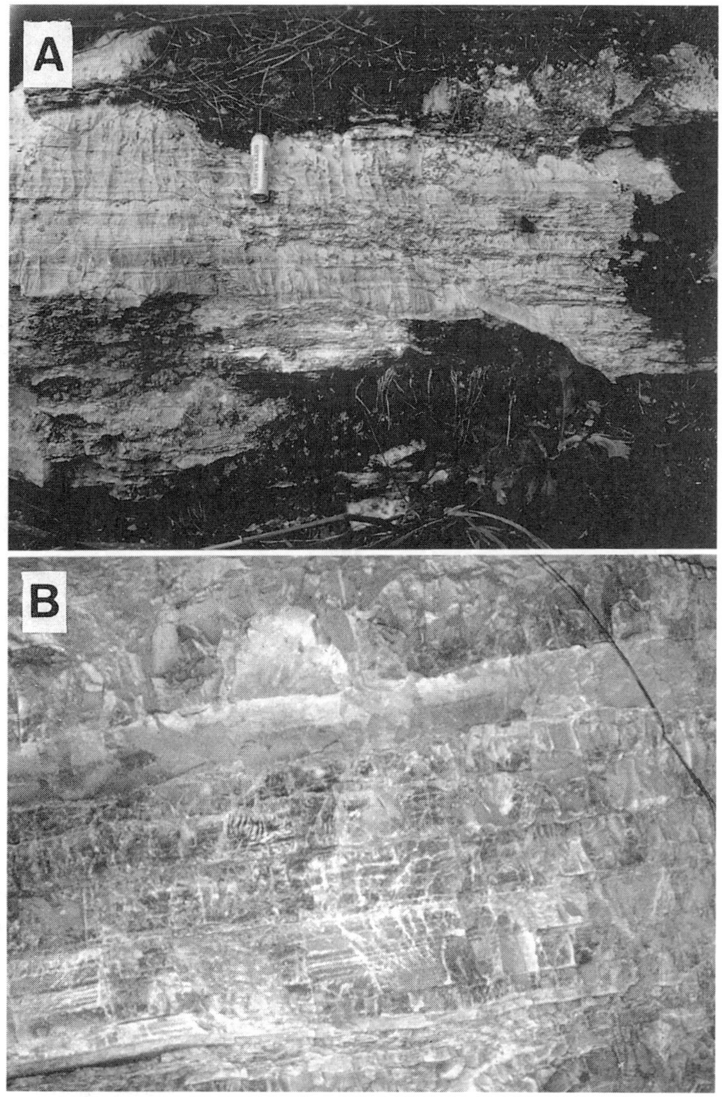

Fig. 6.17. Thin-bedded, fine-grained gypsum deposits. A. Thin to laminar bedded gypsum, seen on outcrop, Marseille Basin (Les Camoins). Marker length is 10 cm. B. Thin to laminar bedded gypsum seen in a mine cut (St. Pierre-les-Martigues). Outcrop is 36 cm, top to bottom.

desiccation polygons on some bedding planes (figure 6.19). Some layers display evidence of bioturbation in the form of numerous small burrows (figure 6.20A), although burrowed beds are not common. In thin section, the burrowed beds are seen to contain a large amount of micritic carbonate in which gypsum is present as discrete lenticular crystals (0.5 mm or less) aligned along the burrows and as diffuse gypsum cement. Commonly these crystals are displaced together with the burrowed matrix by a second generation of larger lenticular gypsum crystals that formed at the center of the burrows (6.20B). In this later generation the crystals are not oriented with respect to the burrow direction.

Fig. 6.19. Desiccation polygons developed in thin-bedded gypsum seen on a bedding plane on the roof of a mine gallery. Roof bolts are 15 cm squares (at St. Pierre-les-Martigues).

Fig. 6.18. Soft-sediment synsedimentary deformation in thin bedded gypsum as seen in a mine cut (St. Pierre-les-Martigues).

Fig. 6.19

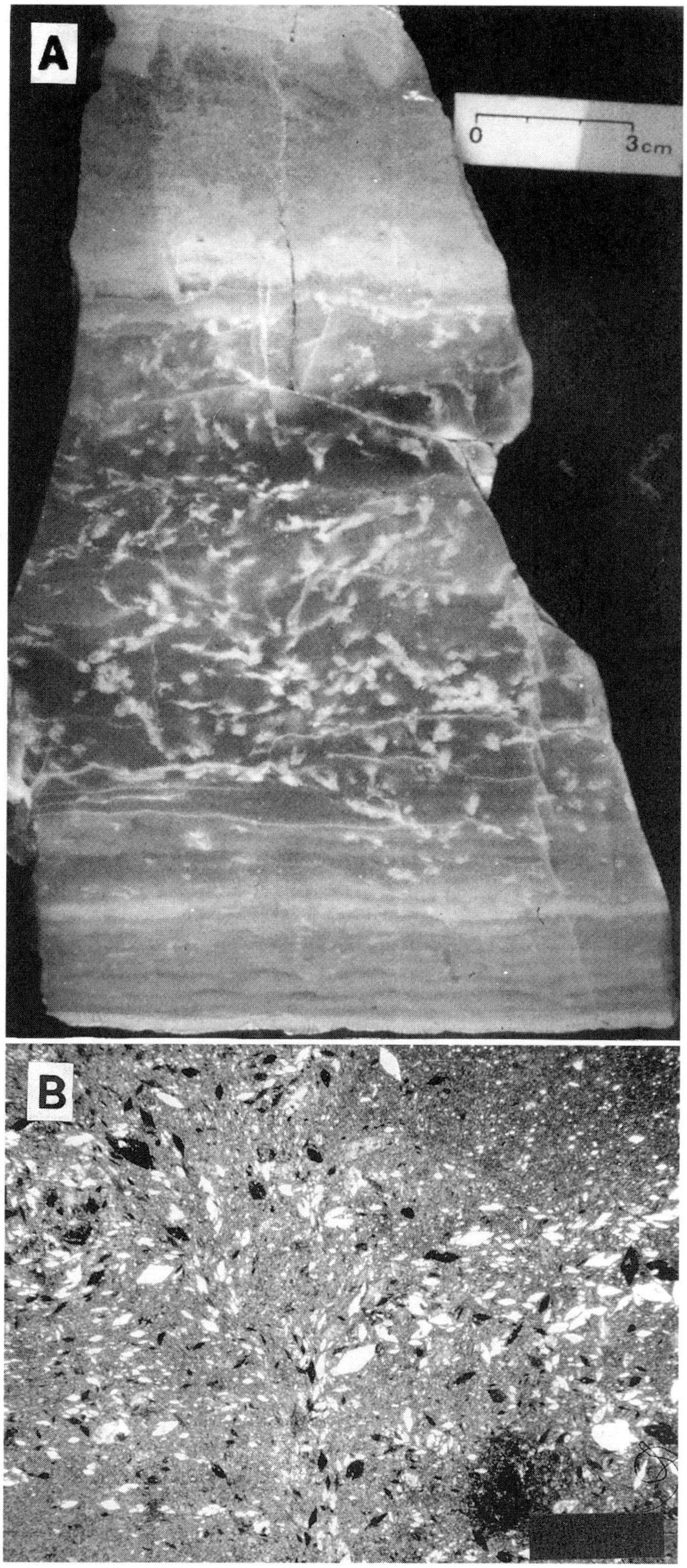

FIG. 6.20. Bioturbated, fine-grained gypsum. A. Hand sample of honey-colored gypsum with white micritic limestone above and below (white) (St. Pierre-les-Martigues; photo courtesy G. Truc). B. Photomicrograph of thin-section illustrating bioturbated fine-grained gypsum. Gypsum crystals are largely lenticular and bi-modal in size, floating in a mixed gypsum-calcite micrite. N.B. vertical burrow-displacement of gypsum crystals (St. Pierre-les-Martigues). Seen with crossed-polars. Scale bar is 0.5 mm.

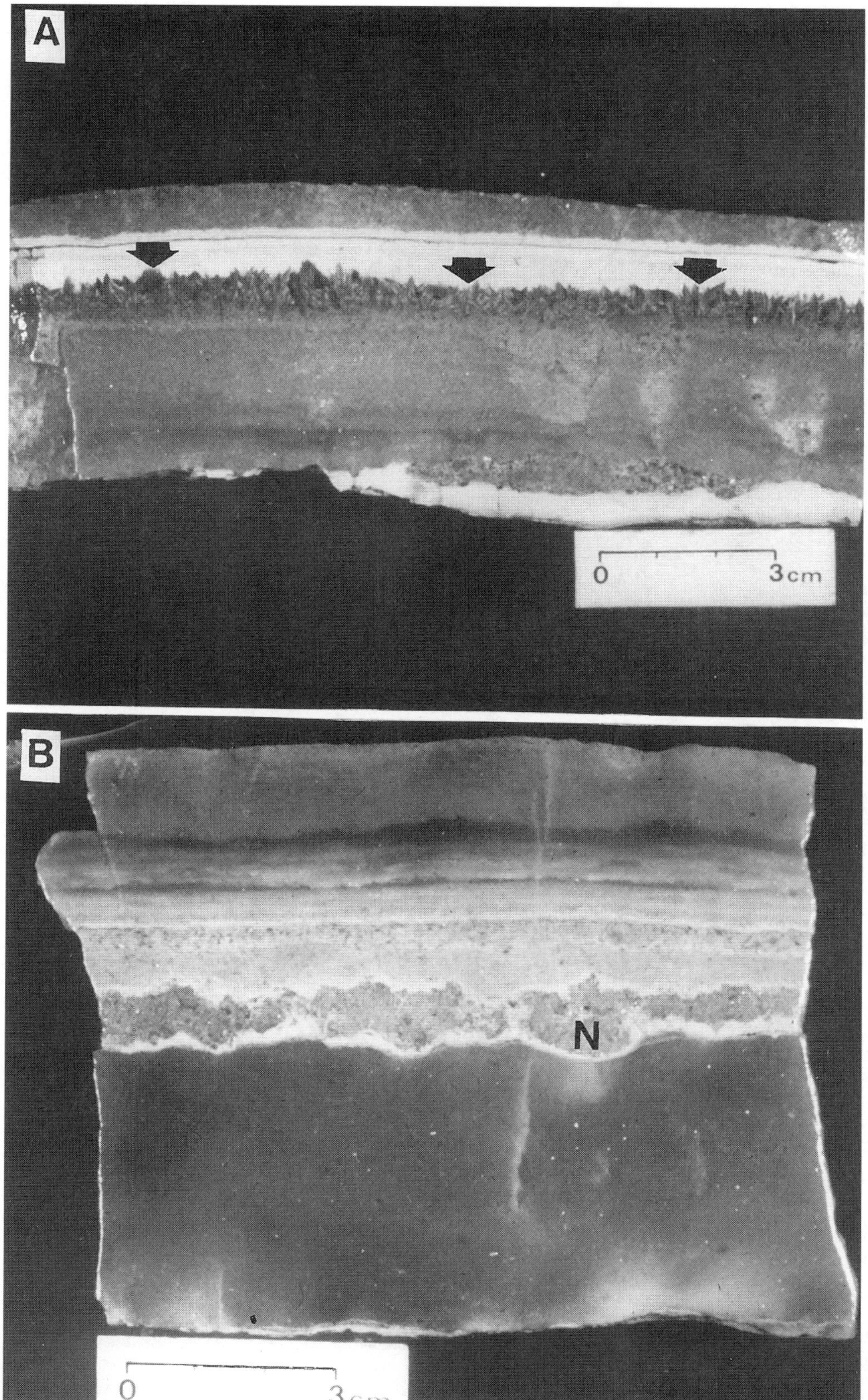

FIG. 6.21. Gypsum crusts developed together with laminar fine-grained gypsum and micrite (St. Pierre-les-Martigues). Photos courtesy G. Truc. A. Hand sample with a thin layer of coarsely crystalline gypsum (at arrows) overlain by white micritic limestone. B. Nodular gypsum crust (N) made up of tiny, interlocking, lenticular crystals, overlain by massive bed of fine-grained gypsum.

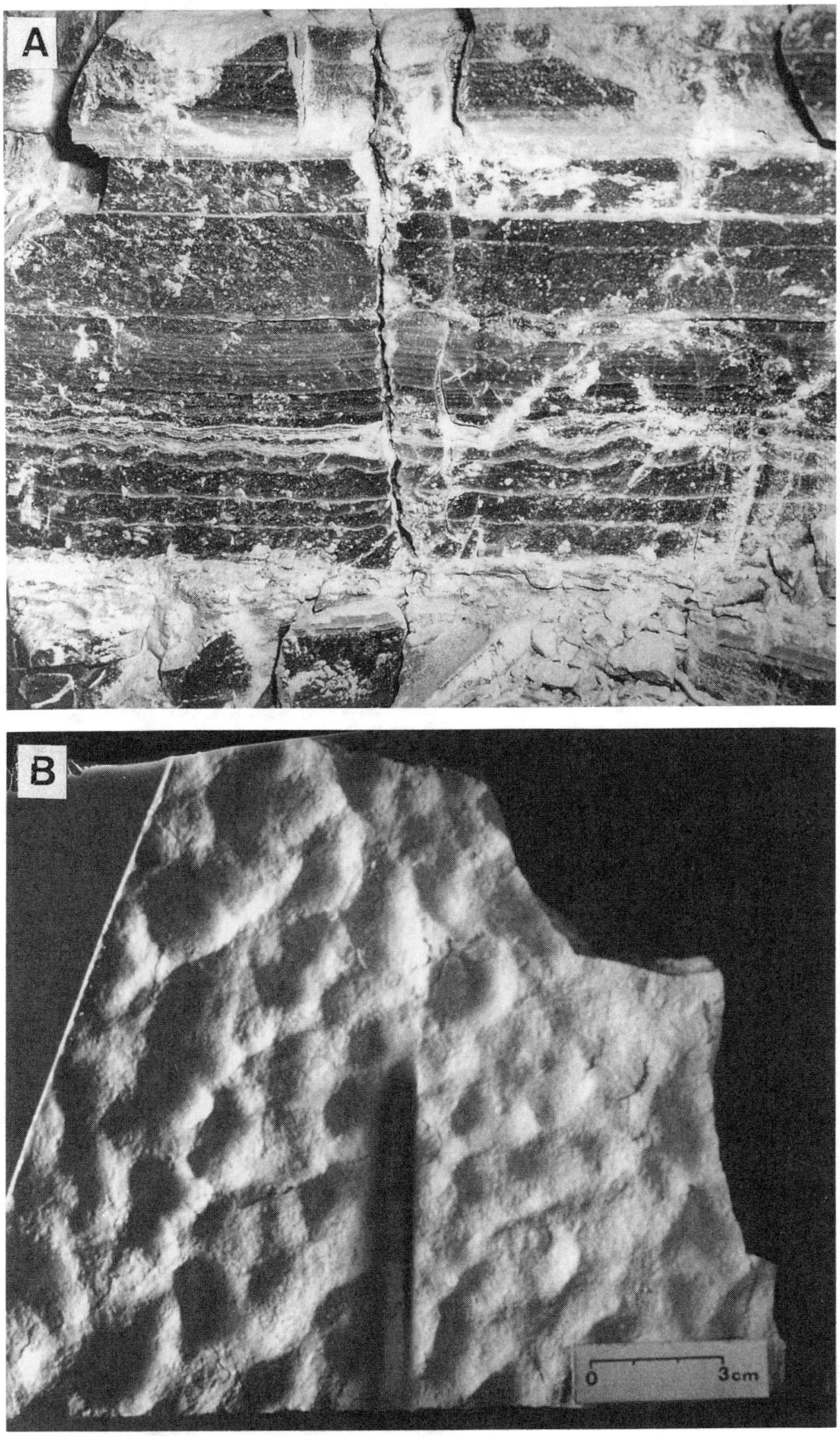

Fig. 6.22. Algally controlled beds of gypsum developed as a product of stabilization by algal mats: A. Irregular to domed gypsum beds seen on mine exposure (St. Pierre-les-Martigues). Rock wall is 0.5 cm, top to bottom. B. Sample of bedding plane showing development of the algal domes. Photo courtesy G. Truc (St. Pierre-les-Martigues).

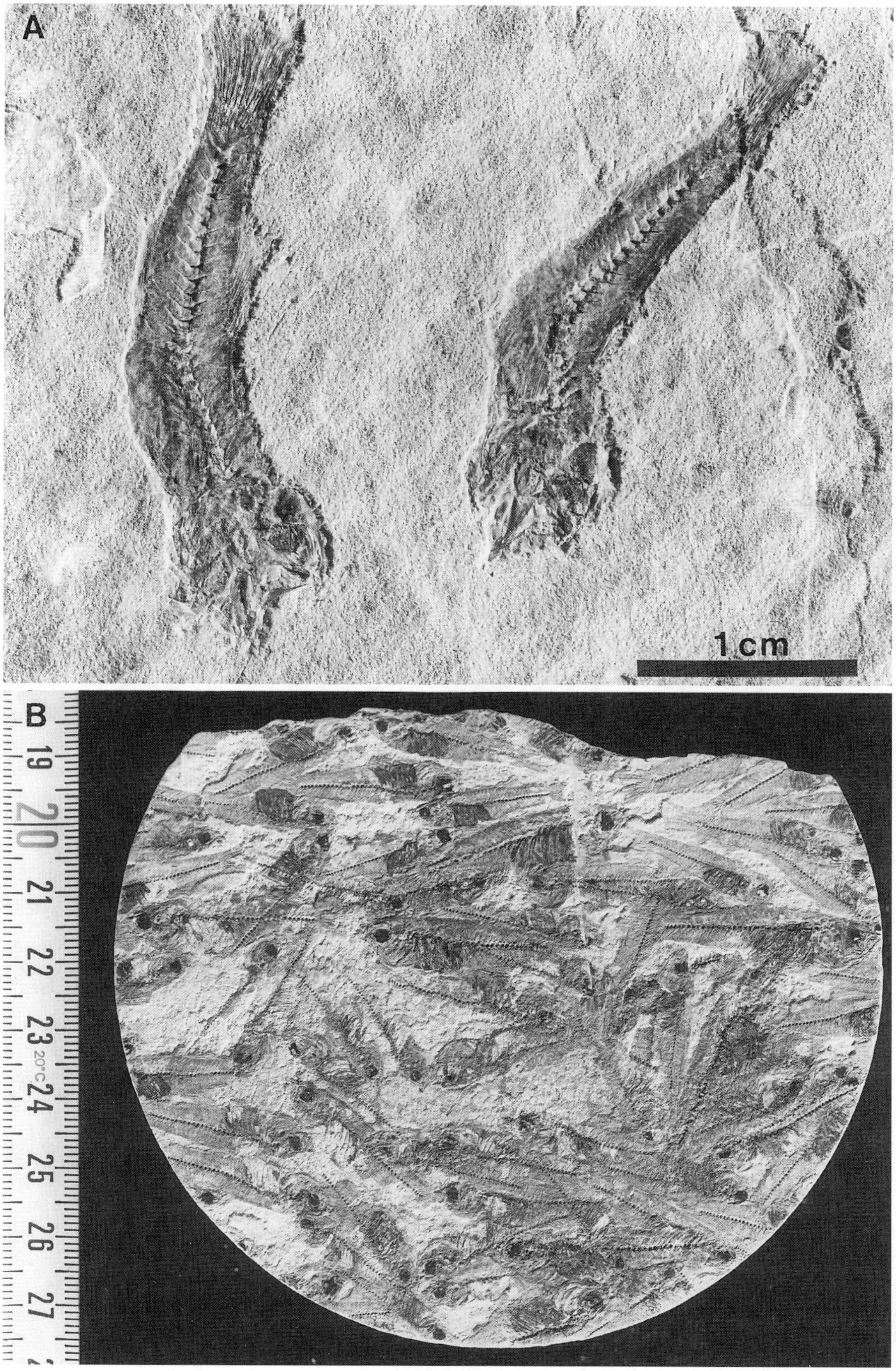

Fig. 6.23

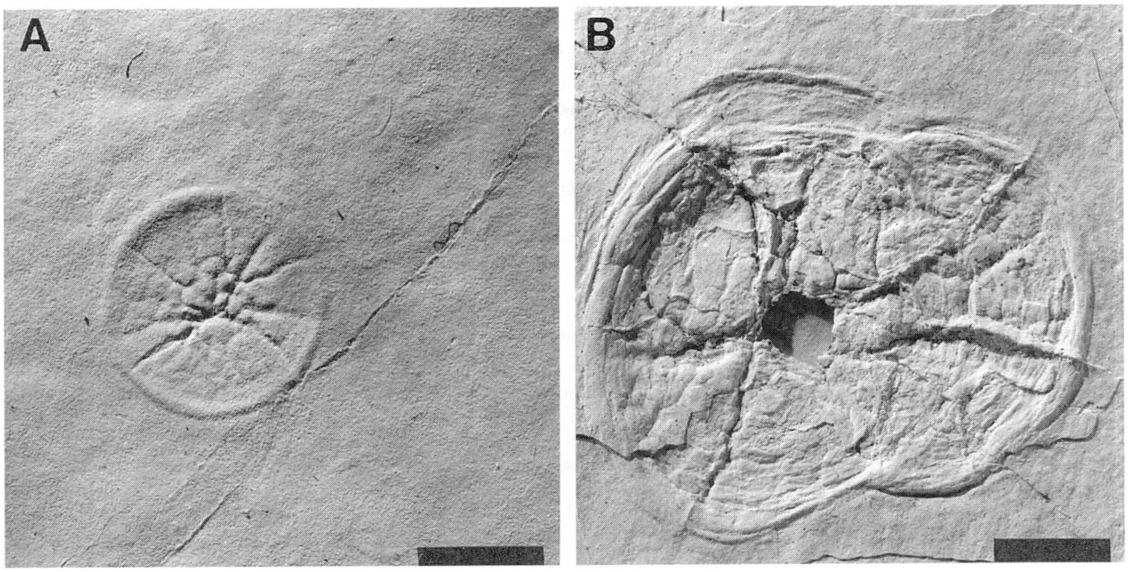

Fig. 6.24. Photographs of characteristic medusa, *Discaloides sp.* (Nel et al. 1987): A. Sample from the Aix Basin. B. Sample from Camoins.

Large gypsum crystals are not common in the sulfates of southern Provence except as late diagenetic fracture filling. A few layers do display thin, grasslike layers of gypsum crystals. These layers are similar to those which are well known in the modern salinas and in many Messinian deposits of Italy, Spain, Cyprus, and Egypt (figure 6.21A). Crusts of nodular gypsum are present at a number of horizons in the gypsum section (figure 6.21B). These nodular crusts are composed of tiny individual lenticular crystals, and are well-known forms that develop in some salina ponds in which thick algal mats are present. These nodules are relatively coherent and may be readily reworked: many beds of gypsum evidence similar nodules, or fragments, when studied in thin section.

In some beds, in particular in the intercalated carbonates, fossils may be noted in the gypsum. The most common biogenic component is that of stromatolitic domes and algally-controlled surfaces (figure 6.22A and B). Because identical structures and domed surfaces are well known in modern analogs there is little question as to the general conditions of their origin. Other fossils are known in laminar carbonates between gypsum beds (figures 6.23 to 6.26). It is unclear if these highly altered forms were washed in or originated within the basin at the time of gypsum formation. Charae oogonia are well preserved and relatively common in the calcareous interbeds. Thus they may well represent an appropriate floral component for fresh water periods within this depositional setting (figure 6.27).

The origin of the gypsum beds of Provence has been the subject of some discussion. It must be pointed out that lenticular gypsum is not the ideal textbook form of gypsum crystallization. It becomes common when trace amounts of certain organic compounds are present (particularly at the

Fig. 6.23. Photographs of characteristic fossils used to determine depositional environments of various sections in southern Provence: A. Fish, *Prolebias cephalotes* (St. Pierre-les-Martigues). Scale bar = 10 mm. B. Fish, *Gobias aries* (Aix-en-Provence). Scale bar = 4 mm.

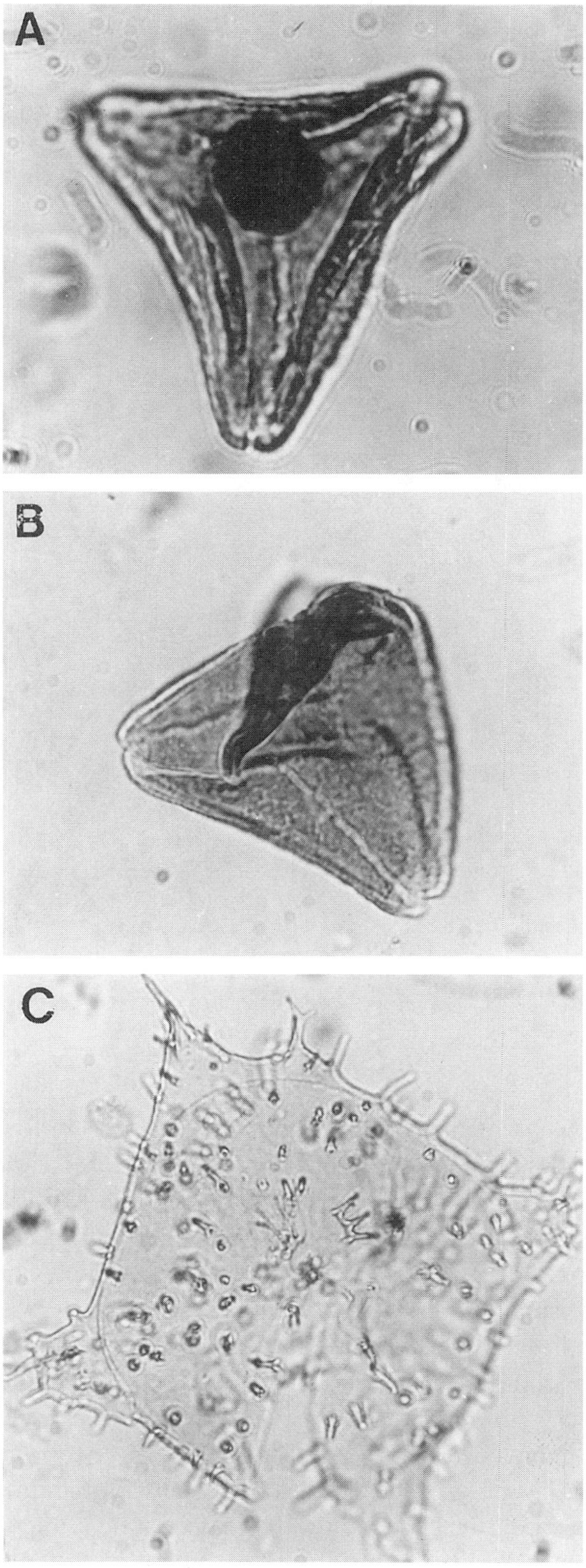

FIG. 6.25. Photographs of characteristic plant pollen and dinoflagellates: A and B. *Boehlensipollis hohli* (Krutzsch 1962). C. *Wetzeliella gochtii* (Costa-Downie 1976). A: field of view (left to right) = 52.5 microns; B: field of view (left to right) = 67.5 microns; C: field of view (left to right) = 90 microns.

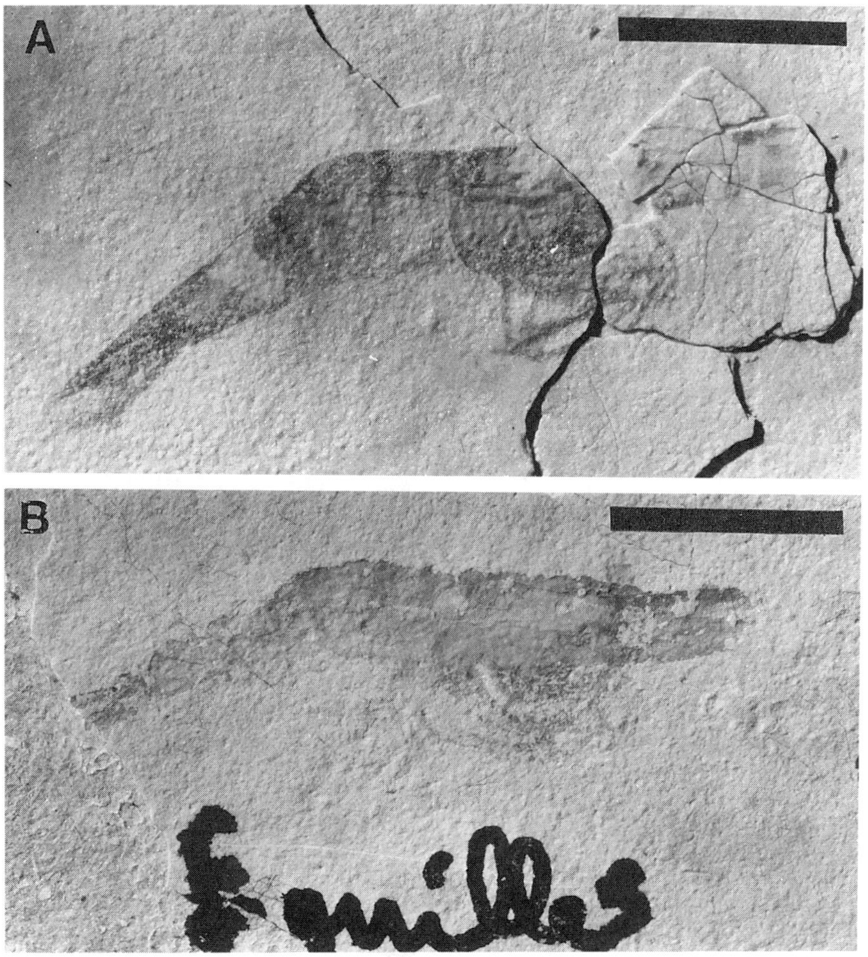

Fig. 6.26. Photographs of characteristic crustacea belonging to the genus *Penneides*. Both A and B are samples from the Aix basin. Scale bars for A and B = 5 mm.

low end of the concentration range of gypsum formation from a solution; Cody and Cody 1988). The debate as to origin arises from the fact that gypsum apparently shows several possible modes of formation. Some layers are clearly direct precipitates of gypsum with crystals forming at the water surface or in the water column. In other cases the laminae are composed of a gypsiferous siltite in which the tiny gypsum crystals apparently grew displacively and replacively very early in the burial history of the layer (syndiagenetic or precocious formation), based on the observation that compaction realignment (parallel to bedding) took place after these crystals formed (figure 6.28A and B). Later-formed gypsum crystals are commonly larger and are not affected by compaction (figure 6.29). Diffuse gypsum cements, either as microcrystalline or more rarely as poikilotopic crystals, are common in many calcareous layers. It seems reasonable that all stages of gypsum formation are present in the beds observed in these basins. Primary precipitates and both early and late diagenetic forms may be present in any series of beds and all of these may develop from the concentrated waters of the formative bodies. Although much of the gypsum is even-bedded and undisturbed, repeated emersion and exposure is demonstrated by the repeated desiccation surfaces known in many areas and at numerous levels

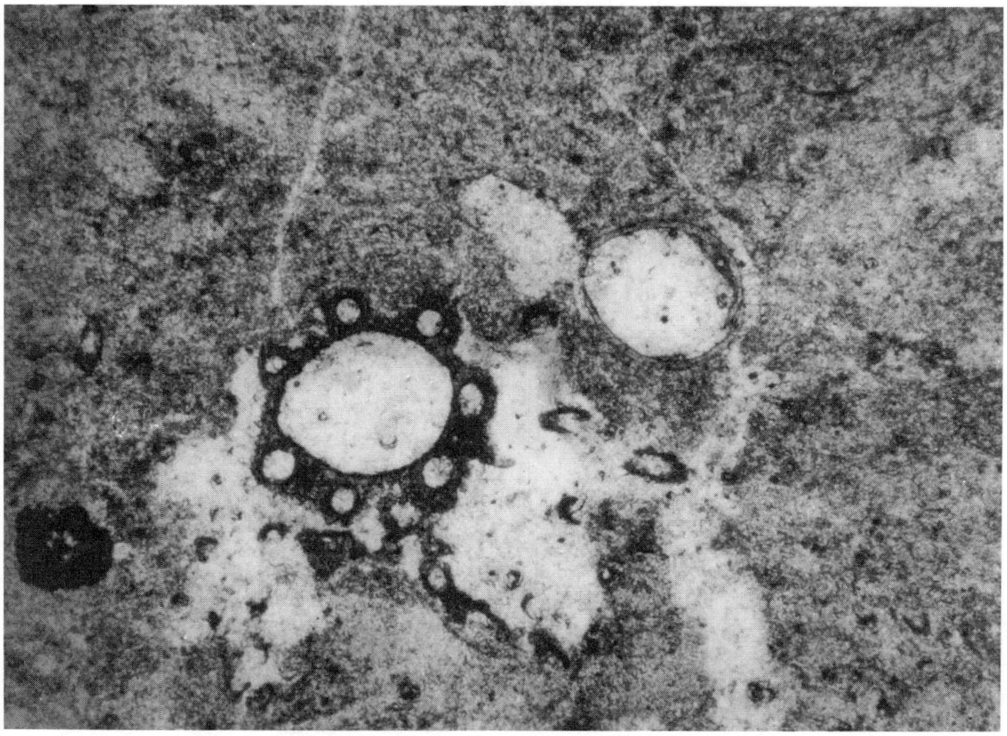

FIG. 6.27. Photomicrograph of thin section showing cross-section of chara and part of a gastropod in a micrite-gypsum matrix. Gypsum is a diffuse cement to the micrite and fossils (St. Pierre-les-Martigues). Seen in plane light; field of view is 3.2 mm, left to right.

in some sections (Collomb 1986). It is also interesting that in these basins the elevated salinities of the water never produced halite, either as preserved deposits or as pseudomorphs or imprints, common in other basins.

5—*Argillites.* Argillite accumulations developed in both fluvial and lacustrine settings. The primary source of the clay influx was from the southern detrital sources, and it is variously composed of smectite, illite, chlorite, and kaolinite. Argillite deposits are commonly interbedded with conglomerates, sands, carbonates, gypsum, and with sparse accumulations of lignite. The beds are rarely well exposed because of their lack of resistance to weathering but are common throughout the Paleogene section (figure 6.30). Attapulgite, a variety of smectite that is typical of soils formed in arid regions (Chamley et al. 1977; Hassouba and Shaw 1980), is found in some of the Provençal calcretes. In those cases where the soil formation is marked, the attapulgite is commonly associated with dolomite.

6—*Lignite.* Lignite is not a common deposit in the Paleogene but it is present in a number of areas and has been mined on a local basis. These coals are apparently swamp deposits with rare tree stumps in situ. Layers of compacted, crushed limnitic gastropods (actual shells, not dissolved or pseudomorphed) are present as are very fine detrital layers (inframillimetric layers of argillites and silt). The water in these swamps must have been relatively neutral or even alkaline, or the gastropod shells would have been destroyed during burial. This suggests that the swamp waters were unsuitable to the more usual levels of bacterial activity, and very little evidence of H_2S productivity is present in the lignite (low pyrite content), and in the carbonate (no pyrite).

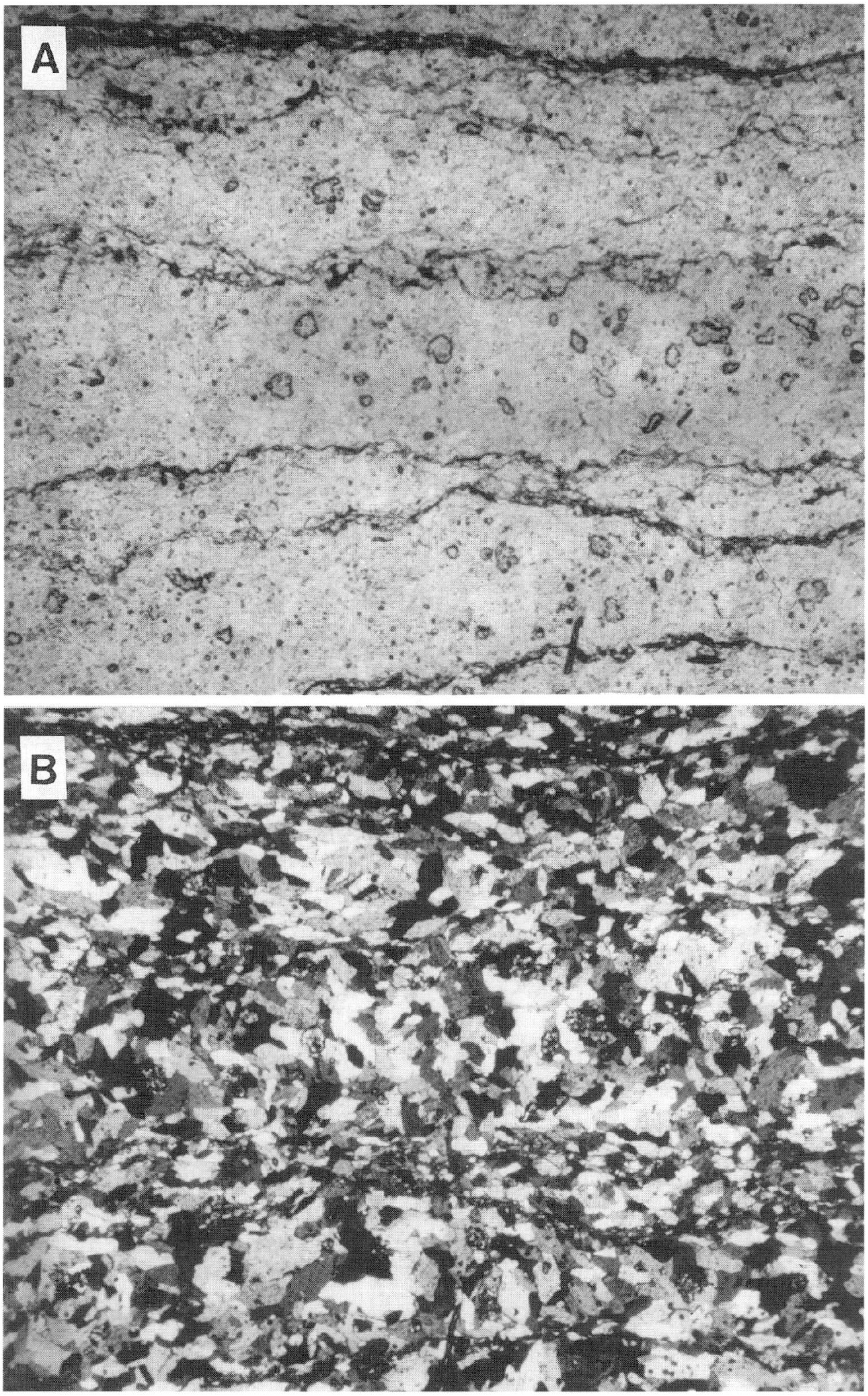

FIG. 6.28. Typical laminar gypsum from St. Pierre-les-Martigues, seen in thin section photomicrographs. A. Seen in plane light, dark horizontal traces are micritic carbonate. B. Identical sample seen with crossed-polars showing interlocking nature of lenticular gypsum crystals. Fields of view are 3.2 mm, left to right.

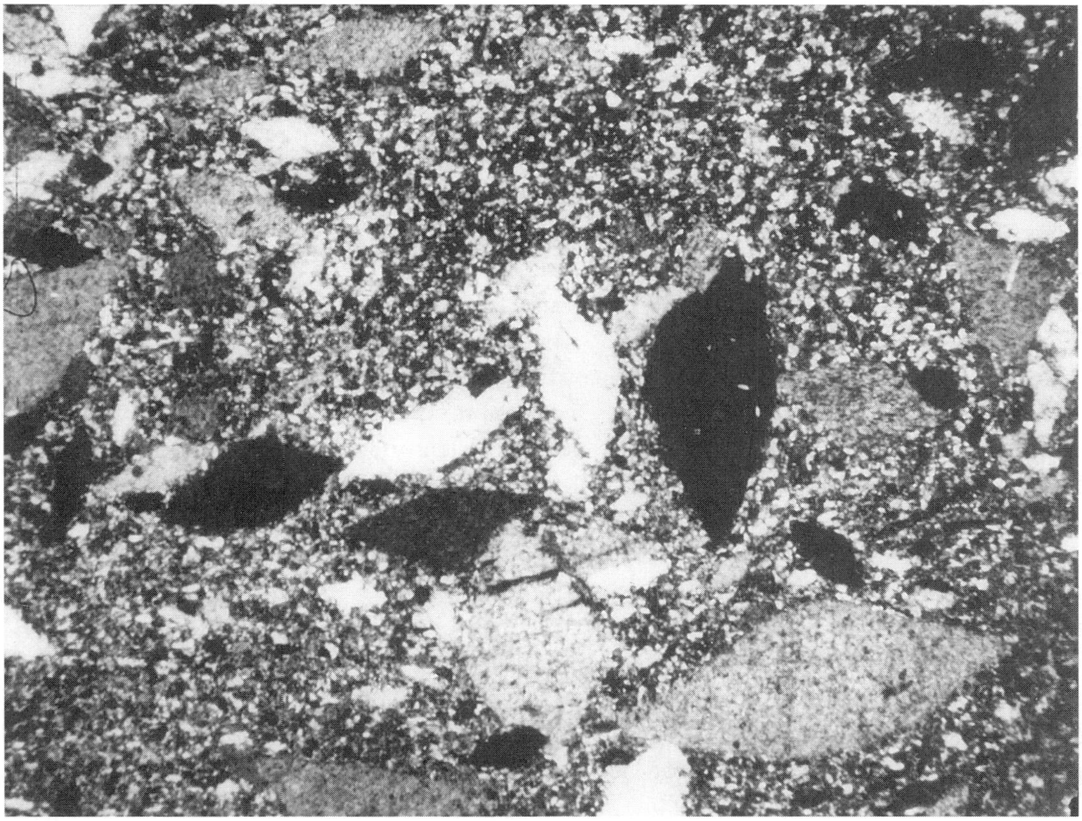

Fig. 6.29. Laminar gypsum from St. Pierre-les-Martigues seen in thin section photomicrograph. In this sample there are two different populations of gypsum. The host matrix is very fine-grained with a second generation of larger, well-formed lenticular crystals. Crossed-polars; field of view 3.2 mm, left to right.

Fig. 6.30. Panoramic view of an exposure in the center of the Marseille Basin showing thin-bedded alternations of argillites with lignite layers. Note conglomerate at top (at arrow). Open-pit mine exposure (St. André).

PALEOBIOLOGICAL ASSOCIATIONS

Paleogene stratigraphy is elaborated with various vertebrates— reptiles and birds for the Eocene and mammals for the Oligocene—as well as charophytes, gastropods, and palynomorphs (table 6.2). Oligocene sections were discussed in the literature by Nury (1987/1988); since then it has become possible to correlate the Provençal series to the Rupelian (Stampian in the Parisian series) and the terminal Chattian (Dutch series). The paleontological basis (floral and faunal) of the stratigraphic subdivisions come principally from Aguilar et al. (1978), Hugueney and Truc (1976), Hugueney et al. (1987), Hugueney and Ringeade (1990), Mein et al. (1971), Vianey-Liaud (1979) for vertebrate and mammal zonation; Feist-Castel (1977) and Riveline (1986) for charophyte zonation; Nury (1987/1988) for gastropod zonation; Châteauneuf (1977), Châteauneuf and Nury (1995) for pollen and dinocyst; Martini (1973) for nannoplankton; and Magne et al. (1987) for absolute age. For the Eocene, however, the correlation is somewhat uncertain. A further review of this sequence is underway, based on the works of Durand (1984); Riveline (1986); and Godinot et al. (1987), and the updated paleobiological information has been included in table 6.2.

ENVIRONMENTS OF DEPOSITION

The Paleogene depositional environments of southern Provence extend across a broad range of settings and include the continental (lacustrine, fluvial, and soil settings), euryhaline water (estuaries and lagoons), and marine shorelines (table 6.3). These assessments are based upon styles of sedimentation together with associated paleontological assemblages. Isotopic analyses of the oxygen and sulfur of the sulfate of some of the gypsum beds have been made by Fontes et al. (1980), and the gypsum has been interpreted as being of largely nonmarine origin, reworked from Triassic sources. The associated paleontologic assemblages, however, suggest an associated marine influx.

1—Continental settings.

The specific continental settings identified on the basis of depositional facies are:

lakes and ponds of variable water depths (±1m) with rare deeper waters. This environmental identification is based on sedimentary criteria (bedding, grain size, styles of diagenesis) and paleontological criteria (assemblages of charae, pelecypods, ostracods, gastropods);

paludial and spring deposits (moss travertines, etc.) based on sediment morphology, petrography, and paleontology (gastropods);

soil horizons and calcretes (sedimentologic and petrographic data);

slow-moving rivers and irregular fluvial torrents (based on bedding, sedimentary structures, and grain size); and

marine shorelines.

Lakes and ponds. The continental lakes and ponds produce sediments that range from massive and bioturbated limestones (containing charae, gastropods, and pelecypods) to finely and regularly laminar limestone or mudstone beds that appear to be abiotic (unburrowed and totally undisturbed). These are generally composed of argillites (with fine silt intercalations) and/or calcareous muds (micritic or pelloidal). The water in the lakes probably was warm and of variable salinity, and contained different amounts of fine terrigenous material. Lake deposits are commonly cut by fluvial channels that are filled by clastic deposits, formed when lake levels became low or fluvial supplies were increased by climatic variations or regional tectonics. It has also been suggested that all the gypsum deposits formed in such lakes, embedded within the carbonate facies, as in the deposits of northern Provence. But significant variation in the paleo-indicators associated with gypsum makes applying the lacustrine interpretation without exception to all occurrences of gypsum problematic.

Paludial and spring deposits. Associated with some lake deposits are apparent paludial carbonates which are irregular in occurrence, have a typical porous appearance and contain thin interca-

TABLE 6.2.
Generalized stratigraphic column for the Paleogene of Southern (Basse) Provence giving the stratigraphic subdivisions employed, together with their paleontologic basis (faunal and floral).

EPOCH	STAGE	AGE	MAMMALIAN REFERENCE UNITS	*CHAROPHYTE ZONES*	*GASTROPOD FAUNAS*	POLLEN, SPORES and DYNOCYSTS	OTHER CHARACTERISTIC DATA
MIOCENE	AQUITANIAN	early Aquitanian	MN 1	*Rantzieniella nitida*	*Potamides girondicus* *Tympanotonos margaritaceus*		NN 1 22.2 MY
OLIGOCENE	CHATTIAN	late Chattian	MP 30	*Chara notata*	*Wenzia ramondi* *Tympanotonos margaritaceus* *Tympanotonos labyrinthum* *Potamides lamarckii*	*Armeria* *Cedrus* *Dicolpollis kockelli*	NP 25 (MARTINI) N 4 (BLOW) 23.7 MY
		middle Chattian					
		early Chattian	MP 26	*Chara microcera*	*Wenzia ramondi minor*	Cold coniferous assemblage (*Abies, Picea, Tsuga*) *Slovakipollis hypophaeoides*	
	RUPELIAN (*sensu strictu* STAMPIAN)	late Rupelian latest			*Potamides lamarckii* *Tympanotonos labyrinthum*	Graminaceae, advent of numerous Chenopodiaceae	
		late Rupelian early	MP 23			*Boehlensipollis hohli* association, *S. hypophaeoides* *Wetzeliella gochti* *Abies, Picea*	
		middle Rupelian late		*Rhabdochara major*	Last *Melanoides tourainei* Impoverished striatelles grouping *Brotia laurae*		
		middle Rupelian early	MP 22			*Boehlensipollis hohli*	
		early Rupelian late					
		early Rupelian earliest	MP 21	*Stephanochara pinguis*	Striatelles grouping (*Melanoides tourainei, M. nysti M. fasciatus, M. acutus, Nystia chastelli, Viviparus soricinensis Pseudamnicola angulifera*)		
EOCENE	PRIABONIAN	late middle Ludian	MP 19-20	*Harrisichara tuberculata, H. lineat*	*Galba longiscata* *Nystia hedonensis*		
	BARTONIAN	Marinesian			*Australorbis mamertensis* *Planorbius vasseuri*		
		Auversien			*Galba longicata, Ischurostoma minutum, Ferrussina subglobosa* *Filhiola genevausi*		
	LUTETIAN				*Australorbis pseudoammonius* *Galba michelini*		
	YPRESIAN	Ilerdian	MP 7 (Palette, Rians) *Ornitholithus arcuatus*	*Nitellopsis thaleri*	*Romanella hopei, Australorbis pseudorotondatus, Bithynia beauduensis, Galba aquensis*		
PALEOCENE	THANETIAN	latest late Thanetian	*Ornitholithus arcuatus*		*Physa columnaris*		
		late Thanetian	*Ornitholithus biroi*	*Sphaerochara edda*	*Physa prisca* *Physa columnaris*		
		early Thanetian		*Dughiella bacillaris*	*Physa prisca*		
	DANO-MONTIAN	Vitrollian ?					

TABLE 6.3.
Cross-reference listing of faunal and floral associations with specific environmental settings.

GROUPS \ MILIEUX	OPEN-MARINE	SHORE-LINE	ESTUARIAN 18‰	LAGOONAL 10‰	FRESH WATER: FLUVIAL	FRESH WATER: LACUSTRINE	CONTINENTAL
Foraminifera	*Globigerina*						
	Asterigerina						
	Cibrononion						
		Miliolodae (*Quinqueloculina*)					
		Cibicides					
		Reussela					
		Rosalina					
		Elphidiidea					
		Nonionidea					
		Ophtalmiidea					
		Bolivinidea					
				Textularia			
Coelenterates	*Discalioides*	*Tarbellastrea*					
		Porites					
Gastropoda		Granulolabium					
			Potamides				
			Tympanotonos				
			Melanoides				
			Melanopsis				
			Nystia				
			Brotia				
			Hydrobiidae				
					Neritidae		
					Viviparus		
					Pseudamnicola		
					Bithynidae		
						Limneidae	
						Planorbidae	
						Physa	
						Ischurostoma	
						Ferussina	
							Pomatiidae
							Helicidae
							Filholiidae
							Romanella
Lamellibranchiata		*Cardium*					
		Divaricella					
		Ostrea					
		Psammobia					
			Polymesoda (*Pseudocyrena*)				
						Sphaerium	
						Pisidium	
Ostracoda		*Cytherella*					
		Hemicyprideis					
		Haplocytheridea					
		Cyamocytheridea					
					Cypris		
					Candona		
Peneidea		indeterminate					
Fish		*Eolates*					
		Perca					
		Mugil					
		Anguila					
		Gobius					
				Prolebias			
				Dapalis			
					Notogoneus		
					Amia		
					Cyprinidae		
Tetrapoda							Rana
							Chelonian
					Crocodilian		
							Birds
							Mammals
Algae		Dasycladaceae					
						Charophytes	
						Botryo-coccaceae	
Dinoflagellates		*Distatodinium*					
		Cordosphaeridium					
		Deflandrea					
			Spiniferites				
			Millioudodinium				
				Wetzeliella			
				Chiropteridium			
Flora							

lated beds composed of charae and gastropods. These deposits are commonly made up of thin broken crusts that are usually filled with coarse, freshwater, sparry calcite cements. These beds are not thick or continuous but their origins may range from moss tufas (Weijermars 1986) to lake margin accumulations.

In depositional settings similar to the palludial deposits, Nury (1988) has described lignite accumulations that are apparent lake-marginal swamps. These accumulations are interbedded with calcareous and argillaceous lake deposits. The lignite apparently accumulated in very quiet water and the details of the preserved plants are exquisitely preserved.

Soil horizons and calcretes. The development of soils, calcretes, and pseudobreccia accumulations apparently occur within arid climates (Wright 1986, 1994). They have developed together with the characteristic *Microcodium* crusts that also have affected most of the exposed Mesozoic limestone surfaces (Bodergat 1974). These limestones represent an association of sedimentary features that point to a warm, arid setting similar to that of modern east-central Spain.

Rivers and fluvial torrents. The fluvial settings, with both slow-moving rivers and torrents, are well defined on the basis of bedform and grain size, and are interbedded, in some cases, with lacustrine deposits. Fluvial development is largely controlled by structure (fault development) and many deposits exhibit synsedimentary fault offsets. Similar river deposits formed by wadis along the Dead Sea and along the coast of the Persian Gulf also exhibit similar forms and variations (D. J. Shearman, personal communication). The slow-moving waters of some shallower rivers formed very clear, somewhat broader, deposits (Provençal sequence), also containing fresh-water pelecypod and gastropod faunas (Nury 1987/1988).

Marine shorelines. Euryhaline settings, particularly those along marine shorelines, range from brackish to hypersaline, but may pass into true marine deposits. While certain species of plants and animals can be carried into the various areas by storms and also by aquatic birds, other marine groups, such as dinoflagellates, coelenterates, peneids, and euryhaline fish cannot be distributed by these vectors. The presence of populations containing substantial numbers of these marine components (Nel et al. 1987), prove the existence of true communication between open, tropical marine waters and the restricted depositional areas. Many of these deposits may have occurred in quiet lagoons and lake-like water bodies but the repeated influx of marine fauna and flora indicate a sporadic, strong marine influence (as in Laguna Madre; Fiske 1959).

A true marine shoreline commonly results in development of a coralline environment in which a thin-bedded succession of biostromes, coral banks, and scattered solitary corals are associated with low energy detrital sediments. In the corraline marine sediment area near Carry a rich paleobiological association, including foraminifers, ostracods, molluscs, dinoflagellates, echinoids, and nanofossils are present together with polyps. Only a few are listed in table 6.3. Complete data may be found in Bureau de Recherche Géologique et Minières (1972). This fossil assemblage suggests that these deposits formed in an open marine setting, on a low-energy rift edge similar to some modern sheltered, fringe-reef environments (Nury and Thomassin 1994).

It is from these quiet euryhaline/marine settings that we find substantial numbers of well-preserved marine assemblages immediately below and then above gypsum deposits. As mentioned earlier, in studies in the Aix basin, the isotope geochemistry suggests that the source of these waters may be nonmarine and chemically reworked from Triassic sources. Each gypsum deposit in Aix, however, is immediately preceded and then followed by a marine influx, based on the fossil content of the sediments. Indeed, it may be that part of the water that enters the lagoons is nonmarine (hence the nonmarine isotopic values), but it remains evident that a clear marine relationship exists in portions of these deposits. In fact no gypsum is found before the sedimentologic onset of evident marine input, and gypsum ends when marine influx is terminated. A similar relationship has been noted in the modern Laguna Madre in Texas (on the Gulf coast), with the existence of marine conditions that persist for a few years, followed by isolation and increasing hypersalinity together with the formation of gypsum,

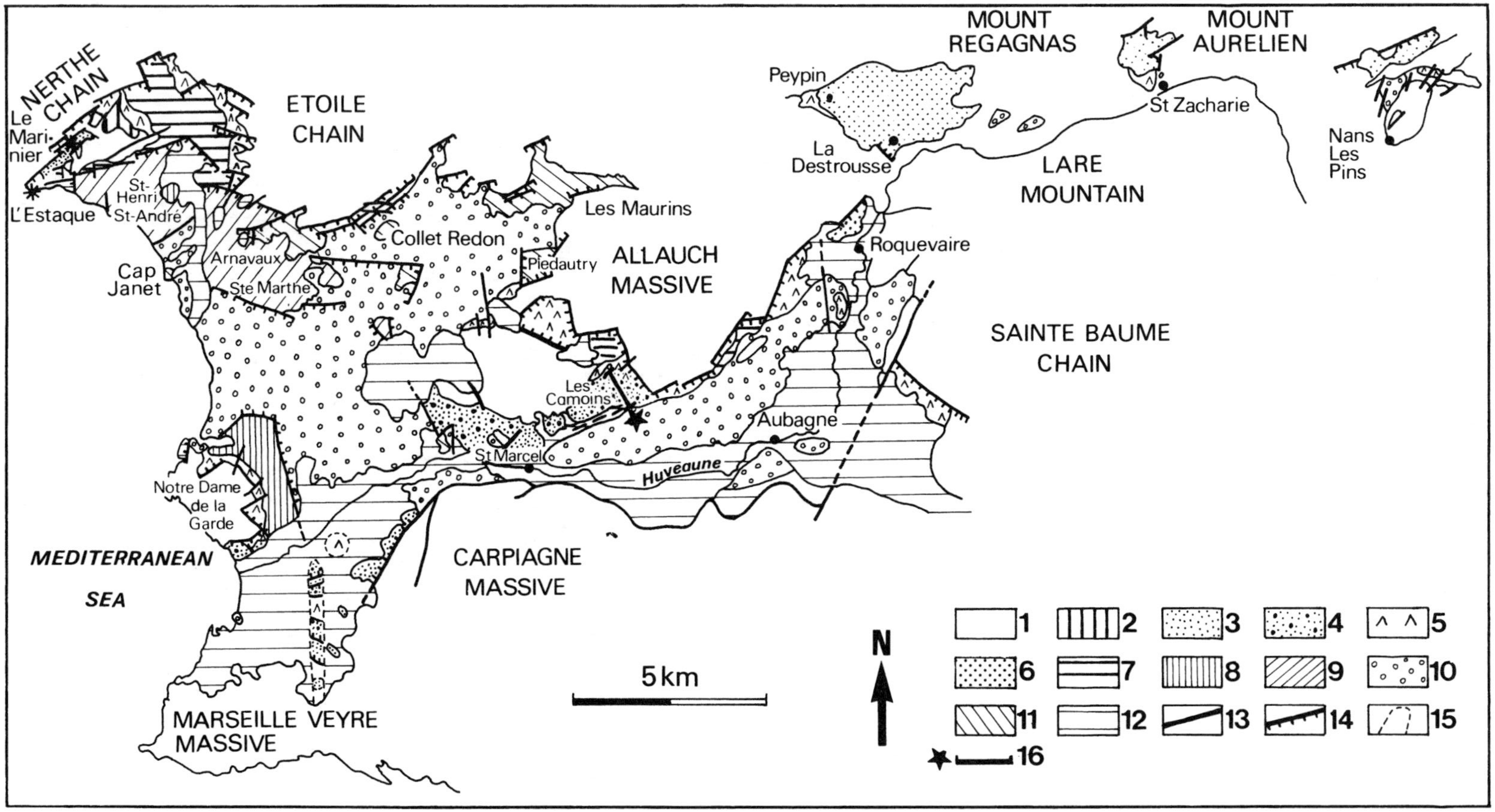

Fig. 6.31. Detailed geological map of the Marseille Basin (Mesozoic through Chattian): Key: 1 = Mesozoic; 2 = earliest early Rupelian; 3 = late early Rupelian; 4 = early middle Rupelian; 5 = late middle Rupelian; 6 = early late Rupelian; 7 = latest late Rupelian; 8 = early Chattian; 9 = middle Chattian; 10 = early late Chattian; 11 = middle late Chattian; 12 = Pleistocene; 13 = fault; 14 = normal fault; 15 = outcrops formed in mines; 16 = position of cross-section given in figure 6.35.

but there has been no suggestion in this latter case of substantial nonmarine control (Fiske 1959).

Marseille Basin

The Marseille Basin contains the most complete Oligocene series in southern (Basse) Provence (figure 6.31; Nury 1987–1988). Eocene deposits are unknown either in outcrops or in bore holes. All of the formations are bounded by normal faults adjacent to Mesozoic limestones and dolomites, and are commonly associated with disorganized pseudobreccias. The sedimentary section contains numerous synsedimentary normal faults, the directions of which are variable through space and time. The lateral formational boundaries are made up of repetitive breccias and megabreccias (with olistoliths). All of the formations are composed of rhythmic and repetitive depositional sequences.

PERIODS OF SEDIMENTATION AND NONDEPOSITION

Sedimentation occurs in the Marseille Basin during three distinct periods that are separated by important lacunae. The areas of sedimentation (depocenters) were gradually displaced throughout the Oligocene (figures 6.32 and 6.33 and table 6.4).

During the earliest Rupelian (Stampian), Striatelles limestone deposits (named for the characteristic gastropod fauna that developed at that time within fluvial and continental settings) are only located in the north of basin and in the Nans-les-Pins area, while lignite is deposited in the St. Zacharie area (figure 6.33A). From late early Rupelian to early late Rupelian, two grabens opened in this region (figure 6.33B to E). In the rather small northern graben, only limestone was deposited, and in the larger southern graben that extended northeastwards as far as Nans-les-Pins, clays, sands, and conglomerates mixed with the bordering limestones and breccias. By the end of early late Rupelian, gypsum graded into the limestones.

From the latest late Rupelian to middle Chattian, the activity of the northern graben became marked but the southern graben apparently ceased subsidence. Deposits of megabreccias and marginal limestones are well developed during the terminal portions of the Rupelian in the northern area. Detrital siliceous sediments are intercalated with limestones and are very fine-grained (mainly clay with only a little sand), but moving upwards through the section the limestones progressively disappear. Fluviatile sediments become coarser upsection (with less clay and more sandstone).

During the late Chattian, detrital fluviatile deposits (yellow and gray clays, sands, and conglomerates) invade the whole area and the Nans-les-Pins region, except for the Peypin—La Destrousse—St Zacharie areas (maximum thickness: 700 m). During the middle part of the late Chattian, red siliceous fluviatile deposits appear, especially in the northern area, while marginal limestones reappear. The latest portion of the late Chattian is missing.

Outcrop studies also show many apparent breaks in sedimentation throughout the section: continental animal burrows cut into clays and soils, and synsedimentary faulting affected early cemented limestone sequences in one area.

MARINE INFLOW (INFLUX)

Variable degrees of marine ingression are evident at different stratigraphic levels in the Marseille Basin, as demonstrated by the presence of mixed faunal and floral assemblages within the sediments. These populations contain foraminifera, Siphonophorids, coelenterates, mollusca, euryhaline fish, and dinoflagellates. The first important ingression, concomitant with the beginning of gypsum deposition, occurs within the first phase of Rupelian sedimentation of the basin (figures 6.33B to D). The advance of the marine environment can be progressively traced northeastward into the southern graben throughout the Rupelian (figures 6.33 and 6.34).

The second phase of apparent marine ingression occurs at the beginning of the second sedimentation

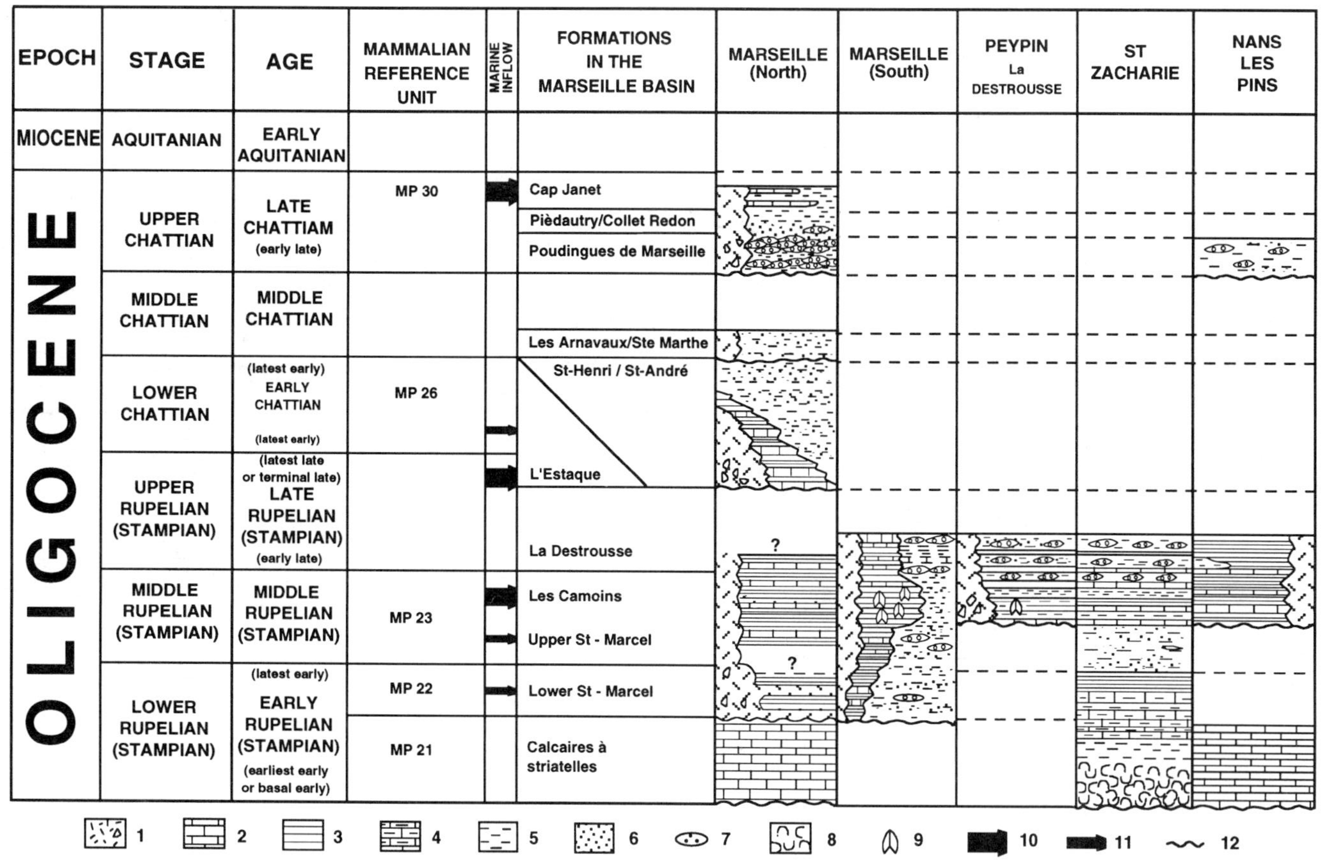

FIG. 6.32. Stratigraphy and lithology in the various subdivisions of the Marseille Basin during the Paleogene. Key: 1 = breccia and megabreccia; 2 = limestone; 3 = laminar limestone; 4 = calcareous marl; 5 = marl; 6 = sand; 7 = conglomerates (in channels); 8 = lignite; 9 = gypsum; 10 = major marine-water inflow; 11 = minor marine-water inflow; 12 = discordance.

TABLE 6.4.
Graphic representation of the distensive stages during the Oligocene, given by basin, together with an approximation of their intensities.

STRATIGRAPHIC SUBDIVISIONS \ BASINS		MARSEILLE NORTH	MARSEILLE SOUTH	PEYPIN LA DESTROUSSE ST.ZACHARIE	AIX SOUTH	AIX NORTH	NERTHE	TOULON NORTH
early Lower Aquitanian								
upper Upper Chattian	upper	+			+	+	+	
	middle	+	+		+	+	+	
	lower	+	+					+
Upper Chattian		+						
middle Upper Chattian		+			+			
lower Upper Rupelian (Terminal Stampian)		+						
upper Middle Rupelian (upper Stampian)			+	+	+			+
Middle Rupelian (middle Stampian)		+	+	+			+	+
Lower Rupelian (lower Stampian)		+	+	+	+		+	+
lowest Lower Rupelian (lowest Stampian)		+		+			+	
Middle Priabonian (middle Ludian)								

period in the northern graben. The third phase of marine ingression appears, also in the northern graben, at the end of the third sedimentary phase and developed as a marine shore-line environment with *Ostrea, Cardium, Psammobia,* found in sediments at the southwestern end of the basin (Cap Janet), and in a maritime marsh (*Potamides*) towards the northeast (at Piédautry).

FLUVIATILE DEPOSITS

Clay and sandstone beds cut by sandstones and conglomerates within channels or beds that fill gullies are common. Channels characterize the Rupelian deposits and these channelized beds become even more common during the late Chattian (Poudingues de Marseille). It seems possible that such channel-fills represent an increase in tectonic activity and also a somewhat wetter climate.

Clays and sandy clays are well developed within two formations: the Formation de St. Marcel (late early Rupelian to early middle Rupelian) and Argiles de St Henri-St André (early Chattian). These are generally micaceous clays principally made of illite, but younger beds begin to contain smectite, a little kaolinite, and some chlorite. Sandstones and conglomerates usually contain a clay matrix and are cemented by carbonate.

Fluviatile paleocurrent directions (based on the imbrication of pebbles, flute-casts, and channel-axes orientations) show that fluviatile materials come from the east and south during almost the whole Oligocene, and afterwards, from the south to western sectors during the middle latest Chattian.

Conglomerate grain size is variable. In general clasts vary in size up to 10 cm and rarely to 30 or 50 cm. Throughout the Rupelian and equally throughout the Chattian, the distribution of the grain size of the various components becomes wider (Formation de La Destrousse and Poudingues de Marseille). This fact seems suggestive of distensive tectonic events in the source areas of the southern crystalline land mass. The proportion of siliceous pebbles within the conglomerate beds varies from 30% to 85%. During the upper Chattian (Poudingues de Marseille), the border conglomerates develop a large component of carbonate peb-

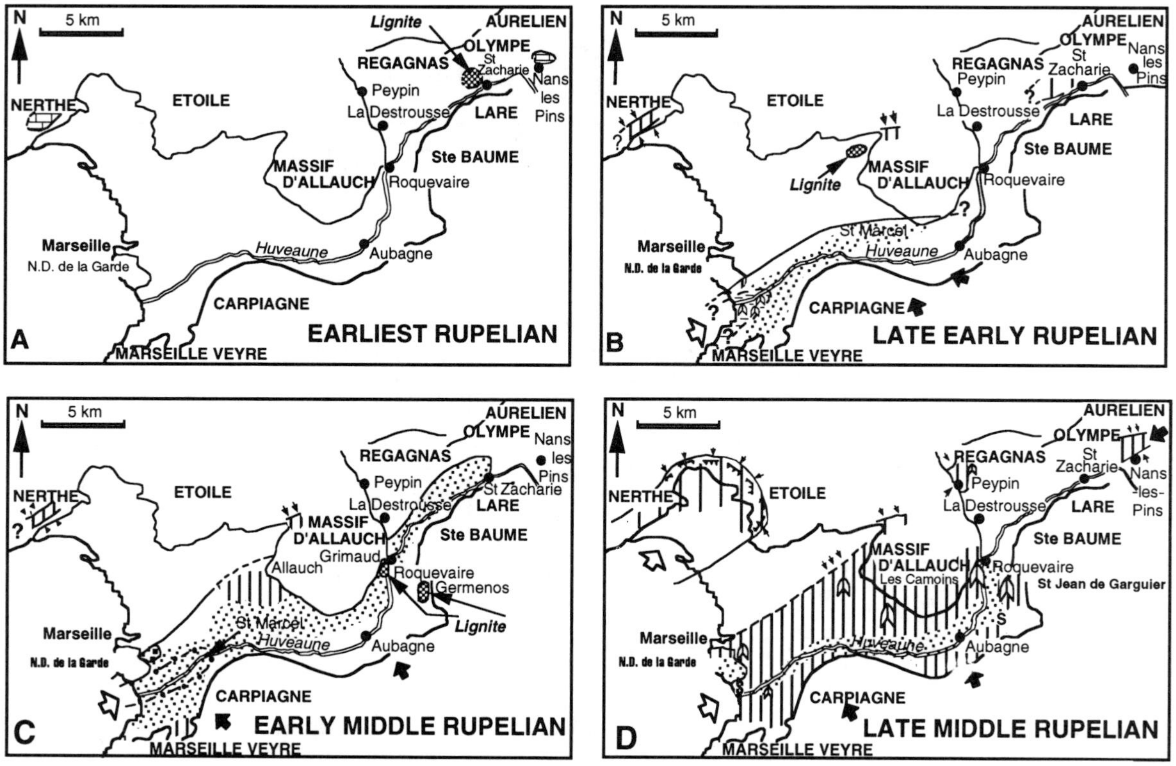

Fig. 6.33. Paleogeographic evolution of the Marseille Basin during the Oligocene: A. earliest Rupelian; B. late early Rupelian; C. early middle Rupelian; D. late middle Rupelian; E. early late Rupelian; F. latest late Rupelian to early Chattian; G. middle Chattian; H. early late Chattian; I. middle late Chattian. Key: 1 = laminar, thin-bedded to massive limestones; 2 = exotic detrital deposits; 3 = *Striatelles* limestone; 4 = marl; 5 = lignite; 6 = normal fault; 7 = mined gypsum; 8 = sparse gypsum; 9 = traces of gypsum; 10 = local detrital input; 11 = fresh water input; 12 = exotic detrital input; 13 = marine water inflow; 14 = boundary, sedimented area; 15 = boundary, suggested marine water inflow; S = Stromatolites.

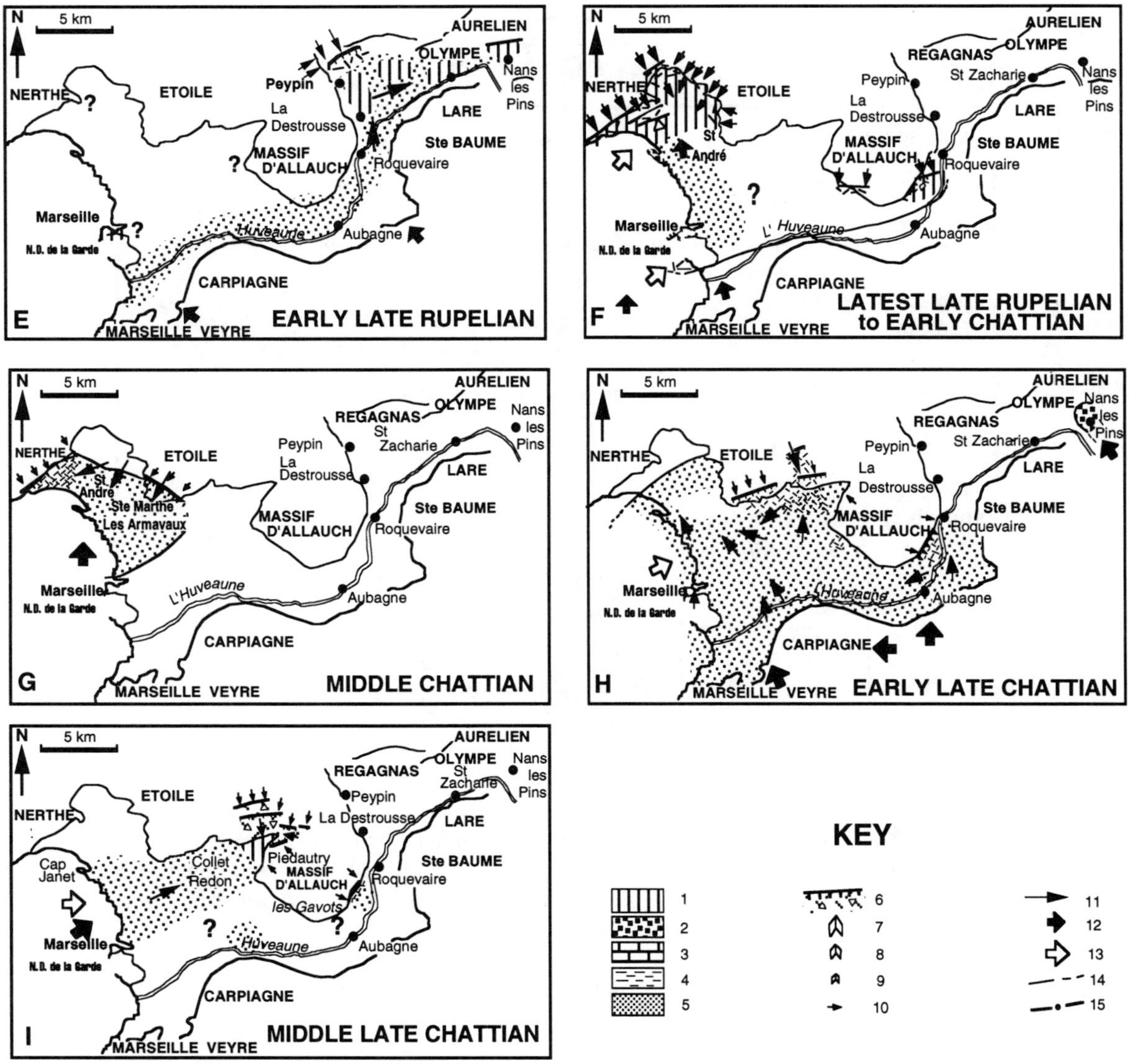

N
5 km
AURELIEN
OLYMPE
Nans les Pins
NERTHE
ETOILE
Peypin
La Destrousse
LARE
MASSIF D'ALLAUCH
Ste BAUME
Roquevaire
Marseille
N.D. de la Garde
Huveaune
Aubagne
CARPIAGNE
MARSEILLE VEYRE
E
EARLY LATE RUPELIAN
N
5 km
AURELIEN
REGAGNAS
OLYMPE
Nans les Pins
Peypin
St Zacharie
NERTHE
ETOILE
La Destrousse
LARE
St André
MASSIF D'ALLAUCH
Ste BAUME
Roquevaire
Marseille
N.D. de la Garde
L' Huveaune
Aubagne
CARPIAGNE
F
LATEST LATE RUPELIAN to EARLY CHATTIAN
MARSEILLE VEYRE
N
5 km
AURELIEN
REGAGNAS
OLYMPE
Nans les Pins
Peypin
St Zacharie
NERTHE
ETOILE
La Destrousse
LARE
St. André
Ste Marthe
Les Armavaux
MASSIF D'ALLAUCH
Ste BAUME
Roquevaire
Marseille
N.D. de la Garde
L'Huveaune
Aubagne
CARPIAGNE
G
MARSEILLE VEYRE
MIDDLE CHATTIAN
N
5 km
AURELIEN
OLYMPE
Nans les Pins
REGAGNAS
Peypin
St Zacharie
NERTHE
ETOILE
La Destrousse
LARE
MASSIF D'ALLAUCH
Ste BAUME
Roquevaire
Marseille
N.D. de la Garde
Huveaune
Aubagne
CARPIAGNE
H
MARSEILLE VEYRE
EARLY LATE CHATTIAN
N
5 km
AURELIEN
REGAGNAS
OLYMPE
St Zacharie
Nans les Pins
Peypin
ETOILE
NERTHE
La Destrousse
LARE
Cap Janet
Collet Redon
Piedautry
MASSIF D'ALLAUCH
Ste BAUME
Roquevaire
les Gavots
Marseille
N.D. de la Garde
Huveaune
Aubagne
CARPIAGNE
I
MARSEILLE VEYRE
MIDDLE LATE CHATTIAN
KEY
1
2
3
4
5
6
7
8
9
10
11
12
13
14
15

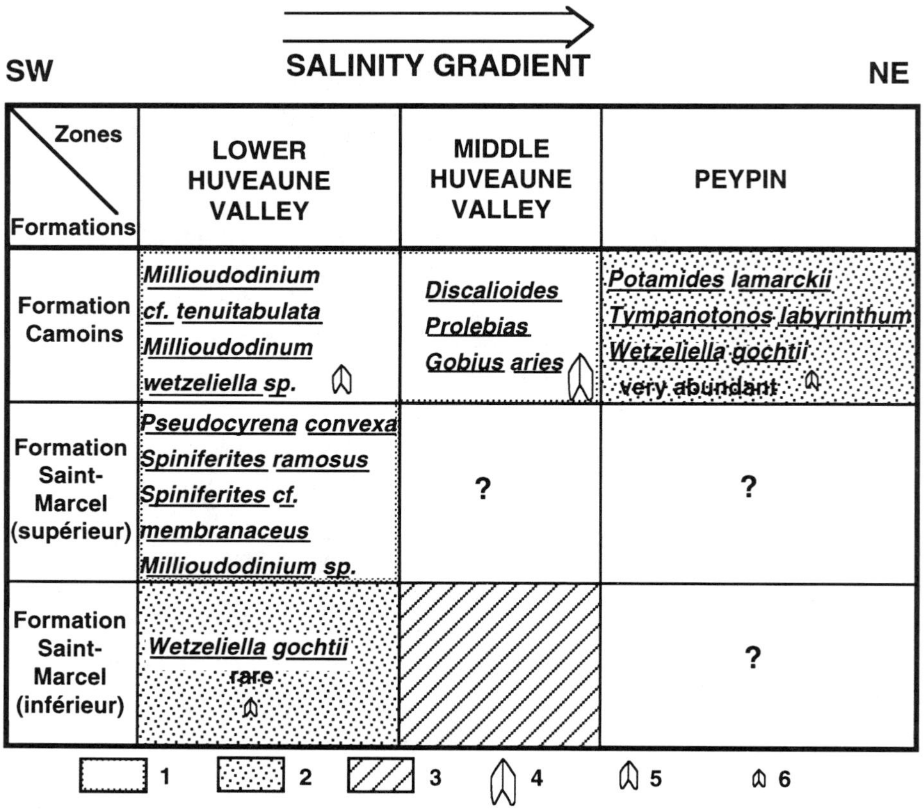

FIG. 6.34. Schematic representation of the progression of marine influence (southwest toward northeast; based on paleontologic and lithologic data) in the Marseille Basin through time and space, as noted in the Huveaune valley, seen through the span of upper early Rupelian through middle Rupelian (early Oligocene). Key: 1 = eroded sections; 2 = environment of fluvial-deltaic deposition; 3 = fresh water environments; 4 = large quantities of gypsum (exploited); 5 = presence of gypsum; 6 = traces of gypsum.

bles (82% to 85% in the >4 cm fraction). This is possibly the consequence of a somewhat wetter climate. Local calcareous detrital input that makes up the breccial components during other periods is in the form of rounded clasts, and is mixed with southern fluviatile input. Heavy mineral groupings include staurolite, kyanite, epidote, tourmaline, zircon, garnet, rutile, and andalusite.

CARBONATES

Pseudobreccia carbonates are widely developed around the borders of the Marseille Basin, especially in the northern Mesozoic border areas. The southern border of the basin has been uplifted by recent tectonics and the equivalent pseudobreccias have been eroded and are absent. Palustrine and travertine limestones, and laminar and thin-bedded carbonates are the principal deposits during the Rupelian.

During the early middle Rupelian (Stampian), fresh-water stromatolite bioherms mark the beginning of the carbonate sedimentation. During the early late Chattian fluviatile phase, stromatolites, reworked into conglomerates, constitute key beds (having a characteristic isotopic signature; Casanova and Nury 1989). Generally, such carbonates reflect a hot and dry climate (table 6.5). Here they are an important component during the Rupelian,

TABLE 6.5.

Oligocene formations of the Marseille Basin showing their relative intensities of distension, the evoloution of their depositional milieus, their vegetation, temperature, and humidity. Key to depositional milieus: S = soils; L = lacustrine and fluvial setting; E = estuaries and lagoon; M = marine shoreline.

STAGE	AGE	VERTEBRATE and MAMMALIAN ZONES	CHAROPHYTE ZONES	FORMATIONS OF THE MARSEILLE BASIN	SEDIMENTARY CYCLES	INTENSITY OF DISTENSION - +	DEPOSITIONAL MILIEUS S L E M	CLIMATE TEMPERATURE HUMIDITY - +
CHATTIAN	late Chattian	MP 30	*Chara notata*	Cap Janet/Piédautry /Collet-Redon Poudingues de Marseille	CYCLE III			
	middle Chattian							
				Amavaux-Ste Marthe	CYCLE II			
	early Chattian	MP 26	*Chara microcera*	St Henri–St André L'Estaque				
RUPELIAN *sensu stricto* / STAMPIAN	latest late Rupelian early							
			Rhabdochara major	La Destrousse	CYCLE I			
	middle Rupellian	MP 23		Les Camoins St Marcel supérieur				
	late	MP 22		St Marcel inférieur				
	early Rupelian earliest	MP 21	*Stephanochara vectensis*	Calcaires à striatelles				

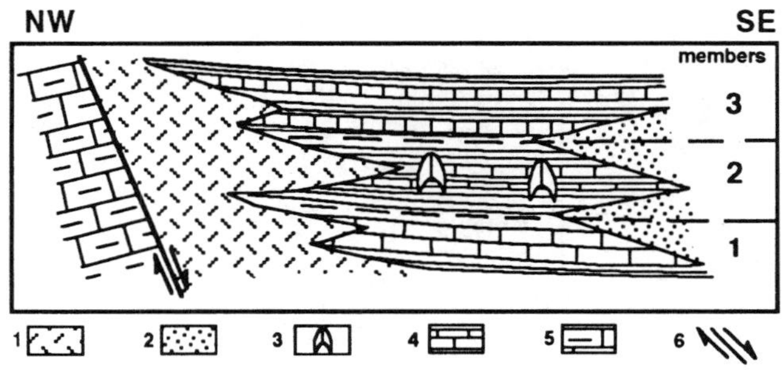

FIG. 6.35. Generalized cross-section of the Marseille Basin showing the relationship of the evaporitic interbeds exposed at the Camoins mine. (For location of this section see indicated line in figure 6.31.) Key: 1 = breccia; 2 = detrital inflow; 3 = gypsum beds; 4 = laminar to massive limestone; 5 = Mesozoic limestone; 6 = fault movement.

and they become less important during the Chattian with a progressive increase in humidity. They reappear with the apparent middle late Chattian dryness.

SULFATES

The sulfates appear as interfingering beds between limestone deposits (figures 6.34 and 6.35). most of the sulfates are found in the formation des Camoins. They are variously formed by precipitation in water column, displacively, or as very early replacive development within existing sediments (synsedimentary). The ubiquitous presence of lenticular gypsum crystals signifies the presence of traces of specific types of organic matter, primary of continental origin (Cody and Cody 1988).

The Formation des Camoins is composed of three members. Member 2 is made up of limestones, calcareous laminites, and marls with gypsum and interbeds or intercalations of sulfur. Above and below the sulfates, Members 1 and 3 are composed of "sterile" calcareous laminites with sporadic influx

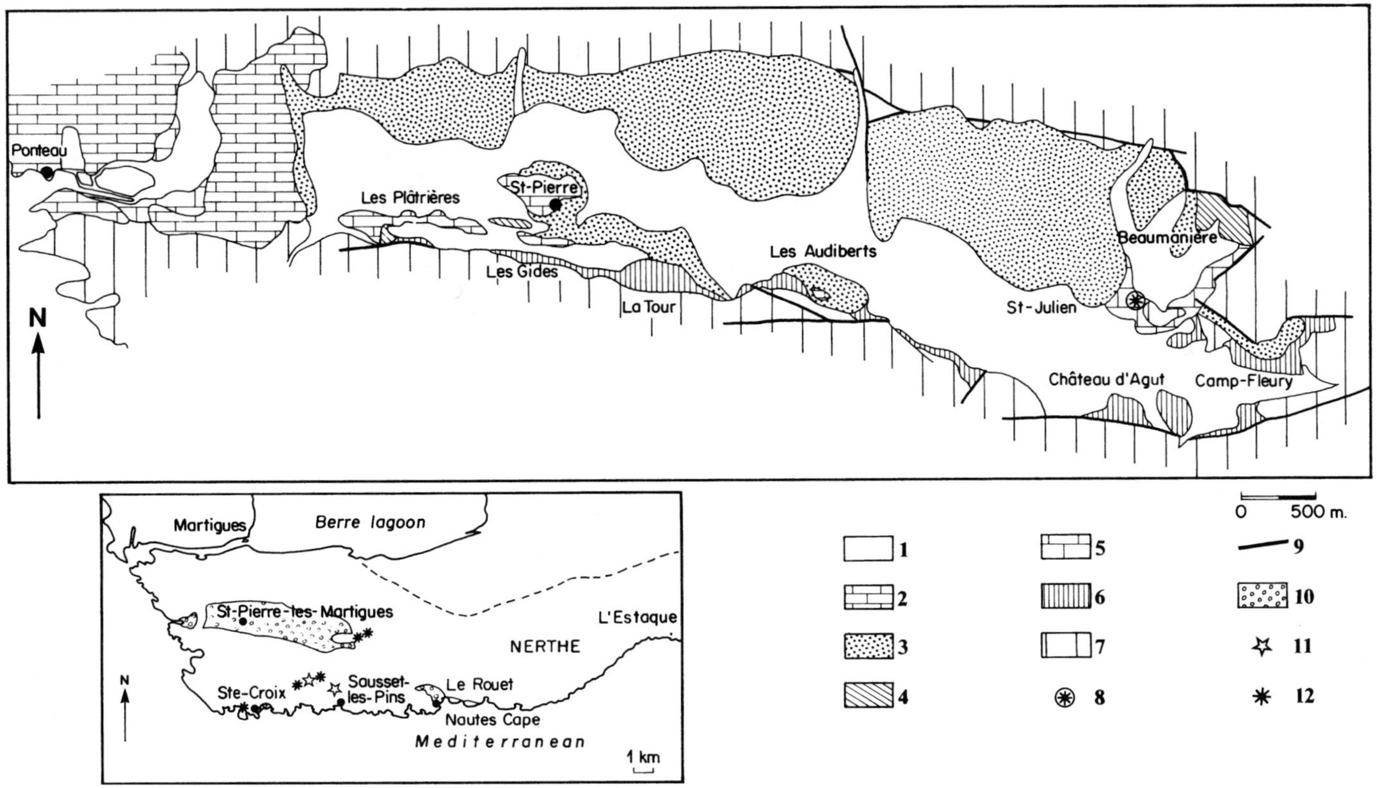

FIG. 6.36. Geological map of the St. Pierre-Nerthe Basin. Key: 1 = Pleistocene; 2 = late Burdigalian; 3 = latest late Oligocene (Poudingues de St. Julien); 4 = early late Rupelian; 5 = late early Rupelian; 6 = earliest early Rupelian (Stampian; Calcaires à striatelles); 7 = Mesozoic; 8 = location of fossiliferous layer dated as late early Rupelian; 9 = faults; 10 = important Oligocene deposits; 11 = isolated Rupelian outcrop; 12 = isolated late Chattian outcrop.

and preservation of delicate marine fauna such as coelenterates, euryhaline fish, and insects (see figures 6.23 to 6.26). In general the milieu was extremely restricted and did not permit molluscs to develop. It is impossible to say from studies of the gypsum if this setting actually received any components of nonmarine sulfate-rich water, but boreholes have shown actualintercalations of nonmarine marls and fine sands. These deposits are probably the consequence of significant fluviatile input which invaded the graben and permitted the detrital particles to mix with the carbonates. Possibly some continentally derived sulfates mixed episodically with the marine water within the confined milieu in a low-energy and possibly stratified water body.

Saint-Pierre-les-Martigues and the Southern Nerthe Chain

Eocene deposits (of either stage usually present in Provence) are unknown in this area. Oligocene deposits overlying the Nerthe Chain are distributed into two depositional areas: St Pierre les Martigues and the southern Nerthe basin (i.e., Rouet and other sporadic outcrops) (figure 6.36). The sedimentary sequence that is present is very thin as compared to that of the Marseille basin series (Nury 1987/1988). The breccia and megabreccia facies are also present and as in Marseille are controlled by normal faults.

PERIODS OF SEDIMENTATION AND NONDEPOSITION

There is a very long sedimentary break between the middle Rupelian (Stampian) and late Chattian depositional periods, as in the Marseille Basin. The sedimentary sequence that is present in these areas is very thin as compared to Marseille. Phases of sedimentation are displaced through time: Rupelian (Stampian) sedimentation occurs principally within the western part of the Nerthe chain while latest Chattian sedimentation appears within the eastern part and afterwards spreads across the whole chain, and is followed without any break by Aquitanian (early Miocene) sedimentation (figures 6.37 and 6.38).

St Pierre-les-Martigues Basin.

In St Pierre (figure 6.38) sedimentation begins with limestones containing lacustrine and fluviatile fauna (Striatelles limestones); above this lays the Formation de Ponteau, containing marls and limestones with meso- to polyhaline fauna marked by *Granulolabium.* After this laminites and thin-bedded limestones with gypsum appear in the western zone, with travertine limestones in the eastern zone. Discordantly above them, a thick sequence of breccias and calcareous conglomerates developed, the components coming from local sources controlled by normal faults just to the north (figure 6.39).

The Nerthe Chain.

We find some rather small late middle Rupelian (Stampian) outcrops scattered throughout the chain but without gypsum. All along the actual sea shore lay Chattian deposits: carbonate breccias (Ste Croix), and carbonate breccias, fluviatile and torrential conglomerates, then shore-line deposits (Le Rouet).

In Rouet we find two Oligocene units (from the basement to the top): a continental unit and a marine one.

The continental unit is the Formation du Poudingues du Rouet: *member 1* is the Grès et Conglomérats du Rouet (inférieurs) constituting a siliceous fluviatile series showing numerous synsedimentary normal faults (principally 090° to 115°, and subordinate 005° to 175°–180° structures); *member 2* is the Argiles et Conglomérats du Rouet (supérieurs), containing a continental fauna and flora (*Helicides, Microcodium*), that constitute a torrentially formed sequence. They lie in discordance above the first member, separated from it by a gullied erosion surface.

The marine unit is composed of three formations: la formation périrécifale du Cap de Nautes, la formation saumâtre de Rousset, and la formation

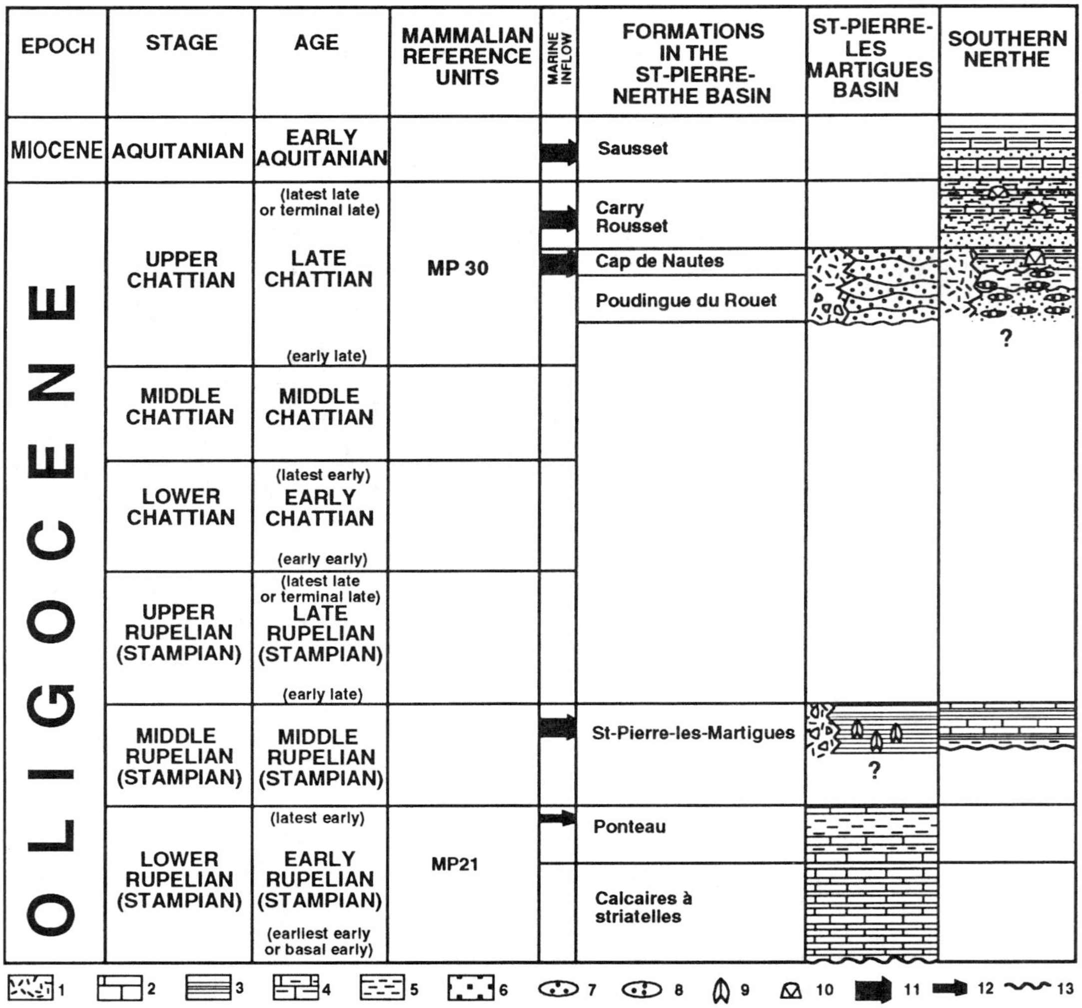

FIG. 6.37. Stratigraphy and lithology in the various subdivisions of the St. Pierre-Nerthe Basin during the Paleogene. Key: 1 = breccia and megabreccia; 2 = limestone; 3 = laminar limestone; 4 = calcareous marl; 5 = marl; 6 = sand; 7= channelized calcareous conglomerates; 8 = channelized mixed siliceous and calcareous conglomerates; 9 = gypsum; 10 = reefs; 11 = important marine inflow; 12 = minor marine inflow; 13 = discordance.

bioclastique de Carry (Nury 1994). The first and third of these are marine but the second is apparently brackish and regressive.

A marine unit lies upon the Formation des Poudingues de Rouet, separated by a trangressive surface. The latest late Chattian age is clearly established by the whole fauna and flora but especially by the presence of mammal reference unit MP 30 (Hugueney et al. 1987; Hugueney and Ringeade 1990) and the nannoplankton zone NP 25 (Martini 1973).

Overlying this Oligocene section lays the formation biodétritique de Sausset that constitutes the earliest Aquitanian deposit.

Every one of these formations has been affected by repeated synsedimentary normal faulting and associated breccias on the northern side of the basin. The outcrops of these deposits illustrate that the deposition of the Argiles et Conglomérats de Rouet is related to the normal faulting that develops from SW towards NE, and the graben opened from the center towards the periphery of the basin.

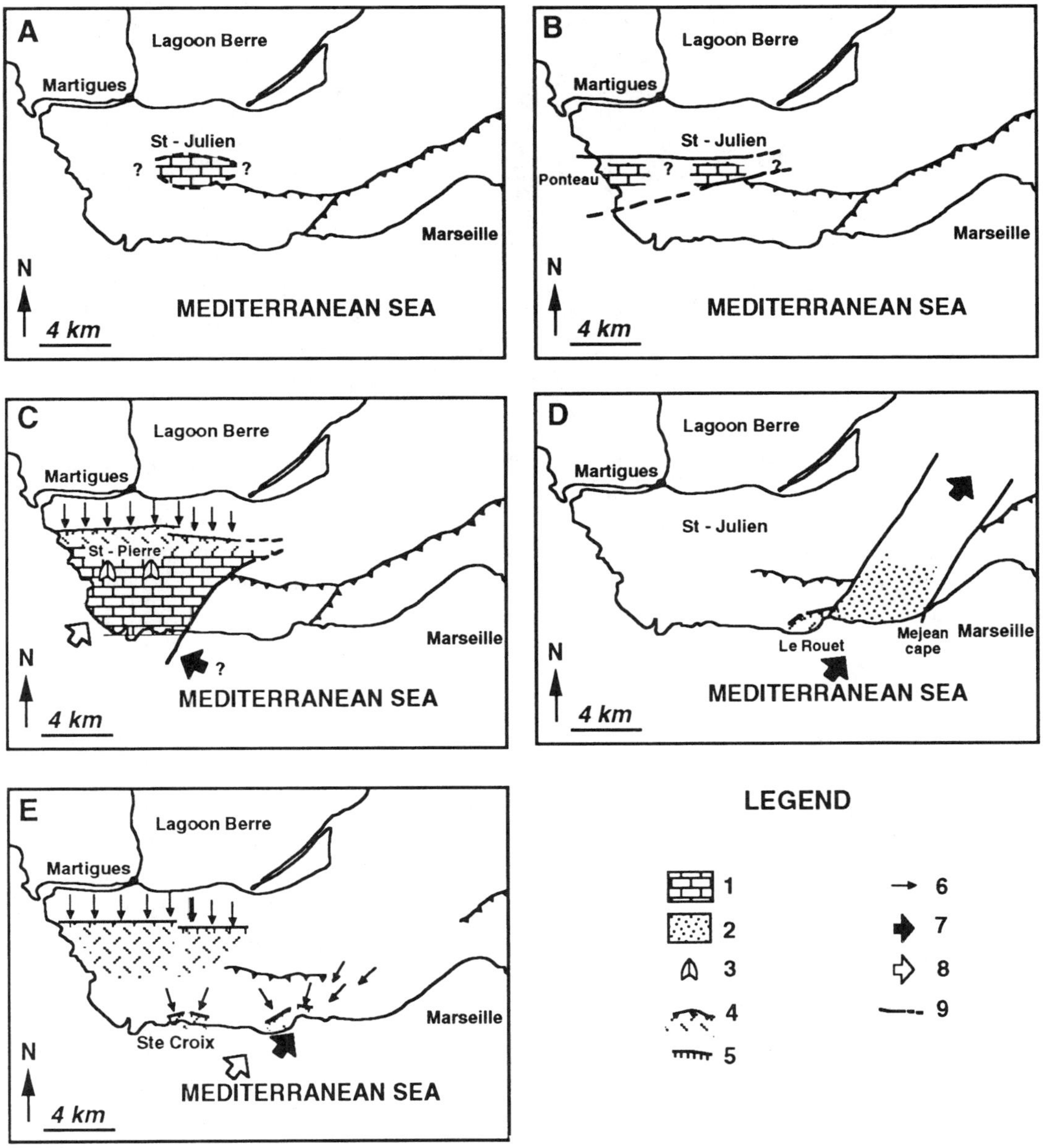

Fig. 6.38. Paleogeographic evolution of the St. Pierre-Nerthe Basin during the Oligocene. A. earliest (or basal) early Rupelian; B. late early Rupelian; C. late middle Rupelian; D. middle late Chattian; E. latest late Chattian. Key: 1 = carbonate deposits; 2 = exotic detrital deposits; 3 = gypsum; 4 = structure related to Pyrenean-Provençal tectonism and associated synsedimentary breccias; 5 = normal faults; 6 = direction of local detrital influx; 7 = direction of exotic detrital influx; 8 = direction of marine influx; 9 = limits of sedimented area.

Repetitive accumulations of clays or sandstone/conglomerates formed in both a continental environment as well as in a shoreline environment. This suggests an uneven, stepped subsidence caused by distensive tectonics. The shoreline deposits probably do not signify the beginning of a generalized transgression, but rather mark the terminal phase of the rifting that allow the Tethys sea water to reach southern Provence.

As in the Marseille Basin many diastems are ob-

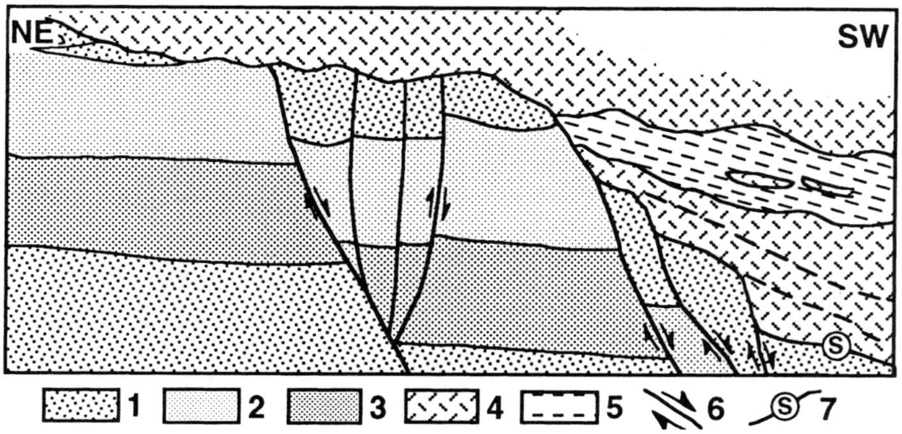

Fig. 6.39. Cross-section illustrating fault development and concomitant sedimentation at the margin of the St. Pierre Basin, represented by the Poudingue du Rouet. Key: Lower section (Grès et Conglomérats inférieurs): 1 = puddingstone conglomerate; 2 = indurated argillaceous sandstone; 3 = upper section (Argiles et Conglomérats supérieurs); 4 = conglomerate; 5 = thin-bedded argillite; 6 = normal faults; 7 = erosion surface.

served cut into earlier conglomerates. These erosional surfaces are affected by *Microcodium* alteration and the development of root-networks, evidence of numerous, short sedimentary breaks.

FLUVIATILE DEPOSITS

Rupelian (Stampian) calcareous beds show only a few pink or green clays interfingering with the gypsum series. Thus it is not certain that siliceous fluviatile influx reached this zone. During the middle late Chattian, siliceous fluviatile input came from south toward the northeast of the Rouet area, mixing calcareous local pebbles with red siliceous materials reworked from the earlier sequences, or perhaps these clasts were carried directly from the southern crystalline land mass. This characteristic red siliceous material does not reach the St Pierre-les-Martigues Basin, filled during this period only by calcareous conglomerates.

In the Formation du Rouet of the Nerthe chain, the fine to coarse red siliceous (quartz) sandstones lie in more or less indurated beds. The heavy mineral groupings in these sandstones contain apatite, garnet, staurolite, chlorite, epidote, zircon, tourmaline, rutile, and andalusite. We find the same mineralogic grouping in the sandstones of the whole series, even extending up into the Miocene beds, the same assemblage that is present in the Marseille Basin. The mineralogy of the conglomerate pebbles is the same in the two continental members but their ratio is different.

	siliceous elements	calcareous elements
member 1	25% to 60%	75% to 40%
member 2	15% to 20%	85% to 80%

During deposition of member 1 of the Formation du Rouet, the siliceous detrital input came from the south and southwest, mixing as in the Marseille Basin with calcareous local drifts coming from the north. The detrital influx of member 2 came from the north, reworking siliceous (and also calcareous) pebbles. But it is possible that some southern siliceous input entered the basin directly.

The local torrential siliceous and calcareous influx continued during deposition of the Formation du Cap de Nautes, with sedimentation along the paleoshoreline. Argillaceous minerals are principally smectites (8 to 10/10), a little illite, and very rarely kaolinite.

CARBONATES

Except in the Rouet area, the sedimentation is almost entirely calcareous: St Pierre les Martigues Basin is almost filled with carbonates and we find the following component facies there:

thin-bedded micritic limestones
travertine and lacustrine limestones
laminated limestones and marls
massives marls
megabreccias, breccias, and conglomerates
reef carbonates

SULFATES

As in the Marseille Basin, gypsum is included within a laminar calcareous series that passes laterally into peripheral breccias. Marine inflow is demonstrated by the presence of hypersaline to mesohaline fauna. Slumped gypsum beds (formed as soft sediment deformation) and convoluted carbonate beds are suggestive of irregular subsidence and/or unstable slopes.

Clearly developed mud-cracks in both the calcareous laminites and in the gypsum deposits (primary laminar and nodular gypsum) characterize an area of very shallow sedimentation. Marine inflow is demonstrated by the repeated intercalation of marine fauna into the sulfate sequence (figure 6.40A). The milieu for the sedimentary gypsum area resembles a consistent maritime marsh. In comparison the Marseille Basin contains variations made up of a mixture of both marine and non-marine features (figure 6.40B). Here, because nearby Triassic outcrops are unknown, any possible chemical reworking of Triassic gypsum could only be the consequence of the normal faulting that may have allowed ground waters to rise to the surface. Otherwise the source of the ions must be at least partially marine in origin.

Marine faunal elements are present beginning in the late early Rupelian (Stampian) part of the section, as shown by the presence of *Globius aries, Granulolabium* (*Pirenella*) *plicata.* This environmental condition persists during the middle Rupelian (Stampian) as demonstrated by the presence of the mollusca *Thedoxus, Hydrobia, Potamides, Melanoides,* and *Polymesoda* (*Pseudocyrena*): Ostracoda (*Halylocytheridea, Candona*) are also included within the gypsum series. In those areas indicated as maritime marshes by their faunal and floral assemblages, both continental and marine-sourced sulfates should be present.

SHORELINE DEPOSITS: MARINE INFLOW

In the Rouet area true littoral marine deposits appear. Marine influx is shown by the presence of mesohaline to hypohaline fauna and flora. The beginning of the inflow is in the late early Rupelian, shown by the presence of the fish (*Gobius aries, Granulolabium (Pirenella) plicata*). The marine influx continues during the middle Rupelian, demonstrated by the presence of molluscs (*Theodoxus, Hydrobia, Potamides, Melanoides, Polymesoda* (*pseudocyrena*)) and the ostracoda *Haplocytheridea.*

In the Rouet area the sea comes from the south, creating a true pararecifal environment (reef and lagoon), marked by a fauna and flora composed of foraminifers, ostracods, mollusks, echinides, and polyps (Andréieff et al. 1972; Nury and Thomassin 1994). The formations are made up of gray, yellow, and black marls, yellow and gray sands and sandstones, conglomerates, and coral reef limestones.

A simplified description of the characteristic sedimentary cycle that was developed in this sequence is: a hardground or a very shallow marine facies, overlain by mud flats, marls, sands, reef-calcarenites, reef limestones, and at the top an erosional break. The succession of these sedimentary cycles probably signifies repeated short phases of irregular subsidence.

During deposition of the Aquitanian Formation

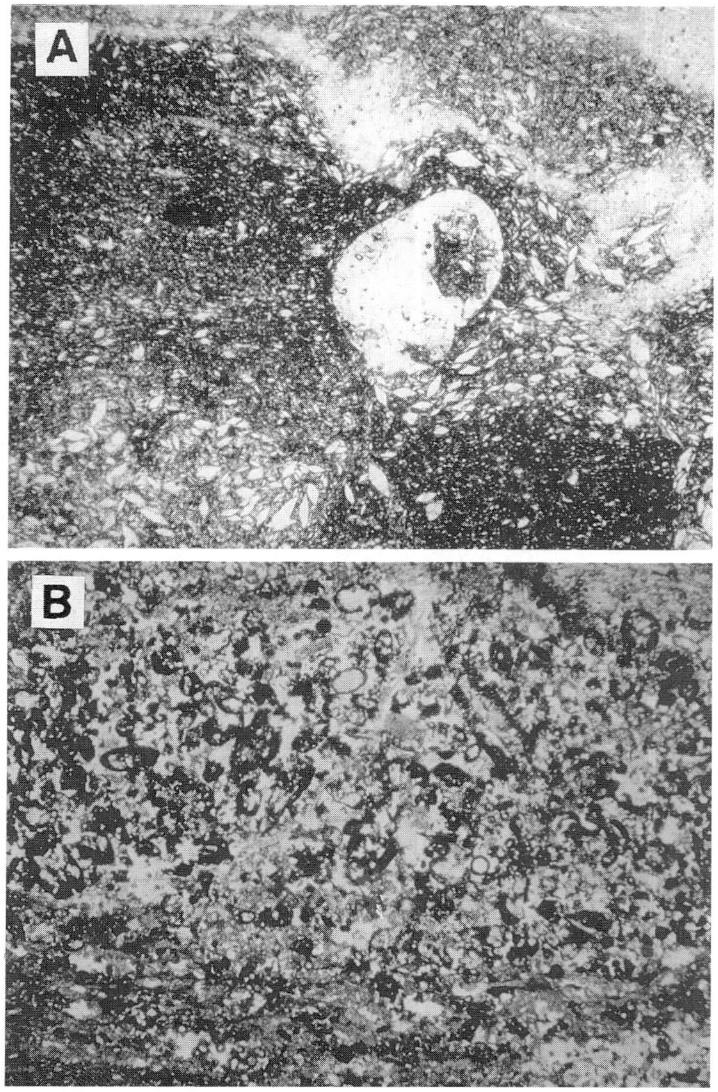

FIG. 6.40. Photomicrographs of fossil remains commonly observed in the gypsum of St. Pierre. A. Small gastropod in a mixed micrite-gypsum matrix (plane light; field of view is 8.2 mm, left to right). B. Fossiliferous packstone composed of diverse faunal elements in a mixed micrite-gypsum matrix; part of the gypsum is a cement composed of large diffuse crystals (0.5 cm) that incorporated the fossils, micrite, and tiny gypsum crystals (plane light; field of view 3.2 mm, left to right).

de Rousset, the depositional environment became brackish, marking a regressive episode, also observed in Aix-en-Provence Basin (Nury 1994).

Aix-en-Provence Basin and Northern Apilles

The Aix-en-Provence Basin contains the most complete Paleogene sedimentary series, with Paleocene, Eocene, and Oligocene deposits. Alone among the basins in Provence, it contains evaporites of Chattian age, but we also may look at the other areas to understand the regional geological story (Durand 1984; Guieu et al. 1987; and Nury 1987/1988). The Oligocene deposits contain breccias starting at the western margin that extend almost to the eastern side of the basin, generated by normal fault movement in the Aix fault.

PERIODS OF SEDIMENTATION AND NONDEPOSITION

Eocene sedimentation seems continuous in a sedimentary area oriented east to west, from the Dano-

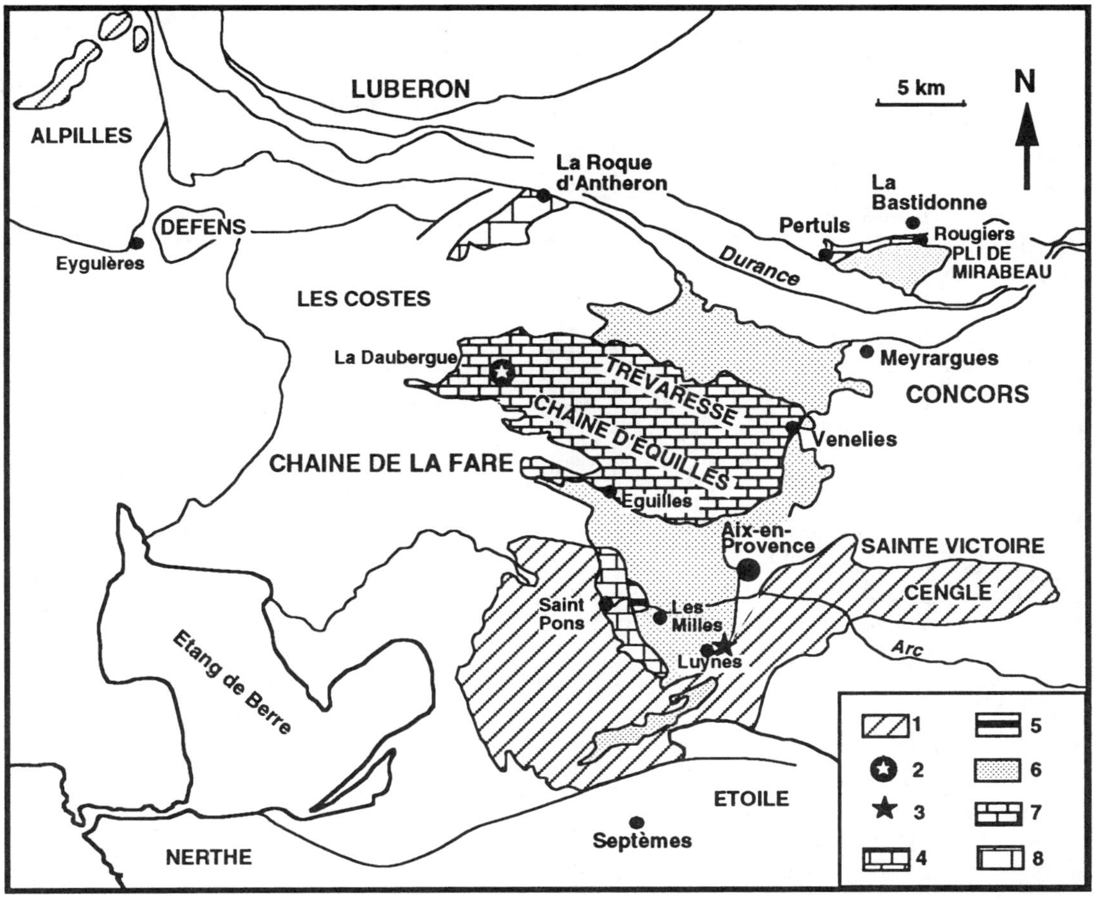

FIG. 6.41. Paleocene, Eocene, and Oligocene deposits from the basins of Aix-en-Provence to Durance. The under- and overlying deposits are not shown. Key: 1 = Paleocene-Eocene deposits (cannot be separated); 2 = location of the de la Daubergue Formation; 3 = location of the Calcaire Luynes; 4 = location of the Calcaire de St. Pons; 5 = location of the Argiles des Milles; 6 = location of the Poudingues de Galice; 7 = location of the d'Aix-en-Provence Formation; 8 = deposits of La Roque d'Antheron, Oligocene deposits.

Montian to late Lutetian in the southern area (Aix-en-Provence), and into late Priabonian in the northern area (Alpilles) (figure 6.41). Deposits are principally constituted of alternating sequences of white to pink lacustrine limestones and gray to red clays containing some conglomerate and sandstone levels. A marked lacuna following this deposition may be correlated with the Pyreneo-Provençal overthrusting that took place at this time.

In the southern area of the basin, the sedimentation begins again during the middle Priabonian ("Ludian" of La Daubergue). Because the outcrops are deeply buried, it is difficult to affirm that this sedimentation is primarily the consequence of distensive tectonics, but it seems very likely.

During the Oligocene, sedimentation occured in the southern portion of the area (Aix-en-Provence) at four discrete periods, and in the north (Pertuis) only during the middle and late Chattian (figure 6.42). During the middle Rupelian, thin-bedded and laminar carbonates containing mesohaline fauna were deposited in Luynes. At this time the Aix-en-Provence Basin was probably connected with the St Pierre-les-Martigues Basin (figure 6.43). After that, a marked sedimentary break extended until the middle Rupelian, when the limestones, cal-

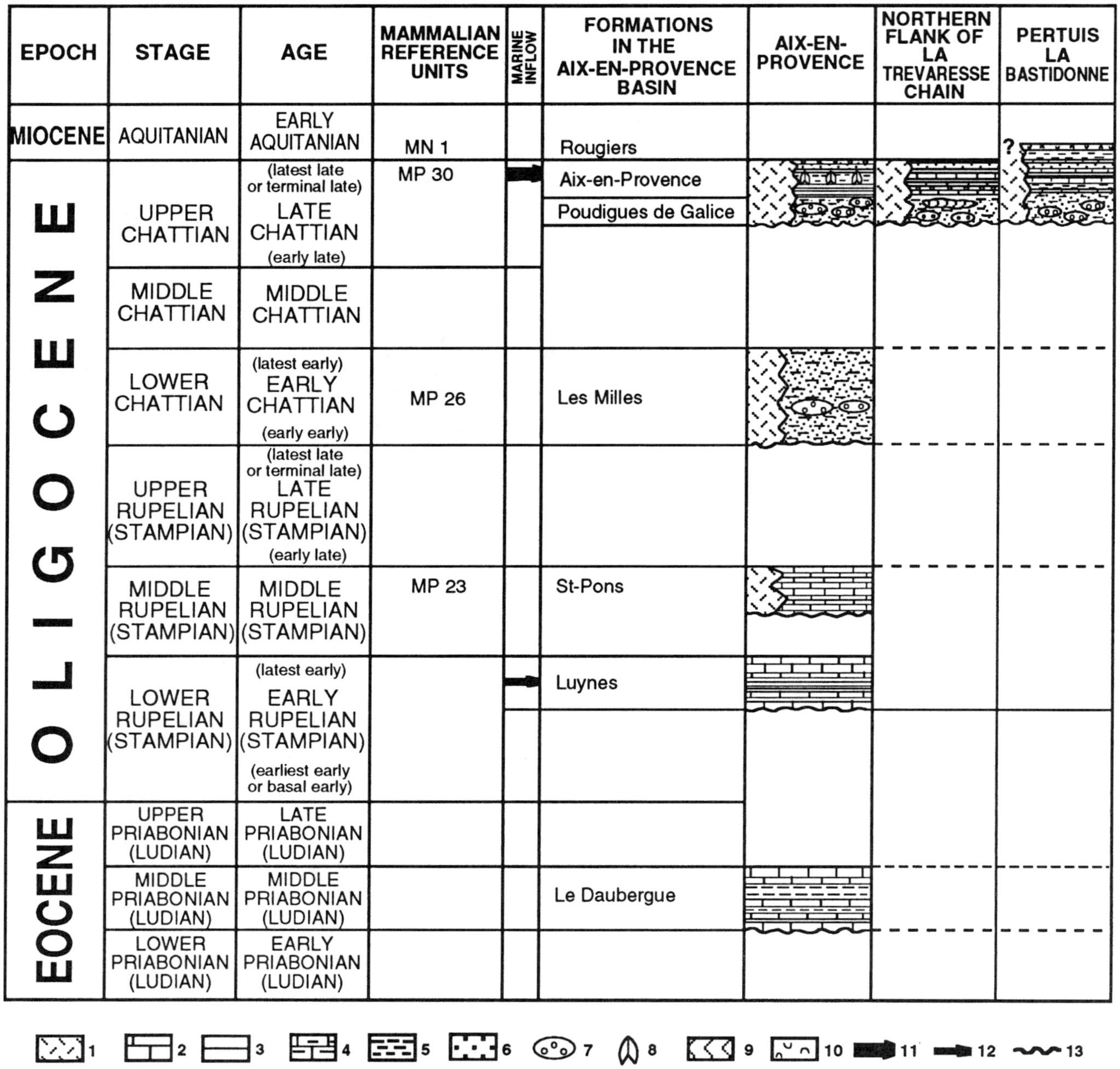

Fig. 6.42. Stratigraphy and lithology in the various subdivisions of the Aix Basin given as successive stages of basin development (tied to A–E in figure 6.43). Key: 1 = breccia and megabreccia; 2 = limestone; 3 = laminar limestone; 4 = calcareous marl; 5 = marl; 6 = sand; 7 = conglomerates (in channels); 8 = gypsum; 9 = calcrete; 10 = lignite; 11 = major marine inflow; 12 = minor marine inflow; 13 = erosion surface.

Fig. 6.43. The Oligocene paleogeography of the Aix Basin given as successive basinal development (A–F): Line A–B in map F marks the position of cross-section given in figure 6.45). A = Eocene; B = late early Rupelian; C = late middle Rupelian; D = early Chattian; E = middle late Chattian; F = latest late Chattian (carbonate member of the Aix Gypsum).

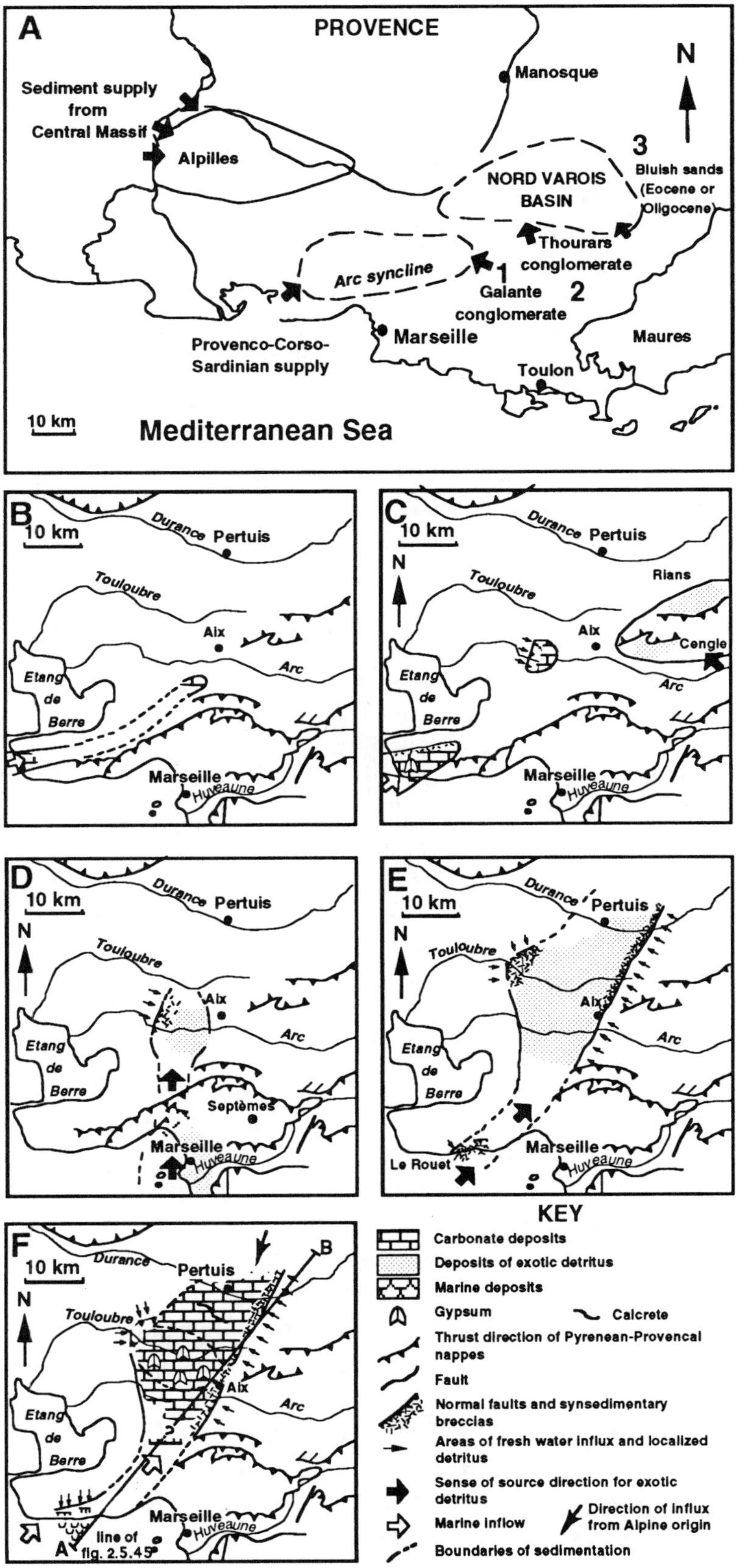

A
PROVENCE
N
Manosque
Sediment supply from Central Massif
Alpilles
3
NORD VAROIS BASIN
Bluish sands (Eocene or Oligocene)
Thourars conglomerate
Arc syncline
1
Galante conglomerate
2
Marseille
Maures
Provenco-Corso-Sardinian supply
Toulon
10 km
Mediterranean Sea
B
10 km
Durance
Pertuis
Touloubre
Aix
Arc
Etang de Berre
Marseille
Huveaune
C
10 km
N
Durance
Pertuis
Touloubre
Rians
Aix
Cengle
Arc
Etang de Berre
Marseille
Huveaune
D
10 km
N
Durance
Pertuis
Touloubre
Aix
Arc
Etang de Berre
Septèmes
Marseille
Huveaune
E
10 km
N
Durance
Pertuis
Touloubre
Aix
Arc
Etang de Berre
Marseille
Huveaune
Le Rouet
F
10 km
N
Durance
Pertuis
B
Touloubre
Aix
Arc
Etang de Berre
Marseille
Huveaune
A
line of fig. 2.5.45
KEY
Carbonate deposits
Deposits of exotic detritus
Marine deposits
Gypsum
Calcrete
Thrust direction of Pyrenean-Provencal nappes
Fault
Normal faults and synsedimentary breccias
Areas of fresh water influx and localized detritus
Sense of source direction for exotic detritus
Direction of influx from Alpine origin
Marine inflow
Boundaries of sedimentation

FIG. 6.43

careous breccias, and conglomerates of Le Calcaire de St Pons were deposited.

After another sedimentary break, during the early Chattian, clays and sands were deposited at Les Milles, coming from the Marseille basin. After this early Chattian sedimentation a well-defined non-depositional period continued until the middle late Chattian, when the sedimentation began again and became very important through the entire late Oligocene. The sedimentary areas developed towards the north (Pertuis) and became much wider through time.

We find two formations (from the basement to the top) (figure 6.44): Poudingues de Galice (fluviatile sandy clays and conglomerates), and Formation d'Aix-en-Provence (carbonate and sulfate-rich beds).

At the beginning of the Aquitanian (Miocene), continental deposition took place in Pertuis (Formation de La Bastidonne, containing Paulhiac mammals; Mein et al. 1971).

Oligocene sedimentation is totally controlled by distensive tectonics with bordering breccias and normal faults (generally 020° to 165°) especially along the Aix fault, an ancient structural normal fault, which is reactivated again principally during the late Chattian. More often than not, the Aix-en-Provence basin looks as though it developed through successive extension of the Nerthe chain or the Marseille sedimentary areas, the exact relationships being buried by more recent deposits, erosion, or deformation.

From the Paleocene to the end of the Oligocene numerous sedimentary breaks are marked by the presence of soils, nodular carbonates, root-marks, etc. They are the consequence of both marked climatic variations and tectonic events.

MARINE INFLOW

The first marine inflow into the basin took place during the late early Rupelian, coming in from the area of Ponteau (St. Pierre-les-Martigues), reached Luynes, and is indicated by the presence of *Granulolabium* (*Pirenella*) *plicatum.* Later, during the middle and latest late Chattian, numerous marine inflows reached the Aix-en-Provence basin as demonstrated by the presence of:

- mesohaline malacofauna such as *Granulolabium, Tympanotonos, Hydrobia, Polymesoda* (*pseudocyrena*)
- polyhaline malacofauna such as *Lucinidae, Veneridae, Psammobia*
- foraminifera such as *Rosalina douvillei, Cibicides, Milliolidae*
- siphonophora coelenterates such as *Discallioides* (Nel et al. 1988)
- Peneides crustacea, a marine to estuarine group
- euryhaline fish such as *Dapalis minutus, D. macrurus, Eolates aquensis, Prolebias cephalotes, Mugil princeps, Perca beaumonti, Anguila, Notogoneus squammoseus, Gobiusaries*
- dinoflagellates such as *Wetzeliella, Chiropteridium, Deflandrea, Spiniferites, Distadodinium.*

The presence of various paleontologic groupings, principally Coderet mammals, permits the correlation of these marine deposits with the Formations de Cap du Nautes/de Rousset and de Carry (Nury 1994), and suggests the probable southern origin of the late marine inflows.

FLUVIATILE DEPOSITS

Considerable siliceous fluviatile input reaches the Aix-en-Provence area during the Paleocene, Eocene, and Oligocene, coming from the southeast, south, and southwest and episodically from the Massif Central (Alpilles, in the Paleogene).

Paleocene and Eocene input. Mottled clays and marls are the principal components of the fluviatile deposits and only rarely interfinger with sands, and were deposited into the subsident basins. The clays are always more or less calcareous and the principal mineral is generally smectite, but is always associated with illite and kaolinite. Illite is very important in the more micaceous beds. Sandstones fill incised channels that were cut into the clays.

We find two conglomerate levels are present: Poudingue de La Galante (at the beginning of Paleocene), and Poudingue de Thouars (probably Thanetian). Both of these conglomerates are composed of Mesozoic limestones and diverse Paleo-

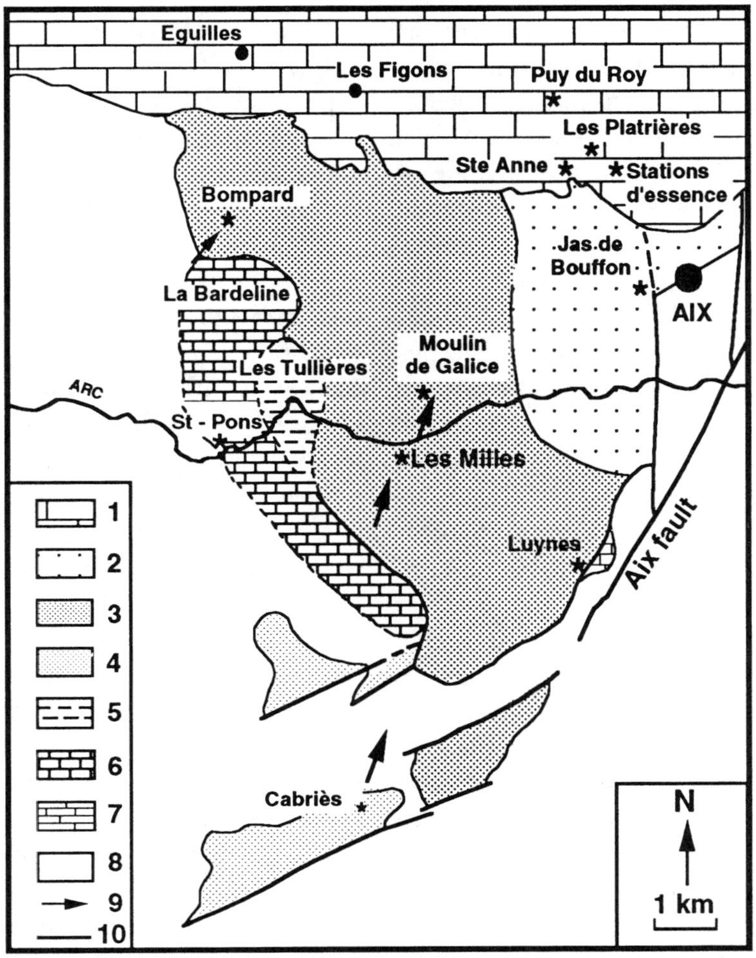

FIG. 6.44. Map showing the depositional limits of the Oligocene formations of the Aix-en-Provence Basin. Key: 1 = Formation d'Aix; 2 = argile et grès de Jas-de-Bouffan (clay and sand); 3 = Formation de Moulin de Galice (clay and conglomerates); 4 = Formation de Cabriès (clay and conglomerates); 5 = Formation de Milles (clays); 6 = Formation de Saint-Pons (limestones); 7 = Formation de Luynes (limestones); 8 = Eocene deposits (undifferentiated); 9 = transport directions of clasts in conglomerates; 10 = faults.

zoic materials (vein quartz, red and green quartzite, phyllite, granite, rhyolite). Interestingly the heavy mineral groupings do not contain andalucite within the assemblage, as in the other basins. This detrital input was apparently the consequence of compressive tectonics that displaced the subsident basin depocenters toward the north and the east.

Oligocene input. Clay, sandy clay, and sparse sandstones make up l'Argile de Les Milles, sedimented during the early Chattian. The clay minerals are smectite (5 to 7/10), illite (2 to 4/10), chlorite (0.5 to 1.5/10). The heavy mineral groupings include rutile, zircon, tourmaline, epidote, staurolite, chlorite, chloritic biotite, kyanite, garnet, and andalucite. The comparison between the Les Milles clays and St Henry-St André clays permits the conclusion that the fluviatile material originated to the south, probably reaching the Aix-en-Provence Basin as an overflow when the Marseille Basin became filled.

The more important of the fluviatile formations, the Poudingues de Galice, is made up of three members: Argiles et Poudingues de Cabries, Argile et Poudingues de Moulin de Galice, and the Argiles et Grès de Jas de Bouffan. The map in figure 6.44 shows that these three members gradually spread laterally from the southwest toward the northeast. The Argile et Poudingues du Moulin de Galice is made up of five expanding conglomerate sheets that cut into the underlying red clays and argillaceous sands. The average dip of the beds is 10° toward the north. The three basic conglomerate sheets are thicker towards the west (by 25 to 10 m). But the maximum thickness of the first sheet is located further to the west than is the maximum thickness of the second sheet, and the second is again more to the west than the third.

The map shown in figure 6.44 also illustrates the gentle discordance between the Poudingues de Galice and the upper Formation d'Aix-en-Provence. These observations suggest that there was considerable tectonic mobility in the sedimentary area, which looks like the development of a half graben that was deeper towards the east and north and migrated through time. This mobility created a progressive displacement of the depocenter and resulted in intraformational discordances between the various members and with the Formation d'Aix-en-Provence. (This movement also can be recognized from the Formation d'Aix-en-Provence because the bordering breccias were clearly developed towards the east.) The conglomerates have a very red matrix that is as red as the interbedded clays: this color probably reflects an arid climate.

The mineralogy of pebbles is rather the same as in the fluviatile Poudingues de Marseille and du Rouet, with a significant quantity of Paleozoic components (64%) showing vein-quartz (18%), quartz-phyllites (1%), quartzite (31%), black radiolarites (8%), and white marbles (6%). In the north of the basin, this Paleozoic material is missing. This observation is in keeping with a southern source, also demonstrated by channel directions and pebble imbrication. The grain size of pebbles is commonly coarse, up to 50 cm, even for the Paleozoic components.

We find the same heavy mineral groupings as before, but at the end of this period, an influx of Alpine minerals appears with noticeable amounts of glaucophane, epidote, and actinote (in the upper sandy members of the Formation d'Aix-en-Provence).

CARBONATES

The calcareous breccias that are present during the whole of the depositional period are indicative of tectonic activity. Paleocene and Eocene limestones are generally white, beige, cream, or pink, deposited in a palustrine or lacustrine milieu, but the Oligocene carbonates are more varied. The early late Rupelian (Stampian) Calcaire de St Pons deposit is an absolutely lacustrine environment, but the calcareous and sulfate composition of the Formation d'Aix-en-Provence is very varied with lacustrine and travertine limestones, massive limestones, stromatolites, laminar and thin-bedded limestones, and laminar marls.

Analysis suggests that deposition took place in a generally marginal to littoral setting with root structures, bioturbation, pedogenetic textures, and mud cracks that signify numerous phases of emersion. However, the northern part of the basin was largely subaerial, characterized by calcrete-forming phases that display typical calcrete features: progressive calcitization of soil horizons (nodular, columnar, to massive beds), and an irregular contact formed between the downward encroachment of the soil development of the lower calcrete boundary.

Mineralogic studies show the progressive development of calcite, attapulgite, and then dolomite within the mineralogy of the calcrete-bearing beds. These calcrete zones apparently signify the development of an arid climate with alternating and climatically contrasting seasons (dry and wet). Perhaps the Paleocene and Eocene limestones, containing dolomite and attapulgite (Durand 1984), are also associated with the calcrete layers. They have not been studied with this idea in mind, thus some sedimentologic details were omitted.

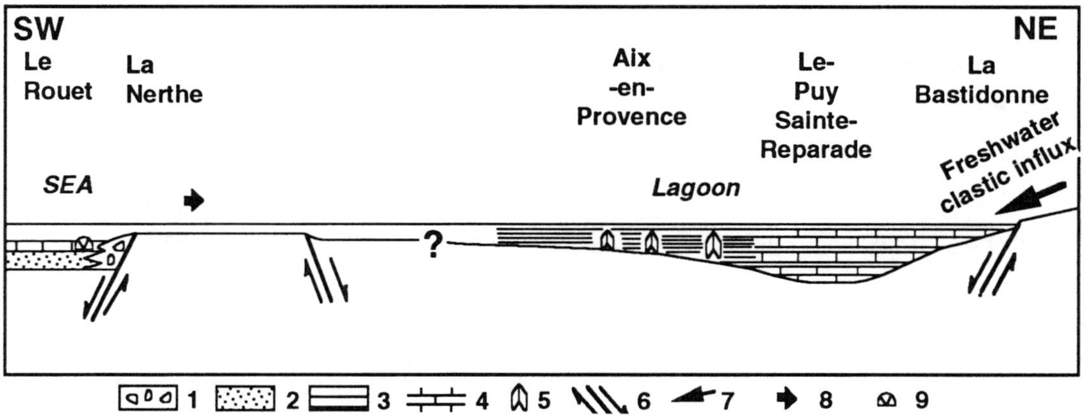

Fig. 6.45. Schematic drawing showing the proposed source-water influx into the Aix Basin from which the carbonates and laminated gypsum of Aix-en-Provence formed (see figure 6.43 for position of section). Key: 1 = breccias; 2 = detrital deposits; 3 = laminites; 4 = limestones; 5 = gypsum; 6 = normal faults; 7 = direction of continental water influx; 8 = direction of marine water influx; 9 = reefs.

SULFATES

Aix-en-Provence gypsums are included within the carbonates at the middle of the Formation d'Aix-en-Provence, especially within member 3:

Member 3 of the Formation d'Aix-en-Provence is composed of alternating beds of:

laminar and thin-bedded limestones with very fine grain size
massive to laminar marls
laminar or thin-bedded (rarely more than 2 m) saccharoidal or lenticular gypsum.

The color of these deposits is white, beige, cream, brown, gray, or green. In the lowest member, the fauna and flora are made up of mesohaline groupings within the rare fossiliferous limestone layers. In the limestones of the middle member we find coelenterates (*Discalioides*) associated with insects, fish, and plants. In the top of this member, there are clear fresh-water groupings. Gypsum is missing in the north of the basin.

Both carbonate and gypsum studies demonstrate that the sedimentary milieu of the Formation d'Aix-en-Provence shows numerous and significant alternations and variations in salinity, passing rapidly from fresh water to solutions from which sulfate precipitation took place over a vertical distance of a few centimeters. Chemical and paleobiological studies suggest a generally estuarine environment.

The sedimentation area of the Aix gypsum seems to have formed in an intermittent maritime marsh, alternating continuously with the open shoreline or isolated and progressively invaded by fresh fluviatile waters (figure 6.45). As in Marseille and St Pierre-les-Martigues, both marine and continental sulfates were apparently mixed. Continental sulfates can easily precipitate from concentrated fresh water, but marine sulfates can develop even more readily within phreatic waters in the same manner as in the modern Senegal mangal areas (Marius 1985). The evolution of the influx and concentration of the water apparently depended both on the climate fluctuations and the regional tectonics that controlled subsidence and the marine inflow coming from the shoreline located toward the Rouet area.

CALCRETE

In the northern area of the Aix Basin during the Oligocene, calcrete zones were formed within red fluviatile deposits (sandy to argillaceous). This was particularly marked during the middle portion of the late Chattian. Such deposits, as in the other basins, suggest alternately contrasting seasons (wet and dry).

Conclusions

In the southern portion of Provence the Paleogene is represented by:

1. continental fluviatile and lacustrine deposits that extend from the Paleocene until Lutetian (middle Eocene) and are present only in the Aix basin;
2. continental, fluviatile, lacustrine, and marine deposits deposited only during the Priabonian (Ludian) to latest Chattian in the other basins; only in Marseille, Aix, and St Pierre les Martigues are there evaporite deposits.

Between these two periods, Provençal tectonic events with marked overthrusting took place. Oligocene deposits that formed in consequence of distensive tectonic events, overlie all of the precedent formations or structures.

TECTONIC CONTROLS

Throughout the Paleogene we can see direct tectonic control of every depositional phenomenon that occurs.

1—Geodynamics.

At the end of the Cretaceous, southern Provence belonged to the eastern lacustrine and palustrine area of the North Pyrenean Sillon. An important phase of subsidence appeared in the Aix syncline and Alpille area created by the "Pyreneo-Provençal" compressive tectonics. Due to the thrust movements, sedimentation stopped in the Aix syncline but apparently goes on in the northern Alpille area until the Priabonian. After the Bartonian phase of thrusting the area of southern Provence remained quiescent. This area was not affected by the important and widespread phase of West European distention during the Priabonian-Bartonian age, except in the Aix basin (La Daubergue).

From the basal Rupelian (Stampian) and continuing until the end of the Oligocene, phases of distensive tectonics created moving grabens. The evidence of extensive tectonics decreases considerably during the Aquitanian (Miocene).

2—Sedimentology.

Tectonic activity is the direct cause of certain types of deposition in the entire region of southern Provence.

Carbonates. The laminar carbonates appear to signify fairly shallow depths, rapid climate variations, and a very confined milieu. The breccias and megabreccias signify normal faulting activity together with associated ground water activity and dedolomitization.

Siliceous detrital input. Mineralogic and sedimentologic studies show that a crystalline landmass which was the major source of the clastic input lay to the south of this area during the Paleogene. Compositionally this source area is the same as the Pyrenees and Sardinia and was sporadically rising and/or subsiding due to tectonic activity. Many fluvial components appear to have traveled distances up to 250 km to 350 km (based on rounding), coming from the east and southeast of Provence until the middle part of the late Chattian when sediment sources shifted to the north and west. The last phases of fluvio-detritus ends with the opening of the Paleo-Mediterranean (Nury 1977; Le Douaran et al. 1984; and Guieu and Arlhac 1986).

During the latest phase of the late Chattian, an Alpine source of detrital material appeared in Provence at the same time as the southern input ceased. This fact demonstrates separation or subsidence of the southern source areas and coincides with the formation of the western portion of the Mediterranean.

Sulfates. The sulfate deposits are primarily intercalated within carbonates in all of Bas Provence. In Aix there is a considerable amount of primary gypsum still present in the sedimentary section and it is rhythmically intercalated with argillites, suggesting that early diagenesis of most of these beds was fairly limited. Displacive gypsum rosettes (rose de sable), however, did form within many associated beds of carbonate mud, and because this is an early form of diagenesis, occurring during desiccation of previously flooded areas, it seems to signify that sporadic drying of the formative lakes and/or lagoons had taken place. In St Pierre much of the gypsum formed as laminar gypsum silt, with rare layers of primary crystalline gypsum, but clear periods of drying also occurred, as demonstrated by the well-preserved desiccation surfaces. In Marseille the gypsum is also laminar and well interbedded within carbonate sequences, but the deposits are not as informative as in Aix or St Pierre. In the Aix area the gypsum-bearing Paleogene deposits contain rare halite imprints or pseudomorphs. Zones of halite do not exist, nor is there any suggestion of their former presence in any of these evaporite basins (no collapse breccias).

The Oligocene sulfate deposits appear to be confined to sedimentation that immediately follows the first appearance of marine faunal and floral elements, especially related to the appearance of pelagic coelenterates (large medusa, up to 16 cm in diameter). In consequence it seems reasonable to suggest that some of the sulfates originate from concentrated marine waters. It has been proposed that most of the Provençal sulfates arise from continental sources, such as leached Triassic sulfates (Fontes et al. 1980). In their study of the stable isotopes of sulfur and oxygen some of the values obtained were indeed appropriate to recycled Triassic sulfates but this geochemical study does not explain many of the other stratigraphical and sedimentological observations made in the basins:

1. Why is gypsum absent in the Bandol and Ollioules basins which are actually nearest to the Triassic gypsum outcrops?
2. In both the St Pierre and the Aix basins there was no known Triassic outcrop. If the sources of Triassic sulfate entered the ground waters of these basins by moving along normal faults, why did this occur in a generally distensive region, under a very arid climate and yet result in no gypsum deposition (until after the first appearance of a marine influx)?
3. Why are there no Oligocene evaporites containing halite in southern Provence when considerable halite is present in the Triassic deposits of Jouques, which was exposed during this period of time (Goy and Busson 1980; Appia-Lamache 1984)? (This point may not be significant since the same arguments for the absence of halite may be made for marine-sourced evaporites.)

Because the first input of marine water (based on faunal evidence) and the first appearance of gypsum deposition took place at nearly the same time, it seems reasonable to assume that a partial marine origin existed for the sulfates of southern Provence, without excluding a partial input of Triassic sulfates which mixed with marine sulfates. In any event tectonic activity is very important to sulfate sedimentation as it provided access to both sulfate sources.

3—Sedimentation: marine versus non-marine.

The depositional environments in Provence are generally shallow lacustrine and palustrine mileux, interrupted by fluvial and marine influx. The marine inflow is restricted to the Oligocene and marine waters reach the interior basins only in association with the most active periods of distention. The presence of nearby marine water at the end of the Oligocene is quite certain because of the access provided by the opening of the western Mediterranean.

Rupelian (Stampian) marine inflow is quite certain because of the presence of marine pelagic fauna of Rupelian (Stampian) age. The sea itself was probably located beyond the southern crystal-

Fig. 6.46. Distribution zones indicated for the Oligocene detrital deposits of southern Provence. The sketches illustrate the changes in detrital transport through time. Key: 1 = normal faults and synsedimentary breccia; 2 = limestone deposits; 3 = deposits of fluvial exotic detritus; 4 = estimated limits of sedimentary areas; 5 = directions of localized sediment influx; 6 = directions of sediment influx from the south; 7 = Directions of influx from Alpine origins.

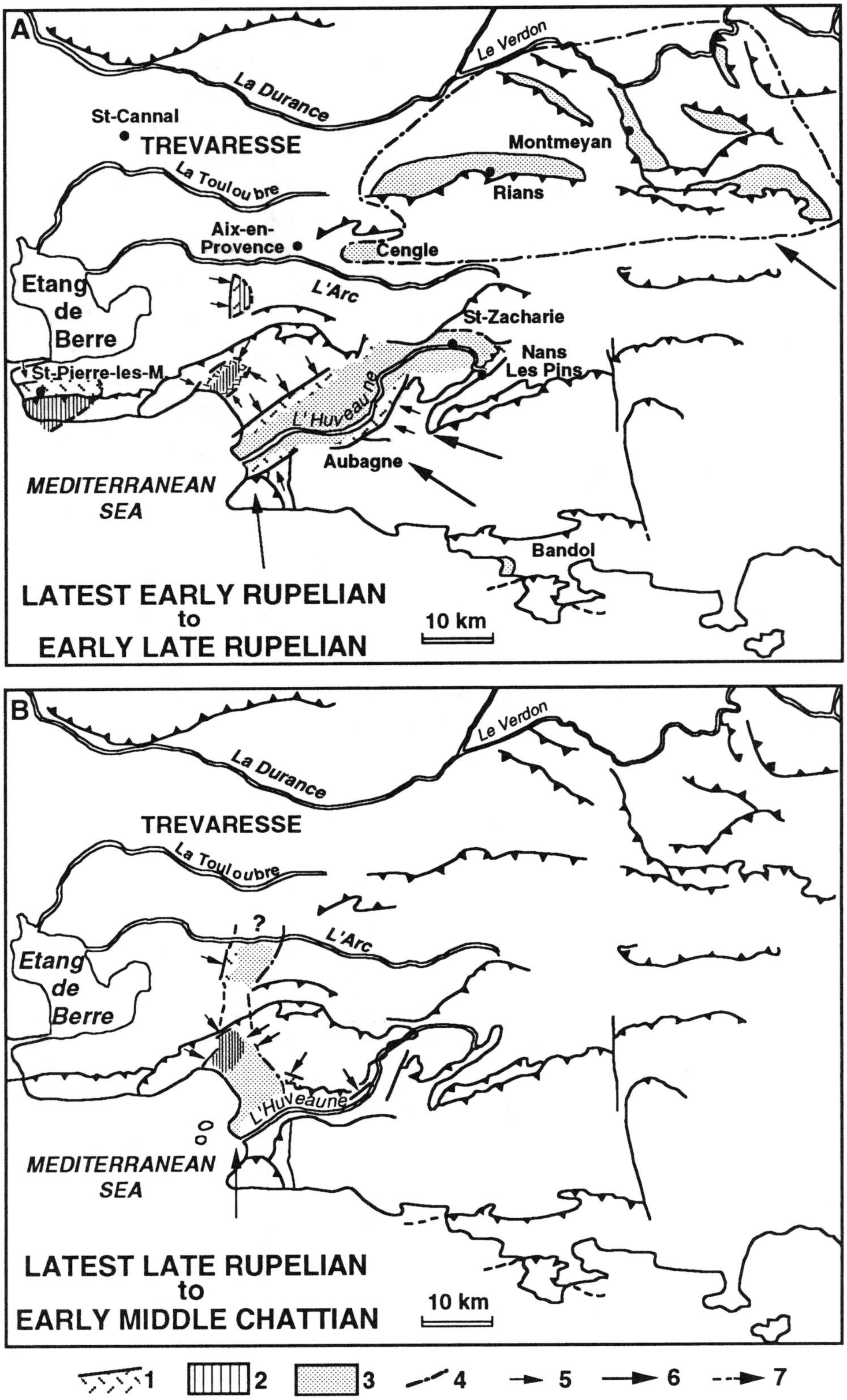
A
Le Verdon
La Durance
St-Cannal
TREVARESSE
Montmeyan
La Touloubre
Rians
Aix-en-Provence
Cengle
Etang de Berre
L'Arc
St-Zacharie
Nans Les Pins
St-Pierre-les-M.
L'Huveaune
Aubagne
MEDITERRANEAN SEA
Bandol
LATEST EARLY RUPELIAN to EARLY LATE RUPELIAN
10 km
B
Le Verdon
La Durance
TREVARESSE
La Touloubre
?
L'Arc
Etang de Berre
L'Huveaune
MEDITERRANEAN SEA
LATEST LATE RUPELIAN to EARLY MIDDLE CHATTIAN
10 km
1
2
3
4
5
6
7

FIG. 6.46

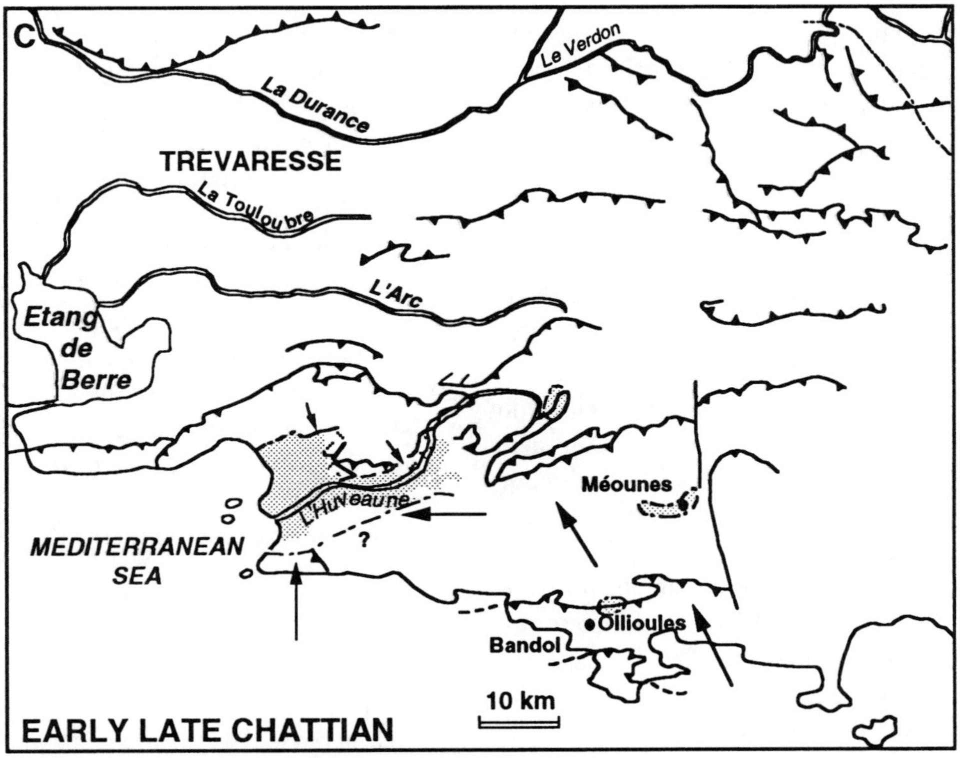
C
Le Verdon
La Durance
TREVARESSE
La Touloubre
L'Arc
Etang de Berre
L'Huveaune
?
Méounes
MEDITERRANEAN SEA
Ollioules
Bandol
10 km
EARLY LATE CHATTIAN

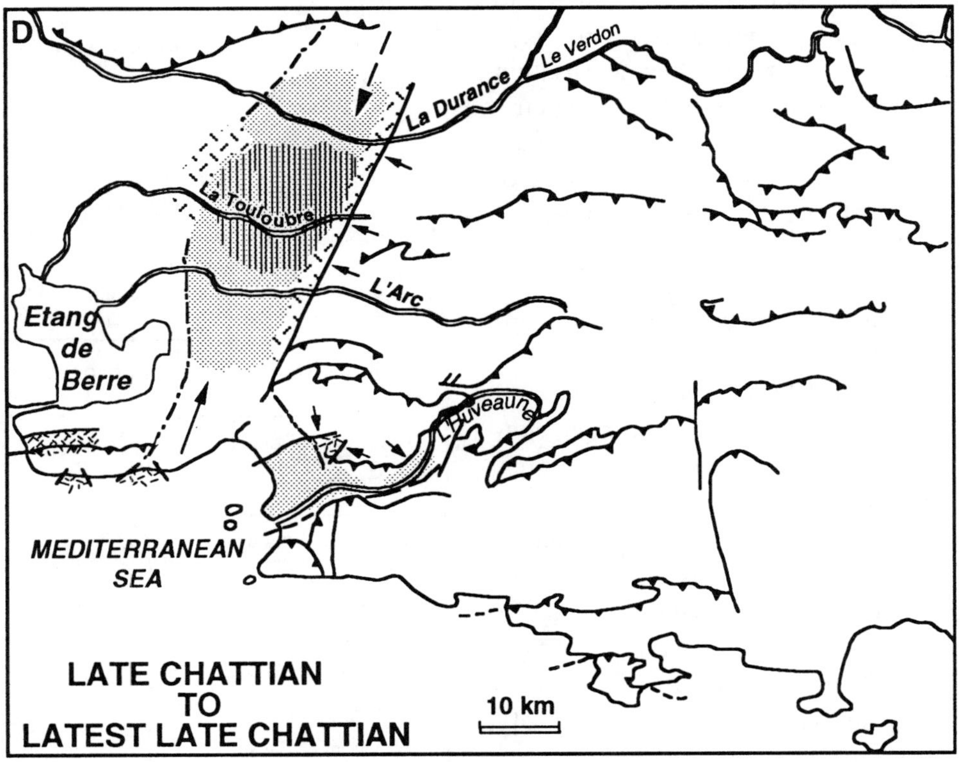
D
Le Verdon
La Durance
La Touloubre
L'Arc
Etang de Berre
L'Huveaune
MEDITERRANEAN SEA
LATE CHATTIAN TO LATEST LATE CHATTIAN
10 km

line land mass. The episodic communication with this distant ocean (the Ligurian Tethys) is the consequence of the marked phases of distensive tectonics.

4—Patterns of fluvial cyclicity and climatic changes through time.

During the whole Oligocene period the climate was dry subtropical with contrasting seasons, not unlike the modern Mediterranean. This is demonstrated by the presence of pedogenic pseudobreccias and calcretes. The dry seasons were quite arid and there were probably long-term dry cycles in the climatic pattern as demonstrated by (1) the rapid consolidation and lithification of sediments; (2) important periods of non-deposition without significant truncation of underlying beds; (3) the presence of evaporites intercalated among the limestones; and, (4) the development of red beds.

There is also evidence of more humid or wetter periods when fluvial sedimentation became important, without evaporites and with less carbonate formation. Small scale (seasonal) climatic variation is also evident in the development of fine millimetric lamination of the lake deposits. Locally it is possible to work out a rough calculation of the large scale variations. In the case of the Poudingues de Marseille (the characteristic conglomerate of the Marseille area), the exposures show a characteristic succession of gullying channel ways filled with conglomerates. The total time estimated for the formation of this formation is 500,000 years. An observed outcrop at Fort Saint Jean, which is 7 m high, contains 18 successive channel cut and fill structures, and because the whole formation is 700 m thick and has approximately the same sedimentary structure throughout, the average time needed for the sedimentation (cut and fill) of one channel is about 278 years. Probably the time period represented by non-deposition is significantly greater than for deposition. Additionally the phases of tectonic activity add to the effects of climatic change.

Summary

The entire Oligocene sedimentary series is very short in terms of geologic time. The fossil groupings show very little or no change. Based on paleontologic studies, the time interval represented by these deposits is probably less than 10^6 years. Moreover the successive sedimentary beds are discordant and the area of sedimentation migrates through time.

These facts show that the non-depositional periods are very important in the stratigraphic record, and they must have provided: the time necessary for the opening of new sedimentary areas (development of accommodation space), and the time necessary for development of the new surface morphology after earlier downcutting.

On the basis of these general observations, we propose an idealized environmental cycle for the evolving distensive environments that provides:

a subaerial period (exposure)
a phase in which there was development of large flooded areas (shallow lakes)
a significant phase of fresh water mixing with marine water inflow
renewed fresh water conditions without apparent marine influx.

TABLE 6.6.
Rates of sedimentation within the Oligocene basin of Marseille.
(Modified from Nury and Raynaud 1986)

FORMATION	RATE OF SEDIMENTATION in mm/1000 years
Poudingues de Marseille Formation (puddingstone)	1400
Les Arnavaux - Ste Marthe Formation St Henri - St André Formation	100
Formations de l'Estaque (bedded and laminar limestone) Formations de Camoins (limestone and laminar gypsum)	135
St Marcel Formation (clays and sands)	90
Calcaires à striatelles Formation (limestone)	40

The intensity of the distensive phase controlled the marine inflow rate into the grabens. Depending on thermal and hygrometric conditions, there were variations in the deposition of evaporites, carbonate beds, or red-beds (table 6.5).

While during the Eocene the structural low of the Aix basin remains in the same place, the Oligocene depocenters gradually migrated during each distensive phase. Migration progressed through time from the south-east toward the north-west (figure 6.46). These basins were developed and filled very rapidly (table 6.6). When one of the small rift valleys became inactive and filled, other valleys opened. The grabens generally maintain a general trend direction of NE-SW but the direction of normal faulting varies locally through time (030° to 100°). This is probably a consequence of two factors: the Ligurian rifting, and the impact of earlier

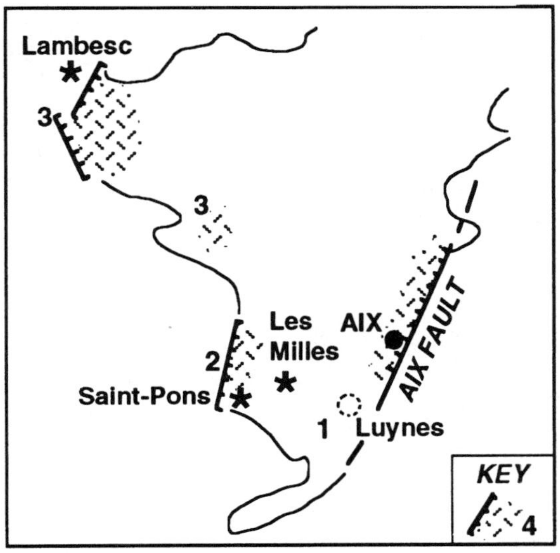

FIG. 6.47. Diagrammatic map representation of the evolving structural scheme for the development of the Aix Basin through the Oligocene. Key: 1 = late early Rupelian; 2 = early late Rupelian to early Chattian; 3 = late Rupelian; 4 = normal faults and synsedimentary breccias.

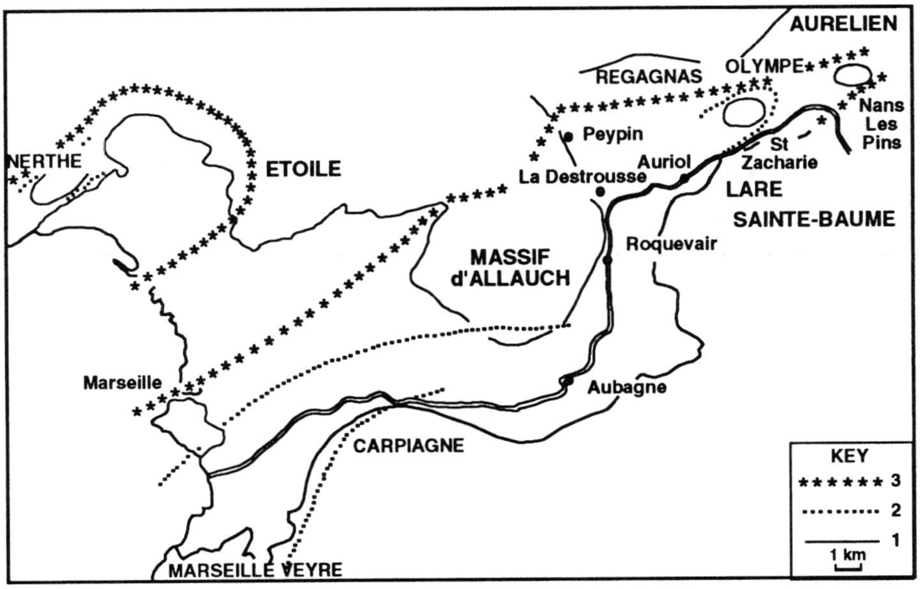

FIG. 6.48. Paleogeographic (map) representation of the opening of the Marseille Basin (north and south areas) from the early Rupelian (Stampian) through the middle Rupelian. Key: 1 = boundaries of basal early Rupelian; 2 = boundaries of latest early Rupelian; 3 = boundaries of middle Rupelian.

rifting upon the Mesozoic crust. The horsts with an NE-SW trend initially acted as barriers preventing siliceous input from the south. As a consequence, there was only carbonate sedimentation in those grabens to the northwest of each horst, at least until the southeastern graben catchments were filled to overflowing.

The grabens opened in a centrifugal manner, that is they opened and progressively widened because of the more and more eccentric normal faulting (figures 6.47 and 6.48). Thus marine water gained greater and greater ingress. The successive distensive phases caused the subsidence of earlier horsts and the uplift of former grabens. This resulted in extensive redistribution of sediment until the next phase of distention, when tectonic adjustment again caused resedimentation (Rouet; Hippolyte et al. 1990).

The most important conclusions of this study are:

1. The migration of the areas of sedimentation (depocenters) takes place because they are bound to distensive tectonics
2. The consistent siliceous detrital input comes from the southern crystalline land mass, and this continued until this source area (i.e. Sardinia and Corsica) became separated from the south of France due to the opening of the Mediterranean
3. The final phase of clastic sedimentation changes over to an Alpine source because of the emplacement of the Alpine chain
4. The continued opening of the Mediterranean during the Oligocene by successive rifting permitted the ingress of marine waters into the southern portion of Provence.

PART 3

Spain: South-Pyrenean Foredeeps and the Ebro Basin

At approximately the same time as eastern France and parts of western Germany were receiving substantial amounts of evaporative deposition in basinal sags, fault basins, and rift areas, the evolving foredeeps of Spain were under the same general climatic regime and also became evaporative. In Spain, within a very different tectonic framework, and with rapid structural evolution, thick sedimentary deposits were formed in marine, alluvial, paludial, lacustrine, and sabkha settings. Because the styles of basinal development were significantly different in the two areas (Spain and France), the sedimentation pattern, under a fairly similar climatic evolution (somewhat offset in time), is necessarily different. However, the areas had many depositional features in common so there are many similarities as well.

The stratigraphical terminology (both marine and nonmarine) used in this section is the international stratigraphic nomenclature, which is marine in origin. Because many of the sections under discussion are continental in origin, a local terminology exists that does not readily translate from one basin to another, or to the marine record. After considerable work Riba et al. (1986) and Cavelier and Pomerol (1986) established a suitable stratigraphic framework, using both international and conventional local terms, that is employed here (table P3.1).

TABLE P3.1.
Marine and continental Cenozoic stratigraphy of Spain.
(Adapted from Riba et al. 1986 and Cavelier and Pomerol 1986)

ERAT	SYSTEM	SUB-SYSTEM	SERIES		STAGE	OTHER UNITS		FACIES and OTHER UNITS
CENOZOIC	TERTIARY	NEOGENE	MIOCENE	UPPER	MESSINIAN or ANDALUSIAN	TUROLIAN continental		CATALONIAN; PONT. s.s.; PONTIAN s.l.; PANNONIAN; SARMATIAN?
					TORTONIAN	VALLESIAN continental		CATALONIAN; VINDOBONIAN; PONTIAN s.l.; PANNONIAN; SARMATIAN?
				MIDDLE	SERRAVALLIAN	ASTARCIAN continental		ARAGONIAN; HELVETIAN; VINDOBONIAN
					LANGHIAN	ORLIANIAN continental		ARAGONIAN; HELVETIAN; VINDOBONIAN
				LOWER	BURDIGALIAN	ORLIANIAN continental		ARAGONIAN
					AQUITANIAN	AGENIAN continental		
		PALEOGENE	OLIGOCENE	UP	CHATTIAN	STAMPIAN s.l.		
					? ?—? —			
				LOWER	RUPELIAN (STAMPIAN) s.s.	STAMPIAN s.l.	LATTORF-IAN s.s.; LATTOR. s.l.	SANNOISIAN
			EOCENE	UPPER	PRIABONIAN		TONGRIAN; LATTOR. s.l.	LUDIAN
					BARTONIAN	BARTON. s.l.	AUVERSIAN = MARINESIAN	
				MIDDLE	? –? – ?– –			
					(BIARRITZIAN)	BARTON. s.l. / LUTET. s.l.		
					—?—?—?—			
					LUTETIAN s.s.	LUTET. s.l.	LEDIAN	
				LOWER	(CUISIAN)	YPRESIAN		
					YPRESIAN	YPRESIAN	SPARNACIAN	
					(ILERDIAN)	YPRESIAN	SPARNACIAN	
			PALEO-CENE	UP	THANETIAN	LANDENIAN		GARUMNIAN
				LOWER	DANIAN	MONTIAN		GARUMNIAN
MESOZOIC	CRETACEOUS		UPPER		MAASTRICHTIAN			GARUMNIAN

7

The Ebro Basin in the Structural Framework of the Iberian Plate

Pere Santanach

The latest Cretaceous through Tertiary strata of the Ebro Basin were deposited within a basinal area that formed, developed, and changed tectonically during their sedimentation. This paper outlines the evolution of the structural framework within which this sedimentation took place. The Iberian plate is bounded by two discrete Alpine orogens, the Pyrenees to the north and the Betics to the south. In general, during the Alpine orogenic cycle this plate did not behave homogeneously throughout these two periods of deformation. Some parts were slightly or not at all deformed, as the Ebro Basin, whereas other areas underwent noticeable deformation. These deformed areas now make up mountain chains such as the Iberian chain and the Catalan coastal range, that cross the Iberian plate (figure 7.1)

The Ebro Basin, located on the northeastern portion of the Iberian plate, was formed during the Pyrenean orogeny as the southern foreland basin in front of the Pyrenees (Zoetemeijer et al. 1990). It is triangularly shaped and is bounded to the north by the Pyrenees, to the southwest by the Iberian chain, and to the southeast by the Catalan coastal range. The basin is asymmetric with a very large amount of tectonism on its northern border and comparatively little on its southern border.

The main areas of sedimentary accumulation in the Ebro Basin are located in the north, along the Pyrenees, where the pre-Tertiary substratum lies at depths ranging from 3,000 to 5,000 m below the surface (Riba et al. 1987). The overall structure of the basin, as seen at the surface, is relatively simple. Tertiary beds are nearly horizontal in most areas of the basin and are deformed significantly only along the margins, especially along the northern margin. The Paleogene basin filling was affected by the Pyrenean thrust and fold belt and is partially covered by Paleogene thrust sheets (Séguret 1972; Puigdefàbregas et al. 1986).

The Iberian chain stretches along the southwestern margin of the basin and consists mainly of an array of north-west trending folds and thrusts. The Tertiary beds of this margin are commonly overthrust by Mesozoic rocks. However, some structures of the chain are unconformably overlapped by upper Tertiary rocks.

Along the southeastern margin of the basin, the Catalan coastal range is characterized by an array of en echelon, northeast-striking faults. In the area where the Iberian chain and the Catalan coastal range join, there is an array of north-vergent, east-trending folds and thrusts. This area, called the Linking Zone (Guimerà 1984), differs from the adjacent areas of the Iberian chain and the Catalan coastal range in having a thicker Mesozoic cover that is cut by low-angle thrusts of large displacement.

In spite of the evident structural differences between the Pyrenees, the Iberian Chain, and the Catalan Coastal Range, several common evolutionary features can be outlined. The fault pattern that con-

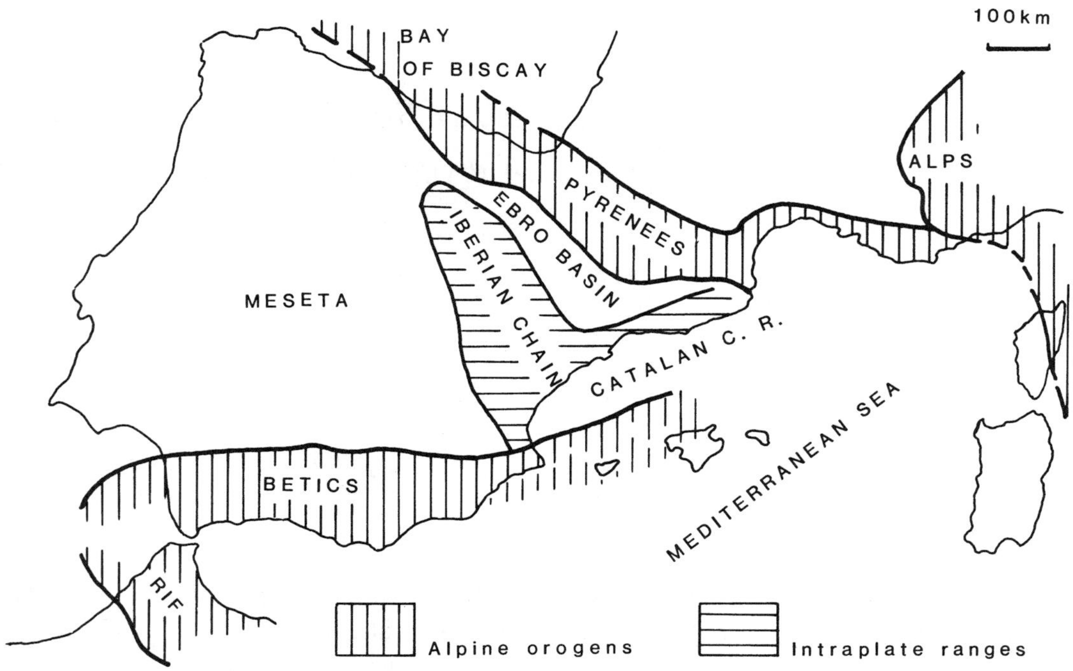

FIG. 7.1. General location of the Ebro Basin among three mountain ranges that make up its borders, given in context of the western Mediterranean structural trends.

trols the Paleogene compressional structure in northeastern Iberia and, hence, the sedimentary infilling of the Ebro Basin, is inherited and probably developed during the late Paleozoic (Arthaud and Matte 1975). These faults were active as extensional normal faults during the Mesozoic, as can be seen in the Mesozoic cover rocks. Because of this fault activity in the Mesozoic, mainly during the Cretaceous, the domains of the Pyrenean, Iberian Chain, and Catalan Coastal Range underwent differental subsidence. However, all subsided to a greater degree than did the Ebro Basin domain (Anadón et al. 1979; Álvaro et al. 1979).

Near the start of the Paleogene, the Ebro Basin began to develop and subside, surrounded by the rising Pyrenees, Iberian Chain, and Catalan Coastal Range (Puigdefàbregas and Souquet 1986). At the Paleogene-Neogene boundary the most easterly part of the Iberian plate developed a structural pattern of sunken and tilted blocks. Grabens located in the eastern Pyrenees, in the Iberian Chain, and along the Mediterranean coast (onshore and offshore) of Iberia belong to this last stage of structural evolution (Fontboté 1954; Vegas et al. 1980).

Development of the Ebro Basin

The development of the Ebro Basin began during the Paleocene as the result of collision between the

FIG. 7.2. Sequential geological maps of the Ebro Basin. A. Age of the basal Tertiary infilling. B. Age of the outcroping Tertiary sediments. C. Subcrop map. (After Riba et al. 1983/87, simplified.)

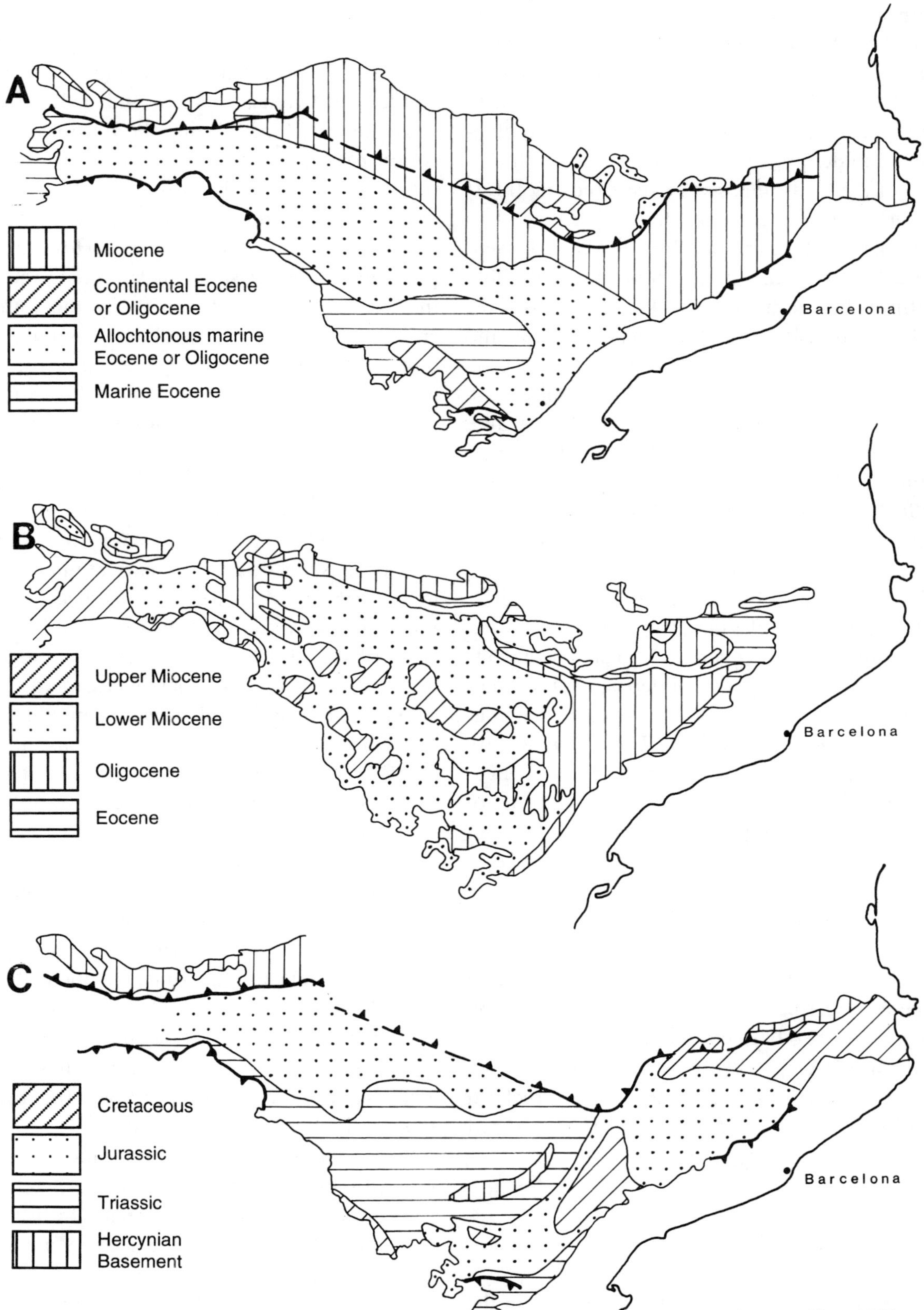
A
Miocene
Continental Eocene or Oligocene
Allochtonous marine Eocene or Oligocene
Marine Eocene
Barcelona
B
Upper Miocene
Lower Miocene
Oligocene
Eocene
Barcelona
C
Cretaceous
Jurassic
Triassic
Hercynian Basement
Barcelona
0 50 100km

FIG. 7.2

European and Iberian plates (Zoetemeijer et al. 1990). The collision occurred first at the eastern end of the Pyrenees causing a tectonic inversion characterized by both north and south vergent thrusts. The Pyrenean foreland basins (Ebro and Aquitanian basins) developed to the north and south as a result of the loads imposed by these moving thrust sheets (Puigdefàbregas and Souquet 1986).

During the Tertiary, the Ebro Basin underwent an areal restriction. The development of the fold and thrust belt in the southern Pyrenees caused the southward translation of the northern border and the axis of the basin by several tens of kilometers from the Paleocene until the Miocene (Riba et al. 1983; Puigdefàbregas and Souquet 1986). Because of the differing structural evolution of the Iberian Chain and the Catalan Coastal Range (Anadón et al. 1985), basinward migration of the southwest and southeast margins were of lesser importance than the shift of the northern one.

The fill of the Ebro Basin consists of a thick stack of Tertiary rocks that range in age from Paleocene to Miocene (figure 7.2). These sediments were laid down in a variety of depositional settings ranging from deep and shallow-marine environments to nonmarine settings. These sedimentary rocks have an asymmetric map pattern, with the Paleogene rocks cropping out mainly in the northern and eastern areas, and the Miocene rocks mainly in the western and central areas (Riba et al. 1987).

Clastic material was provided by the growing border chains, mainly the Pyrenees. The characteristics of the developing sedimentary sequences were controlled principally by the structural evolution of these bordering mountains. However, sea level changes have also been considered a governing factor in the development of some sedimentary sequences (Puigdefàbregas et al. 1986). The structural evolution of the basin-margin ranges is outlined briefly in the succeeding sections.

The Pyrenees

The Pyrenean orogen extends from the Alps in the east to the northwest corner of the Iberian Peninsula in the west. It is about 1,500 km long with an average width of 200 km. Its central part, between the Mediterranean sea and the Bay of Biscay, constitutes the Pyrenean Chain, and in it are found structures developed during the European-Iberian continental collision. At shallow levels, the structure of the Pyrenees displays a fan-like disposition with east-west trending folds and thrusts having opposed polarity about a central axis. However, interpretation of deep seismic reflection data show that the Iberian crust is overthrust by the European crust (figures 7.3 and 7.4). The cross-section based on these data shows a south-directed thrust system mainly involving the upper crust. This thrust system is characterized by a central antiformal stack bounded by imbricate thrust systems (Losantos et. al. 1988; Choukroune et al. 1989; Roure et al. 1989).

To the east, compressive Pyrenean structures are buried by sediments filling basins formed during a period of Neogene extensional tectonics, particularly along the Mediterranean coast in Catalonia, Languedoc (France), and the Gulf of Lion (Vegas et al. 1980). To the west, the Pyrenean orogen lies between the Iberian continental crust and the oceanic crust of the Bay of Biscay (Boillot 1980/81). The emerged area of the orogen constitutes the Basc-Cantabric Pyrenees, which are separated from the Pyrenean Chain (sensu strictu) by the Pamplona fault.

From north to south the main structural units in the central Pyrenean Chain are: (i) the Aquitanian molasse foreland; (ii) the north Pyrenean thrust sheets, which form the north Pyrenean zone; (iii) the north Pyrenean fault zone; (iv) the axial zone, which forms the highest part of the chain and consists mainly of Hercynian basement thrust sheets; (v) the south Pyrenean zone consisting of cover rocks that are thrust southward over the foreland; and (vi) the Ebro Basin foreland.

Thus, east of the Pamplona fault, a northern area with north-directed thrusts (units i and ii) and a southern area with south-directed thrusts (units iv, v, and vi) can be distinguished (Muñoz et al. 1983). The structural history of the southern part of the Pyrenees controlled the morphological and sedimentological evolution of the Ebro Basin.

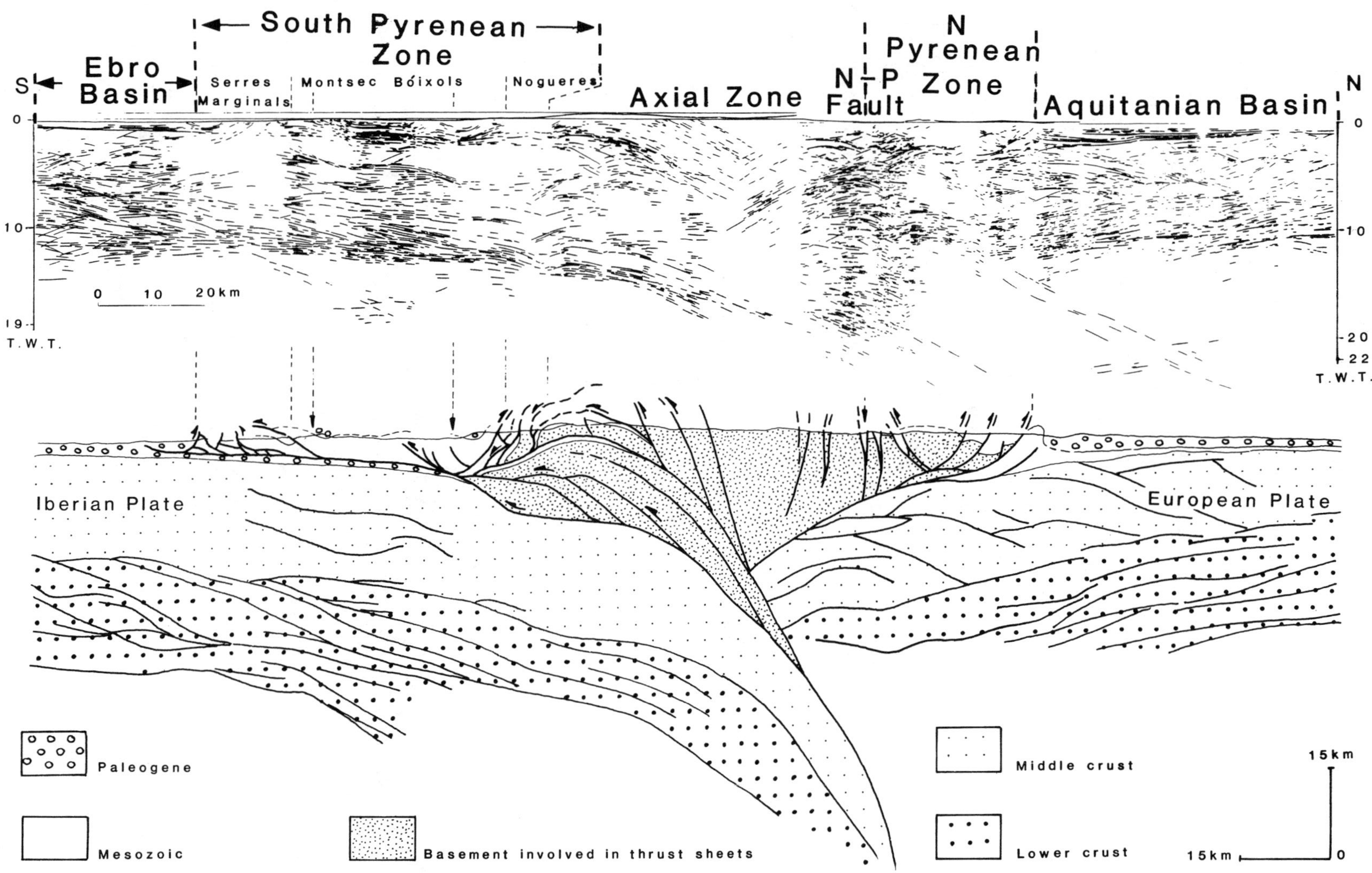

FIG. 7.3. General cross-section of the Pyrenees (after Losantos et al. 1988). At the top a line drawing is given, made from the ECORS Pyrenees profile, extracted from unmigrated seismic data. (See figure 7.4 for location of the profile.)

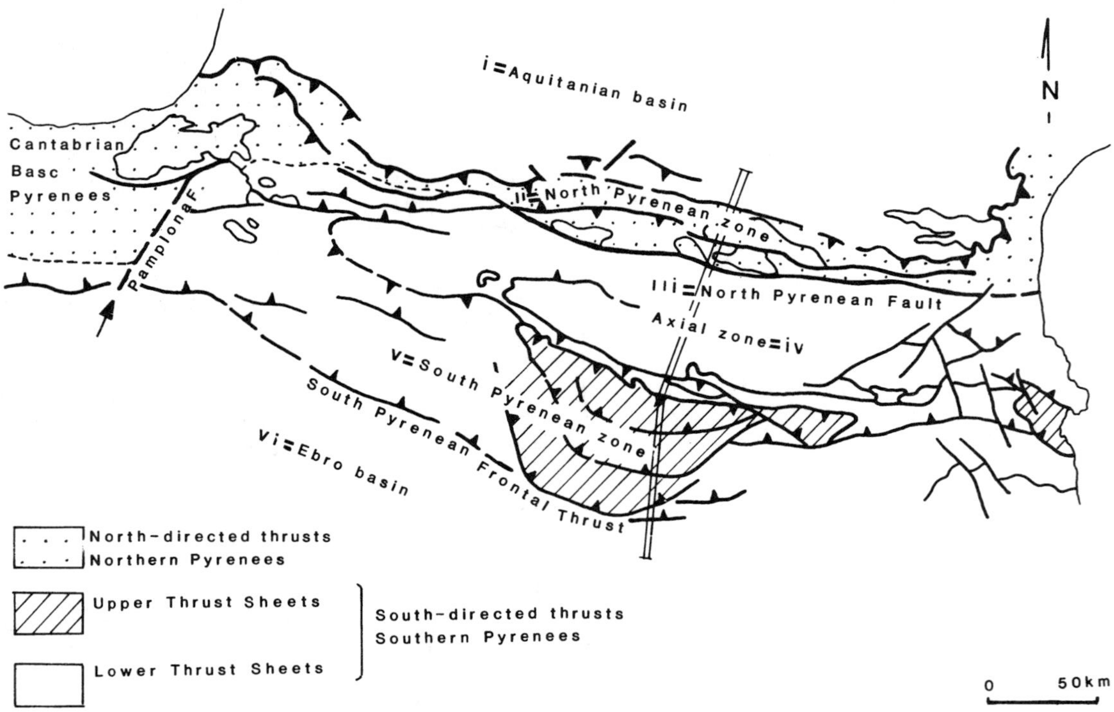

Fig. 7.4. Schematic structural sketch of the Pyrenees. The main structural units and the location of the ECORS profile (figure 7.3) are indicated. (After Muñoz et al. 1987, slightly modified).

The South Pyrenean Structural Units

The boundary between the South Pyrenean units and the Ebro Basin is the South Pyrenean frontal thrust (figures 7.4 and 7.5). In the central and western Pyrenees, this thrust, which is in fact an imbricate structure, outcrops only in Sierra de Alaiz (south of Pamplona), Riglos and Sierra de Guara (north of Huesca), and in the Serres Marginals. Elsewhere it is covered by recent sedimentary rocks or is a blind thrust. The Barbastro anticline and its eastern extensions represent the tip line of the most external thrust of this imbricate (Williams and Fischer 1984; Martínez Peña and Pocoví 1988). In the eastern Pyrenees the Vallfogona thrust, whose characteristics are well studied because of hydrocarbon exploration, constitutes the continuation of the frontal thrust as far as the Mediterranean (Clavell et al. 1988). Between the South Pyrenean frontal thrust and the north Pyrenean fault, different thrust sheets can be distinguished. The sole thrust of the South Pyrenean thrust system ranges from depths of 4 to 5 km before surfacing at the frontal thrust.

The South Pyrenean thrust sheets have been grouped into two sets (figures 7.4, 7.6, and 7.7): the upper thrust sheets and the lower thrust sheets. The upper thrust sheets are formed by cover rocks, mainly Mesozoic in age, whereas the lower thrust sheets involve Hercynian basement plus cover rocks (Muñoz et al. 1983).

UPPER THRUST SHEETS

The Upper Thrust Sheets extend over a large area in the central Pyrenees where the major South Pyrenean central unit has been defined. The Cotiella and Pedraforca are subunits located respectively on its northwestern and northeastern terminations. The main body of the South Pyrenean central unit has been divided into three subunits. From north to south and from structurally highest to lowest these

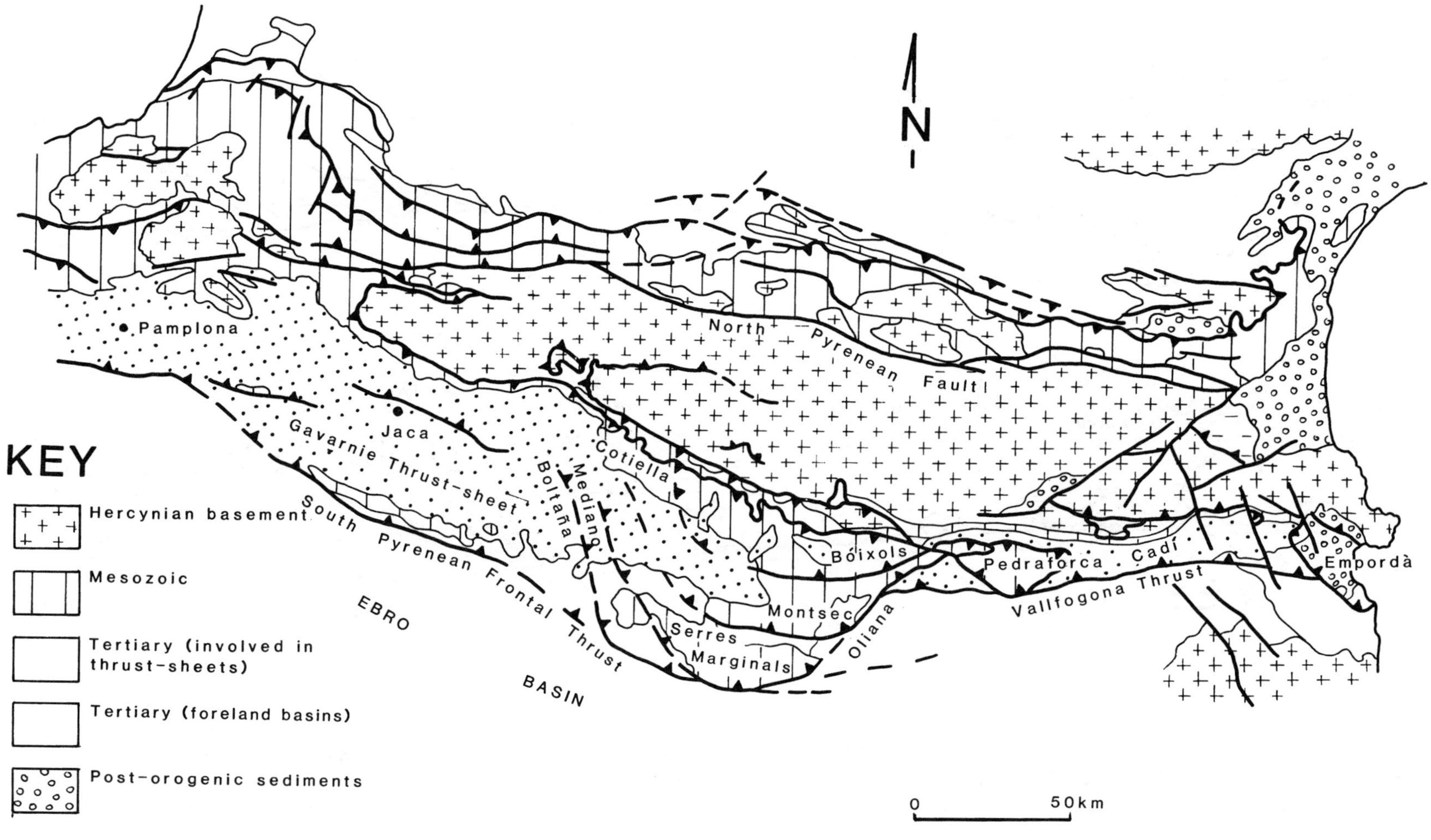

Fig. 7.5. Schematic geological map of the Pyrenees. (After Puigdefàbregas and Souquet 1986, slightly modified.)

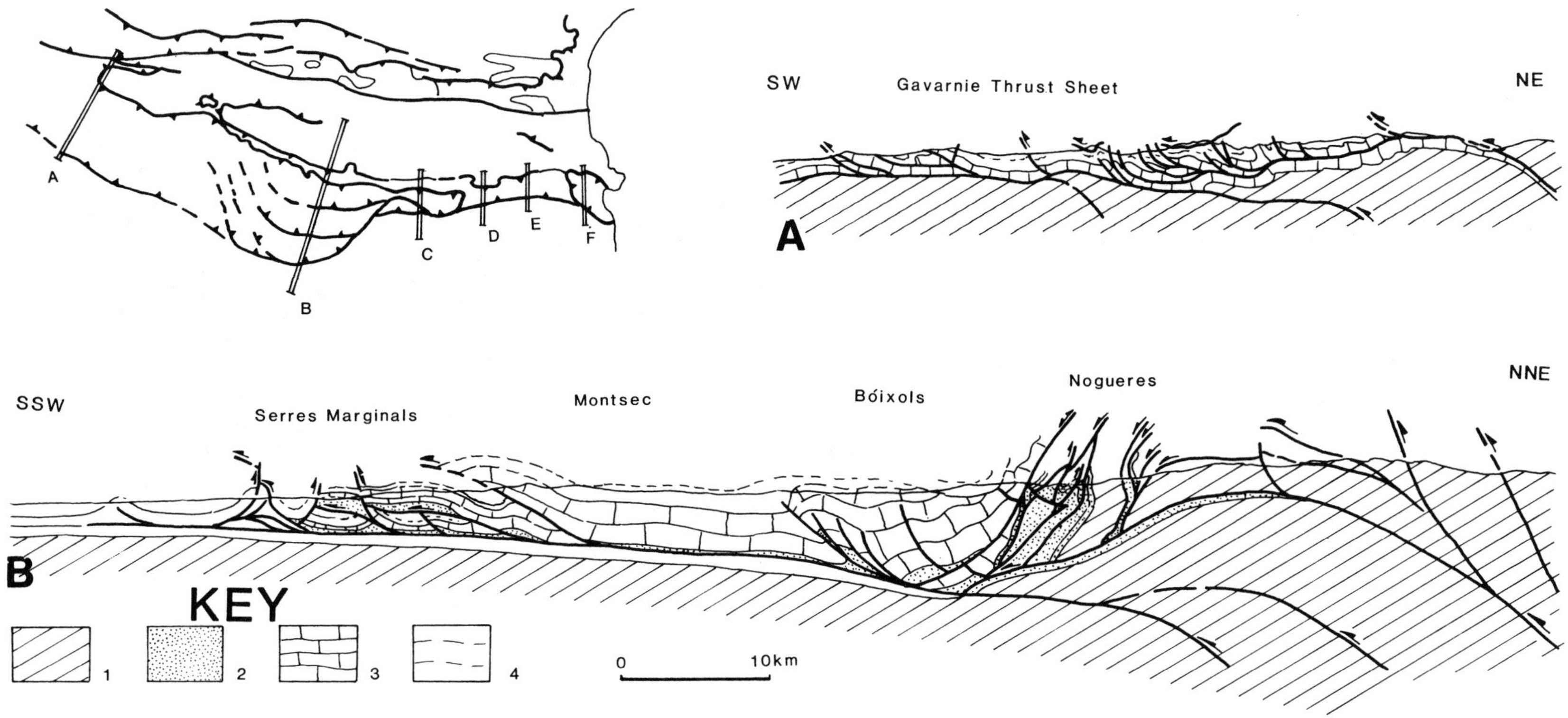

Fig. 7.6. North-south cross-sections of the (A) western, and (B) central southern Pyrenees. Location map given, upper left, showing the positions of the cross-sections 7.6 A and B and 7.7 C–F. Key for A and B: 1 = Hercynian basement, 2 = Triassic (present only in section B), 3 = Mesozoic (in section A includes lower Paleocene limestones), 4 = Tertiary (Cross-section A after Labaume and Séguret 1985; cross-section B after Muñoz, unpublished. Both simplified.)

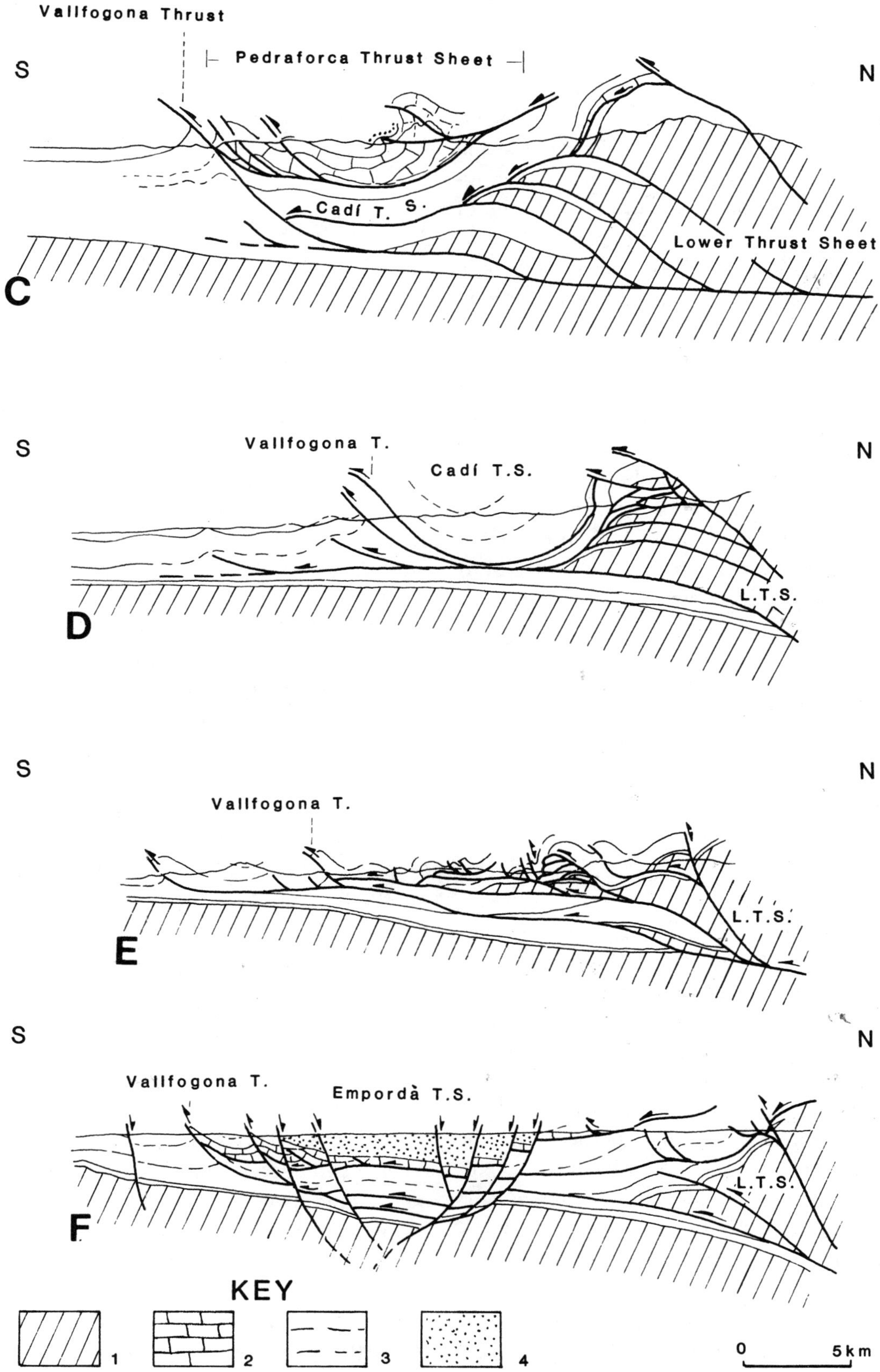

FIG. 7.7. North-south cross-sections of the eastern Pyrenees. Location of the sections in figure 7. 6. Key for C–F: 1 = Hercynian basement, 2 = Mesozoic, 3 = Paleogene, 4 = Post-orogenic Tertiary (Neogene). (Cross-section C and D after Muñoz, 1985; E, after Clavell et al. 1989; F, after Pujadas et al. 1989).

units are the Bóixols, Montsec, and Serres Marginals thrust sheets (Séguret 1972; Garrido-Mejías 1972; Cámara and Klimowitz 1986; Vergés and Martínez 1988). In the easternmost part of the Pyrenees, in the Empordà area, several outcrops of Mesozoic and Paleogene rocks belong to the Empordà thrust sheets. These sheets have been disrupted by Neogene extensional tectonics and today are largely covered by Miocene and Plio-Quaternary sediments (Solé Sabarís 1933; Pujadas et al. 1989).

The internal structure of the Upper Thrust Sheets is characterized by imbricates and associated hanging wall folds. Structures oblique to the east-west general trend are also present and are especially well represented in the Pedraforca thrust sheets and the easternmost part of the Serres Marginals. Some of these structures (Boltaña, Mediano, and Oliana anticlines) formed in the footwall of thrusts synchronously with sedimentation in the South Pyrenean foreland basin. On the basis of these oblique structures, which represent lateral ramps of the imbricate system, a roughly north to south transport direction can be deduced for the South Pyrenean Central Thrust Sheets (Muñoz et al. 1983).

LOWER THRUST SHEETS

These sheets consist of Hercynian basement and cover rocks. The cover succession is characterized by an incomplete, reduced Mesozoic series overlain by Paleogene foreland basin sequences, which were deposited during the southward translation of the Upper Thrust Sheets. The highest sheets are the Cadí Nappe, the Gavarnie Nappe, and the Nogueres Thrust Sheets. The lowermost ones are made up of Hercynian basement rocks overlain locally by pre-Paleogene, post-Hercynian rocks. These sheets constitute most of the region called the Axial Zone (Séguret 1972; Labaume and Séguret 1985; Muñoz et al. 1987; Martínez et al. 1989).

The internal structure of the lower thrust sheets consists of a variety of thrust geometries: imbricates in the most external zones, and duplexes and antiformal stacks of different sizes in the most internal ones.

Displacement and Translational Chronology: South Pyrenean Thrust Sheets

The main South Pyrenean Thrust Sheets have undergone southward translation of at least 10 to 25 km. For the uppermost sheets, such as the Pedraforca, movement of more than 50 km has been suggested. The total shortening for the Southern Pyrenees exceeds 100 km (Séguret 1972; Muñoz et al. 1986; Vergés and Martínez 1988; Cámara and Klimowitz 1986; Williams and Fischer 1984).

The Southern Pyrenean Thrust Sheets were emplaced in a piggy-back sequence between the latest Cretaceous and the Miocene. During this time, younger thrusts developed in the footwall of older thrusts with older thrusts carried on the lower, younger ones. In such an array, the highest units represent the earliest displacements and the lower units represent later displacements. The thrusts propagated toward the foreland with footwall rocks progressively incorporated into the moving Thrust Sheets. During the displacement of the South Pyrenean Thrust Sheets the Ebro Basin developed and its northern edge was progressively incorporated into the Lower Thrust Sheets. The foreland basin depositional sequences have been defined in relation to the thrust movements. Today the emplacement age of most of the Thrust Sheets is well known (Puigdefàbregas et al. 1986).

The Upper Thrust Sheets were emplaced from the latest Cretaceous until the early Lutetian (Bóixols, during latest Cretaceous—early Paleocene; Montsec, Cotiella, and Pedraforca, during latest lower Eocene to early middle Eocene). The Cadí and Gavarnie nappes, as well as the Nogueres Thrust Sheets, and the antiformal stacks that constitute the Axial Zone were emplaced during middle and late Eocene. This emplacement was followed in the Bartonian by the development of several out-of-sequence thrusts. Finally, during the Oligocene and the early Miocene, the Southern Pyrenean Frontal Thrust emerged.

The Cantabrian-Basc Pyrenees

The geodynamic setting of the Cantabrian-Basc Pyrenees differs from that of the Pyrenean Chain proper. Whereas the Pyrenean Chain formed by the continental subduction of the Iberian plate under the continental European plate, the Cantabrian-Basc Pyrenees are related to subduction of the oceanic crust of the Bay of Biscay under the continental Iberian plate (Boillot 1980–81). The transition between these geodynamic settings takes place on several NE-SW trending faults, the Pamplona fault being the most important.

West of the Pamplona fault, a complex array of northeast facing folds and thrusts is developed. The only exception is the Sierra de Cantabria—Montes Obarenes thrust, which shows a significant displacement to the south. This southernmost thrust slice of the Cantabrian-Basc Pyrenees overthrusts continental Tertiary rocks (Oligocene-Miocene) of the Ebro Basin. The geometry of the Sierra de Cantabria thrust is well known from seismic profiles and well data obtained during hydrocarbon exploration (Lamare 1936; Martínez-Torres et al. 1984).

The Iberian Range and the Catalan Coastal Range

The Iberian Range and the Catalan Coastal Range constitute a deformed, Y-shaped, intraplate zone that extends from the Pyrenean orogen to the Betics in southeastern part of the Iberian peninsula (figure 7.8). To the west is the Spanish Meseta and to the east the Mediterranean Sea.

Within this zone eight smaller units, with particular characteristics, can be recognized:

1. the Catalan Coastal Range that stretches along the Mediterranean coast from the Empordà area to the linking zone;
2. the Linking Zone with the Iberian Range;
3. the Aragonian Branch and
4. the Castilian Branch of the Iberian Range, separated by
5. the Tertiary Almazán Basin;
6. the La Demanda—Los Cameros Unit and
7. the Sierra de Altomira, separated from the Castilian Branch by
8. the Tertiary Loranca Basin (Guimerà and Álvaro 1990).

Those units in which structural evolution actually influenced the sedimentation at the southern border of the Ebro Basin are, from east to west, the Catalan Coastal Range, the Linking Zone, the Aragonian Branch, and the Los Cameros—La Demanda Unit. The Castilian branch of the Iberian Range (4) and the Tertiary Almazán basin (5) are not considered here because they are related to the Duero and Tajo basins and are not directly part of the Ebro Basin.

THE CATALAN COASTAL RANGE

The Catalan Coastal Range is composed of Hercynian basement overlain by a sedimentary cover of Triassic to Cretaceous rocks. The upper part of the cover (mainly Jurassic and Cretaceous) is detached from the basement along a décollement within Keuper (uppermost Triassic) lutites and evaporites, while the lower part of the cover remains attached to the basement.

The main structures of the Catalan Coastal Range are nearly vertical basement-involved faults that constitute a right-stepping, en echelon array (figure 7.9, cross-section A). The strike of these faults changes from east-northeast to northeast along the range and is slightly oblique to the trend of the range. During the Paleogene, these faults were active as convergent sinistral strike-slip faults (several kilometers of strike-slip displacement have been estimated), uplifting the southeastern blocks. The vertical uplift is locally greater than 2 km. As a consequence, cover flexures and sporadic folds formed. Some basement slices have overthrust the

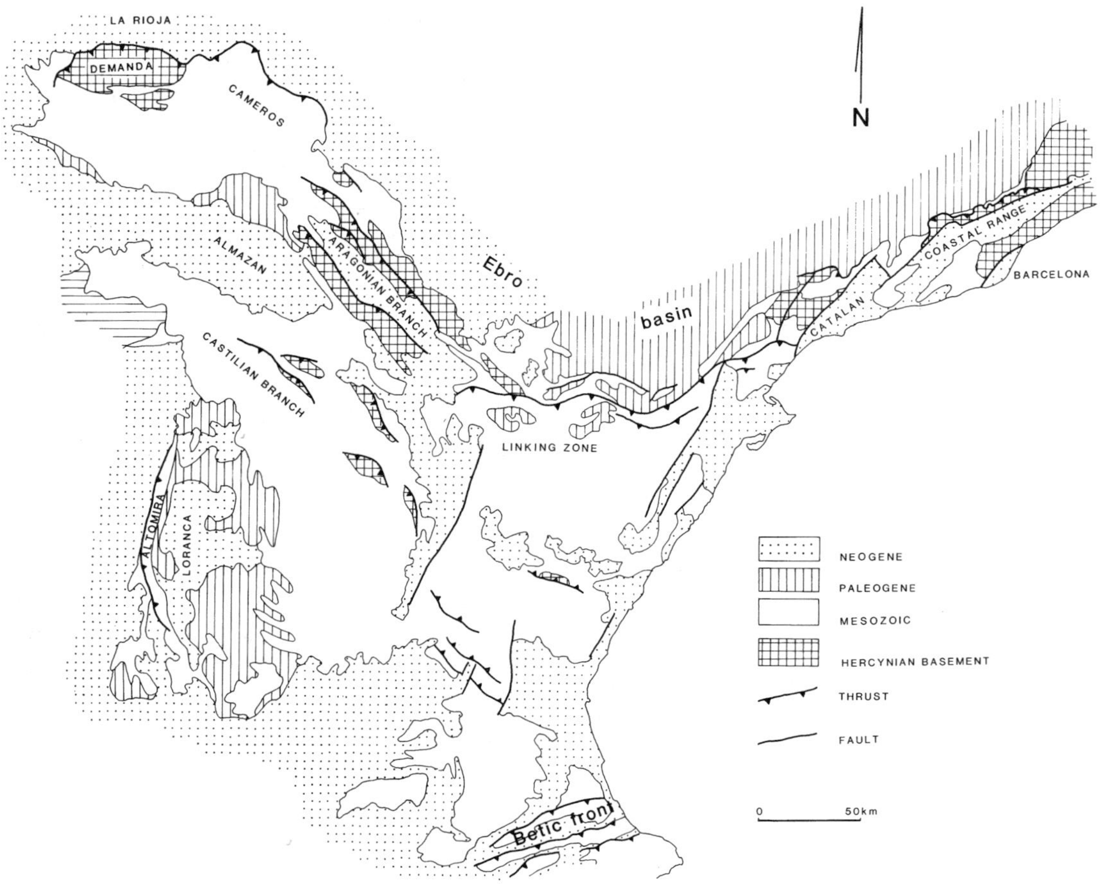

FIG. 7.8. Outline of the structural scheme of the Iberian Range and the Catalan Coastal Range. (After Guimerà and Álvaro 1990, simplified.)

Mesozoic cover and even, in some areas, the Paleogene deposits of the Ebro Basin (Anadón et al. 1985).

The timing of fault activity along the Catalan Coastal Range has been established by studying the evolution of alluvial fan systems that were sourced from uplifted blocks defining the Ebro Basin margin, and by the occurrence of progressive, angular, syntectonic unconformities that developed during deposition of the coarsest sedimentary sequences. Fault movement was diachronous, beginning in the early Eocene in the northern part of the range, but not until late Eocene in the south (figure 7.10). Therefore, because of the en echelon fault array, deformation also migrated basinward toward the northwest. The youngest deformation appears to be of late Oligocene age. In this way, the southeastern margin of the Ebro Basin evolved during Paleogene times, shifting slightly basinward and progressively acquiring its current form and location (Anadón et al. 1985). At the beginning of the Neogene most of these faults were reactivated as normal faults (Fontboté 1954; Anadón et al. 1979–1982).

THE ARAGONIAN BRANCH

In the Aragonian Branch of the Iberian Range, basement-involved faults are also present. They strike to the northwest, and behaved as convergent dextral strike-slip faults during Paleogene compression (Álvaro et al. 1979; Guimerá 1984; Colomer and Santanach 1988). The associated structures are comparable to those of the Catalan Coastal Range, apart from the general en echelon array. Locally, the reverse faults bounding the Paleogene Ebro

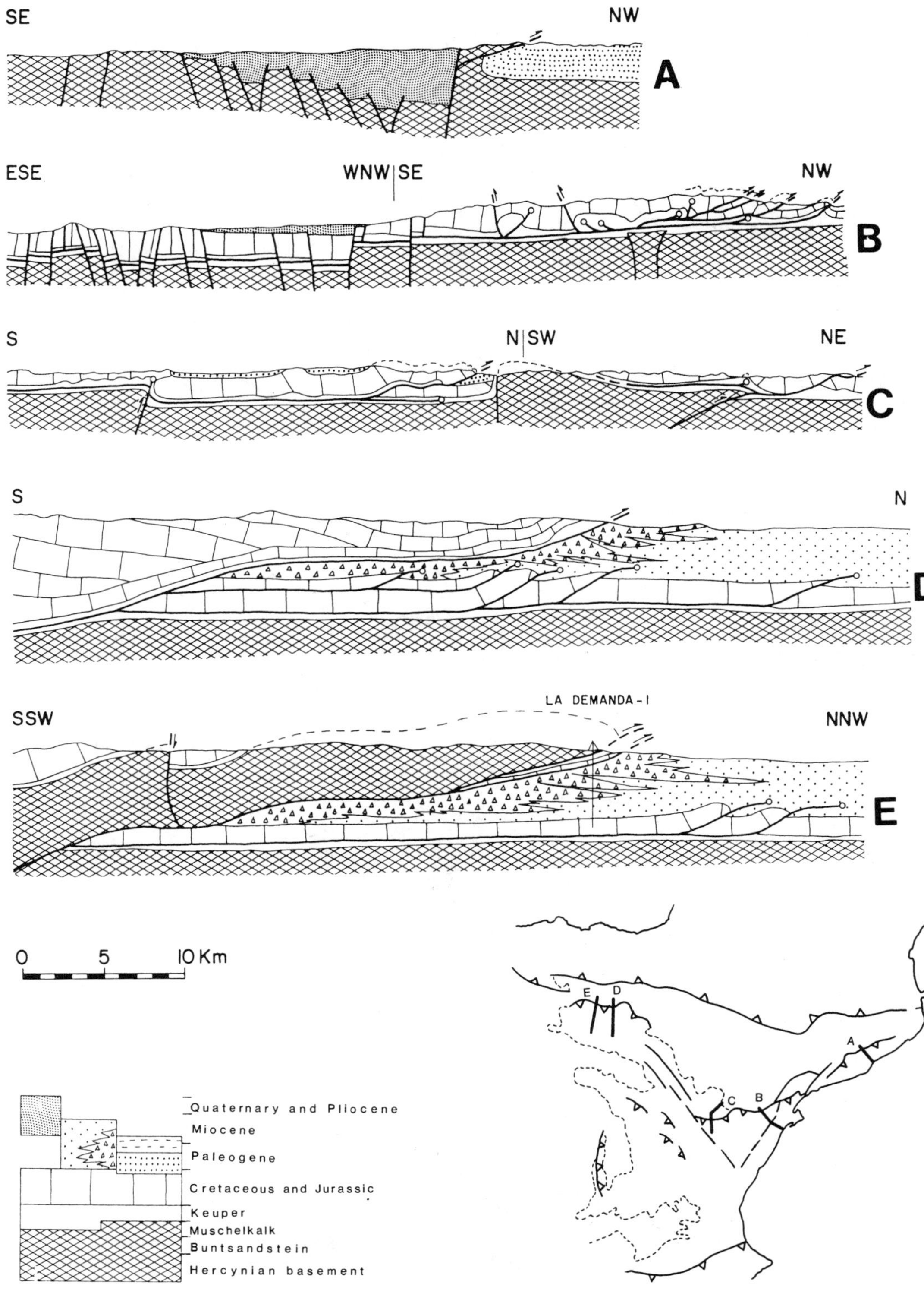

Fig. 7.9. Cross-sections through the Iberian Range and the Catalan Coastal Ranges showing their relationships with the Ebro Basin. (After Guimerà and Álvaro 1990.)

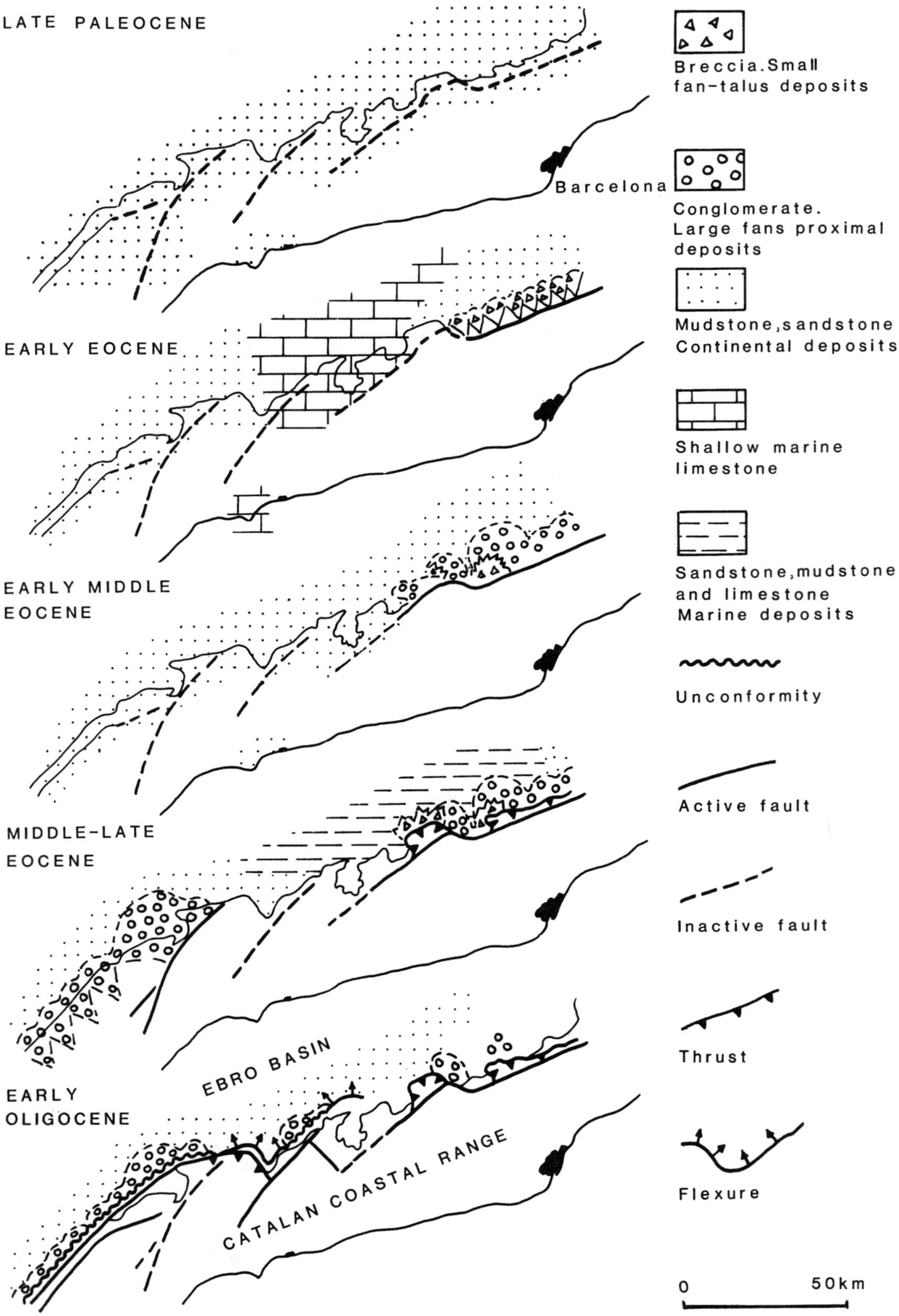
LATE PALEOCENE
Barcelona
EARLY EOCENE
EARLY MIDDLE EOCENE
MIDDLE-LATE EOCENE
EARLY OLIGOCENE
EBRO BASIN
CATALAN COASTAL RANGE
Breccia. Small fan-talus deposits
Conglomerate. Large fans proximal deposits
Mudstone, sandstone Continental deposits
Shallow marine limestone
Sandstone, mudstone and limestone Marine deposits
Unconformity
Active fault
Inactive fault
Thrust
Flexure
0
50km

Fig. 7.10

Basin border are overlapped by younger basinal deposits. As in the Catalan Coastal Range some of the northwest-striking faults have behaved as normal faults since the beginning of the Neogene, causing NW-SE elongated basins, like the Calatayud basin. However these faults maintained a noticeable dextral slip component during the early Miocene.

THE LINKING ZONE

The Linking Zone between the Catalan Coastal Range and the Iberian Range is a triangular shaped unit located where the structures of those units join (Guimerá 1984). The detached cover of the unit is thrust towards the north over the structures of both the Catalan Coastal Range and the Aragonian Branch of the Iberian Range (figure 7.9, cross-sections B and C). This structure is favored by a décollement within the middle Muschelkalk and also within the Keuper (Triassic).

Stratigraphically, this unit is characterized by a thick Mesozoic series, mainly of lower Cretaceous age. The lower Cretaceous is frequently absent or unimportant both in the Catalan Coastal Range and in the Aragonian branch of the Iberian Range (Canérot 1974; Salas 1989).

Its northern front consists of an array of east-trending folds and thrusts. These east-trending cover structures probably have been facilitated by the presence of east-trending basement faults that have controlled the northern border of an early Cretaceous small basin. Most of the thrusts are north-directed, with some showing a displacement of several kilometers affecting the Paleogene infilling of the Ebro Basin. However, there are also some backthrusts (Robles 1975; Guimerà 1984; Guimerà and Álvaro 1990). To the south the intensity of the deformation decreases as thrusts and folds gradually vanish and the Mesozoic beds become subhorizontal.

The east-trending structures of the northern front show structural features (fold interference patterns, bending of fold and thrust axes in relation to the sinistral strike-slip of the Catalan Coastal Range faults) that prove synchronous deformation of this unit with the southern terminations of the Aragonian Branch and the Catalan Coastal Range. Based on the relationships between the frontal thrusts and the related coarse deposits of the Ebro Basin margin (syntectonic angular unconformities), as well as the age of those later rocks, it can be assumed that the frontal thrusts of the linking zone were active from the early Oligocene until the latest Oligocene (Guimerà and Álvaro 1990).

THE LA DEMANDA — LOS CAMEROS UNIT

The La Demanda—Los Cameros unit is the northernmost unit of the Iberian Chain and bounds the southern side of the Ebro Basin at its western end, in the La Rioja area. It shows some similarities with the Linking Zone as well as some significant differences. This unit and the Linking Zone resemble each other in having a thick Mesozoic cover with a substantial lower Cretaceous series and an east-west trending frontal thrust that overthrusts the Ebro Basin sediments. The principal difference with the Linking Zone is that Hercynian basement is involved in the La Demanda—Los Cameros Unit thrusting (figure 7.9, cross-sections D and E). In the western part of the unit (La Demanda), Hercynian basement rocks overthrust the Paleogene and Miocene Ebro Basin fill, whereas in the eastern part (Los Cameros), cover rocks overthrust the Tertiary deposits of the Ebro Basin. The La Demanda—Los

FIG. 7.10. Summary paleogeographic maps, indicated by age, of the southeastern Ebro Basin margin showing the evolution of facies distribution and tectonic activity during the Paleogene. (After Anadón et al. 1985, simplified.)

Cameros Thrust is directed towards the north, and a lateral ramp climbs westward, up from the basement to the Keuper (Triassic).

There is an estimated displacement of the La Demanda—Los Cameros Unit over the Ebro Basin of as much as 35 km. The general geometry and the amount of displacement of this thrust have been shown by seismic profiles made during hydrocarbon exploration. The youngest Ebro Basin rocks affected by this thrust are early Miocene in age (Guimerà and Álvaro 1990).

The Ebro Basin Among Three Mountain Ranges

The development of the Ebro Basin took place mainly during Paleogene time under a northeast-southwest compressional regime resulting from the collision between the Iberian and European plates. This collision began in the latest Cretaceous—early Paleogene in the eastern Pyrenean domain, and propagated to the south and the west during the Paleogene, continuing during the early Miocene at its western end. While compressional conditions controlled the evolution of the western part of the Ebro Basin in early Miocene time, extensional tectonics had already begun to affect the easternmost part of the Iberian plate.

The Ebro Basin is the southern foreland basin of the Pyrenean Chain in which the structural evolution (the development of an important fold and thrust belt) controlled most of its fill and also governed the restriction of the basin on its northern border.

The Catalan Coastal Range and the Iberian Range also arose because of the plate collision. However, because of their northeast and northwest strike of the main faults in the Catalan Coastal Range and the Iberian Range, they behaved mainly as strike-slip faults during Paleogene time. The restrictions of the basin along its southern borders, as well as the detrital influx coming from those growing ranges, were less important than those arising from the northern, i.e. Pyrenean, border. Only the Linking Zone between the Catalan Coastal Range and the Iberian Range, and the La Demanda—Los Cameros Unit, have been thrust significantly over the Ebro Basin itself.

The most marked geographic restriction of the Ebro Basin occurs in La Rioja area. There, the Ebro Basin is reduced in outcrop to a narrow east-trending band about 30 km wide. It is bounded by the Cantabrian Thrust to the north and by the La Demanda—Los Cameros Thrust to the south. In that section the width of the Ebro Basin has undergone a shortening of about 50 km. The current triangular form of the Ebro Basin represents the last stage of an evolving basin located among three mountain ranges, different in size and structural evolution, but which arose at the same time that the basin was filled.

(Accepted for publication April 1992.)

8

Evaporitic Sedimentation in the South Pyrenean Foredeeps and the Ebro Basin During the Tertiary: A General View

FEDERICO ORTÍ

The Ebro Basin deposits discussed here are those Cenozoic sediments deposited either continuously or discontinuously upon Hercynian-Mesozoic basement of the region limited by the Pyrenees, the Iberian chain, and the Catalan Coastal Range (Catalánides). During the Tertiary, a great deal of sedimentation took place in the Ebro Basin. This section of the volume deals with the Tertiary evaporative deposits (both marine and non-marine) of the Ebro Basin, their environments of deposition, and early diagenesis. The associated open-marine and fluvial-clastic sediments are utilized as a framework for studying the evaporites, but are not addressed directly in this presentation.

The Ebro Basin roughly constitutes the southern foreland basin of the Pyrenean chain (figure 8.1). This vast area includes the Paleogene south Pyrenean foredeeps as well as the central Oligocene-Miocene depocenters, the latter forming the actual Ebro river depression. It is in this broad sense that the term "Ebro Basin" is considered throughout this paper. In this basin, a significant part of the Paleogene sediments also have been tectonically incorporated into the structures of the Pyrenees (south Pyrenean zone or also the pre-Pyrenean zone or allochthonous terrains) and, to a lesser degree, into the Catalan Coastal Range.

In the past few decades, many exploratory surveys of the pre-Pyrenean zone and the Ebro Basin depression have been carried out by private companies (for oil, mineral resources, hydrology, etc.), academic, and governmental geological groups. Additionally, a number of specific academic studies were completed which have advanced fundamental stratigraphic and structural knowledge and have explained many specific sedimentological questions. The pioneering and fundamental papers by Riba (1955, 1964, 1976) and coworkers (Riba and Pérez Mateos 1962; Crusafont et al. 1966; Riba et al. 1983) form the framework within which the geology of the Ebro Basin is studied. Academic studies (theses), not commonly mentioned as references in many reviews, have contributed greatly in our understanding of evaporites in the Ebro area and are listed by area in table 8.1. Some of these have been entirely dedicated to the study of evaporites (starred).

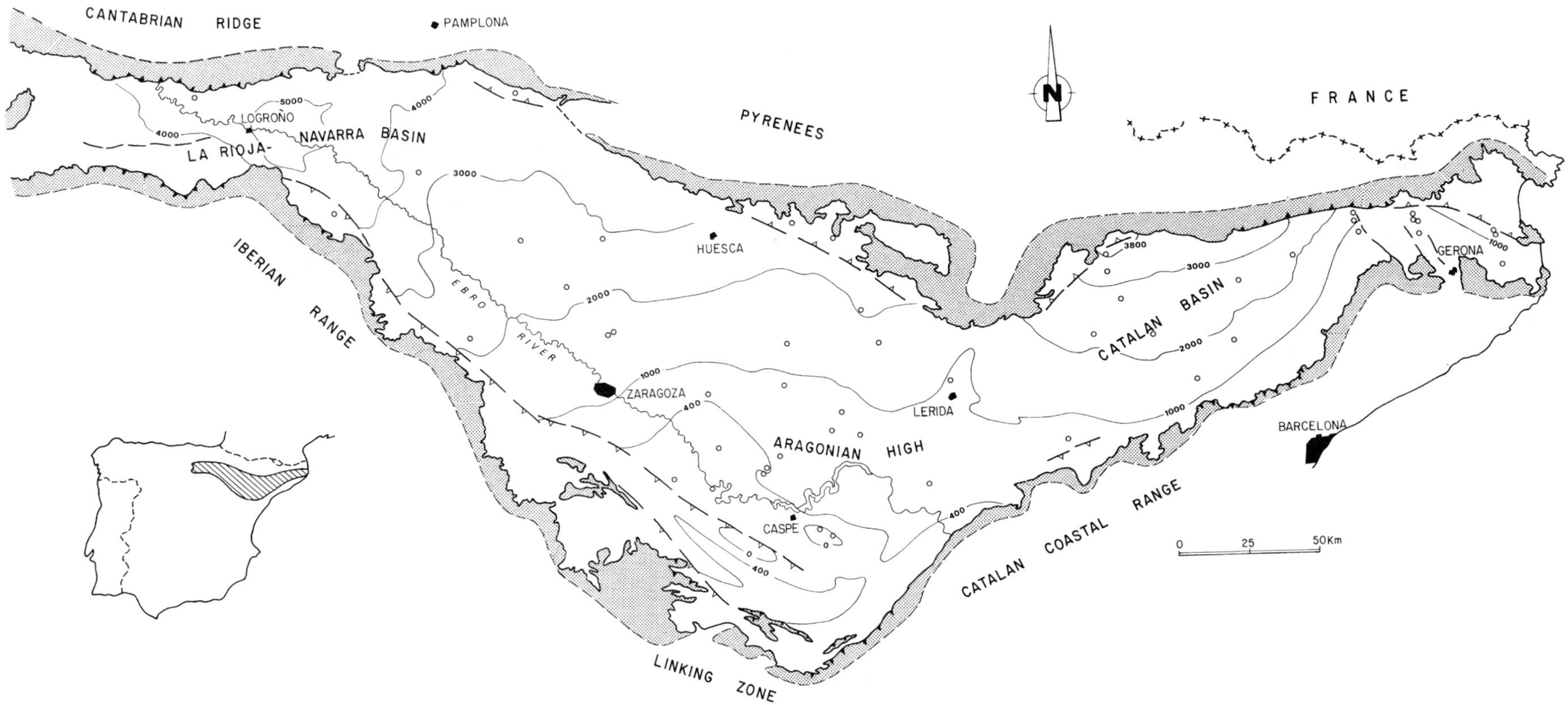

Fig. 8.1. Isobath map of the basement of the Tertiary sediments for autochthonous terrains in the Ebro Basin (simplified after internal report submitted to ENRESA 1987). Shaded area represents Paleozoic and Mesozoic rocks and, along the Pyrenean border, thrusts of Tertiary sediments. Datum: sea level. Open circles represent the position of deep boreholes for petroleum exploration.

TABLE 8.1.
Some of the doctoral theses completed in the last three decades that may be applied to the study of Tertiary evaporites in the Ebro Basin (and also to the Triassic evaporites of the Mesozoic basement) (see Jurado 1989).

REGION	AUTHOR	YEAR	UNIVERSITY
CATALONIA (Eastern Region)			
Pre-Pyrenean Zone	Busquets, P.	1981	Barcelona
Central Depression	Reguant, S	1967	Barcelona
Central Depression	*Pueyo, J. J.	1975	Barcelona
Central Depression	Cabrera, L.	1983	Barcelona
Central Depression	Taberner, C.	1983	Barcelona
Central Depression	Sáez, A.	1987	Barcelona
Coastal Ranges	Anadón, P.	1978	Barcelona
Coastal Ranges	Colombo, F.	1980	Barcelona
ARAGON (Central Region)			
Pre-Pyrenean Zone	Puigfàbregas, C.	1975	Barcelona
Central Depression	*Quirantes, M.	1969	Zaragoza
Central Depression	Birnbaum, S.	1976	Cambridge
Central Depression	Torras, A.	1976	Barcelona
Central Depression	*Mandado, J.	1987	Zaragoza
Central Depression	Jurado, M. J.	1989	Barcelona
NAVARRA (Western Region)			
Pre-Pyrenean Zone	*Rosell, L.	1983	Barcelona
Pre-Pyrenean Zone	*García Veigas, J.	1993	Barcelona
Central Depression	*Salvany, J. M.	1989	Barcelona
ALL REGIONS	*Utrilla, R.	1989	Barcelona

*** = studies focusing entirely on evaporites**

Sedimentation in the Ebro Basin

Sedimentation in the northern (Pyrenean) part of this Tertiary basin was continuous from the upper Cretaceous (Garumn-type facies) onward. In the southern zones the Cenozoic materials overlie very different pre-Tertiary rocks along unconformities or at angular disconformities. Distribution of the Tertiary filling in the Ebro Basin took place in such a manner that the older (Paleogene) sediments crop out in the northern and eastern margins, while the younger (Neogene) ones are exposed in the western and central parts of the basin. Also, a well-defined north-south asymmetry is noted in the distribution of both marine and continental sedimentary units. Thus near the northern and eastern borders, a significant thickness of marine sequences was deposited up until the late Eocene, while on the southern

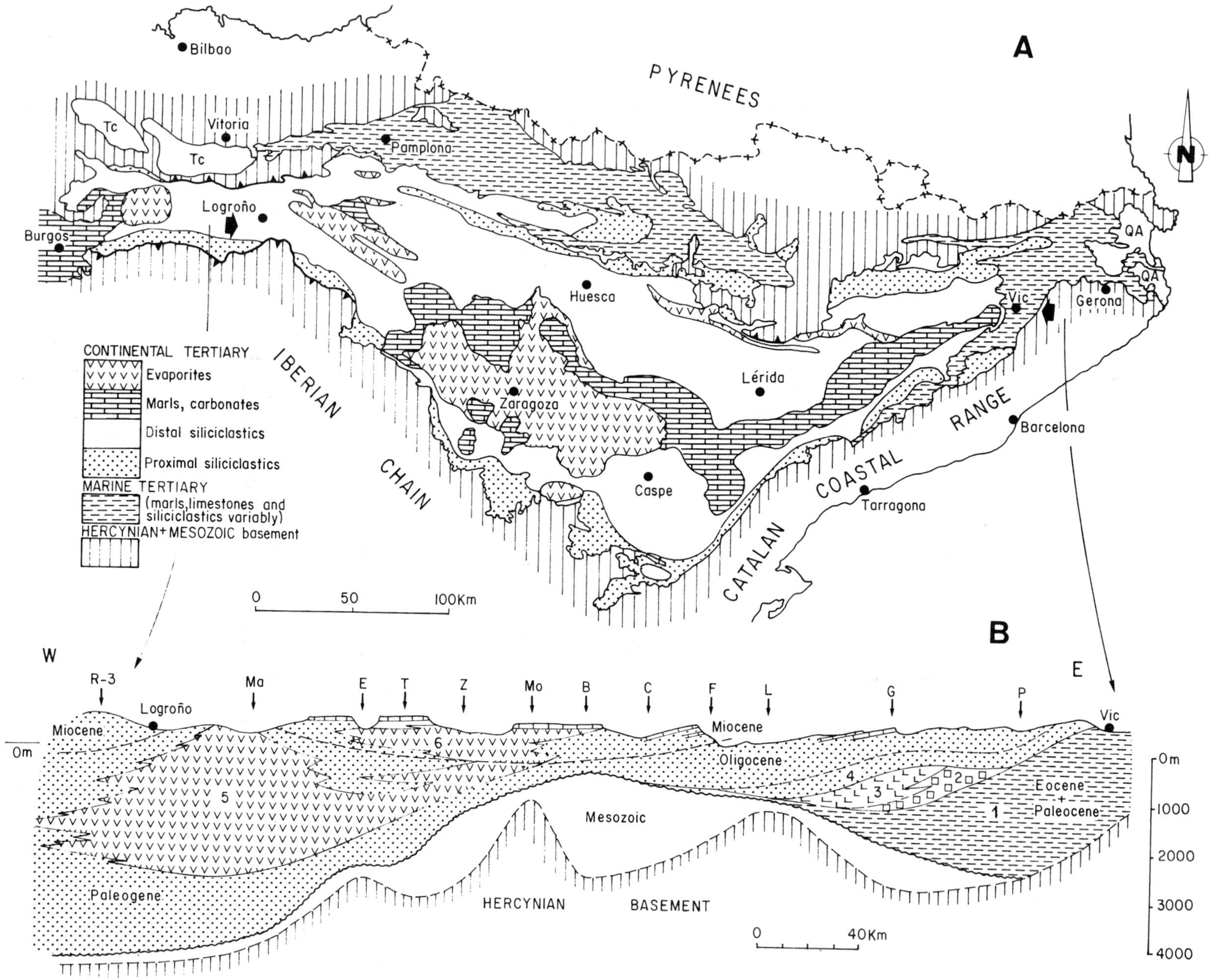

Fig. 8.2. Distribution of the main Tertiary lithofacies in the Ebro Basin. A. Surface distribution of main lithofacies in the continental Tertiary of the Ebro Basin. Tc = undifferentiated Tertiary, QA = Quaternary of the Empordá graben. B. Schematic stratigraphical section of the Tertiary of the Ebro Basin (borehole data projected onto an east-west line). Key: 1 = Paleocene nonmarine and Eocene marine sediments (the basal section comprises both the Tremp (Garumnian) and Pontils formations); 2 = Cardona Saline Formation; 3 = Barbastro Gypsum Formation; 4 = continental Upper Eocene; 5 = Falces Gypsum (at the base) and Lerín Gypsum (at the top) formations; 6 = Zaragoza Gypsum Formation. Boreholes (at arrows): R-3 = Rioja-3; Ma = Marcilla; E = Ejea; T = Tauste; Z = Zaragoza; Mo = Monegrillos; B = Bujaraloz; C = Candasnos; F = Fraga; L = Lérida; G = Guissona; P = Puigreig (modified from Riba, Reguant, and Villena 1983).

margin, marine sediments are entirely lacking throughout the Tertiary.

The maximum sedimentary accumulation occurred along the northern limit of the basin, where the thickness of Cenozoic sediments is as great as 4,000 to 5,000 m, as seen on the isobath map of the base of the Tertiary (figure 8.1). This sedimentary accumulation lies mostly in two subsident subbasins, one in the Navarra—La Rioja zone and the other in Catalonia. They are separated by the structural high of Los Monegros (Aragonian high). In the southern part of the Ebro Basin, Cenozoic sedimentary accumulation is only about a thousand meters thick. In many areas the base of the Tertiary cover is roughly a planar surface tilted towards the north.

A simplified map of the main Tertiary outcropping lithofacies is offered in figure 8.2A. A general cross-section of the lithologies known from boreholes, drawn in an east-west orientation, is presented in figure 8.2B. In the cross-section the relative development of the two major basins is noticeable. Also, it can be seen that marine sediments of Eocene age are restricted to the east, including the potash evaporites of unit 2, while towards the west the lithostratigraphic units are progressively younger, thicker, and of continental origin.

Marine and Continental Evaporites: Regional Distribution

During its Cenozoic history, the Ebro Basin received a rather significant volume of evaporitic sedimentation which has no equivalent in the northern foredeeps of the Pyrenees (in the Aquitanian Basin, France). This deposition records both marine and nonmarine episodes of such variety that it has given the Ebro Basin, among other Tertiary basins in Spain, a very high level of importance for the study of evaporitic environments and facies. Additionally there is substantial economic interest in the mineral deposits of this area.

The marine episodes in the Ebro Basin that originated from Mediterranean and Atlantic inflows are located in the Paleogene south Pyrenean foredeeps and consist of two major phases, one of middle Eocene age (lower Lutetian) and the other of late Eocene (Priabonian). The first marine evaporative phase had a marine connection to the east (with the Mediterranean sea), was restricted to the southeastern Pyrenees (Beuda Gypsum), and is composed of anhydrite and minor amounts of halite. The second developed from Navarra in the west to Catalonia in the east, was linked to the west with the Atlantic Ocean, and constitutes the South Pyrenean Potash Basin (figures 8.3 and 8.4).

Continental evaporitic episodes display wider geographic and temporal distributions than those of marine origin. In the Catalan sector of the Pyrenees continental evaporite sedimentation initiated in the late Paleocene (Tremp Gypsum). Along the contact with the Catalan Coastal Range such sedimentation recurrently took place between the early Eocene and the early Oligocene (Rupelian; or Stampian sensu strictu). Immediately after the second marine evaporite phase (potash basin), a substantial continental evaporite unit developed from Catalonia to Navarra during the late Eocene (Priabonian) and the earliest Oligocene represented by the Barbastro Gypsum—Puente la Reina Gypsum. In Catalonia some minor evaporite deposits (Artesia Gypsum, Camarasa Gypsum, etc.) were formed locally on the thrust sheets immediately after the deposition of the Barbastro Gypsum (Sáez et al. 1991). The Navarrese sector of the Ebro Basin underwent persistent and continuous evaporitic sedimentation throughout the Oligocene that extended into the early Miocene (Falces Gypsum, Lerín Gypsum). The same environment existed in the central sector of the basin, forming the Zaragoza area, with deposition extending from the late Oligocene (?) through the early Miocene (Zaragoza Gypsum). The Cerezo Gypsum is located in the westernmost part of the basin. However, the exact age of this particular unit is still disputed (late Miocene?). Minor evaporitic units are widely distributed in many sectors of the basin throughout the stratigraphical record of the Tertiary (Ortí 1988; Ortí et al. 1989 a, b), or are enclosed within carbonate formations, as in the central Aragonese region during the late Oli-

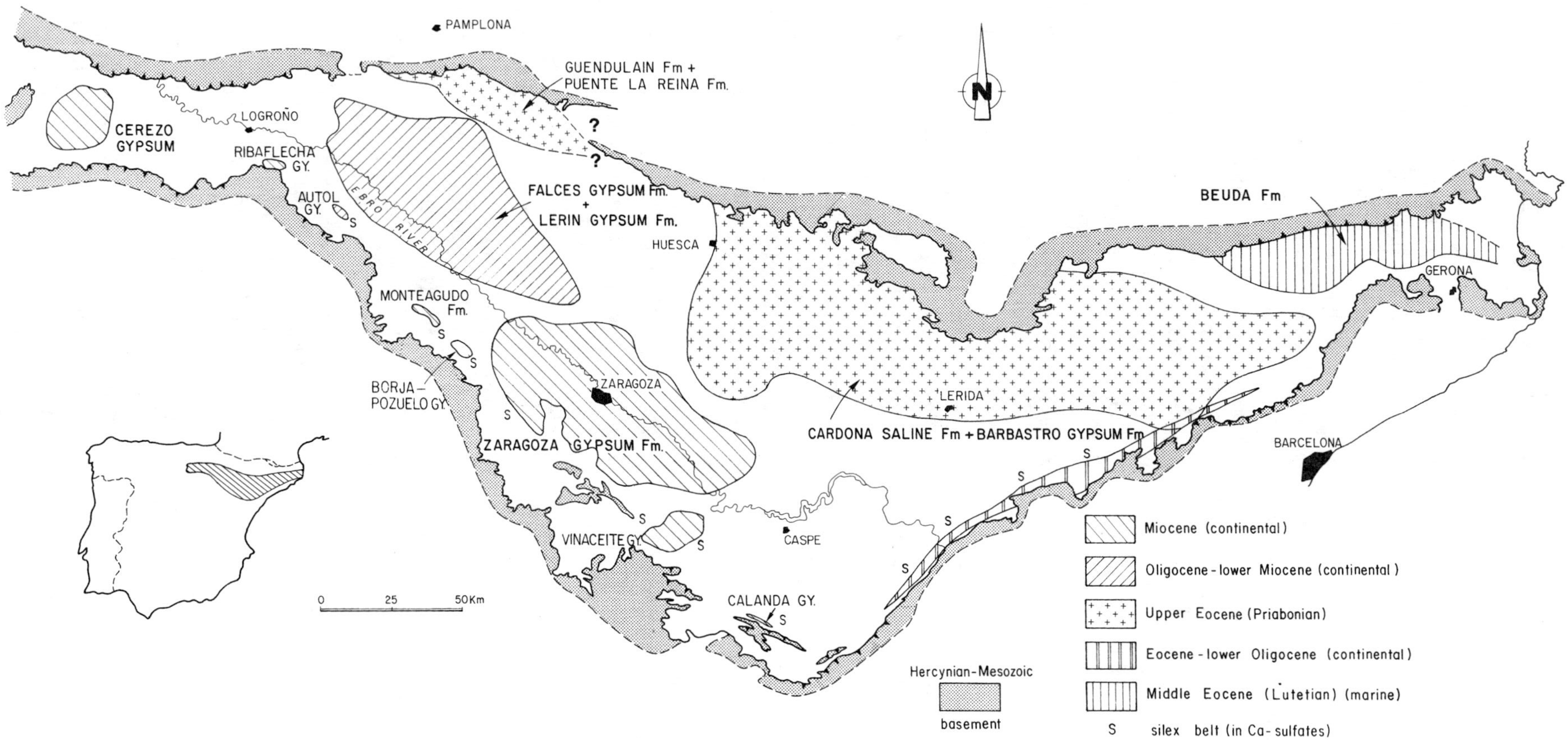

FIG. 8.3. Geographical distribution of major evaporitic areas and formations of the autochthonous Tertiary in the Ebro Basin. In the northern border of the figure the shaded area also includes displaced Tertiary terrain (in part after ENRESA 1987 internal report and Ortí 1990a).

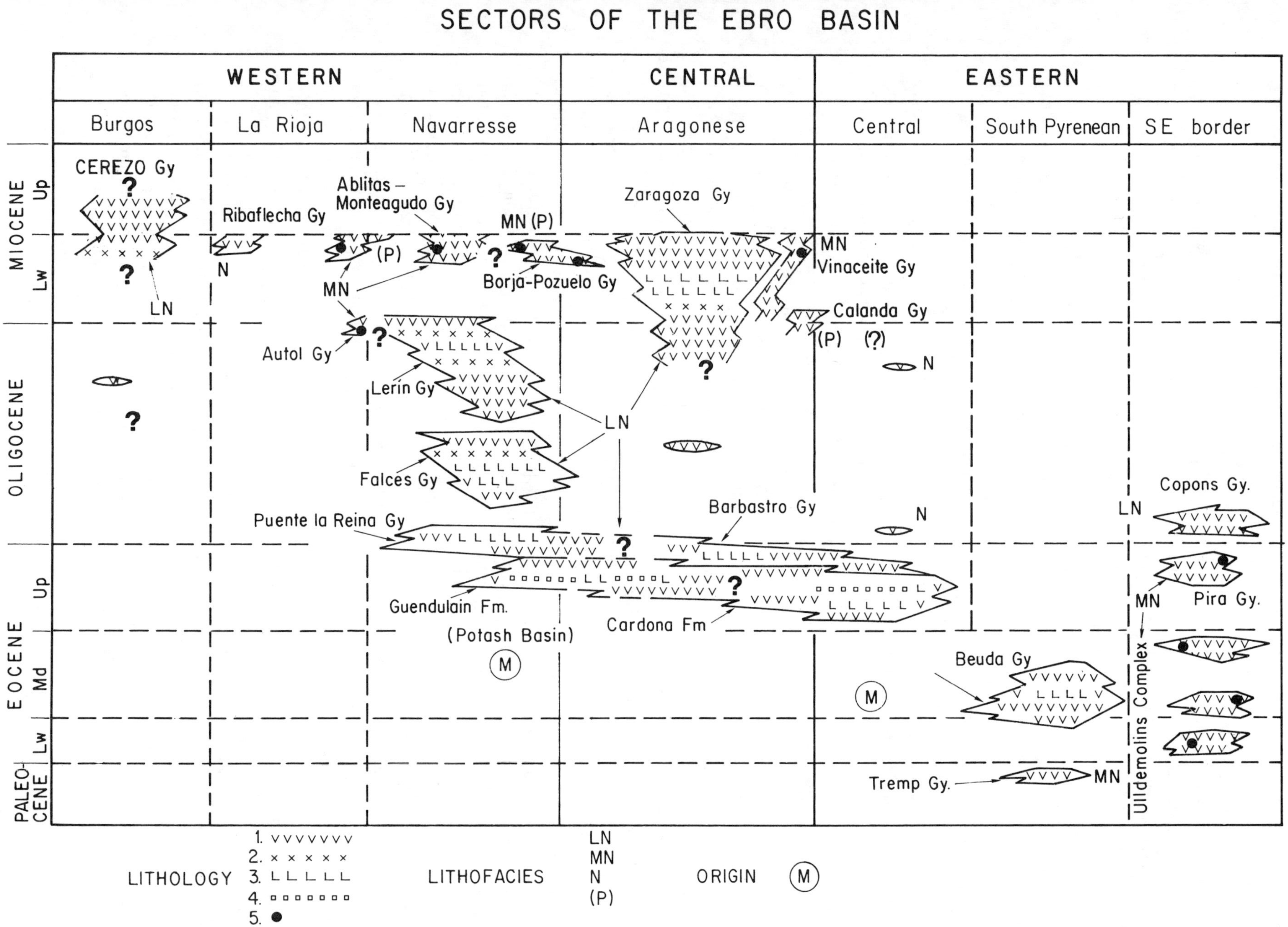

Fig. 8.4. Stratigraphical distribution of the major evaporitic units in the Tertiary of the Ebro Basin (modified from Ortí 1988) (Gy = gypsum in formation and member names). Lithologic key: 1 = Ca-sulfates (anhydrite and secondary gypsum); 2 = Ca-Na sulfates (glauberite and rarely thenardite); 3 = Na-chlorides; 4 = K-Mg salts; 5 = chert nodules. Lithofacies key: LN = laminated to thin-bedded and nodular; MN = microlenticular and nodular; N = nodular; P = some primary gypsum preserved. Origin: M = marine (the other formations are continental).

gocene to early Miocene. Complementary information about the evaporitic sedimentation during the Tertiary in the Ebro Basin is presented in a guide book (Ortí and Salvany 1990) and an overview of the evaporite deposits in Spain (Ortí et al. 1992).

Tectonic Control of the Evaporitic Sedimentation

Evaporitic sedimentation in the Ebro Basin followed its structural evolution faithfully, with each new major tectonic event followed by a new phase of evaporite deposition. This is especially true along the northern (Pyrenean) and southeastern borders (Catalan Coastal Range). Controls exerted by tectonics in the southwestern border mountains (Iberian chain) are also significant in depositional evolution of evaporites.

Sedimentation reflects the various stages of compression during the Eocene-Oligocene in the south Pyrenean zone. This structural control is expressed by continental evaporites that are related to folding phases and marine evaporite deposition related to the emplacement of thrust sheets. In the Catalan Coastal and the Iberian ranges, block uplift with associated continental evaporites is also noted. Similarly, evaporite sedimentation recorded the marked phase of Oligocene subsidence associated with a compressive regime in the Navarrese sector. Finally, evaporite sedimentation also marked Miocene distension and differential subsidence in the central sector of the basin. Throughout the entire structural history of the basin, the large evaporitic formations have commonly occupied the axial depocenters, which progressively shifted towards the south and the west. In the less subsident sectors, evaporites are found only as a second-order component in association with calcareous or argillaceous formations.

Major Depositional Sequences and Evaporites

Marine evaporite formations in the South Pyrenean Zone represent the last episodes of major sedimentary sequences involving marine transgression-regression cycles, and the final infilling of the marine troughs with sulfates and chlorides. These deposits mainly occur as basinal evaporites. In the case of continental sequences, evaporitic sedimentation usually took place during periods of decreased activity on marginal alluvial fan systems, and deposits formed in very shallow lakes developed on distal mud flats.

In contrast to the marine-related evaporitic formations, in the continental formations the depositional relationship with the other major sedimentary sequences is not clearly established, due to the fact that such sequences do not always have a clear regional setting along the passive borders. This is especially the case for the Catalan Coastal Range, where the complex strike-slip movements of basement faults make it difficult to generalize tectono-sedimentary units for the whole chain. In the borderlands of the Iberian chain, where large scale discontinuities are more clearly developed (Pérez et al. 1988; Muñoz 1992), the difficulty lies in accurately correlating them with the thick evaporite deposits that formed in the Navarresse-Aragonese depocenters during Oligocene-Miocene time.

Origin of the Solutes in the Continental Brines

Several chemical and isotopic markers suggest that in the Ebro Basin the major supply of solutes to the continental brines (sulfates and chlorides) came mostly from extensive Triassic and lowermost Liassic evaporites that are present in the surrounding mountains. Important deposits of anhydrite and salt are known from the Pyrenean border (Basque-Cantabrian diapiric zone; central Pyrenean diapiric zone). Anhydrite-rich units bearing halite are also present in the Triassic of the Catalan Coastal Range and the Iberian Chain. These deposits also have been preserved in the Triassic basement of the Ebro Basin (Jurado 1989). The first suggestion of evaporite reworking was offered by Birnbaum and Coleman (1979). Additional supportive isotopic data for the reworking of calcium sulfate from a Mesozoic source in this area has been presented in Utrilla et al. (1987, 1991, 1992) and Utrilla (1989). The general idea of an evaporitic recycling mechanism in this area was reviewed by Ortí et al. (1988). From a hydrologic point of view, this geochemical provenance was demonstrated along the Iberian border (SW) of the basin, where modern groundwaters are significantly enriched in calcium sulfate (Martínez Gil et al. 1988). Such waters are apparent in well-developed springs along that border and also appear in the innermost parts of the Ebro depression feeding several sulfatic lakes.

Theoretically, Eocene marine evaporites from the South Pyrenean area also might have been reprecipitated into the Oligo-Miocene continental evaporites, but apparently this mechanism was not active to any significant extent during that time (Ortí et al. 1988b). However, marked chemical recycling was and is significant in this area. There is a long history of diapir formation and exposure and weathering of Triassic evaporites that began in the upper Jurassic and is still active in the Pyrenean area, as well on other chains in the Iberian peninsula (Brinkmann and Lögters 1968; Serrano et al. 1989). Some small gypsum deposits of Permian age, intercalated among continental red beds in the western Pyrenean zone, have probably contributed only very small amounts of calcium-sulfates (Valero and Gisbert 1994).

Continental Evaporites: Sedimentary Systems and Facies Associations

Evaporitic environments in the continental Tertiary of the Ebro Basin are diverse but in general can be grouped into three main categories (Ortí 1988; Ortí et al. 1989b):

LACUSTRINE WATER BODIES (LOW IONIC CONCENTRATION SULFATIC LAKES).

In these lakes calcium sulfate precipitated mainly as primary gypsum. Anhydrite is locally associated as a secondary component of early diagenetic origin. These lakes were shallow and small, with some intercalated playa deposits. The lacustrine deposits have a limited thickness for each unit are generally less than one hundred meters thick.

These water bodies occupied peripheral positions within the vast endorheic systems of internal drainage. They were either related to proximal alluvial fans or occupied marginal locations in respect to large central evaporitic lakes. Calcium sulfate deposits that formed in these lakes are associated mainly with fine siliciclastic sediments, and to a minor degree to carbonates, marls, chert, and coal. Clastic episodes commonly reworked the gypsiferous sediments of these ponds.

The lake deposits form a rather well-defined belt, Eocene to Miocene age, that almost completely rims the Ebro Basin along its southeastern (Catalánides) and southwestern (Iberian chain) borders. In this belt an associated replacive chert belt is well-developed and primary gypsum is only preserved in the Miocene units (Ortí et al. in press). Other associated sedimentary features are present in the lacustrine deposits, including areas bearing very large anhydrite masses (meganodules), intercalations of thin beds of charophytic carbonates, and intense burrowing in some of the gypsum. Also noted are the lack of chlorides or of other sulfates. These peripheral evaporitic lakes apparently acted as marginal geochemical traps for carbonates, silica, and trace elements. Some minor elements are relatively abundant in these peripheral deposits in comparison with the deposits of the central lakes. Strontium, for example, is present within the crystal lattice of gypsum or exists as celestite micro crystals within the gypsum deposits of the marginal areas while it is uncommon in the basin center.

SALINE MUD-FLATS (SABKHAS)

The formation of sabkha deposits was controlled by the position of the water table and was characterized by the displacive growth of several mineralogical phases, mainly gypsum and anhydrite. These deposits generally developed on the margins of the large saline lakes as wide peripherical mud flats. More rarely these sabkhas were connected to both

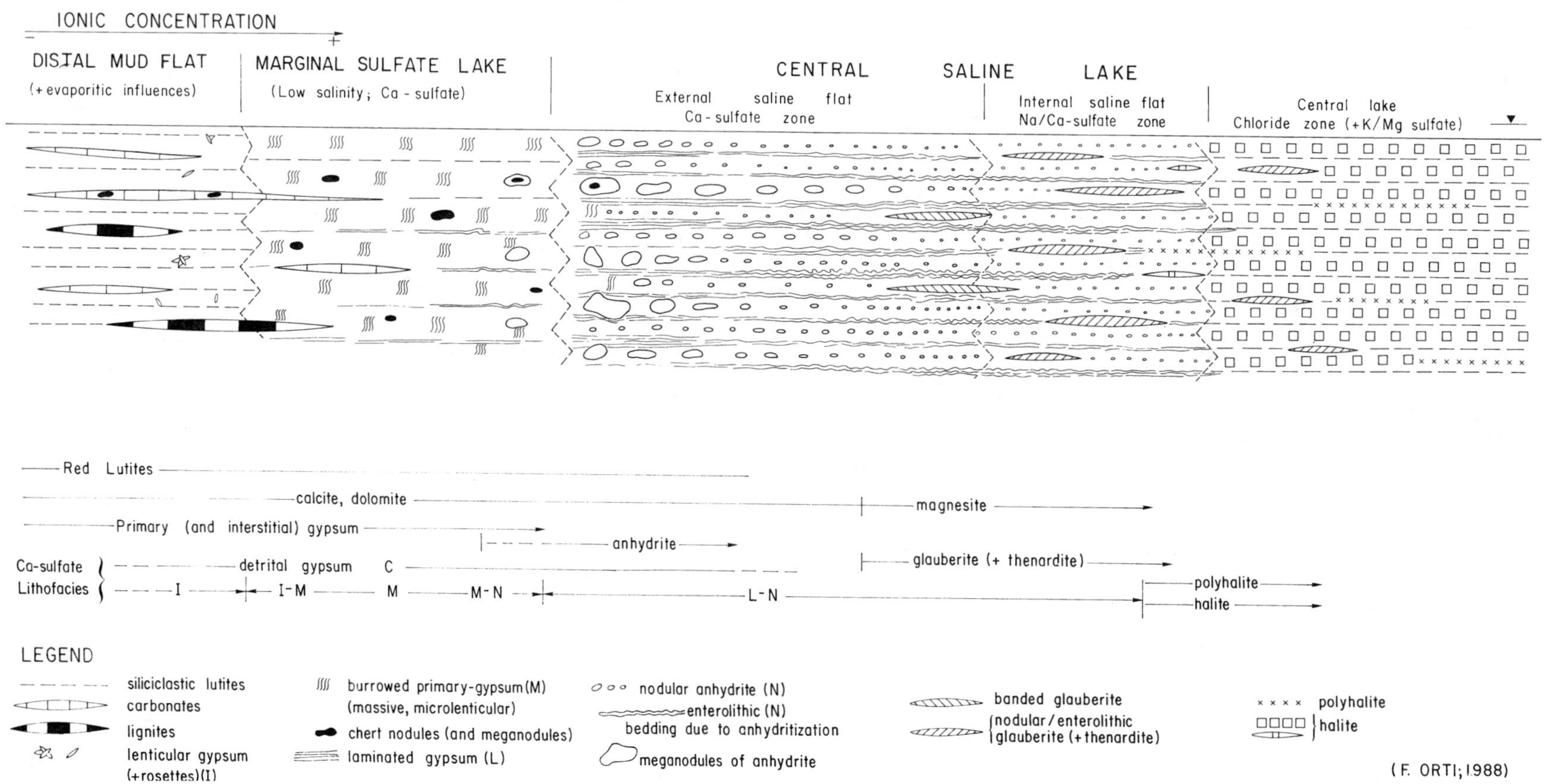

FIG. 8.5. Idealized sedimentologic section, not to scale, of a general evaporative-lacustrine environment and the general facies arrangement, as known from some continental Tertiary basins of Spain (modified from Ortí 1988). Key to calcium sulfate lithofacies: C = clastic; I = macrocrystalline of interstitial origin; M = microlenticular, matrix-poor; N = nodular; L = laminated to thin-bedded.

the peripheral sulfate and the higher salinity central lakes. This was the case in the Azaila-Rodén area. In this area red lutites and carbonates alternate with nodular alabastrine gypsum. These sabkha deposits are commonly about 100 m thick.

SALINE LAKES AT THE BASIN CENTER (HIGH SALINITY)

These lakes were very large and shallow as suggested by the extent of the laminated evaporitic deposits and associated sedimentary features (figures 8.3 and 8.4). Playa episodes were common, as indicated by interstitial and nodular evaporite intercalations, which may represent longer periods of deposition than the flooded phases. The lake deposits exhibit the maximum ionic concentration in the endorheic (internal drainage) system and contain the most highly soluble salts: chlorides, sodium-sulfates, and associated minor amounts of pottasium and magnesium-sulfates. These lake systems generated sedimentary accumulations several hundred to a thousand meters thick, apparently favored by a marked subsidence rate due to tectonism.

FACIES ASSOCIATIONS

In general it seems that there was only a very simple shallow water to sabkha character to the evaporitic sedimentation in the whole of the continental Tertiary of the Ebro Basin. Despite this, the sedimentary facies that developed in these environments were diversified. The most frequent evaporitic lithofacies are calcium sulfates (figures 8.4 and 8.5), with the following characteristics:

Laminated to thin-bedded facies (L). This facies consists of well-layered units displaying internal lamination (mm) or banding (cm). Laminae are planar to undulate and sporadically contorted. Originally they were made up of fine-grained sulfate crystals, generally with idiomorphic or microlenticular habits. They formed in water bodies of very different salinities and in a range of depths. In the most shallow deposits, algal morphologies are frequent (undulate, crenulate, domed, stromatolitic, and as wavy ripples). Cycles having a basal carbonate horizon are also a common feature.

Selenitic facies (S). This facies formed as selenite gypsum crystals (>2mm) arranged in subvertically oriented bands indicative of free growth in a water body. They are very rarely found as primary crystals in a few Miocene formations of the Ebro Basin (Cerezo Gypsum). More commonly they are found as cm-sized pseudomorphs after selenite crystals and selenite domes composed of secondary gypsum that can be found in several of the Paleogene units as in the Artesa Gypsum (Sáez et al. 1991, figure 22).

Microlenticular facies (M). Microlenticular facies formed as matrix-poor or matrix-free deposits. They consist of dm- to m-sized layers of gypsum, gray to green or dark brown, very fine-grained (<1 mm), and displaying lenticular to tabular habits. These gypsum facies are gypsilutites and sometimes gypsarenites which may exhibit the effects of intense bioturbation. Locally, larger crystals are present (up to 1 cm) as elongated prismatic and twinned forms. Gypsum in these facies is primary, partly subaqueous, and was precipitated from water of very low salinity (chloride-poor). This facies is usually intercalated with carbonates (containing charophytes, mollusca) as well as replacive chert nodules and lignitic coals. Halite is totally absent and early anhydrite replacement of the gypsum is very variable.

Nodular facies (N). A general distinction is made among micronodules (<2 cm), macronodules (2 to 50 cm), and meganodules (>50 cm in diameter). Nodules are generally generated as anhydrite during very early diagenesis in playas and saline mud flats, as well as in the border of saline lakes; they may also form extensively in basin center areas, apparently due to sporadic desiccation. Nodular facies in sulfates (anhydrite, gypsum, glauberite) are present in formations of very variable thickness and lateral extent. In general, the nodules have grown interstitially at various depths (cm to meters), either de novo or by replacement of earlier-formed mineral phases (mainly other sulfates and carbonates). The nodules may develop in isolation or grouped in horizons, also passing into enterolithic layering. Vertical elongation of the anhydrite nodules in these horizons is a very common feature in the Tertiary of the Ebro Basin. The meganodules seem to be present only where there is a development of anhydrite facies. The same meganodule development may form as vertical nodular arrangements (in columnar or wall-like structures) of different sizes which cross-cut the primary gypsum layers or the lutitic host sediment. Meganodules normally develop in the sulfate-rich lakes and on the borders of the large central saline lakes. Vertical contorted layers are also a typical feature of anhydrite formation in some stratigraphical units. Nodules made up of primary gypsum (not arising from anhydrite precursors) as those cited in the Calatayud Basin to the south of the Ebro Basin (Rosell and Ortí 1992) were not recognized here.

Macrocrystalline facies of interstitial growth (I). Interstitial sulfate crystals mainly have lenticular habits, although tabular, prismatic, or elongated shapes also are found. They form within a matrix that may be clayey, sandy, calcareous, organic matter, or even gypsiferous in composition. Crystal size ranges from 1 mm to several tens of cms. Such crystals may display zoned growth, especially with depressed crystal faces on the gypsum (skeletal growth), and are frequently arranged in "desert rose" horizons or crusts. All of these forms are normally generated in playas and saline mud flats. It is quite common to find them at the base of gypsum (or glauberite) sequences, marking the beginning of an evaporitic episode.

Gypsiclastic facies (C). This facies represents the product of mechanical reworking, either eolian (gypsilutites and gypsarenites) or aquatic (mainly gypsarenites and gypsrudites). Such deposits are usually present along the border of the evaporitic formations, but locally some gypsarenites are intercalated within the laminated central facies. No examples of gypsum turbidites have been noted in these deposits.

The Evaporative Complex: Facies Models.

An idealized cross-section of the various facies in a typical evaporative complex, based on observations in the Ebro Basin, is given in figure 8.5. The model of the facies complex is not drawn to scale but represents a transect, from border to center, of a large continental evaporite formation. It shows the major facies associations present in the Ebro Basin, and includes the most common types of sulfates and chlorides observed. The most important types of calcium sulfate facies associations displayed by the evaporite formations of the Ebro Basin are outlined in figures 8.4 and 8. 5. Two major groups of evaporite facies associations can be distinguished: laminar-nodular formations and matrix-poor microlenticular formations with meganodules.

Laminated-nodular deposits (L-N). Laminated-nodular deposits occupy central positions in the internal drainage (endorheic) depressions and are the most important both in terms of surface area and volume. They contain variable quantities of halite and sodium-bearing sulfates (more rarely potassium-magnesium sulfates), but chert is lacking and carbonate is present only as a subordinate component. Alternations of laminated to thin-bedded (even algal-like, wavy, or with ripples) and nodular (banded nodular and enterolithic) lithofacies are typical, suggesting continuous changes in the water level of the lakes. Because of the shallow depositional environment, the thick accumulations, which cover large areas, imply both rapid subsidence rates and lateral shifting of the evaporitic depocenters. This category of sediment corresponds to the following formations among others: Barbastro and Puente la Reina Gypsum (Eocene to Oligocene transition), Falces and Lerín Gypsum (Oligocene to lower Miocene), Cerezo and Zaragoza Gypsum (Miocene). Thus this sedimentary association is the most important in the large saline lake category within the central basin area. In the surface the gypsum is almost always secondary in origin, coming from the replacement of either anhydrite or glauberite. A summary of sodium sulfate-bearing evaporite formations in the Teriary basins of Spain was offered by Ortí and Salvany (1991). Articles dealing with glauberite units in the Ebro Basin have been presented by Menduiña et al. (1984), Salvany and Ortí (1992, 1994) and García-Veigas et al. (1994).

Matrix-poor microlenticular formations with meganodules (M-N). These are relatively small units, positioned close to the alluvial fan systems of the basin borders, and displaying variable amounts of burrowing activity in the gypsum. Here salinity of the water was probably low, but with an elevated calcium sulfate content. These deposits also appear to have acted as a sedimentary trap for chert precipitation (Salvany et al. 1994). Locally it is possible to perceive a lateral transition from microlenticular deposits to laminated-nodular types of evaporitic formations towards the basin center. More frequently, toward the centers of basins the waters from these environments (microlenticular deposits) discharged into either the lutitic flats (sabkhas) or more dilute (calcareous) lakes. This facies association is typical of the sulfate lake category. A great number of such formations are located along the southeastern and southwestern borders of the Ebro Basin, as well in the Tremp Basin (allochthonous zone).

CYCLIC SEDIMENTATION

Evaporitic cycles in the calcium-sulfate facies are mainly either sabkha (or salina to sabkha) type or salina (or sabkha to salina) type. The first type displays laminated or microlenticular-bioturbated gypsum at the base and nodular anhydrite at the top. It represents the final infilling of the evaporite lake with lutites: thus the brines were largely confined to the subsurface. The second type has nodular facies at the base and laminated facies at the top. It is the product of the lateral migration of shallow lakes and is present in the Oligocene and lower Miocene of Navarra (Ortí and Salvany 1986b).

Sedimentary Facies in Marine Evaporites

As already mentioned, two evaporitic episodes fed by marine sources developed during the Eocene in the south Pyrenean foredeep basin (figures 8.3 and 8.4):

(a) lower Lutetian phase: Beuda Gypsum
(b) Priabonian phase (also called South Pyrenean Potash Basin). The Priabonian saline formations are subdivided into the Cardona (Catalan Basin) and Guendulain (Navarra Basin)

Both phases display many characteristics of deep basinal evaporites, and developed at the end of well-defined marine sedimentary sequences in which the evaporitic formations roughly mark the transition from marine to continental environments. Each phase corresponds to a single evaporitic cycle with maximum thickness between 100 m (in the Lutetian phase) and 300 m (in the Priabonian phase, in the Catalan Basin). Such evaporitic sequences may have locally turbiditic and anoxic facies at their bases in the preevaporitic stage, and at the top in the postevaporitic stage. Because other papers in this volume are dedicated to both phases, this is a limited and introductory survey.

The major differences between these evaporitic phases are:

1. Geographical distribution: only the eastern part of the chain (in Catalonia) is occupied by the Lutetian phase, while the Priabonian phase extends throughout the basin from Vic to Pampona.
2. Degree of restriction: apparently only an initial chloride stage was developed in the Lutetian phase, while in the Priabonian potash concentration was reached.
3. Structure of the evaporitic units: the Lutetian formation outcrops entirely on allochthonous terrains (though its presence is known also in autochthonous terrains, known by boreholes), while the Priabonian formations are located in both autochthonous (Catalan Basin) and allochtonous (Navarra Basin) terrains.

Our knowledge of the Lutetian phase is rather limited, not only concerning the original basinal limits, but also concerning the volumes and distribution of its chloride deposits as well. Though the latter have been penetrated by deep boreholes, the thickness and geographic extent of the salt throughout the basin is difficult to evaluate.

In both episodes the evaporitic sedimentation initially developed at the top of somewhat deep offshore facies. In the Lutetian phase such open marine facies are composed of a thick sequence of siliciclastic turbidites displaying some local indicators of hypersalinity, while in the Priabonian phase the dominant offshore facies are thick blue marls, which display increasingly euxinic conditions going up section, with a complete lack of any evidence of life near the top. Locally, such preevaporitic conditions give way to a phase of cyclic oxygenated-anoxic deposits just at the top of the thick marl sequence.

The sulfate sedimentation which follows shows rather different patterns for each phase. It is widely distributed in the Lutetian phase and is composed of a thick anhydrite unit (more than 80 m) which extends throughout the turbiditic trough and even laps over the marginal carbonate platforms. In the Priabonian phase the sulfate deposit remains confined to the border areas of the Catalan Basin, forming a narrow belt of sediments up to 30 m thick. This is wedge-shaped and thins out towards the basin center. An equivalent calcium sulfate border deposit is virtually absent in the Priabonian of the Navarra Basin.

In both Priabonian basins the "basal anhydrite" known from the depocenter consists of a thin layer (less than 1 m in Navarra and only a few meters thick in Catalonia) with laminated sulfate lithofacies (no available samples for study exist in the Catalan Basin, but its laminar character appears in old borehole descriptions). Calcium sulfate deposits from the border are almost unknown in the Navarra Basin during the Priabonian.

In the Lutetian phase the vertical evolution of the sulfate horizon displays a shallowing trend: laminated lithofacies are followed upwards by selenitic and finally nodular ones (Ortí et al. 1987–1988). Along the border of the Catalan Basin in the Priabonian, a basal laminar to stromatolitic layer passes upwards into thick layers of selenitic lithofacies, having a truncated upper surface. This discontinuity is finally overlain by some meters of thin-bedded, partly clastic gypsum (Ortí et al. 1985). Such an intra-gypsum discontinuity probably reflects a significant fall of the water level within the basin with associated exposure and dissolution at the borders. In both phases, the detailed study of the sulfate facies is made difficult by the fact that primary gypsum was totally transformed into anhydrite either synsedimentarily or by burial diagenesis. Now, as seen in outcrops and quarries, this anhydrite has not been completely rehydrated into gypsum, thus creating problems for industrial exploitation of the sulfate.

In the two Eocene evaporative phases numerous slumps and intraformational breccias suggest that well-developed gravitational instability existed on the basin borders. In some areas these structures may have affected the phase partially, as in the Lutetian, or totally, as in the Priabonian, where the sulfate horizon and the top of the underlying lutitic series are both involved (Busquets et al. 1985).

Immediately prior to the sulfate stage, a reef episode developed on the border of the Catalan Basin (Priabonian phase) which prograded basinward. Locally, slopes of this reefal system are clearly overlain by the preevaporitic, marly, anoxic facies. In turn, these slopes became overlaid by grainstone layers (made up of Miliolid foraminifera and ooliths) and finally by the first stromatolitic (sulfate) horizons mentioned previously. However, in the allochthonous terrains it is difficult to define such a peripheral reef system either in the Priabonian or in the Lutetian phase.

The chloride stage developed widely during the Priabonian phase. It is characterized by the presence of banded halite and by rhythmic potash facies, either sylvinite and carnallite, as well as by fine preservation of primary crystalline fabrics with no evidence of exposure. The marine origin of all these evaporitic sediments is proved by geochemical (bromine and potassium contents in the Priabonian chlorides and the composition of the fluid inclusions, García-Veigas 1993) and isotopic criteria as well (in both Lutetian and Priabonian sulfates). Nevertheless, some inflow of continental water at the end stage of the evolution of evaporative deposition (potash stage) in the Priabonian chlorides is evident (Ayora et al. 1994).

At the top of both evaporitic sections (sulfate in the Lutetian phase and chloride in the Priabonian) deposition ended with variably grained siliciclastic facies in which marine influences still are noticeable: (a) alternating marls and turbiditic sandstones also displaying anoxic cyclicity, and nearshore sandstones in the Priabonian of the Navarra Basin; and

(b) siliciclastic-calcareous facies of mixed platform origin in the Lutetian phase. Only at the top of the Priabonian of the Catalan Basin is there a marly sequence (transitional member), which also displays an anoxic character devoid of any certain marine influence. Topping all of these facies in both evaporitic phases is a cap composed of coarse clastic materials deposited within continental environments, mainly in the form of alluvial fans.

Primary Versus Secondary Gypsum

The volume of primary gypsum preserved in the evaporite formations of the Ebro Basin is insignificant when compared with the secondary forms. Primary gypsum exists almost exclusively in the microlenticular facies and only rarely in the selenitic facies, and remains restricted to some Miocene formations of the peripheral sulfate belt in the southwestern border of the basin (Ablitas, Borja, Monteagudo, Ribafrecha, Pozuelo, Calanda) and to the Cerezo Gypsum in the central part of the basin. This primary gypsum is indicated in figure 8.4 by the symbol (P). The abundance of secondary gypsum in the surface is related to:

(a) the very shallow character of the Tertiary continental evaporite sedimentation that took place in the central parts of the Ebro Basin. Because the water was extremely shallow it produced highly concentrated, usually chloride-rich, interstitial brines within the laminated sediments, thus favoring their rapid conversion to nodular anhydrite during early diagenesis. In addition, relatively deep burial of the Tertiary sedimentary sequence accounts for later burial conversion of gypsum to anhydrite in the peripheral lake system, in particular for the Paleogene and for some of the Miocene units. Texturally, these secondary gypsum rocks belong to the following petrographic types in order of volume: alabastrine, megacrystalline, and porphyroblastic (Ortí 1977). The alabastrine types, in particular where present as meganodules, have resulted in the development of a substantial alabaster industry in many villages.

(b) the abundace of glauberite formations that have been replaced by secondary gypsum near the surface during exhumation. Petrographic criteria for distinguishing between this type of secondary gypsum (reticulated, acicular fabrics) and that coming from anhydrite have been given in Ortí et al. (1994b).

Comparison with Other Evaporitic Tertiary Basins in Spain

In a general sense, the continental evaporitic sedimentation recorded in the Ebro Basin may be compared readily with other intracratonic Tertiary basins of the Iberian peninsula. Perhaps the most suitable example is the Tajo basin, which also displays several evaporitic episodes in the Paleocene-Eocene to lower-middle Miocene interval. A great variety of facies are present in the Tajo basin, including central basin formations of the laminated to nodular type (L-N) bearing halite and sodium-sulfates (and associated minor potassium-magnesium sulfates), as well as sulfate lakes with bioturbation and chert replacement, or displaying clastic or turbiditic gypsum. A marked chemical recycling process also operated, reworking the Mesozoic evaporites of the surrounding chains by dissolution and feeding them into the endorheic Tertiary drainage systems (Utrilla et al. 1992). The evaporitic formations of these deposits are also of considerable economic value, in particular the deposits of thenardite and glauberite (Ortí et al. 1979; Ordoñez and Garcia del Cura 1994).

Also of great interest is the Calatayud Basin (Zaragoza province) in which a thick evaporite formation of early Miocene age is present. This basin also displays a marginal, matrix-free, microlenticular association (with abundant chert and anhydrite meganodulares) and a central, thick, laminated-nodular association. The latter is formed by secondary gypsum that comes from diagenesis of glauberite and minor anhydrite (Ortí and Rosell 1994). A general section throughout this evaporite formation shows an asymmetric evaporite pattern based on primary and diagenetic lithofacies and a salinity gradient based on the isotopic composition (δ^{18} O, δ^{34} S) of the sulfates (Ortí et al. 1994a).

Climatic Control and Evaporites

The ability of the Ebro Basin, as a whole, to record and preserve evaporite episodes since the Paleogene, including most of the Miocene, is demonstrated in figures 8.3 and 8.4. Several factors have contributed to evaporite formation: climate, geographic isolation, tectonic regime, and solute supply. Doubtless the climatic conditions that have favored some of the major episodes also have affected other geological domains in the vicinity. Thus, in the Eocene-Oligocene transition and the early Oligocene, evaporites were deposited in the rift system of southeast France and in the Paris basin, as well as in Spain. Arid or semi-arid environments prevailed in many areas during the lower-middle Miocene, and extensive evaporite deposits also formed in the Carpathian foreland basin and in the Red Sea during this time.

While climate was the general factor allowing the evaporitic sedimentation, other factors exerted more precise control on depositional conditions. In the case of marine-related depositional episodes the controlling factors were thrust sheet emplacement and general ocean level fluctuations. In the case of the continental deposits, the controls include low levels of activity in the marginal alluvial fan systems and high subsidence rates in the depocenters of the endorheic (internal drainage) systems. Local tectonism is also the reason for the lack of exact coincidence between marine and continental episodes during the Eocene in the Catalan sector of the Ebro Basin.

(Accepted for publication April 1992; revised July 1995.)

9

First Continental Evaporitic Phase in the South Pyrenean Central Area: Tremp Gypsum (Garumn Facies, Upper Paleocene; Allochthonous Zone)

JAVIER GARCÍA VEIGAS

Geological Setting

The Tremp Formation is the oldest Paleogene evaporite deposit and is located within the allochthonous Central South Pyrenean structural unit (Seguret 1972) (figure 9.1). The evaporites are exposed in two different sedimentary basins, the Tremp Basin and the Ager Vale Basin, separated by the Montsec Ridge, which constitutes the front of an allochthonous thrust sheet belonging to the Lower Thrust Sheet sequence and is made up largely of Mesozoic components.

The Tremp Basin formed as a wide synclinal structure composed of Meso-Cenozoic sediments. In its central part, but slightly shifted to the north, the Sant Corneli anticline, which has an east-west trend, developed and acted as a positive high area across the Tremp depression. The Vale of Ager is formed by an assymmetrical syncline whose northern flank was overthrust by the Montsec Ridge.

The Tremp Basin itself began to differentiate from the general Paleogene south Pyrenean basin as a discrete depositional area during the Paleocene-Eocene. In the early Eocene (early Ypresian or Ilerdian; Hottinger and Schaub 1961) it was separated from the "Ager Gulf" by an early uplift of the "Montsec High."

In the zone under consideration, the term Garumn facies refers to the assemblage of continental sediments, stratigraphically intercalated between the underlying marine Maastrichtian (the Areny Sandstone Formation) and the overlying deltaic sediments that display evidence of tidal control (Ager Formation, lower Ypresian, or Ilerdian). The lowermost horizons of the continental South Pyrenean facies are preferentially developed in the eastern sector of the basin (Isona-Conques area), while the upper ones crop out in its western part (Tremp area). Farther to the west, the Paleogene South Pyrenean basin was open to the Atlantic ocean and, as a result, the continental facies of the Tremp area passes laterally westward into transitional (marine-marginal) carbonate sediments. The transition from continental to marine facies is marked by an increase in total sedimentary thickness.

In the south Pyrenean domain those materials composing the Garumn Facies were defined as the Tremp Formation by Mey et al. (1968). Puigdefàbregas and Souquet (1985) consider these uppermost Cretaceous-Eocene sediments (Garumn Facies) the product of the earliest movement of the allochthonous units in the Maastrichtian.

In the Tremp Basin, Rosell (1967) distinguished three different units within the Garumn facies: (A) a lower detrital level bearing coal layers; (B) an intermediate calcareous level; and (C) an upper marly

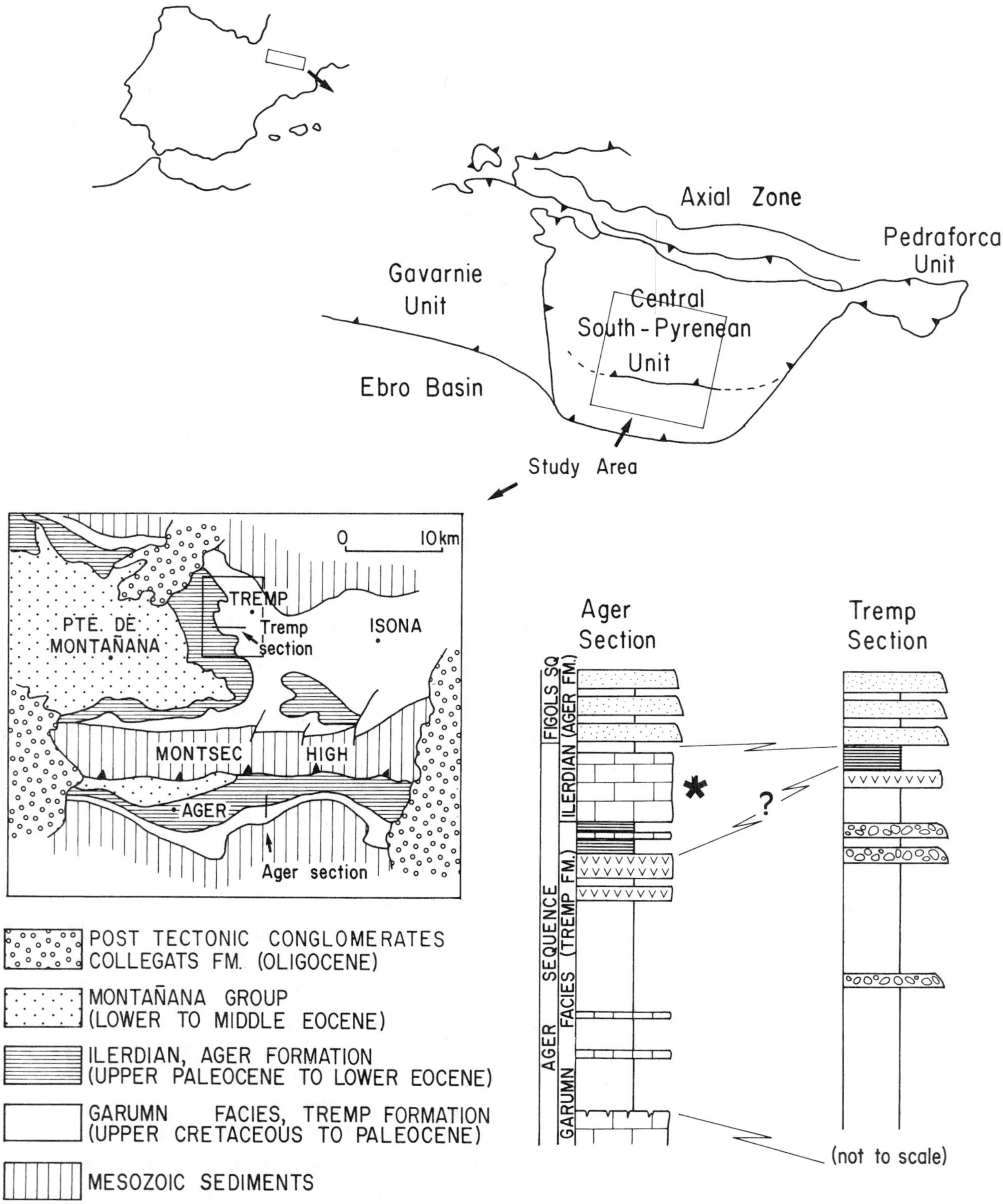

Fig. 9.1. Location map of the south Pyrenean central area under discussion here. Schematic stratigraphic sections and relationships between Tremp basin and the Ager Vale are given at right. Asterisk (*) at right of Ager section indicates Alveolina Limestone.

gypsiferous level. The lower and intermediate levels contain abundant marine fauna and vertebrate remains (Reptilian). Both the A and B levels conformably overlie the Maastrichtian marine sediments. The upper level, however, is azoic and apparently represents deposition throughout the whole of the Paleocene.

In the Ager syncline, Colombo et al. (1986) have differentiated four lithologic units in the Garumn facies: (a) basal limestones of shallow lacustrine environment (90 m); (b) a lower detrital level (180 m); (c) intermediate carbonates of palustrine-lacustrine environment (70 m); and (d) an upper detrital complex of an expansive, overstepping character (500 m) that was formed by deposition within a mud flat environment. These Garumn deposits, about 900 m total thickness, are overlain by a marine, carbonate platform deposit (Alveolina Limestone) of upper Paleocene to lower Eocene age (early Ypresian: Ilerdian). The evaporitic unit under discussion in this paper lies just at the top of the upper sedimentary complex.

This gypsiferous intercalation is included in the Ager sequence defined by Mutti et al. (1985), which includes the upper part of the Tremp Formation and the Alveolina Limestone member of the Ager Formation. This sequence is limited at the base and at the top by the respective regional unconformities. Overlying this sequence is the Figols sequence, which contains all those sediments attributed to the early Ypresian (Ilerdian) in its type-section by Hottinger and Schaub (1961). In the Tremp area the Ager sequence has not been precisely established.

The maximum thickness of the evaporitic facies is developed towards the most central parts of the Tremp and Ager basins. This was caused by a greater subsidence rate in both areas during the uplift of the Montsec high. Stratigraphic correlation of cross sections shows the development of a progressive unconformity in the Montsec area extending from the late Maastrichtian until the Eocene. The evaporitic units are exposed in the uppermost 200 m of the Garumn facies of both basins, and are interlayered within a thick lutitic series. They have been studied in detail by García Veigas (1988).

Evaporitic Facies Analysis

In the evaporitic units of the Garumn facies there is a number of calcium sulfate lithofacies, now present as secondary gypsum in all the studied sections (figure 9.2A). These facies include:

(A) gypsum that is present within infilled bioturbation molds (root structures). It is made up of columnar structures built by coalescent small nodules within a red argillaceous matrix that occupy the space opened by terrestial plant roots. In general, these nodules are composed of fine-grained (alabastrine) gypsum (possibly arid climate soil profiles).

(B) lenticular gypsum rosettes that consist of individual (or aggregates of) lenticular-shaped crystals (now as secondary gypsum pseudomorphs). In general these are no larger than 1 cm (as in a desert-rose crust).

(C) micronodular gypsum formed by millimetric (up to 1 cm) nodules of alabastrine gypsum, very irregularly distributed in a detrital or carbonate matrix.

(D) entrolithic gypsum consisting of contorted horizons of alabastrine gypsum oriented in diverse directions, developed either within an argillaceous and/or gypsiferous-argillaceous matrix or within any or all the above described lithofacies (A to C) as well as in lithofacies E (below). Vertical orientation of these entrolithic structures is common as well as the normally horizontal alignment.

(E) laminated gypsum found in beds that are relatively planar, with fine laminae or beds (mm to cm thick) originally composed of microlenticular (<1mm) gypsum crystals, which may or may not alternate with marly or calcareous laminae. All of this is now alabastrine gypsum.

(F) massive nodular gypsum composed of macronodules (dm- to m-sized), either isolated or arranged in masses, which develop within the laminated facies. These nodules are either replacive or displacive in origin. The secondary gypsum is mainly alabastrine and less commonly megacrystalline or porphyroblastic.

Other lithofacies that are sporadically distributed include:

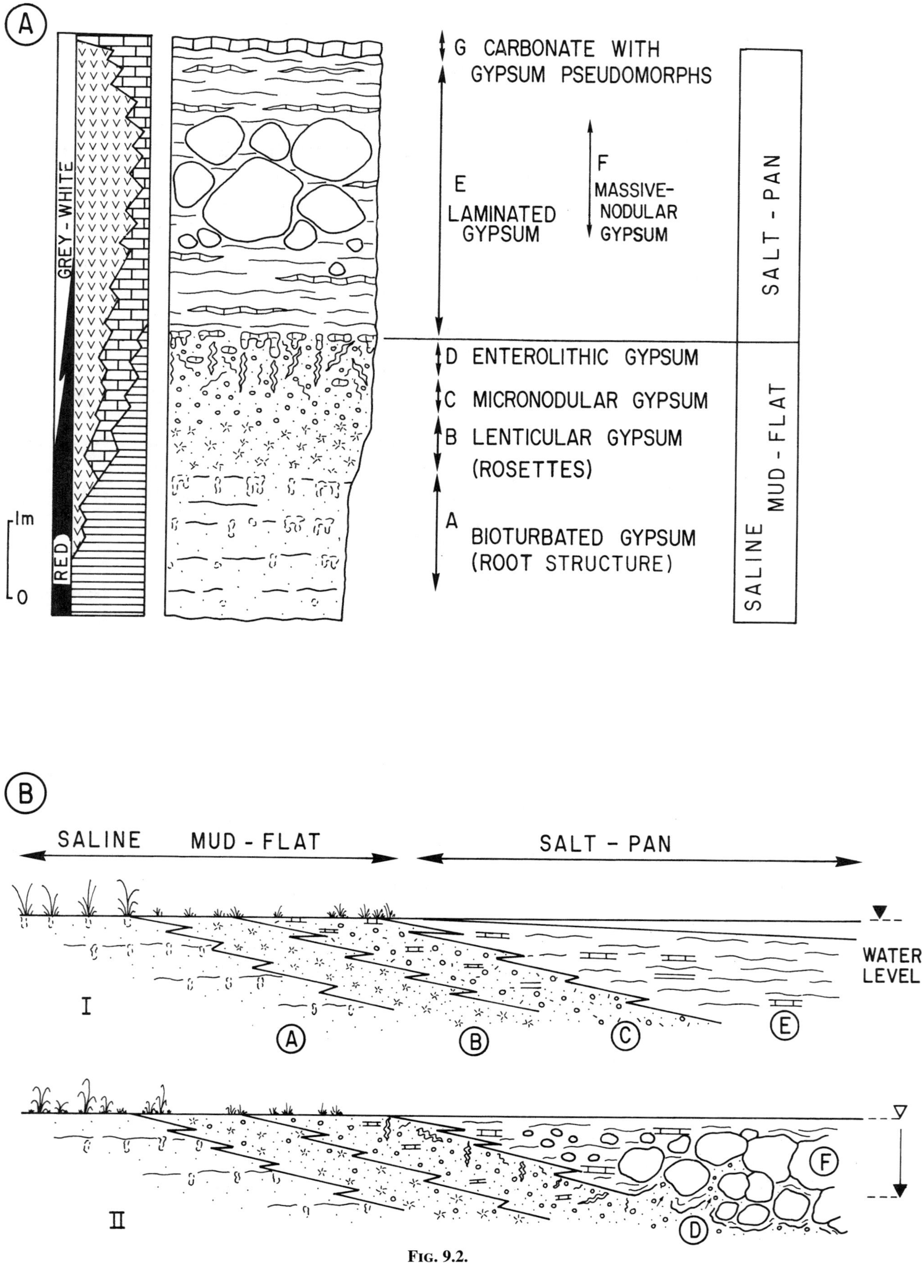
A
GREY - WHITE
RED
1m
0
G CARBONATE WITH
GYPSUM PSEUDOMORPHS
E
LAMINATED
GYPSUM
F
MASSIVE-
NODULAR
GYPSUM
SALT - PAN
D ENTEROLITHIC GYPSUM
C MICRONODULAR GYPSUM
B LENTICULAR GYPSUM
(ROSETTES)
A
BIOTURBATED GYPSUM
(ROOT STRUCTURE)
SALINE MUD - FLAT
B
SALINE MUD - FLAT
SALT - PAN
WATER
LEVEL
I
A
B
C
E
II
F
D

FIG. 9.2.

(G) calcareous horizons that display pseudomorphs of primary lenticular gypsum crystals. These pseudomorphs are commonly noted at the base or at the top of the gypsum layers.
(H) carbonaceous-nodular horizons with nodules of both carbonate and gypsum that develop within a dark clayey matrix bearing carbonaceous plant remains (coals).
(I) detrital gypsum beds of gypsarenites and gypsrudites having a lag-deposit origin. They are present either as siliciclastic or gypsiferous-siliciclastic channel fills. This facies represents the erosion of evaporitic layers that occurred during reactivation of alluvial or fluvial systems.

Calcareous layers associated with these evaporites are scarce in the Tremp area, but are common in the Vale of Ager. They are mainly composed of nodular mudstones developed within the sulfate or the argillaceous layers. In these calcareous sediments the most common lithofacies are: (1) laminated packstones bearing abundant charophyte stalks, (2) burrowed wackestone-packstones, (3) *Microcodium* grainstones, and (4) nodules with septaria structures cemented by celestite.

The evaporitic environment is represented by the development of sulfate units in the central part of both basins. Evaporite precipitation took place in a playa lake complex surrounded by mud flats. These playa deposits are represented by thick mud sequences with intercalations of lacustrine limestones and sandy channels. It is also common to find bioturbated levels caused by root structures that are infilled by gypsum or carbonates. The margins of the basin appear to have been sand flats with numerous braided stream deposits composed of conglomerates and sandstones.

Fig. 9.2. An idealized sedimentary cycle and its probable environmental settings. A. Vertical sequence of a typical evaporitic cycle (layer ST-5). B. Facies model of an evaporitic lacustrine environment (I) and the suggested mode of formation of the nodular anhydrite during early diagenesis (II).

Evaporitic Cycles

A good example of the most common type of depositional cycle developed in the various gypsum layers, particularly in those of the Tremp area, is given in figure 9.2A. The progressive development of an evaporitic environment is demonstrated by the gradual transition from interstitial (A to D) to free or subaquatic (E and G) precipitates. It should be noted that the maximum sediment thickness in any cycle is in the massive-nodular facies. The "G" (calcareous) and "H" (carbonaceous-nodular) facies represent a dilution stage in the lake, but both may be absent in an idealized cycle or even may occupy a different position within it (figure 9.3B). In the sequence, the progressive development of an evaporitic environment is demonstrated by the gradual transition from interstitial (A to D) to free or subaquatic (E and G) precipitates.

The lateral development of layer ST-5 (figure 9.3A) illustrates the facies change in the border zone of a saline lake, and layer SY-2 shows a similar lateral transition in a more central position in the lake. Despite the fact that the limited exposure of the layers usually prevents study in three dimensions, it seems evident that the development of the massive-nodular facies took place towards the central part of the lake. This suggests that such diagenetic growth occurred when the water level dropped below the sediment surface (i.e., the lake desiccated) with associated generation of displacive gypsum and/or anhydrite. The final stage of diagenesis resulted in replacement of the anhydrite by secondary gypsum during uplift. In the laminated facies there are some pseudomorphs of lenticular primary gypsum still preserved, but in most instances they have been destroyed during the ensuing displacive

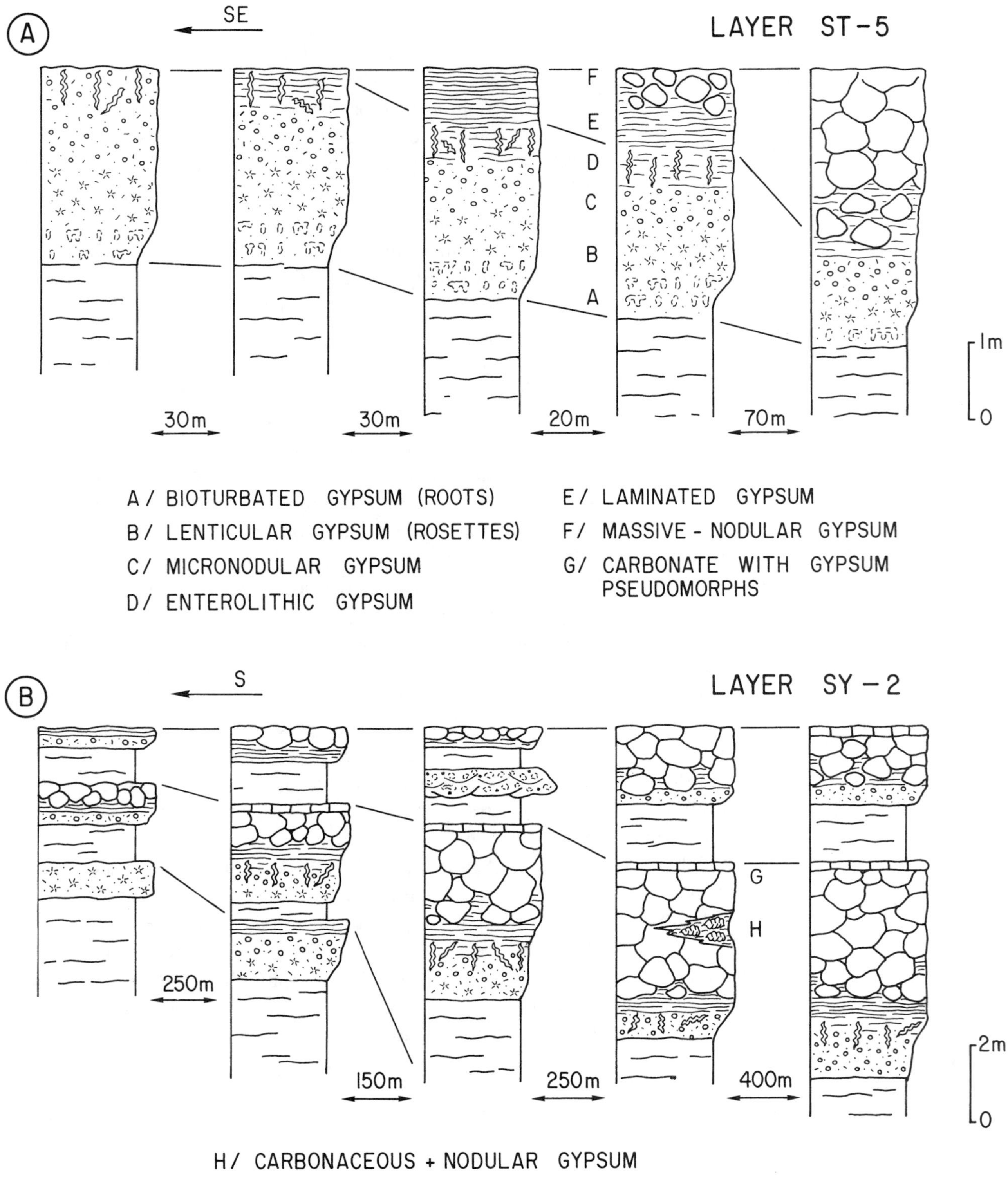

Fig. 9.3. Stratigraphic correlation between laterally related sections illustrating the development of two discrete evaporitic layers. A. Lateral change of facies developed at the margin of the lake (layer ST-5). B. The same change as developed in the lake's center. Location of both sections is indicated in figure 9.4A. Letters A through G refer to facies development within the evaporites; I is rare and is not indicated in the diagram.

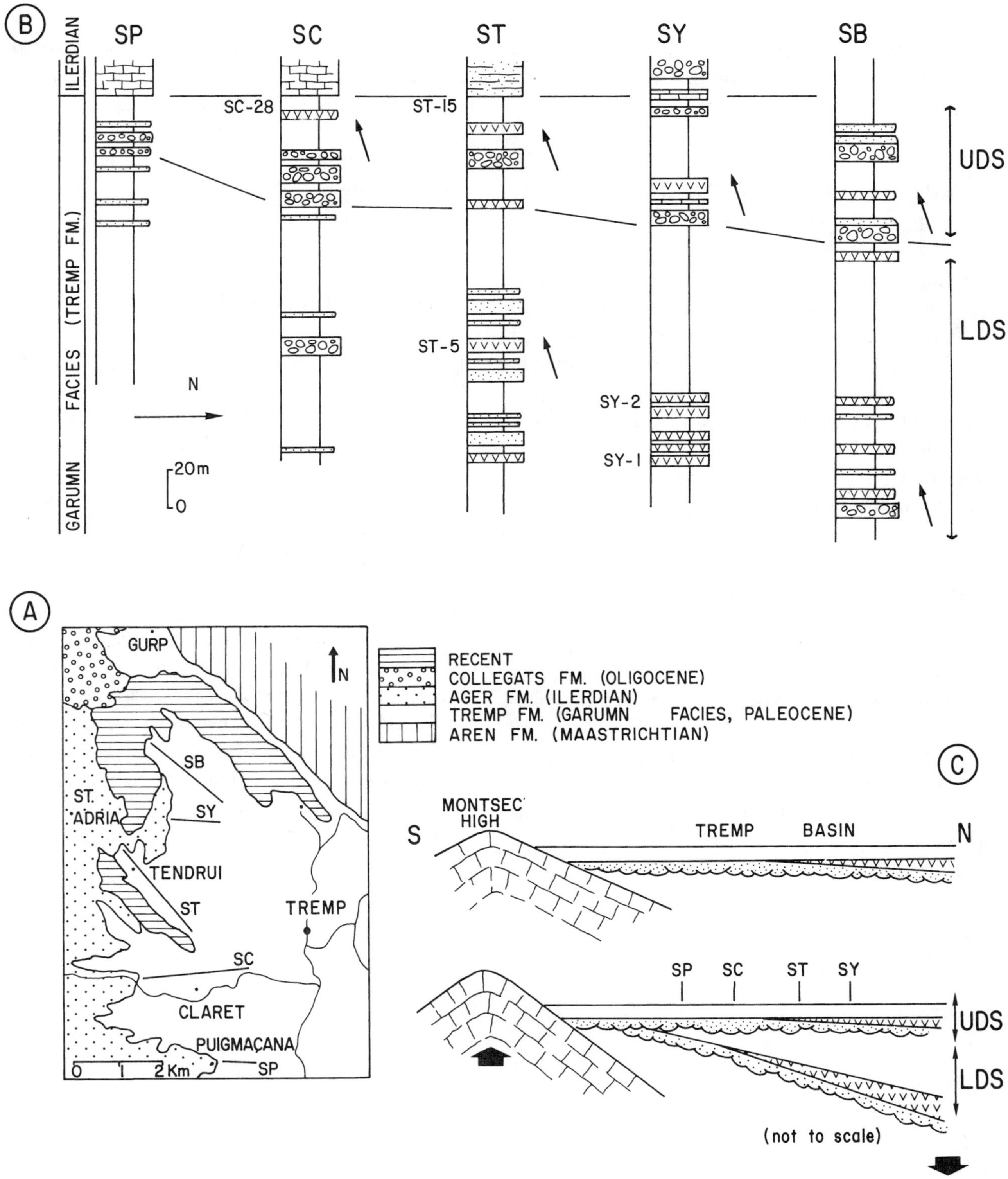

Fig. 9.4. Overview of stratigraphic sections and their development as used in this study. A. Location map of stratigraphic sections represented in B (rectangle indicated on figure 9.1). B. Correlation of critical stratigraphic profiles in the Tremp Basin, involving evaporite development. C. Representative schema for the two evaporitic depositional sequences represented in A, and associated with the uplift of the southern margin of the basin (Montsec high).

development of the nodular anhydrite. There is no clear evidence for formation of primary anhydrite, but it could have been possible.

Depositional Sequences at the Top of the Tremp Formation

In the upper part of the Garumn facies within the Tremp Basin (figure 9.4A) (the "upper marly-gypsiferous level" defined by Rosell 1967) a rather well-defined cyclicity exists, involving fining upward sequences. Each consists of three components (figure 9.4B and C) from bottom to top:

(A) a basal coarse-grained member composed of conglomerates and sandstones formed within paleochannels which probably correspond to tectonically controlled alluvial-fluvial episodes;
(B) an intermediate evaporitic member composed of argillaceous chemical sediments containing the previously described gypsiferous sediments which result from the development of shallow lakes after the stabilization of the basal, coarse-grained sediments; and
(C) an upper lutitic member made up of clays, silts, and marls bearing thin sandy intercalations within thick mud flat deposits developed at the end of a tectonic pulse.

A lower depositional sequence (LDS) and an upper one (UDS) can be distinguished (figure 9.4C). The base of each cycle may be considered isochronous. Each cycle represents the sedimentary response to tectonic reactivation of the detrital systems, implying a progressive unconformity related to the uplift of the Montsec high.

Origin of the Formative Brines

In the Ager Vale the stratigraphic concordance between the transgressive uppermost part of the Garumn facies within which the evaporitic episodes are included and the Alveolina Limestones (the Ager sequence of Mutti et al. 1985) might suggest a marine supply for the shallow saline lakes, in the sense of coastal ponds and salinas. Nevertheless, the isotopic composition of the calcium sulfates of these evaporites shows a marked Triassic affinity, thus demonstrating that a continental, geochemical recycling mechanism was operative (Utrilla 1989; Utrilla et al. 1987). The same isotopic values were obtained from the gypsum of the Tremp area. This is in reasonable concordance with the abundant gypsiferous sections of Triassic materials that were and are outcropping, or known in boreholes, in this area of the central south Pyrenees.

(Accepted for publication April 1992.)

10

First Marine Evaporitic Phase in the South Pyrenean Foredeep: Beuda Gypsum (Middle Eocene: Autochthonous-Allochthonous Zone)

Federico Ortí and Laura Rosell

The structural evolution of the South Pyrenean foredeep is the major factor controlling the development of evaporative depositional environments during the middle Eocene. To understand the formation of these restricted basins and their depositional fill, we first review, on a regional basis, the creation of the basin itself. Beginning with the earliest stages of thrust movement, the southward displacement of South Pyrenean thrust sheets over a stable early Eocene (lower Ypresian: Ilerdian) carbonate platform resulted in the development of the South Pyrenean foreland basin. The southern Pyrenees can be divided into a series of three thrust sheets that were transported toward the south (Muñoz et al. 1986): the Upper Thrust Sheets (Central South Pyrenean thrust sheets; Pedraforca and Montgrí nappes); the Middle Thrust Sheets (Gavarnie and Cadí nappes); and the Lower Thrust Sheets. The thrusts developed in a piggyback sequence, with the Upper Thrust Sheets emplaced first. As the deformation progressed to the south, the axes of successive depocenters also migrated southwards by almost one hundred kilometers, with each sedimentary sequence progressively incorporated into younger thrust sheets.

The Catalan sector of the South Pyrenean foreland basin contains 3,000 m of Paleogene to Oligocene sediments. The most complete successions are exposed in the northern thrust sheets and along the autochthonous border of the Catalan Coastal Range. To establish the tectono-sedimentary evolution of this sector, the depositional sequences are used as defined by Puigdefàbregas et al. (1986) (figure 10.1) who pointed out that in the Southeastern Pyrenees (Catalonia) a noticeable cyclicity is present in the thrust systems. These changes in thrust type correlate well with the major marine evaporite phases recorded in the foreland basin (Beuda and Cardona sequences). Thus the development of the evaporitic events differentiates basin fill into three main depositional cycles that are directly tied to thrusting episodes.

Depositional Cyclicity

FIRST CYCLE

The first depositional cycle (early to early-middle Eocene: Ypresian to earliest Lutetian) coincides with the submarine emplacement of the Upper Thrust Sheets. It is characterized by progressive deepening of the northern margin of the basin due to tectonic loading. The resulting slope and turbidite trough experienced a southward migration (Puigdefàbregas et al. 1986). This cycle is formed

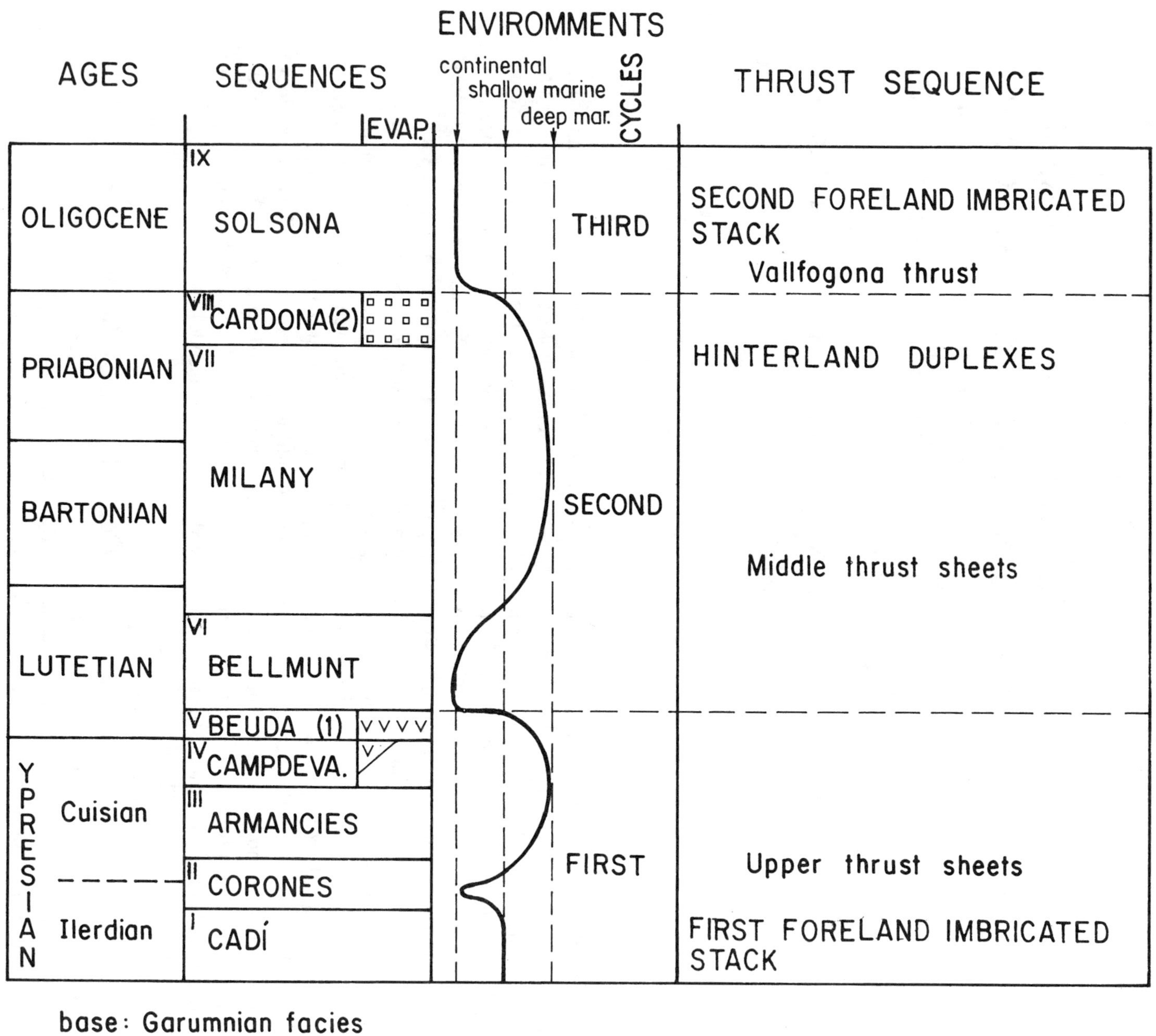

Fig. 10.1. Tectono-sedimentary setting of the southeastern Pyrenean marine evaporite phases (adapted from Puigdefàbregas et al. 1986). Dominant evaporites in the sequences are: 1 = the Beuda Formation: calcium sulfate; 2 = the Cardona Formation: chlorides. Sequence numbers and names (I—IX) are given in figure.

by the following sequences, from bottom to top:

Cadí sequence (I). The Cadí sequence (I) (early Ypresian: Ilerdian) is a transgressive sedimentary sequence that attains a maximum thickness of 300 m. It is made up of the Alveolina Limestones (Cadí and Orpí formations) that were deposited in a shallow marine environment and grade northwards into a slightly deeper, marly environment (Sagnari Formation) (figure 10.2).

Corones sequence (II). The Corones sequence (II) (middle to late Ypresian: Ilerdian and early Cuisian) represents a regressive event, attaining a maximum thickness of 250 m. It includes deltaic sediments in the north, which prograded southward, and a thin cover of terrigenous red beds in the south (Sagnari Formation and Corones Formation). The top of the sequence is characterized by algal carbonates containing hydrocarbons.

Armàncies sequence (III). The Armàncies sequence (III) (late Ypresian: Cuisian) records a sudden deepening of the basin indicated by the presence of marine slope deposits (shales, calcareous turbidites, slope breccias) filling a deep northern trough. At its southern margin a shallow carbonate platform developed (Penya Formation). This platform probably had a gradual transition southwards to the red beds of the Pontils group (figure 10.2A).

Campdevànol sequence (IV). The Campdevànol sequence (IV) (late Ypresian: Cuisian) marks a sudden clastic influx of terrigenous turbidites attaining a thickness of 900 m (figure 10.2B). The deep turbidite trough originated in front of the moving Pedraforca and Montgrí nappes and had west-oriented, longitudinal paleocurrents, roughly parallel to the axis of the basin. The sparse, restricted faunal content and other paleosalinity indicators, such as the presence of sporadic levels of calcitized anhydrite nodules and a few intercalations of gypsum, suggest an hypersaline environment. At the same time, some evaporitic deposition took place on the shallow carbonate platform, located at the southern margin, but these carbonate-sulfate facies are only known from boreholes (figure 10.2B).

Beuda sequence (V). The Beuda sequence (V) (early Lutetian) is an evaporative sequence that was deposited as the basin was constricted during the final emplacement stage of the upper thrust sheets. The marine evaporites developed at the top of this first cycle are described later in this paper. The spatial distributions of the various units are shown in a north-south general section in figure 10.2A and in figure 10.3.

SECOND CYCLE

The second depositional cycle (middle to upper Eocene: early Lutetian-Priabonian) coincides with the development of an antiformal stack behind the leading edge of the previously emplaced thrust sheets. Rejuvenation of topographic relief resulted in the establishment of prograding deltas at the same time that the basin depocenter migrated southwards. This depositional cycle is formed by the following sequences, from bottom to top:

Bellmunt sequence (VI). The Bellmunt sequence (VI) (Lutetian) is characterized by a coarse-grained deltaic complex up to 1,000 m thick, that prograded southwards in tandem with a retreating carbonate platform. The deltaic complex includes prodelta marls (Banyoles Formation), delta-front sandstones (Barcons Formation), and delta plain to alluvial conglomerates and sandstones (Bellmunt Formation). The associated carbonate platform (Tavertet Formation) is a high-energy, nearshore belt composed of *Nummulites* (the various lithostratigraphic units of this sequence have been omitted in figure 10.4A).

Milany sequence (VII). The Milany sequence (VII) (Bartonian—early Priabonian) consists of another deltaic complex, up to 1,200 m thick. This sequence starts with a transgressive event indicated by either glauconitic sandstones (Folgueroles Formation) or a nearshore complex of carbonates and sandstones (Collbàs Formation). This southwards progradation of the coarse-grained delta complex includes prodelta marls (Igualada Formation), delta front sandstones (Milany Formation), and delta plain to alluvial fan conglomerates (Sta. Magdalena Formation). In the Catalan Coastal Range, northwards progradation of fan deltas occurred during the deposition of this sequence (Montserrat Formation). At the final stage of both northern and southern deltaic progradation a coral reef belt

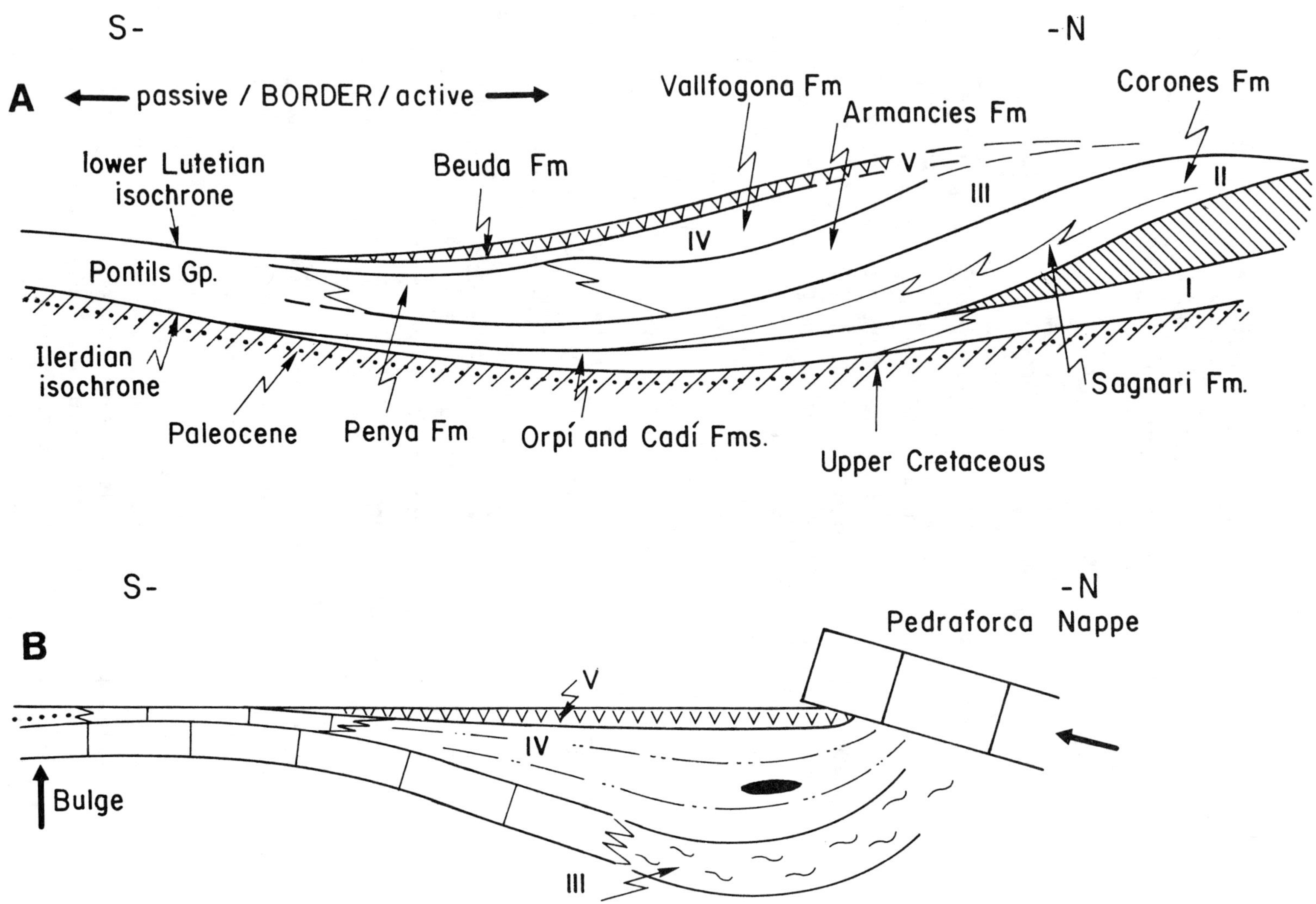

Fig. 10.2. Cross-sectional views of the lithostratigraphic-tectonic development for the first sedimentary cycle of the Eocene sequences in the south Pyrenean foredeep. Sequence numbers given in figure 10.1. A. Lithostratigraphic units and depositional sequences for the first sedimentary cycle of the Eocene (adapted from Puigdefàbregas et al. 1986). B. Schematic representation of the tectono-sedimentary evolution of the foreland basin during the Beuda sequence (V) (adapted from Puigdefàbregas et al. 1986). Some Triassic olistoliths (in black) are contained in the Vallfogona Formation of sequence (IV).

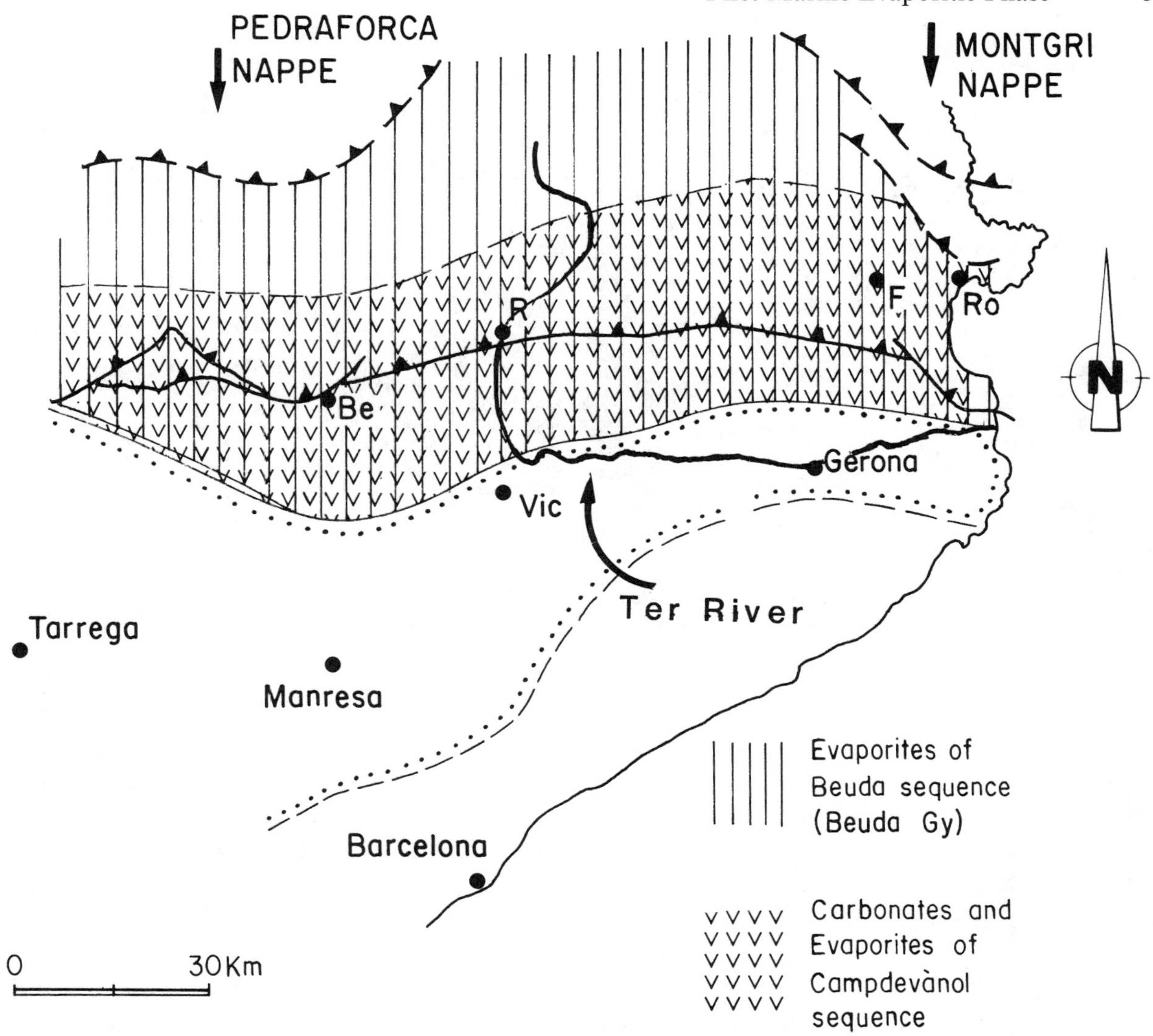

Fig. 10.3. Palinspastic reconstruction of the assumed original facies distribution for evaporites in the Campdevnol and Beuda sequences (simplified from Puigdefàbregas et al. 1986). This schematic map takes into account the displacement of the south Pyrenean thrust sheets. The position of Pedraforca-Montgrînappe corresponds to the end of the Beuda sequence. Modern position of the Vallfogona thrust is indicated. Be = Berga; R = Ripoll; F = Figueres; Ro = Roses.

(Tossa Formation) developed over the outer edge of the delta-front sandstones.

Cardona sequence (VIII). The Cardona sequence (VIII) (late Priabonian) consists of a distinctive evaporite cycle developed in the center of the residual basin. These sediments are the consequence of the final stages of the emplacement of the Cadí nappe and its westerly equivalents, which resulted in the confinement of the last marine foreland basin. Marine evaporites belonging to the top of the second cycle will be discussed in the following paper (chapter 11). The spatial distribution of these units in a north-south general section is shown in figure 10.4A, and a schematic model explaining the tectonic and sedimentary position of the Cardona sequence appears in figure 10.4B.

THIRD CYCLE

The third depositional cycle (late Priabonian—Oligocene) (figure 10.1) was caused by significant

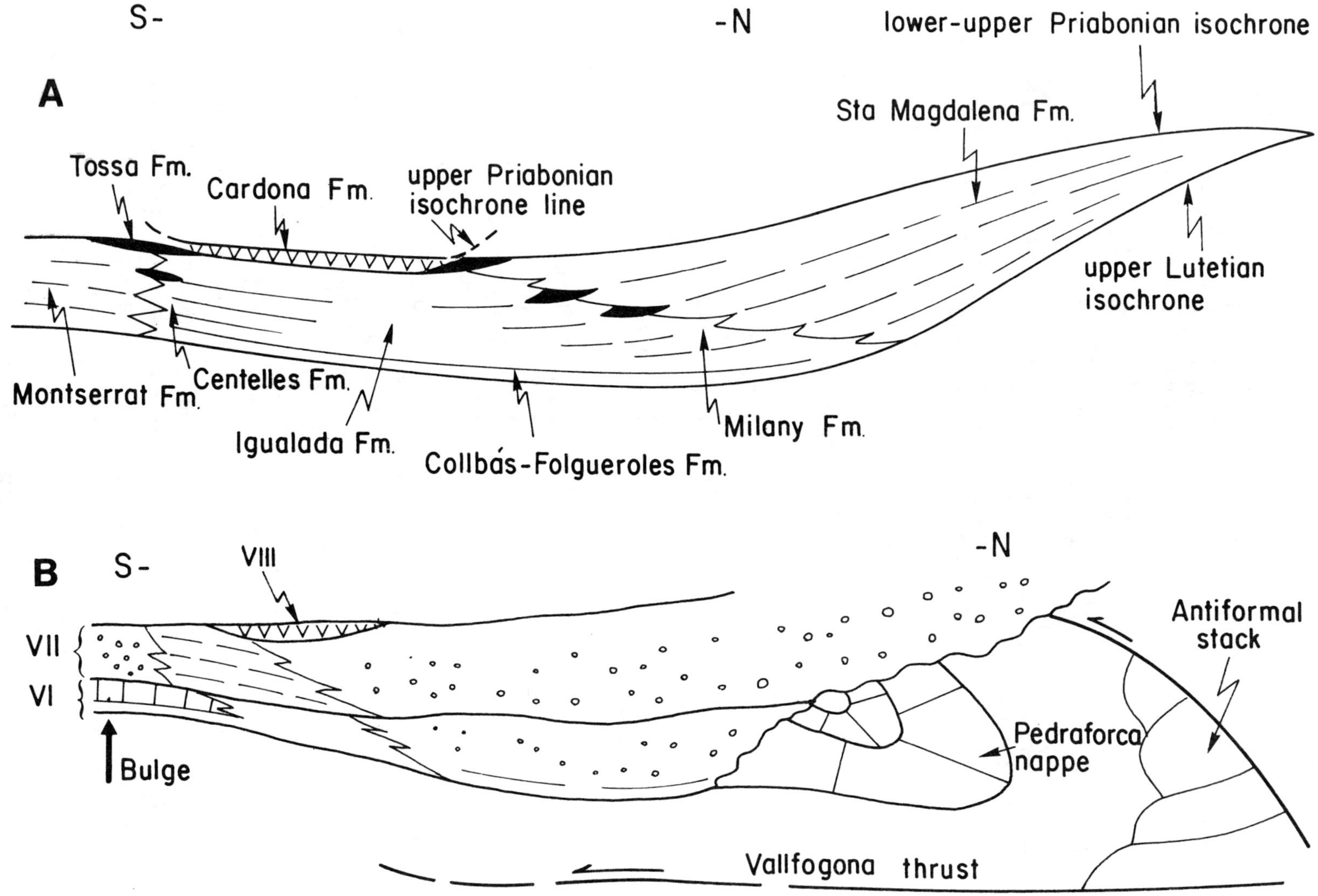

Fig. 10.4. Cross-sectional views of the lithostratigraphic-tectonic development of the Milany (VII) and Cardona (VIII) sequences. A. Lithostratigraphic units in the Milany and Cardona sequences (upper part of the second sedimentary cycle of the Eocene, adapted from Puigdefàbregas et al. 1986). B. Schematic representation of the tectono-sedimentary evolution of the foreland basin during the Cardona sequence (VIII) (adapted from Puigdefàbregas et al. 1986). VIII = Cardona sequence; VII = Milany sequence; VI = Bellmunt sequence (antiformal stack of Freser valley and major deformation of Pedraforca nappe occurred mostly during the Milany sequence).

southward thrusting that carried all other thrust sheets and incorporated the northern parts of the foreland basin sequences. These older thrust sheets, displaced by the emergent Vallfogona thrust, supplied clastics to alluvial fans that were laterally connected with lacustrine areas in the center of the basin. This cycle has only a single depositional sequence (Puigdefàbregas et al. 1986).

Solsona sequence (IX). The Solsona sequence (IX) (Oligocene) consists of a succession of alluvial and lacustrine deposits that are considered the last major stage in the development of the South Pyrenean foreland basin. Alluvial fans developed on the basin margins and progressive unconformities developed along the northern, most active, zone due to displacement on the Vallfogona thrust. The center of the basin was flooded with fluvial waters coming from active alluvial fans. Newly formed lacustrine areas were filled by thin turbidites, carbonates, evaporites, and intercalated alluvial deposits, including lignite.

In the complex tectonosedimentary evolution of the South Pyrenean foredeep, evaporitic events of marine origin apparently coincide with the development of flat thrust trajectories, whereas basinward progradation of clastic systems correlates with ramping episodes (Puigdefàbregas et al. 1986). Also involved in the sedimentary pattern is the effect of eustatic or worldwide sea level fluctuation as an additional factor controlling marine evaporite deposition during the Eocene (Vail et al. 1977).

Geological Setting

The presence of marine evaporites during the late Ypresian (Cuisian) to Lutetian in the southeastern Pyrenees is noted in the upper part of the first cycle of sedimentation (Campdevànol and Beuda sequences) (Puigdefàbregas et al. 1986). A palinspastic reconstruction of the original deposition area of these evaporites is given in figure 10.3.

The middle Eocene evaporitic deposits of the southern Pyrenees are exposed in the Cadí thrust sheet (Cadí nappe) between the villages of Vilada to the west and Olot to the east (figure 10.5A). The deposits consist mainly of a relatively continuous and thick (up to 80 m) calcium sulfate horizon. This evaporitic deposit developed towards the top of the first sedimentary cycle (lower to lower-middle Eocene: Ypresian to earliest Lutetian) of the south Pyrenean areas and is referred to as the Beuda sequence (Puigdefàbregas et al. 1986). In the central area of the basin, near Ripoll, the above-mentioned cycle is defined by the lithostratigraphic units shown in table 10.1.

These sediments were deposited within a compressive tectonic regime in a sedimentary trough that trended east-west, whose genesis and evolution was controlled by the emplacement of the allochthonous units of the southeastern Pyrenees. The character of the lithostratigraphic successions in table 10.1 is as follows: units 4 and 5 are turbiditic and represent the greatest maximum depth in the marine trough, while units 2 and 3 received significant fresh water input, based on the presence of lacustrine limestones. Units located at the base and at the top of the cycle are continental in origin.

In their zone of surface exposure, the evaporitic materials are enfolded within the Ripoll syncline, which is the major structure of the Cadí nappe, emplaced during the Late Eocene to Oligocene (figure 10.5B). A detailed study of this sequence was presented by Ortí et al. (1987–1988).

PRE-EVAPORITIC SEDIMENTS (VALLFOGONA FORMATION: LATE YPRESIAN, CUISIAN)

The basal unit of the evaporite sequence is a siliciclastic turbidite unit known as the Vallfogona Formation (Estévez 1970), which corresponds to the Campdevànol Sequence of Puigdefábregas et al. (1986). It is about 900 m thick and its general sedimentologic features were described mainly by Kromm (1967) and Busquets (1981). It consists of alternating decimetric layers of gray lutite and fine-grained sandstones. Olistoliths of clayey sulfatic Triassic masses are intercalated within the turbidites of this formation. This turbiditic unit is also referred to as the Ripoll Turbidite Sequence, described by Costa (1989).

Two main lines of evidence suggest restriction of the basin with concomitant stratification of the water body that began before evaporite deposition.

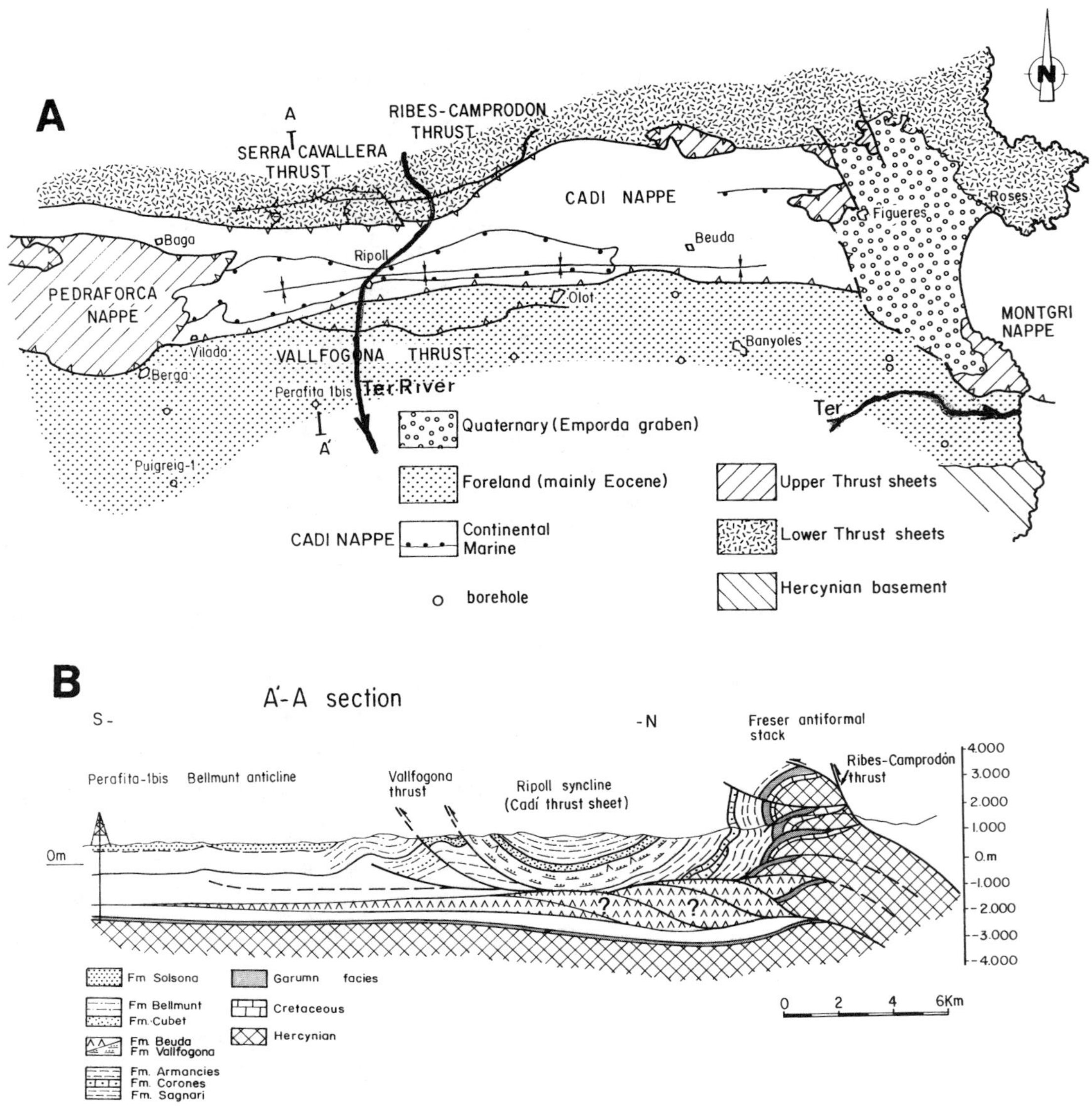

FIG. 10.5. Map and cross-sectional view of the southeastern Pyrenees. A. Schematic geological map of the southeastern Pyrenees showing the position of the Cadí nappe (simplified from Clavell et al. 1988). Small circles correspond to deep boreholes (names of boreholes not given in all cases). The approximate location of the section AA′ is shown in figure B. The La Garrotxa volcanic zone is omitted. B. North-south geological section (AA′) of the southeastern Pyrenees (ENRESA 1987, slightly modified from Muñoz et al. 1986). This interpretation assumes a large structural accumulation of the Beuda anhydrites at the base of the Cadí nappe.

Only the lowermost third of this turbiditic sequence, in outcrops along the northern flank of the Ripoll syncline, shows any sign of bioturbation or benthonic activity. Paleontologic studies carried out on these turbidites utilizing nannoplankton, palynomorphs, fish, and plant remains, suggest there was an intermittent input of continental water into this marine basin (Busquets et al. 1987). Taken together, the excellent preservation of the plant remains and the absence of any benthonic activity suggest the existence of a reducing or oxygen-poor environment on the floor of the turbidite trough, although surficial waters were well oxygenated and produced a diverse fauna and flora. It seems that

TABLE 10.1.
Stratigraphic units of the first sedimentary cycle of the Eocene in the area near Ripoll.
(Puigdefàbregas et al. 1986)

SECOND CYCLE	BELLMUNT SANDSTONE AND LUTITE FM. (CONTINENTAL)
FIRST CYCLE	7. COUBET SANDSTONE AND LUTITE FM. 6. BEUDA GYPSUM FM. 5. VALLFOGONA SANDSTONE AND LUTITE FM. 4. ARMANCIES MARL FM. 3. CORONES LIMESTONE FM. 2. SAGNARI MARL FM. 1. CADI LIMESTONE FM.
BASE	GARUMN FACIES (CONTINENTAL)

stratification of water in the basin existed during much of the pre-evaporitic phase. But gradually this deep basin actually became evaporative, as evidenced by the presence of sporadic laminated, dark gypsum layers intercalated between siliciclastic turbidites. These layers are sporadic and localized and may display features suggestive of turbiditic emplacement (such as graded bedding).

EVAPORITIC SEDIMENTS (BEUDA FORMATION; EARLIEST LUTETIAN)

The Campdevànol Gypsum and Anhydrite Formation (Busquets 1981) or its lateral equivalent to the east of the Ter river, the Beuda Gypsum and Anhydrite Formation, crops out linearly on both flanks of the Ripoll syncline. A number of quarries exist, particularly in the northern flank, where the sulfate horizon locally reaches a thickness of one hundred meters. The formation is made up, in surface exposures, of secondary gypsum containing scattered, irregularly distributed, masses of fine-grained anhydrite. Only minor amounts of lutites, carbonates, and sandstones are associated with the calcium sulfates. Vestiges of halite are absent at the surface, but the existence of salt has been recorded in deep boreholes. The distribution of both the Beuda Gypsum and the underlying turbidites (Vallfogona Formation) is shown in figure 10.6.

In the subsurface this evaporitic unit becomes rapidly anhydritic and extends laterally to the east of the Beuda meridian. Towards the south it is also known in the autochthonous zone of the foreland as seen in boreholes data. According to drilling data, it is not present to the west of the Puigreig meridian (Van Romphey 1965). Also we suggest that it was originally deposited north of the present outcrop region. Some boreholes have cut through tectonically thickened sections of these sulfates beneath the thrust sheets.

Lithofacies.

In a vertical sequence of this formation, the following main lithofacies may be distinguished:

(a) *Banded-Laminated Gypsum* (figure 10.7) is dominant in the lower two thirds of the unit. In general, lamination is planar, with the thickness of each lamina or band varying between a few mm and one decimeter. Locally, this banding disappears in some layers, giving way to massive beds. Interlaminae are made up of dark clay or carbonate in very thin layers. The carbonates usually develop at the base of gypsum layers. No macroscopic pseudomorphs of primary gypsum crystals are present in outcrop samples of this lithofacies.

(b) *Nodular Sulfate* is dominant in the upper third of Beuda section, where two variants of this lithofacies may be distinguished:

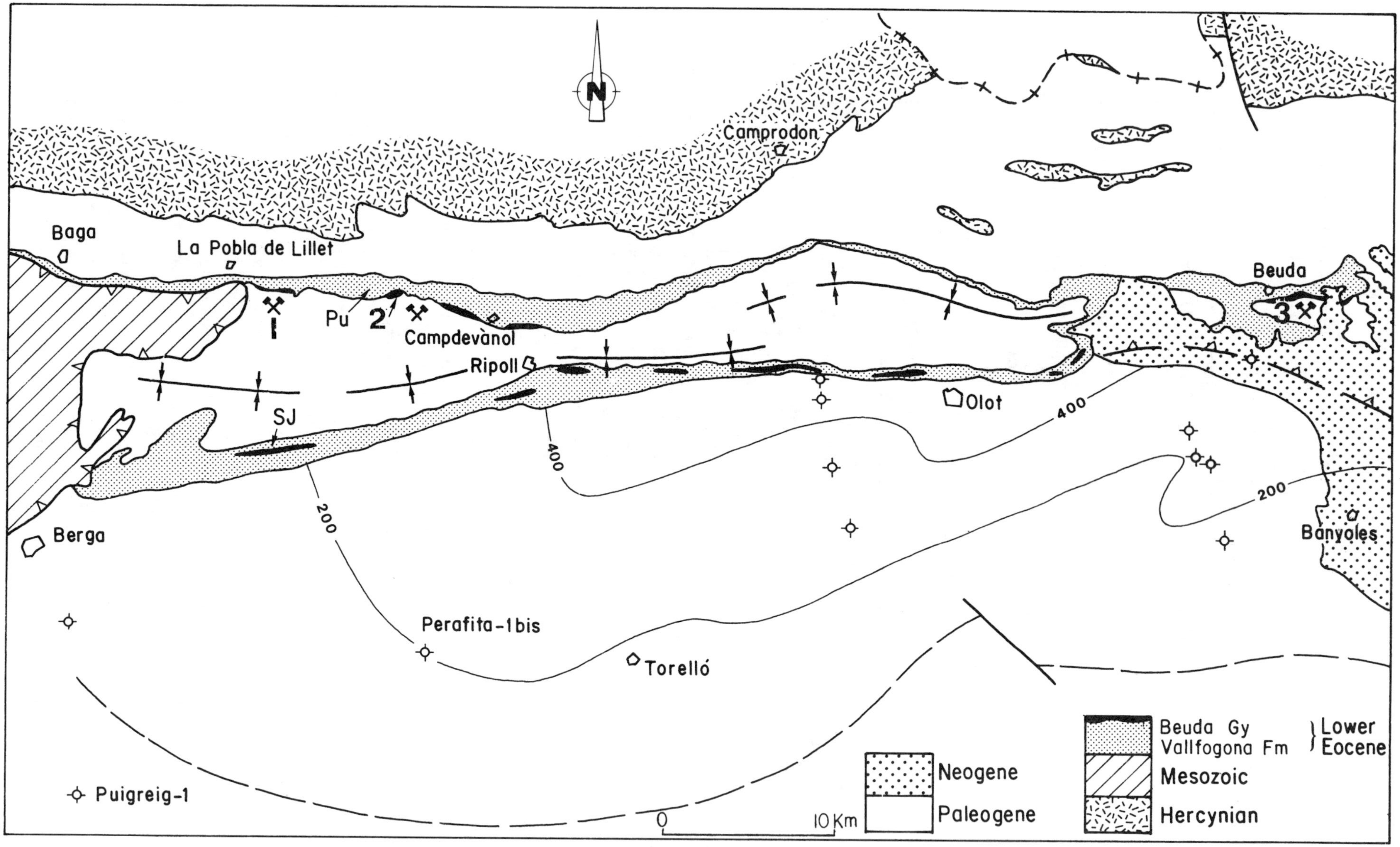

Fig. 10.6. Schematic outcrop map of the Vallfogona Formation (turbidites) and the Beuda Formation (evaporites) in the Ripoll-Beuda area. Some isopach lines of the Beuda Formation are indicated in the autochthonous foreland zone (simplified from ENRESA 1987; internal report). Gypsum quarries: 1 = Guixer; 2 = Gusachs; 3 = Beuda. Turbidite sections: Pu = Puigbò; SJ = Sant Jaume de Frontanyà. The names and positions of some of the deep boreholes are indicated (Puigreig-1, Perafita-1 bis).

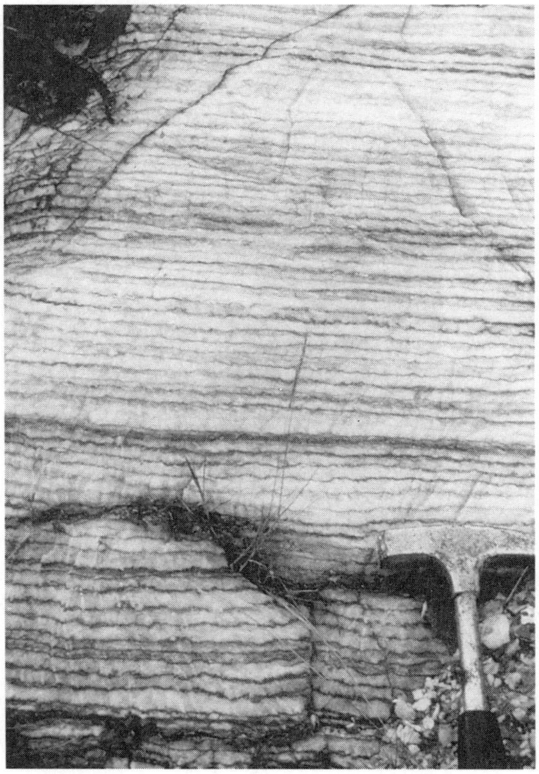

Fig. 10.7. Banded laminated lithofacies of the Beuda Gypsum in the Gusachs quarry (layers are 0.5 cm to 1 cm in thickness).

Nodular-mosaic gypsum: The nodular-mosaic is made up of a population of closely packed small nodules (between a few mm and some cms) giving way to nodular layers displaying typical nodular mosaic (chicken-wire) structures. Some small pseudomorphs of primary gypsum are noted.

Macronodules: Macronodules reach tens of cms in size and are generally depressed within the stratification plane. In the Beuda section, individual macronodules are intercalated within the laminated facies (up to 40 cm in diameter), displacing the regular, even layers (figure 10.8).

(c) *Pseudomorphic Tabular-Prismatic Sulfate* (figure 10.9) is dominant in the upper third of Guixer section, where gypsum layers are composed of centimetric-order pseudomorphs of tabular-prismatic primary gypsum (between 0.5 and 4 cm). Such crystal pseudomorphs are not twinned but locally they may be oriented in a particular direction. In general, transport indices are absent as well as any competitive growth features. We infer from this that the crystals grew displacively within the lutitic host sediment. Sporadically wavy ripples are also present in these layers.

The secondary gypsum is mainly of the fine-grained alabastrine variety. Locally, some porphyroblastic gypsum is present, as well as megacrystalline masses. The latter is mainly found in association with fractures. In many quarries residual masses of anhydrite hydrated along fractures and cleavage planes are still preserved. Crystalline fabrics of this anhydrite are randomly oriented and microgranular.

The vertical succession of lithofacies is interpreted as a shallowing upward sequence. Laminated facies resembling typical basinal deposits formed in the deepest water, while the pseudomorphic crystalline facies formed by intermittent flooding of the coastal zone in shallow water. Nodular layers, in particular those displaying mosaic fabrics, correspond to the most permanent subaerial exposures of the sulfate deposit. In general the water depths represented by the evaporite facies are uncertain. The laminated facies called "basinal" is simply subaqueous and in a central position in the sedimen-

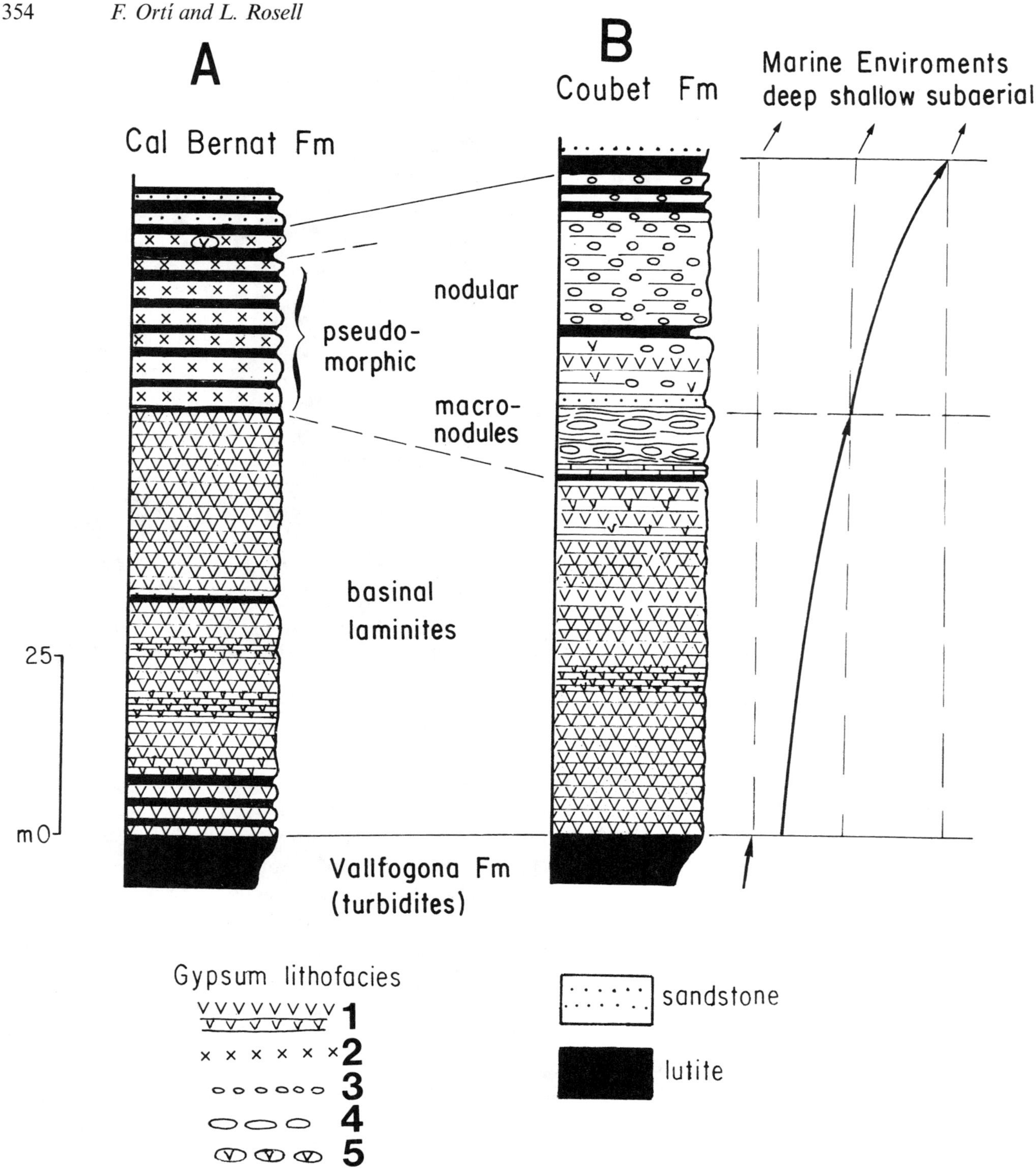

Fig. 10.8. A comparison between two representative sections of the Beuda Formation: A. the Guixer quarry, and B. the Beuda quarries. In both cases a shallowing-upwards tendency is interpreted (modified from Ortí et al. 1987/88). Key to gypsum lithofacies: 1 = banded to laminated; 2 = pseudomorphic of selenitic forms; 3 = nodular bedded; 4 = macronodules; 5 = isolated nodules.

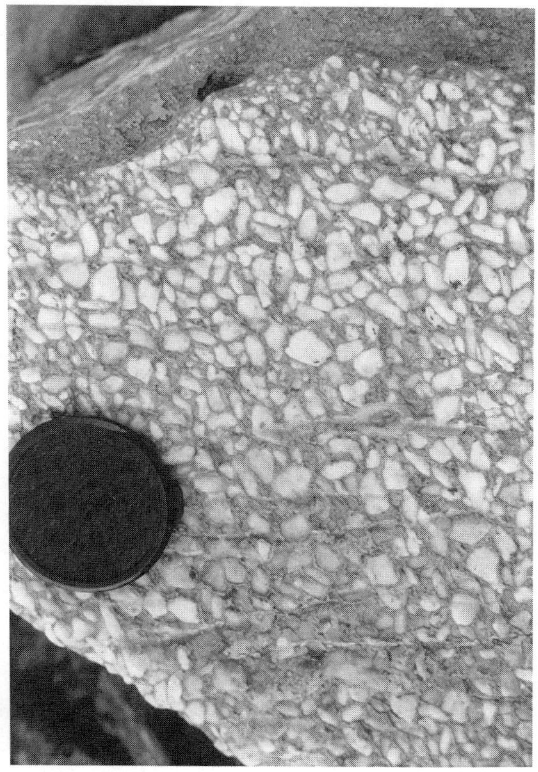

Fig. 10.9. Pseudomorphic tabular-prismatic lithofacies of the Beuda Gypsum in the Guixer quarry. Top view of a single bed.

tary pattern; the term is not a quantitative indicator of depth.

Macronodules interlayered within the laminated facies of the Beuda section are likely of subaquatic diagenetic origin, perhaps in connection with periods of major water drops or oscillations. In the German Zechstein, the sudden appearance of this facies (termed "Flaser anhydrite"), present as thick horizons interlayered with the basinal anhydritic laminites, has been interpreted by Richter-Bernburg (1985) as the result of periods of warm (shallower) water entering into the deep basin. In the south Pyrenean potash basin, another similar deposit of nodular anhydrite (basal anhydrite; Priabonian) was assigned a subaqueous origin by Rosell and Ortí (1980).

THE POST-EVAPORITIC SEDIMENTS (COUBET FORMATION; LUTETIAN)

The sediments forming the Coubet Formation (or Cal Bernat Formation) belong to a diversified spectrum of environments, very variable in depth, that represent the final infilling of the marine trough. On both flanks of the Ripoll syncline the general evolution of that formation is quite similar (figure 10.10).

Lithofacies.

From bottom to top, two different depositional units have been distinguished in this formation (Ortí et al. 1987-88): (a) sandstones and lutites (lower unit), formed in a restricted marine environment; and (b) mixed calcareous-siliciclastic platform sediments (upper unit), formed in an open marine environment.

The most common sediments in the lower unit are dark, very finely laminated marls, intercalated with sandstone beds, the latter having an average thickness of 20 to 30 cm. Such sandstone layers become more abundant above the lowermost 30 m of the formation. They are medium-grained, organized in fining upward sequences, and exhibit sedimentary structures typical of turbiditic deposits.

No signs of bioturbation were found in these tur-

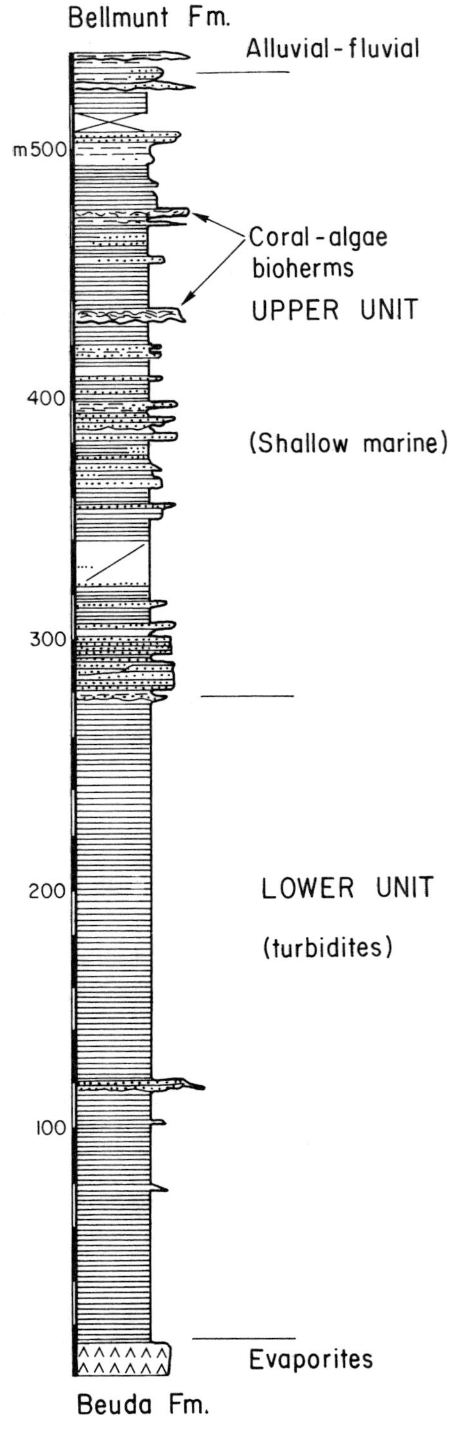

FIG. 10.10. Stratigraphical section of Coubet Formation, in St. Jaume de Frontanyà (southern flank of the Ripoll syncline; locality is indicated in figure 10.6) (slightly modified from Ortí et al. 1987-88, drawn by C. Taberner).

bidites. The dark marls contain significant amounts of perfectly preserved carbonaceous plant remains. They also contain well-defined anoxic cycles marked by the presence of centimeter-thick ferruginous crusts intercalated within the lutites. These crusts are abundant mainly in the lower part of the unit. The iron in them is now oxidized but some framboidal aggregates of pyrite can be still seen (using the SEM), which would suggest that the primary iron-rich minerals were originally formed under reducing conditions. Fewer ferruginous crusts are found going upward in the sequence, accompanied by the parallel increase in the number and frequency of sandy turbiditic layers. Iron crusts are completely absent in the transition to the overlying unit, which is characterized by the presence of carbonates with abundant benthonic foraminifera.

The upper unit of the Coubet Formation is mainly composed of siliciclastic layers whose facies associations have been interpreted as having formed in shallow marine environments (delta fans, delta front fans, beaches). Carbonate sediments are found in close association with these terrigenous materials, overlain by bioherms constructed by corals, red algae, and oysters (Busquets et al. 1986).

The non-bioturbated lutitic zones of the Vallfogona, Beuda, and Coubet formations are dark and similar in facies and in stratigraphic position to laminites underlying evaporite deposits elsewhere in the world. Many of these subevaporitc lutitic laminites contain elevated amounts of trace elements such as copper, lead, and zinc. In the three aforementioned formations, 154 samples were selected to analyze for these elements. Additionally in the Coubet formation 92 samples were also selected for study of their nickel, manganese, and iron contents (Ortí et al. 1987-88). In general it was found that these formations contain amounts of these trace elements similar to the pre-evaporitic laminites in other basins.

Discussion

The turbiditic sediments of the Vallfogona Formation were apparently deposited in a marine trough under reducing (euxinic) bottom conditions. Surficial waters were oxygenated and even relatively diluted, rich in nannoplankton and dinoflagellates. However, based on the average trace element content of lutitic beds, it seems that the bottom conditions were restricted, resembling some black-shale facies deposited in other marine basins during preevaporitic or intra-evaporitic intervals.

To explain the increase in siliciclastic deposition, Puigdefàbregas et al. (1986) have proposed that a relative worldwide water-level fall produced a general regression (based on the general sealevel curves of Vail et al. 1977), resulting in exposure of the basin margins and erosion. This produced an increased input of terrigenous turbidites into the basin (Vallfogona Formation). The same lowered sealevel caused the restriction in bottom conditions (the black-shale facies), and on the south side of the trough the earlier carbonate platform was influenced by shallow evaporitic ponds. Local geologic variations, such as regional uplift, faulting, and thrusting may also have controlled or added to the clastic deposition.

The first certain vestiges of restricted sedimentation within this marine trough are found toward the top of the Vallfogona Formation, where the depositional sequence first becomes intercalated with some gypsum layers and finally becomes purely lutitic. The establishment of expansive evaporative environments in the basin (Beuda Gypsum Formation) and on the adjacent carbonate platform seems to have been a rapid and apparently continuous process. This is suggested by the limited thickness (a few meters) of the alternation of lutite with gypsum that forms the base of the evaporite unit.

All the studied sections suggest that there was a filling of the sedimentary trough with evaporites, with a tendency toward progressive basin shallowing at the end of that phase. Nevertheless, emersion probably was only sporadic, because a restricted marine sedimentary sequence overlies the evaporites. These post-evaporitic marine sediments reach thicknesses up to 500 m locally and they contain evidence of a progressive upwards reoxygenation of the water column (Coubet Formation). This was accomplished through the renewed entrance of repeated turbidity flows into the basin and represents the end of bottom-water anoxic conditions. The development of a new shallow carbonate platform, with the development of algal-coral reefs, testifies to the re-establishment of open marine conditions. Finally, the complete infilling of the marine basin is shown by the generalized cover by the continental sediments of the Bellmunt Formation.

(Accepted for publication April 1992.)

11

Second Marine Evaporitic Phase in the South Pyrenean Foredeep: The Priabonian Potash Basin (Late Eocene: Autochthonous-Allochthonous Zone)

Laura Rosell and Juan-José Pueyo

Geological Setting

The late Eocene (Priabonian) evaporites which contain a considerable accumulation of potassic minerals, including sylvite and carnallite together with halite and anhydrite, occurs within a narrow and elongated basin which extends from Navarra in the west to Catalonia in the east, a distance of more than 300 km. This evaporite accumulation represents the last marine deposition episode in the South Pyrenean foreland basin.

The distribution of the potash formation is known from mine, borehole, and geophysical data (figures 11.1 and 11.2). While the Catalan Potash Basin is located in the autochthonous zone, its Navarrese equivalent occurs within allochthonous terrains (Del Valle 1978). Continuity of the evaporite formation between both potash subbasins has not been proved because intermediate boreholes are lacking. However, the petrological and geochemical similarities between them (Pueyo 1975a; Rosell 1983) and the geometric possibility of subsurface continuity, if structural modifications due to South Pyrenean thrust sheet emplacement are considered, seem to support the idea of continuity (Puigdefàbregas 1975).

Although the supposition that the evaporative period was related to the Pyrenean ridge uplift (which would have confined a residual branch of the sea) has been previously discussed in several papers, it has been more precisely defined recently. This refined definition is the result of greater knowledge about the Pyrenean structural evolution. According to Puigdefàbregas et al. (1986) confinement of the basin, leading to the development of the Priabonian evaporite deposition, is a direct consequence of the final stages of the emplacement of the Cadí thrust sheet and its western equivalents. Priabonian evaporites are considered by these authors as the top of the second depositional cycle, developed in the eastern South Pyrenean foreland basin during the Paleogene (the Cardona sequence).

During the evaporitic event, concentrated brines accumulated mainly in two small subbasins (Catalonian and Navarrese), where the potash stage of evaporite precipitation was reached. At the end of evaporitic deposition, lacustrine and fluvial environments were established and the continental infilling of the basin took place. The saline sequence in the center of both potash subbasins is quite similar and represents a complete evaporite macrocycle, but contains no intermediate magnesium sulfate beds (following the classification of Braitsch 1971). Only the thickness of each saline unit varies from

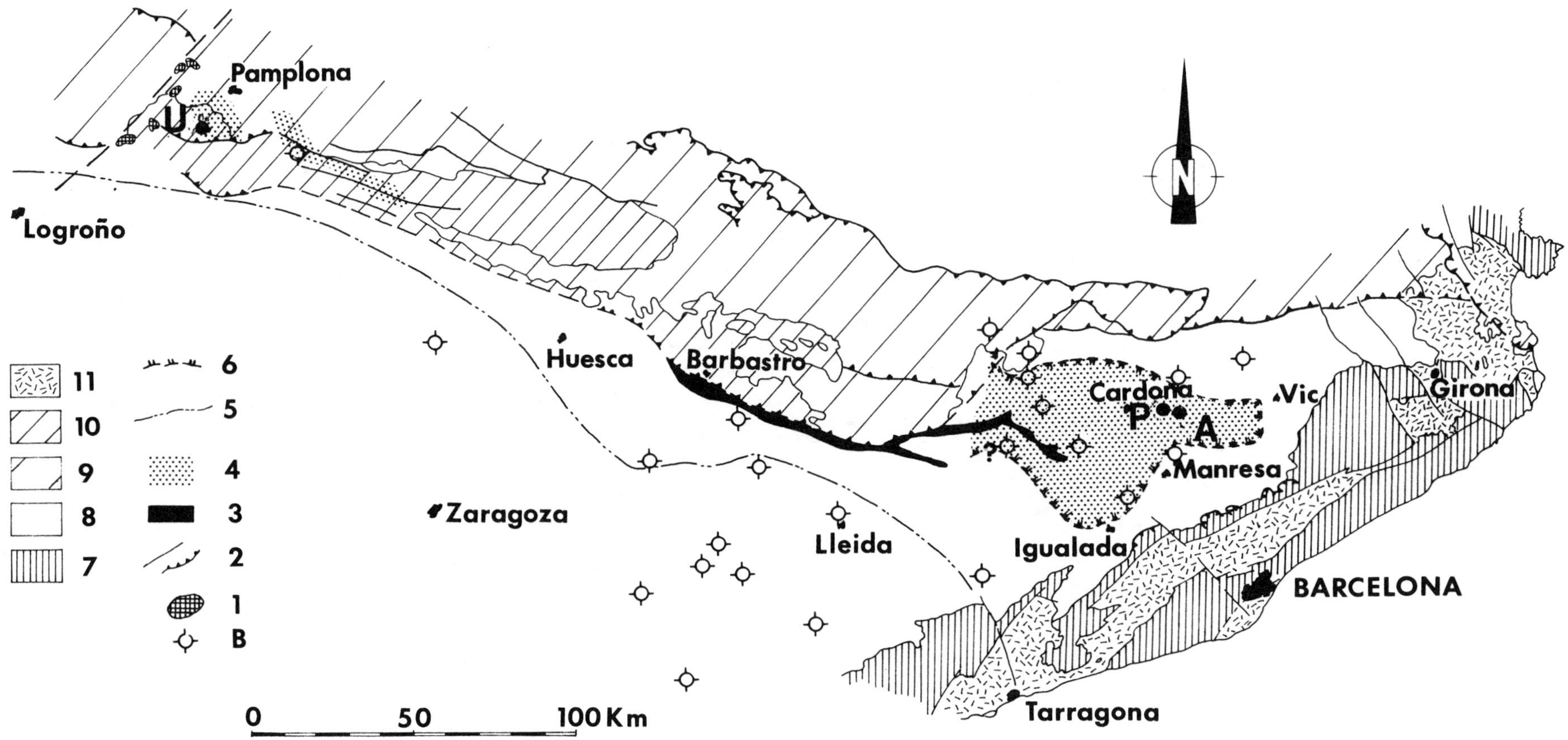

Fig. 11.1. Distribution of the upper Eocene evaporite formations in the South Pyrenean Potash basin (modified from Rosell and Anadón, in Rosell 1983). Key: 1 = Triassic diapirs; 2 = fault, overthrust; 3 = Barbastro gypsum; 4 = limits of known potash evaporite formation; 5 = marine Eocene boundary; 6 = outer boundary known of the Catalan Potash Basin; 7 = pre-Tertiary terrains; 8 = autochthonous Tertiary in the Ebro Basin; 9 = pre-Oligocene (allochthonous); 10 = Oligocene (allochthonous); 11 = Neogene and Recent; B = borehole; P = Puigreig 3; A = Avinyó; U = Uterga.

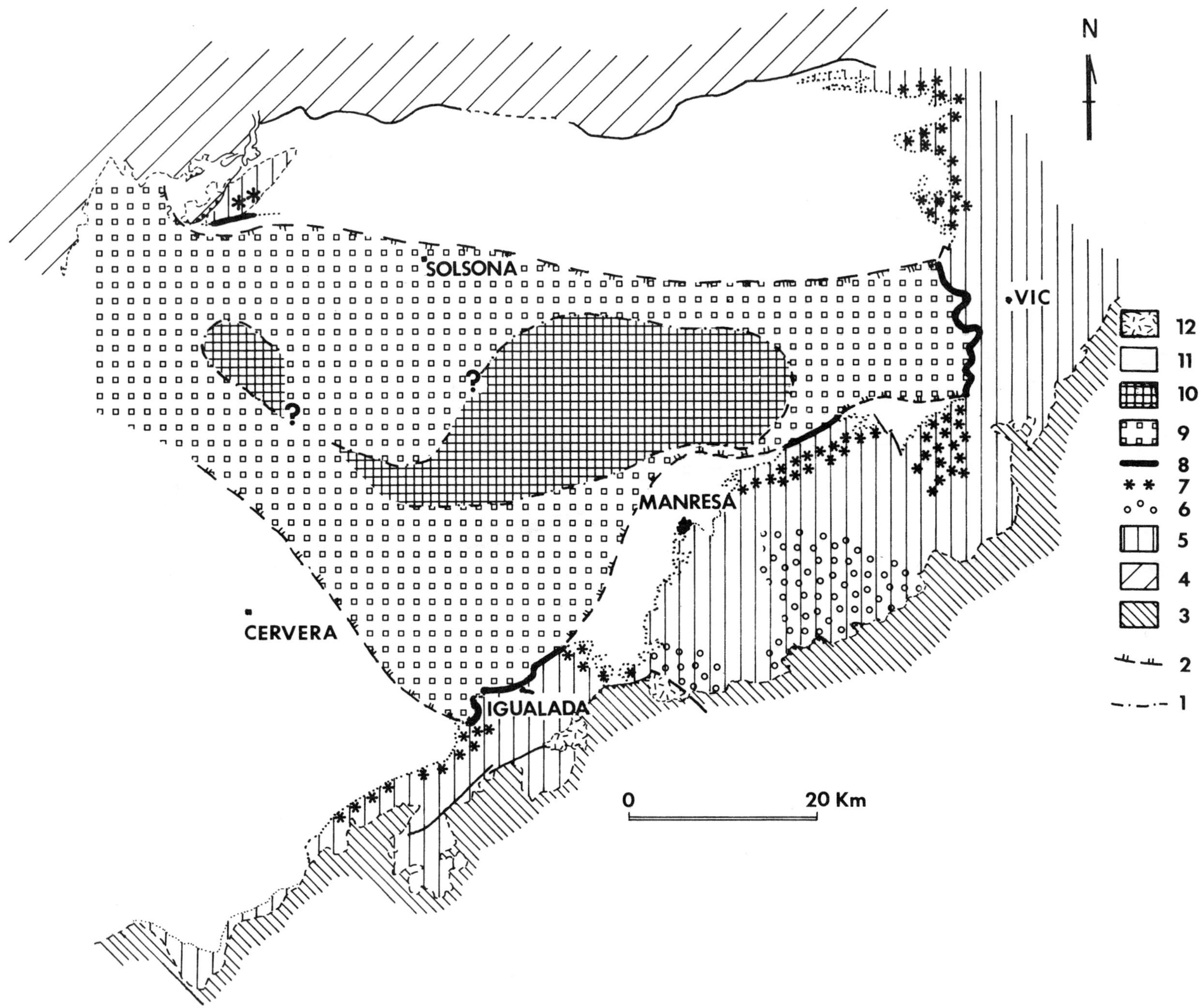

Fig. 11.2. Facies distribution in the Catalan Potash Basin (adapted and slightly modified from Rosell and Anadón, in Rosell 1983). Key: 1 = potash salt boundary; 2 = evaporite formation boundary; 3 = Catalanides Range (Paleozoic and Mesozoic); 4 = allochthonous thrust sheets (Mesozoic and Tertiary); 5 = pre-evaporitic Paleogene; 6 = fan-deltas; 7 = reefs; 8 = calcium-sulfate belt; 9 = halite zone; 10 = potash zone; 11 = post-evaporitic Paleogene; 12 = Recent.

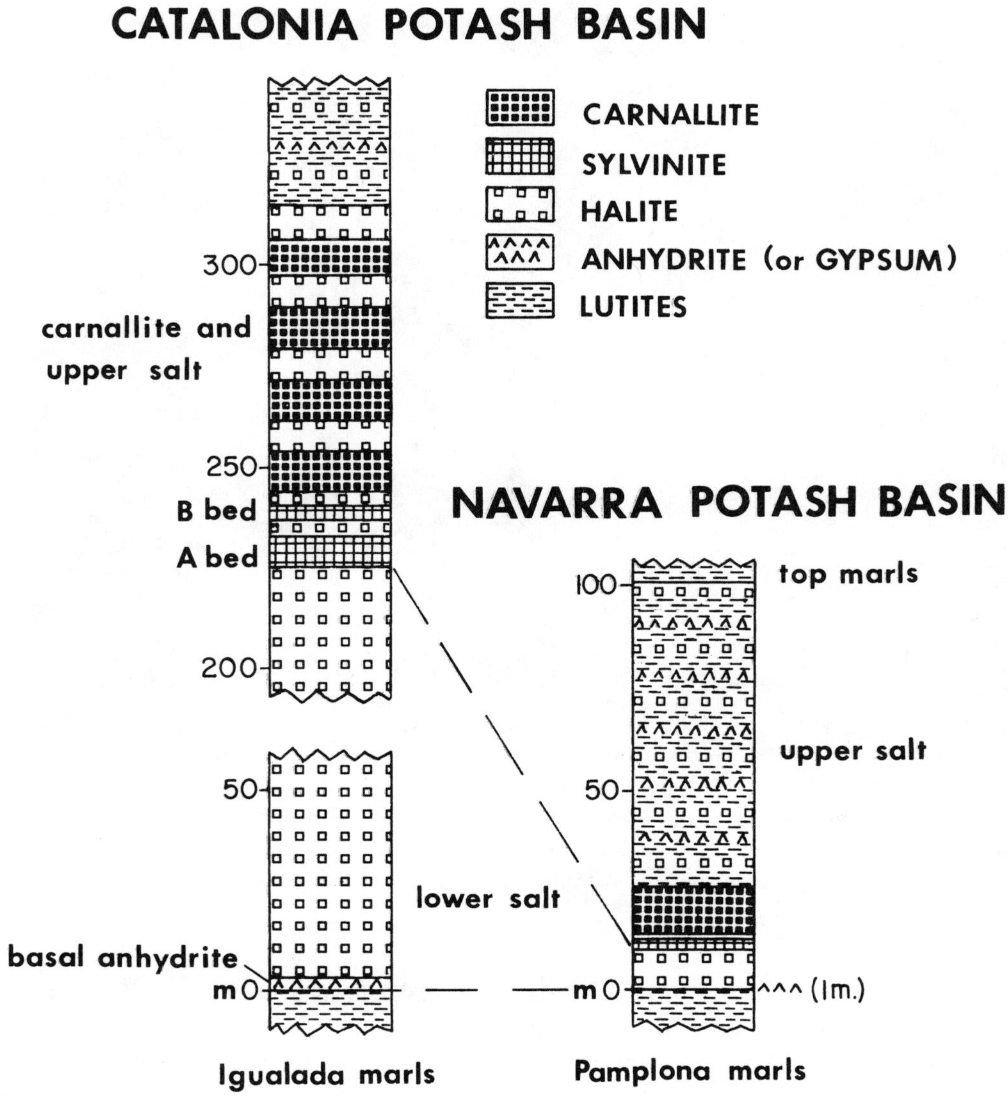

FIG. 11.3. General stratigraphic diagram depicting the evaporite macrocycles as developed in the Catalonian and Navarrese Potash basins (adapted from Rosell 1983).

Fig. 11.4. Cardona salt mountain: A. General view of the western part of the Riu Salat salt outcrop. Quaternary clays at the top were formed as a residue from the dissolution of the rising diapiric structure. B. Detail illustrating residual materials that correspond to the salt levels intercalated among potash beds.

one subbasin to the other, and the total thickness is an average three times greater in Catalonia (more than 300 m) than in Navarra (100 m) (figure 11.3).

In the Catalan Basin, where outcrop conditions are better than in Navarra, three facies belts are observed in an approximately concentric arrangement (figure 11.2). From the margin toward the center the facies are:

(a) a peripheral reef (carbonate) belt (Tossa Formation and its equivalents)

(b) a calcium sulfate belt (Odena Gypsum and its equivalents)

(c) a central saline (chloride) body (Cardona Saline Formation). This chloride body is exposed in the Cardona diapir (Salt Mountain, figure 11.4)

In the Navarra Basin, the saline formation and the overlying detrital unit have been named the Guendulain formation (Puigdefàbregas 1975). Detailed information concerning the South Pyrenean Potash Basin is synthesized mainly from the studies listed in table 11.1.

TABLE 11.1.
References for the South Pyrnenean Potash basin (through 1991). G, S, B and G = general aspects, stratigraphy, borehole, and geophysical data; P, G, S = petrology, geochemistry, and sedimentology.

AUTHORS	YEAR OF PUBLICATION	TOPICS
Busquets et al.	1985	P, G, S
Del Valle	1929, 1932	G, S, B and G
Font and Montoriol	1968	Mineralogy
García Siñeriz	1941	G, S, B and G
Marín	1922, 1923, 1926, 1932, 1945	G, S, B and G
Masachs	1952	G, S, B and G
Mir	1942	Mineralogy
Montoriol	1968	Mineralogy
Montoriol and Font	1969	Mineralogy
Montoriol and Traveria	1969	Mineralogy
Ortí et al.	1984, 1985	P, G, S
Ortí Cabo and Palou	1977	P, G, S
Pueyo	1975 a & b, 1977	P, G, S
Pueyo and San Miguel	1974	P, G, S
Ramirez and Riba	1975	P, G, S
Riba	1975	G, S, B and G
Ríos	1963, 1968	G, S, B and G
Rojas et al.	1973	G, S, B and G
Rosell	1983	P, G, S
Rosell and Ortí Cabo	1980, 1981-82	P, G, S
Rosell and Pueyo	1984	P, G, S
Solans and Montoriol	1968	Mineralogy

Facies Description

This analysis is arranged by depositional zones [(a), (b), (c)], beginning at the base of the central depositional area and moving outward.

POTASSIC ZONE

1—Basal marine marls.

The basal portion of the potassic sequence consists of gray-bluish marine marls: the Igualada-Vic Marls in Catalonia and the Pamplona Marls in Navarra.

In the Catalan Basin this marly formation is made up of the more distal sediments of a series of fan deltas (Montserrat, Sant Llorenç de Munt, etc.) developed in the southeastern margin of the basin. In the central part of the basin, in the chloride zone, these marls are only known from borehole data. In some boreholes carbonate layers 10 m thick are intercalated within the marls.

The Pamplona Marls, which locally have a turbiditic character, have been interpreted as slope deposits (Puigdefàbregas 1975). Analysis of their

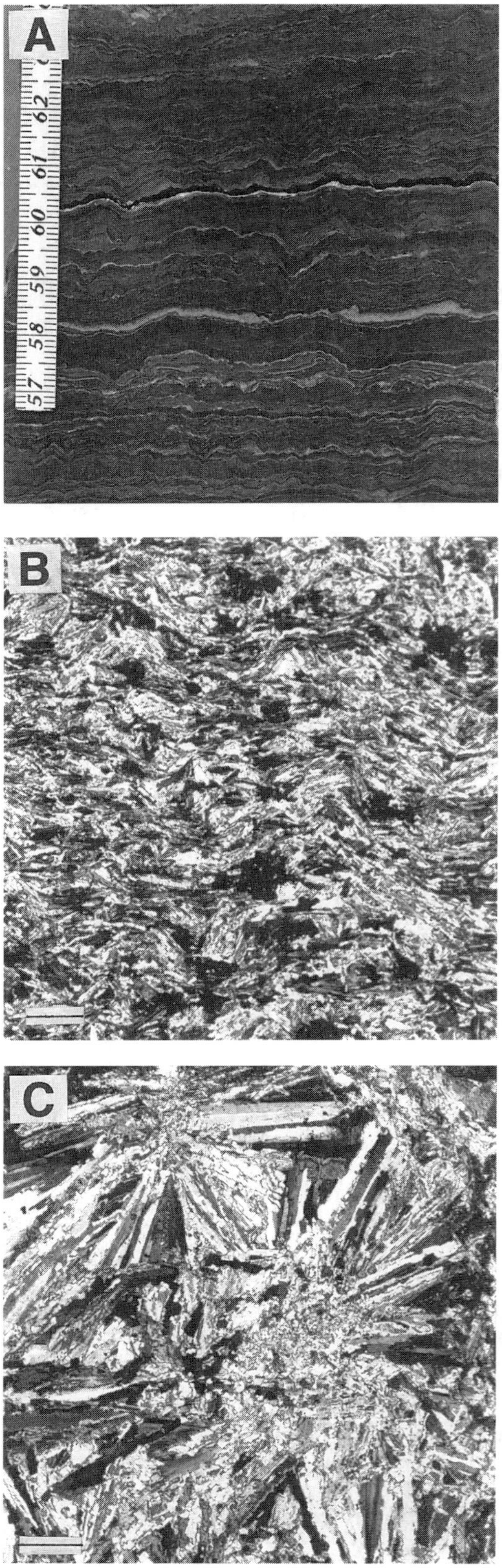

FIG. 11.5. The basal laminated lithofacies in the calcium sulfate belt of the Navarrese Potash Basin. A. Slabbed core. The wavy appearance of laminations suggests a stromatolitic origin. Scale in cm. B. Thin section photomicrograph of anhydrite prismatic-crenulated fabric from the wavy laminated lithofacies. Crossed polars. Scale bar, 250 microns. C. Thin section photomicrograph of anhydrite fibro-radiate fabric. In the center of radial aggregates note cores of microgranular anhydrite. Crossed polars. Scale bar, 250 microns.

faunal content reveals a deep marine origin with a possible shallowing upward evolution in their upper part (Brouwer, in Puigdefàbregas 1975). The uppermost layers of the marls, close to the base of the saline sequence, show some features characteristic of anoxic facies: sections with perfectly preserved millimetric lamination and the absence of organic remains (fossils) except for carbonized plant fragments.

2—Basal Sulfate Unit.

The lower part of the saline formation consists of an anhydrite bed. In the Catalan Basin, where it is only known from borehole data, this bed is several meters thick (4 to 5 m) and is made up of a laminated sulfate lithofacies. In the Navarrese Basin the basal sulfate bed is exposed in some mine galleries. It consists of a gray anhydrite bed 0.6 to 1 m thick which is mainly laminated. Nodular structures are noted locally.

These laminated lithofacies are made of alternations of irregular anhydrite and carbonate laminae whose morphology suggests a possible stromatolitic origin (figure 11.5A). Anhydrite pseudomorphs after gypsum crystals (and also after halite) are frequent in both lithofacies. The anhydrite has a prismatic-crenulated fabric in the laminated lithofacies (figure 11.5B), is fibrous-radial in nodular lithofacies, and also shows microgranular crystalline fabrics (figure 11.5C). The host sediment accompanying this anhydrite consists of non-fossiliferous carbonate, siliciclastic lutites, and fine-grained arenites.

Diagenetic analcime ($Na(AlSi_2O_5) \cdot H_2O$) having diverse crystal morphologies is present within the various anhydrite fabrics. This mineral is probably generated by reaction of concentrated interstitial brines with clay minerals present in the host sediment (Rosell and Ortí 1980).

3—Lower Salt Unit.

This unit is composed of halitic layers directly overlain by potash beds. It shows similar characteristics in both potassic subbasins, with only a pronounced difference in thickness: 130 to 200 m in the Catalan Basin and only 10 m in the Navarrese Basin.

The Lower Salt Member displays a banded halite lithofacies with alternations of dark gray and clear, pink, or white centimetric bands, which range in thickness from 2 to 10 cm (figure 11.6A). The dark gray bands are predominantly formed by coarse hopper crystals with scattered clayey fragments. Pink or white bands are formed by hopper crystals (smaller than the coarse ones in the gray salt) and by clear (inclusion-free) crystals (figure 11.6B). Clay fragments are absent here. The crystal sizes range in general from 1 to 3 mm. Hopper crystals range from 7 to 10 mm. Chevron- or cornet-shaped textures have not been observed in this facies.

The uppermost portion of the Lower Salt Member, just below the sylvinite beds, shows slightly different features: clayey horizons are much more noticeable, crystal size is smaller, and the overall appearance is very similar to that of halite layers within the sylvinite beds. The microstructure of this halite unit is a coarse crystalline mosaic of anhedral halite crystals that display inclusion-rich nuclei enclosed by inclusion-free halite (clear halite) (figure 11.6C). These "mixed" crystals are commonly associated with smaller clear halite ones which may form well-defined horizons.

Anhydrite and polyhalite are present in the Lower Salt Member, with the latter being slightly more dominant. They may occur together or separately. These anhydrite crystals are equant or slightly elongated perpendicular to cleavage. Polyhalite crystals exhibit various habits and are often twinned (figure 11.7). Sulfate crystals may also be found associated with zones of insolubles but are also scattered throughout the halite, either inside the halite crystals or at their margins (figure 11.8). The insoluble component is made up of clay and magnesitic micrite.

4—Sylvinite (lower potash) Unit.

The Sylvinite Unit directly overlies the Lower Salt Unit. Its thickness ranges from 5 to 20 m in the Catalonia basin and averages 2 m in Navarra. The Sylvinite Unit is made of sylvinite beds (a mixture

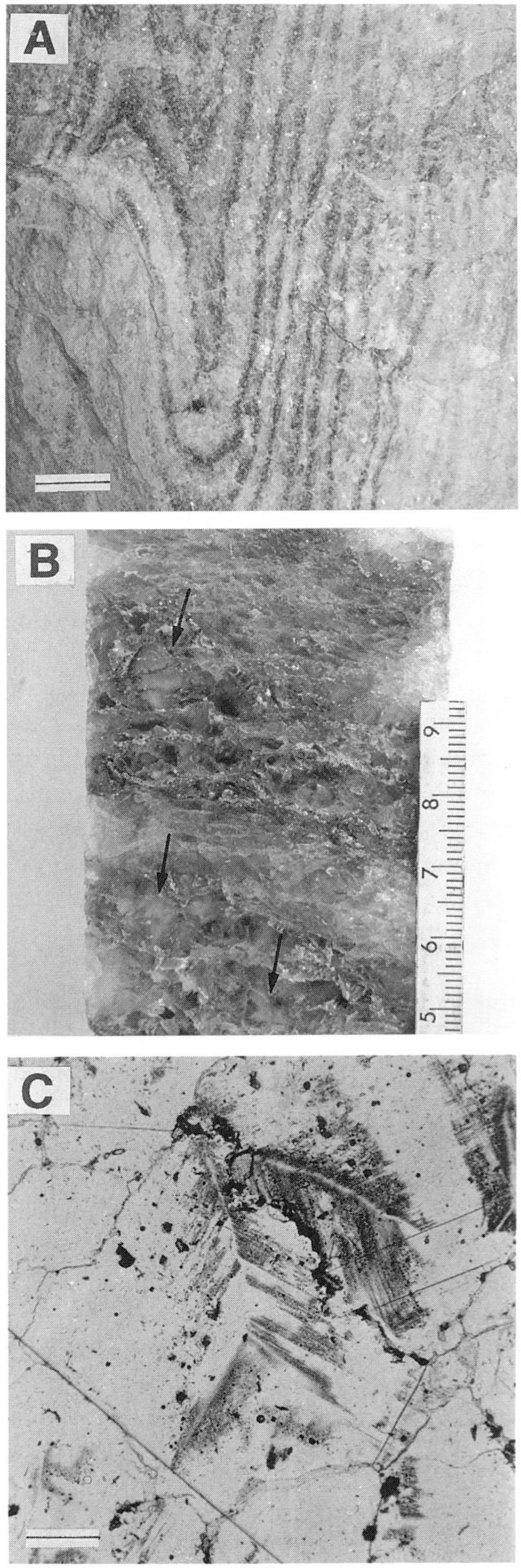

FIG. 11.6. Lithofacies and textures in the Lower Salt Unit: A. Alternation of clear (halite) and dark (insoluble-rich) layers mark a fold. Cardona mine, Catalan Potash Basin. Scale bar, 30 cm. B. Detail seen in a borehole slab: abundant hopper crystals with cloudy centers (at arrows). Navarrese Potash Basin. Scale bar in cm. C. Thin section photomicrograph of large halite hopper crystals partially dissolved and overgrown by clear halite. Plane light. Scale bar, 1 mm.

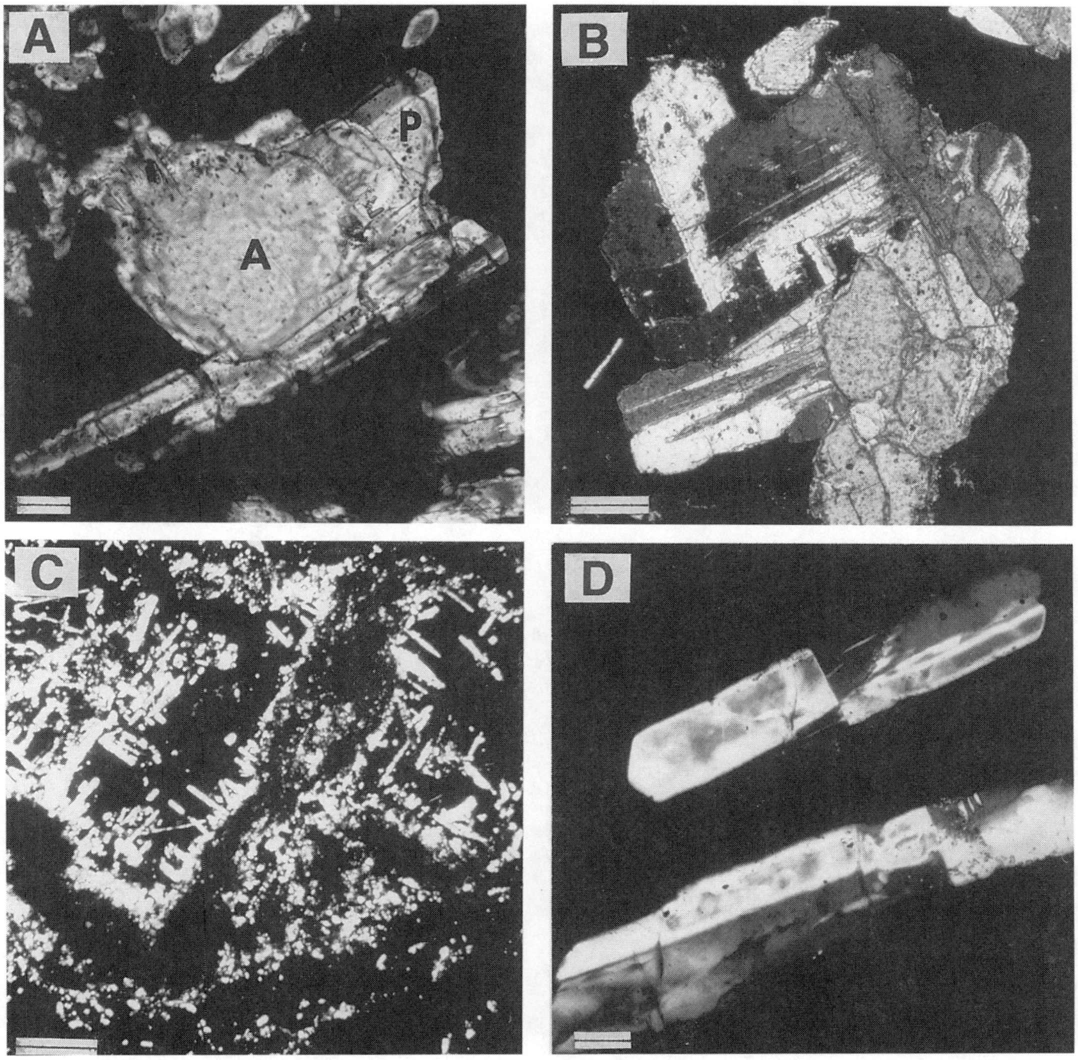

Fig. 11.7. Polyhalite mineralogy and morphologic variations: A. Thin-section photomicrograph of an anhydrite crystal (A) partially replaced by polyhalite (P). Halite layers in the sylvinite unit, Navarrese Potash Basin. Crossed polars. Scale bar, 50 microns. B. Thin-section photomicrograph showing an aggregate of polyhalite crystals with polysynthetic twinning. Sallent mine, Catalan Potash Basin. Crossed polars. Scale bar, 0.3 mm. C. Thin-section photomicrograph showing acicular polyhalite crystals developed perpendicular to the cube faces of halite. Puigreig drill core, Catalan potash basin. Crossed polars. Scale bar, 1 mm. D. Thin-section photomicrograph of two polyhalite crystals displaying characteristic twinning. Halite layers in the sylvinite unit, Navarrese Potash Basin. Crossed polars. Scale bar, 50 microns.

of halite and sylvite) alternating with halite beds (figure 11.9A). In the sylvinite beds the complete sedimentary cycle is: insolubles → halite → sylvite or sylvite/halite → halite → insolubles (figure 11.9B). However, variations of this sequence may exist, and sometimes more than one sylvite layer occurs in any single cycle. The thickness of individual sylvinite cycles ranges from less than 10 to 30 cm in the Catalonia basin and from 4 to 14 cm in the Navarrese Basin.

The macroscopic appearance of the various sylvinite layers is not homogeneous and variations mainly related to the color (pink or red) and grain sizes are observed. The smaller the crystal size the

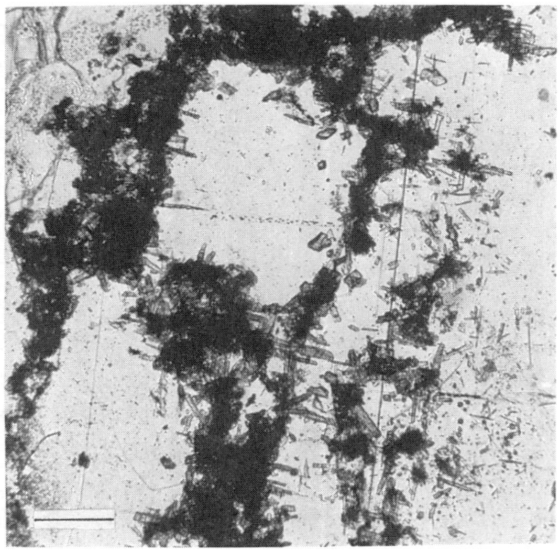

Fig. 11.8. Lithofacies of the Lower Salt Unit seen in thin section photomicrograph, showing halite rich in insoluble material. Acicular sulfate crystals are disposed perpendicularly to the cube faces of halite crystals. Sallent mine, Catalan Potash Basin. Scale bar, 1 mm.

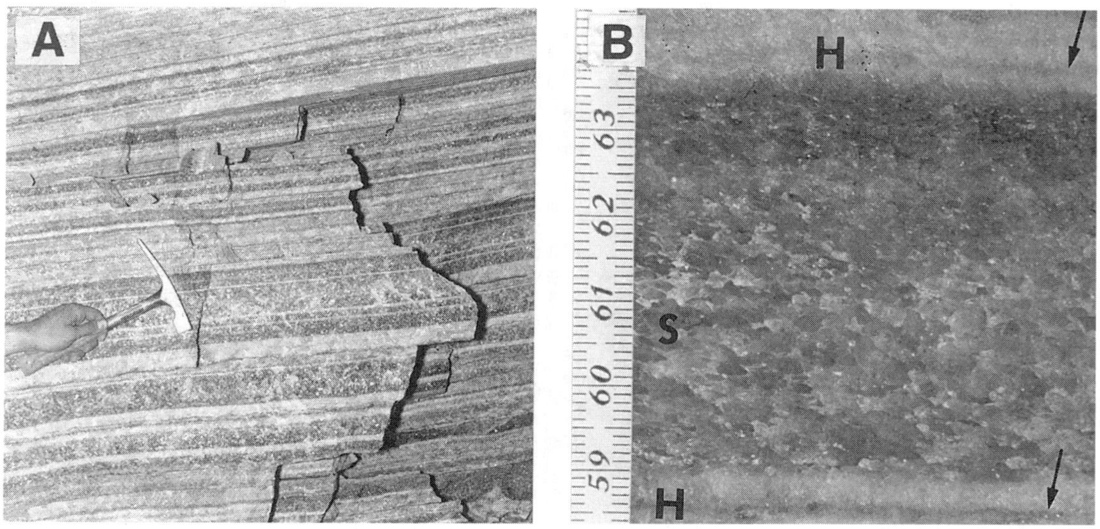

Fig. 11.9. Sylvinite lithofacies and textures. A. Sylvinite with alternating halite (clear) and sylvite or sylvite-halite (dark) layers. Sallent mine, Catalan Potash Basin. Hammer for scale. B. Detail of a sylvinite cycle. Halite (H) with insoluble horizons (arrows) at the bottom and the top, and a sylvite layer (S) in the central part. Crystal size decreases toward the top of the sylvite layer along with an intensification of its red color. Navarrese Potash Basin. Scale in cm.

more intense the reddish color. Halite beds alternating with sylvinite beds range in thickness from 2 to 12 cm and show halite-insoluble cycles (figure 11.9A). Generally these halite beds are pink while halite in the sylvinite cycles is normally white.

In thin section, the sylvite layers generally show a granular mosaic texture suggesting that some recrystallization has probably affected individual sylvite crystal boundaries (figure 11.10A—D). They also commonly display a size gradation with the coarser crystals often occupying a central position within the layer, although they also may occur at the base of some layers. In the Catalan Basin sylvite layers are sometimes completely recrystallized, forming a continuous layer made up of large interlocking crystals (3 cm to 4 cm).

In the halite layers of the sylvinite cycles, as well as in the intermediate beds (figure 11.11A), the

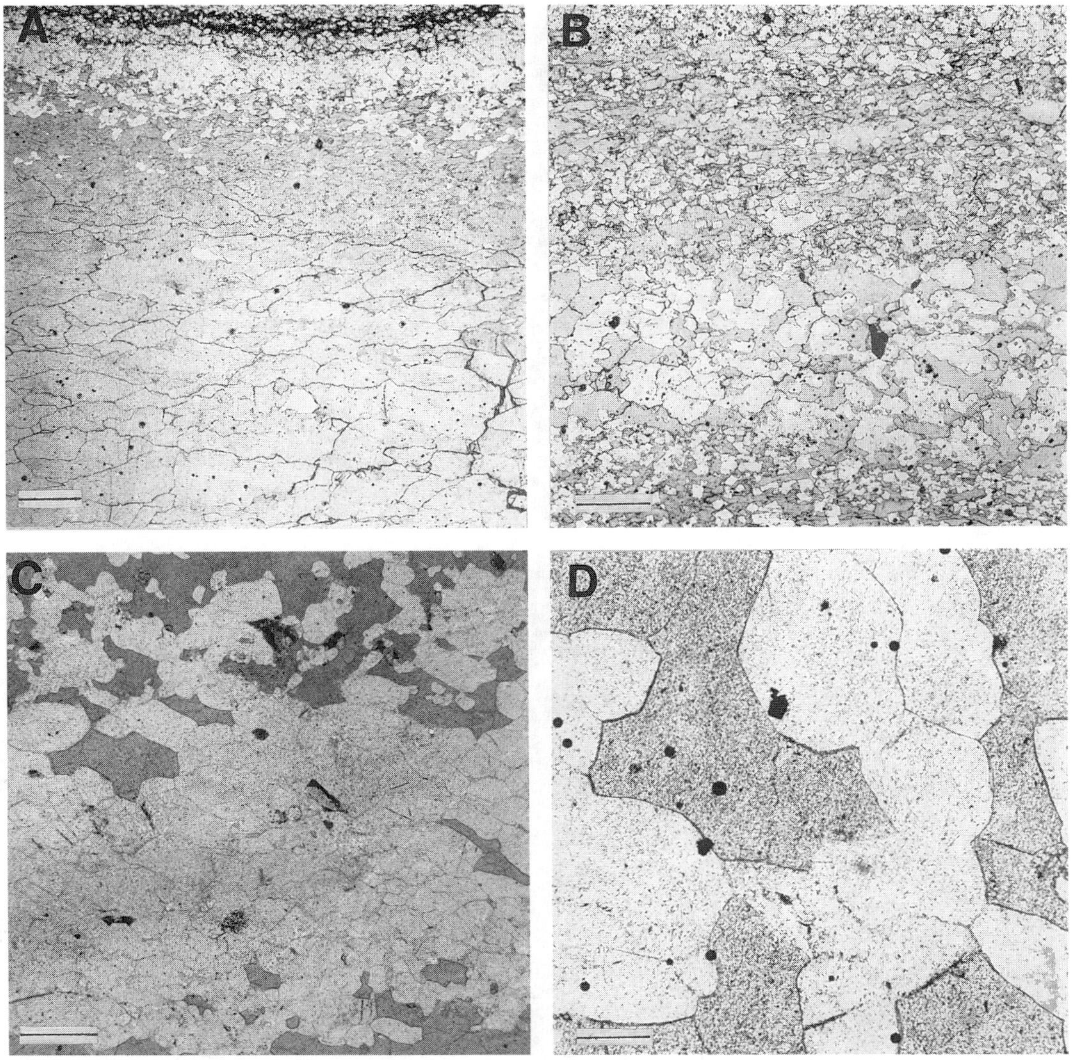

FIG. 11.10. Thin section photomicrographs of sylvinite microstructure: A. The upper part of a sylvinite cycle. The sylvite layer (gray), noticeably monomineralic, displaying decreasing crystal size toward the upper part of the cycle; halite layers are white and insoluble horizons are black. Navarrese Potash Basin. Plane light. Scale bar, 2 mm. B. The microstructure of a sylvinite layer made up of sylvite and halite crystals. Crystal size is similar for both minerals although sylvite has probably suffered recrystallization. Halite crystals are dominantly euhedral except in the central band. Navarrese Potash Basin. Plane light. Scale bar, 2 mm. C. Detail of a halite layer (white) located within the basal part of a sylvinite cycle. Halite layer shows a sharp lower contact while the upper one is gradual. Catalan Potash Basin. Plane light. Scale bar, 1.5 mm. D. Detail of (C) showing the typical aspect of the boundaries between halite (white) and sylvite (gray) crystals. Plane light. Scale bar, 0.3 mm.

clear halite textural type predominates. However, alternation of clear and cloudy halite types also occurs (figure 11.11B). In the Catalan Basin clear halite crystals frequently are graded in size and large halite cubes may grow at the base of most cycles (figure 11.11C). Synsedimentary dissolution structures may also occur, although they are uncommon (figures 11.11D and 11.12A). In the Navarrese Basin noticeable accumulations of hopper crystals may develop associated with the insoluble horizons, and are particularly well-developed just above them.

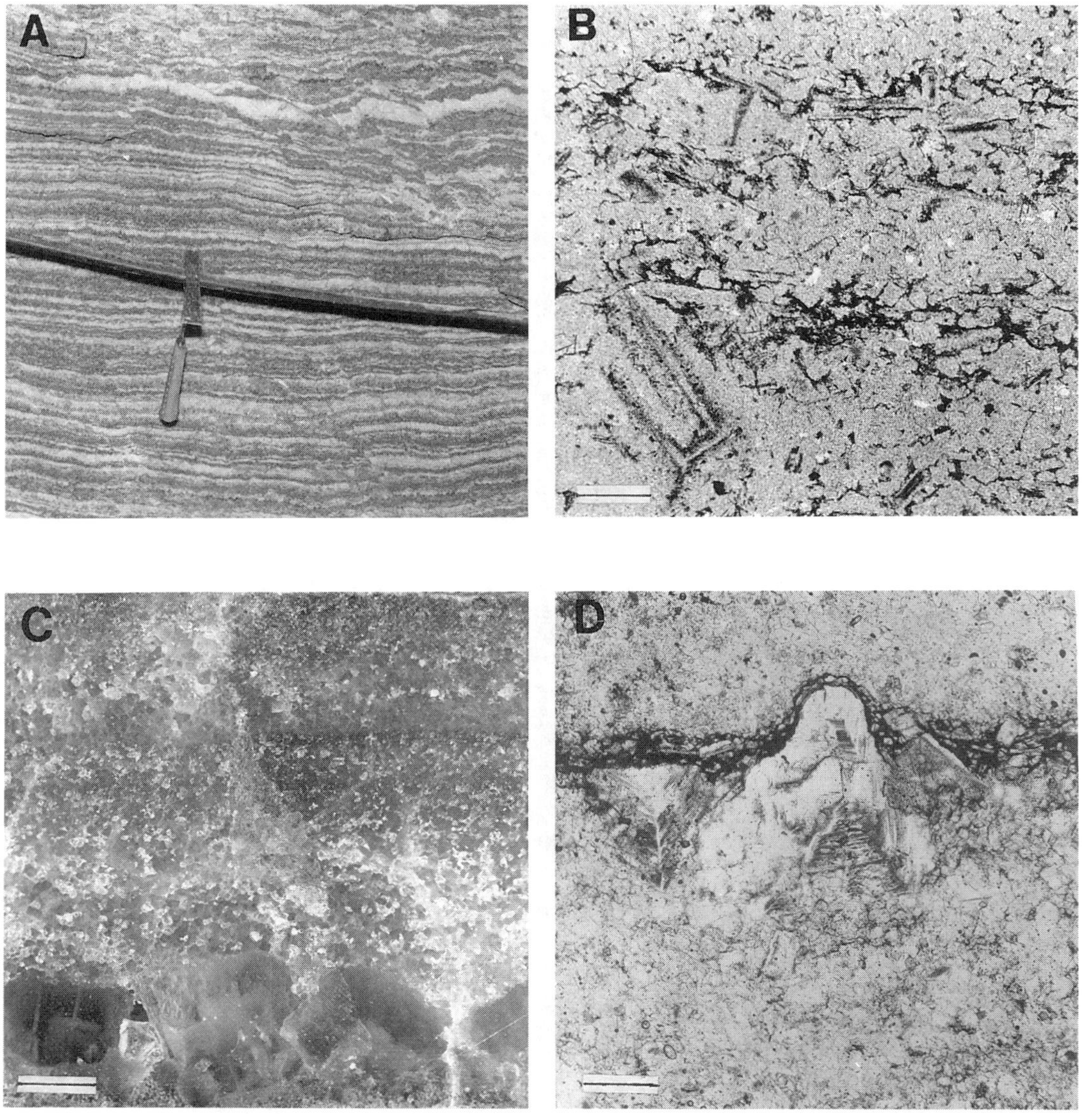

Fig. 11.11. Halite in the Intermediate Salt of the Catalan Potash Basin (Sallent mine). A. Halite layering between A and B sylvinite beds (see figure 11.3). Detrital material (dark) is composed dominantly by clays. Hammer for scale. B. Thin section photomicrograph of a detail of cloudy (hopper) and clear halite alternations under plane light. Sallent mine. Scale bar, 1 mm. C. Slabbed core showing basal growth of halite cubes in a salt layer in Sallent mine. Scale bar, 5 mm. D. Thin-section photomicrograph showing a synsedimentary dissolution structure, which are rare. The dissolution surface is directly overlain by a new layer of primary halite (hopper crystals). Plane light. Scale bar, 2 mm.

Fig. 11.12. Detail of halite cyclicity. A. Cycles in the Intermediate Salts (halite in the sylvinite unit) of the Catalan Potash Basin. (a) Cycles showing grain-size gradation; sulfate minerals (anhydrite, polyhalite) in black. (b) Cycles showing dissolution upper surface. The basal portion of each new cycle contains an accumulation of tabular hopper crystals (see also figure 11.11D). B. Types of cycles (a, b, c, d) in the halite of the carnallite unit and in the upper salts of the Navarrese Potash Basin. Insol = insolubles; H = halite; H in = inclusion-bearing halite; H t = clear halite; An = anhydrite; Dol = dolomite.

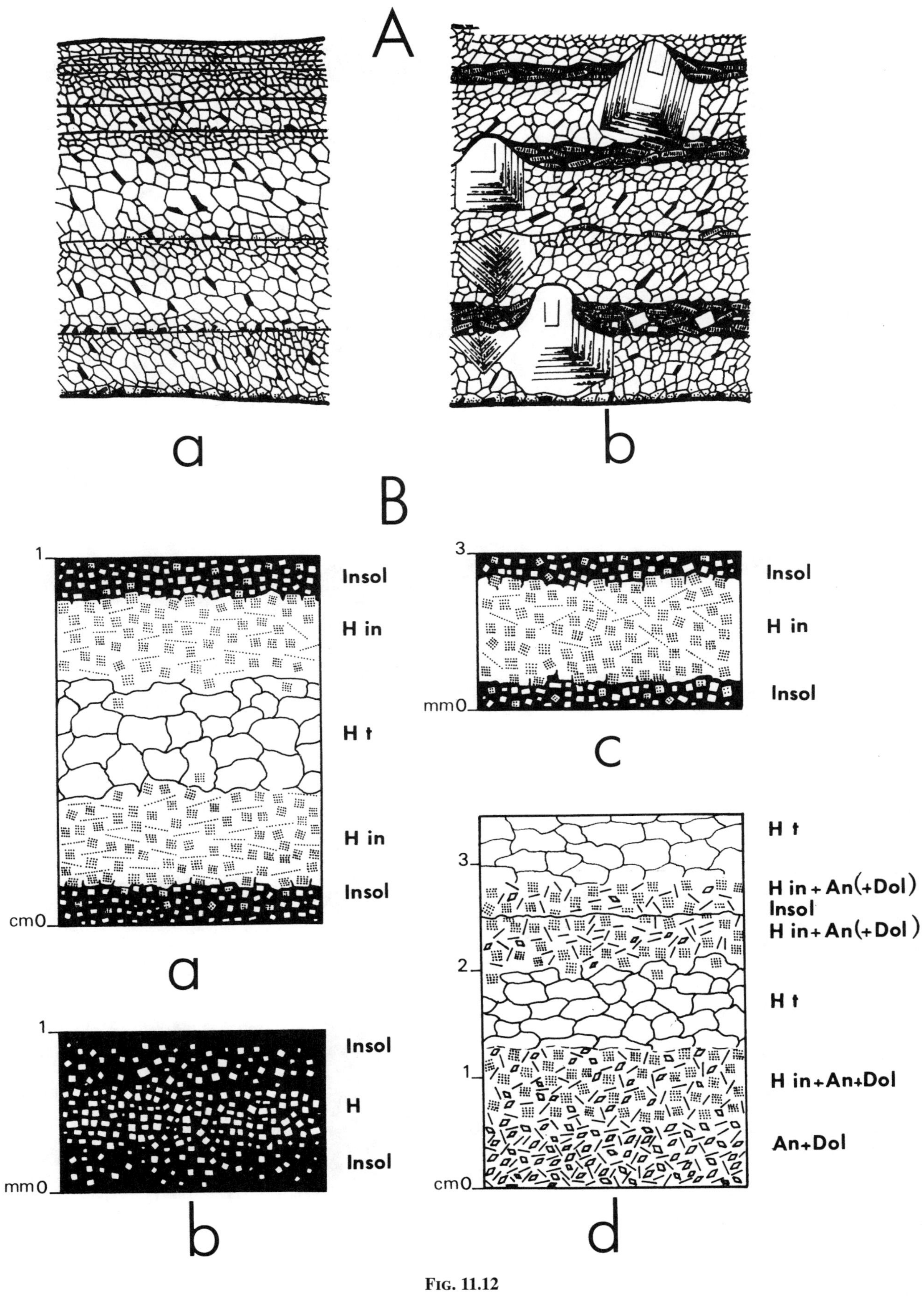
A
a
b
B
1
Insol
H in
H t
H in
Insol
cm0
a
1
Insol
H
Insol
mm0
b
3
Insol
H in
Insol
mm0
c
H t
3
H in + An(+Dol)
Insol
H in + An(+Dol)
2
H t
1
H in + An + Dol
An + Dol
cm0
d

FIG. 11.12

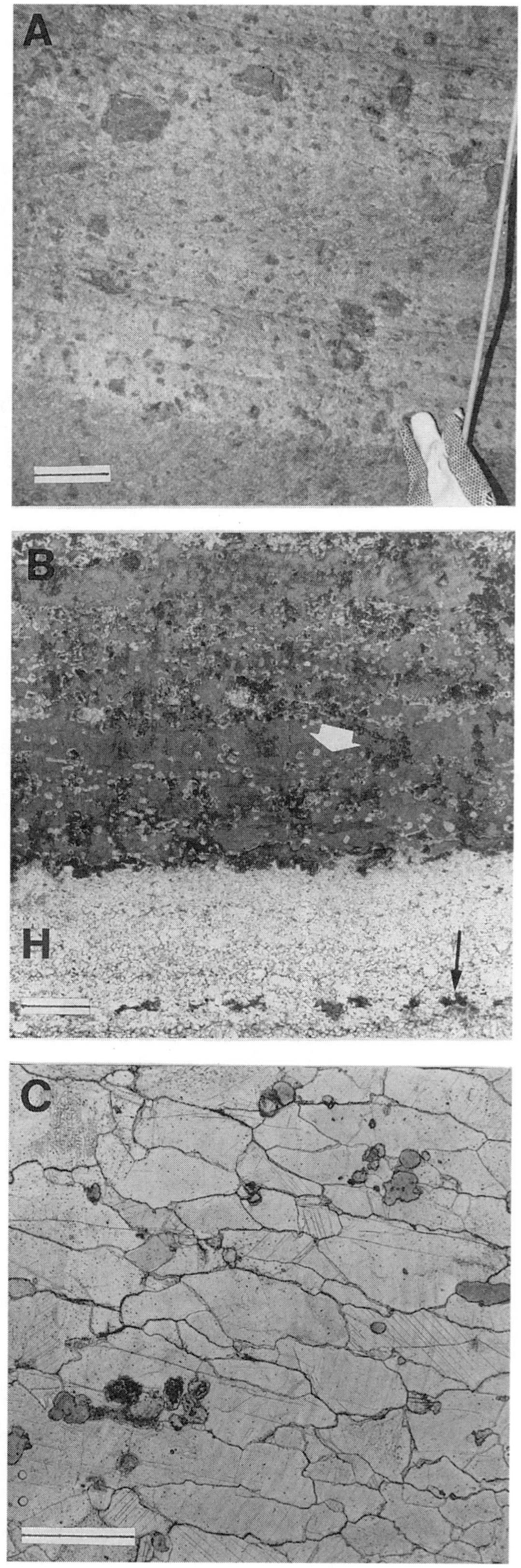

Fig. 11.13. Carnallite lithofacies and textures. A. Brecciated carnallite lithofacies made of irregular carnallite fragments (arrows) in a pink, fine-grained matrix. Subiza mine, Navarrese Potash Basin. Scale bar, 8 cm. B. Thin section photomicrograph of carnallite microstructure corresponding to a banded lithofacies. Halite layer (h) in the lower part is rich in inclusion-bearing crystals (hoppers) and shows a discontinuous carnallite horizon (black arrow). Dark carnallite layer in upper part of the photograph exhibits euhedral halite crystals (white arrow). Navarrese Potash Basin. Plane light. Scale bar, 2 mm. C. Thin section photomicrograph showing carnallite microstructure with noticeable schistosity. Dark crystals are sylvite. Catalan Potash Basin. Plane light. Scale bar, 5 mm.

Anhydrite and polyhalite are present throughout the sylvinite unit although they are relatively scarce in the sylvite layers. The insoluble fraction is made of clay and magnesitic micrite.

In the strongly tectonized zones of the Catalan Basin (Cardona dome), sylvinite beds lose their primary textural features and display a homogeneous appearance ("silvinita pintada"). In other areas in the Catalan Basin, anhydritic "hartsalz" zones developed locally. The hartsalz is made of halite, sylvite, anhydrite, and other sulfate minerals. This lithofacies forms dark, fine-grained lenticular masses within the sylvinite beds. However, clay and sulfate minerals can be locally more abundant and sylvite may disappear progressively in the anticlinal area causing a halitic lithofacies ("impoverished zones" of Borchert 1969).

5—Carnallite (upper potash) Unit.

The Carnallite Unit overlies sylvinite beds, although in the Navarrese Basin an intermediate halite unit up to 1 m thick exists. The Carnallite Unit is made up of carnallite beds alternating with halite. It has a total thickness of 40 to 80 m in Catalonia and 12 m in Navarra. In the Navarrese Basin eight carnallite beds may be distinguished, although in places some of them may be locally lacking. Carnallite is mainly composed of a brecciated or massive lithofacies (figure 11.13A) although nodular or banded varieties may also occur (figure 11.13B and C). Halite beds are formed by thin halite layers 1 m to 5 m thick, alternating with clay partings.

Halite in this unit displays the two petrographic types already described: clear halite (inclusion-free) and cloudy halite (inclusion-bearing), the latter being formed by the remains of primary hopper crystals. Different types of inclusion zonation may be distinguished, probably corresponding to different primary crystalline habits (figures 11.14A to E). In these halite layers three distinct types of cycles may be observed (figures 11.12B, 11.14E and F):

Type A: insolubles → inclusion-bearing halite → clear halite → inclusion-bearing halite → insolubles
Type B: halite → insolubles
Type C: insolubles → inclusion-bearing halite → insolubles

Anhydrite is the only sulfate mineral that exists in the carnallite unit, mainly developing with prismatic morphologies. Dolomite crystals also occur with several different habits. Clay and magnesitic or dolomitic micrite constitutes the insoluble fraction.

Altered sylvinite: In both potassic basins some of the carnallite beds have been replaced by a secondary sylvinite lithofacies that has been interpreted as having originated by the "solution metamorphism" of carnallite (Braitsch 1971; Borchert and Muir 1964; Pueyo 1975a; Rosell 1983). This term must be qualified, because true metamorphic temperatures were never achieved. Indeed textures identical to those in true metamorphic rocks are apparent, but these form at much reduced temperatures in salts.

In the Navarrese Basin the lower beds of the carnallite unit are more frequently replaced but recrystallization and transformation may affect the whole unit. The altered sylvinite occurs mainly as isolated lenticular bodies associated with dissolution zones of the sylvinite unit or of the whole potash unit. These zones seem to be located preferentially in the proximity of major faults.

Generally the appearance of altered sylvinite is chaotic: clay horizons are found as scattered fragments and it is practically impossible to recognize the original depositional cyclicity. However in some zones of the Navarrese mine altered sylvinite displays well-developed cycles in which clay horizons, halite and sylvinite layers are still noticeable (figure 11.15A). These cycles are apparently inherited from the original halite-carnallite precursors. Altered sylvinite is made up of coarse red (sylvite) and transparent crystals of sylvite and halite associated with anhydrite (figure 11.15B). Hexagonal hematitic inclusions as well as goethite fibers are present, also inherited from halite-carnallite precursors.

6—Post-potassic Unit.

In the Catalan Basin a weakly defined halite unit is generally present overlying the potassic beds.

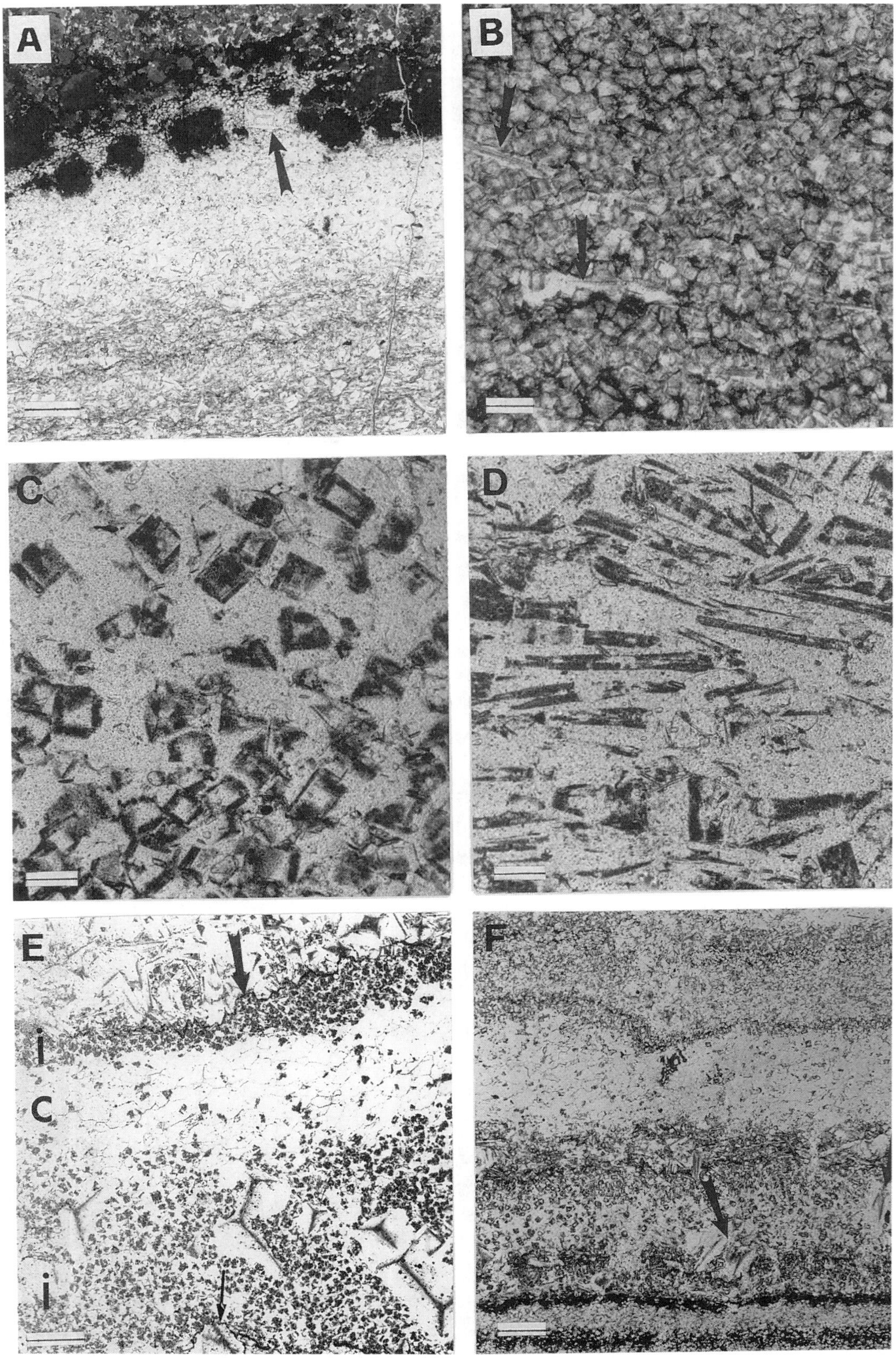

FIG. 11.14

However the uppermost (last) halite bed of the carnallite-halite cycles is commonly thicker than the other halite beds of the unit. This uppermost bed is overlain by the upper gray lutites, which include gypsum and halite intercalations and, locally, sporadic carbonate horizons. The thickness of this unit averages 35 m (borehole data) although locally it may be significantly thicker (for instance in the cores of anticlines). This unit is overlain by the Súria Formation in both the Súria and Cardona zones, composed mainly of tabular sandstones and red lutites with lacustrine limestone layers. This formation has been interpreted as terminal alluvial fan deposits prograding from the Pyrenean margin to the basin center (Sáez 1987; figures 11.16 and 11.17). In the Súria zone a gypsum unit with laminated and less frequently nodular lithofacies intercalates between the upper gray lutites and the Súria Formation detrital materials. It is 10 to 20 m thick (borehole data) but becomes gradually thicker towards the west (Barbastro gypsum).

In the Navarra Basin the Carnallite Unit is overlain by a thick (50 to 75 m) saline unit (Upper Salt Unit). It consists of halite layers, 1 cm to 10 cm thick, alternating with laminated lutite layers, 0.5 cm to 2 cm thick. Thin anhydrite beds occur associated with the halite and become more common toward the top of the unit.

Halite layers, in which the grain size increases from the bottom to the top of the unit, show cycles very similar to those already described for the halite of the carnallite unit. However, a new type of halite (Type D) exists here, characterized by the presence of anhydrite and dolomite. The complete sedimentary cycle in this unit is constituted by: dolomite/anhydrite → dolomite/anhydrite/inclusion-bearing halite → clear halite → inclusion-bearing halite/anhydrite (/dolomite) (figure 11.12B). In the Upper Salt Unit the base of each cycle is usually formed by accumulations of large hopper crystals of halite.

Anhydrite is the only sulfate mineral present in this unit. Clay minerals and dolomite micrite form the insoluble fraction.

In the Navarrese Basin a lutite unit (Top Marl Unit) occurs overlying the upper salt. It is mainly made of variegated and finely laminated lutites with anhydrite intercalations that locally display enterolithic morphologies. These top marls are locally rich in marine nannoplankton and their thickness averages 50 m. This unit is overlain by a siliciclastic de-

FIG. 11.14. Halite microfacies and textures of the Navarrese potash basin seen in thin-section photomicrographs (plane light).

A. Halite layer rich with inclusion-bearing crystals, especially in the lower third of the photograph, in a banded carnallite lithofacies. At the base of carnallite layer (black), nodular carnallite is present and a large hopper crystal is noticeable (arrow). Scale bar, 2 mm.

B. Inclusion-rich halite crystals with cubic inclusion zonation from a halite layer in the carnallite unit. Halite cubes are randomly oriented ("cumulates," Lowenstein 1982). Two elongated sheets of connected halite crystals (at arrows) are sunken hopper rafts. Scale bar, 0.5 mm.

C. Thin-section photomicrograph of halite inclusion-bearing crystals with cubic zonation. Halite layer in the carnallite unit. Scale bar, 250 microns.

D. Thin-section photomicrograph of flat, elongate halite crystals (inclusion-bearing). Upper Salt Unit section taken perpendicular to bedding. Scale bar, 250 microns.

E. Thin-section photomicrograph of halite cyclicity showing an a-type cycle: inclusion-bearing halite crystals (i), clear halite (c), inclusion-bearing halite crystals (i), insoluble film (arrow). The base of a new cycle is visible at the top of the photograph. The base of a cycle contains larger hopper crystals. Insoluble films correspond to dissolution horizons. Upper Salt Unit. Scale bar, 2 mm.

F. Thin-section photomicrograph of halite cyclicity showing a- and c-type cycles. Inclusion-bearing crystals mainly show cubic zonation. Large hopper crystal residues are present (at arrows). Halite layers are seen in the Carnallite Unit. Scale bar, 2 mm.

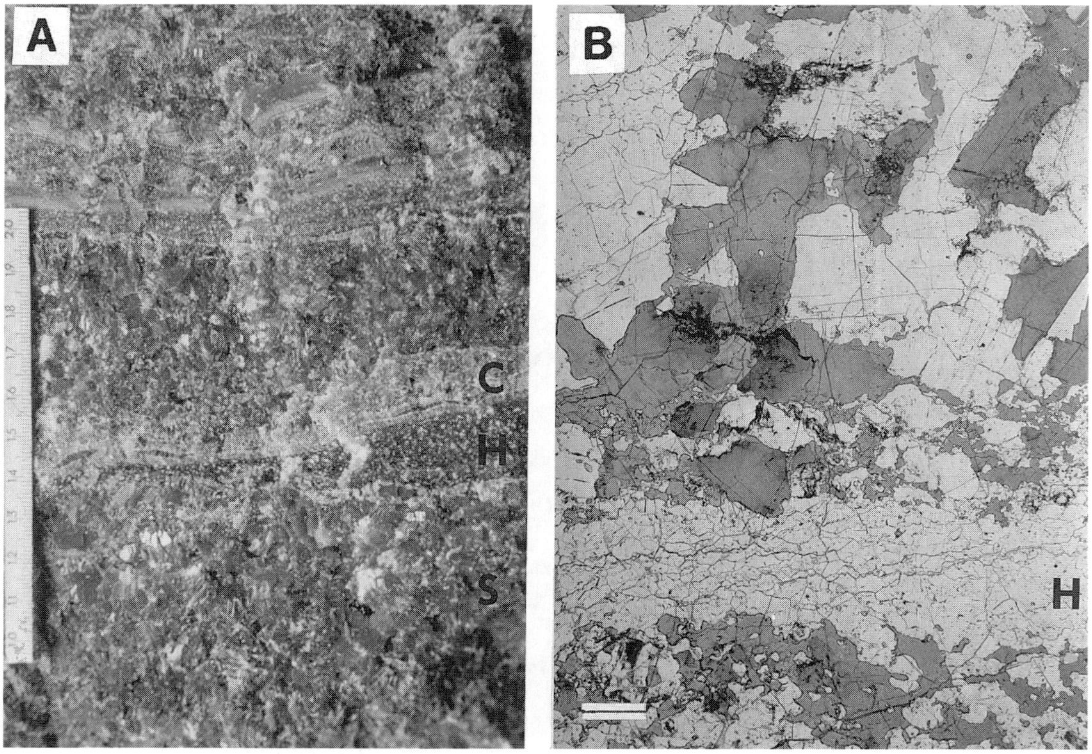

Fig. 11.15. Altered sylvinite lithofacies and microfacies, Navarrese Potash Basin. A. Slab sample showing well developed cyclicity may be observed with alternating sylvinite (S), halite (H), and clay-anhydrite (C) layers. Scale in cm. B. Thin-section photomicrograph showing microstructure. Coarse dark (red sylvite) and transparent crystals (sylvite, halite) characteristic of this rock type are visible. Halite layer (H) made of smaller crystals is in the lower half of the photograph. Plane light. Scale bar, 2 mm.

Fig. 11.16. General view of Súria Formation components in the village of Súria. In the right part of the photograph an anticline (Mig Món anticline) nucleus is visible. Anticlinal structure brings the saline formation near the surface and the mine is located in the anticlinal zone.

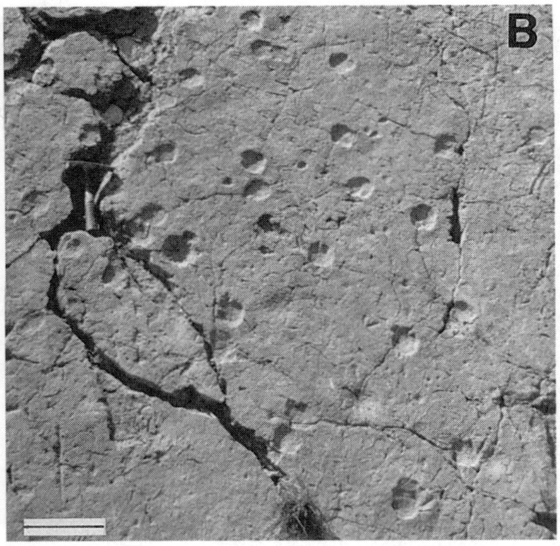

FIG. 11.17. Lacustrine sediments that overlie the Cardona Saline Formation showing: A. Bird tracks. Scale bar, 10 cm. B. Mammal tracks, northern flank of the Cardona anticline, under the Cardona castle. Scale bar, 20 cm.

trital sandstone (Galar or Liédena Sandstone), locally containing nummulitic foraminifera that represent clear evidence of marine influence.

HALITE ZONE: LATERAL EVOLUTION OF THE CHLORIDE BODY

In the marginal areas of the Catalan Basin, where the characteristics of the saline sequence are known from two boreholes, the potassic member is missing. However the saline formation is made of a halitic succession having features similar to those of the Lower Salt Member in the potassic zone of the basin. Thus the concentric facies distribution in the Catalan Basin is also present in the chloride facies.

The chloride succession is made up of a banded halite lithofacies with white milky transparent and gray bands. Halite crystal size usually ranges from 3 to 5 mm, but in clay-rich layers, larger crystals up to 1.5 cm may exist. Hopper crystals are present in both types of layers (figures 11.18A and B). In this zone anhydrite is the only sulfate that is present accompanying the halite (Ortí et al. 1984).

BASIN MARGIN ZONE

1—The peripheral reef belt.

Along the Catalan Basin margin a reef belt occurs including several stages of reef development that intercalate with deltaic deposits. This alternation of facies corresponds to the different stages of deltaic progradation. The youngest barrier reefs outcrop continuously from Igualada to Vic (Tossa Formation, Collsuspina Formation, and Sant Martí Xic Formation) and represent the last open marine period of sedimentation. This last reef stage has some evidence of karstification and dolomitization at the reef-tops, probably related to the sea level fall during the evaporite event. A more detailed description and discussion concerning this reef belt can be found in Salas (1979), Taberner (1982), and Busquets et al. (1985).

2—Calcium sulfate belt.

Along the eastern and southeastern border of the Catalan Potash Basin a calcium sulfate belt up to

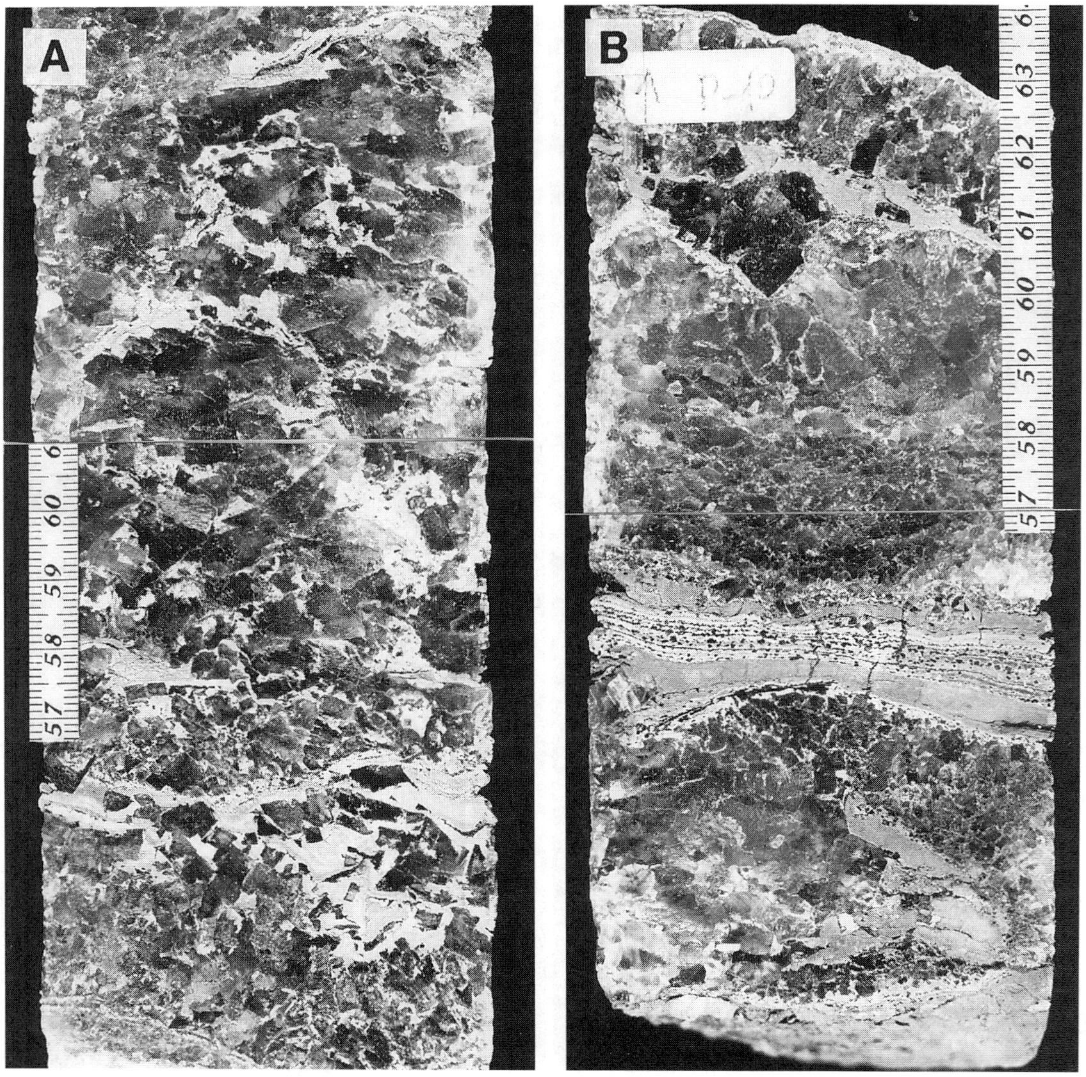

FIG. 11.18. Drill cores of the lower salt unit in the Puigreig area. A. Halite is coarse grained with poorly developed layering. B. In the right core a band displays well defined layering. In this band, matrix is composed of clays, sulfates (anhydrite), and minor amounts of carbonates (dolomite and magnesite). Scale in cm.

30 m thick crops out in several areas (Odena—Artés—La Noguera Gypsum). In the northern border of the basin, this unit appears in isolated outcrops near Oliana (Caus 1973). It consists of secondary gypsum and residual anhydrite masses and overlies the blue marine marls that display features characteristic of anoxia in their uppermost part (Reguant 1967; Taberner 1983; Taberner and Santisteban 1984). The gypsum belt thins toward the toe of the reef talus as well as toward the center of the basin, where it is the equivalent of the basal anhydrite unit.

Three units constitute the succession in the calcium sulfate belt (Busquets et al. 1985):

Basal stromatolites. In the Igualada area (Odena Gypsum) this unit is 1 to 2 m thick bed and has a gypsum-carbonate lithology. The lamination is millimetric as are as the pseudomorphs after primary gypsum that are noticeable within it. In thin section, remains of algal filaments can be identified in

Fig. 11.19. Gypsum structures in the Catalan Potash Basin. A. Basal, thin, regular gypsum-carbonate layers (stromatolite) of the calcium sulfate belt. Vic area. Scale bar, 20 cm. B. Secondary gypsum nodules displacing basal stromatolite lamination. Calcium sulfate belt, Vic area. Scale bar, 20 cm.

these carbonates, and celestite accompanies secondary gypsum in significant proportions. The calcareous character of this stromatolite facies varies laterally.

In the Vic area (in the La Noguera Gypsum) this layer shows similar morphologic features (figure 11.19A), but locally contains intercalated gypsum nodules that displace the laminations (figure 11.19B). In the Artés area basal stromatolites have not been found in the quarries and the lower gypsum unit lies directly on the blue marine marls.

Lower unit is up to 20 m thick and is of great industrial interest. It is made of fine-grained, white-gray, alabastrine, secondary gypsum, with carbonate laminations, consisting of pelletal micrite, irregularly distributed throughout the whole unit. Locally carbonate may dominate in the lowermost few meters. This lithofacies varies from laminated to banded although locally it displays a massive character. Towards the top of the lower unit, secondary gypsum pseudomorphs after primary gypsum crystals become more noticeable. They are cm-sized and display prismatic gypsum and pseudohexagonal habit and are commonly twinned (figure 11.20A and B). These crystals deform the laminations but also may cut them. Under the microscope the alabastrine gypsum shows a considerable amount of relic anhydrite.

Anhydrite masses up to several meters long are included within this unit, especially in the lower half. Anhydrite displays a fibro-radiate fabric and is roughly spherulitic. Spherulites are mm-sized and are either isolated or in veined arrangements that reach some tens of centimeters in length. Dark-colored, cm-sized porphyroblasts of secondary gypsum are scattered in the anhydrite masses. Hydration of anhydrite into secondary gypsum occurs mainly along a typical vein pattern. Secondary gypsum frequently displays a fibro-radiate fabric also, possibly inherited from anhydrite.

In the Vic area, the lower gypsum unit displays large spheroidal or conically-shaped bodies, either adjacent and connected to each other or isolated and surrounded by blue lutites. These gypsum structures resemble the primary "supercones" (figure 11.21A) described by Dronkert (1977) in the Messinian gypsum of the Sorbas basin (Almeria, Spain). Unfortunately in our case the inner structure of these bodies, now preserved as alabastrine secondary gypsum, is not clearly distinguishable. In the Artés area similar isolated supercones are noted locally and intercalate with blue clay levels.

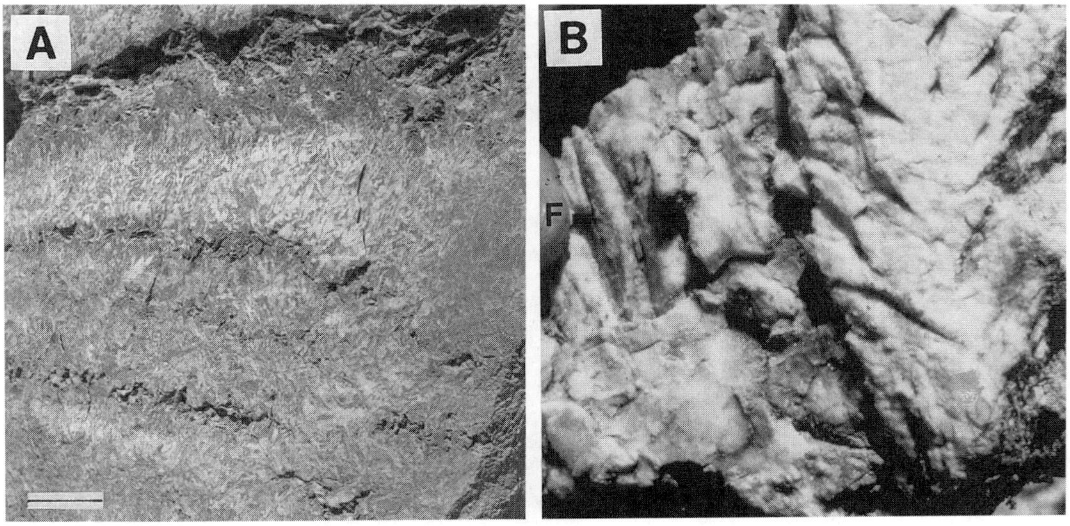

Fig. 11.20. Gypsum structures, Catalan Potash Basin. A. Lithofacies of the lower gypsum unit showing pseudomorphs after selenite crystal forms. Abandoned quarry in Artés area. Scale bar, 15 cm. B. Alabastrine secondary gypsum pseudomorphs after twinned primary gypsum crystals. Calcium sulfate belt, Vic area. Scale, finger (F) at left.

Fig. 11.21. Gypsum structures in the Catalan Potash Basin. A. Large cone-shaped bodies ("supercones") forming the base of the lower gypsum unit. Supercones now consist of secondary alabastrine gypsum and depress and overlay blue marine marls. Calcium sulfate belt, Vic area. B. Upper gypsum unit showing laminated lithofacies (U). Contact with the lower unit (L) is an erosional surface (reinforced black line). Calcium sulfate belt, Odena quarry.

Upper unit is 5 to 7 m thick, gray, and made up of laminated lithofacies that consist of alabastrine and megacrystalline secondary gypsum. The laminae show evidence of current deposition such as low-angle cross-bedding. Clastic textures are evident in thin section but the grainy appearance may be attributed in part to transport, and partly to later tectonic deformation of the section. In the Igualada area the contact between the two gypsum units is an erosional surface (figure 11.21B). Anhydrite is absent in outcrop but is always present as tiny inclusions under the microscope.

In the Vic area this unit has a thickness between 1 and 5 m and displays diffuse carbonate lamination that resembles the algal morphologies, especially at the base of the unit. Laterally this unit probably passes into a stromatolite zone. Contact with the lower unit is probably erosional in the southern half of this area, but in the northern half a thin lutitic horizon preferentially occurs between both gypsum units. In the Artés area the upper unit reaches some 5 m in thickness and is also separated from the lower one by a thin clay horizon.

Geochemistry of the Chloride Facies

Major (Na, K, Ca, Mg) and minor (Li, Rb, Sr, Fe, Br) element distribution in the South Pyrenean Potash Basin has been studied (Pueyo 1975b, 1977; Rosell and Pueyo 1984). Mean values and standard deviations for each depositional unit in both the Navarrese and Catalan Potash basins are shown in table 11.2. Because analyses in the Catalan Basin were run ten years before those in Navarra, some minor elements reported in Navarra were not run in Catalonia. The data for Navarra is listed here although a complete comparison between the basins is not possible.

MAJOR ELEMENTS

Distribution of the major elements (Na, K, Ca, Mg) is very similar in both potassic basins:

Calcium.

The values for this element are similar for the Lower Salt and Sylvinite members in both basins. In these units the calcium is present in anhydrite and polyhalite. Elevated values are present in the carnallite and upper salt of the Navarra Basin due to the abundance of anhydrite. In the Catalan Basin, however, calcium is most common in the "impoverished" lithofacies, apparently generated by the processes of "solution metamorphism."

Magnesium.

Magnesium is present in the lower salt wherever there is polyhalite. It is also present in some of the residual brines. The lower values present in the Catalan Basin seem controlled by the effects of diapirism. In the halite from the sylvinite unit, higher values are noted in Navarra and are probably related to the abundance of polyhalite. In the Upper salts of the Navarra Basin there is no polyhalite but the magnesium is present in admixed dolomite. Differences in magnesium within the carnallite in both basins seem related to the purity of the samples analyzed in Navarra.

Potassium.

The potassium content follows that of magnesium within the lower salt, and for in the same reasons. Similarly the high potassium content is caused by the presence of polyhalite within the halite of the Sylvite Unit in Navarra. In the halite of the carnallite unit and from the upper salt there is a dearth of potassium minerals in Navarra.

MINOR ELEMENTS

Bromine.

The bromine content observed in the Lower Salt Member corresponds to the theoretical values obtained for marine halite before K-Mg salt precipitation (Valyashko 1956; Holser 1966a and b; Braitsch 1971). Catalan Basin samples average 130 ppm. However most of the samples analyzed come from the Cardona diapir and probably belong to the lower stratigraphic parts of the deposit, so they have been affected by structurally induced alteration. In the marginal area of this basin bromine contents for the Puigreig-3 borehole samples yield an average of 100 ppm. In the detailed bromine profile of these data, a progressive increase toward the top of the sequence may be observed from 60–70 ppm to 120–140 ppm (figure 11.22A). Bromine content of Avinyó-3 borehole samples have an average value of 122 ppm.

In the Navarrese Basin bromine content of the

TABLE 11.2.

Chemical analyses of various elements of salts taken from the South Pyrenean Potash Basin. (A) Mean values and standard deviations of analyzed elements in saline lithofacies of the Navarra Potash Basin (from Rosell 1983); (B) Mean values and standard deviations of analyzed elements in saline lithofacies of the Catalan Potash Basin. (From Pueyo 1975a)

A

ROCK TYPE	# samples	Ca		Mg		K		Na	
		χ	σ	χ	σ	χ	σ	χ	σ
Lower Salt	27	9020	3865	2640	1755	4240	2564		
Halite (sylvinite unit)	16	8905	3791	4895	2803	12290	9076		
Halite (carnallite unit)	5	18995	4843	3540	641	875	276		
Upper Salt	13	26865	9558	5650	4253	270	165		
Sylvinite	26	6750	4614	2020	1361			111225	57686
Altered Sylvinite	7	17665	2597	4085	1425			159610	47941
Carnallite	17	13160	4921	62820	8026			80795	38605

ROCK TYPE	# samples	Li		Rb		Sr		Fe		Br	
		χ	σ	χ	σ	χ	σ	χ	σ	χ	σ
Lower Salt	27	0.8	0.65	-		240	118	150	154	195	22
Halite (sylvinite unit)	16	1.3	0.95	-		110	45	680	425	265	53
Halite (carnallite unit)	5	0.9	0.15	-		50	8	605	144	280	14
Upper Salt	13	1.0	0.35	-		75	32	1195	800	250	19
Sylvinite	26	0. 6	0.50	4.5	2.8	55	30	385	275	1970	297
Altered Sylvinite	7	1.9	0.70	2.7	1.2	50	10	675	178	740	151
Carnallite	17	1.2	0.75	110	82	75	35	490	315	2885	345

B

ROCK TYPE	Ca		Mg		K	
	χ	σ	χ	σ	χ	σ
Lower Salt	6925	4067	1475	2333	3585	5111
Halite (sylvinite unit)	10120	4676	3400	1577	6765	6177
Sylvinite	5900	7492	3200	2499	209600	101200
Carnallite	3700	3768	44200	19297	131000	57555
Impoverished	18300	13210	11700	8075	38000	36522

ROCK TYPE	Rb		Sr		Fe		Br	
	χ	σ	χ	σ	χ	σ	χ	σ
Lower Salt	-	-	60	58	40	46	130	50
Halite (sylvinite unit)	-	-	105	54	695	606	275	58
Sylvinite	8 *[20-40]	7.1	90	54	305	240	1065 *[2100-2600]	431
Carnallite	145 *[~300]	61	70	-	310	218	1610 *[2000-3000]	631
Impoverished	-	-	475	555	190	242	330	-

Data given in ppm * calculated values for individual minerals - Below sensitivity

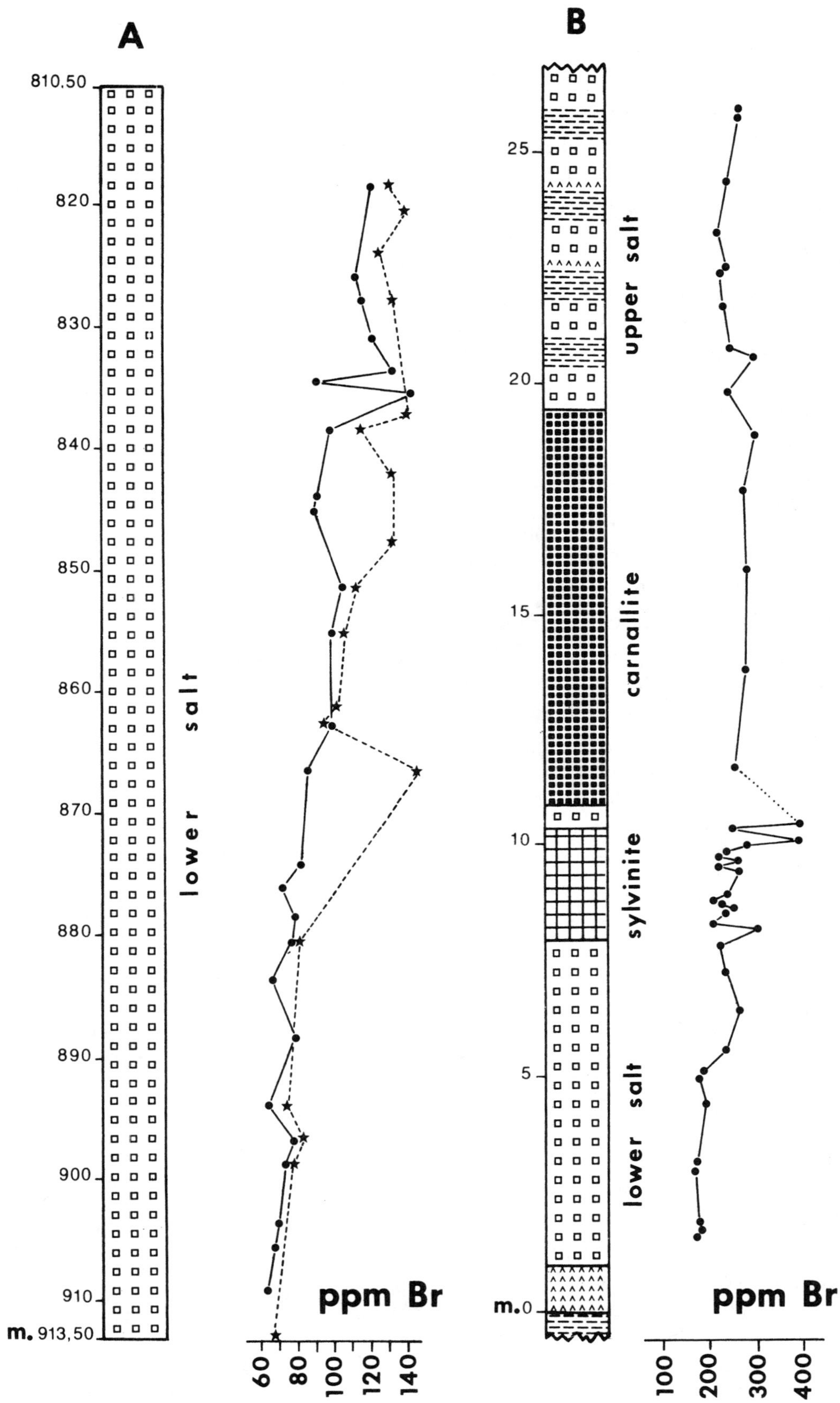

FIG. 11.22. Vertical bromine profiles from halite sections: A. Marginal halite zone, Puigreig borehole, Catalan Potash Basin. B. Potassic zone, Beriain mine, Navarra Potash Basin. These analyses were made without separation of fluid inclusions from carefully chosen samples.

Lower Salt Member is comparatively higher: 195 ppm on average. Samples from the lower parts of the unit near the basal sulfate member also show extremely high bromine values: 180 ppm (figure 11.22B). However, in a marginal (non-potassic) area of the Navarrese Basin, the Uterga borehole samples show bromine content closer to theoretical marine values: ranging from 65 ppm (in halite close to the basal anhydrite) to 90 ppm. To account for these facts an asymmetrical basin arrangement has been proposed in which potash beds occur offset from their basal halitic member (Ortí et al. 1985). In such a basin type, the most concentrated brines would accumulate at one side of the basin margins where potash precipitation takes place.

The bromine content of halite accompanying potash beds fits well with calculated theoretical marine values (Valyashko 1956; Braitsch 1971) and is very similar in both potash basins: 275 ppm on average in Catalonia and 265 (sylvinite unit) to 280 ppm (carnallite unit) on average in Navarra. The bromine content of sylvite and carnallite also fits with the calculated theoretical marine values, but it must be emphasized that altered sylvinite in the Navarra Basin displays much lower values (740 ppm on average) than the primary sylvinite (1970 ppm on average).

Rubidium.

Rubidium occurs in solid solution, substituting for potassium within potassium minerals, especially in carnallite. A detailed discussion of its distribution in salt minerals and its partition coefficient may be found in McIntire (1963), Braitsch (1966, 1971), Wardlaw (1968), Kühn (1968, 1972) and Schock and Puchelt (1971).

Kühn (1968) and Wardlaw (1968) suggested that dissolution and recrystallization processes lead to an enrichment in this element in the case of carnallite, or to an impoverishment in it in the case of sylvite. However, altered sylvite after carnallite should have higher rubidium contents than primary sylvite.

In the South Pyrenean Potash Basin both sylvite and carnallite show values matching those proposed as characteristic of a primary origin by Kühn (1968), although slightly lower in the case of sylvite. Altered sylvite shows extremely low values, even lower than those of primary sylvite, which for the moment are not explainable. Combined evaluation of Br and Rb as a genetic characterization method for sylvite and carnallite, proposed by Kühn (1972), has been applied to the South Pyrenean Potash Basin, but no coherent results have been obtained.

Iron.

In most saline formations this element is related to hematite and goethite inclusions in potassic lithofacies, and these minerals mainly occur as plates and fibers respectively. Discussion of the origin of these iron minerals in carnallite and sylvite may be found in Wardlaw (1968) and Braitsch (1971). Among the potash lithofacies, altered sylvinite is the chloride component that shows the highest iron content and, also, the larger crystal sizes. High iron contents displayed by the upper salts in the Navarrese Basin seem to be related to the presence of ferroan dolomite.

Strontium.

This element, besides being found in celestite, which may be present within halite beds, is also included diadochically in other compounds, especially replacing Ca in calcium sulfate minerals. It is also present in calcium-free sulfates and chlorides, mainly within halite (Kühn 1968; Braitsch 1971). In the Catalan Basin strontium is especially enriched in the "impoverished" lithofacies developed in the sylvinite zone, associated with folding and fracturing.

Pronounced differences in the respective strontium contents of the lower salt units from the two potash subbasins may be related to an impoverishment that developed in association with accumulation of insoluble phases. This accumulation is related to the diapiric processes in the Cardona zone.

Discussion

Severe basinal restriction and aridity are two necessary constraints that permit evaporite deposition to take place in modern environments (Perthuisot 1980). In the South Pyrenean Potash Basin the restriction was caused by the tectonic events related to the uplift of the Pyrenean range. Moreover, sea level changes that occurred at the end of the Eocene would also account for basinal confinement (Vail and Hardenbol 1979; Vail 1981; Taberner 1983).

In addition, the evolution of regional relief and increase in exposed land surface that accompanies a sea level fall would have caused a change in the paleogeographic setting, leading to an increase in the continentality of the basin. The high degree of continentality could have led to the development of the very low relative humidities needed for potash mineral precipitation (Kinsman 1976). However complete and permanent disconnection from the sea does not seem probable because evidence of continued marine influence (presence of marine nanoplankton) is found in the Navarrese Basin toward the top of the evaporite macrocycle.

The evaporitic event is preceded by the deposition of anoxic marine marls in a stratified water body with a lower anoxic water mass. According to Busquets et al. (1985) these anoxic marls postdate the last phase of reef growth and progradation. In the Navarrese Basin a depth decrease toward the top of basal marine marls is revealed by microfauna analyses (Brouwer, in Puigdefàbregas 1975).

The depositional phase of calcium sulfate produced a calcium sulfate sedimentary belt that is limited to the basin margin. Only a thin anhydrite unit is present in the basin center. Scarcity of calcium sulfate in the basin center may be related to the anoxic conditions prevailing in the bottom waters, which could have favored bacterial sulfate-reduction processes. Although lateral relations between the marginal calcium sulfate belt and basin center evaporites are not well known, Ortí et al. (1984) have suggested that the erosional surface at the top of the lower gypsum unit may have been caused by a water level fall accompanying potash deposition in the basin center.

Chloride stage precipitation apparently took place in the basin without interruption, as significant erosional surfaces have not been found. Halite shows two textural types, clear and cloudy. These facies have previously been described in many papers (Dellwig 1955; Wardlaw and Schwerdtner 1966; Shearman 1970). Cloudy halite exhibits three distinct varieties: (a) large hopper crystals (up to 1 or 2 cm); (b) square-shaped crystals (cubes, equant hoppers of Lowenstein 1982); (c) elongate crystals (tabular).

Two types of relationship between clear and cloudy halite are described in ancient halite rocks. Patchy disposition is characterized, according to Lowenstein and Hardie (1985), by the existence of "sharp and commonly curved boundaries" that separate clear halite from remnants of cloudy halite. The most widely accepted explanation for this relationship was proposed by Shearman (1970) who interpreted clear halite as a syndepositional cement filling dissolution voids in chevron halite (also see Curial and Moretto in this volume).

In other halite rocks, as in the South Pyrenean Potash Basin, clear and cloudy halite form bedded bands or laminae that show a well-defined cyclicity. This type of relationship is described by Dellwig (1955) in Silurian Salina Salt of the Michigan basin and by Geisler (1979) in the Lorraine Keuper salts. Dellwig (1955) proposed a seasonal origin for these alternations, clear halite precipitating from the bottom brines due to supersaturation during the cooler seasons (i.e., slower crystal growth). This view was used to explain halite cyclicity in the South Pyrenean Potash Basin by Solans and Montoriol (1968), Pueyo (1975a) and Rosell and Ortí (1981–82).

The potash lithofacies of the South Pyrenean Potash Basin present three problems that are worthy of further discussion: the origin of the sylvinite unit; the origin of the altered sylvinite; and the origins of possible brine types from which the sylvinite-carnallite potash succession formed without magnesium sulfate minerals.

Discrepancies between theoretically predicted and actual potash-bearing evaporite sequences, particularly the general deficiency in $MgSO_4$ in the latter, have been explained classically by syndepositional alteration of seawater composition: by bacterial sulfate reduction (Borchert 1969; Neev and Emery 1967; Braitsch 1971); by gypsum precipitation resulting from inflow of calcium-rich solutions; or by removal of magnesium by dolomitization (Valyashko 1958, cited in Braitsch 1971; Shearman 1966; Braitsch 1971). Braitsch (1971) shows that in the stages of maximum $MgSO_4$ impoverishment, calculated from a Jänecke ternary diagram, primary sylvite-halite and carnallite-halite paragenesis develops. Mineral sequences calculated by Harvie et al. (1980) including calcium-bearing phases seem to be in better accord with observed natural sequences although primary sylvite formation is not explained.

Although sylvite is the most common potassium mineral in ancient marine evaporite formations, the specific conditions needed for its precipitation (extremely low or high temperatures, metastable equilibrium conditions, and significant impoverishment in $MgSO_4$) have led many authors to consider sylvite only as a secondary diagenetic mineral, formed by incongruent dissolution of carnallite. A different view, proposed by Hardie (1984), supposes a nonmarine origin for sylvite-carnallite bearing sequences that develop from essentially hydrothermal parent waters.

A primary origin by accumulation of crystals precipitated at the air-brine interface was suggested by Lowenstein and Spencer (1990) for the Oligocene sylvite of the Rhine graben based on petrographic and fluid inclusion evidence. In the same paper a syndepositional origin as a pore-filling cement was suggested for sylvite in the Permian Salado and Devonian Prairie formations.

In the South Pyrenean Potash Basin a series of observations also suggests a primary origin for the sylvinite unit (Pueyo 1975a; Rosell and Ortí 1981-82). These characteristics are:

(a) the well-defined layering of each sylvinite cycle and observable lateral continuity (at least 10 km). As mentioned, the sylvinite cycles display pronounced stratigraphic variations of the internal distribution of their components (insolubles, halite, sylvite) and in their thickness. Some cycles (individually) are very characteristic, for example cycle 8 or cycle 16 in Navarra, and may be used as marker horizons. These depositional features plus the absence of erosional surfaces suggest a primary subaqueous origin in a perennial brine body; and

(b) the pronounced textural and geochemical differences between the Sylvinite Unit and the altered sylvinite lithofacies strongly suggest totally different origins for these cases. For the latter lithofacies a secondary diagenetic origin seems reasonable.

Brine cooling is the mechanism generally invoked to account for sylvite precipitation due to the strong temperature coefficient of KCl solubility (Sturmfels 1943; Solans and Montoriol 1968; Braitsch 1971; Pueyo 1975a; Rosell and Ortí 1981-82; Sonnenfeld 1984; Lowenstein and Spencer 1990). Such cooling could be produced by a seasonal drop in the temperature of the brine; or the influx of warmer surface layers along the basin floor (Wardlaw and Schwerdtner 1966) or to the cooler central parts of the basin (Braitch 1971). Richter-Bernburg (1972) also suggested that alternations of halite and sylvite were deposited in the deeper areas of the basin where the bottom water was at a lower temperature.

Such a potash precipitation mechanism would explain, according to Braitsch (1971), the noticeable monomineralic character of most of the sylvite compounds of sylvinite cycles, because isothermal evaporation would produce much higher NaCl content. The origin of parent waters has been discussed by Hardie (1984), who suggested that the usual application of bromine data from salts to distinguish marine from nonmarine chlorides may be faulty because continental waters may have bromine content overlapping seawater values. However most examples of nonmarine halites studied by Hardie are characterized by a wide range of variation in their bromine content (see fig. 6 of Hardie 1984). In this study, the agreement of our values with theoretically calculated ones for marine halite, sylvite, and carnallite; the evolution of the vertical bromine profiles (see figure 11.22); and the homogeneity in both the values and vertical profiles observed in the whole basin, seem more consistent with parent waters of a largely marine origin.

After the period of potash deposition the sedimentary evolution of the sequences differs in Catalonia and Navarra. A well-developed Upper Salt Unit is lacking in Catalonia. However the end of the evaporite macrocycle is represented in both subbasins by fine detrital sedimentation: an upper sequence of gray lutites in Catalonia and an upper sequence of marls in Navarra. During deposition of the Upper Gray Lutite a transition from marine to nonmarine conditions took place in the Catalan area (Sáez 1987). In Navarra, however, distinct marine influences have been found locally in the Liédena Sandstone.

Although it is not possible to estimate the precise water depth, two facts must be emphasized: there was a decrease in water depth throughout the evaporite event from basal marine marls to very shallow detrital sediments at the top of the evaporite macrocycle (Súria Formation, Liédena Sandstone), and desiccation periods are not apparent. Significant erosional surfaces are unknown in any of the evaporite units.

In the South Pyrenean Potash Basin, the characteristics of the sedimentary facies and their distribution fit the "basin-center evaporite type" models proposed by Schmalz (1969) and Busson (1979, 1988).

In comparing the two major potash basins described in this volume, the Alsace Basin and the South Pyrenean Basin, several features should be emphasized. First, the geological setting is quite different in the two basins. The Alsace Basin is located in a rift system, the Tertiary Rhine graben, and the evaporite sequence overlies continental deposits. The South Pyrenean Potash Basin lies in the southern foreland basin of the Pyrenean range and the evaporite sequence overlies marine deposits and occurs at the transition from marine to nonmarine conditions.

Yet potassic paragenesis is nearly identical in both basins, with sylvite and carnallite and an absence of $MgSO_4$ salts. However there is a noticeable petrologic difference within the sylvinite cycles. In the Alsace Basin chevron (and cornet-shaped) textures in the halite layers are common, but these textures are absent in the South Pyrenean Basin. Finally, as we have already seen, a primary depositional origin of sylvite from a perennial brine body has been proposed in both basins.

(Accepted for publication June 1992.)

12

Continental Evaporitic Sedimentation on the South East Border of the Ebro Basin (Catalánides) During the Paleocene to Lower Oligocene

FEDERICO ORTÍ

Geological Setting

As mentioned by Santanach in this volume, the southern margins of the Ebro Basin are formed by the Iberian Range to the southwest, the Catalan Coastal Range to the east and the "Linking Zone" between them. These chains display characteristic structural features and, as these developed, so did sedimentation in the Ebro Basin. At the southeastern border of the Ebro Basin continental evaporite sedimentation took place during the early Paleogene along the area that now makes up the Catalan Coastal Range (Catalánides), from the region of Gandesa in the southwest to Manresa in the northeast, a distance of about 150 km (figure 12.1). These evaporites are intercalated within several fine-grained detrital units having a Catalánide provenance. The evaporitic units range in age from the late Paleocene to early Oligocene (Stampian, s.l.). Along this borderland the lowermost Paleogene sediments were deposited upon a Hercynian-Mesozoic basement.

Three paleogeographic domains have been distinguished in the Catalan Coastal Range by Anadón et al. (1979–82): (a) a northern domain, northeast of the Llobregat river; (b) a central domain located between the Francolí and Llobregat rivers; and (c) a southern domain, to the southwest of the Francolí river (figure 12.1B). Two factors that are of great importance in the study of the evaporites present within this range are:

1. the two periods of Eocene marine transgression where the sea entered from the north (Atlantic domain); the sediments that resulted are present in the central and northern areas of the Catalan Coastal Range but are lacking in the south. These two phases of transgression resulted in deposition of the Orpí Formation and the Santa Maria Group
2. the pre-Paleogene basement in the northern area which contains almost no Triassic evaporites due to earlier erosion or non-deposition, whereas in the central and southern areas there are rather significant volumes of Middle (Muschelkalk) and Upper (Keuper) Triassic evaporites overlying the Hercynian basement.

The sedimentological history of the southeastern border of the Ebro Basin and the Catalan Coastal Range during the Paleogene is briefly reviewed here (Anadón et al. 1979; Colombo 1980). During the Paleocene to early Eocene the mountain chain behaved as a rigid block with little noticeable tectonic activity. From the early to middle Eocene the base-

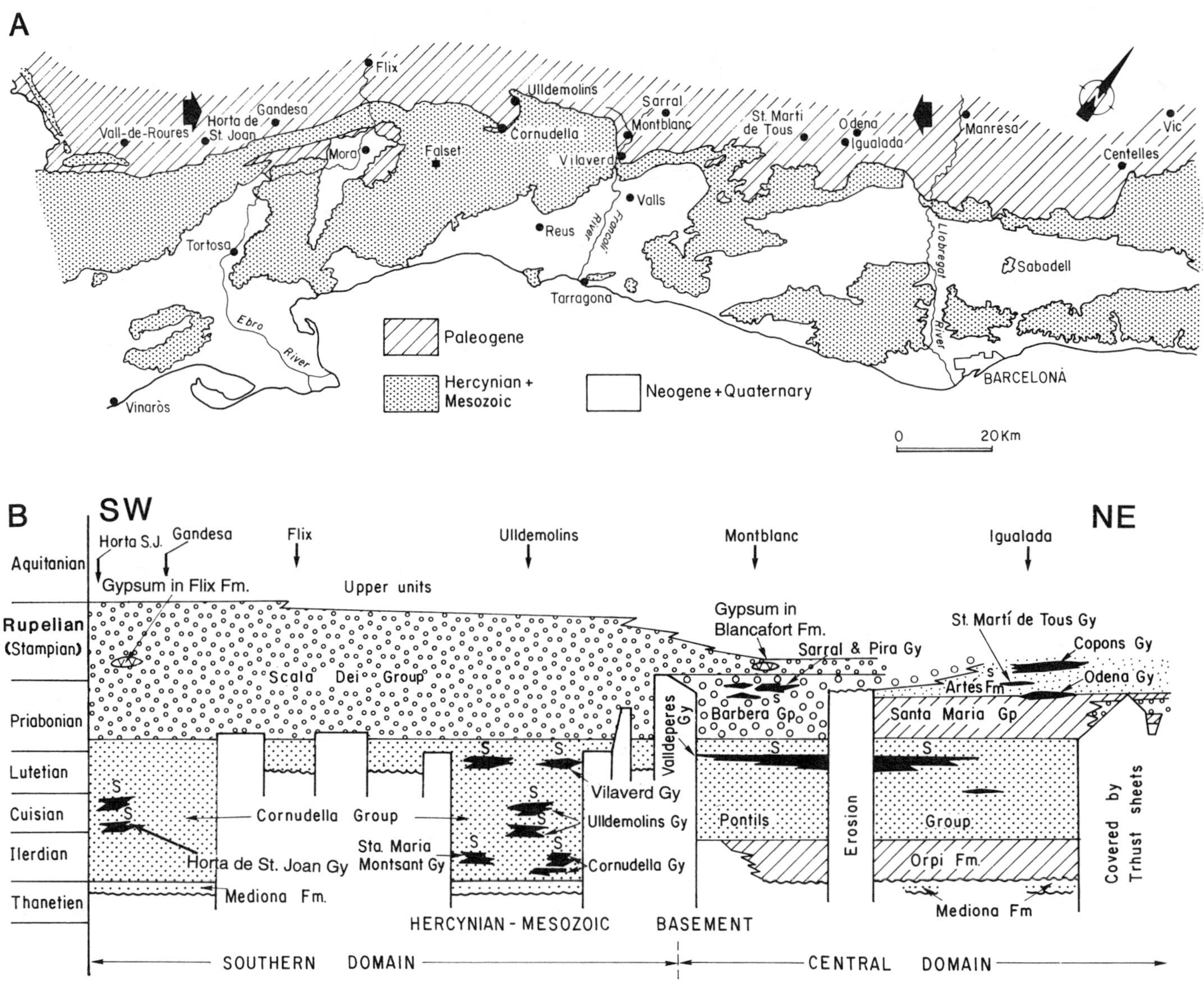

Fig. 12.1. Geological and stratigraphic distribution of the Paleocene-lower Oligocene continental evaporites in a part of the Ebro Basin (Catalánides).

A. Schematic geological map of the contact zone between the Catalan Coastal Range and the Paleogene of the Ebro Basin.

B. Stratigraphic distribution of the major evaporitic units in the lower Paleogene along the southeastern border of the Ebro Basin (lithostratigraphy and chronostratigraphy adapted from Anadón et al. 1979–82) (not to scale). Only gypsum layers thicker than 10 m are shown (in black). Key: S = abundant chert nodules and meganodules; s = sporadic chert nodules.

ment was faulted and a number of small basins were created with deposition controlled by block relief. During the middle to late Eocene a well-defined basin border developed and the associated sedimentary systems reflected significant amounts of marginal tectonism. This geodynamic activity continued during the Oligocene with variable intensity.

Thus, for a part of the Paleogene, large volumes of coarse clastic materials were deposited by the alluvial systems developed along the basin border and were controlled by the active basement faults. Movement on various faults was initiated at different times and resulted in the diachronous development of major alluvial fan systems (the conglomerates of Montserrat, Sant Miquel del Montclar, Montsant) (Anadón and Marzo 1986). Syntectonic unconformities are present in the coarse-grained facies, demonstrating a clear relationship between the fault movements and sedimentation, as well as local changes in the Ebro Basin's southeastern borders during the Paleogene. Nevertheless, these localized unconformities are restricted to the marginal areas and die out rapidly towards the interior of the basin, where depositional sequences are more continuous and apparently unaffected by synsedimentary deformation and erosion.

Only the most significant lithostratigraphic Paleogene units that formed in the central domain are considered here. These units, defined by Anadón and Marzo (1986) are, from base to top (figure 12.1B):

Continental unit: The Mediona Lutite Formation (late Paleocene), at the base of the Paleogene section is 3 m to 35 m thick and is comprised of sandstones and red lutites formed in alluvial plain environments, with some shallow calcareous lacustrine episodes.

First marine transgression: The Orpí Limestone Formation (Ypresian in part: Ilerdian) consists of foraminiferal grainstones of shallow marine origin, mainly composed of *Alveolina* and *Orbitolites* foraminifera.

Regressive units: The Pontils Group (late Ypresian (Cuisian)—early Bartonian) was deposited in continental to transitional environments, and consists of several formations. Some of the formations (including the Valldeperes Dolostone and Gypsum Formation, with considerable amounts of evaporites: up to 120 m thick), were deposited in mud flats and playa lakes. Minor amounts of gypsum are present in other units, such as in the Carme Lutite Formation.

Second marine transgression: The Santa Maria Group (Bartonian—middle Priabonian) consists of three marly and calcareous formations reaching almost 900 m in thickness, with an evaporitic unit, the Odena Gypsum Member of the Cardona Saline Formation (up to 40 m thick), at the top.

Regressive unit: The Artés Molasa Formation (late Priabonian—early Oligocene) consists of alluvial to fluvial sandstones together with a red lutitic complex (mud-flats).

In the southern domain, where time equivalent deposits from the marine transgressions are absent, somewhat different stratigraphic units have been distinguished (Colombo 1986). In some a significant phase of evaporitic deposition is present, especially in the Cornudella Group, the lateral equivalent to the Pontils Group. Evaporites are also present in the Barberà Group.

Evaporitic Sedimentation

The most significant sedimentological feature near the southeastern border of the Ebro Basin is the development of the spectacular alluvial fans controlled by the movements of major fault zones. However there is considerable evidence that evaporitic environments existed along that border. These may pre-date or post-date the alluvial progradation.

In general, based on sedimentological evidence, it appears that these evaporitic deposits developed in very shallow lakes and playa complexes on the lutitic plains of the alluvial and alluvial-fluvial systems. In rare instances these clastics grade laterally into shallow lacustrine carbonates. Some of these evaporitic units have a remarkable lateral extent, although the individual lithological bodies composing them are discontinuous. The only evaporative material that is present is calcium sulfate, with chert and some carbonate. Evidence of the former presence of halite is practically absent. All the gypsum on outcrop is secondary, after earlier anhydrite.

These formations commonly crop out along a southwest to northeast trend and many of them have been quarried for the plaster industry. Due to the general areal tilting (10° to 30° NW), transverse sections and exposures are lacking. From study of the few exposures present, these units seem to thin towards the center of the basin and rapidly die out to the northwest. The general morphology of these formations is usually lenticular and they are elongate in the direction of the basinal border (southwest-northeast). The age of these evaporites has only been approximated and is based on the ages established for their detrital and carbonate host units.

From bottom to top, the most important evaporitic units are (figure 12.1B):

Intercalations in the Pontils and Cornudella Groups (late Ypresian p. p. (Cuisian) to early Bartonian). Evaporite intercalations in the central domain in these groups are stratigraphically located between the sediments deposited during the two periods of marine transgression. Maximum calcium sulfate deposition occurs in the Cornudella Group in the Montsant area (southern domain), with the following depositional units: Cornudella Gypsum, Santa Maria del Montsant Gypsum, Ulldemolins Gypsum, and the Vilaverd Gypsum (Colombo 1986). To the northeast, the Valldeperes Gypsum (Pontils Group) occupies the same stratigraphic position as the Vilaverd Gypsum. Near the southwestern limit of the depositional area, the Horta de Sant Joan Gypsum developed in the Cornudella Group. Only a small amount of gypsum is present locally in the Carme Formation (lower part of the Pontils Group). Sparse quantities of coarse detrital sediments reflect the absence of tectonic activity.

These gypsiferous units all display massive bedding and are overall quite similar to each other. They were originally composed of a microlenticular gypsum lithofacies (see figure 8.2M of Ortí in this volume) and display limited development of nodular facies of sabkha-like origin formed either by nodules or meganodules (figure 12.2A). In the Valldeperes Gypsum nodular development is common.

Fluidized soft sediment deformation structures are common in these matrix-free gypsum deposits and, as a result, gypsiferous patches rich in clay or carbonate are irregularly scattered within the gypsum layers. Many pseudomorphs after lenticular crystals of primary gypsum are present in the clayey zones.

Nodular chert is also a common feature in these layers. Such silica nodules are usually located within the matrix-free gypsum layers (figure 12.2B) but are also present within the gypsum nodules. Pseudomorphs of lenticular and rosette gypsum forms are preserved within the chert (figure 12.2C). Bioturbation structures in the gypsum are frequently replaced by silica (figure 12.2D). Chert meganodules (>50 cm diameter) are common.

As an example of the depositional patterns common in these gypsum sections, the general stratigraphic column of the Cornudella Group at Horta de Sant Joan is given in figure 12.3. Details of one of these units and some small-scale depositional cycles developed within these gypsum deposits are illustrated in figure 12.4. Some mineralogical and petrological investigations of these evaporitic units were carried out by Estop (1974) and Ortí et al. (1987). Their studies demonstrate the generally secondary character of the gypsum in these units, resulting from the hydration of earlier anhydrite, the common presence of small celestite crystals, and the very early diagenetic formation of the chert.

Intercalations in the Barberà group (Bartonian?-Priabonian). This continental group, developed in the abovementioned central domain, is the lateral equivalent of the marine Santa Maria Group to the northwest. Gypsum intercalations within the Barberà group are the lateral equivalents of the Montblanc Lutite Formation in the localities of Pira and Sarral. These lutites grade laterally into thick conglomeratic units. The gypsiferous units are similar to those intercalated in the Pontils and Cornudella groups in that they also bear chert and locally have nodular sabkha-like deposits. The latter facies is readily seen in the quarries near Sarral as vertical

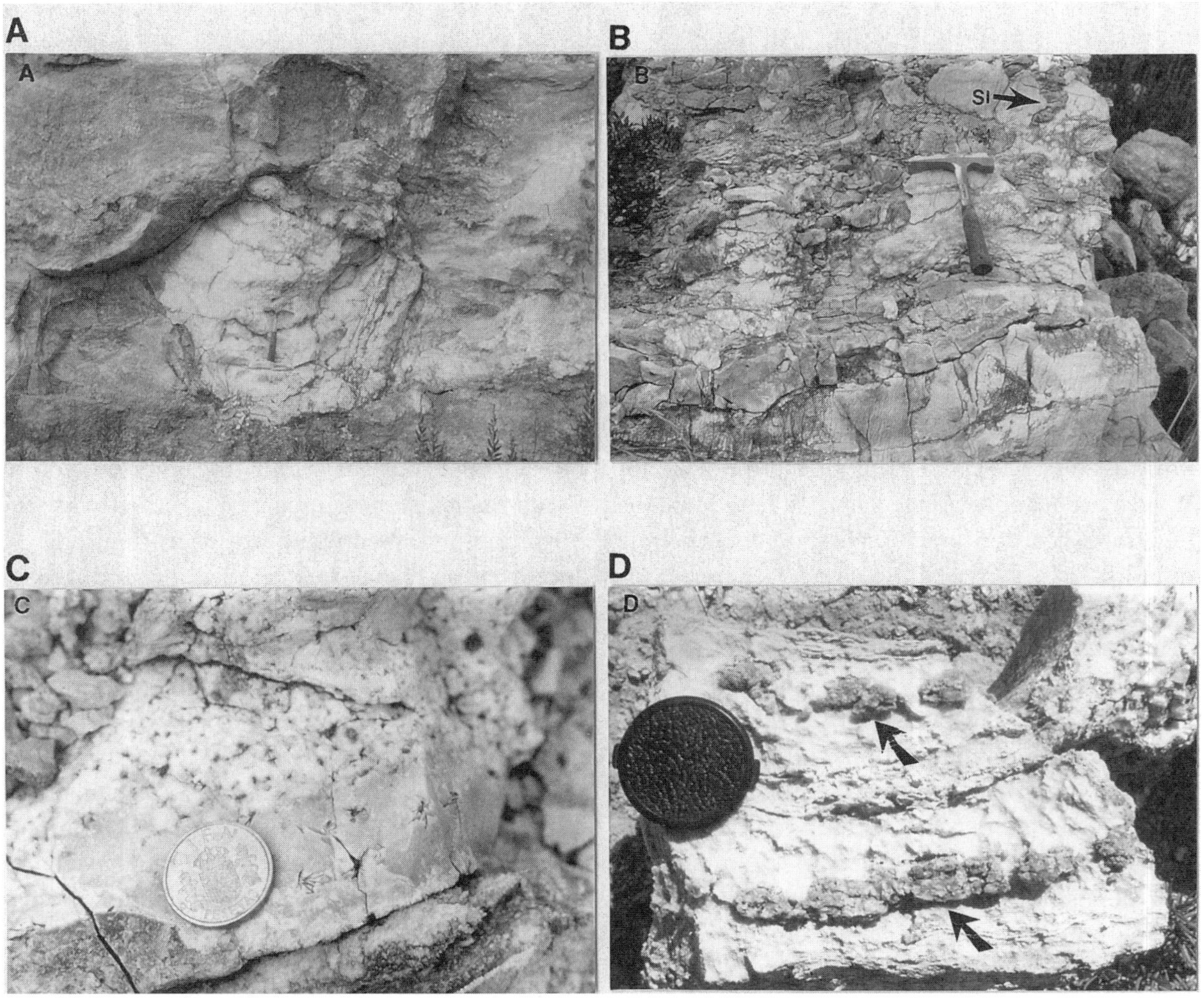

Fig. 12.2. Outcrop photographs of various physical features of gypsum (now alabastrine secondary gypsum at outcrop). A. Meganodule of alabastrine secondary gypsum both replacing and displacing the matrix-free microlenticular gypsum layers in the Vilaverd Gypsum of the Cornudella Group (see hammer for scale). B. Silica nodules (Sl) replacing matrix-free microlenticular gypsum (Vilaverd Gypsum). C. Pseudomorphs and moldic porosity of lenticular and rosette gypsum inside a chert nodule (diameter of coin, 2.2 cms) (Vilaverd Gypsum). D. Silicified burrows in the secondary gypsum of the Sta. Maria del Montsant Gypsum (Cornudella group).

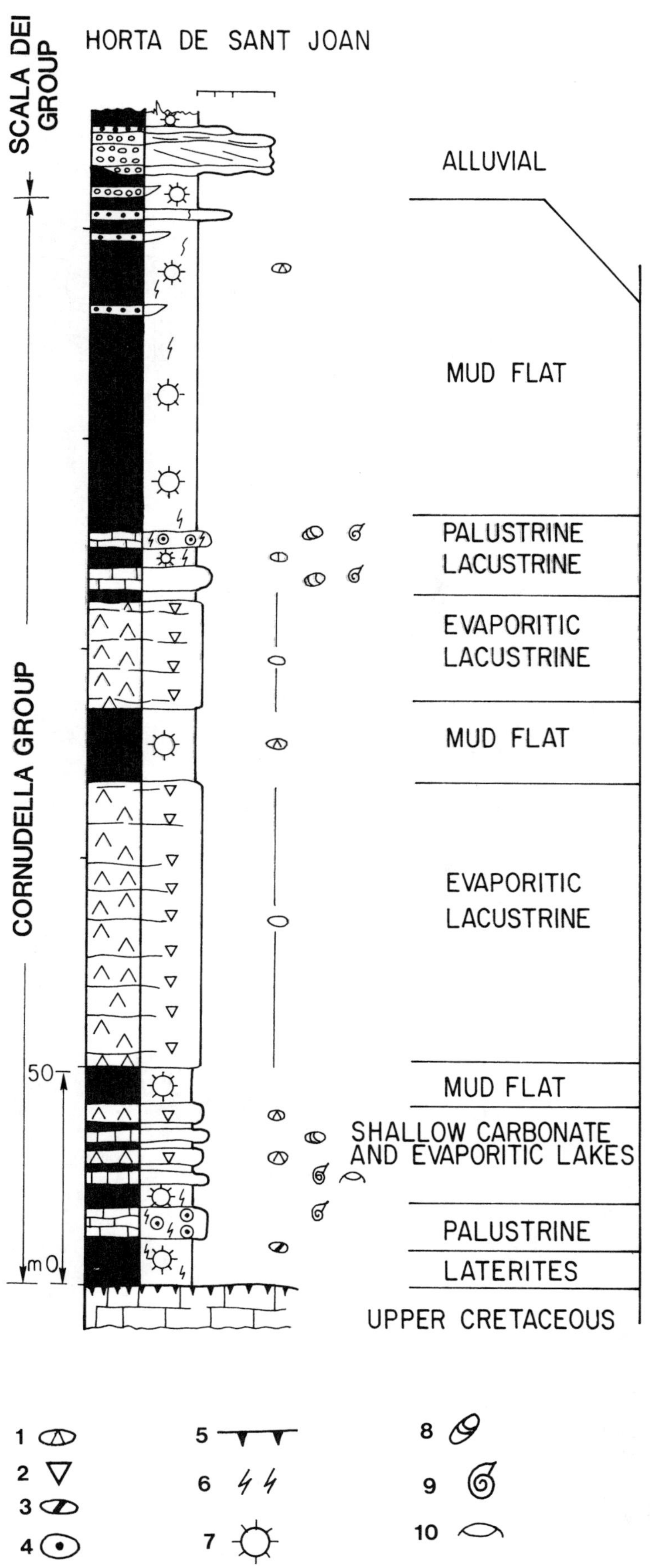

Fig. 12.3. Stratigraphic section and depositional environments of the Cornudella Group in Horta de Sant Joan (after Cabrera et al. 1985). Key: 1 = evaporite nodules; 2 = chert nodule; 3 = ferruginous nodules; 4: = pisolites; 5 = hard ground; 6 = burrowing; 7 = mottling; 8 = Charophytes; 9 = Gastropods; 10 = Ostracods.

Fig. 12.4. *Following page:* Stratigraphic section of part of the Cornudella Group: A. Detail of Vilaverd Gypsum unit in the quarries close to Vilaverd. B. Characteristic small-scale cycles of the gypsum beds (a), and some other depositional variants (b, c, d).

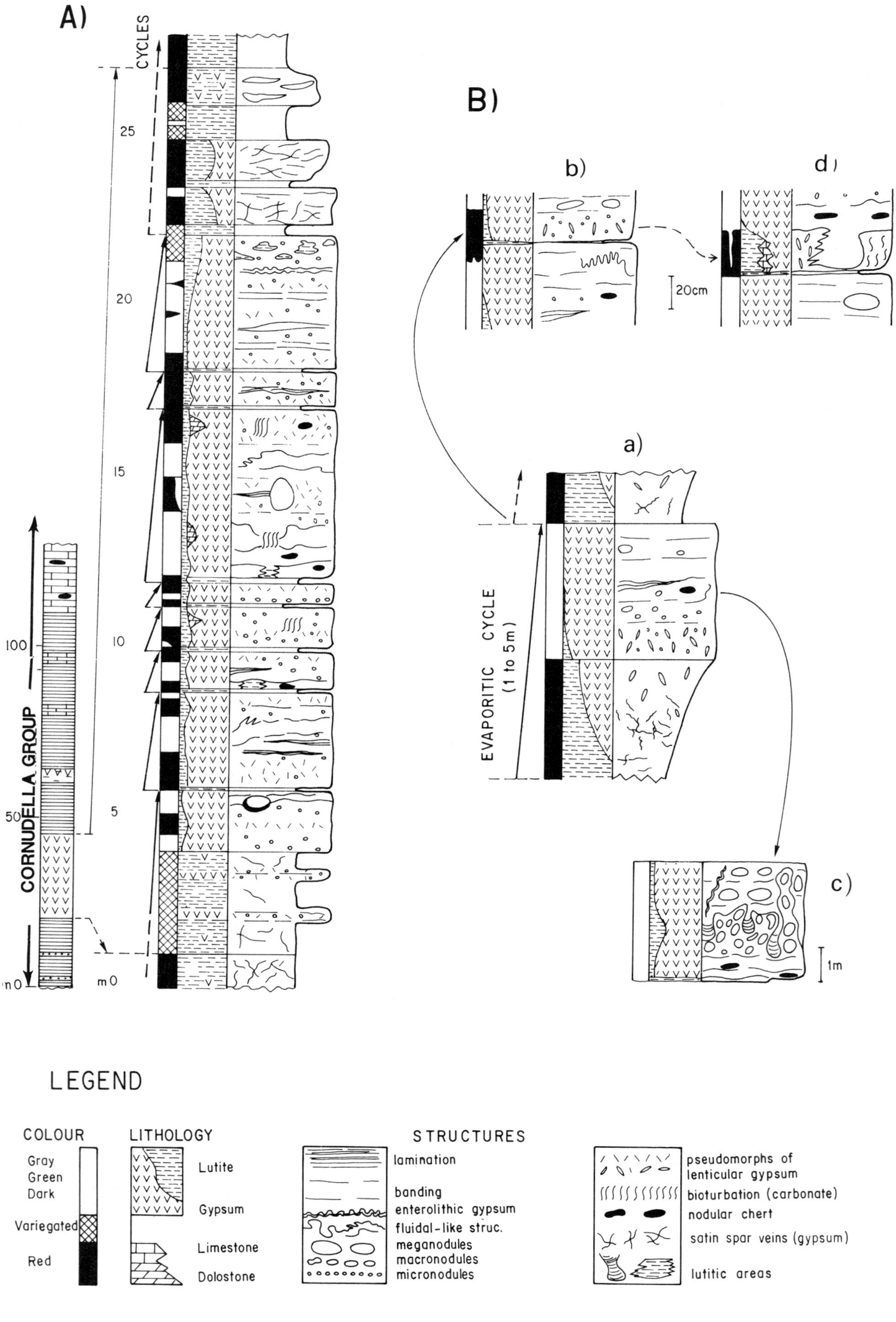
A)
CYCLES
25
20
15
10
5
m0
100
50
m0
CORNUDELLA GROUP
B)
b)
d)
20cm
a)
EVAPORITIC CYCLE (1 to 5 m)
c)
1m
LEGEND
COLOUR
Gray
Green
Dark
Variegated
Red
LITHOLOGY
Lutite
Gypsum
Limestone
Dolostone
STRUCTURES
lamination
banding
enterolithic gypsum
fluidal-like struc.
meganodules
macronodules
micronodules
pseudomorphs of lenticular gypsum
bioturbation (carbonate)
nodular chert
satin spar veins (gypsum)
lutitic areas

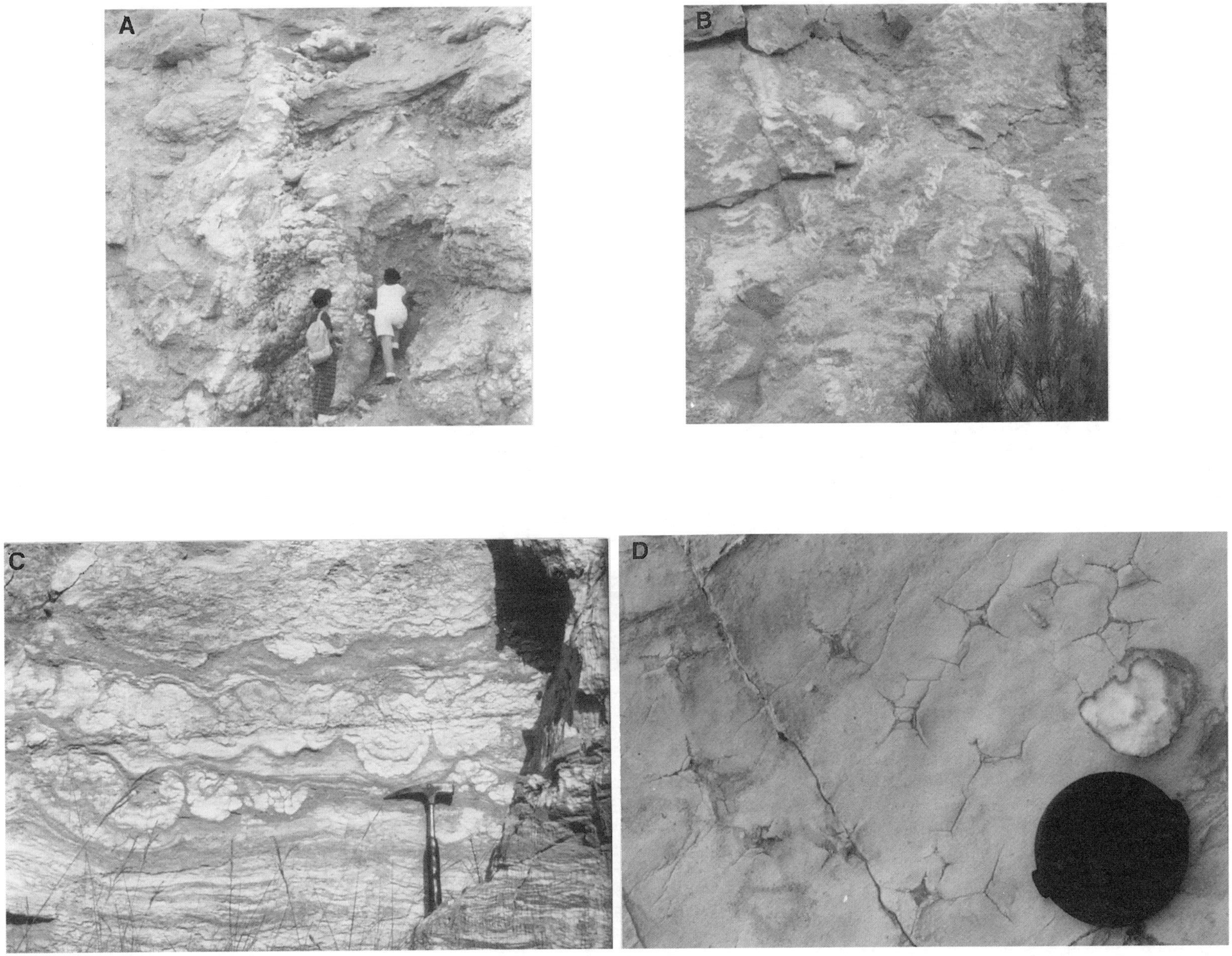

Fig. 12.5. Outcrop details of the secondary gypsum units: A. Large scale flow structure in the nodular sabkha of Sarral Gypsum (Barberà Group). Vertical columns of nodules mixed with lutitic zones give a chaotic appearance. Quarry near the village of Ollés. B. Detail of outcrops is the Ollés quarry, displaying the abundant subvertical enterolithic layers developed within a massive microlenticular gypsum matrix. C. Alternations of laminated and nodular lithofacies in the Copons Gypsum in quarry near Clariana (Artés Formation). D. Halite casts in Copons Gypsum in quarry near Clariana (Artés Formation).

columns of nodular masses of earlier anhydrite, apparently emplaced by soft sediment flow and deformation (figure 12.5A). They are accompanied by almost vertical enterolithic layers of calcium sulfate (gypsum after anhydrite) (figure 12.5B).

Intercalations in late Eocene (Priabonian)—early Oligocene (Rupelian/Stampian). These evaporitic intercalations post-date the second marine transgression developed in the central domain. In the southern part of the basin where the conglomeratic Scala Dei Group is present, gypsum interbeds are present that are lateral equivalents of the Blancafort Formation, a lutitic unit located to the northeast of the Montsant Conglomerates. The Montsant Conglomerate unit was assigned to the Rupelian (Stampian) by Colombo (1986). Some gypsum also is present in the Flix Sandstone and Lutite Formation (also Rupelian/Stampian).

In the northern portion of the central domain, thick evaporitic units, the Sant Martí de Tous Gypsum and the Copons Gypsum, are intercalated within the Artés Formation, and are considered to be Rupelian (Stampian) (Saéz 1987). These gypsiferous units are characterized by the absence of chert, except for Sant Martí de Tous Gypsum, where some silica nodules are noted. The Copons Gypsum and the Sant Martí de Tous Gypsum are made up of nodular-laminated gypsum lithofacies (L-N, of Ortí in this volume) and represent sedimentation within more concentrated evaporitic lakes than the other units (figure 12.5C). Abundant halite casts are present in the Copons Gypsum unit (figure 12.5D).

Relationships Between Marine and Continental Evaporitic Episodes

Based on the results of isotopic studies of the sulfate, the lower Paleogene gypsiferous intercalations originated from the chemical recycling of marine Triassic evaporites present in the Catalan Coastal Range (Utrilla 1989). The gypsum deposits developed essentially in the southern and central domain of the basin border where the Triassic evaporites have been preserved within the Mesozoic cover of the chain.

Although correlations are somewhat uncertain between the marine and continental units, in general, Ypresian (Cuisian) to early Lutetian marine evaporites of the southeastern Pyrenees might have a temporal coincidence with some of the continental evaporites in the Cornudella and Pontils groups of the southern and central domains of the Catalan Coastal Range (Campdevànol and Beuda sequences of Puigdefàbregas et al. 1986; also see figures 8.4 and 10.1, this volume). It is more difficult to recognize a stratigraphic equivalent to the middle Priabonian marine evaporites of the potash basin, although some Priabonian evaporites are present in the Barberà Group in the southern domain of the Catalan Coastal Range.

Chert-Evaporite Relationships

The existence of abundant chert, not only in the Eocene gypsum units of Catalonia that we deal with in this study, but also in the peripheral sulfatic lake system of the Miocene of Aragón and Navarra (see figures 8.3 and 8.4, this volume), suggest that much of the chert developed in low salinity lakes as an evaporitic by-product that precipitated when the waters rich in silica coming from the marginal alluvial systems mixed with lacustrine, sulfate-rich water.

Generally the growth of some isolated and very large meganodules of anhydrite (in the gypsum units of the Cornudella Group) displaced both the microlenticular gypsum masses and chert nodules that were present within them during their growth. This observation suggests that those silica nodules formed before the anhydrite meganodules, during very early diagenesis. Also, commonly, evidence of burrowing is well preserved within the chert nodules, but later alteration of gypsum to anhydrite has destroyed any other vestiges of bioturbation within the gypsum layers.

Carbonates associated either laterally or vertically with the gypsum units considered here also contain chert locally. The same observation is made for other carbonate units in association with sulfates of Eocene, Oligocene, and Miocene age throughout the Ebro Basin.

(Accepted for publication April 1992.)

13

Continental Evaporitic Sedimentation in Navarra During the Oligocene to Lower Miocene: Falces and Lerín Formations

JOSEP M. SALVANY

Structural Setting

Tertiary deposits of continental origin (Oligocene-Miocene) in the western sector of the Ebro Basin are found in the geographical area known as the Ribera of Navarra and La Rioja (north and south of the Ebro River, respectively). They are bounded by some very well-defined structural elements (figure 13.1). To the south, the Tertiary sediments are in contact with the Mesozoic of the Cameros Chain (Iberian Range) along major reverse faults having several hundred meters of slip. Along these structures the Tertiary rocks are tilted almost vertically and are even overturned. To the north, the contact also is along reverse faults. The affected sediments are mainly made up of the marine Eocene of the Pamplona Basin, which separates the Ribera zone from the Pyrenees. Riba (1964) called those structural lines confining the continental Tertiary "master faults," and pointed out their important synsedimentary role in the configuration and later subsidence of the basin during the Oligocene and Miocene. Across these master faults a series of transverse structures is superposed to the first set, which acted as wrench-faults and have a normal orientation.

In the inner part of the basin the Tertiary sediments show only gentle fold structures with axes trending ESE-WNW. These folds disappear towards the west and the southeast, giving way to a flat (tabular) landscape almost unaffected by any tectonism. On those folds that are present, the axes are sometimes deformed or display sudden periclinal ends, because they commonly involve evaporitic formations. Deformation intensity decreases towards the south and the age of the folds becomes progressively younger in that direction as well.

Sedimentation during the Oligocene and Miocene took place under a compressive regime controlled by the uplift and structural evolution of the Pyrenees and the Iberian ranges. Between these ridges, the sedimentary troughs gradually shifted toward the south as a result of the predominance of the Pyrenean deformation over the Iberian one. This continuous shift is exactly paralleled by the migration of the evaporitic formations that generally occupied the basinal depo-axis (figure 13.2). The same observation is demonstrated by the shift in clastic provenance (Riba and Pérez Mateos 1962). In the Eocene to early Oligocene, the paleocurrents came generally from the south (Iberian Range), while during the late Oligocene, and particularly during the Miocene, they displayed preponderantly Pyrenean provenance.

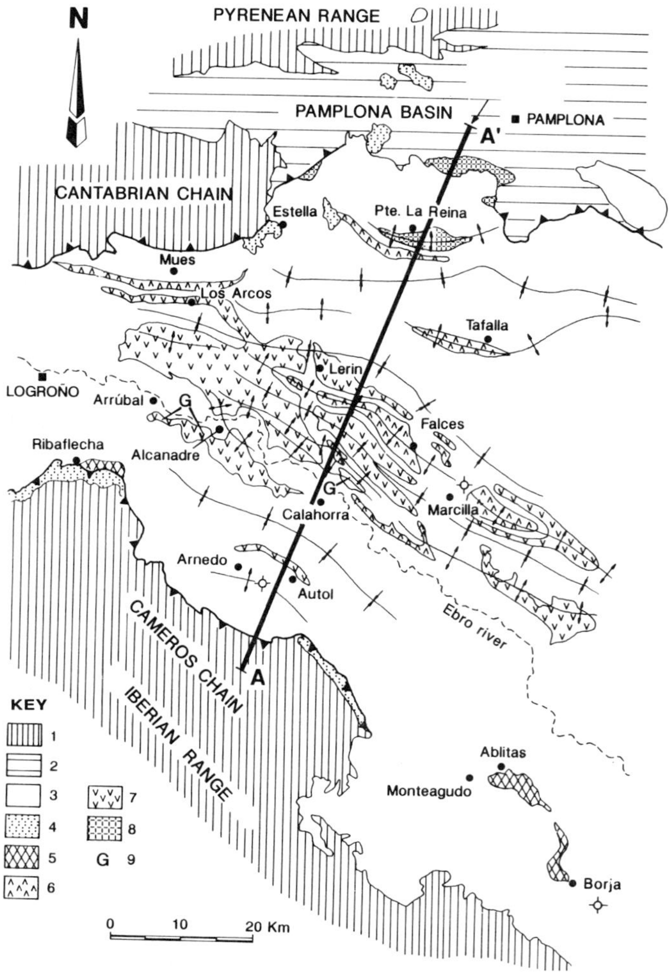

FIG. 13.1. Geological map of the Navarra-La Rioja sector of the Ebro Basin. The position of the outcropping evaporitic formations and the deep boreholes is shown. Section A–A′ is given in figure 13.2. Key: 1 = Mesozoic; 2 = marine Tertiary; 3 = continental Tertiary, detrital and calcareous; 4 = Triassic diapirs (Keuper); 5 = Monteagudo and Ribafrecha Gypsum formations; 6 = Falces Gypsum Formation; 7= Lerín Gypsum Formation; 8 = Puente La Reina Gypsum Formation; 9 = glauberite in the Lerín Formation.

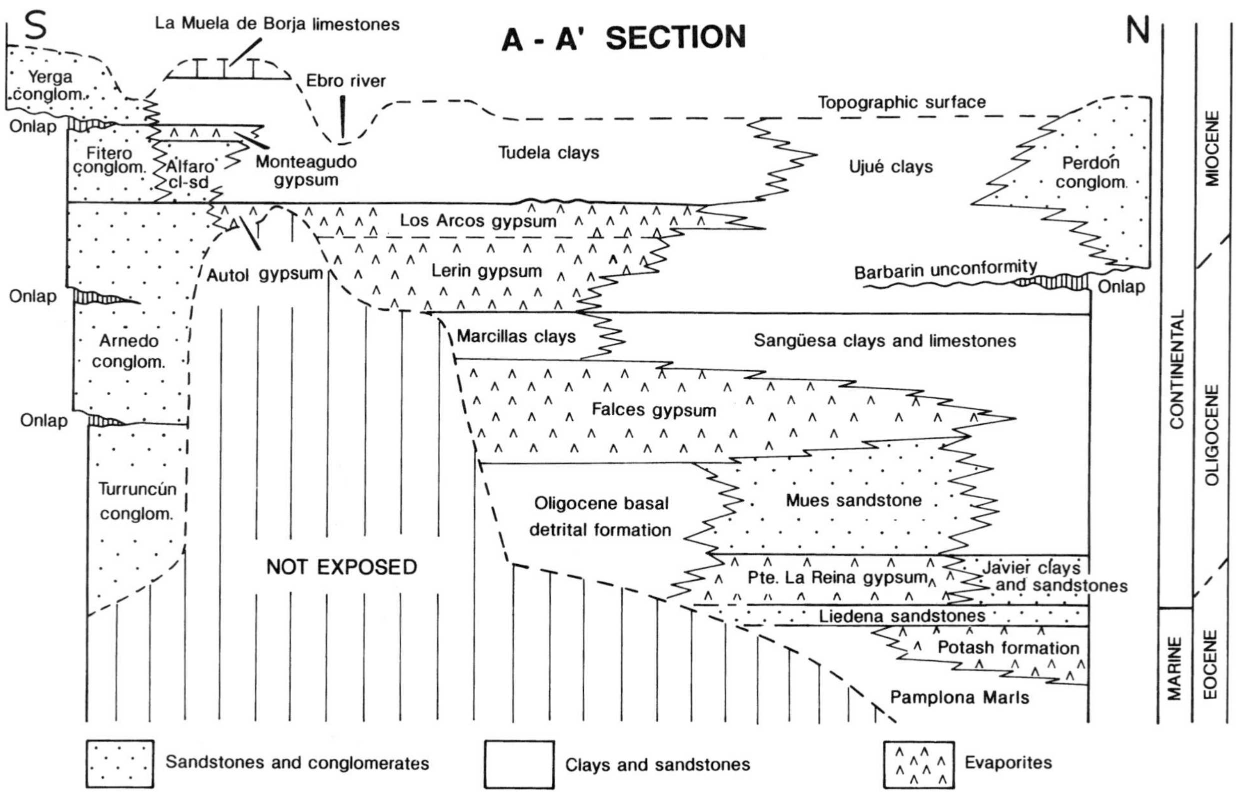

FIG. 13.2

Stratigraphical Setting

In the continental Tertiary sequence a marked distinction is seen between those coarse-clastic materials (conglomerates and sandstones) of the basin borders and the fine-detrital and chemical-evaporitic sediments of the central areas (clays, sulfates, and chlorides). The former are represented by proximal alluvial fans, while the latter consist of playa lake deposits initiated on the distal argillaceous plains of those fans. Calcareous lacustrine facies are of lesser importance and are variably developed depending on the sectors and history of basin evolution.

BASIN CENTER

The general stratigraphic characteristics of the basin center are known from studies by Riba and Pérez Mateos (1962), Riba (1964), Crusafont et al. (1966), Castiella et al. (1978), Solé (1972), Ortí et al. (1986), Ortí and Salvany (1986b), Salvany (1989a, 1989b), and Riba and Jurado (1992). The continental Tertiary sequence is up to 4,000 m thick and consists mainly of alternating detrital and evaporitic formations displaying very complex lateral facies distribution. During the Oligocene three major evaporitic units were deposited: the Puente la Reina Gypsum, the Falces Gypsum, and the Lerín Gypsum (Castiella et al. 1978). Other minor evaporitic formations were deposited during the Miocene at the southern border of the basin. The Puente la Reina Gypsum directly overlies the top of the Eocene marine sediments (Guendulain Formation). This unit constitutes the basement of the continental Teriary in the northern part of the Navarrese sector of the Ebro Basin. The Puente la Reina Gypsum is poorly known because of meager outcrops and deformed exposures. The total thickness is estimated to reach between 300 and 400 m. The evaporites are largely composed of calcium sulfate (secondary gypsum with anhydrite having both nodular and laminated lithofacies); however, some halite is present.

Overlying the Puente la Reina Gypsum is a thick detrital unit, the Mués Sandstone Formation, that reaches thicknesses between 2,100 and 2,400 m in the Los Arcos sector (Solé 1972). This formation and its laterally equivalent units represents a prolonged period in which alluvial and calcareous lacustrine facies were deposited. The age of this detritic phase is attributed to the middle Oligocene, and may be due to a relatively humid period.

The Falces Formation constitutes the second major evaporitic unit, in which the depositional area clearly expanded and overstepped a great part of the basin. This thick unit overlies the Mués Formation in the Los Arcos sector but to the south is deposited on the "Oligocene basal detrital formation" (ENRESA 1987). In this southern zone the base of the Falces Formation constitutes a planar base that is the décollement horizon for the whole fold structure of the Navarrese Oligocene and Miocene.

The uppermost of the three gypsum formations, the Lerín Gypsum, is separated stratigraphically from the Falces Formation by a detrital unit whose thickness and facies development varies in each sector of the basin. In the central part this unit constitutes a monotonous argillaceous sequence 300 to 400 m thick, with some sandstone layers and minor amounts nodular gypsum (Marcilla Clay Formation). To the north and northeast these clays change laterally into more detrital argillites (Sangüesa Formation), an intercalation of common sandstone and limestone layers reaching thicknesses of 650 m with a 1,250 m maximum.

The Lerín Formation is considered to lie at the Oligocene-Miocene boundary (figure 13.2), but this age assignment is somewhat uncertain because of the lack of vertebrate remains within these

---

FIG. 13.2. Lithostratigraphic schema, without scale, of the detrital and evaporitic formations of the Navarra-La Rioja sector of the Ebro Basin, drawn in a north-south profile.

facies. Nevertheless, several paleontologic findings (Cuence et al. 1992) have suggested an age that is entirely Miocene for the Los Arcos unit and all the sediments overlying the Lerín Formation.

The Miocene materials overlying the Lerín Formation are a thick clayey sequence built up of red argillites, the Alfaro and Tudela Clay formations, which extensively crop out in the southern half of the Ribera de Navarra and La Rioja. Both represent the fine detrital sedimentation in the central area of the western Ebro Basin. The Alfaro Clay, which is well developed to the southwest of the Ribera, contains numerous sandstone layers and represents a fluvial system coming from the west. The Alfaro Clay passes laterally into the Tudela clay (argillites with some limestones and sandstones), well developed in the southeast. The Tudela Clay, with thin layers of interfingering lacustrine carbonate, was deposited by a fluvio-lacustrine system that represents an intermediate facies assemblage between the Alfaro Clays and the Zaragoza Formation that lies at the center of the Ebro Basin (figure 8.3, this volume). The Alfaro amd Tudela Clays contain small, local intercalations of gypsum, the only evaporitic vestiges within these detrital and calcareous formations. To the north, the Alfaro and Tudela formations pass laterally into the Ujuè Clays that represent the distal alluvial facies coming from the Pyrenees (figure 13.2).

The Tertiary sequence ends with limestone units (La Muela de Borja Formation) that are algal and bioclastic limestones, intensively burrowed, containing gastropods and ostracods and are a few tens of meters thick (maximum 30 to 40 m).

BASIN BORDERS

The southern border of the basin is well known because of several significant studies (Bomer, 1954; Riba, 1955; Muñoz et al. 1986/1987; Pérez et al. 1988; Muñoz 1992; and Riba and Jurado 1992). They all point out the importance of the tectonic control exerted by the Iberian Range on the detrital sediments during the Paleogene and Neogene. In the Arnedo sector six tecto-sedimentary units have been identified by Muñoz et al. (1986/1987). These tecto-sedimentary units ("Unidad Tecto Sedimentaria" or UTS) are limited by sedimentary breaks that appear as unconformities on the basin borders and/or as significant changes in lithologies in the general sedimentary trends in the basin center (figure 13.3).

In these sequences, conglomeratic horizons represent maximum development of alluvial systems, implying significant clastic input due to orogenic activity. Deposition of evaporites and/or carbonates marks periods of lesser tectonic activity allowing maximum expansion of lacustrine environments. Clastic sequences generally coarsen upward marking the uplift phases of the development of the Iberian Range. However, in some sequences fining upwards trends were developed.

Passing toward the center of the basin there is a very rapid lateral change from the clastic facies into evaporites. In only a few kilometers a marginal conglomeratic sequence passes into a sandy and argillaceous sediment, and almost immediately develops into argillaceous-evaporitic sediments. At the same time, angular unconformities present on the borders attenuate towards the basin and completely disappear in its center, where the contacts between the various formations always seem concordant. In the entire evaporitic assemblage, only the youngest member can be directly correlated with detrital equivalents at the borders.

The most significant of the unconformities noted in this section was developed near the top of the Monteagudo Gypsum. The Miocene sequence overlying it extends broadly to the south, even covering the Mesozoic section of the Cameros Chain. At its contact with the lower Miocene conglomerates, an angular unconformity is noted.

The stratigraphic characteristics of the northern border of the basin appear more complex and are poorly known. Nevertheless, articles by Riba and Pérez Mateos (1962) and Riba (1964, 1992) have explained the tecto-sedimentary behavior of that border and identified an important angular unconformity (Barbarin discordance) inside the detrital sequence that separates the Falces and the Lerín formations. This unconformity disappears towards the basin center but displays a noticeable continuity along the northwestern border of the basin, over

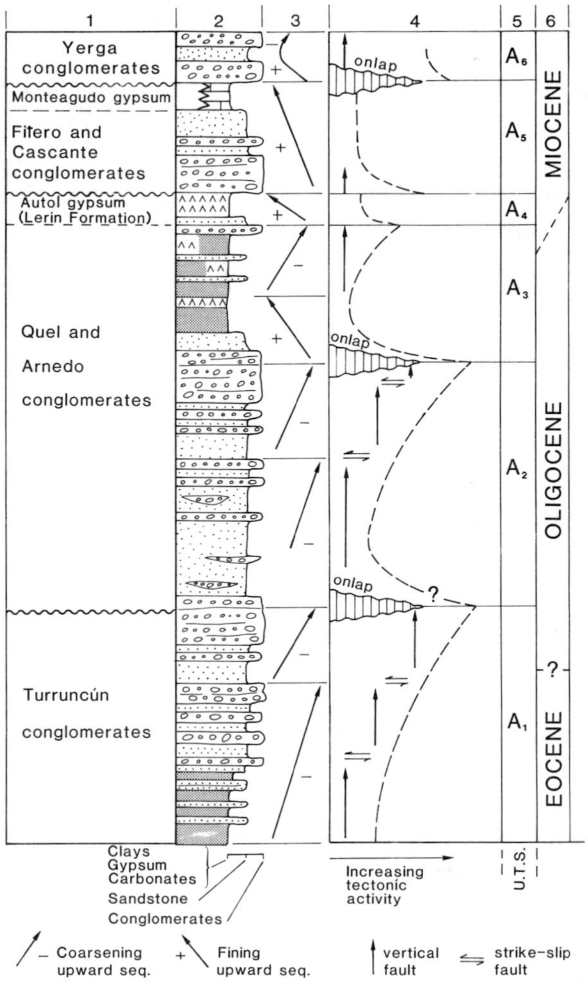

FIG. 13.3. Lithostratigraphic sequence of Paleogene and Neogene materials of the southern border of the Ebro Basin in the Arnedo (La Rioja) sector, modified from Muñoz et al. (1986/87). Key: 1 = lithostratigraphic units; 2 = lithologies; 3 = sedimentary cycles; 4 = tectonic evolution of the basin; 5 = tectono-sedimentary units (U.T.S.); 6 = chronostratigraphic units.

tens of kilometers. This Oligo-Miocene unconformity separates a lower folded sequence from a less disrupted but folded upper one, which onlaps the Mesozoic and marine Eocene materials of the south Pyrenean basin (Ujué Formation and the Perdón Conglomerates).

Falces Gypsum

Although the Falces Gypsum Formation displays sedimentological and petrological characteristics identical to the overlying Lerín Gypsum Formation, it is poorly known because of limited area of outcrop. The Falces Gypsum Formation is a very thick evaporitic unit that always occupies the highly deformed cores of diapiric anticlines while the Lerín Gypsum normally forms their flanks, which are far less affected by tectonism.

Because of its position within diapiric structures it is quite difficult to estimate the true thickness of the formations. In Desojo (near Los Arcos), where this unit is completely exposed, the beds are almost vertical and the gypsum is between 750 and 950 m thick (Solé 1972). On the core of the Falces anticline, the Marcilla borehole was drilled through the Falces Formation, and a diapiric thickness of 2,800 m was present (IGME 1987). The true sedimentologic thickness may be about one thousand meters.

At the surface the Falces Gypsum is a monotonous sequence of secondary gypsum layer and common gray argillite fine intercalations. The gypsum lithofacies are composed mainly of the laminated-nodular (L-N) association (see chapter 8, this volume). In the Marcilla borehole the calcium-sulfate is always anhydrite, with significant amounts of intercalated halite. Some glauberitic beds (see chapter 14, this volume) have been recognized in cores of the Marcilla borehole, displaying features similar to the glauberite of the Alcanadre deposits of the Lerín Formation (Salvany and Ortí 1994).

In the Marcilla borehole, which lies at the center of a diapiric anticline, the following lithologic members were described, from base to top (ENRESA 1987):

1. lower anhydrite member (296 m thick)
2. lower anhydrite-halite member (333 m thick)
3. lower member with anhydrite and gray clays (171 m thick)
4. intermediate anhydrite-halite member (389 m thick)
5. intermediate anhydrite member (357 m thick)
6. upper member with anhydrite and gray clays (215 m thick)
7. upper anhydrite member (1004 m thick)

Carbonates are almost completely lacking in the Falces Formation. The clays present are a mixture of illite, chlorite, and kaolinite.

Lerín Gypsum

This formation is a complex evaporitic formation made up of alternating gypsiferous and clayey units displaying significant lateral facies changes (Salvany 1989). Generally it is made up of five major gypsum units of great lateral continuity, separated by argillaceous units of thicknesses similar to the gypsiferous ones. The gypsum units show a clear thickening upward tendency, and each unit expands over the underlying one and shifts toward the west (figure 13.4 and 13.5).

The most important gypsum level in the Lerín Formation is the uppermost one (Los Arcos Gypsum) located in the central part of the basin, reach 250 m thick and more than 100 km long. To this unit belong the glauberite deposits of Alcanadre-Arrúbal (chapter 14 this volume). Toward the southern border of the basin this gypsum deposit passes into the Autol Gypsum, characterized by the presence of alabaster meganodules scattered within a micronodular matrix that was formed by replacement of the microlenticular primary gypsum. The characteristics of the Autol Gypsum are very similar to those of the Monteagudo and Ribafrecha gypsums (see Ortí and Salvany in this volume).

Clay horizons interlayered between these five gypsum units also contain minor gypsum layers and a small amount of sandstone. The development of these gypsum layers varies from region to region. In the marginal part of the formation they are rare (Allo, Sartaguda, Mendavia, and Villafranca units), but in the central part they give way to gypsum-clay alternations (Lodosa and Peralta units).

The gypsum and clay units formed as couplets made up of a basal gypsum unit and an overlying clay-gypsum unit that represent an "evaporitic megacycle" (Salvany 1989). The megacycles show the broadest areal coverage by the evaporites at the base; moving up through each sedimentary cycle, the detrital (clayey) phase invades the evaporite area resulting in reduction of the evaporative area. During this later period the detrital sediments cover a very broad area, extending from the margin to the center of the basin. The most saline facies (laminated gypsum halite and glauberite) are found near the center of the basal gypsiferous units. Toward the top and the margins these megacycles are solely composed of nodular anhydrite with rare laminated gypsum (now as secondary alabastrine gypsum), within an argillaceous matrix.

These megacycles may be interpreted as both the result of changes in the rate of subsidence or due to long-term climatic fluctuations. (1) When the subsidence was slow, the alluvial facies were restricted

FIG. 13.4. Lithostratigraphic subdivison of the Lerín Formation: A. Northwest-southeast section; B. Plan view showing the mega-sequence structure. Key: Numbers 1 through 5 indicate correlation of megacycle succession indicated in figure 13.4A. Arrows indicate migration direction of depocenter.

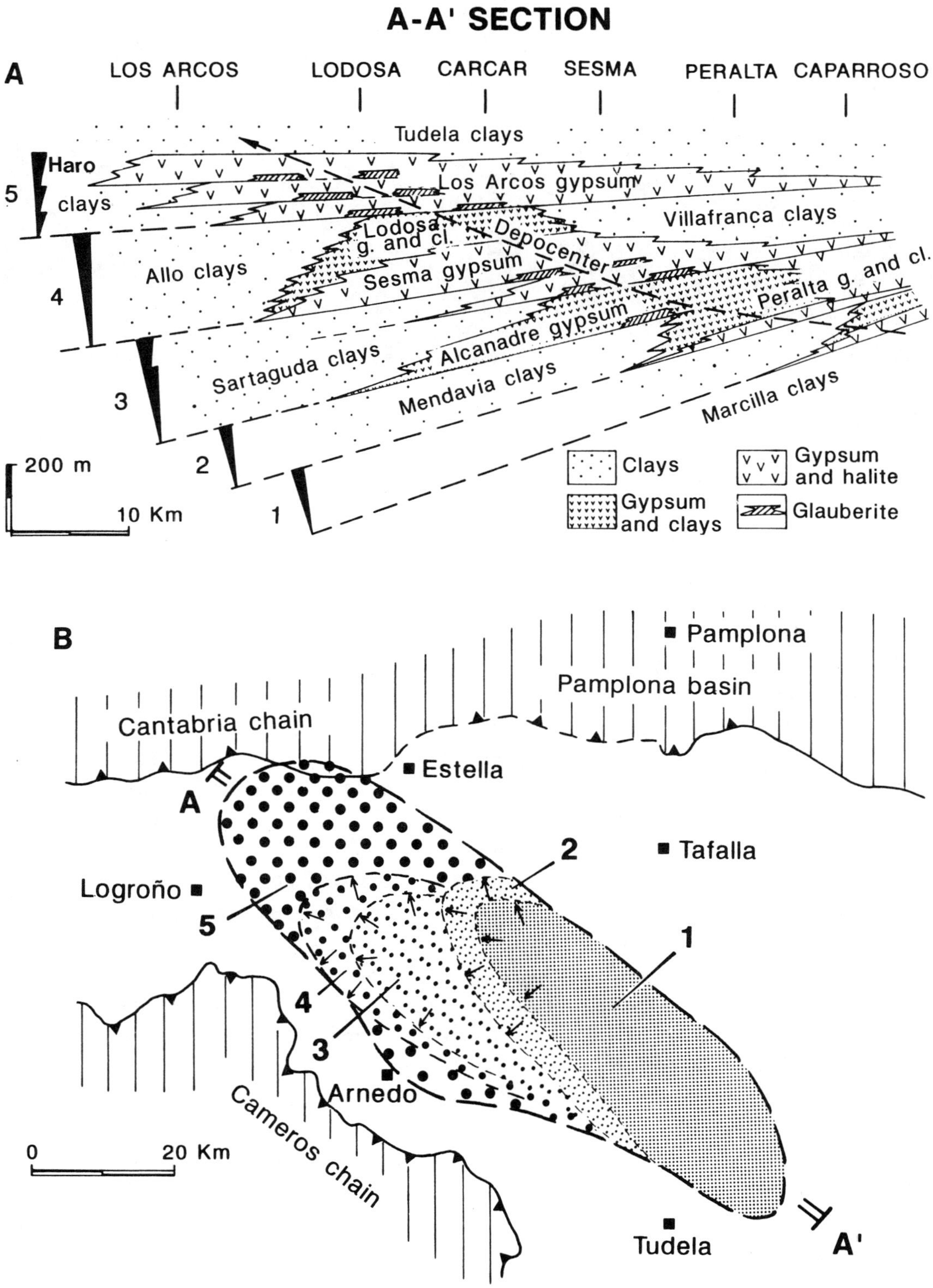
A-A' SECTION
A
LOS ARCOS
LODOSA
CARCAR
SESMA
PERALTA
CAPARROSO
Tudela clays
Haro clays
5
Los Arcos gypsum
Villafranca clays
Lodosa g. and cl.
Depocenter
Allo clays
4
Sesma gypsum
Peralta g. and cl.
Alcanadre gypsum
Sartaguda clays
3
Mendavia clays
Marcilla clays
200 m
2
10 Km
1
Clays
Gypsum and halite
Gypsum and clays
Glauberite
B
Pamplona
Pamplona basin
Cantabria chain
Estella
A
Tafalla
2
Logroño
5
1
4
3
Arnedo
Cameros chain
0
20 Km
Tudela
A'

FIG. 13.4

Fig. 13.5. Panoramic view of Lerín Gypsum Formation in the village of Peralta (Navarra), as seen on the southern flank of Falces anticline (see figure 13.1 for location). Key: A = Los Arcos Gypsum; B = Villafranca Clays; C = Sesma Gypsum; D = Peralta Gypsum and Clay (these units are located in figure 13.4).

to the basin margins, and the evaporites became well developed toward the center of the basin, with little detrital influx. When the rate of subsidence increased, the marginal alluvial systems grew and their distal systems expanded to the basin center, preventing the development of evaporites. (2) During a dry period, the alluvial facies were restricted to the basin margins with evaporites toward the basin centers. However during a wetter period there was a greater influx of water, building extensive alluvial systems and flooding the basin center with less concentrated water.

Therefore from the Marcilla Clays to the Los Arcos Gypsum the sedimentary evolution shows a progressive increase in the evaporite/detrital ratio, which may be the result of either progressive tectonic quiescence or deposition during a drier portion of the climatic cycle.

Sedimentary Facies

GYPSUM/ANHYDRITE

Three calcium sulfate lithofacies (anhydrite at depth and secondary meteoric-sourced gypsum after anhydrite at surface) characterize most of the gypsum units:

Nodular and enterolithic facies. This is the dominant lithofacies in the Falces and Lerín formations. Nodules display various forms, and their sizes range from 1 or 2 cm to 20 or 30 cm in diameter (figure 13.6). Layers of nodular facies may reach thicknesses up to one meter, in which the host matrix is clay or dolomicrite. These appear to be typical continental sabkha facies. They are present in all stratigraphical sections, and particularly in the marginal part of each gypsiferous unit, where they may be the only facies present.

Laminar facies. In this facies, gypsum or anhydrite forms layers having a millimetric lamination, but some morphologic variations are present (figure 13.7). These laminar layers may accumulate into thicker horizons (up to several meters) in which nodular levels, mainly of the enterolithic type, may be intercalated. The sedimentary structures are perfectly preserved and display such features as wave-ripple lamination and undulations of likely algal origin. No pseudomorphs of selenitic gypsum are found in this facies.

Lenticular facies. Some gypsum layers are made of aggregates of lenticular gypsum crystals, up to 2

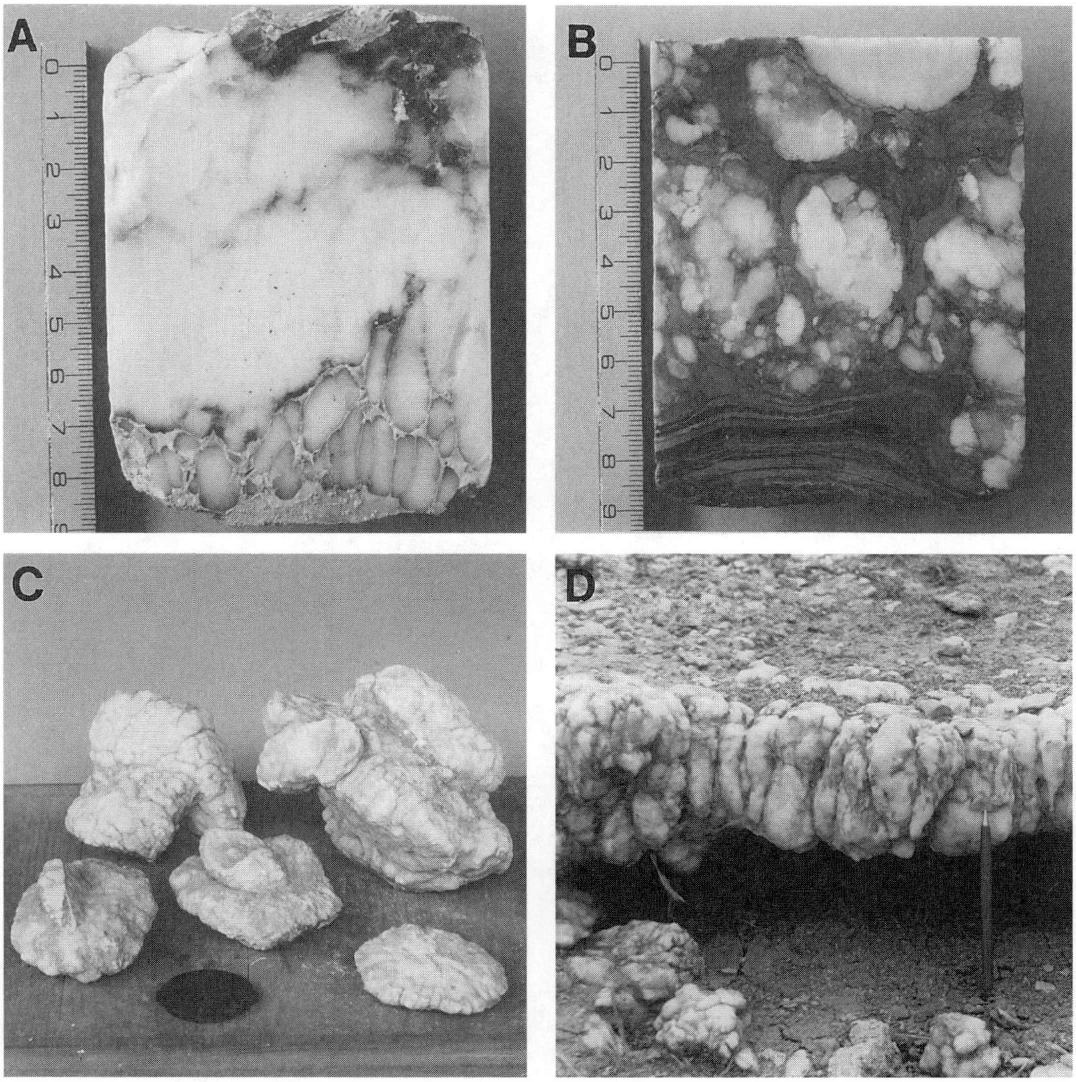

Fig. 13.6. Photographs of various morphologies of nodular lithofacies in the Falces and Lerín formations. A. Massive nodular anhydrite; B. Nodular anhydrite within an abundant calcareous (dolomicrite) matrix; C. Single or composite gypsum nodules, which are arranged scattered within clay horizons; D. Vertically elongated fabric in the horizons of gypsum nodules intercalated in the clay layers.

or 3 cm long, preserved as pseudomorphs and composed of polycrystalline secondary gypsum. These layers are usually intercalated within the laminated lithofacies. Such aggregates are also present within the carbonate beds where the individual pseudomorphs are generally smaller than 1 cm.

HALITE

Halite, a common mineral of the Falces and Lerín formations, is preferentially associated with the major gypsum horizons, but it is never seen on exposure because of its high solubility. In boreholes

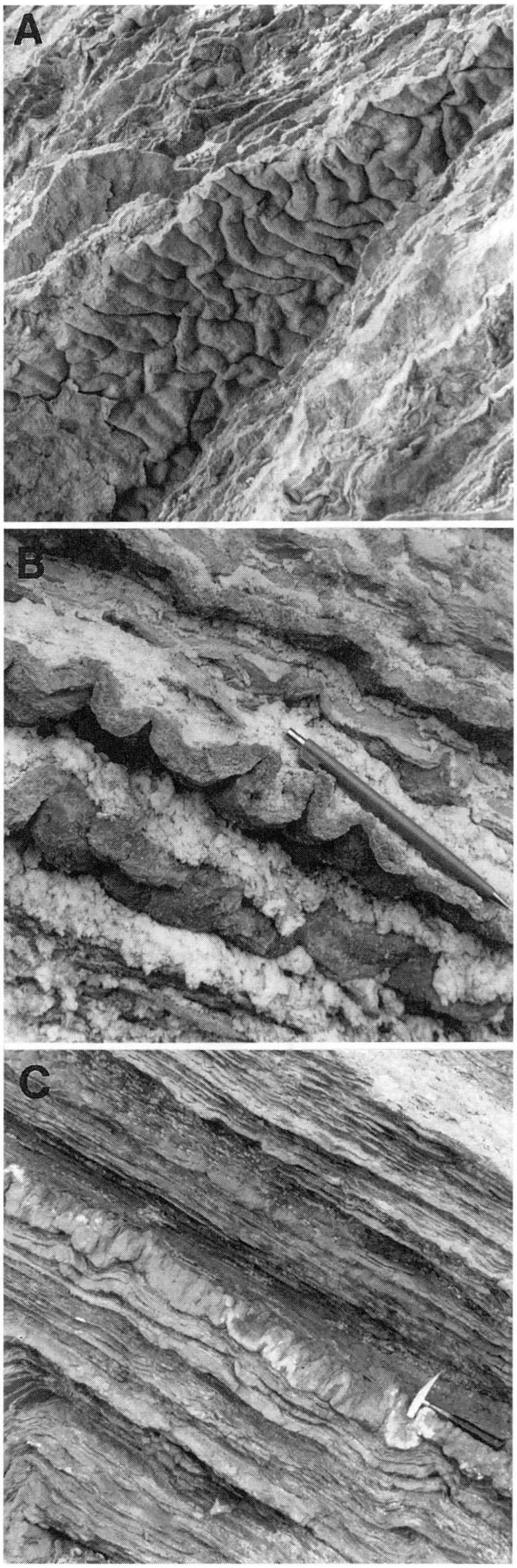

FIG. 13.7. Various morphologies of laminated gypsum in the Falces and Lerín formations. A. and B. Two aspects of the laminated gypsum layers showing undulating morphologies that apparently originate from stromatolitic control, intercalated within the major horizons of laminated gypsum (note pencil); C. Characteristic appearance in outcrop of a laminated gypsum layer within the Lerin Formation. Intercalations of enterolithic horizons are a normal feature of these laminated gypsum beds (note hammer).

this mineral forms beds several meters thick displaying an internal banding in which alternation with gypsum-anhydrite and argillaceous layers is common. Bromine content is extremely low in the halite (between 2 and 10 ppm), appropriate to its continental origin.

CARBONATE

There is very little carbonate in the Falces and Lerín formations mainly as dolomite. In some areas the carbonate thickens and is present as beds less than 1 meter thick within the gypsum sequences, but, in general, the carbonate intercalations are only a few centimeters thick. Such layers may be featureless mudstones (argillaceous dolomicrite) that locally contain evaporites (lenticular and micronodular gypsum). In other places the carbonate is a matrix for gypsum nodules, or forms as millimetric alternations with laminated gypsum. In glauberite deposits, carbonate is present as dolomite and magnesite.

CLAYS

Clays, in general, display a remarkable mineralogical homogeneity, regardless of the coloration they exhibit. They are composed, variously, of illite, chlorite, and kaolite, to which micritic calcite and/or dolomite is variably associated. In the Autol Gypsum (at the southern edge of the basin), the smectite content is high (Inglès et al. 1994).

SANDSTONE

Sandstones when present are always very well sorted and fine-grained. They display tabular layers with a thickness of less than half a meter or are developed within channels (1 to 1.5 m deep). Climbing ripples, flat pebbles, and load casts are frequent. These layers are intercalated inside the principal argillaceous horizons. The mineralogical composition is mixed but is mostly siliciclastic (quartz, feldspars, micas, metamorphic, and igneous rock fragments).

Lithologic Cyclicity

The lithologic assemblage composing the evaporitic units of the Falces and Lerín formations are usually arranged in sedimentary cycles that may be single or complex and have been called "elemental cycles" by Salvany (1989). Such cycles are well developed in the mixed argillaceous-gypsiferous units such as in the Peralta and Lodosa units (figure 13.8). In these units the cycles are diversified, but in general may be considered to fall into two principal patterns:

1. *Sabkha type* (figure 13.9A). This cyclic pattern is particularly well-developed in the peripheral zones of each evaporitic unit. A lower clayey horizon is present, which includes sandstone intercalations and gypsum nodules. The nodules are frequently scattered inside the clays in an irregular manner, but also give way to nodular horizons a few cm thick. The upper horizon is made up of a well-defined nodular gypsum layer with variable proportions of clay matrix. Such cycles are interpreted as the result of sedimentation and syngenetic evolution on an exposed mud flat with the deposition of interstitial evaporites and where the water table rarely reaches the surface (Salvany 1989).

2. *Evaporitic lake type* (figure 13.9B). This pattern of sedimentation is developed in the central part of the evaporitic units. The typical lake cycle involves three horizons: (a) a lower clayey horizon in which sandstone layers and scattered gypsum nodules are intercalated; (b) an intermediate carbonate member, which is formed by dolomitic beds either interlayered with gray argillites or a calcium sulfate horizon made up of a nodular lithofacies; and (c) a calcium sulfate horizon, in which the lower part is nodular and the upper part is mostly laminated with some enterolithic intercalations. Commonly the calcareous member (b) is lacking or limited to matrix material distributed between the gypsum nodules. This cycle (the salina cycle of Ortí and Salvany, 1986) is the result of the progressive establishment of a shallow evaporitic lake on a mud flat (figures 13.10 and 13.11).

These sedimentation cycles are also present within the gypsiferous and clayey units, but in these cases they can be difficult to recognize. In the gyp-

Fig. 13.8. Details of the Lerín Formation seen on outcrop. A. View of a sabkha cycle type in the Peralta unit; Key: A = clays with gypsum nodules that are either scattered or form discrete layers, B = nodular gypsum giving way to well-stratified discrete layers. B. View of an evapoiritc lake cycle type in the Lodosa unit (40 m, base to top); Key: A = clays with some sandstone intercalations; B = clays with calcerous intercalations bearing some gypsum nodules; C = dolomitic bed with intercalations of nodular gypsum; D = laminated and nodular gypsiferous beds.

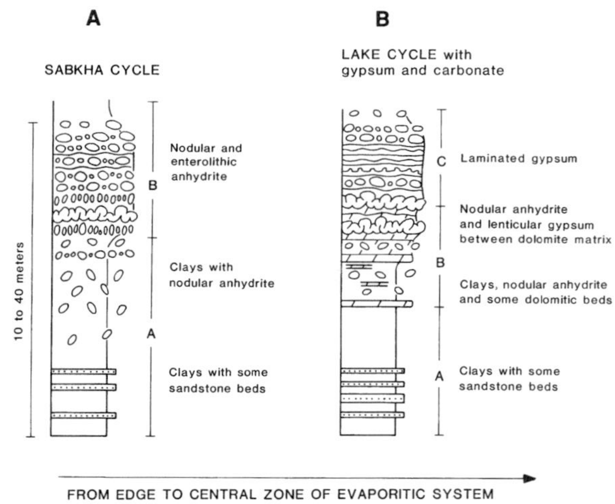

Fig. 13.9. Idealized evaporitic cycles recognized in the Falces and Lerín Gypsum formations.

sum units such elemental cycles are only composed of gypsum and no clear clay horizons are present. The gypsum units appear as thick sequences of elemental cycles showing nothing but a monotonous succession of laminar and nodular gypsum. The clayey units are made of a succession of clayey horizons with minor development of gypsum facies and/or rare nodular gypsum layers.

The cyclic succession pattern formed by the lithofacies development within the Lerín and Falces evaporites are illustrated in figures 13.10 and 13.11. The evaporites developed within the playa lake environment are made up by the development of three depositional zones with the following facies associations:

the internal zone, with facies composed of laminar gypsum and halite, developed within areas of shallow

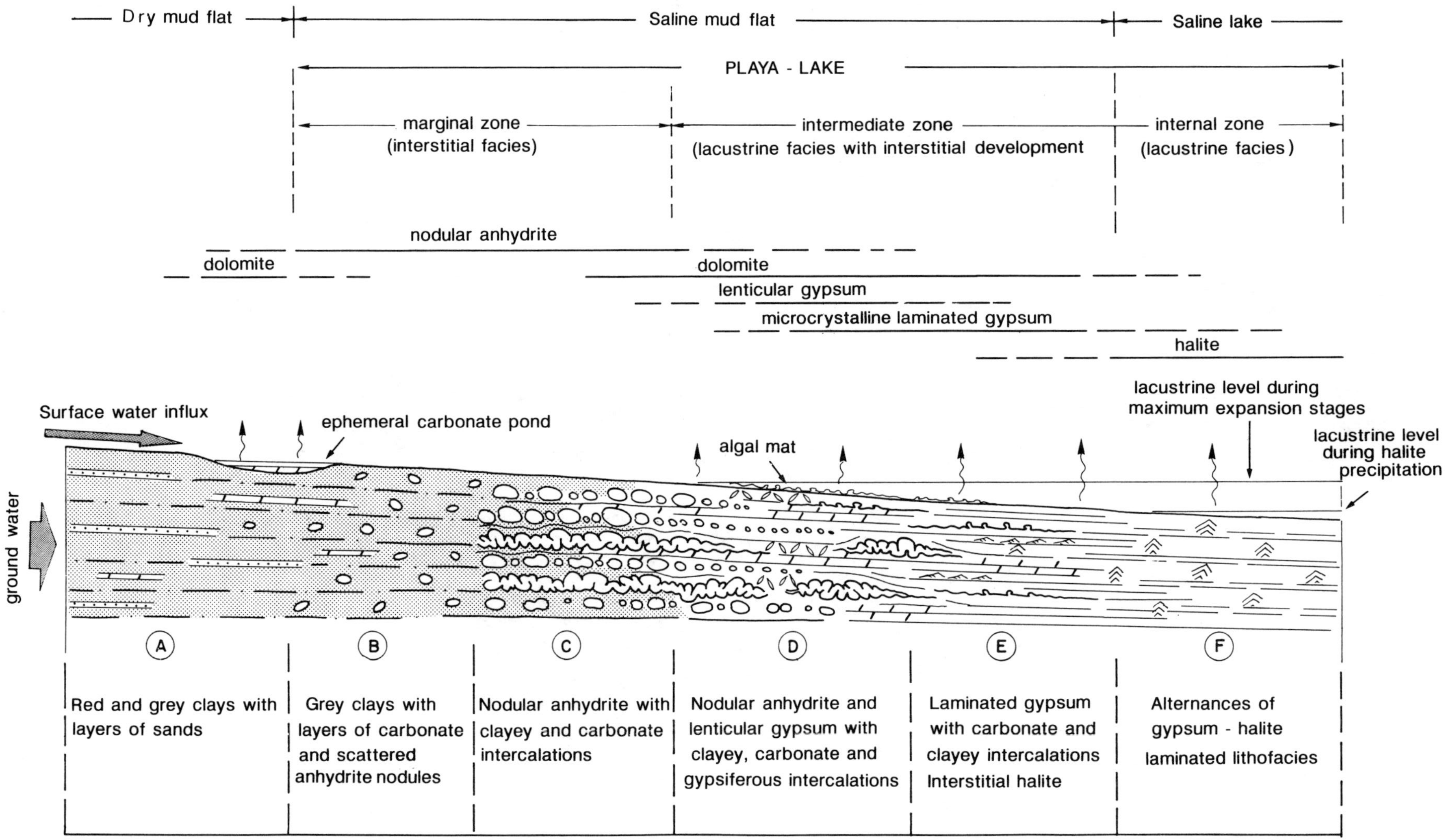

Fig. 13.10. Interpretative playa lake environment during a single depositional stage, developed from the facies distribution in the Falces and Lerín Gypsum formations.

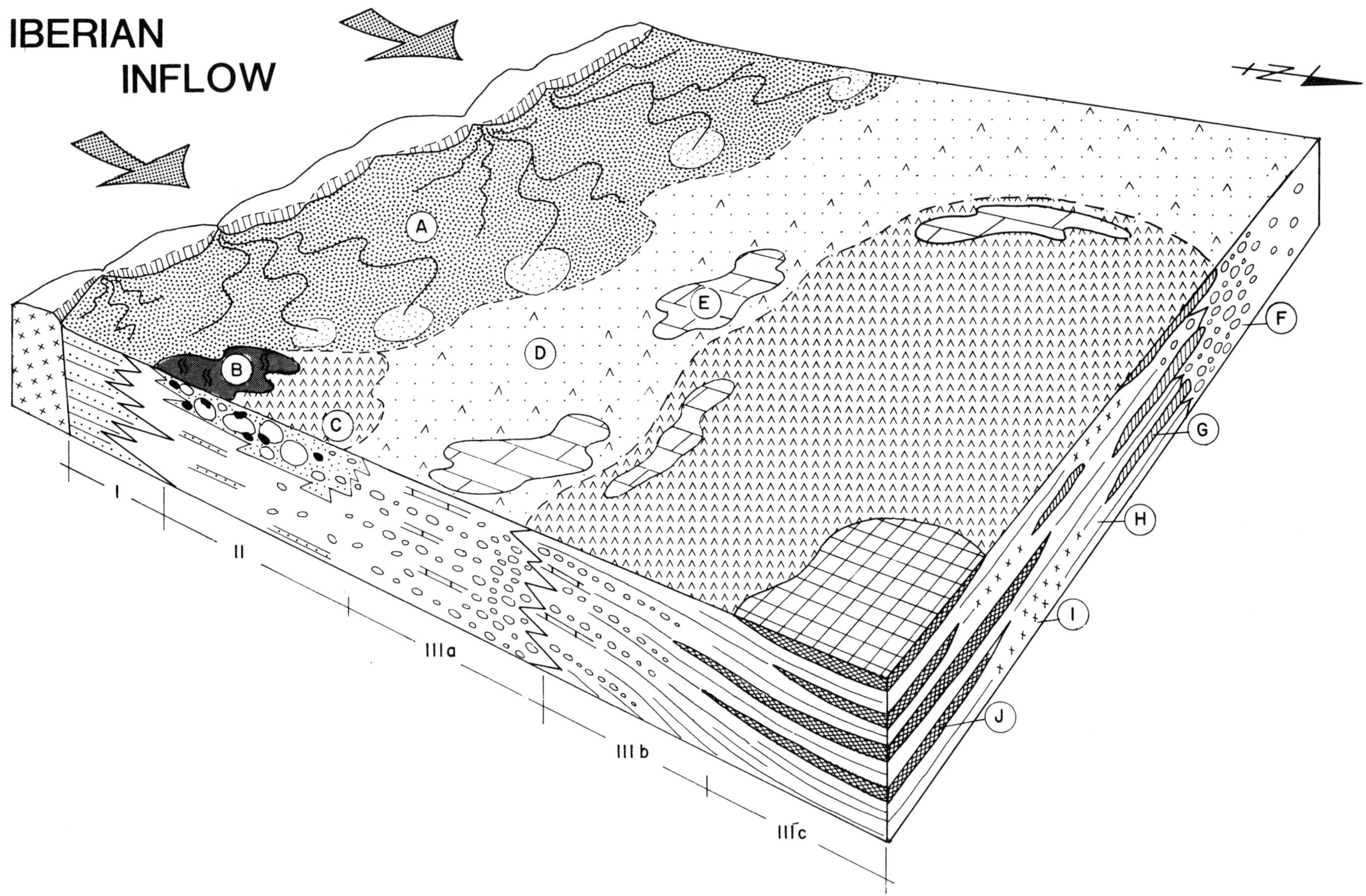

Fig. 13.11. Three-dimensional diagram showing detrital and evaporitic facies distribution for the Lerín Formation (Autol and Los Arcos units) from the southern border toward the center of the Ebro Basin. Key: I = alluvial fan zone; II = distal alluvial plain zone with marginal evaporites (Autol Gypsum unit); III = playa lake zone (IIIa = marginal zone, IIIb = intermediate zone, IIIc = internal zone). A = conglomerates, sandstones, and clays; B = lenticular gypsum with burrows and chert nodules; C = nodular and meganodular anhydrite with chert nodules; D = clays with interstitial nodular anhydrite and carbonates; E = carbonate; F = nodular anhydrite; G = glauberite; H = laminated gypsum; I = polyhalite; J = halite.

lacustrine sedimentation. During the high salinity stages halite is precipitated, and during phases of lower salinities the predominant precipitate is gypsum.

the intermediate zone, with the facies composed of laminar gypsum developed during the lacustrine phases, and nodular anhydrite or lenticular gypsum during the sabkha stages.

the marginal zone, with an argillaceous or calcareous matrix containing only interstitial anhydrite with no gypsum. This zone can be very wide, ringing a small central playa lake, but in other cases it is restricted to the margins of a broad playa lake. This variation is dependent on the basinal water budget.

During the phases of maximum stability and brine evolution of the evaporitic system, glauberite and polyhalite deposits are also formed, as noted previously. These sulfates are described by Salvany and Ortí in this volume.

In summary, the pair of gypsum deposits, the Falces and Lerín units, represent a very long period of evaporitic conditions during the middle to late Oligocene and early Miocene in the Navarra Basin. Throughout this time span the sedimentologic characteristics of the central saline lakes and their saline mud flats remained virtually unaltered. The major evaporitic units represent stages of significant stabilization of the evaporitic environment. Here, the alluvial systems were confined to the basin borders and were represented in the central area only by thin argillaceous intercalations. In the gypsum units the lithofacies dominance was controlled by the position of the evaporitic lakes. Thus in the basin center laminated facies dominated (either gypsum and/or halite), but the same unit at the margin was almost exclusively made up by nodular anhydrite lithofacies. The shifting of the sedimentary trough was the result of basement fault movement which took place under the compressive regime during the Oligocene and continued into the early Miocene.

(Accepted for publication April 1992; revised July 1995.)

14

Glauberite Deposits of the Lerín Formation (Lower Miocene: Alcanadre Zone, La Rioja-Navarra)

Josep M. Salvany and Federico Ortí

During the late lower Miocene the Alcanadre-Arrúbal area (La Rioja) received deposition of considerable amounts of glauberite within the Los Arcos unit of the Lerín Formation (see figures 13.1 and 13.4, this volume). In these sediments, the mineralogic paragenesis consists of gypsum-anhydrite, glauberite, polyhalite, halite, and magnesite, and the gypsum is always secondary. No indication of the presence of thenardite or mirabilite exists, as is usual in many other deposits of this type in ancient evaporite formations and many modern lakes.

Glauberitic deposits (or beds) are also known in some other continental evaporites of the Ebro Basin, both in Oligocene and Miocene sediments. In all these deposits the glauberite beds are always associated with the thickest gypsum horizons and are located in the paleogeographical centers of the basins. In general these sediments tend to crop out in unfavorable exposures (steep slopes with collapse structures) having little additional exposure in the third dimension. It is difficult to estimate their volumetric importance but based on borehole evidence they are ubiquitous in the subsurface.

The glauberite deposit of Alcanadre-Arrúbal has been known for a long time and has been discussed in the literature by numerous authors (Rios 1963; Riba 1964; Ortí et al. 1979; Ordoñez et al. 1982; Ortí et al. 1986b; Ortí and Salvany 1986b; Salvany and Ortí 1987, 1994; Salvany 1989). The glauberite is seen clearly in the cliffs along the Ebro river where the whole unit crops out (Los Arcos Unit) and is up to 200 m thick. It consists of a monotonous gypsiferous sequence, with argillaceous intercalations, in which several glauberite layers are particularly well-developed at the base of some of the thickest gypsum horizons. In this outcrop four main glauberite layers are noticeable ranging from 2 to 4 m in thickness. There are many other thinner beds as well. The principal layers were exploited in the past, although today the mines at this site are abandoned (figure 14.1). In 1983 and 1984, Unión Salinera de España S.A. drilled four boreholes with continuous coring: the Alcanadre-1 (118 m deep), Rioja-1 (112 m), Rioja-2 (85 m), and Rioja-3 (109.5 m). All the boreholes cut through several layers of glauberite, halite, and polyhalite in the Los Arcos Unit. These layers have been studied petrographically and are reported here.

Petrography

GLAUBERITE

Glauberite is mainly made up of nodular and/or enterolithic lithofacies in beds 10 to 30 cm thick (figure 14.2A and B). These beds are intercalated between the gypsum layers and may also occur in

Fig. 14.1. Glauberite beds (white) intercalated within the laminated gypsum portion of the Los Arcos Unit (Lerín Formation). The Alcanadre deposit (La Rioja).

groups of beds, creating glauberite horizons meters thick. Much less commonly the glauberite lithofacies is banded and massive, with cm-thick tabular beds in which large crystals of glauberite are present (up to 1 cm). These contain variable amounts of clayey matrix with a reddish or gray tint. On the surface all these beds weather to a white powder of sodic sulfate composition. Usually, glauberite in the nodular, banded and massive lithofacies is coarsely crystalline (from less than 1 to 15 mm), with sub- to euhedral habits. Very fine-grained glauberite textures (30 to 500 microns) are also present.

Under the microscope the glauberite shows an unoriented fabric with typical rhombohedral sections (figure 14.3A), approximately interpenetrating, and some calcareous or clayey intercrystalline matrix. In some samples there is marked interpenetration, and the glauberite exhibits an almost anhedral, equigranular fabric (figure 14.3B). The petrographic study reveals the existence of three genetic varieties of glauberite:

1. *"Primary" (interstitial) glauberite.* In this variety the crystals apparently nucleated and grew directly from a brine within a soft host sediment of calcareous or clayey nature. Growth is displacive and may pass into nodular, enterolithic, and also banded lithofacies.
2. *Secondary glauberite.* This variety of glauberite forms by the interstitial brine reaction with a calcium-sulfate precursor mineral (anhydrite and/or gypsum) in a very early stage of diagenesis. Relic anhydrite inclusions are commonly present within the glauberite. However primary gypsum relics within these glauberite crystals have never been found.
3. *Poikilitic secondary glauberite.* In this variety, large anhedral glauberite crystals replaces polyhalite poikilitically (figure 14.3E). In the process polyhalite pseudomorphs (spherulites) are well preserved. This variety of glauberite develops in a relatively more advanced stage of the early diagenesis than type 2. This is because polyhalite is itself a secondary mineral, and comes from the early replacement of the earlier types 1 and 2 of the glauberite facies. This replacement sequence is relatively simple to recognize because pseudomorphs of both polyhalite and glauberite are noticeable when studying poikilitic glauberite under plane polarized light (figure 14.3F).

POLYHALITE

Polyhalite (figure 14.2C and D) frequently displays a laminated lithofacies inherited from primary gypsum, early diagenetic anhydrite, or glauberite

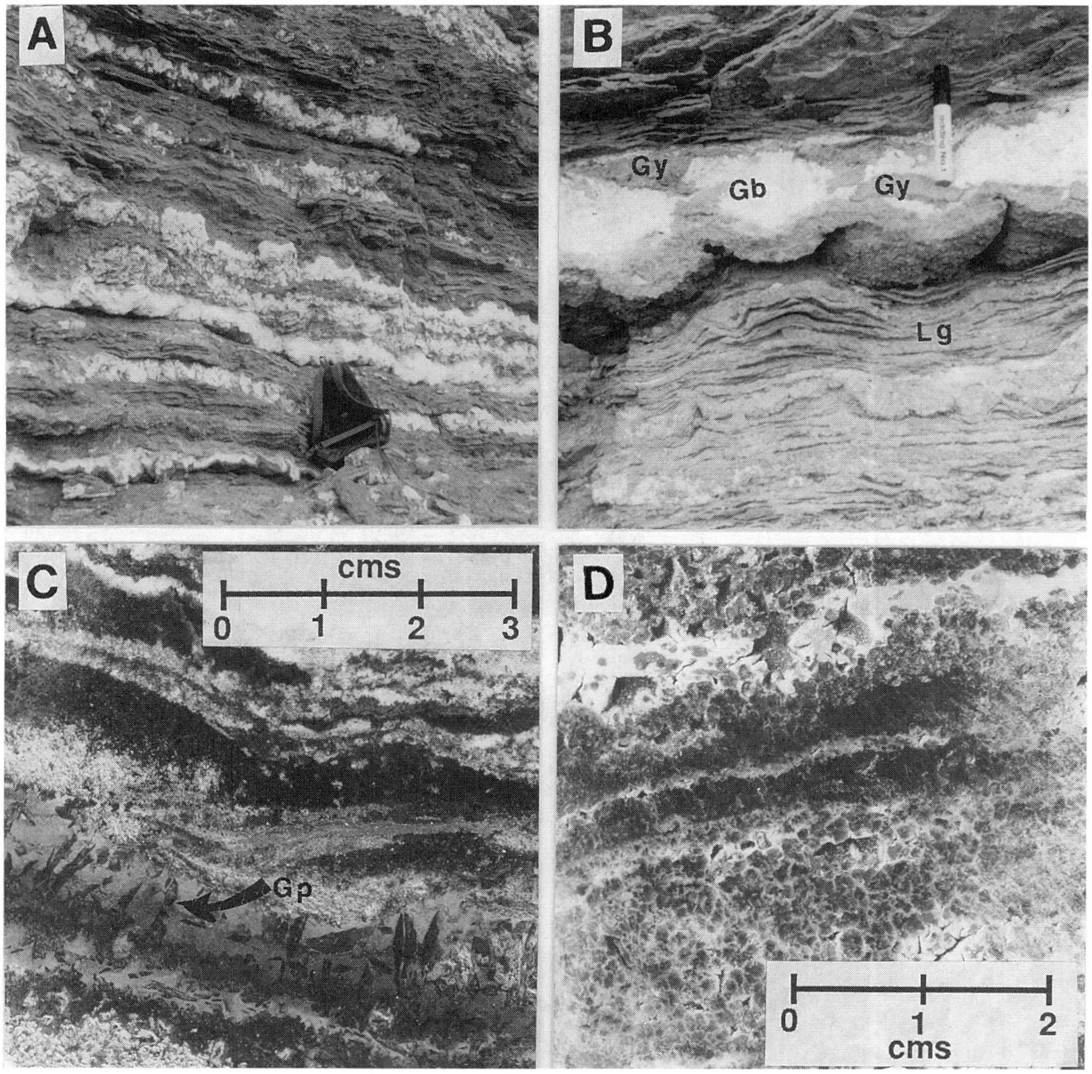

FIG. 14.2. Evaporative lithofacies of the Lerín Formation. A. Nodular and enterolithic layers of glauberite (white) alternating with laminated gypsum horizons (gray) in Los Arcos Unit. Alcanadre deposit. B. Detail of a nodular layer of glauberite (Gb) which has been partially replaced by secondary gypsum (Gy) from the borders towards the center of the layer. Original lithofacies has been strictly preserved. Lg = laminated gypsum layers. C. Detail (polished slab of a core sample) of pseudomorphs (Gp) of polyhalite after glauberite, in a level of laminated polyhalite. D. Detail of a spherulitic layer of polyhalite (polished slab of a core sample). Individual spherulites reach up to 1–2 mm and are distributed within an abundant calcareous (magnesite) matrix.

FIG. 14.3. Photomicrographs of microstructures found in the evaporites of the Alcandre deposits. A. Primary fabric in glauberite, composed of very characteristic rhombohedral sections of glauberite crystals, with abundant carbonate matrix (gray-black). B. Secondary crystals of glauberite bearing relic inclusions of anhydrite. C. Halite crystals displaying polyhalite replacement on their borders (see arrow). D. Spherulitic polyhalite. E. Polyhalite layer (Ph), partially replaced (retrodiagenetically) by tabular anhydrite crystals (Ah) and by anhedral poikilitic glauberite (Gb). F. Spherulitic polyhalite pseudomorphs after an idiomorphic glauberite precursor (central part of figure); at present all is secondary glauberite that replaced polyhalite in retrodiagenesis.

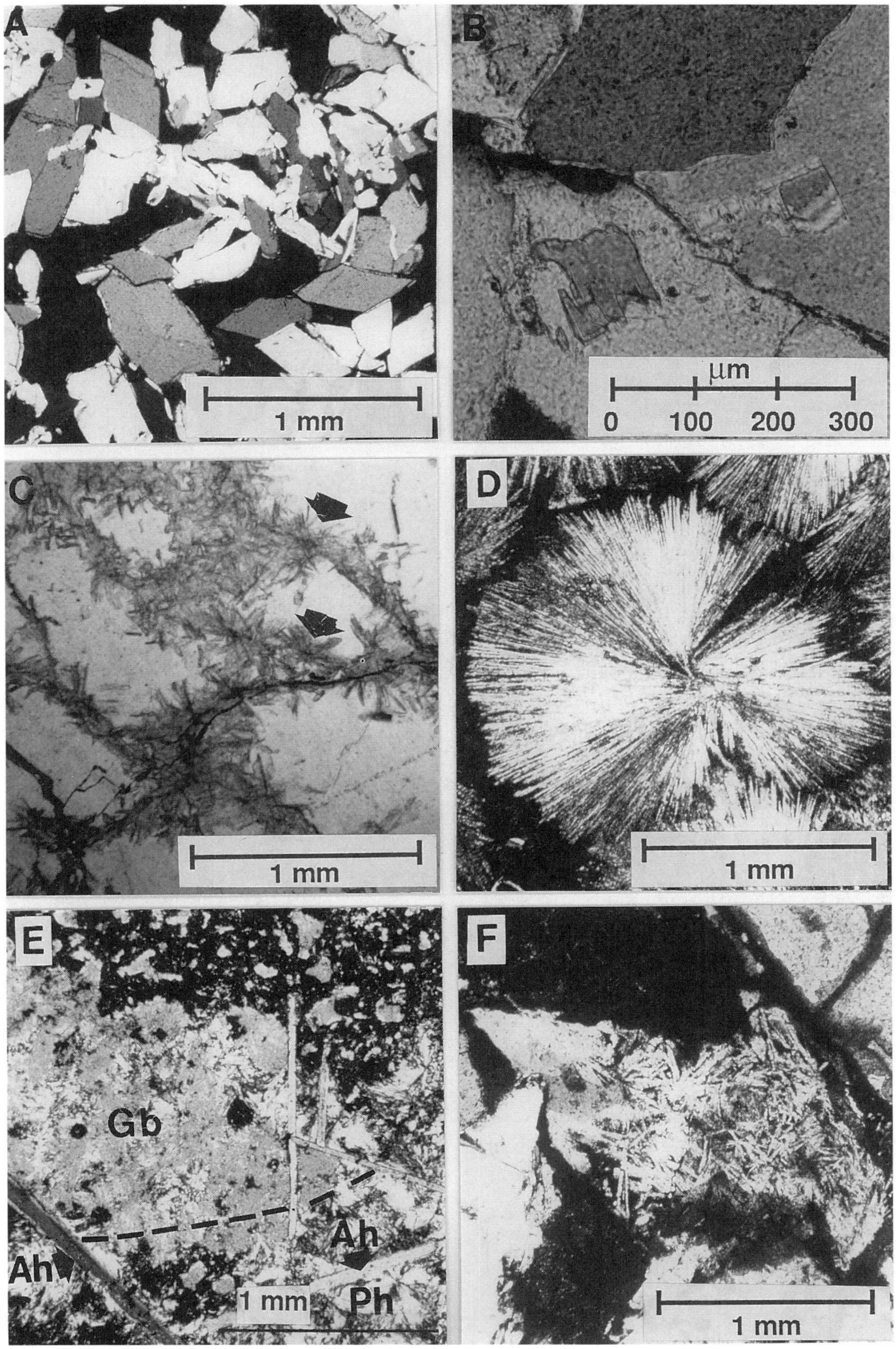

FIG. 14.3

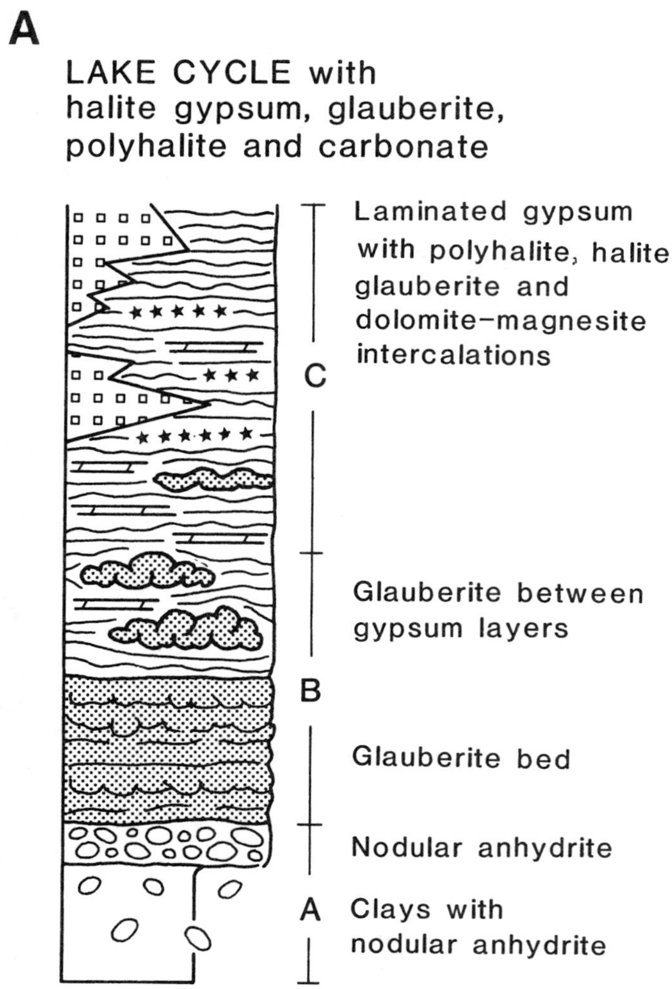

FIG. 14.4. A. Evaporitic cycle as recognized within the Alcanadre glauberite deposit of the Lerín Formation (from outcrop data). B. Interpretative playa lake environment during a single depositional stage, developed from the facies distribution in the Los Arcos Unit (Lerín Formation). Notice that the development of facies from the margin to the playa center parallels that of figure 13.10 (this volume) but is mineralogically different (Salvany and Ortí 1994).

lithofacies (types 1 and 2), and commonly all these minerals are preserved as pseudomorphs of polyhalite. The fabrics observed are fibrous and spherulitic, with sizes between 200 microns to 1 or 2 mm (figure 14.3C). In some instances the spherulites grow within a carbonate matrix without any apparent mineral precursor, or as a cement in cavities. The development of the polyhalite is mainly related to the halite precipitation. It is common to observe both minerals growing paragenetically within a single layer (figure 14.3D).

MAGNESITE

Magnesite is the carbonate mineral usually accompanying glauberite and polyhalite. It always displays a micritic texture.

Genesis of the Glauberite Deposits

To explain the origin of the glauberite deposits two different processes which have operated simultaneously must be take into account:

SEDIMENTARY PROCESSES

Glauberite layers are always related to the main gypsiferous horizons in the various units of the Lerín as well as in the Los Arcos Formation. Usually such layers developed towards the base of thick, laminated gypsum horizons, and the following idealized sequence apparently developed (from several meters to several tens of meters thick). From base to top the sequence is made up of the following components (figure 14.4A):

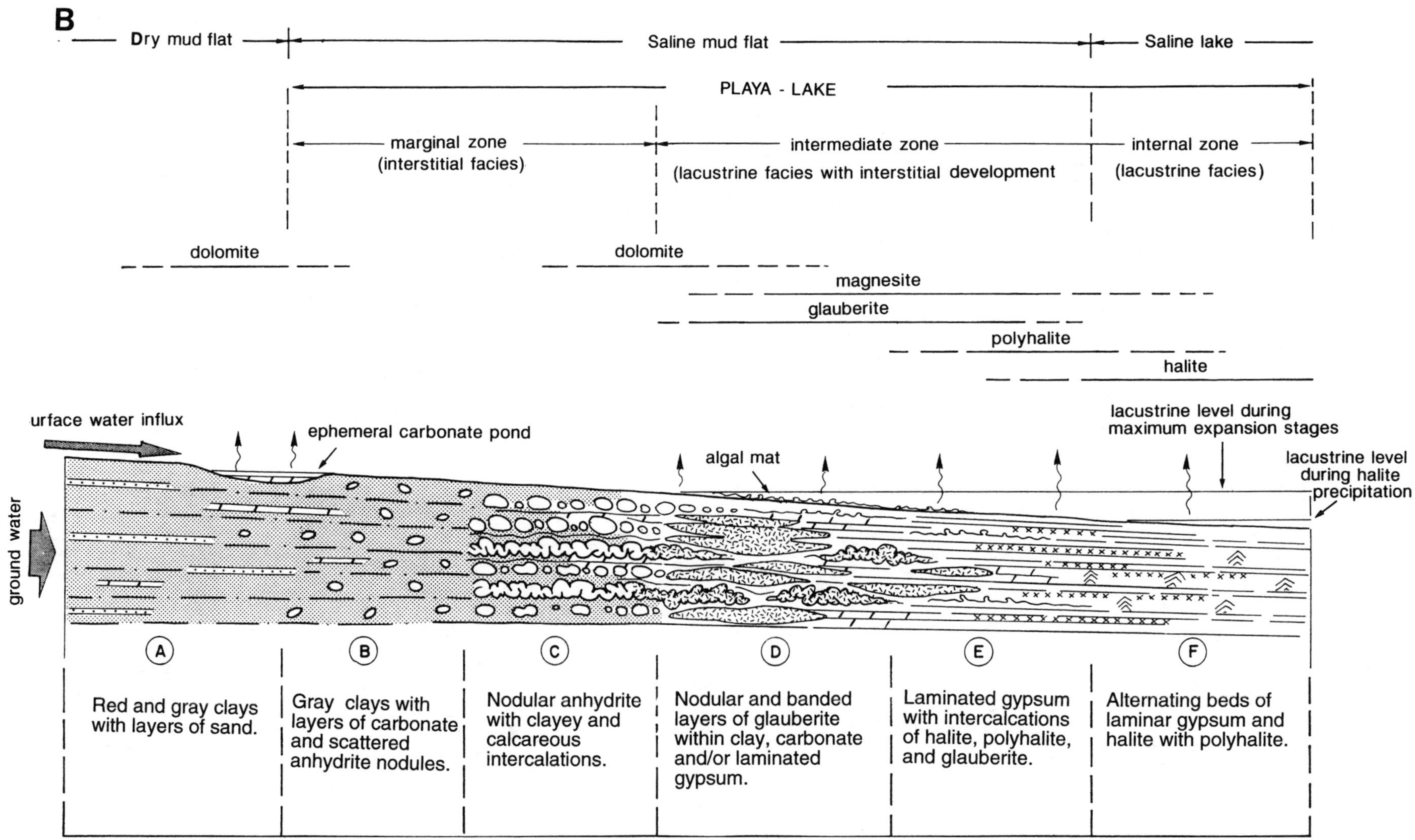
B
Dry mud flat
Saline mud flat
Saline lake
PLAYA - LAKE
marginal zone
(interstitial facies)
intermediate zone
(lacustrine facies with interstitial development
internal zone
(lacustrine facies)
dolomite
dolomite
magnesite
glauberite
polyhalite
halite
urface water influx
ground water
ephemeral carbonate pond
algal mat
lacustrine level during maximum expansion stages
lacustrine level during halite precipitation
A
B
C
D
E
F
Red and gray clays with layers of sand.
Gray clays with layers of carbonate and scattered anhydrite nodules.
Nodular anhydrite with clayey and calcareous intercalations.
Nodular and banded layers of glauberite within clay, carbonate and/or laminated gypsum.
Laminated gypsum with intercalcations of halite, polyhalite, and glauberite.
Alternating beds of laminar gypsum and halite with polyhalite.

Fig. 14.4B

(A) *argillaceous lower horizon* containing thin sandstone beds at the base and gypsum nodules at the top. Some carbonate beds can also be intercalated within the argillites.

(B) *glauberite intermediate horizon* is a glauberitic horizon up to a few meters thick, in which several glauberitic lithofacies may alternate. Usually, however, a banded lithofacies is developed within the lower part just above the argillite, and layers of nodular and enterolithic glauberite occupy the central and upper part of the horizon.

(C) *gypsiferous-chloride upper horizon* displays both laminated and laminated-nodular gypsum lithofacies. Individual layers of nodular or enterolithic glauberite and banded halite are intercalated within the laminated gypsum. Some carbonate laminae may alternate with the gypsiferous ones. Individual crystals of glauberite, arranged in single layers, may also be present within the gypsum laminae. It is mainly in such zones that polyhalite is developed, replacing both the laminated gypsum and the nodular glauberite, and halite as well. This C horizon is the thickest one, reaching 10 to 40 m or more.

This sedimentary cycle, composed of the A, B, and C lithofacies, constitutes a very restricted stage of evaporitic development. Facies distribution in this highly concentrated playa lake is composed of concentric depositional belts similar to that in the evaporitic lake setting. The detailed relationship of the various facies belts that are to be found in one of these glauberitic lakes is illustrated in figure 14.4B.

DIAGENETIC PROCESSES

Brine evolution has played a fundamental role in the development of these saline lakes. Mineral phases originated either as primary precipitates from the brines or as secondary minerals, the latter being a reaction product between the interstitial brine and a mineral precursor. In the first case the precipitation would be a free growth on the lake floor, commonly the case for gypsum and halite. Intrasedimentary crystallization is the case for all the other sulfates, but only in part for gypsum and halite.

Under progressive evaporation conditions, with increasing ionic concentration the mineral sequence of precipitation commonly follows this pattern: (1) carbonates → (2) gypsum (+ anhydrite) → (3) glauberite (+ gypsum/anhydrite) → (4) halite (+ glauberite, + anhydrite, + polyhalite).

1. *Carbonate* precipitates from slightly concentrated brines, probably as calcite or aragonite. The reaction of these phases with the interstitial brines would cause progressive alteration to dolomite and magnesite.
2. *Gypsum* precipitates within shallow lakes as small crystals (gypsiltites or gypsarenites), forming laminar lithofacies. In the border zones of the basins, coarse crystals of gypsum precipitate interstitially within previously existing sediments, within a clayey or calcareous host matrix. A part of this gypsum may be converted into anhydrite in very early diagenesis, giving way to nodular and enterolithic layers that alternate with the less altered laminar lithofacies.
3. *Glauberite* precipitates interstitially as variably sized idiomorphic crystals, either as a direct product or replacing both gypsum and anhydrite. It mainly displays nodular and enterolithic lithofacies. No certain evidence has been found to support a primary origin for the glauberite as a free-growth precipitate on a lake floor (such as competitive upward-directed crystalline fabrics formed in open water).
4. *Halite* precipitates during the highest brine concentration, either as a primary product or as a cement filling porosity in the sediments. In association with halite, sulfates such as glauberite also may precipitate. The sulfate polyhalite characterizes the chloride stage of these lakes, replacing any of the earlier minerals (gypsum, anhydrite, glauberite) and depleting the interstitial brines of potassium and magnesium.

Some retrodiagenetic processes take place when the brine of these lakes becomes diluted by rain or periods of flooding. Polyhalite is apparently replaced by poikilitic glauberite, as well as by anhydrite in phases of reduced concentration. Also glauberite may be replaced by anhydrite. If dilution progresses even further, glauberite crystals will dissolve, leaving casts that are later infilled and cemented (pseudomorphed) by halite or polyhalite in the next stage of concentration.

During burial the most significant diagenetic process seems to be the transformation of any residual primary gypsum into anhydrite.

Exhumation of lake deposits produces new diagenetic (late diagenetic) reactions. The most important is the replacement of any sulfate (not only anhydrite) by secondary gypsum. If the process takes place slowly and in the subsurface, all replacement commonly occurs isovolumetrically, and pseudomorphs of precursor sulfates are clearly preserved in the secondary gypsum. But rapid hydration at the surface may destroy the most soluble minerals (halite, glauberite, polyhalite), and only anhydrite is replaced isovolumetrically by alabastrine or megacrystalline secondary gypsum. An example of replacement textures is illustrated in figure 14.2B, where a nodular layer of glauberite is being delicately replaced by secondary gypsum. It is also noted that in abandoned mine galleries the remaining glauberite layers may alter into secondary gypsum on exposed surfaces. In these near-surface mines the halite intercalations have already been leached away and are only present in the deeper boreholes.

(Accepted for publication April 1992; revised July 1995.)

15

Continental Evaporitic Sedimentation in the Ebro Basin During the Miocene

Federico Ortí and Josep M. Salvany

As in other continental Tertiary basins of Spain, evaporitic sedimentation continued in the Ebro Basin during the Miocene. The distribution of Miocene deposits follows a similar pattern to those of the Paleogene (see figure 8.3, this volume). Many of these deposits have also been affected by synsedimentary diagenesis and modification. These changes are well developed and clearly documented. It is for this particular reason that we include the Miocene sediments in what is essentially a discussion of Paleogene sedimentation.

The Miocene evaporitic deposits are made up of sediments that formed within:

(a) small lacustrine systems adjacent to basin borders, in which the ionic concentration of the water is low;
(b) large central lake systems, in which the salinity reaches the chloride precipitation stage;
(c) continental sabkha zones which may be related to either of the other bodies of water (a) or (b).

The first group (a) includes the assemblage of small, basin-margin formations developed close to the Iberian Ridge border. These formations are present in the areas of Zaragoza, Navarra, and La Rioja provinces. Other evaporative areas include Borja, Ablitas-Monteagudo, Pozuelo, as well as minor deposits in such regions as Ribafrecha and Calanda, and are roughly time-equivalent. These deposits are characterized by the preservation of primary gypsum, which is of the microlenticular matrix-free type. In these deposits gypsum bioturbation is ubiquitous, nodular chert is common, and halite vestiges are absent. The bedding is thick and massive. Sedimentologically these deposits are equivalent to those of the Paleogene age that developed along the Catalánide border.

The second group (b) includes the extensive system of the Zaragoza Formation and its peripheral units in the central area of Aragón, and the Cerezo Gypsum in Burgos province. In both Aragón and Burgos the sequences are very thick and are dominated by the laminated-nodular facies association, which intercalates with sodium-calcium-sulfates in Cerezo and sodium-calcium-sulfates and halite in Zaragoza.

The third group (c) includes the deposits of the Mediana, Gelsa, and Azaila type, characterized by the development of extensive nodular anhydrite in a clayey or calcareous matrix. The laminated gypsum facies is almost entirely absent. In this grouping, the interstitial and displacive growth of early diagenetic anhydrite was the dominant formative process.

Central Evaporitic Systems of the Ebro Basin During the Early-Late Miocene

At the beginning of the Miocene the axis of maximum subsidence in the Ebro Basin shifted towards the south, trended NW-SE, and occupied approximately the same position as does the present-day Ebro river. During the Oligocene to Miocene transition, a broad "pre-Aquitanian" unconformity developed throughout the western part of the basin. In this new arrangement, the single evaporitic depocenter of the Navarra zone (the Falces Gypsum and Lerín Gypsum) became differentiated into two subsident zones and evaporites collected within both, although not synchronously. Of the two zones, one was located to the west in the Burgos region (Cerezo Gypsum, upper (?) Miocene) and the other, more important, to the southeast in the central part of Aragón (Zaragoza Gypsum, lower Miocene). Thus the Oligocene evaporitic systems of Navarra continued throughout the Miocene within the central and western parts of the Ebro Basin (see figures 8.3 and 8.4, this volume).

ZARAGOZA GYPSUM FORMATION

The Zaragoza Gypsum Formation is present at outcrop over a nearly rectangular zone trending northwest-southeast (figure 15.1A). The evaporitic basin is about 80 km long and about 40 km wide (Quirantes 1969). Maximum thickness in outcrop reaches 100 meters. Exploratory boreholes (for oil and mining) have penetrated and drilled into several halite layers bearing sulfates about 800 m below the surface. Several rivers cut through the formation, resulting in impressive, almost vertical cliffs (Ebro, Gállego, Jalón, Huerva, etc.), and facilitate its study. The age of the Zaragoza Gypsum is somewhere between the base of the Miocene and the middle Aragonian (early Miocene).

Laterally, the Zaragoza Formation passes into other time-equivalent evaporitic sulfate units that may be clearly differentiated from the Zaragoza itself, such as the Borja-Pozuelo Gypsum to the west and the Vinaceite Gypsum to the south. A schematic representation of this transition is shown in figure 15.1B. This formation passes laterally to several siliciclastic deposits and into some minor calcareous units. Several general studies on the overall deposition of the Zaragoza Formation have been carried out (Quirantes 1969; Birnbaum 1976; Mandado 1987).

Within the Zaragoza Formation there are four main intercalated halitic bodies, in particular in the areas of Remolinos, Torres de Berrelén, and Zaragoza (Torrescusa and Klimowitz 1990). These halite deposits are largely composed of banded halite (figure 15.2), which exhibit primary salt textures (hopper and chevron structures) in which some nodular anhydrite is noted (Ortí and Pueyo 1977; Fernández-Nieto and Galan 1979; Ortí 1990b). The bromine content of the salt is commonly between 1 and 15 ppm, which clearly demonstrates its continental origin. These halite facies formed in a very shallow saline lake affected by weak dilution episodes, in which nodules of anhydrite grew within thin pelitic interlayers, or within the salt itself. A peripheral belt of nodular anhydrite, partly replacing the halite, surrounds the central salt body. Synsedimentary replacive processes did not significantly affect the salt towards the inner part of the lake, in which perfectly preserved depositional structures are present. However both early and late diagenetic dissolution phenomena can be observed locally in the mine galleries, especially in some areas where the banding has been destroyed and replaced by patches of transparent, inclusion-free, halite crystals (Ortí and Pueyo 1977). The clear halite crystals, of centimetric to metric size, display the lowest bromine content of any of the halite ("sal de compás").

A 106 m exploratory borehole drilled inside the gallery of the Remolinos mine cut salt layers with glauberite crystals and halite/anhydrite pseudomorphs after glauberite that extended for 80 m; glauberite continued to the bottom of the hole. The geochemical study of this salt (Garcîa Veigas 1993; Garcîa Veigas et al. 1994) reveals low bromine, magnesium, and potassium content. Fluid inclusion studies of the same salt also contain very low bromine, magnesium, and potassium—falling below the limit of detection. The SO_4^{-2} content is significant only in samples associated with glauberite beds. These data agree with the mineral paragenesis in the Zaragoza Gypsum Formation (gypsum/anhydrite, glauberite, thenardite, and halite), where no magnesium or potassium minerals are present. This is in strong contrast to the polyhalite bearing sodium sulfate deposits of Alcanadre (Lerín Gypsum Formation) (see chapter 14, this volume).

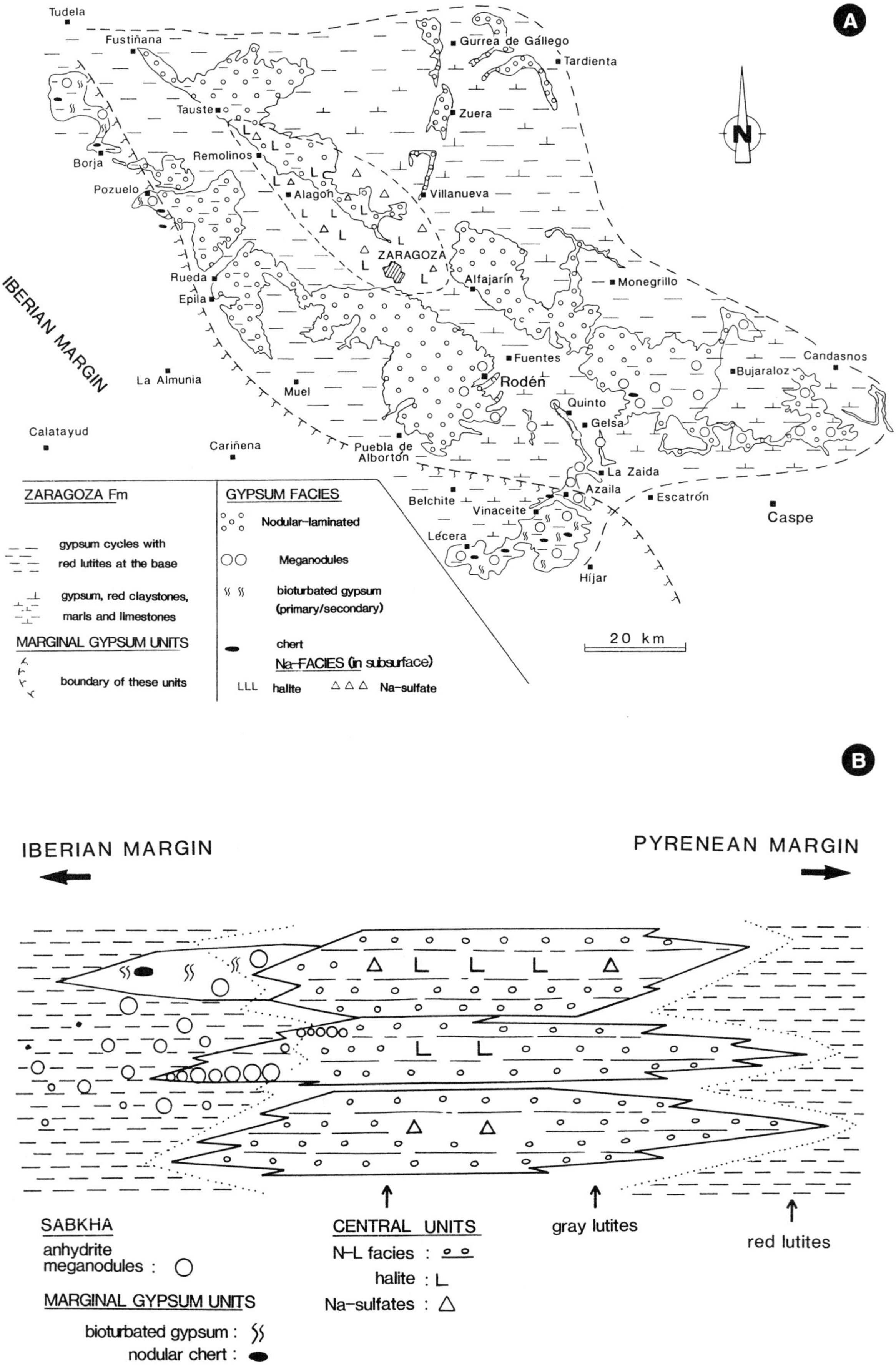
A
Tudela
Fustiñana
Gurrea de Gállego
Tardienta
Tauste
Zuera
Borja
Remolinos
Pozuelo
Alagón
Villanueva
ZARAGOZA
Rueda
Epila
Alfajarín
Monegrillo
IBERIAN MARGIN
Fuentes
Candasnos
Bujaraloz
La Almunia
Rodén
Muel
Quinto
Gelsa
Calatayud
Cariñena
Puebla de Alborton
La Zaida
Azaila
Escatrón
Belchite
Vinaceite
Caspe
Lécera
Híjar
20 km
ZARAGOZA Fm
gypsum cycles with red lutites at the base
gypsum, red claystones, marls and limestones
MARGINAL GYPSUM UNITS
boundary of these units
GYPSUM FACIES
Nodular-laminated
Meganodules
bioturbated gypsum (primary/secondary)
chert
Na-FACIES (in subsurface)
LLL halite
△△△ Na-sulfate
B
IBERIAN MARGIN
PYRENEAN MARGIN
SABKHA
anhydrite meganodules :
MARGINAL GYPSUM UNITS
bioturbated gypsum :
nodular chert :
CENTRAL UNITS
N-L facies :
halite : L
Na-sulfates : △
gray lutites
red lutites

FIG. 15.1

Fig. 15.2. Bedding characteristic of the Zaragoza Halite. Decimetric-order banding of hopper chevron halite seen in the Mina Real of Remolinos (Zaragoza province).

Further from the basin center, a huge saline mud flat (sabkha) developed over an area of more than 50 km^2, in which nodular anhydrite (and probably also glauberite) of early diagenetic origin alternates with a laminated gypsum facies. The original laminated gypsum was completely transformed into anhydrite and glauberite during very early diagenesis. Today it crops out on the surface as secondary gypsum. As happens in the Lerín and Falces formations, borehole data show that metric-order glauberitic layers are also intercalated within the salt across a wide zone (in the area of Utrebo-Zaragoza and in Remolinos), thus demonstrating the same paragenetic assemblage. Mandado (1987) also cited the existence of some thenardite layers.

Units peripheral to the Zaragoza Formation such as the Vinaceite Gypsum (zone between Vinaceite and Azaila) are characterized by the presence of primary microlenticular gypsum (preserved), some large nodules of alabastrine secondary gypsum, and chert nodules. These evaporitic units constitute type (a) (small lacustrine systems). The thickness of the evaporites reaches up to 100 m in many outcrops. Other similar units are also developed regionally. Locally, in the transition between the (a) and (b) types, some sabkha-type units develop, as in the area near Rodén and Azaila, which is characterized by the presence of large alabastrine nodules within argillaceous beds of red clays (figure 15.3A and B) that are very actively being exploited.

Fig. 15.1. The Zaragoza Gypsum Formation. A. Lithologic map showing the facies distribution in the central part of the outcropping area of the Zaragoza Gypsum Formation. B. Cross-sectional scheme, not to scale, illustrating the lateral transitions between the central halite/glauberite/anhydrite units of the Zaragoza Gypsum Formation and the marginal gypsum units with lower concentration facies and chert (i.e., Pozuelo to Remolinos cross-section). The model also includes some sabkha occurrences with extensive development of meganodules of anhydrite (i.e., from Azaila to the north). (Taken from Ortí 1990b.)

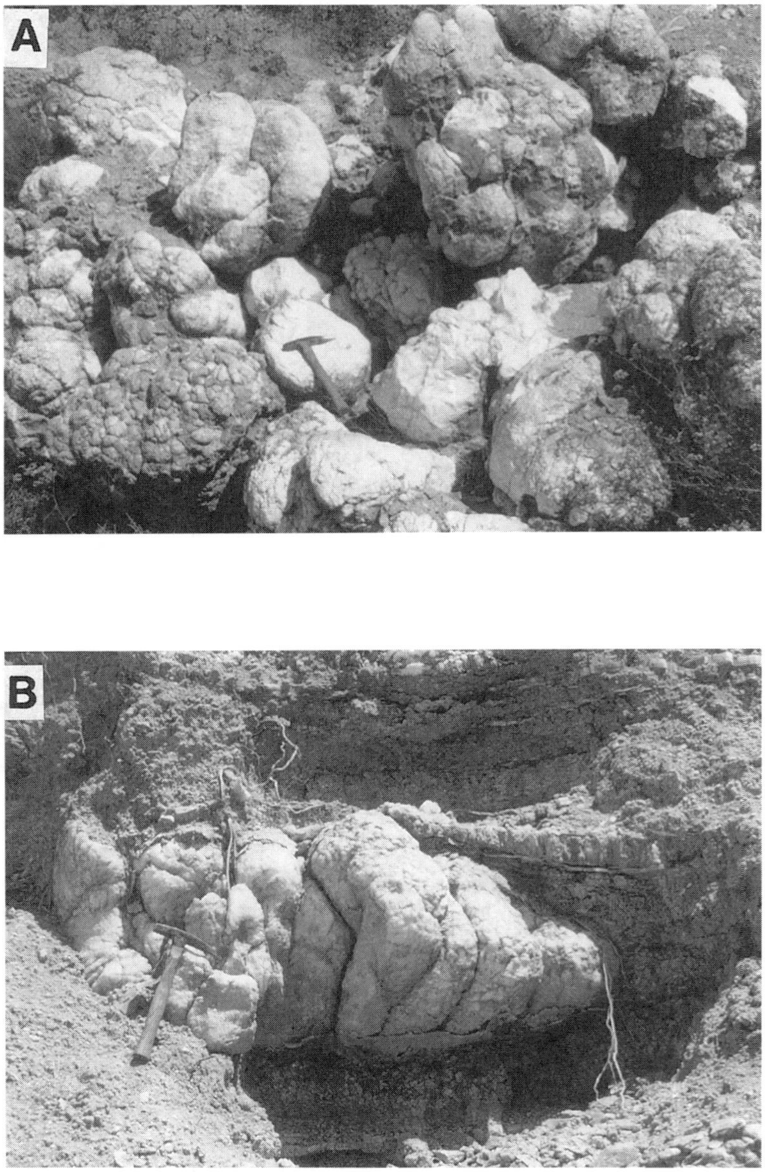

Fig. 15.3. Anhydrite meganodules of probable sabkha origin in the Miocene near Rodén, now preserved as alabastrine secondary gypsum. A. Spheroidal nodules. B. Vertically aligned elongate gypsum nodules that together form a lenticular mass. Hammer for scale in both photographs.

CEREZO GYPSUM UNIT

In the western border of the Ebro Basin there exists a narrow, tectonically-controlled zone (30 to 40 km wide) trending east-west, called the Bureba corridor, which is considered to have been the connection between the Ebro and the Duero basins. In this graben, subsidence during the Tertiary was locally as much as 5000 m. During the Miocene significant amounts of evaporitic sedimentation took place in the westernmost part of the Bureba area in what is called the Cerezo Gypsum Unit (see figures 8.3 and 8.4, this volume). The Cerezo Gypsum reaches an outcrop thickness of 150 m, but data about its continuity in the subsurface have not been published. Portero et al. (1979) assigned this evaporitic unit to the early Miocene, while Riba et al. (1983), Anadón (1990), and Riba and Jurado (1992) have argued that it is late Miocene. This unit passes laterally into detrital and calcareous facies and is not overlain by any younger deposits. The gypsum facies, which are mostly formed by secondary gypsum with only minor amounts of primary selenitic crystals at the top of the sequence, are defined by an alternation of

FIG. 15.4. Sabkha cycles in the Cerezo Gypsum Unit seen in the Villalómez quarries, Province of Burgos (1–5 m per cycle). In each cycle laminated gray lower horizon (L, lake stage) is overlain by an upper massive nodular layer (white alabastrine; S, sabkha stage). Almost all calcium sulfate now present is secondary gypsum (see hammer for scale).

FIG. 15.5. Polished slab of the glauberite rock of Cerezo de Rio Tirón deposit. Note the presence of a fine-grained anhydritic matrix (A) between the coarse glauberite crystals.

laminar-nodular facies, which locally exhibit well-defined cyclic trends. Many of these depositional cycles in the marginal zone are of the sabkha type (figure 15.4). These facies are rather similar to those of the Zaragoza Formation (lower Miocene) and Lerín and Falces formations (Oligocene–lower Miocene of Navarra). The secondary gypsum comprising this formation ranges from alabastrine to megacrystalline and also contains typical features proving that a glauberite precursor was present.

The Cerezo Gypsum Unit contains intercalated glauberite layers that crop out in the Belorado—Cerezo de Rio Tirón area. Boreholes in that locality show an alternation of glauberite and anhydrite beds. At least six major layers of glauberite are known in the Cerezo, with individual thicknesses of several meters. Anhydrite is mainly composed of fine-grained nodular textures while glauberite is banded or massive, with a range in crystal size of millimeters to one centimeter (figure 15.5). More rarely, the glauberite displays fine-grained textures with nodular and enterolithic layering. Neither halite nor thenardite layers have been found: a complete investigation of the sequence from boreholes has never been published. The geology of this deposit in Cerezo was discussed by Menduiña et al. (1984). Two intensely exploited glauberite mines, in Belorado and Cerezo de Rio Tirón, are currently active.

Primary Gypsum Deposits of the Peripheral Systems: Ablitas Gypsum

Although it is a fairly thin deposit, the Ablitas-Monteagudo Gypsum Unit constitutes an interest-

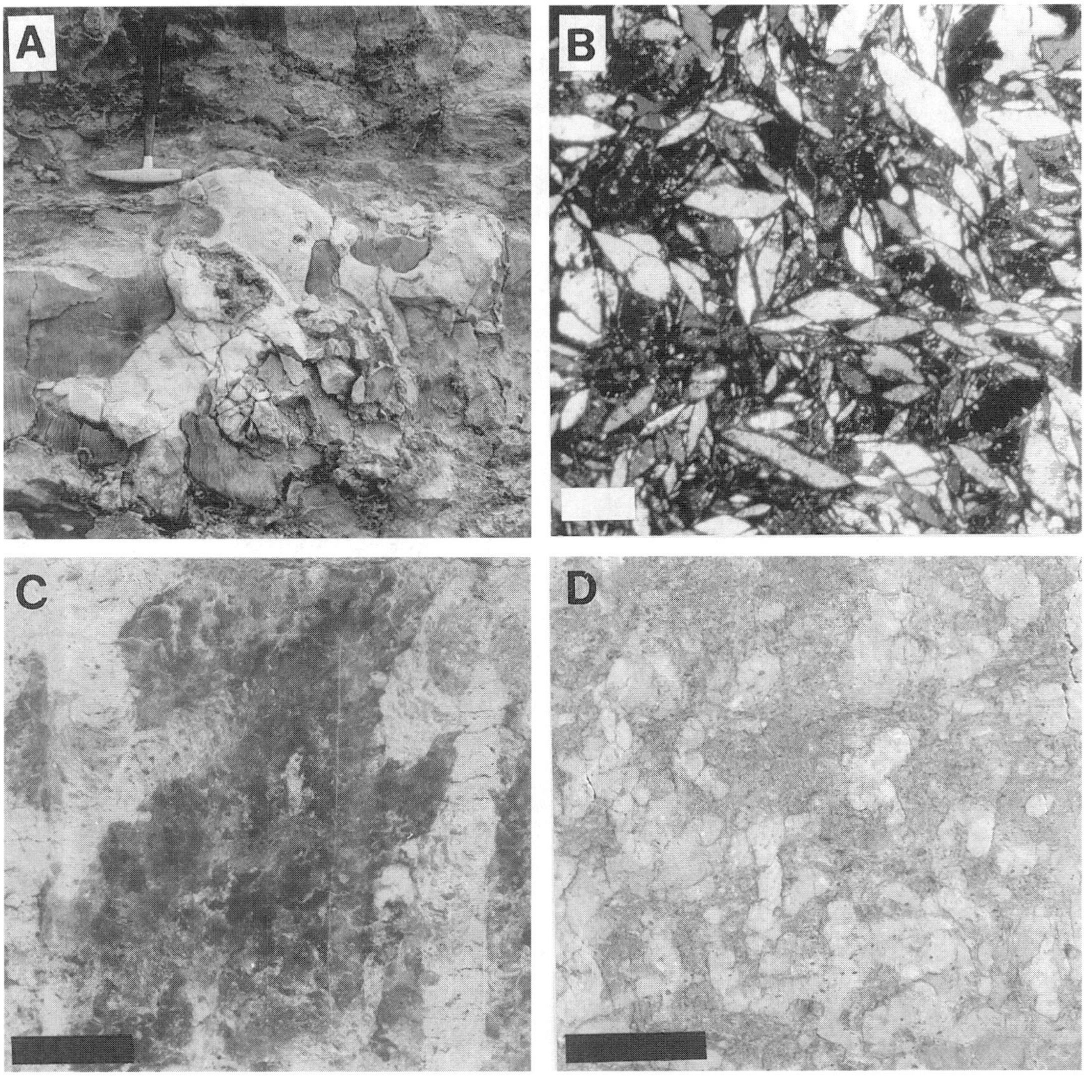

Fig. 15.6. Aspects of the Ablitas Gypsum, both primary and diagenetic features. A. Chert meganodule replacing the primary microlenticular gypsum in the Ablitas quarry. B. Photomicrograph of the microlenticular primary gypsum (scale bar, 100 microns). C and D. Two aspects of burrowing affecting the microlenticular primary gypsum (polished slab). Burrows are filled by gypsum and also by carbonate (scale bars, 1 cm).

ing evaporitic formation that displays very well- defined sedimentary and petrologic characteristics (similar to those described by Ortí in chapter 12 of this volume for the gypsum of the Cornudella Group) (Castiella et al. 1978; González and Galán 1984; Ortí and Salvany 1986a; Ortí et al. 1986a; Mandado 1987; Pérez et al. 1988; Salvany 1989; and Salvany et al. 1994b). This formation is composed of gypsum with associated carbonate, clay, and chert nodules (figure 15.6A–D). No vestiges of halite or sodium-sulfates are known. All these features are rather similar to those in the gypsum units present in the areas to the southeast, such as Borja, Pozuelo de Aragón, Vinaceite, Calanda, and to the west at Autol and Ribafrecha (figure 15.7; and see figures 8.3 and 8.4, this volume).

Gypsum is mainly alabastrine (secondary), but towards the western part of the formation it progressively changes to the microlenticular primary facies. The alabastrine secondary gypsum displays

Fig. 15.7. View of Ablitas Gypsum Unit near the village of Ablitas. Gypsum layers form the top of the tabular relief and are intercalated within the lutites of the Alfaro Formation.

Fig. 15.8. Meganodule of secondary gypsum replacing the microlenticular primary gypsum as seen in the Ablitas quarries. The nodule, displaying a thick envelope of satin spar gypsum veins (at arrow), is formed by a single crystal of secondary gypsum (megacrystalline variety) showing cleavage surfaces. This replacement does not result in deformation of enclosing sediment but is clearly cross-cutting.

nodular lithofacies of two major types: (a) nodules and meganodules (large white alabastrine nodules) of very pure calcium-sulfate varying between 20 cm and 1 to 2 m in diameter, and (b) micronodules, in general having a greenish tint and variable sizes (between less than 1 cm and 2 to 4 cm). Meganodules may occur as individuals or be arranged in layers. In the first case they are commonly surrounded either by micronodules, or by microlenticular primary gypsum, which they replace (figure 15.8).

The primary gypsum forms beds with a thickness of about 1 meter made up of two major lithofacies:

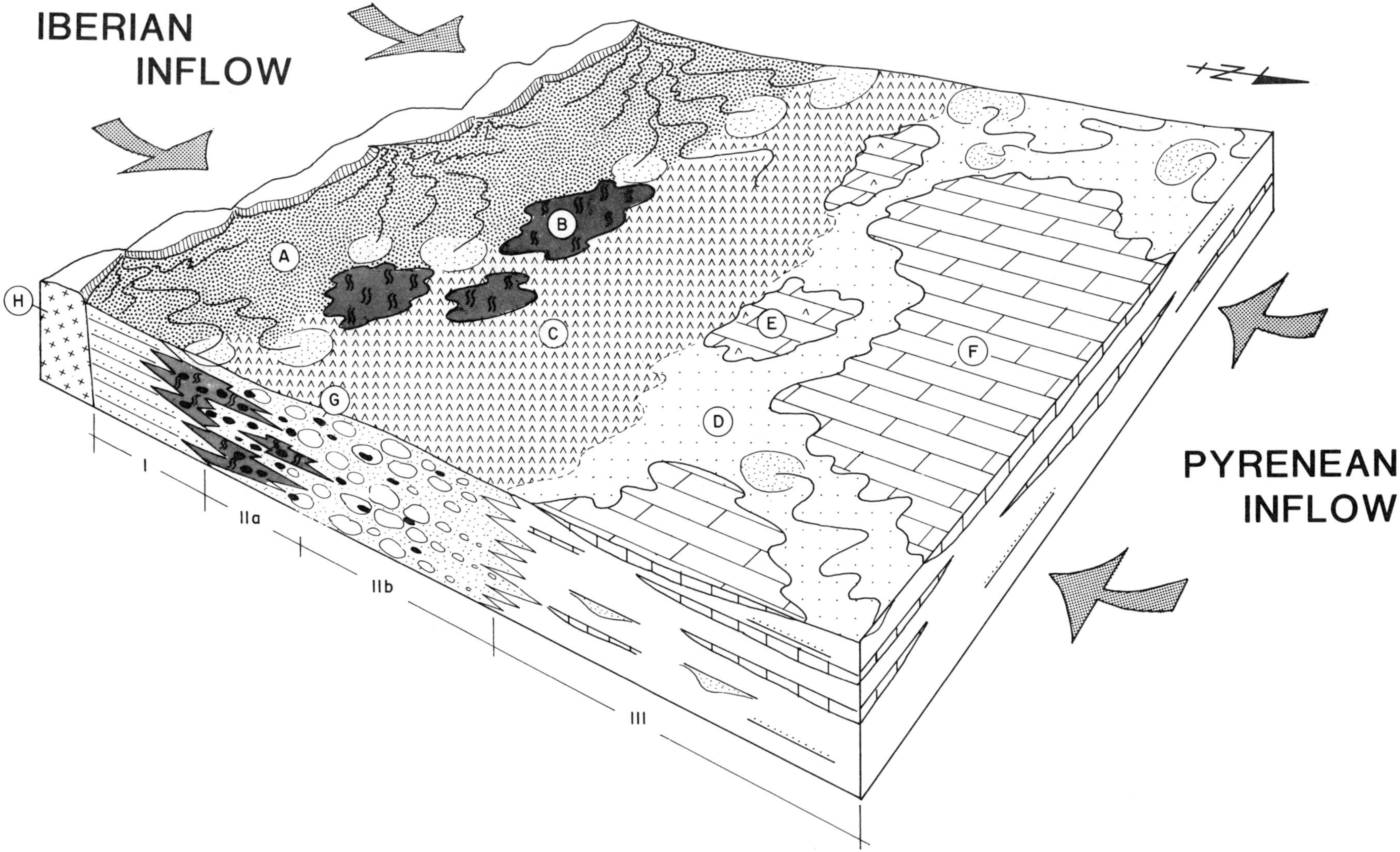

FIG. 15.9. Three-dimensional diagram of the Monteagudo-Ablitas Gypsum Unit showing the distribution of the detrital (I), evaporitic (II a and b), and carbonate (III) facies; drawn from the southern border of the basin (Iberian Range) towards the center. Key: A = alluvial fans with conglomerates, sandstones, and clays; B = a peripheral gypsiferous shallow lake displaying bioturbation and chert nodules (black); C = zone of anhydritization containing meganodules of both anhydrite (white) and chert; D = distal lutitic plain with carbonates and sandstones; E = shallow calcareous lake with gypsum intercalations (rosettes); F = shallow calcareous lake with well-developed bioturbation; G = chert nodules; H = Paleozoic-Mesozoic basement (Iberian Range) (taken from Salvany 1989).

(a) fine-grained (100 to 500 microns), brownish gypsum with a massive appearance, composed of tiny crystals with a microlenticular habit. This lithofacies may contain an abundant matrix of irregularly distributed carbonate, and is probably the result of primary, very shallow subaquatic precipitation.

(b) medium-grained (<1 mm to 2–3 cm), greenish gypsum, with crystals having a spatulate-lenticular twinned habit.

Both lithofacies are intensively burrowed (figure 15.6C and D). Burrows are usually filled with gypsum of the same type, but sometimes also with carbonate and chert.

The micronodular secondary gypsum apparently results from the early diagenetic transformation of primary gypsum into anhydrite, caused by water level fluctuations in the formative sulfatic lake. Meganodules apparently originate in a more advanced stage of diagenesis, perhaps due to groundwater circulation of concentrated solutions (Salvany et al. 1994).

In this sequence, the carbonate mineral is calcite, with minor amounts of dolomite. Texturally it is a mudstone to wackestone with abundant bioclasts of charophytes, ostracods, and gastropods. The clays are mainly illite and chlorite, but smectites and minor amounts of kaolinite are also present. In this formation magnesian smectites and sepiolite are also known (González and Galán 1984). Chert nodules are well developed in the various types of gypsum, although they occur preferentially in the primary gypsum. Meganodules of chert (>50 cm in diameter) are common. Both gypsum and carbonate are replaced by the chert. Because of the position of these evaporites within the detrital host argillaceous formations, and the central distribution of the gypsum facies within the formation, the Ablitas-Monteagudo Gypsum may be interpreted as having formed within a playa lake setting adjacent to the toe of the alluvial fans along the southern border of the Ebro Basin. Such fans grade to the north into a broad lutitic plain in which fluvial and calcareous lacustrine facies developed. During the early Miocene, clastic sediments having a Pyrenean provenance were most common in this part of the Ebro Basin, while those with a detrital composition having an Iberian range provenance were restricted to a narrow southern belt. In the transition zone between the two sedimentary systems, deposition of evaporites of the Ablitas Gypsum took place (figure 15.9).

In several areas near the southern border of the basin, other restricted Miocene evaporite formations with similar petrologic and sedimentologic characteristics to those of the Ablitas-Monteagudo Gypsum developed (see figures 8.3 and 8.4, this volume). These include, among others, the Borja Gypsum and the Pozuelo de Aragón Gypsum in the province of Zaragoza, which are closely related to the Ablitas Unit, and the Autol Gypsum. However, the latter is older and has been completely altered into secondary gypsum. Finally, the Ribafrecha Gypsum, close to Logroño, is a formation in structural contact with the Triassic evaporites of the Iberian Range border. The sediments display a playa lake character, with a clear overprint by the nodular facies (meganodules) that developed by replacement under a ground water regime. Nodular chert is present and some primary gypsum is preserved.

(Accepted for publication April 1992; revised July 1995.)

PART 4

Conclusions: Paleogene Evaporites from France and Spain

Georges Busson

Similarities and Differences

OUTCROP

The outcrops of evaporitic sediments are far more extensive within the Ebro basin of Spain than in most of the French basins, despite the fact that most of southwestern Europe was arid during the Paleogene and large volumes of these deposits were formed in both areas. The preservation of the Spanish deposits is far better than in France because continued regional aridity has preserved the original evaporites. Additionally the Ebro evaporites are located in settings sometimes tectonically favored, within thrust sheets that have conveniently brought many of them to the surface, for instance where they are now well exposed in the northern margin of the Pyrenean foreland.

SUBSURFACE

The potassic-bearing areas of Spain, however, are known solely from the subsurface. Most of the French Paleogene evaporites are best known from boreholes with outcrop of evaporites being largely limited to the area of Provence. Boreholes whose target was halite are numerous: some are scattered within basins and others, for instance in the boreholes in Tersanne and the Etrez area, are represented by a very dense network of drilling. Seismic lines are rare in many areas of the Spanish evaporites, and most of those are proprietary, but in France the seismic and borehole data is plentiful and generally freely accessible. For these reasons the two parts of the book necessarily have been presented in very different formats.

TECTONIC CONTROL OF THE EVOLUTION OF THE VARIOUS EVAPORITE BASINS: A DISPARITY

France.

In France during the Paleogene the evolving basins, including Alsace, Bresse, and Valence were subjected to diverse styles of tectonism and degrees of subsidence, although most were the product of distensive deformation. Accordingly each received different amounts and types of evaporite (mainly halite) deposition (table P4.1). Other small basins evolved within France in the same time period but are not covered in this book, including the basins of Vauvert, Camargue, Limagne, and Riom.

All of the most subsident French basins evolved during the Paleogene development of the West European Rift and lie along a structurally linear trend. They were controlled by regional distension and were marked by strong phases of subsidence and associated evaporite accumulation during the Middle-Upper Priabonian (Ludian) and Rupelian (Stampian). In the Upper Oligocene (Chattian)

TABLE P4.1.

Tabulation of the evaporites present within the depositional sections in the various basins that developed in France during the Paleogene.

STAGE	PARIS BASIN	ALSACE		BRESSE	VALENCE		MORMOIRON	LE PUY	RHONE VALLEY Aix-Marseille	LOIRE AND ALLIER BASINS
		1000 km^2 area of halite 700 m halite thickness 400 km^3 volume halite		8000 km^2 total 2200 km^2 $CaSO_4$ 800 km^2 halite area, 1400 m thick						
CHATTIAN				Upper Evaporite Formation ?	T 2 (>500 m)	U. Evaporite Formation				
RUPELIAN / *STAMPIAN*	Fontainebleau Sandstone. Marls with oysters (5.4 m)	Cyrena Marls (110 m) Melettea Beds (290 m max) Marls with foraminifera (5 to 18 m) top of Salt Series	NP 23	Intermediate Marl Fm. BRECCIA	T 3 (500 m)	Varigated Marls Layers with cyrenia and potamides Layered marls w/ foraminifera INTERMEDIATE MARL FORMATION			Gypsum	
	Sannois Limestone (6.30 m) "Caillasse" of Orgemont	Marls without salt (>80 m) Salt V (365 m) Salt IV (>230 m)	NP 22	E 1 (163 m) E 2 (>160 m) E 3 (146.8 m)	T 4 (275 m) T 5 T 6	LOWER SALT FMTN.				
lower RUPELIAN (Lower Stampian) (Sannoisian)	Romainville green clays (5.25 m)	Fossiliferous Zone (60-80 m)		E 4 Upper Part (230 m)	SUBSALT FORMATION (500 m)	possible gypsum	chemical complex of Bleauvac (carbonate and gypsum)	gypsum	La Fayette Limestones (Apt basin) Limestone with *Brotia laurae* (Mormoiron) Limestones with *Striatelles* (Aix, St. Pierre-les-Martigues)	mainly gypsum and anhydrite with halite in one borehole in Riom
middle-upper PRIABONIAN (middle-upper Ludian)	Loamy clays w/ Cyrenia Pantin white marls (6.85 m) Argenteuil blue marls (8.80 m) 1st to 3rd Gypsum Masses [26.5 m] + Mg clays	Salt III (middle salt) (>220 m) Lower Salt (>900 m)	Salt II (175 m) MVL 2 Salt I (400 m) MVL I	E 4 Lower Part E 5 (600 m)						
BARTONIAN (Marnesian)	4th Gypsum Mass	Subsalt Series		E 6						
upper LUTETIAN	Gypsum	Siderolithic (22.4 m)	marine facies	?						

evaporites reappear to a limited extent, except in the Valence basin.

Spain.

In Spain the depositional basins containing evaporites in the south Pyrenean trough were formed within evolving foreland basins due to the loading of the crust by advancing thrust sheets. Each major tectonic phase seems directly related to a phase of marine evaporite deposition (e.g. Beuda Gypsum and the chlorides of the Cardona Formation) (also see Burbank et al. 1992).

Clastic sedimentation was apparently reciprocal to evaporite deposition, and originated from the three developing ranges that surrounded the Ebro Basin (the Pyrenees, Catalan Range, and Iberian Chain); however, most of the clastic input came from the Pyrenees. The Pyrenean thrust sheets gradually migrated southward over a distance of 10 to 25 km and caused a comparable shift in the evolution of the depocenters. The north to south shift of the thrust sheets began in the northern and northeastern portions of the Ebro Basin during the latest Cretaceous and continued in several stages into the early Paleogene. This period of tectonism resulted in Eocene evaporite formations that are restricted to the northern and northeastern side of the basin, and were sourced by waters that were largely marine. Younger and thicker stratigraphic units, continental in origin, developed to the south and west of the area in what is now the central and western portions of the Ebro Basin (during Oligocene and Miocene times).

The motions and the amount of crustal loading that originated from the north resulted in the evolution of the depocenters and the thickness and distribution of sediment in both main subbasins (Catalonia and Navarra-Rioja). The northern part of the section is up to 4,000 m thick and is continuous and marine until the upper Eocene. To the south, the entire Cenozoic sequence is no more than 1,000 m thick, lies unconformably over a pre-Tertiary sequence, and is completely continental in origin. The Tertiary surface is nearly planar and tilted toward the north. The evolving sedimentary depocenters, partly filled by evaporites, gradually migrated southward over a distance of 100 km from the late Cretaceous through the early Neogene.

Additionally, in the north, the centers in which Paleogene sedimentation took place gradually became enclosed within younger and younger Pyrenean thrust sheets. Associated block uplift and differential subsidence enhanced the formation of continental evaporitic deposits. Each phase of evaporite formation was probably preceded by a phase of tectonism and generally was linked to Paleogene compression and later to Miocene distension. Puigdefàbregas et al. (1986) has suggested that the Upper Eocene marine evaporites (the Cardona) were directly tied to the emplacement of the Cadí nappe (and its western counterparts) which restricted access to the open seas. The basinal subsidence that resulted from the emplacement of this load was responsible for the development of the great marine-sourced evaporite deposits of Catalonia and Navarra.

POST-DEPOSITIONAL TECTONICS

Except for the halite deposits in the Vauvert (Camargue) the Paleogene evaporites of France are relatively undeformed, and because they are not deeply buried they are very well preserved. In the Mulhouse Basin there is, however, considerable salt tectonics in one portion of the basin, while the remaining part is little disturbed. In this latter case individual salt beds can be readily traced over the whole of the mining concession.

The Paleogene deposits of Spain within the several subbasins of the main Ebro Basin are present within the foreland area in front of the Pyrenees. The compression produced by the collision of the European and Iberian plates that formed these mountains was active from the late Cretaceous through the early Miocene with strongly marked phases during the latest Cretaceous through the early Neogene, and produced both folding and thrusting. Thus the entire section was affected both during and after deposition. The Ebro Basin sequence is principally deformed along its northern

TABLE P4.2.
Tabulation of the main evaporative formations present in the Ebro Basin (Paleocene through Miocene). Note, C = Continental, and M = Marine.

FORMATION	REG.	AGE	LOCATION	SIZE	THICKNESS	MAIN LITHOLOGIES	FORMATION BOTTOM	FORMATION TOP	COMMENTS
Cerezo Gypsum	C	L. Miocene or U. Miocene	Bureba corridor west border of Ebro Basin Burgos area	30 X 40 km	150 m in outcrops	secondary laminar-nodular gypsum			Tertiary subsidence up to 5000 m
Zaragosa Gypsum	C	Earliest Miocene to M. Miocene	Central part of Aragon	80 X 40 km	200-300 m ?	primary gypsum			Several layers of halite (up to 100m thick) w/.glauberite + thenardite intercalations. Br = 1 to 15 ppm
Lerin Gypsum	C	Upper Oligocene	Ribera of Navarra and La Rioja			Alternating gypsiferous and clayey levels		Miocene red lutites	Common evaporites in entire Oligocene; from central saline lakes and assoc. mud flats
Falces Gypsum	C	Oligocene	Ribera of Navarra and La Rioja		759 - 900 m	laminar and nodular gypsum	2000 m thick continental ss.	continental detrital	Occurrences of glauberite
Puenta la Reina Gypsum	C	Lower Oligocene	Navarra		300 - 400 m in outcrop	Laminar and nodular gypsum	Guendulain Fm.	Mues SS. Fm.	
Barbastro Gypsum	C	Upper Priabonian	Navarra and Catalonia		30 - 60 to 300 m in outcrops	Laminar and nodular gypsum	Top Gray Lutites	Tora Formation	Conformity evident above the intermediate layers between marine and continental facies
Potash Basin	M	U. Priabonian	Catalonia and Navarra	300 km	300 m in Catalonia 100 m in Navarra	Anoxic marls, carbonates, $CaSO_4$, K and Na chlorides			No continuity between Navarra and Catalonian subbasins but correlation suggests continuous deposition
CATALANIDES:	C	U. Paleocene to L. Oligocene	S.E. Ebro Basin	150 km					Evaporite pinch out toward basin center. Source: Triassic evaporites (Catalanides)
* Odena Gypsum		Bartonian to Priabonian			40 m	$CaSO_4$, cherts, carbonates, massive-nodular facies			In Santa Maria Group
* Valldeperes Dolostone & Gypsum		Cuisian to Lower Bartonian			120 m	gypsum with intercalated fine detrital layers			In Pontils Group
Beuda Gypsum (& Coubet Gy.)	M	Lower Lutetian (U.Ypresian and Lutetian)	Southeastern Pyrenees		40 m	Anhydrite	Vallfogona sequence	restricted marine sequence	small amount of halite in subsurface
Tremp Gypsum	C	Upper Paleocene	Central-south Pyrenean area					*Alviolina* limestone	

border. It is separated from the Catalan Range by a series of faults. Other than this the Tertiary deposits are only mildly distorted.

Generally speaking, the greatest disparities between these two areas arise from the differences in tectonic control. The French basins may present an E-W longitudinal symmetry (perpendicular to their main trend), as in the Haute-Alsace between the Vosges and the Black Forest, but the development of the Spanish basins is wholly asymmetric. This difference comes about because the Spanish basins are primarily compressional while those in France are largely rift-controlled.

Ages and Main Features of Principal Evaporite Formations

The general features of the evaporite formations of Spain and France are summarized in tables P4.1 and P4.2.

LITHOFACIES DISTRIBUTION: SPAIN

1. Continental Evaporites.

Most of the continental evaporites appear to have been deposited within very shallow lakes that were located on distal mudflats, during times when marginal alluvial fans were only weakly active (see Ortí in this volume).

A. THE DEPOSITIONAL SYSTEMS

(1) *Shallow sulfate lake systems with playa stages.* These lakes were small or lay at the margins of large central chloride-rich lakes that were the product of interior drainage systems. In particular such deposits developed from the Eocene through the Miocene in a well-defined belt along the SE and SW borders of the Ebro Basin. These marginal systems interfingered with fine siliciclastic sediments and apparently acted as geochemical traps for carbonates, silica (chert), and strontium compounds (carbonate and sulfate). The calcium sulfate lithologies include microlenticular and meganodular gypsum beds that contain only small amounts of matrix. Some coal deposits are also present.

(2) *Large lakes at the basin center (chloride rich).* In addition to calcium sulfate and halite the sediments of these basins contain sodium sulfate and small amounts of potassium and magnesium sulfate salts. These deposits may be very thick (up to 1,000 m) and some playa stage deposits are interbedded within the lake sediments.

3. *Different lithological groupings of $CaSO_4$ within the sequence.* At least five different lithological groupings composed of $CaSO_4$ have been clearly identified in the lake and playa deposits of Spain, most of which originated in the form of gypsum (based on morphologies) but are now variously pseudomorphed by secondary gypsum, anhydrite, chert, and/or calcite (table P4.3):

Laminated sulfate facies (gypsum and/or anhydrite) with or without algal morphologies (probably subaqueous).

Microlenticular facies are fairly fine grained, sometimes very fine-grained, with little or no matrix (usually gypsum but may be pseudomorphed by polycrystalline anhydrite or chert). Microlenticular sulfates are commonly interbedded with micritic carbonates that bear Charophytes and Mollusca. These were very shallow subaqueous deposits, formed in chloride-poor phases of lake development.

Micro-, macro-, and meganodular facies formed during early diagenesis within playas or associated sabkhas. These deposits developed during dry periods by interstitial precipitation of gypsum, anhydrite, or other sulfates (e.g. glauberite).

Gypsiclastic facies were formed due to mechanical reworking of gypsum, and were usually subaerial but also may have been subaqueous.

Composite lithologies: laminated-nodular alternations are present in the central parts of the lake areas and are commonly interbedded with halite. This composite facies is very important in both area and volume and is due to repeated changes in lacustrine water levels (subaqueous to sabkha stages). The microlenticular and nodular facies types are usually limited to marginal deposition in many of the same lakes. These latter facies also may be interbedded with cherts and carbonates.

TABLE P4.3.
Comparison of many of the lithologic features (basin centers versus the margins) observed in evaporative formations of the Ebro Basin (Paleocene through Miocene).

FORMATION	CENTRAL SEQUENCE	MARGINAL SEQUENCE
CEREZO GYPSUM (lower or upper Miocene)	laminar to nodular gypsum cycles of the sabkha type glauberite layers present	detrital and calcareous facies
ZARAGOSO FORMATION (lower-middle Miocene)	primary gypsum with halite beds	nodular anhydrite, alternating with primary laminar gypsum and carbonates, alabastrine gypsum and cherts
LERIN GYPSUM (upper Oligocene)	both laminar and nodular gypsum with intercalations of lutite, halite, glauberite, and polyhalite	sabkha cycles
FALCES GYPSUM (Oligocene)	both laminar and nodular gypsum with intercalations of halite and polyhalite	
PUENTA la REINA GYPSUM (lower Oligocene)	laminar gypsum, nodular gypsum and halite layers	nodular gypsum in sabkha cycles
BARBASTRO GYPSUM (upper Priabonian)	laminar and nodular gypsum and halite layers	nodular gypsum in sabkha cycles
TREMP GYPSUM (upper Paleocene)	massive, nodular gypsum	

In some instances a regular facies sequence developed with microlenticular gypsum at the bottom and nodular sulfate at the top and apparently was produced by the terminal filling of an evaporative lake with the evolution of a sabkha at the top (the final brines being restricted to the subsurface).

2. Marine Evaporites

A. BEUDA GYPSUM AND THE COUBET GYPSUM.

These formations are made up of layers containing beds of laminated gypsum, pseudomorphs of crystalline gypsum, and nodular lithologies. These deposits appear to have come from sporadic marine flooding in fairly deep marine basins, but at the top contain sediments formed during interbedded phases of shallowing or desiccation. Only gypsum and anhydrite are known in outcrop but halite is found in boreholes.

Under the marine gypsum there are nonbioturbated laminated lutites that contain substantial quantities of trace elements such as Cu, Pb, and Zn (Vallfogona Sequence), possibly formed as a restricted marine deposit (deeper water than during gypsum formation). Above the evaporites there is also a restricted marine sequence. This position of the gypsum suggests deposition within a dense brine having a basinal reflux that would have carried the chlorides out of the system. Above and below the gypsum the deposits were probably deposited in stable stratified waters.

B. THE SOUTH PYRENEAN POTASH BASIN.

A structural sill separates the two subbasins of the South Pyrenean Basin, but sedimentary continuity is suggested by the stratigraphic similarity of the two sections despite the apparent disparity in thickness (Catalonia = 300 m; Navarra = 100 m).

FACIES PATTERNS.

Three concentric facies belts are present and are more clearly observed in Catalonia than in Navarra because of the favorable nature of the exposures:

The *peripheral facies belt* (Tossa Formation and its equivalents) is open marine, calcareous, and represents deposition in the last open marine period in the whole area of Catalonia. This belt also contains several deltaic accumulations interbedded with the carbonate trend. The reef sequence is karstified and dolomitized due to the postdepositional lowering of the basinal water level.

The *intermediate depositional belt* is largely composed of $CaSO_4$ (Odena Formation and its equivalents). This sequence is well represented by a deposit about 30 m thick at the east and southeastern border of the Catalan Basin. It overlies blue marine marls that are anoxic at the top. The sulfate deposits thicken toward the reef core and thin toward the center of the basin, beneath the central chloride filling. The three microfacies that make up the vertical section of this depositional belt are: (a) basal stromatolites (1 to 2 m) composed of gypsum and carbonates; (b) fine-grained alabastrine gypsum that pseudomorphs selenitic gypsum beds containing irregularly scattered carbonate laminae (total of 20 m); (c) current- and cross-bedded, laminated gypsum (5 to 7 m), now entirely composed of secondary gypsum (megacrystalline and alabastrine). This depositional progression suggests deposition under very shallow to emergent conditions.

The *central chloride region* that represents the depocenter section shows the following sequences:

(1) Catalonia. This chloride series is represented by the Cardona Formation in Catalonia. It begins at the base with anoxic marine marls (blue to gray) that are assumed to be distal sediments of a deltaic fan sequence. In some boreholes these anoxic marls contain carbonate beds. In marls, the fauna begins as typical of deep water that shallows upward. Typically near the top of this unit the sediment becomes anoxic with millimetric laminations but with a lack of appreciable organic remains.

The basal evaporite section is made up of laminated anhydrite with irregularly alternating laminae of carbonate (4 to 5 m thick). The morphlogy of these beds suggests a possible stromatolitic origin.

The lower salt unit is 130 to 200 m thick, composed of banded centrimetric layers of halite having dark gray and clear, pink, or white alternations. Small amounts of anhydrite, polyhalite, magnesian clay, and micrite are noted.

The sylvinite unit (lower Potash) is 5 to 20 m thick, with individual cycles of 10 to 30 cm composed of cyclic sets consisting of either insoluble—halite—sylvite or sylvite—halite—sylvite. This unit has fine-scale lateral continuity, cycle by cycle, over a distance of more than 10 km, suggesting subaqueous formation under a perennial water body.

The carnallite unit is 40 to 80 m thick and is made up of carnallite beds alternating with halite beds.

The upper salt unit is weakly defined, with intercalations of lutite and gypsum.

The top marl unit is composed of gray, gypsiferous lutite and locally with carbonate (35 m thick). It is overlain by the Barbastro Gypsum.

(2) Navarra. The basin center sequence begins with marls that are locally turbiditic and are interpreted as slope deposits.

Here the basal evaporites are represented by only one bed of gray laminated anhydrite (0.6 to 1 m).

The lower salt unit is relatively thin as compared to Catalonia (10 m), with the halite being present in centimetric layers (2 to 10 cm).

The sylvinite unit is a total of 2 m with sylvinite layers alternating with halite in beds 1 to 14 cm thick.

The carnallite unit is 12 m thick and is made up of carnallite beds alternating with halite beds.

The upper salt unit is up to 50 to 75 m thick and is composed of halite alternating with dolomite. The halite also contains intercalations of laminated lutites (0.5 to 2 cm) and thin anhydrite layers. The section becomes less halitic and more anhydritic toward the top of the unit.

The top marl unit is 50 m thick and is laminated and locally rich in calcareous nannoplankton (a fairly open marine assemblage).

C. DISCUSSION.

The concentric lithofacies pattern noted in the Spanish marine Paleogene extends only in part into the potash deposits. These deposits appear somewhat displaced with regard to the underlying halite depocenter. The concentric pattern of facies shown

TABLE P4.4.
Comparison of the clay paragenesis observed in the Bresse and Valence basins during the deposition of the Paleogene evaporites.

<table>
<tr><th>BRESSE</th><th>VALENCE</th><th>ASSOCIATED FEATURES</th><th>OBSERVATIONS</th></tr>
<tr><td colspan="2">Illite + small % kaolinite and chlorite</td><td>detrital</td><td></td></tr>
<tr><td colspan="2">Illite + smectite + very small % chlorite</td><td></td><td>smectite may be very low or beyond illite</td></tr>
<tr><td>Illite + illite-smectite, small % kaolinite and chlorite</td><td></td><td>aggradation characterized by the development of swelling minerals (smectite + smectite-illite)</td><td>associated with carbonates by magnesian aggradation on the finest illite particles</td></tr>
</table>

by these basins differs from the ideal Schmalz depositional pattern (1969) that predicted that carbonates and sulfates continued with the same thickness from the margin to the depocenter. On the other hand it corresponds exactly to the pattern proposed by Busson (1978, 1988).

LITHOFACIES DISTRIBUTION: FRANCE

The lithofacies present in the various basins of France vary considerably from basin to basin, so they will be discussed area by area, as well as by facies groupings.

A. Distribution of siliciclastics.

Siliciclastic sediments cover large areas of the French basins and are common throughout the section. They are interpreted as distal deposits formed from fluvially scoured alluvial fans and plains carried into the basins by temporary flows moving along wadis. The argillaceous deposits found in all of the French basins are commonly laminated, generally contain well-preserved lignitic plant remains, and are commonly pyritiferous. Where present in the depocenter areas, red argillites are uncommon and are associated with the rare desiccation surfaces. At the margins the argillites are commonly red and desiccation surfaces are numerous. A comparison of the clay components is presented in table P4.4.

(1) Valence Basin (N 20° Trough): This basin is about 30 km from north to south and is a few km wide. The east flank has a gentle slope beginning at the Montmiral high, and on the other side of this high lies the Royans subbasin. To the west of this area there is a steep contact with the Vivarais mountains marked by a synsedimentary fault. To the south lies the actual Valence area with 1000 m of detrital sediment below the T4, equivalent to the subsalt unit and part of the lower salt formation. Here most of the detrital materials come from the south and west (Vivarais), with ubiquitous anhydrite or gypsum nodules and halite phenoblasts within these sediments, suggesting that free or interstitial supersaturated brines existed within the areas receiving the clastic deposits.

(2) Bourg-en-Bresse Basin: This basin is located in the southern part of the Saône graben (200 km x

30 to 60 km). The salt deposits occupy an area of about 60 km x 18 km with the steep side of the basin towards the east (the Jura) and a gradual slope upward toward the west, to the Limonest high (total area covered by the basin = 8,000 to 9,000 km^2; $CaSO_4$ = ~ 2,200 km^2; halite = ~ 800 km^2). Further to the west of the Limonest high there is a small western subbasin. In the Bresse area most of the detrital material originated from a southern detrital source. In some instances the influx of clastics, and their associated diluted waters, displaced the zone of halite accumulation toward the northeast (Etrez).

(3) Haute-Alsace: The deposits of the Haute Alsace Basin cover a total area of 7,000 km^2 with interrelated sediments, and an area of 2,000 km^2 that is covered by evaporites. Of this evaporite region about 500 km^2 contains potassic salts. In Alsace the deposits of alluvial fans ("coastal conglomerates") cover only a small area and are closely tied to the nearby Vosgian highlands. They apparently represent the lateral equivalents of the evaporite deposits of the depocenter areas.

B. Carbonates.

Both calcite and dolomite have been found only in remarkably small quantities within the depocenters of the French evaporite basins. In many instances calcite is present as tiny rhombs (or rosettes) making up sporadic laminae in rhythmic salt deposits. In other instances calcite is present along the walls of gypsum crystals or as pseudomorphs of gypsum. Such finely disseminated calcite also lines the walls of halite phenoblasts in other parts of the section.

Massive carbonates are rare, and when present in the Bresse Basin they are usually found in beige layers a few decimeters thick. In some instances the beds are undulate, suggestive of stromatolitic origins. In the Valence Basin there is additional diversity in the calcite layers with pelletal carbonates (T5), oolites, laminar, and well-developed, domed stromatolite forms (T5 Hauterives, T4 Tersanne). Oolitic limestones in the upper part of the intermediate marl of Valence are similar to those that form in the modern analog of the Laguna Madre (Texas). There are also finely laminar limestones (infra- to finely millimetric) that show no bioturbation, coprolites, desiccation cracks, or evaporitic crystals. These laminar beds are sometimes argillaceous and variably rich in organic matter. In some of the finely laminar beds fish and/or marine nannoplankton occurrences have been noted (Valence, Horizon B of the T4, and in the lower part of the Intermediate Marl Formation).

Dolomite is largely restricted to the Valence Basin and is present only within a distinctly continental sequence.

C. Evaporites

(1) CALCIUM SULFATE.

Deposits of calcium sulfate are rare within the main depocenters of the Paleogene of France, but they are relatively common in the peripheral areas and in some of the smaller basins (particularly in Provence). In the depocenters themselves the cumulative percentage of sulfate is far below that expected from normal seawater, but outside of these areas calcium sulfate is intercalated with carbonate, present in quantity, generally without halite. In several levels within the depocenters there are thin anhydrite beds commonly associated with organic-rich laminae, however the deposits in marginal areas do not show such restricted lithology. The deposits of calcium sulfate in France are far smaller than in the time equivalent basins of Spain. The microfacies noted include:

Subaqueous bottom-growth gypsum ("grass-like" shapes) with internal millimetric to centimetric laminae; now present as secondary alabastrine gypsum.

Displacive tabular to curved-face crystals or clusters that are commonly early diagenetic forms (synsedimentary); the morphology is largely controlled by the trace amounts of diverse organic compounds entering the basin from the continental ground-waters.

Irregular nodular shapes that are commonly very early diagenetic forms developed just below the water-sediment interface. Such nodular forms apparently continued to increase in volume within the host matrix and coalesced, forming enterolithic, "chickenwire," or massive textures.

TABLE P4.5.
Comparison of the salt accumulations (thickness and purity) in the main Paleocene salt beds of the Mulhouse, Bresse, and Valence basins.

STRAT. LEVEL	MAXIMUM THICKNESS	MAXIMUM % HALITE
	MULHOUSE BASIN	
SALT I	greatest thickness 400 m Great Salt Bank 200 m	50 % in the thickest beds
SALT II	175 m	70 %
SALT III	> 220 m	40 %
SALT IV	> 230 m	> 65 %
SALT V	365 m	64 %
	BRESSE (Etrez)	
E 5	600 m	
E 4	230 m	67 %
E 3	146.8 m	
E 2	> 160 m	
E 1	163 m	
	VALENCE BASIN	
T 6	> 200 m	85 %
T 5	260 m	65 %
T 4	275 m	80 %
T 2	> 500 m	30 %

(2) SALTS (HALITE AND POTASSIC SALTS).

The general distribution of halite in the Paleogene of France is reviewed in table P4.5. The microfacies that are present include:

rhythmic halite beds having 15 to 20 thin repetitive layers per meter (represented in the T6 of Tersanne and the E2 and 3 of Bresse) and probably the product of cumulate crystal precipitation within the water column;

cloudy halite layers (milky due to fluid inclusions), the product of rapid primary crystal growth either at the brine/air or the brine/sediment interface (present in almost all of the basins);

clear halite develops during very early or syngenetic precipitation. The simplest form is present as clear overgrowths rimming cloudy halite cores. These clear rims have the same bromine content as the cloudy crystals they enclose and are in crystallographic continuity with the original primary crystals. In other cases clear halite fills fractures, solution cavities, and desiccation cracks and is commonly preceeded by partial infill by other materials (carbonate, anhydrite, celestite, pyrite, glauconite, or calcareous argillites);

potassium chloride (sylvite) is reviewed in table P4.6 in which the upper Eocene of Catalonia is compared to the lower Oligocene of Alsace.

TABLE P4.6.
A comparison of significant sedimentary characteristics of the sylvinite deposits of Alsace (Rhine graben) and the Ebro basin (Spain).

TOPIC	ALSACE	CATALONIA
AGE	lower Rupelian (Stampian)	upper Priabonian
AREA	Lower Layer: 500 km ^{2} Upper Layer: 120 km ^{2}	
THICKNESS	Lower Layer: 6 m Upper Layer: 2 m	sylvinite unit { 40 to 80 m in Catalonia; 12 m in Navarre carnallite unit { 5 to 20 m in Catalonia; 12 m in Navarre] 600 km ^{2}
ORIGIN OF THE IONS	Possibly marine	Marine
POSITION W/ REGARD TO MAXIMUM THICKNESS	Lower Layer: follows isopercentage lines Upper Layer: migrates west	
COMMENTS	Very regular concentric pattern for total thickness, maximum halite, and boundary of potash layer not in keeping with hydrothermal source	
ELEMENTARY CYCLES	Insoluble components, anhydrite and halite (halite having chevron and cornet forms), sylvinite *N.B.	Insoluble components, halite, sylvite or sylvite/halite
COMMENTS	Regularly increasing marine paragenesis is not compatible to a possible hydrothermal origin Magnesium sulfate is absent in both basins	
LATERAL CONTINUITY	30 km	less than 10 km within the sylvite
COMMENTS	This degree of continuity suggests deposition in a perennial saline water body	

* In Alsace, the presence of deposits of argillaceous anhydrite above the sylvinite suggests that depth of the saline but perennial water body protected the underlying potassic salts during refreshment

Secondary Sulfates Present Within the Evaporite Deposits

A. PRIMARY AND EARLY DIAGENETIC LITHOLOGIES

Primary gypsum morphologies are rare in the Paleogene deposits and almost always are overprinted by later diagenesis: the original forms are recognizable only through their gross morphologies. The relics of these primary facies are commonly quite evident with shapes suggestive of their original grass-like, subaqueous, bottom growth. Although they are now present as secondary alabastrine gypsum, anhydrite, calcite and/or chert pseudomorphs, the internal laminae, and even saw-tooth crystal layers, curved-face crystals or crystal clusters are clearly evident to the eye. Early diagenetic (subbottom) nodular to massive sulfates, originally developed as simple displacive nodules, massive layers or enterolithic beds are also present within argillaceous, muddy or calcareous matrices. These also have undergone alteration and/or several stages of replacement. Commonly this transformation has taken place early in the burial history of the sediments, and occasionally a layer or two in a sequence remains unchanged or slightly overprinted while the layers just below are altered.

B. LATE DIAGENETIC LITHOLOGIES

Late diagenetic anhydrite may be found as anhydrite cements in porous limestones and argillites. Similarly it may appear as porphyroblasts replacing the gypsum of several adjoining lenses. Gypsum and anhydrite veins also form in consequence of the release of overpressured fluids that fill fractures and crosscut beds made up of carbonate, other sulfates, and halite. The mobility of calcium sulfate is in part due to the high chloride content in the formational brines which permits about three times as much sulfate to dissolve in brines than in non-saline waters (Bock 1961).

C. SECONDARY MINERALS

Accumulations of glauberite and celestite make up parts of a number of deposits in both France and Spain. The glauberite, which is very common in the continental deposits of Spain, appears to be the result of fairly early diagenesis (synsedimentary) while in some cases the celestite may be much later in origin. Secondary gypsum and anhydrite commonly replace each other and also the glauberite throughout the section. Celestite makes up about 7% of the total volume of the rock in some areas. In some cases it seems to have formed as a result of bacterial breakdown of some of the anhydrite. However celestite may also form as the product of aragonite inversion to calcite and also due to anhydrite hydration to gypsum. In these cases of alteration, the amount of strontium that can fit into the new crystal morphology is far lower than in the original minerals.

Organic Matter

A general evaluation of the sedimentologic and organic content present in the evaporitic series of Haute-Alsace is shown in table P4.7. From this table it is evident that in Salt III, for instance, the areas richest in organic content also have the purest halite (highest isopercentages). The areas having the thinnest layers of intercalated impurity are the richest in TOC. At the scale of individual beds, the highest TOC is just at the top of the argillo-anhydrite interbeds, and the lowest TOC is just at the top of the halite.

In the Bresse Basin, the maximum TOC is found within the breccias at the top the salt deposits. These layers are rich in lignitic remains (type 3). Organics are also common in other levels, such as the barren intermediate series. Elsewhere in this sequence type 2 organics are present.

In contrast to the French evaporite sequences, most of the Spanish deposits contain very little organic matter and in general are far more oxidized.

TABLE P4.7.
The evolution, through time, of the clay types and of the organic matter present in the evaporative sequence in Haute-Alsace.

MEMBER NAME	CLAY PARAGENESIS	BROMINE CONTENT	TYPE OF ORGANIC MATTER	FAUNA AND FLORA	COMMENTS
Foraminifera Marls			Type 2 O.M.; higher TOC		
Marls without salt	Illite, chlorite, and kaolinite		TOC minimum		
Salt V	Illite, kaolinite, and chlorite		Decreasing TOC	disappearance of marine fossils	more and more abundant, green or red-brown shades. Rounded quartz and desiccation structures.
Salt IV	Kaolinite and illite	70 ppm (200 to 340 ppm in halite: 2800 to 3900 ppm in sylvinite)	TOC decreases, but is Type 3 TOC at maximum. Type 2 at base	coccoliths present from the base up to Unit 12 together with very small foraminifera	
Fossiliferous Zone	Illite and kaolinite (up to 40%)		TOC decreasing Type 3, rich in higher plant remains	restricted nannoplankton: numerous dinokysts and other marine algal remains (ex. for the upper 2 m of the Fossiliferous Zone) numerous fossils typical of marine and brackish water (Hydrobia, Mytilus, forams etc.)	
Salt III		90 - 150 ppm (concentration appropriate for potash in some cases)	Type 2 TOC maximum ranging between 3 % and 40 % (highest in the upper part)	disppearance of lacustrine fossils, passing upwards into marine or brackish-water forms (Cerithium, Mytilus, and diverse dinokysts and achritarchs; nannoplankton in Bank 8)	
Lower Salt	Well crystallized illite with a little chlorite	30-40 ppm with 70 ppm in Salt II	Type 3	mainly lacustrine biota	
Sub-saliferous			poor in O.M. but higher than in the Siderolithic		
Siderolithic			very low content of O.M.		

Facies Patterns on a Basinal Scale

VALENCE BASIN (N 20° TROUGH)

In this area the depocenter remained topographically lower than the rest of the basin throughout the depositional period and contains laminites and rhythmic salt and displays an apparent lack of desiccation throughout the section.

T6 shows a clearly established sedimentological relationship from the center (halite rich) to the border facies. T5 shows a restriction of the halite to a more central area than during T6 deposition, and is intercalated with fairly thick accumulations of sandstone, shale, limestone, and anhydrite. These marginal facies include carbonates to the east of the halite and detrital sediments to the south and west. There are no carbonate or detrital intercalations within the central halite area and the general sediment distribution shows a concentric but asymmetric pattern.

In T4 there is onlap of halite after a basal argillite that passes up into pure halite. The halite, however, contains some limestone horizons in which coccoliths and fish remains are present in Member B.

Intermediate Marl formed in a non-restricted to evaporitic environment and characterizes the Lower Salt Formation and Upper Evaporite Formation. It is a largely marly succession in which laminated limestone and anhydrite beds occur. It is unusually homogenous across the basin and shows no internal facies variations.

Upper Evaporite Formation: here the evaporites are restricted to the central trough. Facies limits apparently correspond to the isopach lines. Even in the depocenters, great amounts of carbonates, sulfates, and clays interfinger with the halite. Glauberite is present in the lower part of the section. There is a weakly established concentricity of lithofacies from the margin to the center (carbonates, sulfates, to halite) and the entire section becomes increasingly silicic upwards.

BRESSE BASIN

This basin contains a thick evaporative section (1,400 m). The percentage of halite within the section is quite high and may exceed 90%. Laterally the carbonates are proportionally more abundant toward the east and become thinner and pinch out toward the depocenter. The argillaceous detrital input came into the basin from the south and was largely restricted to that area at the same time as the evaporites were forming in the depocenter.

To the north of the main Bresse Basin lies the Louhans Basin, which contains carbonates and sulfates, and may have possibly played a role as an evaporative annex for the Bourg-en-Bresse Basin.

HAUTE-ALSACE

This basin contains a 1,700 m salt section in the depocenter and thins to about 400 m thick on the Mulhouse horst, where the deposits are largely carbonates and $CaSO_4$.

In the Lower Salt the greatest isopachous thickness is coincident with the highest mapped isopercentages within the deposits. In the Middle and Upper Salt the maximum isopercentages are offset with respect to greatest thicknesses of the accumulation, and are found atop a platform in the basin center. It is suggested that this latter is due to the thinning of the zones of insoluble materials between the halite beds.

CONCLUSIONS

The French evaporite basins are characterized by their general asymmetry (as opposed to a concentric pattern of basin center evaporites), their small size, and the trends of both their detrital and marine input. There is also a direct relationship between salt sedimentation and differential subsidence.

Ionic Sources and Depositional Environments

SPAIN

In the Spanish basins the distinction between ionic influx of marine and continental-sourced water is quite clear.

1. Marine-sourced evaporites.

Only two formations are clearly established as having marine affinities—the Beuda Formation (lower Lutetian) and the Cardona Saline Formation (potash) (upper Priabonian). In the latter the encircling formations are wholly marine (based on fauna and flora) and the bromine content of the salt itself is very high, as is appropriate for marine-sourced deposits.

2. Continental-fed evaporites.

All of the other formations studied in this report are apparently sourced by continental water. This interpretation is based on geochemistry, particularly the isotopic markers, and they indicate a Triassic-Liassic source for the SO_4^{-2} ions that are present within the evaporite deposits (Utrilla et al. 1987; Utrilla 1989; Ortí et al. 1988). The source areas apparently outcropped in the Basque-Cantabrian diapiric areas in the Catalan and Iberian chains, and are also present in the basement of the Ebro Basin itself. There is no input from upper Priabonian marine evaporites to the uppermost series. It must be noted that some of the continental evaporites lie conformably just above or below marine formations, or grade laterally into marine or shoreline formations (for example in the Tremp Formation).

FRANCE

In the French Paleogene basins the ion sources and depositional environments remain the subject of much greater controversy than in Spain. The Upper Evaporite Formation in the Valence Basin and parts of those in Provence are deposits that can be unquestionably assigned to a wholly continental and recycled origin. In the next section we shall review the arguments for the sources and environments for the other formations in the main basins of Alsace, Bresse, and Valence.

1. Petrographic and geochemical data

A. *Bromine:* There is rather close agreement between bromine content of the salt and the paleontological and organic matter assemblages contained within these evaporite sections.

In Alsace the section is relatively continental at the base (Salt I and Salt II), becoming more marine in the overlying Salts III and IV, and returning to an increasingly continental character in Salt V.

In Valence the coincidence of a high bromine content of T4, Salt A, and the marine microfauna present in T4, member B below, and the Intermediate Marl Formation above, is clear. However, in the T5, the existence of continentally recycled Paleogene salt could explain the low bromine values.

In Bresse the bromine levels in E4 and E5 appear to correspond to a marine influx, but the low bromine in E1 and E3 also seems to be the result of a continental reworking of other Paleogene salts.

B. *Paragenesis and purity of salt:* The commonly observed paragenetic relationships within many of the salt formations in France is appropriate to marine sources and does not contain the characteristic non-marine mineral assemblage noted in the upper evaporite formation of Valence or the Ebro Basin. Similarly the presence of hundreds of meters of very pure, rhythmic salt deposits seems more typical of continuing marine influx rather than the interruptions by terrigenous intercalations so typical of repetitive continental sourcing.

2. Paleobiological Data.

In Haute Alsace, marine flora and fauna with a diverse assemblage including calcareous nannofossils, foraminifera, dinoflagellates, acritarchs, pelecypods are present in the Salt III, the fossiliferous zone, and lower part of Salt IV. This is in good agreement with other data from the same basin. In Valence there are calcareous nannofossils in the B level of the T4 and the base of the Intermediate Marl. Only the Bresse Basin contains no marine organic remains, but this could be because the basin was simply too saline, as in the modern, marine-fed analog within Lac Assal.

3. Paleogeographic data

Continental origin: A continental origin for much of the evaporite section of Alsace is not in contradiction with the proximity of existing Permian salt and potash basins. The area of the three basins is also underlain and surrounded by Triassic evaporites, providing plentiful potential sources. In addition to the closeness of appropriate soluble source rocks, the Paleogene was a period of marked tectonism in this region, so that parts of these older evaporites may have been exposed and available for ready chemical recycling.

Marine origin: Contemporary marine deposits (Paleogene) are known to have existed only a short distance from each of the major rift basins, although marine deposits did not extend to the rifts themselves (as in the relationship between the modern Indian ocean and Lac Assal at another scale). The maximum onlap of these Tethyan extensions also are coincident with the main evaporite stages in the rift basins (middle-upper Ludian and lower Stampian). Because proven marine input did exist at some levels in both Alsace and Valence and is suggested for Bresse, it seems that the proximity of the Tethyan extensions may well be indicative of a sporadic throughgoing connection that fed the rift areas. These Tethyan marine sources that came from the eastern side of the basins also may be the reason that both the Bresse and Valence basins have greater amounts of carbonate and sulfate deposits on the eastern sides while the non-marine influx, marked by a substantial clastic influx, was from the south and west. Purely continental inflow into a closed basin would have a far lower likelihood of developing this asymmetry.

4. Other data.

During this period the formation of evaporites extended far beyond the borders of the actual rift basins, and included the Paris basin, several small Atlantic basins, the Massif Central basins, and Languedoc. This broad distribution is certainly not consistant with the idea of a marine origin for all the deposits, but it is equally difficult to suggest that all of these evaporites are purely the product of chemically recycled evaporites from existing Permo-Triassic deposits.

Climate

Broad regional aridity is a necessary but not sufficient condition for evaporite deposition. It certainly constrained deposition in the disparate basins during the Paleogene, prevented their rapid infill due to clastic influx, and permitted the basins to reach an appropriate concentrative balance for evaporite precipitation. This is significant because the various basins of France and Spain had such diverse structural histories and rates of subsidence yet they all received evaporite deposition during the same time periods. Sporadic wetter stages or localized wet uplands resulted in the development of marginal alluvial fan deposits associated with all of the basins, but such periods were not prolonged or well developed because many of the evaporite depocenters remained continuously saliferous throughout the middle upper Priabonian (Ludian) and lower Rupelian (Stampian), a testimony to continuously elevated aridity. The broad distribution of the major palynomorph groups supports these climatic generalizations (pinaceae and cupressaceae). Widely divergent climatic inferences based on the presence of other pollen and spore groups (Schuler 1990) suggesting general regional rainfall of 1,500 to 2,000 mm/year during the deposition of the evaporite sediments seem difficult to accept. This is not to say that the palynological observations are wrong, but that there must be other explanations for their presence within the evaporitic sequences. Two reasonable alternatives to the presence of the observed high-rainfall flora are: (a) the palynomorphs come from surrounding mountain areas that were wetter; or (b) the palynomorphs came from clay interbeds within the evaporites and represent short climatic excursions during otherwise dry periods.

Water Depths During Deposition of Evaporites

In spite of a superficial resemblance to deposits formed in shallow salt pans, all authors in this volume support the idea that some parts of these evaporite successions formed under perennial subaqueous conditions.

SHALLOW WATER

The most convincing argument for shallow-water formation is that the "elementary" cycles of halite, only a few centimeters to decimeters thick, are made up of repeated, identical, rhythmic halitic layers with dissolution-truncated upper surfaces which would be most difficult to generate in deep water bodies. When each new influx entered the basin it appears that the brine became diluted right down to the bottom, and this is very difficult to achieve in a deep basin. Similarly it may explain the strong variations in bromine content within each of the "elemental" cycles, that change regularly from high to low, going from the basal to the uppermost layer of halite in each cycle. This is very difficult to achieve in any but shallow water bodies. The third point of argument is found in the petrology of the rocks themselves, in which some of the halite layers display distinctive shallow-water morphologies, comparable with those presented by some authors (Shearman 1970, for example).

DEEP WATER

The argument in favor of a deep water depositional environment is based on an analysis of regional stratigraphy. There is a complete mineralogical differentiation between the central and marginal deposits. For example in the area of Haute-Alsace there are 1,600 m of evaporites at the basin center, mainly as halite and potassic salts, while only 400 m of carbonates and sulfates were deposited on the Mulhouse horst. The lack of carbonate and sulfate in the basin center, the lack of colonization by benthic organisms (for example the Membranipora), the absence of migration of mammals restricted to the horst areas, all give evidence for a bathymetric differentiation and even for the perennial character of a stratified system. This differentiation appears to have been permanent throughout the upper Priabonian (Ludian) and lower Rupelian (Stampian).

In the central saline area the salts are extremely pure and it seems that these salts were precipitated within a very short time period. The presence of thick rythmic salt beds gives clear evidence of the long-term stability of the conditions of sedimentation, unlikely in very shallow water. Additionally, fine-scale correlation of laminar salt beds over long distances suggests the existence of stable stratified water bodies (maintained through geologically significant time periods) that probably existed independant of basin-bottom morphology. Laminated shales interbedded within salts that correspond to periods of reducing conditions demonstrate the long-term stability of this stratified system, which is more likely in a deep water system. Finally, there is a general lack of desiccation structures within the depocenters of these basins, and such features should be present, were the basin truly shallow.

References

Aguilar, J.-P., J. Michaux, D. Nury, J. Sudre, F. Touraine, and M. Vianey-Liaud. 1978. Découverte d'un gisement trés riche (Mammifère et Mollusques principalement) à Venelles (Bouches-du-Rhône), premier niveau repère important dans la série du "Gypse d'Aix." *C. R. Acad. Sci. Paris,* D, 287: 439–42.

Alabouvette, B., G. Berger, and C. Cavelier (coordinators). 1984a. Miocène post-Aquitanien: Puissance et Faciès. In Synthése geologique du Sud-Est de la France. Map No. 3, *Bur. Rech. Geol. Min. Mém.* 125, Orléans, France.

Alabouvette, B., C. Cavelier, J. J. Châteauneuf, J. Gaudant, and J. M. Schlund. 1984b. In S. Debrand-Passard, S. Courbouleix, and M. J. Lienhart, eds., Paleogene, Synthése geologique du Sud-Est de la France, pp. 415–21. *Bur. Rech. Geol. Min. Mém.* 125, Orleans, France.

Allard, P., E. Gannat, N. Laplaiche, A. Lefavrais-Raymond, and P. Marie. 1959. Sur un niveau à *Microcodium* à la base du Tertiaire de Bresse. *C. R. somm. Bull. Soc. Géol. France:* 6: 150–51.

Álvaro, M., R. Capote, and R. Vegas. 1979. Un modelo de evolución geotectónica para la Cadena Celtibética. *Acta Geològica Hispànica* 14: 122–77.

Anadón, P. 1978. *El Paleógeno continental anterior a la transgresión biarritziense (Eoceno medio) entre los ríos Gaià y Ripoll (Provs. de Tarragona y Barcelona).* Ph.D. Thesis, University Barcelona. 267 pp.

——. 1990. Los Yesos de Cerezo (Mioceno Superior continental, provincia de Burgas). In F. Ortí and J. K. Salvany, eds., *Formaciones evaporiticas de la Cuenca del Ebro y cadenas periféncas, y de la zona de Levante.* pp. 127–31, ENRESA, University Barcelona.

Anadón P., Ll. Cabrera, J. Guimerà. and P. Santanach. 1985. Paleogene strike-slip deformation and sedimentation along the southeastern margin of the Ebro basin. In K. T. Biddle and N. Christie-Blick, eds., *Strike-slip Deformation, Basin Formation, and Sedimentation,* pp. 303–18. S.E.P.M. Special Publication No. 37.

Anadón, P., F. Colombo, M. Esteban, M. Marzo, S. Robles, P. Santanach, and L. L. Sole Sugrañes. 1979. Evolución tectonoestratigráfica de los Catalánides. *Acta Geològica Hispànica* 14: 242–70.

Anadón, P. and M. Marzo. 1986. Sistemas deposicionales eocenos del margen oriental de la Cuenca del Ebro: sector de Igualada-Montserrat. *XI Congreso Español de Sedimentologia, G.E.S. Barcelona, Spain. Excursion Guidebook:* 4.1–4.59.

Anadón, P., M. Marzo, O. Riba, A. Sáez, and J. Vergès. 1989. Fan-delta deposits and syntectonic unconformities in alluvial fan conglomerates of the Ebro Basin. In M. Marzo and C. Puigdefàbregas, eds., *Guidebook Series of the 4th International Conference on Fluvial Sedimentology,* 100 pp.

Andreieff, P., R. Anglada, G. Carbonnel, F. Catzigras, C. Cavelier, J.-J. Chateauneuf, E. Colomb, C. Glintzboeckel, C. Jacob, J. Lay, A. l'Homer, L. Lezaud, C. Lorenz, H. Mercier, and A. Parfenoff. 1972. Contribution à l'étude de la Aquitanien: La coupe de Carry-Le-Rouet (Bouches-du-Rhône, France). *Bull. Bur. Rech. Géol. Min.* (2), section 1, no. 4, Ve Congrès de Néogène Méditerranéen, III: 1–135.

Angelier, J. and F. Bergerat. 1983. Systèmes de contrainte et extension intra-continentale. *Bull. Centres Rech. Explor. Prod. Elf-Aquitaine* 7: 137–47.

Ansorge, J., D. Emter, K. Fuchs, J. P. Lauer, E. Peterschmidt, and C. Prodehl. 1973. Seismic refraction investigations in the central and southern part of the Rhine Graben. *Approaches to Taphrogenesis.* Schweizerbart, Stuttgart. 460 pp.

Appia-Lamache, C. 1984. *Le Trias évaporitique du Sud-Est de la France: du bassin profond salifère à la bordure cévenole détritique et sulfatée (Largentière, Lodevois)*. Thèse 3ème cycle. Univ. Paris-Sud (Orsay). 159 pp.

Arbey, F. 1980. Les formes de la silice et l'identification des évaporites dans les formations silicifiées. In Les évaporites: mécanismes, diagenèse, et applications. *Bull. Cent. Rech. Explor. Prod. Elf Aquitaine* 4: 310–65.

Arthaud, F. and Ph. Matte. 1975. Les décrochements tardihercyniens du sud-ouest de l'Europe: Géometrie et essai de reconstitution des conditions de la deformation. *Tectonophysics* 25: 139–71.

Arthaud, F., F. Megard, and M. Seguret. 1977. Cadre tectonique de quelques bassins sédimentaires. *Bull. Centres Rech. Explor. Prod. Elf-Aaquitaine* 1: 147–88.

Arthaud, F., M. Ogier, and M. Seguret 1980/1981. Géologie et géophysique du golfe du Lion et de sa bordure nord. *Bull. Bur. Rech. Geol. Min.* 3: 175–93.

Arthaud, F. and M. Seguret. 1981. Les structures pyrénéennes du Languedoc et du Golfe du Lion (Sud de la France). *Bull. Soc. Géol. France,* 23: 51–63.

Arthurton, R. S. 1973. Experimentally produced halite compared with Triassic layered halite-rock from Cheshire, England. *Sedimentology* 20: 145–60.

Aubouin, J. and G. Mennessier. 1963. Essai sur la structure de la Provence. In *Livre à la mémoire du Professeur P. Fallot*. Mém. h. s., Soc. Géol. France 2: 45–98.

Augier, C. 1980. Diagraphies et matrices complexes. *Bull. Cent. Rech. Explor. Prod. Elf Aquitaine* 4: 143–77.

Ayora, C., J. Garcia Veigas and J.-J. Pueyo.1994. X-ray microanalysis of fluid inclusions and its application to the geochemical modeling of evaporite basins. *Geochimica et Cosmochimicha Acta,* 58:43–55.

Baranyi, I., H. Lippolt, and W. Todt. 1976. Kalium-Argon-Altersbestimmungen an tertiären Vulkaniten des Obberheingraben-Gebietes: II. Die Alterstraverse vom Hegau nach Lothringen. *Oberrhein. geol.* 25: 41–62.

Barbreau, A., J. Bodelle, P. Escalier des Orres, B. Mohammadioun, J. Vogt, and C. Weber. 1976. Essai de confrontation de la sismicité historique et de données structurales nouvelles en Provence occidentale et centrale. *Bull. Soc. Géol. France* 5: 1111–18.

Baudrimont, A. F. and P. Dubois. 1977. Un bassin mésogéen dans le domaine péri-alpin: le Sud-Est de la France. *Bull. Centres Rech. Explor. Prod. Elf-Aquitaine,* 1: 261–308.

Bellaiche, G., F. Irr, and M. Labarbarie. 1976. Découverte de sédiments marins fini-oligocène-aquitanien au large du massif des Maures (canyon des Stoechades). *C.R. Acad. Sci. (Paris)* 283: 319–22.

Bellon, H. 1981. Chronologie radiométrique (K-Ar) des manifestations magmatiques autour de la Méditerranée occidentale entre 33 et 1 Ma. In F. C. Wezel, ed., *Sedimentary Basins of Mediterranean Margins.* pp. 341–60. Tecnoprint, Bologna.

Bergerat, F. 1977a. La fracturation de l'avant-pays jurassien entre les fossés de la Saône et du Rhin. Analyse et essai d'interprétation dynamique. *Rev. Géogr. phys. Géol. dyn.* 4: 325–37.

——. 1977b. Le rôle des décrochements dans les liaisons tectoniques entre le fossé de la Saône et du Rhin. Analyse et essai d'interprétation dynamique. *C.R. Somm. Soc. Géol. France,* 4: 195–98.

——. 1981/1982. Le couloir rhodanien au Paléogène: analyse de la fracturation et interprétation cinématique régionale. *Rev. Géol. dyn. Géogr. phys.* 23: 329–43.

——. 1983. Paléocontraintes et évolution tectonique paléogène du fossé rhénan. *C.R. Acad. Sci. (Paris)* 297: 77–80.

——. 1985. *Déformations cassantes et champs de contrainte tertiaires dans la plate-forme européenne.* Thèse Doct. Etat, Mém. Sc. Terre Univ. Pierre et Marie Curie, Paris 85–07: 317 pp.

——. 1986. Apports et résultats de l'analyse microstructurale dans l'étude de la Bresse chalonnaise (fossé de la Saône): tectonique cassante et paléo-champs de contrainte au tertiaire. *Bull. Inf. Géol. Bass. Paris* 23: 17–22.

Bergerat, F., M. Cazes, B. Damotte, S. Guellec, J. L. Mugnier, F. Roure, and C. Truffert. 1989. Les structures distensives en Bresse d'après les données du profil sismique Jura-Bresse (programme ECORS). *C.R. Acad. Sci. Paris* 309: 325–32.

Biju-Duval, B., J. Letouzey, and L. Montadert. 1978. Structure and evolution of the Mediterranean basins. In K. J. Hsü and L. Montadert et al. eds., *Report Deep Sea Drilling Project, Leg 42 A,* (I): pp. 951–84. Washington, U.S. Government Printing Office.

Birnbaum, S. J. 1976. *Nonmarine evaporite and carbonate deposition, Ebro Basin, Spain.* Ph.D. Thesis, Cambridge University, 148 pp.

Birnbaum, S. J. and M. Coleman. 1979. Source of sulphur in the Ebro Basin (Northern Spain). Tertiary nonmarine evaporite deposits as evidenced by sulphur isotopes. *Chemical Geology* 25: 163–68.

Blanc-Valleron, M.-M. 1981. Les Facies Calcaires du Lutêtien a *Planorbis pseudoammonius* dans le Bas-Languedoc: Argilogenése et Silicifications Associées aux Encroûtements Calcaires. Unpubl. Thesis, l'Université Louis Pasteur, Institut de Geologie Strasbourg, 43 pp.

——. 1986a. Enseignements géologiques tirés de l'étude systématique de la matière organique d'un bassin salifère. Le bassin potassique de Mulhouse (Alsace, France). *Comte Rendu de Académie Sciences, Paris,* (II), 302: 825–30.

——. 1986b. La sédimentation salifère paléogène du bassin potassique de Mulhouse: premiers apports de l'étude géochimique de la matière organique contenue dans les passées argileuses. *Docum. Bur. Rech. Geol. Min.,* 110: 111–26.

——. 1990/91. *Les formations paléogènes évaporitiques du bassin potassique de Mulhouse et des bassins plus septentrionaux d'Alsace.* Thèse Doct. Sci., Univ. Strasbourg, 1990 et Doc. *Mém.Bur. Rech. Geol. Min.* No. 204, 1991, 350 + 136 p.

Blanc-Valleron, M.-M., A. Foucault, and E. Gannat. 1989. Le contrôle climatique, orbital et solaire de la sédimentation évaporitique: exemple de la base du Salifère supérieur (Oligocène inférieur) du bassin de Mulhouse (Sud du fossé rhénan, France). *Comte Rendu de Académie Sciences, Paris,* (II), 308: 435–41.

Blanc-Valleron M.-M. and E. Gannat. 1985. Cartographie de subsurface du Salifère supérieur du bassin potassique de Mulhouse (Oligocène, Alsace). *Bull. Soc. géol. France* 6: 823–36.

Blanc-Valleron, M.-M., J. P. Gely, M. Schuler, F. Dany, and M. Ansart. 1991. La matière organique associée aux évaporites de la base du Sel IV (Oligocène inférieur) du bassin de Mulhouse (Alsace, France). *Bull. Soc. géol. France, Paris,* 162: 113–22.

Bless, J. M., R. Conil, P. Defourny, E. Groessens, L. Hance, and M. Hennebert. 1980. Stratigraphy and thickness variations of some Struno-Dinantian deposits around the Brabant massif. *Meded. Rijks Geol. Dienst.* 32(1): 56–65.

Bock, E. 1961. On the solution of anhydrous calcium sulfate and of gypsum in concentrated solutions of sodium chloride at 25°, 30°, 40°, and 50°C. *Canadian Journal of Chemistry* 3: 1746–51.

Bodergat, A. M. 1974. *Les Microcodiums, milieux et modes de développment.* Thèse de spécialité, Lyon, no. 388, 132 pp.

Bodeur, Y. 1976. Evaluation de l'amplitude du décrochement cévenol par le décalage des faciès récifaux portlandiens des environs de Ganges (Hérault). *C. R. Acad. Sci. Paris:* 282: 961–63.

Boeke, H. E. 1908. Über das Krystallisationsschema der Chloride, Bromide, Jodide von Natrium, Kalium und Magnesium sowie über das Vorkommen des Broms und das Fehlen von Jod in den Kalisalzlagerstätten. *Zeitschrift fur Krystallographie* 45: 346–91.

Boigk, H. and H. Schöneich. 1974. The Rhinegraben: geologic history and tectonic activity. In J. H. Illies and K. Fuchs, eds., *Approaches to Taphrogenesis.* Inter-Union Commission on Geodynamics, Scientific Report 8, pp. 60–71. Schweizerbart, Stuttgart.

Boillot, G. 1980/81. De la subduction à la collision: l'exemple des Pyrénées. *Bulletin Bur. Rech. Geol. Min. Mem.* 2: 93–101.

Bomer, B. 1954. Trois aspects du contact entre Monts Celtibériques occidentaux et Bassin de l'Ebre. *Bulletin Asso. Géogr. Fran.* 239/240: 35–41.

Borchert, H. 1969. Principles of oceanic salt deposition and metamorphism. *Bulletin Geological Society America* 80: 821–64.

Borchert, H. and R. O. Muir. 1964. *Salt Deposits: The Origin, Metamorphism, and Deformation of Evaporites.* Van Nostrand, London, 338 pp.

Braitsch, O. 1966. Bromine and Rubidium as indicators of environment during sylvite and carnallite deposition of the Upper Rhine Valley Evaporites. In: J. L. Rau, ed., *Second Symposium on Salt,* pp. 293–301, Northern Ohio Geological Society, Cleveland, Ohio.

——. 1971. *Salt Deposits: Their Origin and Composition.* Springer-Verlag, New York, 297 pp.

Braitsch O. and A. G. Herrmann. 1963. Zur Geochemie des Broms in salinaren Sedimenten Teil. I: Experimentelle Bestimmung der Br Verteilung in verschiedenen natürlichen Salzsystemen. *Geochimica et Cosmochimica Acta* 27: 361–91.

Brinkmann, R. and H. Lögters. 1968. Diapirism in Western Pyrenees and Foreland, Spain. In J. Braunstein and G. D. O'Brien, eds., *Diapirism and Diapirs,* pp. 275–92. American Association Petroleum Geologists, Memoir 8.

Brookins, D. G. 1989. *Eh-pH Diagrams for Geochemistry.* Berlin, Springer-Verlag. 176 pp.

Brousse, R. and H. Bellon. 1983. Réflexions chronologiques et pétrologiques sur le volcanisme associé au développement des rifts de France. *Bull. Centres Recherch. Explor. Prod. Elf-Aquitaine* 7: 409–24.

Burbank, D. W., J. Vergés, J. A. Muñoz, and P. Bentham. 1992. Coeval hindward- and forward-imbricating thrusting in the south-central Pyrenees, Spain: Timing and rates of shortening and deposition. *Bull. Geol. Soc. Amer.,* 104: 3–17.

Bureau de Recherche Géologique et Minières. 1972. Contribution à l'étude de la Aquitanien: La coupe de Carry-Le-Rouet (Bouches-du-Rhône, France). *Bull. Bur. Rech. Géol. Min.* (2), section 1, no. 4, Ve Congrès de Néogène Méditerranéen, III: 1–135.

Burg, P. and A. Etchecopar. 1980. Détermination des systèmes à contraintes liés à la tectonique cassante au coeur du Massif Central français: la région de Brina (Haut-Allier). *C.R. Acad. Sci. Paris* 290: 397–400.

Busquets, P. 1981. *Estratigrafia i sedimentología del Terciari pre-pirinenc entre els rius Llobregat i Freser-Ter.* Ph.D. Thesis, University Barcelona, 190 pp.

Busquets, P., I. Matias, and N. Solé de Porta. 1987. Aportaciones sedimentológicas y micropaleontológicas al conocimiento de las formaciones luteciens es del manto del Cadí. Prepirineo catalán. *Revista Investigacions Geològiques* (Barcelona), 42/43: 77–88.

Busquets, P., F. Ortí, J. J. Pueyo, O. Riba, L. Rosell, A. Saez, R. Salas, and C. Taberner. 1985. Evaporite deposition and diagenesis in the Saline (Potash) Catalan Basin, upper Eocene. In *VI European Regional Meeting, I.A.S. Lérida, Spain, Excursion Guidebook:* 11–59.

Busquets, P., S. Reguant, C. Santisteban, and C. Taberner. 1986. Depósitos de plataformas mixtas siliciclásticas-carbonáticas del Eoceno medio-superior en el área de Vic-Ripoll-Berga. In *XI Cong. Esp. Sedimentología, G.E.S., Barcelona. Excursion Guidebook:* 5.1–5.28.

Busson, G. 1974. Sur les evaporites marines: sites actuels ou recents de depots d'evaporites et leur transposition dans les series du passe. *Rev. Geogr. phy. Geol. dyn.*, ser. 2, 16: 189–208.

——. 1978. L'unité des faciès confinés en milieu de plateforme carbonatée. In *Livre jubilaire J. Flandrin,* pp. 87–112. Docum. Lab. Géol. Fac. Sci. Lyon, Special Publication no. 4, Lyon.

——. 1979. Couches laminées riches en matière organique et précédant les roches salines: les enseignements d'un enchainement de faciès. In *Milieux de sedimentation de schistes bitumineux et de black shales,* pp. 5–18. Docum. Lab. Géol. Fac. Sci. Lyon no. 75, Lyon.

——. 1980. Les grandes cuvettes evaporitiques en milieu detritique: comment elles se creusent, comment elles se remplissent. *Bull. Cent. Rech. Explor. Prod. Elf Aquitaine* 4: 557–88.

——. 1988. Relations entre les types de dépôts évaporitiques et la présence de couches riches en matière organique (roches-mères potentielles). *Revue Institut Française du Pétrole,* 43: 181–215.

Busson, G., A. Cornée, N. Dulau, J. C. Fontes, D. Geisler, D. Gouleau, J. Jaccard, J. C. Landry, D. Noël, J. P. Perthuisot, C. Pierre, C. Poumot, J. Tetard, J. C. Thomas, M. Thomas, N. Trauth, and L. Zaninetti. 1982. Données hydrochimiques, biologiques, isotopiques, sédimentologiques et diagénétiques sur les marais salants de Salin-de-Giraud (Sud de la France). *Géologie Méditerranéene* IX, 4: 591 pp.

Busson, G., M. J. Lienhart, and C. Appia. 1984. Domaine rhodanien, Bas-Dauphine, Basse-Ardeche, Bas-Rhone, et Languedoc. In S. Debrand-Passard, S. Courbouleix, and M. J. Lienhart, eds., *Chapitre Trias, Synthese Geologique du Sud-Est de la France,* pp. 95–99. Bur. Rech. Geol. Min. Mém. no. 125, Orleans.

Busson, G. and D. Noël. 1972. Sur la constitution et la genèse de divers sédiments finement feuilletés ("laminites") à alternances de calcaire et de matière organique ou argileuse. *C. R. Acad. Sci.* 274 (D): 3172–75.

Busson, G. and J. P. Perthuisot. 1977. Interet de la sebkha el Melah (sud tunisien) pour l'interpretation des series evaporitiques anciennes. *Sedim. Geol.*, 19: 139–64.

Butler, G. P. 1969. Modern evaporite deposition and geochemistry of coexisting brines, the Sabkha, Trucial Coast, Arabian Gulf. *Journal of Sedimentary Petrology* 39: 70–81.

——. 1970. Holocene gypsum and anhydrite of the Abu Dhabi sabkha, Trucial Coast: An alternative explanation of origin. In J. L. Rau and L. F. Dellwig, eds., *Third Symposium on Salt* (1), pp. 120–52. Northern Ohio Geol. Soc., Cleveland.

Butler, G. P., P. M. Harris, and C. G. St.C. Kendall. 1983. Recent evaporites from the Abu Dhabi coastal flats. In C. R. Handford, R. G. Loucks, and G. R. Davies eds., *Depositional and Diagenetic Spectra of Evaporites: A Core Workshop,* SEPM Core Workshop No. 3: 33–64. Tulsa, Oklahoma.

Cabrera, L. 1983. *Estratigrafía y sedimentología de las formaciones lacustres del tránsito Oligoceno-Mioceno del SE de la Cuenca del Ebro.* Ph.D. Thesis, University Barcelona, 443 pp.

Cabrera, L., F. Colombo, and S. Robles. 1985. Sedimentation and tectonics interrelationships in the Paleogene marginal alluvial systems of the SE Ebro Basin: Transition from alluvial to shallow lacustrine environments. In J. Rosell and M. Milà, eds., *VI European Regional Meeting, I.A.S., Lérida, Spain. Excursion Guide-Book:* pp. 394–492.

Cámara, P. and J. Klimowitz. 1986. Interpretación geodinámica de la vertiente centro-occidental surpirenaica. *Estudios geológicos* 41: 391–404.

Canérot, J. 1974. *Recherches géologiques aux confins des Chaînes Ibérique et Catalane (Espagne).* Trabajos de Tesis, ENADIMSA, ser. 5, 4: 517 pp.

Casagrande, L. 1989. Evolution tectono-sédimentaire mio-pliocène du bassin du Comtat Venaissin (vallée du Rhône). *Bull. Bur. Rech. Geol. Min., Géologie de la France* 3: 13–20.

Casagrande, L., J. Andrieux, and J. L. Morel. 1989. Le massif de Suzette (Vaucluse): L'inversion tectonique d'un graben oligocène. *Bull. Bur. Rech. Geol. Min. Mem., Géologie de la France* 3: 3–12.

Casanova, J. and D. Nury. 1989. Biosédimentologie des stromatolites fluvio-lacustres du fossé oligocène de Marseille. *Bull. Soc. géol. France* (8)5: 173–84.

Casas, S. E. and T. K. Lowenstein. 1989. Diagenesis of salt pan halite: comparison of petrographic features of modern, Quaternary, and Permian halites. *Journal Sedimentary Petrology* 59: 724–39.

Castillo-Herrador, F. 1974. Le Trias évaporitique des bassins de la vallée de l'Ebre et de Cuenca. *Bull. Soc. Géol. France* 6: 666–76.

Caus, E. 1973. Aportaciones al conocimiento del Eoceno del anticlinal de Oliana (provincia de Lérida). *Acta Geológica Hispánica* 8: 7–10.

Cavelier, C. 1979. La limite Eocène-Oligocène en Europe occidentale. *Sci. Géol., Mém.*, Strasbourg, 54, 280 p.

——. 1984. Paléogène. In S. Debrand-Passard, ed., *Synthèse géologique du Sud-Est de la France.* Bur. Rech. Geol. Min. Mém., Orléans, no. 125(1): 389–468.

Cavelier, C. and Pomerol, C. 1986. Stratigraphy of the Paleogene. *Bull. Soc. géol. France,* Paris, (8)2: 255–65.

Cerchi, A. and L. Montadert. 1982. The Oligo-Miocene rift of Sardinia: a model for the early history of the Western Mediterranean Basin. *Nature* 298: 736–39.

Chamley, H., G. Giroud d'Argoud, and C. Robert. 1977. Genèse des smectites messiniennes de Sicile: implications paléoclimatiques. *Géologie Méditerranéenne,* IV, 4: 371–78.

Charollais, J., P. A. Hochuli, H. J. Oertli, K. Perch-Nielsen, M. Toumarkine, F. Roegl, and J. L. Pairis. 1980. Les Marnes a foraminiferes et les Schistes a *Meletta* des chaines subalpines septentrionales (Haute-Savoie, France). *Eclogae Geol. Helv.* 73: 9–69.

Châteauneuf, J.-J. 1977. Étude palynologique de l'Oligocène du Bassin de Marseille. *Géologie Méditerranéenne,* IV, 4: 37–46.

Châteauneuf, J.-J. and D. Nury. 1994. La flore de l'Oligocène de Provence méridionale: Implications stratigraphiques, climatiques. *Géologie de la France* 2: 43–55.

——. 1995. La flore de l'Oligocène de Provence méridionale: Implications stratigraphiques, environmentales et climatiques. *Géologie de la France,* n°. 2.

Chauve, P., R. Enay, P. Furk, C. Sittler, and J. B. Edel. 1980. Vosges, Fossé rhénan, Bresse, Jura. In C. Lorenz, coordinator, *Géologie de pays européens,* 28ème CGI. Dunod éd., Paris: 353–430.

Chauve, P. and Perriaux, J. 1988. Le Jura. In J. Debelmas: *Géologie de la France* 2: les chaînes plissées du cycle alpin et leur avant-pays. Doin éd.: 443–64.

Choignard, J. 1965. *Stratigraphie et sédimentologie du Paléogène de Bresse: Résultats et interprétation.* Unpubl. thesis, University of Dijon (France), 142 pp.

Choukroune, P. and Team ECORS. 1989. The Ecors Pyrenean deep seismic profile reflection data and the overall structure of an orogenic belt. *Tectonics* 8: 23–39.

Chowns, T. M. and J. E. Elkins. 1974. The origin of quartz geodes and cauliflower cherts through the silicification of anhydrite nodules. *Journal Sedimentary Petrology* 44: 885–903.

Clavell, E., A. Martínez, and J. Vergés. 1988. Morfologia del basement del Pirineu oriental: evolució i relació amb els mantells de corriment. *Acta Geològica Hispànica* 23: 129–40.

Cody, R. D. 1979. Lenticular gypsum: occurences in nature, and experimental determinations of soluble green plant material on its formation. *Journal Sedimentary Petrology* 49: 1015–28.

Cody, R. D. and A. M. Cody. 1988. Gypsum nucleation on crystal morphology in analog saline terrestrial environments. *Journal Sedimentary Petrology* 58: 247–55.

Collomb, M. 1986. Les évaporites paléogène du châinon de la Nerthe (St.-Pierre-les-Martigues, B.d.R.). Unpublished thesis, Université Claude Bernard, Lyon I, 35pp.

Colombo, F. 1980. *Estratigrafía y sedimentología del Paleógeno continental del borde meridional occidental de los Catalánides.* Ph.D. Thesis, University Barcelona, 609 pp.

Colombo, F., J. L. Cuevas, and L. Mercade. 1986. Las facies Garumnienses del flanco sur del sinclinal de Ager. Análisis sedimentológico. *XI Cong. Esp. Sedimentologia, G.E.S. Barcelona, Abstracts book:* p. 50.

Colomer, M. and P. Santanach. 1988. Estructura y evolución del borde sur-occidental de la fosa de Calatayud-Daroca. *Geogaceta* 4: 29–31.

Conrad, G. 1969. L'evolution continentale post-hercynienne du Sahara algerien (Saoura, Erg Chech-Tanezrouft, Afnet-Mouydir). *Edit. Centr. Nation. Rech. Scient.* no. 10, 517 p., Paris.

——. 1975. Modele de sedimentation evaporitique continentale, actuelle, en zone aride nord-saharienne (Algerie): Comparison avec le Quaternaire. *Ninth Cong. of Sedim.* 2: 29–33, Nice.

Contini, D. and N. Theobald. 1974. Relations entre le fossé rhénan et le fossé de la Saône: Tectonique des régions sous-vosgiennes et préjurassiennes. In J. H. Illies and K. Fuchs, eds., *Approaches to Taphrogenesis.* Schweitzerbart Verlag: Stuttgart, pp. 310–321.

Cornée, A. 1982. Bactéries des saumures et des sédiments des marais salants de Salin-de-Giraud (Sud de la France). *Géol. Médit* IX, 4: 369–89.

——. 1984. Etude préliminaire des bactéries des saumures des sédiments des salins de Santa Pola (Espagne). Comparaison avec les marais salants de Salin-de-Giraud (Sud de la France). *Revista del Investigaciones Geologicas,* Barcelona 38/39: 109–22.

Costa, J. M. 1989. *Turbidites de Ripoll. Relació amb llurs plataformes.* Ph.D. Thesis, University Autónoma of Barcelona, Bellaterra, 154 pp.

Costa, L. I. and C. Downie. 1976. The distribution of the dinoflagellate *Wetzeliella* in the Paleogene of northwestern Europe. *Paleontology* 19: 591–614.

Courjault-Rade, P. 1982. *Environnement sédimentaire des minéralisations (Pb, Zn) liées aux strates de la plateforme carbonatée du Cambrien inférieur de la Montagne Noire: sédimentologique Aspect et métallogénique.* Thèse 3ème cycle, Univ. Toulouse II.

Courtot, C., E. Gannat, and E. Wendling. 1972. Le bassin potassique de Mulhouse et les environs. Etude du Tertiaire. *Sci. géol. Bull.* 25: 69–91.

Cravatte, J., P. Dufaure, M. Prim, and S. Rouaix. 1974. Les sondages du Golfe du Lion; stratigraphie et sédimentologie. *Notes Mém. C.F.P.* 2: 209–74.

Crusafont, M., J. Truyols, and O. Riba. 1966. Contribución al conocimiento de la estratigrafía del Terciario continental de Navarra y La Rioja. *Not. y Com. del IGME* 90: 53–76.

Cuenca, G., J. I. Canudo, C. Laplana, J. A. Andrés. 1992. Bio y cronoestratigrafía con mamíferos en la Cuenca Terciaria del Ebro: ensayo do síntesis. *Acta Geologica Hispanica* 27 (1/2): 127–43.

Curial, A. 1986. Les dissolutions dans la partie supérieure du Salifère paléogène de Bresse (SE de la France): chronologie, extension et mécanismes. *Rev. Géol. dyn. et Géogr. Phys.* 27: 225–35.

——. 1986/87. *La sédimentation salifère et suprasalifère du Paléogène bressan (France): comparaisons entre les données diagraphiques et lithologiques. Etude diagraphique du champ d'Etrez et synthèse de bassin.* Thèse Doctorat, Univ. Lyon, Document U.A. 1209, 1986, Paris, 11: 251 p. and Docum. Lab. géol. Lyon, 1987, no. 100: 192 p.

——. 1987. Le détritisme durant la sédimentation salifère paléogène en Bresse méridionale. *Géologie Alpine* 13: 391–95.

——. 1988. Multiple Applications of Dipmeter Curves for Analyzing Evaporite Formations: Examples from Paleogene of Bresse Trough, France. *American Assoociation Petroleum Geologists Bulletin* 72: 1323–33.

Curial, A., D. Dumas, and R. Moretto. 1988. Evolution comparée et corrélation entre deux bassins salifères paléogènes du rift ouest européen: les fossés de Bresse et de Valence (France, Sud-Est). *C. R. Acad. Sci.* 306: 655–61.

Curial, A. and R. Moretto. 1985. Le Paléogène salifère de Bresse (département de l'Ain, SE de la France): étude diagraphique et sédimentologique. *Documents du GRECO 52 (CNRS, Paris)* 5: 147–50.

——. 1988. La matière organique du bassin salifère paléogène de Bresse: caractérisation par pyrolyse Rock-Eval, relation avec l'uranium et détection par spectrométrie du rayonnement gamma naturel. In G. Busson (ed.), *Evaporites et hydrocarbures, Mém. Mus. nat. Hist. nat.* (C), 55: 71–94.

Curial, A., R. Moretto, D. Dumas, and G. Dromart. 1990. Organic Matter and Evaporites in the Paleogene West European Rift: The Bresse and Valence Salt Basins (France). In A. Y. Huc (ed.), *Deposition of Organic Facies, A.A.P.G. studies in Geology,* 30: 119–32.

Curnelle, R. 1983. Evolution structuro-sédimentaire du Trias et de l'Infra-Lias d'Aquitaine. *Bull. Centre Rech. Explor. Prod. Elf-Aquitaine* 7: 69–99.

Curnelle, R., P. Dubois, and J. C. Seguin. 1982. The Mesozoic-Tertiary evolution of the Aquitaine basin. *Phil. Trans. R. Soc. London* 305: 63–84.

Debrand-Passard, S., S. Courbouleix, and M. J. Lienhardt. 1984. Synthèse Géologique du Sud-Est de la France. 1: Stratigraphie et paléogéographie. *Bur. Rech. Geol. Min. Mém.* 125: 615 p.

de Brun, P. and C. Châtelet. 1923. Sur la découverte du Bartonian sur le revers nord les Alpilles (BdR). *Bull. Geol. Soc. France* (4), 23: 109–12.

Degens, E. T. and D. A. Ross. 1974. *The Black Sea: Geology, Chemistry, and Biology.* American Association Petroleum Geology Memoir #20, 633 pp., Tulsa.

de la Peña J. A., J. M. Garcia Ruiz, R. Marfil, and M. Prieto. 1982. Growth features of magnesium and sodium salts in a recent playa lake of La Mancha (Spain). *Estudios geol.* 38: 245–57.

Dellwig, L. F. 1955. Origin of the Salina Salt of Michigan. *Journal Sedimentary Petrology* 25: 83–110.

——. 1963. Environment and mechanics of deposition of the Permian Hutchinson salt member of the Wellington shale. In *First Symposium on Salt,* Cleveland, pp. 74–85.

——. 1968. Significant features of deposition in the Hutchinson salt, Kansas, and their interpretations. In R. B. Mattox, W. T. Holser, H. Odé, W. L. McIntyre, N. M. Short, R. E. Taylor, and D.C. Van Siclen, eds., *Saline Deposits.* Spec. Paper *Geol. Soc. Amer.* 88: 421–27.

Del Valle, J. 1978. *Mapa Geológico de España, E 1:50000, 2a ser. 1a ed., Hoja no 141 "Pamplona" y Memoria Explicativa.* I.G.M.E. Serv. Pub. Min. Industria. Madrid.

De Magnée, I., A. Delmer, and M. Cordonnie. 1986. La dissolution des évaporites du Dinantien et ses conséquences. *Bull. Soc. Belge Géol.* 2/3: 213–20.

Demarcq, G. 1970. Etude stratigraphique du Miocène rhodanien. *Bur. Rech. Geol. Min. Mém.* 61: 257 pp.

Demarcq, G., A. Lefavrais-Raymond, and J. Perriaux. 1984. Néogène Bresse (Dombes et Lyonnais). *Bur. Rech. Geol. Min. Mém.* 125: 483–87.

Destombes, J. P. 1961. Description géologique du bassin oligocène de Manosque-Forcalquier (Lubéron oriental). *Bull. Serv. Carte géol. France* 266: 461–568.

Dewey, J. F., M. L. Helman, E. Turco, D. W. H. Hutton, and S. D. Knott. 1989. Kinematics of the western Mediterranean. In M. P. Coward, D. Dietrich, and R. G. Park, eds., *Alpine Tectonics.* pp. 265–83, Geological Society Special Publication 45. Oxford: Blackwells.

Doebl, F. 1967. The Tertiary and Pleistocene sediments of the northern and central part of the Upper Rhine graben. In J. P. Rothe and K. Sauer, eds., *The Rhine Graben Progress report. Abh. Geol. Ladesamt Baden-Wurttemberg* 6: 48–62.

——. 1970. Die tertiären und quartären Sedimente des südlichen Rheingrabens. In J. H. Illies and S. Mueller, eds., *Graben Problems,* Schweitzerbart. éd., Stuttgart, pp. 56–66.

Doebl, F., C. Müller, M. Schuler, C. Sittler, and H. Weiler. 1976. Les Marnes à Foraminifères et les Schistes à Poissons de Bremmelbach (Bas-Rhin). Etudes sédimentologiques et micropaléontologiques. Reconstitution du milieu au début du Rupélien dans le Fossé Rhénan. *Sci. Géol., Bull., Strasbourg,* 29: 285–320.

Dorel, J., D. Pradoux, and D. Fourvel. 1990. Sismicité du bassin oligocène d'Ambert dans le Massif Central français. *C. R. Acad. Sci. Paris* 310: 749–56.

Doudoux, B., G. Mercier de Lepinay, and M. Taredy. 1982. Une interprétation nouvelle de la structure des massifs subalpins savoyards (Alpes Occidentales): nappes des charriges oligocènes déformations superposées. *C.R. Acad. Sci., Paris,* 295(1): 63–68.

Doveton, J. H. 1986. *Log Analysis of Subsurface Geology: Concepts and Computer Methods.* New York, Wiley, 273 pp.

Dronkert, H. 1977. The evaporites of the Sorbas basin. *Revista del Instituto de Investigaciones Geológicas, Dip. Prov. Barcelona* 32: 55–76.

Dubois, P. and P. Umbach. 1974. A propos du Trias de deux bassins sédimentaires français: le bassin de Paris et le bassin du Sud-Est. *Bull. Soc. Géol. France* 6: 696–707.

Duclos, P. 1967. *Géologie et minéralisation uranifères de la plaine tertiaire du Forez (Massif Central français).* Thèse Univ. Clermont-Ferrand and rapport CEA-R 3117: 104 pp.

Dumas, D. 1984. Les évaporites du bassin oligocène de Narbonne-Sigean (Aude, Sud de la France). Conditions de dépôt et de diagénèse. *Rev. Géol. dyn. Géogr. phys.* 25: 229–42.

——. 1986. La sédimentation paléogène dans le fossé salifère de Valence (SE de la France): géométrie des dépôts et aperçu des mécanismes sédimentaires. *Géologie de la France* 4: 399–416.

——. 1988. *Le Paléogène salifère du bassin de Valence (Sud-Est de la France): Géométrie et sédimentologie des dépôts, synthèse de bassin.* Thèse Doc. Univ. Claude Bernard, Lyon: 293 pp.

Duplaix, S. and S. Guillaume. 1962. Application de la méthode des minéraux lourds à l'étude des formations crétacées et tertiaires du Jura. *C. R. Acad. Sci.* 254: 1843–45.

——-. 1963. Etude stratigraphique et minéralogique de formations tertiaires du Jura. *Rev. Géogr. Phys. et Géol. dyn.* 2(5): 37–54.

Duplaix, S., S. Guillaume, and A. Lefavrais-Raymond. 1965. Le Tertiaire de la Bresse, stratigraphie et minéralogie, comparaison avec les régions voisines. *Rev. Géogr. Phys. et Géol. dyn.* 2(7): 135–48.

Durand, J.-P. 1984. Paléogène: Descriptions régionales Basse Provence. In S. DeBrand et al., *Synthèse du Couloir Rhodanien. Bur. Rech. Geol. Min. Mém.* No. 125: 426–29.

Duringer, Ph. 1988. *Les conglomérats des bordures du rift cénozoïc rhénan. Dynamique sédimentaire et contrôle climatique.* Thèse Doct. Sci. Nat., Strausbourg, 278 pp.

Edel, J. B. 1978. Contribution du magnétisme et de la gravimétrie à la connaissance du socle varisque dans l'Est de la France et le Sud de l'Allemagne. *Sci. géol. Bull.* 31: 45–55.

Edel, J. B., K. Fuchs, C. Gelbke, and C. Prodehl. 1975. Deep structure of the Southern Rhine Graben area from seismic refraction investigation. *J. Geophys.* 41: 333–56.

Emre, T. and G. Truc. 1978. Mise en évidence d'un contact discordant Oligocène-Trias dans le massif de Suzett. Implications tectoniques et conséquences sur l'origine des évaporites ludiennnes du bassin de Mormoiron (Vaucluse). *Géol. alpine* 54: 17–23; 87–139.

ENRESA 1987. *Inventario nacional de formaciones geológicas favorables para el almacenamiento de residuos radiactivos de alta actividad (Fase 11): Region del Ebro.* Madrid (Internal report) 64 pp.

Espitalie, J. 1979. Charakterisierung der organischen Substanz und ihres Reifegrades in vier Bohrungen des mittleren Oberrhein-Grabens sowie Abschätzung der paläogeothermischen Gradienten. *Fortschr. Geol. Rheinl. Westfalen, Krefeld,* 27: 87–96.

Espitalie, J., G. Deroo, and F. Marquis. 1985. La parolyse Rock-Eval et ses applications. *Rev. Inst. France Pétrole,*

Rueil-Malmaison 40 (Part I): 563–79; 40 (Part II): 755–84.

——. 1986. La parolyse Rock-Eval et ses applications. *Rev. Inst. France Pétrole, Rueil-Malmaison,* 40 (Part III): 73–89.

Estévez, A. 1970. La estructure de la Garrotxa (Gerona) en el sector compredido entre el Cóme Negre y St. Joan les Fonts. Relaciones entre zócalo y cobertera. *Cuad. Geol. Univ. Granada,* 1/2: 122–32.

——. 1973. *La vertiente meridional del Pirineo Catalán al N del curso medio del rio Fluvià.* Ph.D. Thesis, University Granada, 514 pp.

Estop, E. 1974. *Estudio de la distribución de especies minerales y elementos menores en yacimientos de yeso de la provincia de Tarragona.* Graduation Thesis, University Barcelona. 129 pp.

Eugster, H. P. 1984. Geochemistry and sedimentology of marine and non marine evaporites. *Eclogae Geol. Helv.* 77: 237–48.

Eugster, H. P. and L. A. Hardie. 1978. Saline Lakes. In A. Lerman ed., *Lakes: Chemistry, Geology, Physics.* Springer-Verlag, Berlin: 237–93.

Fabre, P., J. Medus, and J. L. Pairis. 1986. Caractérisation de l'Eocène et de Oligocéne marins dans le chaines méridionales à l'Ouest de Gap (haut Alpes). *Eclog. Geol. Helv.* 79: 719–30.

Feist-Castel, M. 1977. Étude floristique et biostratigraphique des Charophytes dans les séries du Paléogène de Provence. *Géologie Méditerranéenne* (IV)2: 109–38.

Fernández-Nieto, C. and E. Galan. 1979. Mineralogia de los depositos de sales de Remolinos (Zaragoza). *Soc. Esp. Mineralogía,* Extra 1: 51–65.

Fertl, W. H. 1981. Openhole crossplot concepts: a powerful technique in well log analysis. *Journal Petroleum Technology* (March 1981): 535–49.

Fischer, M. W. 1984. Thrust tectonics in the North Pyrenées. *J. Structural Geology* 6: 721–26.

Fisk, H. N. 1959. Padre Island and the Laguna Madre flats, coastal South Texas. In R. J. Russel, ed., *2nd Coastal Geography Conference,* National Academy of Science National Research Council, pp. 103–51.

Fleury, E. 1909. *Le Sidérolithique suisse. Contribution à la connaissance des phénomènes d'altération superficielle des sédiments.* Mém. Soc. fribourgeoise Sci. nat., Fribourg, 6, 262 p.

Fontboté, J. M. 1954. Las relaciones tectónicas de la depresión del Vallès-Penedès con la cordillera prelitoral catalana y con la depresión del Ebro. *Boletín de la Real Soc. Española de Historia Natural,* vol. Homenaje al Prof. E. Hernández Pacheco, pp. 218–310.

Fontes, J. C., A. Filly, J. Gaudant, and P. Duringer. 1991. Origine continentale des évaporites paléogènes de Haute Alsace: arguments paléoécologiques, sédimentologiques et isotopiques. *Bull. Soc. géol. France,* 162(4): 725–37.

Fontes, J.-C., J. Gaudant, and G. Truc. 1980. Données paléoécologiques, teneurs en isotopes lourds et paléohydrologie du bassin gypsifère oligocène d'Aix-en-Provence. *Bull. Soc. géol. France,* 22(3):491–500.

Fourniguet, J. 1977. Mise en évidence de mouvements actuels verticaux dans le Sud-Est de la France par comparaison de nivellements successifs. *C. R. Somm. Soc. Géol. France* 5: 266–68.

——. 1984. Neotectonique. In Synthése geologique du Sud-Est de la France. *Bur. Rech. Geol. Min. Mém.* 125: 564–65.

——. 1987. Géodynamique actuelle dans le Nord et le Nord-Est de la France. Apport des comparaisons de nivellements. *Bur. Rech. Geol. Min. Mém.* 127: 90 pp.

Fourniguet, J., J. Vogt, and C. Weber. 1981. Seismicity and recent crustal movements in France. *Tectonophysics* 71: 195–216.

Franzen, J. L. 1968. *Revision der Gattung Palaeotherium CUVIER 1804 (Palaeotheriidae, Perissodactyla, Mammalia).* Thèse Doct., Univ. Freiburg, 181 p.

Fredet, J. M. 1987. *Tectonique et Sedimentation en Domaine Continental: Evolution du Bassin Paleogene d'Alès (Gard).* Ph.D. Dissertation, Université Claude Bernard, Lyon 1. 244 pp.

Freeman, T. 1962. Quiet water oölites from Laguna Madre, Texas. *Journal Sedimentary Petrology* 32: 475–83.

Gao, G., S. D. Hovorka, and H. Posey. 1990. Limpid dolomite in Permian San Andres halite rocks, Palo Duro Basin, Texas Panhandle: characteristics, possible origin, and implications for brine evolution. *Journal Sedimentary Petrology,* 60: 118–24.

García Veigas, J. 1988. *La formación evaporítica de la facies Garumniense de la Conca de Tremp (Lleida): petrología, geoquímica y sedimentología.* Graduation Thesis, University Autónoma of Barcelona, Bellaterra, 175 pp.

——. 1993. *Geoquimica de inclusiones fluidas en formaciones salinas. Microanlisis Cryo-SEM-EDS.* Ph.D. Thesis, Univ. Barcelona, 259 pp.

García Veigas, J., F. Ortí, and C. Fernández-Nieto. 1994. Modelo hidroquimico de sedimentacion de glauberita-halita: sondeo PURASAL, Formacion Yesos de Zaragoza (Mioceno Inferior, cuenca del Ebro). *Geogaceta,* 16: 136–39.

Garrido-Mejías, A. 1972. Sobre la colocación del manto de Pedraforca y sus consecuencias: una nueva unidad tectónica independiente, "el manto del Montsec." *Bol. Geol. Min.* 83: 242–48.

Gaudant, J. 1981. Contribution à la Paléoichthyologie continentale à la reconstitution des paléoenvironnements cénozoïques d'Europe occidentale: approche systématique, paléoécologique, paléogéographique, et paléoclimatologique. Thèse d'Etat, Univ. Pierre et Marie Curie, Paris: 368 pp.

Gaudant, J. and Guillevin, Y. 1979. Découverte de *Dopalis* (Smerdis) *macrurus* (póisson téleostéen) dans l'Oligocène du bassin de Valence (Drôme): implications stratigraphique et paléogéographique. *Bull. Mus. nat. Hist. nat., Paris,* 4 (C), 1: 25–33.

Gebelein C. D. and P. Hoffman. 1973. Algal origin of dolomite laminations in stromatolitic limestones. *Journal Sedimentary Petrology* 43: 603–13.

Geisler, D. 1979. Problèmes posés par la géochimie du brome dans les niveaux salifères du Keuper inférieur lorrain. *Sciences de la Terre,* 23(2): 23–38.

——. 1981. Genèse et évolution des gypses des marais salants de Salin-de-Giraud (Camargue). *Bull. Minéral.,* 104: 625–29.

Geisler-Cussey, D. 1982–1984. De la mer au sel: les faciès superficiels de marais salants de Salin-de-Giraud (Sud de la France). *Géologie Méditerranéenne, Marseille* (IV), 9, 4: 521–49.

——. 1985. *Approche sédimentologique et géochimique des mécanismes générateurs de formations évaporitiques actuelles et fossiles. Marais salants de Camargue et du Levant espagnol, Messinien méditerranéen et Trias lorrain.* Thèse Doct. ès Sci., University Nancy I: 297 pp.

——. 1986. Approche sédimentologique et géochimique des mécanismes générateurs de formations évaporitiques actuelles et fossiles. Marais salants de Camargue et du Levant espagnol, Messinien méditerranéen et Trias lorrain. *Documents GRECO 52 (CNRS Paris)* 10, and *Mém. Sci. de la Terre* 48: 268 pp.

Geisler-Cussey, D., J. Hilly, C. Marchal, R. Moretto, and H. Rashid. 1985. Typology of the halitic dissolutions observed in evaporitic series. *6th European Regional Meeting of Sedimentology.* Lérida, Spain, abstract: 582–85.

Gélard, J. P. 1978. La fracturation de la Bourgogne méridionale. Essai d'interprétation et implications tectoniques régionales. *Rev. Géogr. Phys. Géol. Dyn.* 1: 5–28.

Gennesseaux, M. and D. Lefebvre. 1980. Le Golfe du Lion et le Paleo-Rhône messinien. *Géologie Méditerranéene* 1: 71–80.

Girod, M. and N. Girod. 1977. Contribution de la pétrologie à la connaissance de l'évolution de la Méditerranée occidentale depuis l'oligocène. *Bull. Soc. Géol. France* 3: 481–88.

Glasbergen, P. 1985. The origin of groundwater in Carboniferous and Devonian aquifers at Maastricht. *Geol. Mijnbouw* 64: 123–29.

Godinot, M., J.-Y. Crochet, J.-L. Hartenberger, B. Lange-Badre, D. E. Russel, and B. Sige. 1987. Données nouvelles sur les mammifères de Palette (Eocène inférieur, Provence). *Münch. Geowiss. Abh.* (A) 10: 273–88.

Goemaere, E., J. Thorez, and R. Dreesen. 1985. A propos des milieux évaporitiques supratidaux dans les psammites du Condroz (Famennien supérieur, Belgique). In "Evaporites pré-permiennes en Europe" (Bruxelles, 1985). meet. Greco 52, *Soc. belge Géol. and Groupe de Contact Sédimentol.,* abstract, 2 p.

Goguel, J. 1948. Essai d'interprétation de la prospection géophysique de la Bresse et du bas-Dauphiné. *Pub. B.R.G.G.* 6: 79 pp.

——. 1957. Gravimétrie et fossé rhénan. *Verh. Koninkl. Nederl. Geol. Mijnbouwk. Genoot.,* Geol. Ser. 18: 125–47.

González, I. and E. Galán. 1984. Mineralogía de los materiales terciarios del área de Tarazona-Borja-Ablitas (Depresión del Ebro). *Estudios Geológicos* 40: 115–28.

Gornitz, M. V. and B. C. Schreiber. 1981. Displacive halite hoppers from the Dead Sea: some implications for ancient evaporite depositions. *Journal Sedimentary Petrology* 51: 87–94.

Gouleau, D. 1982. Chimie des eaux interstitielles des sédiments des marais salants de Salin-de-Giraud (Sud de la France). In G. Busson, ed., *Géologie Méditerranéenne. Annales de l'Université de Provence* 9: 349–67.

——. 1984. Relations entre saumures libres et saumures interstitielles du marais salant de "Bras del Port," Santa Pola (Espagne). Géochimie des ions majeurs et de la silice dissoute. *Revista del Instituto de Investigaciones Geologicas,* Barcelona 38: 55–78.

Gouleau, D. and D. Noël. 1984. L'importance des diatomées dans le cycle de la silice dissoute des saumures libres et des eaux interstitielles dans le marais salant de Salin-de-Giraud (SE de la France). *Rev. Géol. dyn. et Géogr. Phys.* 25: 177–86.

Gourvernet, C., G. Guieu, and C. Rousset. 1979. *La Provence. Guides géol. régionaux.* 2d ed. Masson & Cie., Paris: 229 pp.

Goy, G. and G. Busson. 1980. *Inventaire des évaporites de sous-sol: les bassins profonds du Trias de Sud-Est français, du Paléogène de Bresse et de Valence.* C. R. Action concertée: Valorisation des ressources du sous-sol. Muséum national d'Histoire naturelle, Paris: 167 pp.

Grappin, C., J. M. Jaffrezo, and J. C. Touray. 1975. Conditions de dépôt du sel gemme dans le bassin de Manosque: les données du dosage des traces de Br à diverses échelles de prélèvement (5 mg–100 mg). *9th Intern. Congr. Sedim., Nice* 2: 69–72.

Grolier, J. and G. Tchimichkian. 1963. Connaissances nouvelles sur la géologie du socle de la Limagne d'après les sondages de la Régie autonome des pétroles. *Bull. Soc. géol. France* 7: 930–37.

Guieu, G. and P. Arlhac. 1986. Hypothèse d'un bombment crustal golfe du Lion-Sud Provence entre le Crétacé spérieur et le Miocène. Mécanismes, conséquences. *C. R. Acad. Sci.* 303: 1691–96.

Guieu, G. and J. Rousset. 1980. La Provence. In Introduction à la Géologie du Sud-Est (France). Livret guide, *26è congr. inter.*, pp. 65–79.

Guieu, G., J. Philip, J.-P. Durand, D. Nury, and C. Redondo. 1987. Le détrisme provençal du Crétacé moyen à l'Oligocène dans son cadre paléogéographique structural et géodynamique. *Géol. Alpine, Grenoble Mem. H. S.* 13: 247–71.

Guillevin, Y. 1979. Eléments de pétrographie des evaporites oligocènes de bassins de Bresse et de Valence. Illustration et interprétation de quelques séquences. In *Dépôts évaporitiques comité des Techiniciens de la Chambre syndicale des Hydrocarbures.* Editions Technip, Paris: 41–48.

——. 1981. Petrographical data of Oligocene evaporites in the Bresse and Valence basins (Eastern France). In *Evaporite Deposits, Illustration and interpretation of some environmental sequences.* Ed. Technip, Paris: 41–48.

Guimerà, J. 1984. Paleogene evolution of deformation in the northeastern Iberian Peninsula. *Geological Magazine* 121: 413–20.

Guimerà, J. and M. Alvaro. 1990. Structure et évolution de la compression alpine dans la Chaine Ibérique et la Chaine Cotière Catalane. *Bull. Soc. Géol. France,* 6: 339–48.

Handford, C. R. 1981. Coastal sabkha and salt pan deposition of the lower Clear Fork Formation (Permian), Texas. *Journal Sedimentary Petrology* 51: 761–78.

——. 1988. Depositional interaction of siliciclastics and marginal marine evaporites. In B. C. Schreiber, ed., *Evaporites and Hydrocarbons.* Columbia University Press, New York: 139–81.

Handford, C. R. and P. E. Fredericks. 1980. *Facies Patterns and Depositional History of a Permian Sabkha Complex: Red Cave Formation, Texas Panhandle.* Bureau of Economic Geology, University of Texas at Austin, Geology Circular no. 80–9, 38 pp.

Haq, B. U., J. Hardenbol, and P. R. Vail. 1987. The chronology of fluctuating sea level since the Triassic. *Science* 235: 1156–67.

Hardie, L. A. 1968. The origin of the Recent non-marine evaporite of the Saline Valley, Inyo County, California. *Geochimica, Cosmochimica Acta,* 32: 1279–1301.

——. 1984. Evaporites: marine or non-marine? *Amer. Jour. Science* 284: 193–240.

Hardie, L. A., T. K. Lowenstein, and R. J. Spencer. 1985. The problem of distinguishing between primary and secondary features in evaporites. In B. C. Schreiber and L. Harner, eds., *Sixth Symposium on Salt:* pp. 11–39. Cleveland: Northern Ohio Geological Society.

Hardie, L. A., J. P. Smoot, and H. P. Eugster. 1978. Saline lakes and their deposits: A sedimentological approach. In A. Matter and M. E. Tucker, eds., *Modern and Ancient Lake Sediments.* International Assoc. Sedimentologists, Special Publication 2: 7–41.

Harvie, C. E., J. M. Weare, L. A. Hardie, and H. P. Eugster. 1980. Evaporation of seawater: calculated mineral sequences. *Science* 204: 498–500.

Hassouba, H. and H. F. Shaw. 1980. The occurrence of palygorskite in Quaternary sediments of the coastal plain of north-west Egypt. *Clay Minerals,* 15: 77–83.

Helman, M. L. 1989. *Tectonics of the Western Mediterranean.* D. Phil. Thesis, University of Oxford. 290 pp.

Herry, M. 1983. Etude géométrique par diagraphies différées du bassin évaporitique keupérien de Bresse: Jura. *Sci. de la Terre* 25: 107–39.

Herrmann A. G., D. Knake, J. Schneider, and H. Peters. 1973. Geochemistry of modern seawater and brines from salt pans: main components and bromine distribution. *Contr. Mineral. and Petrol.* 40: 1–24.

Hippolyte, J.-C., J. Angelier, F. Bergerat, D. Nury, and G. Guier. 1993. Tectonic-stratigraphic record of paleostress time changes in the Oligocene basins of the Provence, southern France. *Tectonophysics* 226: 15–35.

Hippolyte, J.-C., J. Angelier, F. Bergerat, D. Nury, and S. Raynaud. 1990. Evolution tectono-sédimentaire d'un bassin faillé: le baissin oligocène de Marseille. *C. R. Acad. Sci. Paris,* 310 (II): 53–58.

Hirn, A. 1976. Sondages sismiques profonds en France. *Bull. Soc. géol. France* 7: 1065–71.

Holdoway, K. A. 1978. Deposition of evaporites and red beds of the Nippewalla Group, Permian, Western Kansas. *Kansas Geol. Surv. Bull.* 215. 1–43.

Holliday, D. W. 1967. Secondary gypsum in Middle Carboniferous Rocks of Spitsbergen. *Geological Magazine* 104: 171–77.

——. 1970. The petrology of secondary gypsum rocks: a review. *Jour. Sed. Petrol.* 40: 734–44.

Holser, W. T. 1966. Bromide geochemistry of salt rocks. In J. L. Rau, ed., *Second Symposium on Salt* 2: 248–75. Cleveland: Northern Ohio Geol. Soc.

——. 1979. Mineralogy of evaporites. In R. G. Burns ed., *Marine Minerals.* Mineral. Soc. Am. Short Course Notes 6: 211–94.

Horn, P., H. J. Lippolt, and W. Todt. 1972. Kalium-Argon-Altersbestimmungen an tertiären Vulkaniten des Obberheingrabens. *Ecl. geol. Helv.* 65: 131–56.

Hovorka, S. D. 1983a. Sedimentary structures and diagenetic modifications in halite and anhydrite, Palo Duro Basin. In *Geology and Geohydrology of the Palo Duro Basin, Texas Panhandle.* Bureau of Economic Geology, Austin, Circular 83/4: 49–57.

——. 1983b. Dissolution and recrystallization fabrics in halite and the timing of their development, Palo Duro Basin. In *Geology and Geohydrology of the Palo Duro Basin, Texas Panhandle.* Bureau of Economic Geology, Austin, Circular 83/4: 58–65.

——. 1988. Alteration of gypsum to anhydrite: examples from the Permian Basin. *Geological Society of America* Abstract, p. A263.

——. 1992. Halite pseudomorphs after gypsum in bedded anhydrite: clue to gypsum-anhydrite relationships. *Journal Sedimentary Petrology* 62: 1098–1111.

Hottinger, L. and H. Schaub. 1961. División en pisos del Eoceno. Introducción de los pisos Ilerdiense y Biarritziense. *Not. y Com. del IGME* 61: 199–223.

Hugueney, M., J.-P. Berger, and M. Weidmann. 1987. Présence de mammifères oligocènes dans le parastratotype de l'Aquitanien, *Bull. Soc. Fribourg Sc. Nat.* 76: 129–135.

Hugueney, M. and M. Ringeade. 1990. Synthesis on the "Aquitanian" lagomorph and rodent faunas of the Aquitanian basin (France). In E. H. Lindsay et al. (eds.), *European Neogene Mammal Chronology,* pp. 139–56. Plenum Press, New York.

Hunt, C. B., T. W. Robinson, W. A. Boles, and A. L. Washburn. 1966. Hydologic basin, Death Valley, California. *U. S. Geological Survey Paper* 494B: 138 pp.

Hugueney, M. and G. Truc. 1976. Découvertes récentes de mammifères et de mollusques dans des formations d'âge oligocène terminal et aquitanien du Sud-Est de la France. *Géobios* (Lyon), 9: 359–62.

IGME. 1987. *Contribución de la exploración petrolífera al conocimiento de la geología de España.* Public. IGME, 465 pp., Madrid.

Illies, J. H. 1967. An attempt to model the Rhinegraben tectonics. *Abh. geol. Landesamt Baden-Würtemberg* 6: 10–12.

——. 1972. The Rhine Graben rift system. Plate tectonics and transform faulting. *Geophys. Surv.* 1: 27–60.

——. 1974. Taphrogenesis and plate tectonics. In J. H. Illies and K. Fuchs eds. *Approaches to Taphrogenesis.* Schweitzerbart. éd., pp. 433–60, Stuttgart.

——. 1975. Recent and paleo intra plate tectonics in stable Europe and the Rhine-graben rift system. *Tectonophysics* 29: 251–64.

——. 1977. Ancient and recent rifting in the Rhine Graben. *Geol. in Mijnbouw* 56: 329–50.

——. 1981. Mechanism of graben formation. *Tectonophysics* 73: 249–66.

Illies, J. H. and H. Baumann. 1982. Crustal dynamics and morphodynamics of the Western European rift system. *Z. Geomorph.-N.F.-Suppl.* 42: 135–65.

Illies, J. H. and G. Greiner. 1978. Rhine graben and Alpine system. *Bull. Geol. Soc. Amer.* 89: 770–82.

Inglès, M., M. Muñoz, A. Pérez, and J. M. Salvany. 1994. Relación entre la mineralogía y los ambientes sedimentarios en el Terciario continental del sector occidentalde la Cuenca del Ebro. *II Congreso del grupo Español del Terciario.* Jaca. Resúmenes de Comunicaciones: 247–50.

Jauzein, A. 1974. Les données sur le système $CaSO_4 \cdot 2H_2O$ et leurs implications géologiques. *Rev. Géogr. Phys. et Géol. dyn.* 2(6): 151–60.

Jung, J. 1946. *Géologie de l'Auvergne et de ses confins bourbonnais et limousins.* Mém. pour servir à l'explication de la Carte géol. France: 372 pp.

Jurado, M. J. 1989. *El Triásico del subsuelo de la Cuenca del Ebro.* Ph.D. Thesis, University Barcelona, 259 pp.

Kahle, C. F. 1974. Ooids from Great Salt Lake, Utah as an analogue for the genesis and diagenesis of ooids in marine limestones. *Journal Sedimentary Petrology* 44: 30–39.

Keizer, J. and W. J. Letsch. 1963. Geology and the Tertiary in the Netherlands. *Verh. K. Ned. Geol. Mijnbouwkd. Genoot., Geol. Ser.* 21/2: 147–72.

Kendall, A. C. 1984. Evaporites. In R. G. Walker, ed., *Facies Models.* Geoscience Canada, Reprint series 1: 259–96.

Kessler, P. 1909. Die tertiären Küstenkonglomerate in der mittelrheinischen Tiefebene mit besonderer Berüksichtigung des elsässischen Vorkommen. *Mitt. geol. Landesanst Els.-Lothr., Strasbourg,* 7: 167–290.

Kiener, M. 1984. Etudes sédimentologiques et corrélations par diagraphies: le sommet de la série salifère et sa couverture dans le sondage EZ 08 (Paléogène de Bresse). *C.R. de la Journée sur le Sel,* pp. 207–31.

Kieken, M. 1980. Eocène supérieur. In *Géologie du bassin d'Aquitaine.* Ed. Bur. Rech. Geol. Min. Mem.: 1–24.

Kinsman, D. J. J. 1966. Gypsum and anhydrite of Recent Age, Trucial Coast, Persian Gulf. In J. L. Rau, ed. *Second Symposium on Salt,* Northern Ohio Geological Society, 1: 302–26.

——. 1969a. Modes of formation, sedimentary associations, and diagnostic features of shallow water and supratidal evaporites. *American Assoociation Petroleum Geologists Bulletin* 53: 830–40.

——. 1969b. Interpretation of Sr^{+2} concentrations in carbonate minerals and rocks. *Journal Sedimentary Petrology,* 39: 486–508.

——. 1976. Evaporites: relative humidity control of primary mineral facies. *Journal Sedimentary Petrology* 46: 273–79.

Kromm, F. 1967. Le Flysch de Vallfogona et son contexte paléogéographique. *Ac. Soc. Linn. Bordeaux, sér. B.* 104: 11 pp.

Kühn, R. 1968. Geochemistry of the German potash deposits. *Geological Society America,* Special Paper 88: 427–504.

——. 1972. Combined evaluation of Br and Rb contents for the genetic characterization of carnallites and sylvite rocks. *Geology of Saline Deposits* pp. 77–89. Proceedings Hannover Symposium.

——. 1979. Diagenese in Evaporiten. *Geol. Rundschau* 68(3): 1066–75.

Kühn, R. and H. Roth. 1979. Beiträge zur Kenntnis der Saltzlagerstätte am Oberrhein. *Z. geol. Wiss.*, Berlin, 7: 953–66.

Labaume, P. and M. Séguret. 1985. Evolution of a turbitidic foreland basin and analogy with an accretionary prism: example of the Eocene south-Pyrenean basin. *Tectonics* 4: 661–85.

Lamare, P. 1936. Recherches géologiques dans les Pyrénées basques d'Espagne. *Mémoire Soc. géol. France,* XII(1–4): 463 pp.

Landry J. C. and Jaccard J. 1982. Chimie des eaux libres dans le marais salant de Salin-de-Giraud (Sud de la France). *Géologie Méditerranéene* 9(4): 329–48.

——. 1984. Chimie des eaux libres dans le marais salant de Santa Pola, Salina de Bras del Port. *Revista del Instituto de Investigaciones Geologicas,* Barcelona, 38/39: 37–53.

Larroque, J. M. 1987. Analyse de la déformation de la série salifère du bassin potassique de Mulhouse et évolution du champ de contraintes dans le sud du fossé rhénan au Tertiaire et à l'actuel. Thèse Doct. Univ. Sci. Techn. Languedoc, Montpellier: 206 pp.

Larroque, J. M. and M. Ansart. 1985. Les déformations liées à la tectonique distensive oligocène du bassin potassique de Mulhouse: cas du secteur minier. *Bull. Soc. géol. France,* 6: 837–47.

Lauer, J. P., A. Roche, and W. Wimmenauer. 1967. Structure, evolution, and paleomagnetism of the Kaiserstahl volcano. *Abh. geol. Lamdesant Baden-Wurttemberg* 6: 134–38.

Le Douaran, S., J. Burrus, and F. Avedik. 1984. Deep structure of the northwestern Mediterranean Basin: results of the two ships seismic survey. *Marine Geology* 55: 325–44.

Lefavrais-Raymond, A. 1962. Contribution à l'étude géologique de la Bresse d'après les sondages profonds. *Mém. Bur. Rech. Géol. Min.* 16: 170 pp.

Lefavrais-Raymond, A., E. Gannat, and C. Cavelier. 1984. Bresse. In Synthèse géologique du Sud-Est de la France, *Mém. Bur. Rech. Géol. Min.* 125: 403–6.

Lenguin, M., M. P. Mouline, P. Roger, and J. Szulc. 1990. Continental sedimentation in the Paleogene Narbonne Basin. Main facies and relationships with extensive deformation. *IXth Congress R.C.M.N.S., Barcelona. Global Events and Neogene Evolution of the Mediterranean,* abstracts: p. 213.

Leroy, C. 1984. Utilisation des méthodes sismiques et de forages profonds pour la reconnaissance structurale d'un bassin salifère. *C.R. de la Journée sur le Sel* (Mar), Ecole Polytechnique, Palaiseau, 163–76.

Letouzey, J. and P. Tremolieres. 1980. Paleo-stress fields around the Mediterranean since the Mesozoic derived from microtectonics: comparisons with plate tectonic data. *26ème Congr. Geol. Int., C5. Mém. Bur. Rech. Géol. Min:* 261–73.

Lienhardt, G. and P. Rat. 1974. Le groupe d'étude "Bresse" et les tableaux stratigraphiques proposés pour le Miocène, le Pliocène et le Villafranchien du remplissage bressan. *C.R. somm. Soc. Géol. France* 4: 95–100.

Lippolt, H. J., W. Todt, and P. Horn. 1974. Apparent potassium-argon ages of Lower-Tertiary Rhine graben volcanics. In J. H. Illies and K. Fuchs, eds., *Approaches to Taphrogenesis.* Schweitzerbart éd., pp. 213–21, Stuttgart.

Llwellyn, P. G. 1968. Dendritic halite pseudomorphs from the Keuper Marl of Leicestershire, England. *Sedimentology* 11: 293–97.

Llewelyn, P. G., S. A. Mahmoud, and R. Stabbins. 1968. Core anhydrite from the Anhydrite Series, carboniferous Limestone, Hathern borehole, Leicestershire, West Cumberland. *Trans. Inst. Min. Metall.* 77B: 21–25.

Logan, B. W. 1987. The MacLeod Evaporite Basin, Western Australia: Holocene environments, sediments, and geological evolution. *Amer. Assoc. Petrol. Geol.,* Memoir no. 44, 140 pp.

Logan, B. W., R. Rezak, and R. N. Ginsburg. 1964. Classification and environmental significance of algal stromatolites. *Journal Geology* 72: 68–83.

Loreau, J. P. 1982. Sédiments aragonitiques et leur genèse. *Mèm. Mus. nat. Hist. nat., Paris,* Série C, XLVII: 312 pp.

Losantos, M., X. Berástegui, J. A. Muñoz, and C. Puigdefàbgregas. 1988. Corte geológico cortical del Pirineo central (perfil ECORS): Evolución geodinámica de la cordillera pirenaica. *II Congreso Geológico de España, Simposios* 7–16.

Lowenstein, T. K. 1982. Primary features in a potash evaporite deposit, the Permian Salado Formation of West Texas and New Mexico. In C. R. Handford, R. G. Loucks, and G. R. Davies, eds., *Depositional and Diage-*

netic Spectra of Evaporites: a Core Workshop, pp. 276–304, SEPM Core Workshop 3, Calgary, Canada.

Lowenstein, T. K. and L. A. Hardie. 1985. Criteria for recognition of salt-pan evaporites. *Sedimentology* 32: 627–44.

Lowenstein, T. K. and R. J. Spencer. 1990. Syndepositional origin of potash evaporites: petrographic and fluid inclusions evidence. *American Journal Science* 290: 1–42.

Maget, P. and C. Maiaux. 1980. Trias-faciès évaporatiques. In C. Megnien, ed., *Synthèse géologique du bassin de Paris, II (Atlas).* Mém. Rech. géol. min. Carte TRS, Orléans.

Magne, J., Y. Gourinard, and M. J. Wallez. 1987. Comparaison des étages du Miocène inférieur définis par stratotypes ou par zones paléontologiques. *Strata,* sér. 1, 3: 95–107.

Maiklem, W. R., D. G. Bebout, and R. F. Glaister. 1969. Classification of anhydrite. A practical approach. *Bulletin Canadian Petroleum Geology* 17: 194–233.

Maïkovsky, V. 1941. Contribution à l'étude paléontologique et stratigraphique du bassin potassique d'Alsace. *Mém. Serv. Carte géol. Als.-Lorr.*, Strasbourg, 6, 192 pp.

——. 1952. Le pétrole dans le Haut-Rhin: le sondage de Staffelfelden IV-DP 25. *Bull. Soc. industr. Mulhouse* 1: 25–34.

Malecha, A. 1966. The basin of southern Bohemia. In J. Svoboda et al., eds., *Regional Geology of Czechoslovakia.* Part I, The Bohemian Massif. Geol. Survey Czechoslovakia, Prague: 581–600.

Malkovsky, M. 1966. The North Bohemian Basin. In J. Svoboda et al. eds., *Regional Geology of Czechoslovakia.* Part I, The Bohemian Massif. Geol. Survey Czechoslovakia, Prague: 532–42.

Mandado, J. 1987. *Litofacies yesíferas del Sector Aragonés de la Cuenca Terciara del Ebro. Petrogénesis y Geoquímica.* Ph.D. Thesis, University Zaragoza, 443 pp.

Marchal, C. 1983. Le gîte salifère keupérien de Lorraine-Champagne. *Sci. de la Terre* 44: 139 pp.

Marfil R., E. Bermejo, and J. A. de la Peña. 1975. Sedimentacion salina actual en las lagunas de la zona Corral Rubio-La Higuera (provincia de Albacete). *Estudios geol.* 31: 543–53.

Marius, C. 1985. Mangrove du Sénégal et de la Gambie: Ecolgie, Pédologie, Géochmie, Mise en valeur et aménagement. Edition de l'Ordstrom. *Collection Travaux et Documents No. 193,* Paris, 355 pp.

Martínez, A., J. Vergés, E. Clavell, and J. Kennedy. 1989. Stratigraphic framework of the thrust geometry and structural inversion in the southeastern Pyrenees: La Garrotxa area. *Geodinamica Acta* 3: 185–94.

Martínez Gil, F. J., J. A. Sánchez, J. L. De Miguel, and J. San Roman. 1988. El drenaje subterráneo de la Cordillera Ibérica en la Cuenca del Ebro como proceso de movilización y transporte de sustancia en disolución. Sus implicaciones en el aporte de sulfatos. In A. Pérez and A. Muñoz, eds., *Sistemas lacustres neógenos del margen ibérico de la Cuenca del Ebro.* III Reunión Grup. Esp. Trab. PICG 219; Zaragoza, pp. 57–80. Excursion Guide-book.

Martínez Peña, M. B. and A. Pocoví. 1988. El amortiguamento frontal de la estructura de la cobertera surpirenaica y su relación con el anticlinal de Barbastro-Balaguer. *Acta Geològica Hispànica* 23: 81–94.

Martínez-Torres, J. M., L. Eguiluz, R. Ramon Lluch, and H. Llanos. 1984. Geometría del Pirineo occidental y Arco Vasco. *I Congreso Español de Geología* III: 207–17.

Martini, E. 1973. Nannoplankton-Massenvorkommen in den Mittleren Pechelbronner Schichten (Unter-Oligozän). *Oberrhein. geol. Abh.*, Karlsruhe, 22: 1–12.

Masse, J. P. 1980. Sur l'amplitude du décrochement sénestre de la faille de Nîmes, déduite des variations de faciès du Barrémien-Bédoulien. *8 Réunion Annuelle des Sciences de la Terre* Marseille (University of Provence) volume of abstracts: 245.

——. 1983. Relations possibles entre fossés d'effondrement, décrochements et bombements dans les rifts intracontinentaux. *Bull. Centres Rech. Explor. Produt. Elf Aquitaine* 1: 149–54.

Mauffret, A. and M. Gennesseaux. 1989. Compression, décrochement et distension sur le poutour méditerranéen nord-occidental. *C. R. Acad. Sci. Paris* 308: 961–67.

Mauffret, A., M. Labarbarie, and M. Montadert. 1982. Les affleurements des séries sédimentaires pré-pliocènes dans le bassin méditerranéen nord-occidental. *Marine Geology* 4: 159–75.

Mauffret, A., L. Montadert, M. Lavergne, and C. Willm. 1978. Geological and geophysical setting of DSDP, Site 372 (Western Mediterranean). In K. J. Hsü, L. Montadert et al., eds., *Report Deep Sea Drilling Project, XLII* (I): 889–97. U.S. Govt. Printing Office, Washington, D.C.

Mauffret, A., J. P. Rehault, M. Gennesseaux, G. Bellaiche, M. Labarbarie, and D. Lefebre. 1981. Western Mediterranean basin evolution: from a distensive to a compressive regime. In F. C. Wezel, ed., *Sedimentary Basins of Mediterranean Margins.* CNRS Italian Project of Oceanography. pp. 67–81. Technoprint, Bologna.

Maurin, A. F. 1984. Halocinèse et sédimentation dans la Mésogée. *Strata* 1: 3–5.

Maury, R. C. and J. Varet. 1980. Le volcanisme tertiaire et quaternaire en France. *26ème CGI Géologie de la France. Bur. Rech. Geol. Min. Mém.* 107: 137–59.

McDonald, G. J. F. 1953. Anhydrite-gypsum equilibrium relations. *American Journal Science* 251: 884–98.

McIntire, W. L. 1963. Trace element partition coefficients: a review of the theory and application to geology. *Geochimica et Cosmochimica Acta* 27: 1209–64.

Mein, P., G. Truc, and G. DeMarq. 1971. Micromammifères et Gastéropodes continentaux des biozones de Paulhiac et de la Romieu dan le Miocène de la Bastidonne et de Mirabeau (Vaucluse, Sud-Est de la France). *Comte Rendu de Académie Sciences, Paris,* D, 273: 566–68.

Menduiña A. J., S. Ordóñez, and M. A. Garcia Del Cura. 1984. Geología del yacimiento de glauberita de Cerezo de Río Tirón (provincia deBurgos). *Bol. Geológico y Minero* 95: 33–51.

Mey, P. H., P. J. C. Nagtegaal, K. J. Robert, and J. J. A. Hartevelt. 1968. Lithostratigraphic subdivision of post-hercynian deposits in the south-central Pyrenees, Spain. *Leidse Geologische Mededelingen* 41: 221–28.

Michel, P., G. Appert, and J. Lavigne. 1951. Sondages RAP du Mont Myon et du Revermont. *Bull. Soc. Géol. France* 6(49): 819–21.

Milliken, K. L. 1979. The silicified evaporite syndrome: Two aspects of silicification history of former evaporite nodules from southern Kentucky and northern Tennessee. *Journal Sedmentary Petrology* 49: 245–56.

Morange, A., F. Heritier, and J. Villemin. 1971. Contribution de l'exploration pétrolière à la connaissance structurale et sédimentaire de la Limagne, dans le Massif Central. In *Symposium J. Jung,* Plein Air Service édit., Clermont-Ferrand: 295–308.

Moretto, R. 1985. Sédimentologie de la série salifère du Paléogène de la Bresse (France). *Bull. Soc. Géol. France, Paris* 8: 849–55.

——. 1986/87. Etude sédimentologique et géochimique des dépôts de la série salifère paléogène du bassin de Bourg-en-Bresse (France). Thèse Université de Nancy (1986). *Documents U. A. 1209 (CNRS Paris)* 12 (1987) and *Mém. Sci. de la Terre* 50 (1987). 252 pp.

——. 1988. Observations on the incorporation of the trace elements in halite of oligocene salt beds, Bourg-en-Bresse Basin, France. *Geochimica et Cosmochimica Acta* 52: 2809–14.

Müller, Gerhard. 1964. *Ein Beitrag zur Geochimie des Stronthums in Ca-Sulfat Gesteinen.* Thèse Univ. Saarbrück, 166 pp.

Müller, German. 1962. Zur Geochemie des Sr in ozeanen Evaporiten unter besonderer Berücksichtigung der sedimentären Cölestinglagerstätter von Hemmelte-West (Süd-Oldenburg). *Geologie (série II),* 35: 1–90.

Muñoz, M. 1992. Analisis tectosedimentario del Terciario del sector accidental de la cuenca del Ebro (Comunidad de la Rioja). *Ciencias de la Tierra* 15, Ediciones del Instituto de Estudios Riojanos, Lograno, 347 pp.

Muñoz, A., G. Pardo, and J. Villena. 1986/87. Análisis tectosedimentario del Terciario de la depresión de Arnedo (prov. de La Rioja). *Acta Geològica Hispànica* 21/22: 427–35.

Muñoz, J. A., A. Martínez, and J. Vergés. 1986. Thrust sequences in the Spanish Eastern Pyrennees. *Journal Structural Geology* 8: 399–405.

Muñoz, J. A., C. Puigdefàbregas, and J. M. Fontboté. 1987. El ciclo alpino y la estructura tectónica del Pirineo. *Libro Jubilar de J. M. Rios* II: 185–205, I.G.M.E., Madrid.

Muñoz, J. A., F. Sabat, and P. Santanach. 1983. Cisaillements alpins et hercyniens dans le versant méridional de la zone axiale des Pyrénées entre le Freser et le Ter. *C. R. Acad. Sci. Paris* 296: 1453–58.

Mutti, E., J. Rosell, G. Allen, F. Fonnesu, and M. Sgavetti. 1985. The Eocene Baronía tide-dominated delta-shelf system in the Ager Basin. *VI European Regional Meeting, IAS, Lérida (Spain)* Excursion Guidebook: 577–600.

Naiman, E. R., A. Bein, and R. L. Folk. 1983. Complex polyhedral crystals of limpid dolomite associated with halite, Permian upper Clear Fork and Glorieta Formations, Texas. *Journal Sedmentary Petrology* 53: 549–55.

Neev, D. and K. O. Emery. 1967. The Dead Sea. *Geological Survey of Israel Bulletin* 41: 147 pp.

Nel, A., G. A. Gil, and D. Nury. 1987. Découverte d'empreintes attribuables à des Coulentérés Siponophores chondrophorides dans l'Oligocène de Provence. *C. R. Acad. Sci. Paris* 305(Ser. II): 637–41.

Neugebauer, H. J. 1978. Crustal doming and the mechanism of rifting. Part I: rift formation. *Tectonophysics* 40: 159–86.

Neugebauer, H. J. and Braner, B. 1978. Crustal doming and the mechanism of rifting. Part II: rift development of the upper Rhinegraben. *Tectonophysics* 40: 1–20.

Neugebauer, H. J. and P. Temme. 1981. Crustal uplift and the propagation of failure zones. *Tectonophysics* 73: 33–51.

Noël, D. 1982. Les diatomeés des saumures des marais salants de salin-de-Giraud (Sud de la France). In G. Busson, ed., *Géologie Méditerranéene Marseille* IX(4): 447–70.

Nurmi, R. D. 1978. Use of well logs in evaporite sequences. In W. E. Dean and B. C. Schreiber, eds., *Marine Evaporites.* SEPM Short Course 4, pp. 144–76. Tulsa, Okla.

——. 1988. Geologic interpretation of well logs and seismic measurements in reservoirs associated with evaporites. In B. C. Schreiber, ed., *Evaporites and Hydrocarbons,* pp. 405–59. Columbia University Press, New York.

Nury, D. 1970. Données nouvelles sur le bassin oligocène de

Saint-Pierre-les-Martigues (Bouches-du-Rhône). *Paléobiologie continentale* 1: 15–18.

——. 1977. Le bassin oligocène de Marseille, stratigraphie, évolution paléogéographique. *Géologie Méditerranéene* 1: 25–36.

——. 1984. Basse Provence-Oligocène. In S. Debrand-Passard (ed.), *Synthèse géologique du Sud-Est de la France.* Mém. Bur. Rech. Geol. Min. Mém. no. 125(1): 429–33.

——. 1987/88. L'Oligocène de Provence méridionale. Stratigraphie, dynamique sédimentaire, reconstitutions paléogéographiques. Thèse Doct. Etat, Univ. Provence (Aix-Marseille): 378 pp.

——. 1994. Relations géométriques entre carbonates et évaporites: Exemple de l'Oligocène terminal de la région d'Aix-en-Provence/Marseille (France). *Géologie Méditerranéenne, Marseille,* XXI. n°. 1–2.

Nury, D. and B. A. Thomassin. 1994. Paléoenvironment tropicaux, marins et Lagunaires d'un littoral abrité (fonds meubles à bancs coralliens, lagune évaporitique) à l'Oligocène Terminal, en Basse-Provence (région d'Aix-en-Provence/Marseille, France). *Géologie Méditerranéene, Marseille,* XXI, n°. 1–2.

Nury, D. and Raynaud, S. 1986. Etude tectono-sédimentaire du Fossé oligocène de Marseille (Bouches-du-Rhône). Evolution postérieure. *Bull. Bur. Rech. Geol. Min. Mem., Géol. de la France* 4: 377–97.

Nury, D. and C. Rousset. 1985/86. Dépôts oligocènes de la région de Toulon (Var, France). *Géologie Méditerranéene* 3/4: 175–84.

Odin, G. S. and C. Odin. 1990. Echelle numérique des temps géologiques. *Géochroniques* 35: 12–21.

Ordóñez, S. and M. A. Garcia del Cura. 1994. Deposition and diagensis of sodium-calcium sulfate salts in the Tertiary saline lakes of the Madrid Basin, Spain. In R. W. Renault and W. M. Last, eds., *Sedimentology and Geochemistry of Modern and Ancient Salt Lakes:* 229–238.

Ordóñez, S., J. Menduiña, and M. A. Garcia Del Cura. 1982. El sulfato sódico natural en España. *Tecniterrae* (Madrid) 8(46): 16–32.

Ortí, F. 1977. Aproximación sal estudio de las microestructuras de las rocas de yeso secundario y a su origen. *Revista Instituto Investigaciones Geologicas, Dip. Prov. Barcelona* 32: 87–152.

——. 1982. Caracteristicas deposicionales y petrologicas de las secuencias evaporiticas continetale en las cuencas terciaras peninsulares. *Temas Geologico-Mineros, J. G. M. E.*, Madrid, Part 2: 485–506.

——. 1988. Sedimentación evaporítica continental durante el Terciario en la Península Ibérica: aspectos generales. *II Congreso Geología de España, SGE, Granada.* Simposios: 509–18.

——. 1990a. Introducción a las evaporitas de la Cuenca Terciaria del Ebro. In: F. Ortí and J. M. Salvany (eds.), *Formaciones evaporíticas de la Cuenca del Ebro y cadenas periféricas, y de la zona de Levante,* ENRESA–University Barcelona, 62–66.

——. 1990b. Observaciones sobre la Formación Zaragoza y unidades evaporíticas adyacentes (Mioceno continental). In: F. Ortí and J. M. Salvany (eds.), *Formaciones evaporíticas de la Cuenca del Ebro y cadenas periféricas, y de la zona de Levante,* ENRESA–University Barcelona, 117–19.

Ortí, F., P. Busquets, L. Rosell, C. Taberner, R. Utrilla, and M. Quadras. 1987/88. La fase evaporítica del Eoceno medio (Luteciense) en la cuenca surpirenaica catalana. Nuevas aportaciones. *Rev. Inv. Geològ.* 42–43.

Ortí, F., G. Busson, A. Cornée, N. Dulau, D. Geisler-Cussey, D. Gouleau, J. Jaccard, J. C. Landry, C. Mosser, D. Noël, C. Pierre, C. Poumot, J. J. Pueyo, J. C. Thomas, M. Thomas, G. Truc, R. Utrilla, R. Casal, and L. Zaninetti. 1984. Introduction to the sedimentology of the coastal salinas of Santa Pola (Alicante, Spain). *Revista del Instituto de Investigaciones Geológicas,* Barcelona, 38/39: 235 p.

Ortí, F. and J. J. Pueyo. 1977. Asociación halita bandeada-anhidrita nodular del yacimiento de Remolinos, Zaragoza (sector central de la Cuenca del Ebro). Nota petrogenética. *Rev. Inst. Invest. Geol., Dip. Prov. Barcelona* 32: 167–202.

——. 1980. Polihalite diagenética en una sequencia evaporitica continental (Mioceno, Cuenca del Tajo, España). *Rev. Inst. Invest. Geol., Dip. Prov. Barcelona* 34: 209–22.

Ortí, F., J. J Pueyo, D. Geisler-Cussey, and N. Dulau. 1984. Evaporitic sedimentation in the coastal salinas of Santa Pola (Alicante, Spain). *Revista del Instituto de Investigaciones Geologicas,* Barcelona 38/39: 169–220.

Ortí, F., J. J. Pueyo, and L. Rosell. 1985. La halite du bassin potassique sud-pyrénéen (Éocène supérieur, Espagne). *Bulletin Societé Géologique France* (8)I: 863–72.

Ortí, F., J. J. Pueyo, L. Rosell, M. Inglès, R. Utrilla, and C. Pierre. 1988. Estudio sedimentológico, geoquímico e isotópico de las formaciones evaporíticas peninsulares. El reciclaje de evaporitas marinas mesozoicas, eocenas y messinienses en las terciarias continentales. *Memoria final Proyecto CAICYT* no. 2153/83, 388 pp.

Ortí, F., J. J. Pueyo, and A. San Miguel. 1979. Petrogénesis del yacimiento de sales sódicas de Villarrubia de Santiago (Terciario continental de la cuenca del Tajo). *Bol. Geolog. y Minero* 94: 347–73.

Ortí, F., J. J Pueyo, and G. Truc. 1984. Las salinas maritimas de Santa Pola (Alicante, Espana). Breve introduccion al estudio de un medio natural controlado de sedimentacion evaporitica somera. *Revista del Instituto de Investigaciones Geologicas,* Barcelona 38/39: 9–29.

Ortí F. and L. Rosell. 1994. Origin glauberitico del cuerpo central de los Yesos de Calatayud. *II Congreso del Grupo Espanol del Terciario (G.E.T.), Jaca, Sept. 1994, Simposio IGCP,* 324: 255–58.

Ortí F., L. Rosell, L. Fallick, and R. Utrilla. 1994. Yesos de Calatayud: aplicacion del estudio de facies y geoquimica de sulfatos al conocimiento de un sistema evaporitico. *Geogaceta* 15: 74–77.

Ortí, F., L. Rosell, and A. Lascorz. 1994. Fabricas cristalinas del yeso secundario de reemplazamiento de glauberita: aplicacion enprospeccion de sulfato sodico. *Geogaceta* 17: 49–52.

Ortí, F., L. Rosell, and J. J. Pueyo. 1984. Cuenca evaporítica (potásica) surpirenaica del Eoceno Superior. Aportaciones para una interpretación deposicional. *Libro Homenaje a Luis Sánchez de la Torre, Publicaciones de Geología,* no. 20, pp. 209–31, Universidad Autónoma de Barcelona.

Ortí, F., L. Rosell, J. J. Pueyo, M. Inglès, and R. Utrilla. 1987. Investigación de las formaciones evaporíticas en Cataluña. Composición y génesis. I: Catalánides. *Mem. Ajut Inv. CIRIT,* 42 pp. (unpublished)

Ortí, F., L. Rosell, and J. M. Salvany. 1992. Depositos evaporiticos en Espana: aspectos geologicos y recursos. In: J. García Guinea and J. Martinez Frias (Coords.), *Recursos minerales de Espana,* C.S.I.C., Madrid: 1171–1209.

Ortí, F., L. Rosell, J. M. Salvany, and M. Inglès. in press. Chert in continental evaporites of the Ebro and Calatayud basins (Spain): distribution and significance. In *Siliceous Rocks and Culture,* VI Inter. Flint Symposium, Univ. Granada-CSIC, 14 pp.

Ortí, F., L. Rosell, J. M. Salvany, M. Inglès, A. Permanyer, and J. García Veigas. 1989. Sedimentología y diagénesis como criterios de prospección de recursos en el Terciario evaporítico de la Cuenca del Ebro. *G.E.S. XII Congreso Español de Sedimentología* Simposios: 253–62, Bilbao.

Ortí, F., L. Rosell, R. Utrilla, M. Ingles, J. J. Pueyo, and C. Pierre. 1988. Reciclaje de evaporitas en la Península Ibérica durante el ciclo alpino. *II Congr. Geológ. España,* SGE, Granada. Communications 1: 421–24.

Ortí, F. and J. M. Salvany. 1986a. *Programa de investigación de las formaciones evaporíticas en Navarra.* Serv. Rec. Hidrológicos y Geología, Gobierno de Navarra. Vols. I and II (internal report).

——. 1986b. Evaporite cycles in the continental Tertiary of Navarra (Ebro Basin, Spain). *7th European Reg. Meet.* I.A.S., Kraków. Abstract book, 2 pp.

——. 1991. Depositos de glauberita en Espana: aspectos sedimentologicos y petrologicos generales. In J. J. Pueyo (Coord.), *Genesis de formaciones evaporiticas. Modelos andinos e ibericos* Estudi General, 2, Univ. Barcelona: 193–230.

Ortí, F., J. M. Salvany, J. M. Quadras, and J. J. Pueyo. 1986. Glauberite growths in playa-lake environments (Tertiary of the Ebro Basin, Alcanadre-Arrúbal zone). *II Inter. Symp. on crystal growth in sedimentary environments* Granada: Abstracts book, 2 pp.

Ortí, F., J. M. Salvany, L. Rosell, and M. Ingles. 1989. Sistemas lacustres evaporíticos del Terciario de la Cuenca del Ebro. *Geogaceta* 5: 103–4.

Ortí, F., J. M. Salvany, L. Rosell, J. J. Pueyo, and M. Ingles. 1986. Evaporitas antiguas (Navarra) y actuales (Los Monegros) de la cuenca del Ebro. *XI Congr. Español de Sedimentología,* GES Barcelona. Excursion Guidebook 2, 40 pp.

Pairis, J. L., J. Charolais, and C. Kerckhove. 1984. Evolution des Alpes au cours du Paléogène. In S. Passard-Debrand, S. Courbouleix, and M. J. Lienhart, (eds.), Ch. Paléogène. Synthèse Géologique Sud Est de la France. *Mém. Bur. Rech. Géol. Min.* Orleans, 125: 452–53.

Pelet, R. 1985. Sèdimentation et évolution de la matière organique. *Bull. Soc. Géol. France, Paris* 1(7): 1075–86.

Pérez, A., A. Muñoz, G. Pardo, et al. 1988. Características de los sistemas lacustres en la transversal Tarazona-Tudela (sector navarro-aragonés de la cuenca del Ebro)." *II Congr. Geológ. España,* SGE, Granada. Symposia: 519–27.

Perthuisot, J-P. 1975. La Sebkah el Melah de Zarzis. Genèse et évolution de un basin paralique. *Tra. Lab. Géol. Ecole norm. sup.* Paris, 9: 252 pp.

——. 1980. Sites et processus de la formation d'évaporites dans la nature actuelle. *Bull. Cent. Rech. Explor. Prod. Elf-Aquitaine* 4: 207–33.

——. 1982. Introduction générale à l'étude des marais salants de Salin-de-Giraud (Sud de la France). Le cadre géographique et le milieu. *Géologie Méditerranéene* IX(4): 309–27.

Perthuisot, V. and N. Guilhaumou. 1983. Les diapirs triasiques du domaine vocontien: phases diapiriques et hydrothermales en domaine périalpin. *Bull. Soc. géol. France* 7: 397–410.

Philip, J., P. Masse, and L. Machour. 1987. L'évolution paléogéographique et structurale du front de chevauchement nord-toulonnais (Basse-Provence occidentale, France). *Bull. Soc. Géol. France* 3: 541–50.

Pierre, C. 1982. *Teneurs en isotopes stables (^{18}O, ^{2}H, ^{13}C, ^{34}S) et conditions de genèse des évaporites marines: application à quelques milieux actuels et au Messinien de la Méditerranée.* Thèse d'Etat, Université de Paris-Sud, Orsay, 266 pp.

——. 1983. Polyhalite replacement after gypsum at Ojo de Liebre Lagoon (Baja California, Mexico): an early diagenesis by mixing of marine brines and continental waters. In B. C. Schreiber and L. Harner, eds., *Sixth International Symposium on Salt.* Salt Institute, Alexandria, Virginia, 1: 257–65.

Pierre C., L. A. Ortlieb, and A. Person. 1984. Supratidal evaporitic dolomite at Ojo Liebre Lagoon: mineralogical and isotopic arguments for primary crystallization. *Journal Sedimentary Petrology* 54: 1049–61.

Platt, J. and R. L. M. Vissers. 1989. Extensional collapse of thickened continental lithosphere: a working hypothesis for the Alboran Sea and Gibralter Arc. *Geology,* 17: 540–43.

Plaziat, J. C. 1984. *Le domaine pyrénéen de la fin du Crétacé à la fin de l'Eocène.* Thèse Doct. Etat Univ. Paris-Sud (Orsay): 1362 pp.

Portero, J. M., A. Hernández, J. Ramírez Del Pozo, M. J. Aguilar, and C. Leal. 1979. Mapa Geológico de España, a escala 1:50.000. Memoria y Hoja núm. 168, Briviesca. *MAGNA, IGME,* Madrid. 36 pp.

Poutoukis, D. 1991. Hydrochimie, teneurs isotopiques et origine des saumures associées aux gisements de potasse d'Alsace. Thèse Univ. Paris-Sud, 140 pp.

Pueyo, J. J. 1975a. *Estudio petrológico y geoquímico de los yacimientos potásicos de Cardona, Súria, Sallent y Balsareny (Barcelona, España).* Ph.D. Thesis, University Barcelona, 352 pp.

——. 1975b. Distribución del estroncio en los yacimientos potásicos catalanes. *Rev. Inst. Inv. Geol. Dip. Prov. Barcelona* 30: 5–10.

——. 1977. El bromo y el rubidio como indicadores genéticos en las evaporitas de la Cuenca Potásica Catalana. *Rev. Inst. Inv. Geol. Dip. Prov. Barcelona* 32: 77–86.

——. 1978/79. La precipitacion evaporitica actual en las lagunas s ladas del area: Bujaraloz, Sastago, Caspe, Alcaniz y Calanda (provincias de Zaragoza y Teruel). *Revista del Instituto de Investigaciones Geologicas* Barcelona 33: 5–56.

Pueyo J. J. and M. Ingles Urpinell. 1986. Substrate mineralogy, pore brine composition, and diagenetic processes in the playa lakes of Los Monegros and Bajo Aragon, Spain. *I International Symposium on Geochemistry of the Earth's Surface,* Grenada.

Puigdefàbregas, C. 1973. Miocene point-bar deposits in the Ebro Basin, Northern Spain. *Sedimentology* 20: 133–44.

——. 1975. La sedimentación molásica en la cuenca de Jaca. *Mon. Inst. Est. Pirenaicos* 104: 188 pp. C.S.I.C.

Puigdefàbregas, C., J. A. Muñoz, and M. Marzo. 1986. Thrust belt development in the eastern Pyrenees and related depositional sequences in the southern foreland basin. *Special Publication Association Sedimentology* 8: 229–46.

Puigdefàbregas, C. and P. Souquet. 1985. Cyclicity, basin evolution and tectonic control in the Mesozoic and Cenozoic of the Pyrenees. *Terra cognita* 5: 119 pp.

——. 1986. Tecto-sedimentary cycles and depositional sequences of the Mesozoic and Tertiary from the Pyrenees. *Tectonophysics* 129: 173–203.

Pujadas, J., J. M. Casas, J. A. Muñoz, and F. Sàbat. 1989. Thrust tectonics and Paleogene syntectonic sedimentation in the Empordà area, southeastern Pyrenees. *Geodinamica Acta* 3: 195–206.

Purser, B. H. 1973. *The Persian Gulf: Holocene carbonate sedimentation and diagenesis on a shallow epicontinental sea.* Springer-Verlag, Berlin. 471 pp.

Quirantes, J. 1969. *Estudio sedimentológico y estratigráfico del Terciario continental de los Monegros.* Ph.D. Thesis, University Granada, 200 pp.

Rat, P. 1974. Le système Bourgogne-Morvan-Bresse (articulation entre le Bassin Parisien et le domaine péri-alpin). In J. Debelmas, ed., *Géologie de la France* II: les chaînes plissées du cycle alpin et leur avant-pays. Doin éd.: 480–500.

Rat, P., G. Cattaneo, and G. Doret 1975. Décrochements dans le seil de Bourgogne. *C.R. somm. Soc. Géol. France* 1: 26–28.

Raup, O. B. 1966. Bromine distribution in some halite rocks of Paradox Member, Hermosa Formation, in Utah. In J. L. Rau, ed., *Second Symposium on Salt* 1: 236–47. Northern Ohio Geol. Soc., Cleveland.

Rauscher R. and M. Schuler. 1988. Les dinokystes, des témoins d'influences marines dans le Paléogène d'Alsace. *Bull. Cent. Rech. Explor. Prod. Elf-Aquitaine,* 12(1): 405–25.

Rebelle, M. 1986. *Sédimentologie, géochimie et palynologie du bassin évaporitique du Zechstein à partir de données de sub-surface (Mer du Nord, Hesse, R.F.A.).* Thèse Doct. Sci., Univ. Paris-Sud: 303 pp.

Reguant, S. 1967. *El Eoceno marino de Vic.* Ph.D. Thesis, University Barcelona, 617 pp.

Rehault, J. P. 1981. *Evolution tectonique et sédimentaire du bassin Ligure (Méditerranée occidentale).* Thèse Doct. Etat, Univ. P. et M. Curie, Paris VI: 132 p.

Rehault, J. P., G. Boillot, and A. Mauffret. 1984. The Western Mediterranean Basin Geological evolution. *Marine Geology* 55: 447–77.

Rehault, J. P., J. Mascle, and G. Boillot. 1984. Evolution geodynamique de la Mediterranee depuis l'Oligocene. *Mem. Soc. Geol. It.* 27: 85–96.

Renaut, R. W. 1993. Zeolitic diagenesis of late Quaternary fluviolacustrine sediments and associated calcrete formation in Lake Bogoria basin, Kenya Rift Valley. *Sedimentology* 40: 271–301.

Renaut, R. W. and J-J. Tiercelin. 1994. Lake Begoria, Kenya Rift Valley: a sedimentological overview. In R. W. Renaut and W. M. Last, eds., *Sedimentology and Geochemistry of Modern and Ancient Saline Lakes.* SEPM Special Publication #50: 101–123.

Rey, R. 1966. *Essais de corrélation entre les bassins oligocènes de l'Europe occidentale à de gastèropodes continentaux.* Thèse d'état, Univ. Rennes, 87 pp.

——. 1971. Biostratigraphie des bassin tertiaires du Massif Central. In J. Jung (ed.), Symposium: *Géologie, géomorphologie et structure profonde du Massif Central français* pp. 309–29. Plein Air Serv. Ed., Clermont-Ferrand.

Riba, O. 1955. Sobre la edad de los conglomerados terciarios del borde Norte de las sierras de la Demanda y Cameros. *Not. y Com. IGME* 39: 39–50.

——. 1964. Estructura sedimentaria del Terciario continental de la Depresión del Ebro en su parte riojana y navarra. *Aport. al XX Congr. Geogra. Intern., Londres,* pp. 127–38. Madrid.

——. 1976. Tectogenèse et sédimentation: deux modèles de discordance syntecttonique pyrénées. *Bull. Bur. Rech. Geol. Min. 2ieme sér.* 4: 383–401.

——. 1992. Las secuencias oblicuas en el borde Norte de la Depresion del Ebro en Navarra y la Discordancia de Barbarin. *Acta Geologica Hispanica* 27(1/2): 55–68.

Riba, O. and M. J. Jurado. 1992. Reflexiones sobre geología de la parte occidental de la Depresión del Ebro. *Acta Geologica Hispanica* 27(1/2): 177–93.

Riba, O. and J. Pérez Mateos. 1962. Sobre una inversión de aportes sedimentarios en el borde Norte de la Cuenca del Ebro (Navarra). *Inst. Edaf. Sec. Petro. Sedim.* pp. 201–21, II Reunión del GES, Sevilla 1961. Madrid.

Riba, O., S. Reguent, and M. Tarradell. 1986. Una Taula dels Temps Geològics. *Instituto d'Estudis Catans* 81: 127 pp.

Riba, O., S. Reguant, and J. Villena. 1983/1987. Ensayo de síntesis estratigráfica y evolutiva de la Cuenca Terciaria del Ebro. *Instituto Geología y Minero de España* (Madrid), 2: 131–59.

Richter-Bernburg, G. 1955. Über salinare Sedimentation. *Zeitschr. Deutsch. Geol. Gesellsch.* 105: 593–645.

——. 1972. Sedimentological problems of saline deposits. In *Geology of Saline Deposits,* Proc. Coll. Hanover, UNESCO édit., Earth Sciences 7: 316 pp.

——. 1985. Zechstein Anhydrite. Fazies und Genese. *Geologisches Jahrbuch* A85: 3–82.

Rider, M. H. 1986. *The geological interpretation of well logs.* Blackie (Halsted Press), Glasgow, 175 pp.

Ríos, J. M. 1963. Materiales salinos del suelo español. *Memoria IGME* 64, 166 pp. Madrid.

Riveline, J. 1986. Les charophytes du Paléogène et du Miocéne inférieur d'Europe occidentale: Biostratigraphie des formations continentales. *Cahiers de Paléontologie, CNRS,* Paris, 227 pp.

Robles, S. 1975. Estudio geológico del Mesozoico del bloque del Cardó y sectores adyacentes (Provincia de Tarragona). *Acta Geològica Hispànica* 10: 59–66.

Roedder, E. 1984. The fluids in salt. *American Mineralogist* 69: 413–439.

Rosell, J. 1983. *Estudi petrològic, sedimentològic i geoquímic de la formació de sals potàssiques de Navarra (Eoceno superior).* Ph.D. Thesis, University Barcelona, 321 pp.

Rosell, L. and F. Ortí. 1980. Presencia de analcima y observaciones diagenéticas en la anhidrita basal de la cuenca potásica de Navarra (Eoceno superior, cuenca del Ebro, España). *Revista Instituto Investigaciones Geologicas Dip. Prov. Barcelona* 34: 223–35.

——. 1981/82. The Saline (Potash) formation of the Navarra Basin (Upper Eocene, Spain). Petrology. *Revista Instituto Investigaciones Geológicas, Dip. Prov. Barcelona* 35: 71–121.

——. 1992. Geochemical characteristics and facies analysis of a Miocene lacustrine gypsum deposit (Calatayud Basin, Zaragoza, Spain). *IGCP-324 GLOPALS, Annual Meeting,* Madrid 1992, Abstracts: 39–41.

Rosell, L. and J. J. Pueyo. 1984. Características geoquímicas de la formación de sales potásicas de Navarra (Eoceno superior). Comparación con la cuenca potásica catalana. *Acta Geològica Hispànica* 19: 81–95.

Rosendahl, B. R., D. J. Reynolds, P. M. Lorber, C. F. Burgess, J. McGill, D. Scott, J. J. Lambiase, and S. J. Derksen. 1986. Structural expressions of rifting: lessons from Lake Tanganyika, Africa. In L. E. Frostick, R. Renaut, I. Reed, and J.-J. Tiercelin, eds., *Sedimentation in the African Rifts* pp. 29–43, Geological Society Spec. Publ. No. 25.

Rosset, C. 1964. Les formations du bassin oligocène du Sigean-Portel (Aude) et leurs chronologie. *C. R. somm. Soc. Géol. France* 10: 415–16.

Rouchy, J. M. 1987. Pétrographie des sulfates et des sédiments associés: modalités de dépôt et de diagénèse. Annexe: Les déformations de type halocinétique dans le Lodévois et leurs conséquences. In G. Busson and J. M. Rouchy, eds., *Etude pétrographique et géochimique des grandes accumulations évaporitiques du Trias et du Paléogène français.* Unpublished rep.: 201–28.

Rouchy, J.-M., E. Groessens, and A. Laumontais. 1984. Sédimentologie de la formation anhydritique viséenne du sondage de Saint-Ghislain (Hainaut, Belgique). Implications paléogéographiques et structurales. *Bull. Soc. Géol. Belgique* 93: 105–45.

Rouchy, J. M., A. Laumondais, and E. Groessens. 1986. The

Lower Carboniferous (Visean) evaporites in northern France and Belgium: Depositional, diagenetic and deformational Guides to reconstruct a disrupted evaporitic basin. In T. Peryt, ed., *Evaporite Basins.* Lecture Notes in Earth Sciences 13. Springer-Verlag, pp. 31–67, Berlin-Heidelberg.

Roure, F., P. Choukroune, X. Berástegui, J. A. Muñoz, A. Villien, P. Matheron, M. Bareyt, M. Séguret, P. Cámara, and J. Déramond. 1989. ECORS deep seismic data and balanced cross sections: geometric constraints on the evolution of the Pyrenees. *Tectonics* 8: 41–50.

Roure, F., M. Seguret, and A. Villier. 1988. Structural style of the Pyrenees: a view from seismic reflection to surface studies. *A.A.P.G. Mediterranean Basins Conference* field-trip guidebook: 109–33.

Rousset, D. and R. Bayer. 1990. Nouvelles cartes gravimétriques du sud du Fossé Rhénan. *C.R. Acad. Sci. Paris* 310: 199–206.

Ruhland, M. 1982. Structure et compartimentage du fosse rhénan méridional. Rôle des accidents issus du socle des Vosges et de la Forêt Noire sur les champs de failles. *Trav. Lab. Sci. Terre,* Marseille B 19: 3–4.

——. 1974. Le rôle des décrochements dans le socle vosgien et en bordure du Fossé rhénian. In J. H. Illies and K. Fuchs, eds., *Approaches to Taphrogenesis.* Schweizerbart, Stuttgart: 167–71.

Saéz, A. 1987. *Estratigrafía y sedimentología de las formaciones lacustres del tránsito Eoceno-Oligoceno del NE de la Cuenca del Ebro.* Ph.D. Thesis, University Barcelona, 353 pp.

Saéz, A., J. Vergès, C. Taberner, J. J. Pueyo, J. A. Muñoz, and P. Busquets. 1991. Eventos evaporíticos paleógenos en la cuenca de antepaís Surpirenaica: causas climáticas-causas tectónicas? *I Congreso Grupo Español del Terciario,* Libro-Guía de Excursione 5 (F. Colombo, Ed.), 85 pp.

Salas, R. 1979. *El sistema arrecifal del Eoceno superior de la cuenca de Igualada (Barcelona).* Tesi de Llicenciatura (Graduation thesis). University Barcelona. 196 pp.

——. 1989. *El Malm i el Cretàci inferior entre el Massís de Garraf i la Serra d' Espadà. Anàlisi de conca.* Thesis, University Barcelona, 345 pp.

Salvany, J. M. 1988. Aspectos petrológicos y sedimentológicos de los yesos de Ablitas y Monteagudo (Navarra): Mioceno de la Cuenca del Ebro). *I Encuentro nacional de estudios sobre el Moncayo,* Tarazona, Sep.-Oct. 1988, Centro de Estudios Turiasonenses, CSIC, Inst. Fernando el Católico: 123–46.

——. 1989a. *Las formaciones evaporíticas del Terciario continental de la Cuenca del Ebro en Navarra y La Rioja: litoestratigrafía, petrología y sedimentología.* Ph.D. Thesis, University Barcelona, 397 pp.

——. 1989b. Los sistemos lacustres evaporiticos del sector Navarro-Riojano de la Cuenca del Ebro durante el Oligoceno, Mioceno inferior. *Acta Geológica Hispánica* 24: 231–41.

Salvany, J. M. and F. Ortí. 1987. La paragénesis de sulfatos de Ca y Na en el Mioceno continental de Alcanadre-Arrúbal (La Rioja) y San Adrián (Navarra). *Bol. Soc. Esp. Mineralogía* 10: 47–48.

——. 1992. El yacimento glauberítico de Alcandre: procesos sedimentarios y diagenéticos (Mioceno Inferior, Cuenca del Ebro). In J. García Guinea and J. Martinez Frias (Coords.), *Recursos minerales de Espana,* CSIC, Madrid: 1251–74.

——. 1994. Miocene glauberite deposits of Alcanadre, Ebro Basin Spain: Sedimentary and dagenetic processes. In R. W. Renaut and W. M. Last, eds., *Sedimentology and Geochemistry of Modern and Ancient Saline Lakes.* SEPM Special Publication #50: 203–15.

Salvany, J. M., A. Muñoz, and A. Pérez. 1994. Nonmarine evaporitic sedimentation and associated diagenetic processes of the southwestern margin of the Ebro Basin (Lower Miocene), Spain. *Jour Sed. Research* A64(2): 190–203.

Schiradin, J. 1954. Les formations littorales et côtières du Sannoisien de la Moyenne Alsace. *Bull. Serv. carte géol. Als.-Lorr.* 7: 35–67.

Schmalz, R. 1969. Deep-water evaporite deposition: a genetic model. *Bull. Amer. Assoc. Petrol. Geol.* 53: 798–823.

Schock, H. H. and H. Puchelt. 1971. Rubidium and cesium distribution in salt minerals—I. Experimental investigations. *Geochimica et Cosmochimica Acta* 35: 30–317.

Schreiber, B. C. 1978. Environments of subaqueous gypsum deposition. In W. E. Dean and B. C. Schreiber, eds., *Marine Evaporites.* S.E.P.M. Short Course 4: 43–73. Oklahoma City.

——. 1986. Arid shorelines and evaporites. In H. G. Reading, ed., *Sedimentary environments and facies* (2d ed.), Blackwell Scientific Publishers, pp. 189–228, Oxford.

——. (ed.). 1988. *Evaporites and Hydrocarbons.* Columbia University Press, New York: 475 pp.

Schreiber, B. C. and D. Walker. 1988. Halite pseudomorphs of gypsum in cyclic evaporites G.S.A. 20(7): 7 (abs).

——. 1992. Halite pseudomorphs after gypsum: a suggested mechanism. *Journal Sedimentary Petrology* 62: 61–70.

Schnaebele, R. 1948. Monographie géologique du champ pétrolifère de Pechelbronn. *Mém. Serv. carte géol. Als.-Lorr.* 7: 254 p.

Schuler, M. 1988. *Environnements et paléoclimats paléogènes. Palynologie et biostratigraphie de l'Eocène et de l'Oligocène inférieur dans les fossés rhénan, rhodanien et de Hesse.* Thèse Doct. Etat, Univ. Strasbourg, 339 pp.

——. 1988/90. Environnements et paléoclimats paléogènes.

Palynologie et biostratigraphie de l'Eocène et de l'Oligocène inférieur dans les fossés rhénan, rhodanien et de Hesse. Thèse Doct. Sci., Strasbourg, 1988 et Doc. *Bur. Rech. Geol. Min. Mém.* 190, 503 p.

Schuler, M., M. M. Blanc-Valleron, J. P. Gely, and M. Ansart. 1991. Matière organique et palynofaciès de l'Oligocène inférieur évaporitique du bassin de Mulhouse (France). *Palynosciences* (Montpellier), 1: 41–58.

Séguret, M. 1972. Etude téctonique des nappes et séries decollés de la partie centrale du vesant Sud des Pyrénées. *Pub. University Sci. et Tec. du Languedoc Pub. Ust. sér. Géol. Struct.* Montpellier, 167 pp.

Seguret, M. and M. Daignières. 1986. Crustal-balanced cross-sections of the Pyrenees: Discussion. *Tectonophysics* 129: 303–18.

Serra, O. 1979. Diagraphies différées, base de l'interprétation. Tome 1, Acquisition des données diagraphiques. *Bull. Cent. Rech. Explor. Prod. Elf-Aquitaine* 1: 316 pp.

——. 1980. Aspects diagraphiques des évaporites. *Bull. Cent. Rech. Explor. Prod. Elf-Aquitaine* 4: 411–31.

——. 1985. Diagraphies différées, Base de l'interprétation. Tome 2: Interprétation des données diagraphiques. *Bull. Cent. Rech. Explor. Prod. Elf-Aquitane, Pau* 7: 631 pp.

——. 1986. Evaporites et diagraphies différées. In *Les séries à évaporites en exploration pétrolière* pp. 13–71. Paris, Editions Technip.

Serra, O. and A. Curial. 1985. Progrès récents dans la reconnaissance des accumulations salines par diagraphies. *Bull. Soc. Géol. France* 8: 797–806.

Serrano, A., W. Martinez Del Olmo, and P. Camara. 1989. Diapirismo del Trías salino en el dominio Cántabro-Navarro. *Libro homenaje a R. Soler,* A.G.G.E.P., Madrid, pp. 115–21.

Shearman, D. J. 1963. Recent anhydrite, dolomite, and halite from the coastal flats of the Arabian shore of the Persian Gulf. Proc. *Geol. Soc. London* 1607: 63–65.

——. 1966. Origin of marine evaporites by diagenesis. *Transactions Institute Mining and Metallurgy* 75B: 208–15.

——. 1970. Recent halite rock, Baja California, Mexico. *Inst. Mining Met. Trans.* 79B: 155–62.

——. 1978. Halite in sabkha environments. In W. E. Dean and B. C. Schreiber, eds., *Marine Evaporites.* SEPM Short Course 4: 30–42. Tulsa, Okla.

——. 1985. Syndepositional and late diagenetic alteration of primary gypsum to anhydrite. In B. C. Schreiber and L. Harner, eds., *Sixth Internat. Symp. on Salt,* Salt Institute, Alexandria, Virginia, I: 41–50.

Sichère, M. C. 1976. *Dosage du Chlore et du Brome dans les évaporites.* Thèse doctorat Université, Univ. Paris VI: 101 pp.

Siedlecka, A. 1976. Silicified Precambrian evaporite nodules form Northern Norway: a preliminary report. *Sedimentary Geology* 16: 161–75.

Sittler, C. 1964. Relations sédimentologiques entre divers oligocènes de l'Est de la France. *Mém. Bur. rech. géol. min.* Paris 28: 531–44.

——. 1965. Le Paléogène des fossé rhénan et rhodanien. Etudes sédimentologiques et paléoclimatiques. *Mém. Serv. carte géol. Als.-Lorr., Strasbourg* 24: 392 pp.

——. 1967. Le soubassement et le remplissage sédimentaire du Fossé rhénan au niveau du bassin de Pechelbronn et du sénil d'Erstein. Coupes géologiques à travers le Fossé rhénan. *Abh. geol. Landesamt Baden-Württemberg* 6: 69–80.

——. 1985. Les hydrocarbures d'Alsace dans le contexte historique et géodynamique du fossé rhénan. *Bull. Cent. Rech. Explor. Prod. Elf-Aquitaine,* 9: 335–71.

Smith, D. B. 1971. Possible displacive halite in the Permian Upper Evaporite Groupe of northeast Yorkshire. *Sedimentology* 17: 221–232.

Smoot, J. P. and B. Castens-Seidel. 1994. Sedimentary features produced by efflorescent salt crusts, Saline Valley and Death Valley, California. In R. W. Renault and W. M. Last, eds., *Sedimentology and Geochemistry of Modern and Ancient Saline Lakes.* SEPM, Special Publication 50, Tulsa: 73–90.

Solans, J. and J. Montoriol. 1968. Estudio termoquimico de la zone potasica de los yacimentos de Súria y Balsareny (Barcelona). *Acta Geológica Hispánica* 3: 102–4.

Solé, J. 1972. *Formación de Mués, litofacies y procesos de sedimentación.* Graduation Thesis, University Barcelona, 46 pp.

Solé Sabarís, Ll. 1933. La geologia dels voltants de Figueres i la tectònica de l'Empordà. *Bulettí Institució Catalana Història Natural* 33: 250–57.

Sonnenfeld, P. 1984. *Brines and Evaporites.* Academic Press, Orlando, 613 pp.

Sonnenfeld, P. and P. P. Hudec. 1985. Origin of clay films in rock salt. *Sediment. Geol.* Amsterdam, 44: 113–20.

Southgate, P. N. 1982. Cambrian skeletal halite crystals and experimental analogues. *Sedimentology* 29: 391–407.

Soyer, R. 1960. Les zones gypseuses antéludiennes préservées de l'Ile-de-France. *Bull. Soc. Géol. France* 10: 145–51.

Stanley, D. J. and E. Mutti. 1968. Sedimentological evidence for an emerged land mass in the Ligurian Sea during the Paleogene. *Nature* 218: 32–36.

Stoertz, G. E. and G. E. Ericksen. 1974. Geology of salars in Northern Chile. *Geologic Survey Professional Paper.* Wahington, D.C., No. 811, 65 pp.

Sturmfels, E. 1943. Das Kalisalzlager von Buggingen (Südbaden). *Neues Jahrb. Mineral. Abh. Beil. Bd.* 78A: 131–216.

Szulc, J. R. P., M. P. Mouline, and M. Lenguin. 1991. Evolution of lacustrine systems in the Tertiary Narbonne Basin, northern Pyrenean foreland, southeast France. *Special Publication International Association Sedimentologists* 13: 279–90.

Taberner, C. 1983. *Evolución ambiental y diagenética de los depósitos del Terciario inferior (Paleoceno y Eoceno) de la cuenca de Vic.* Ph.D. Thesis, University Barcelona, 1400 pp.

Taberner, C. and C. Santisteban. 1984. La sedimentación del tránsito Eoceno-Oligoceno en la cuenca catalana (sector de Vic). Su significación geodinámica. *Libro Homenaje a Luis Sánchez de la Torre,* Publicaciones de Geologia n°. 20, Universitat Autònoma de Barcelona, pp. 283–318.

Tchimichkian, G., J. Reulet, and A. Vatan. 1958. Etude pétrographique des matériaux mollassiques de quelques sondages profonds de Bresse. *Eclogae. Geol. Helv.* 51: 1093–1114.

Teichmüller, R. 1974. Die tektonische entwicklung der Niederrheinischen Bucht. In J. H. Illies and K Fuchs (eds.), *Approaches to Taphrogenesis.* Schweizerbart Verlag, Stuttgart, pp. 269–85.

Tempier, C. 1987. Modèle nouveau de mise en place des structures provençales. *Bull. soc. Géol. France* (8)3: 533–40.

Thomas, B., H. Gennesseaux, and L. Lecca. 1988. La structure de la marge occidentale de la Sardaigne et la fragmentation de l'île au Cenozoïque. *Marine Geol.* 83: 31–41.

Thomas J. C. 1984. Formations benthiques à Cyanobactéries des salins de Santa Pola (Espagne): Composition spécifique, morphologie et caractéristiques biologiques des principaux peuplements. *Revista del Instituto de Investigaciones Geologicas Barcelona* 38/39: 139–58.

Thomas J. C. and D. Geisler. 1982. Peuplements benthiques à Cyanophycées des marais salants de Salin-de-Giraud (Sud de la France). *Géologie Méditerranéene* IX(4): 391–411.

Thomas M. 1982. Approche géochimique du système sédimentaire des marais salants de Salin-de-Giraud (Sud de la France). *Géologie Méditerranéene* IX(4): 487–500.

Tiercelin, J. J. and A. Vincens, eds. 1987. Le Demi-graben de Baringo-Bogoria, Rift Gregory, Kenya. 30,000 ans d'histoire hydrologique et sédimentaire. *Bulletin des Centres de Recherches Exploration Production Elf-Aquitaine* 11: 249–540.

Tobien, H. 1968. Mammifères éocènes du Bassin de Mayence et de la partie orientale du Fossé Rhénan. *Mém. Bur. Rech. géol. min.* 58: 297–307.

Torras, A. 1976. *Sedimentogia de los limos yesiferas de la depresion media del Ebro.* Ph.D. Thesis, University Barcelona, 220 pp.

Torrescusa, S. and J. Klimowitz. 1990. Contribución al conocimiento de las evaporitas miocenas (Fm. de Zaragoza) de la Cuenca del Ebro. In F. Ortí and J. K. Salvany, eds., *Formaciones evaporiticas de la Cuenca del Ebro y cadenas periféncas, y de la zona de Levante* pp. 120–2, ENRESA, University Barcelona.

Toulemont, M. 1987. Les gypses lutétiens du bassin de Paris. Sédimentation, karstification et conséquences géotechniques. *Lab. Ponts et Chaussées, Rap. des Lab.* GT-24 Paris: 319 pp.

Touraine, F. 1971. L'Oligocène nord-varois; sa position stratigraphique et sa signification dans la tectonogénèse subalpine (état des connaissances actuelles). *Annales Guébhard, Neuchâtel* 47: 394–421.

Trauth, N., E. Gannat, J.-M. Schlund, and G. Truc. 1984. Evaporites et faciès associés au Paléogène. In Synthèse géologique du Sud-Est de la France. *Mém. Bur. Rech. Géol. Min.* 125: 456–61.

Triat, J.-M. and G. Truc. 1974. Evaporites paléogènes du domaine rhodanien. *Rev. Géogr. Phys. et Géol. dyn.* 2(16): 235–62.

——. 1983. Le rôle des failles N50 dans la sédimentation des temps mésozoïques et cénozoïques et dans l'évolution tectonique du bassin du S. E. (France). *Bull. Centres Rech. Explor. Prod. Elf-Aquitaine* 1: 425–32.

Trichet, J. 1967. Essai d'explication de dépot d'aragonite sur des substrats organiques. *Comte Rendu de Académie Sciences, Paris* 265D: 1464–67.

Trümpy, R. 1980. *Geology of Switzerland: a guide-book.* Schweizerische Geologishe Kommission. Wepf and Co., Basel: 8–93.

Tucker, M. E. 1976. Quartz replaced anhydrite nodules ("Bristol Diamonds") from the Triassic of the Bristol District. *Geological Magazine* 113: 569–74.

——. 1981. *Sedimentary Petrology: An Introduction.* Blackwells, Oxford, 252 pp.

Usdowski, E. 1973. Das geochemische Verhalten des Strontiums bei der Genese und Diagenese von Ca-Karbonat und Ca-Sulfat Mineralen. *Contrib. Miner. Petrol.* 38: 177–95.

Usiglio, J. 1849. Analyse de l'eau de la Méditerranée sur les côtes de France. *Ann. Chimie et Phys.* (3)27: 32–107.

Utrilla, R. 1989. *Les composicions isotopiques ($\delta^{18}O$, $\delta^{34}S$) del sulfat, com a indicadores de l'origen de les evaporites del Mesozoic i del Cenozoic de la Peninsula Ibérica i les Illes Balears.* Ph.D. Thesis, University Barcelona: 276 pp.

Utrilla, R., F. Ortí, C. Pierre, and J. J. Pueyo. 1991. Composición isótopica de las evaporitas terciarias continentales de la Cuenca del Ebro: relación con los ambientes deposicionales. *Revista de la Sociedad Geológica de España* 4(3/4): 353–60.

Utrilla, R., C. Pierre, F. Ortí, and J. J. Pueyo. 1992. Oxygen and sulphur isotope compositions as indicators of the origin of Mesozoic and Cenozoic evaporites from Spain. *Chemical Geology (Isotope Geoscience Section)* 102: 229–44.

Utrilla, R., C. Pierre, F. Ortí, L. Rosell, M. Ingles, and J. J. Pueyo. 1987. Estudio isotópico de los sulfatos en formaciones evaporíticas mesozoicas marinas y terciarias continentales. Aplicación a la Cuenca del Tajo. *II Congr. Geoquimica* Soria, Communications: 91–94.

Vacl, J. 1966. The Cheb Bassin. In J. Svoboda et al., eds., *Regional Geology of Czechoslovakia. Part I, The Bohemian Massif.* Geol. Survey Czechoslovakia, Prague: 516–31.

Vail, P. R. 1981. Cenozoic sea-level changes: Introduction to colloquium 3. Geology of Continental margins. *26th International Geological Congress Paris.*

Vail, P. R. and J. Hardenbol. 1979. Sea level changes during the Tertiary. *Oceans* 22: 71–79.

Vail, P. R., R. M. Mitchum, and S. Thompson. 1977. Seismic stratigraphy and global changes of sea level. In C. E. Payton, ed. *Seismic Stratigraphy: Applications to Hydrocarbon Exploration:* pp. 83–97. Memoir Association Petroleum Geologists 6.

Valyashko, M. G. 1956. Geochemistry of bromine in the processes of salt deposition and the use of bromine as genetic and prospecting criterion. *Geochemistry* 6: 570–89.

Valero Garcés, B. L. and J. Gisbert Aguilar 1994. Permian saline lakes in the Aragón-Béarn basin, Western Pyrenees. In R. W. Renaut and W. M. Last, eds., *Sedimentology and Geochemistry of Modern and Ancient Saline Lakes.* SEPM Special Publication 50: 267–90.

Van Rompey, C. 1965. L'Eocène Nord Catalan en rélation avec les permis "Berga," "Montserrat" et "Olot." *SEPE* (Internal Report).

Vatin-Perignon, N. and R. Michel. 1966. Présence d'un fossé d'effondrement volcano-tectonique dans la partie centrale du Cantal. *Bull. Soc. Géol. France* 7: 308–10.

Vegas, R., J. M. Fontboté, and E. Banda. 1980. Widespread Neogene rifting superimposed on Alpine regions on the Iberian Peninsula. In *Evolution and Tectonics of the Western Mediterranean and Surrounding Areas* pp. 109–28. Instituto Geográfico Nacional, Special publication. Madrid.

Vergés, J. and A. Martínez. 1988. Corte compensado del Pirineo oriental: Geometría de las cuencas de antepaís y edades de emplazamiento de los mantos de corrimiento. *Acta Geológica Hispánica* 23: 95–105.

Vergès, J. and J. A. Muñoz. 1990. Thrust sequences in the Southern Central Pyrenees. *Bull. Soc. Geol. France* 2: 265–72.

Viallard, P. 1987. Un modèle de charriage epiglyptique: la nappe des Corbières orientales (Aude, France). *Bull. Soc. Géol. France* 3: 551–60.

Vianey-Liaud, M. 1979. Evolution des Rongeurs à l'Oligocène en Europe occidentale. *Palaeontographica* (Stuttgart), A166: 136–236.

Villemin, T. 1986. Tectonique en extension, fracturation et subsidence: le Fossé rhénan et le bassin de Sarre-Nahe. Thèse Univ. Pierre et Marie Curie, Paris. Mém. Sci. Terre: 1–270.

Vincent, P. M., M. Aubert, P. Boivin, J. M. Cantagrel, and J. F. Lenat. 1977. Découverte d'un volcanisme paléocène en Auvergne: les maars de Menat et leurs annexes. Etude géologique et géophysique. *Bull. Soc. Géol. France* 10: 1057–70.

Wardlaw, N. C. 1968. Carnallite-sylvite relationships in the Middle Devonian Prairie Evaporite Formation, Saskatchewan. *Bulletin Geological Society America* 79: 1273–94.

Wardlaw, N. C. and W. M. Schwerdtner. 1966. Halite-anhydrite seasonal layers in Middle Devonian Prairie evaporite formation, Saskatchewan, Canada. *Bulletin Geological Society America* 77: 331–42.

Weijermars, R., C. W. Mulder-Blanken, and J. Wiegers. 1986. Growth rate observation from the moss-built Checa travertine terrace, central Spain. *Geol. Mag.* 123: 279–86.

West, I. M. 1973. Vanished evaporites. Significance of strontium minerals. *Journal Sedimentary Petrology* 43: 278–79.

West, I. M., A. Brandon, and M. Smith. 1968. A tidal-flat evaporitic facies in the Visean of Ireland. *Journal Sedimentology Petrology* 4: 1079–93.

Williams, G. D. and M. W. Fischer. 1984. A balanced section across the Pyrenean orogenic belt. *Tectonics* 3: 773–80.

Wimmenauer, W. 1967. Igneous rocks of the Rhinegraben. *Abh. geol. L.Amt. Baden-Württ.* 6: 144–48.

Winnock, E. 1974. Le bassin d'Aquitaine. In J. Debelmas, *Géologie de la France,* 1: Vieux massifs et grands bassins sédimentaires, pp. 259–92, Doin Ed., Paris.

Wright, V. P. 1986. The role of fungal biomineralisation in the formation of early Carboniferous soil fabrics. *Sedimentology* 33: 831–38.

——. 1994. Paleosols in shallow marine sequences. *Earth-Science Reviews* 35: 367–95.

Zarkhov, M. A. 1984. *Paleozoic Salt-bearing Formations of the World.* Springer, New York: 427 p.

——. 1981. *History of Paleozoic Salt Accumulation.* Springer, New York: 308 p.

Zen, E.-A. 1965. Solubility measurements in the system $CaSO_4$-NaCl-H_2O at 35°, 50° and 70° and one atmosphere pressure. *Journal Petrology* 6: 124–64.

Ziegler, P. A. 1980. Northwestern Europe: subsidence patterns of Post-Variscan basins. 26ème Congr. Géol. intern., Géologie de l'Europe. *Bur. Rech. Geol. Min. Mém.* 108: 249–80.

——. 1982. *Geological Atlas of Western and Central Europe.* Shell Intern. Petrol. Maastsch. B. V., The Hague: 130 p.

Zoetemeijer, R., P. Desegaulx, S. Cloetingh, F. Roure, and I. Moretti. 1990. Lithospheric dynamics and tectonic-stratigraphic evolution of the Ebro Basin. *Journal of Geophysical Research* 95: 2701–11.

Contributors

Georges Busson
Laboratoire de Geologie du Museum National
43, rue de Buffon
75005 Paris
France

B. Charlotte Schreiber
Lamont Doherty Earth Observatory
Columbia University
Palisades, New York 10964
United States

Marie-Madeline Blanc-Valleron
Laboratoire de Geologie du Museum National
43, rue de Buffon
75005 Paris
France

Alain Curial
DIASTRATA
20 bis, rue Songieu
69100 Villeurbanne
France

Gilles Dromart
Ecole Normale Superieure de Lyon
Department of Earth Sciences
46, Allée d'Italie
69364 Lyon Cedex 07
France

Daniel Dumas
3, rue Richan
Fr 69004 Lyon
France

Dominique Geisler-Cussey
IPRA—Faculte des Sciences
Universitaire de Pau
Avenue de l'Université
64000 PAU,
France

Robert Moretto
DIASTRATA
20 bis, rue Songieu
69100 Villeurbanne
France

Denise Nury
Université de Provence
Aix-Marseille 1
Centre de Saint Charles
Laboratoire de Geologie Structurale et Appliquee
3, Place V. Hugo (Case 28)
F-13331 Marseille Cedex 3
France

Jean-Marie Rouchy
Laboratoire de Geologie du Museum Nation
43, rue de Buffon
75005 Paris
France

Monique Schuler
Centre de Sédimentologie et de Géochimie de la Surface de Strasbourg
1, rue Blessig
F-67084 Strasbourg Cedex
France

Federico Ortí
Universitat de Barcelona
Facultat de Geologia
Departament di Geoquimica, Petrologia i Prospecció Geològica
Zona Universitària de Pedralbes
08071 Barcelona
Spain

Juan-José Pueyo
Universitat de Barcelona
Facultat de Geologia
Departament de Geoquimica, Petrologia i Prospecció Geològica
Zona Universitària de Pedralbes
08071 Barcelona
Spain

Laura Rosell
Universitat de Barcelona
Facultat de Geologia
Departament de Geoquimica, Petrologia i Prospecció Geològica
Zona Universitària de Pedralbes
08071 Barcelona
Spain

Josep M. Salvany
Universitat Politècnica de Catalunya
Departament d'Enginyeria del Terreny i Cartogràfica
E.T.S. Enginyers de Camins, C.iP.
c/o Gran Capità, s/n
08034 Barcelona
Spain

Pere Santanach
Universitat de Barcelona
Facultat de Geologia
Departament de Geoquimica, Petrologia i Prospecció Geològica
Zona Universitària de Pedralbes
08071 Barcelona
Spain

J. García Veigas
Universitat de Barcelona
Facultat de Geologia
Departament de Geoquimica, Petrologia i Prospecció Geològica
Zona Universitària de Pedralbes
08071 Barcelona
Spain

Index